Plant Protein and Proteome Altlas–Integrated Omics Analyses of Plants under Abiotic Stresses

Plant Protein and Proteome Altlas–Integrated Omics Analyses of Plants under Abiotic Stresses

Special Issue Editors

Tingyun Kuang
Xuchu Wang
Xiaochun Qin
Shaojun Dai
Pingfang Yang
Ling Li

MDPI • Basel • Beijing • Wuhan • Barcelona • Belgrade

Special Issue Editors
Tingyun Kuang
Chinese Academy of Sciences
China

Xuchu Wang
Hainan Normal University
China

Xiaochun Qin
University of Jinan
China

Pingfang Yang
Chinese Academy of Sciences
China

Ling Li
Mississippi State University
USA

Shaojun Dai
Shanghai Normal University
China

Editorial Office
MDPI
St. Alban-Anlage 66
4052 Basel, Switzerland

This is a reprint of articles from the Special Issue published online in the open access journal *International Journal of Molecular Sciences* (ISSN 1422-0067) in 2019 (available at: https://www.mdpi.com/journal/ijms/special_issues/protein_omics).

For citation purposes, cite each article independently as indicated on the article page online and as indicated below:

LastName, A.A.; LastName, B.B.; LastName, C.C. Article Title. *Journal Name* **Year**, *Article Number, Page Range.*

ISBN 978-3-03921-960-5 (Pbk)
ISBN 978-3-03921-961-2 (PDF)

Cover image courtesy of Pingfang Yang.

Contents

About the Special Issue Editors . . . . **ix**

Preface to "Plant Protein and Proteome Altlas–Integrated Omics Analyses of Plants under Abiotic Stresses" . . . . **xi**

Xuchu Wang
Protein and Proteome Atlas for Plants under Stresses: New Highlights and Ways for Integrated Omics in Post-Genomics Era
Reprinted from: *Int. J. Mol. Sci.* **2019**, *20*, 5222, doi:10.3390/ijms20205222 . . . . **1**

Shanshan Li, Juanjuan Yu, Ying Li, Heng Zhang, Xuesong Bao, Jiayi Bian, Chenxi Xu, Xiaoli Wang, Xiaofeng Cai, Quanhua Wang, Pengcheng Wang, Siyi Guo, Yuchen Miao, Sixue Chen, Zhi Qin and Shaojun Dai
Heat-Responsive Proteomics of a Heat-Sensitive Spinach Variety
Reprinted from: *Int. J. Mol. Sci.* **2019**, *20*, 3872, doi:10.3390/ijms20163872 . . . . **8**

Lingyun Yuan, Jie Wang, Shilei Xie, Mengru Zhao, Libing Nie, Yushan Zheng, Shidong Zhu, Jinfeng Hou, Guohu Chen and Chenggang Wang
Comparative Proteomics Indicates That Redox Homeostasis Is Involved in High- and Low-Temperature Stress Tolerance in a Novel Wucai (*Brassica campestris* L.) Genotype
Reprinted from: *Int. J. Mol. Sci.* **2019**, *20*, 3760, doi:10.3390/ijms20153760 . . . . **29**

Shuping Wang, Yingxin Zhang, Zhengwu Fang, Yamin Zhang, Qilu Song, Zehao Hou, Kunkun Sun, Yulong Song, Ying Li, Dongfang Ma, Yike Liu, Zhanwang Zhu, Na Niu, Junwei Wang, Shoucai Ma and Gaisheng Zhang
Cytological and Proteomic Analysis of Wheat Pollen Abortion Induced by Chemical Hybridization Agent
Reprinted from: *Int. J. Mol. Sci.* **2019**, *20*, 1615, doi:10.3390/ijms20071615 . . . . **51**

Satheeswaran Thangaraj, Xiaomei Shang, Jun Sun and Haijiao Liu
Quantitative Proteomic Analysis Reveals Novel Insights into Intracellular Silicate Stress-Responsive Mechanisms in the Diatom *Skeletonema dohrnii*
Reprinted from: *Int. J. Mol. Sci.* **2019**, *20*, 2540, doi:10.3390/ijms20102540 . . . . **69**

Xinbo Wang, Yanhua Xu, Jingjing Li, Yongzhe Ren, Zhiqiang Wang, Zeyu Xin and Tongbao Lin
Identification of Two Novel Wheat Drought Tolerance-Related Proteins by Comparative Proteomic Analysis Combined with Virus-Induced Gene Silencing
Reprinted from: *Int. J. Mol. Sci.* **2018**, *19*, 4020, doi:10.3390/ijms19124020 . . . . **89**

Quanliang Xie, Guohua Ding, Liping Zhu, Li Yu, Boxuan Yuan, Xuan Gao, Dan Wang, Yong Sun, Yang Liu, Hongbin Li and Xuchu Wang
Proteomic Landscape of the Mature Roots in a Rubber-Producing Grass *Taraxacum Kok-saghyz*
Reprinted from: *Int. J. Mol. Sci.* **2019**, *20*, 2596, doi:10.3390/ijms20102596 . . . . **105**

Yajing Wang, Xinying Zhang, Guirong Huang, Fu Feng, Xiaoying Liu, Rui Guo, Fengxue Gu, Xiuli Zhong and Xurong Mei
iTRAQ-Based Quantitative Analysis of Responsive Proteins Under PEG-Induced Drought Stress in Wheat Leaves
Reprinted from: *Int. J. Mol. Sci.* **2019**, *20*, 2621, doi:10.3390/ijms20112621 . . . . **124**

Jianqiang Huo, Dengjing Huang, Jing Zhang, Hua Fang, Bo Wang, Chunlei Wang, Zhanjun Ma and Weibiao Liao
Comparative Proteomic Analysis during the Involvement of Nitric Oxide in Hydrogen Gas-Improved Postharvest Freshness in Cut Lilies
Reprinted from: *Int. J. Mol. Sci.* **2018**, *19*, 3955, doi:10.3390/ijms19123955 . . . . **143**

Xuan Wang, Tinashe Zenda, Songtao Liu, Guo Liu, Hongyu Jin, Liang Dai, Anyi Dong, Yatong Yang and Huijun Duan
Comparative Proteomics and Physiological Analyses Reveal Important Maize Filling-Kernel Drought-Responsive Genes and Metabolic Pathways
Reprinted from: *Int. J. Mol. Sci.* **2019**, *20*, 3743, doi:10.3390/ijms20153743 . . . . **162**

Yuying Wang, Ling Qiu, Qilu Song, Shuping Wang, Yajun Wang and Yihong Ge
Root Proteomics Reveals the Effects of Wood Vinegar on Wheat Growth and Subsequent Tolerance to Drought Stress
Reprinted from: *Int. J. Mol. Sci.* **2019**, *20*, 943, doi:10.3390/ijms20040943 . . . . **188**

Tingting Jia, Jian Wang, Wei Chang, Xiaoxu Fan, Xin Sui and Fuqiang Song
Proteomics Analysis of *E. angustifolia* Seedlings Inoculated with Arbuscular Mycorrhizal Fungi under Salt Stress
Reprinted from: *Int. J. Mol. Sci.* **2019**, *20*, 788, doi:10.3390/ijms20030788 . . . . **211**

Yuan Wang, Yuting Cong, Yonghua Wang, Zihu Guo, Jinrong Yue, Zhenyu Xing, Xiangnan Gao and Xiaojie Chai
Identification of Early Salinity Stress-Responsive Proteins in *Dunaliella salina* by isobaric tags for relative and absolute quantitation (iTRAQ)-Based Quantitative Proteomic Analysis
Reprinted from: *Int. J. Mol. Sci.* **2019**, *20*, 599, doi:10.3390/ijms20030599 . . . . **235**

Hui Liu, Fen-Fen Wang, Xian-Jun Peng, Jian-Hui Huang and Shi-Hua Shen
Global Phosphoproteomic Analysis Reveals the Defense and Response Mechanisms of *Jatropha Curcas* Seedling under Chilling Stress
Reprinted from: *Int. J. Mol. Sci.* **2019**, *20*, 208, doi:10.3390/ijms20010208 . . . . **249**

Xiaoyun Zhao, Xue Bai, Caifu Jiang and Zhen Li
Phosphoproteomic Analysis of Two Contrasting Maize Inbred Lines Provides Insights into the Mechanism of Salt-Stress Tolerance
Reprinted from: *Int. J. Mol. Sci.* **2019**, *20*, 1886, doi:10.3390/ijms20081886 . . . . **276**

Guowu Yu, Yanan Lv, Leiyang Shen, Yongbin Wang, Yun Qing, Nan Wu, Yangping Li, Huanhuan Huang, Na Zhang, Yinghong Liu, Yufeng Hu, Hanmei Liu, Junjie Zhang and Yubi Huang
The Proteomic Analysis of Maize Endosperm Protein Enriched by Phos-tag[tm] Reveals the Phosphorylation of Brittle-2 Subunit of ADP-Glc Pyrophosphorylase in Starch Biosynthesis Process
Reprinted from: *Int. J. Mol. Sci.* **2019**, *20*, 986, doi:10.3390/ijms20040986 . . . . **299**

Haixia Guo, Huihui Guo, Li Zhang, Yijie Fan, Yupeng Fan, Zhengmin Tang and Fanchang Zeng
Dynamic TMT-Based Quantitative Proteomics Analysis of Critical Initiation Process of Totipotency during Cotton Somatic Embryogenesis Transdifferentiation
Reprinted from: *Int. J. Mol. Sci.* **2019**, *20*, 1691, doi:10.3390/ijms20071691 . . . . **316**

Zhongyuan Lin, Cheng Zhang, Dingding Cao, Rebecca Njeri Damaris and Pingfang Yang
The Latest Studies on Lotus (*Nelumbo nucifera*)-an Emerging Horticultural Model Plant
Reprinted from: *Int. J. Mol. Sci.* **2019**, *20*, 3680, doi:10.3390/ijms20153680 . . . . **350**

Li-Qin Li, Cheng-Cheng Lyu, Jia-Hao Li, Zhu Tong, Yi-Fei Lu, Xi-Yao Wang, Su Ni, Shi-Min Yang, Fu-Chun Zeng and Li-Ming Lu
Physiological Analysis and Proteome Quantification of Alligator Weed Stems in Response to Potassium Deficiency Stress
Reprinted from: *Int. J. Mol. Sci.* **2019**, *20*, 221, doi:10.3390/ijms20010221 . . . . **363**

Huihui Guo, Haixia Guo, Li Zhang, Zhengmin Tang, Xiaoman Yu, Jianfei Wu and Fanchang Zeng
Metabolome and Transcriptome Association Analysis Reveals Dynamic Regulation of Purine Metabolism and Flavonoid Synthesis in Transdifferentiation during Somatic Embryogenesis in Cotton
Reprinted from: *Int. J. Mol. Sci.* **2019**, *20*, 2070, doi:10.3390/ijms20092070 . . . . **381**

Lei Dong, Lei Qin, Xiuru Dai, Zehong Ding, Ran Bi, Peng Liu, Yanhui Chen, Thomas P. Brutnell, Xianglan Wang and Pinghua Li
Transcriptomic Analysis of Leaf Sheath Maturation in Maize
Reprinted from: *Int. J. Mol. Sci.* **2019**, *20*, 2472, doi:10.3390/ijms20102472 . . . . **404**

Zhuang Yang, Wen Li, Xiao Su, Pingfei Ge, Yan Zhou, Yuanyuan Hao, Huangying Shu, Chonglun Gao, Shanhan Cheng, Guopeng Zhu and Zhiwei Wang
Early Response of Radish to Heat Stress by Strand-Specific Transcriptome and miRNA Analysis
Reprinted from: *Int. J. Mol. Sci.* **2019**, *20*, 3321, doi:10.3390/ijms20133321 . . . . **417**

Ye Jin, Lin Liu, Xuehong Hao, David E. Harry, Yizhi Zheng, Tengbo Huang and Jianzi Huang
Unravelling the MicroRNA-Mediated Gene Regulation in Developing Pongamia Seeds by High-Throughput Small RNA Profiling
Reprinted from: *Int. J. Mol. Sci.* **2019**, *20*, 3509, doi:10.3390/ijms20143509 . . . . **439**

Jichao Tang, Zhigui Sun, Qinghua Chen, Rebecca Njeri Damaris, Bilin Lu and Zhengrong Hu
Nitrogen Fertilizer Induced Alterations in The Root Proteome of Two Rice Cultivars
Reprinted from: *Int. J. Mol. Sci.* **2019**, *20*, 3674, doi:10.3390/ijms20153674 . . . . **456**

Qi Song, Fang Lv, Muhammad Tahir ul Qamar, Feng Xing, Run Zhou, Huan Li and Ling-Ling Chen
Identification and Analysis of Micro-Exon Genes in the Rice Genome
Reprinted from: *Int. J. Mol. Sci.* **2019**, *20*, 2685, doi:10.3390/ijms20112685 . . . . **476**

Meichao Ji, Kun Wang, Lin Wang, Sixue Chen, Haiying Li, Chunquan Ma and Yuguang Wang
Overexpression of a S-Adenosylmethionine Decarboxylase from Sugar Beet M14 Increased *Araidopsis* Salt Tolerance
Reprinted from: *Int. J. Mol. Sci.* **2019**, *20*, 1990, doi:10.3390/ijms20081990 . . . . **490**

Rendi Ma, Wangyang Song, Fei Wang, Aiping Cao, Shuangquan Xie, Xifeng Chen, Xiang Jin and Hongbin Li
A Cotton (*Gossypium hirsutum*) *Myo*-Inositol-1-Phosphate Synthase (*GhMIPS1D*) Gene Promotes Root Cell Elongation in *Arabidopsis*
Reprinted from: *Int. J. Mol. Sci.* **2019**, *20*, 1224, doi:10.3390/ijms20051224 . . . . **504**

Qi Jia, Defeng Kong, Qinghua Li, Song Sun, Junliang Song, Yebao Zhu, Kangjing Liang, Qingming Ke, Wenxiong Lin and Jinwen Huang
The Function of Inositol Phosphatases in Plant Tolerance to Abiotic Stress
Reprinted from: *Int. J. Mol. Sci.* **2019**, *20*, 3999, doi:10.3390/ijms20163999 . . . . **521**

About the Special Issue Editors

Tingyun Kuang, born in 1934, is the Academician of Chinese Academy of Sciences, chairman of the Chinese Botanical Society. Since the 1960s, she has been mainly engaged in research on photosynthesis. She has published more than 400 papers. She is the chief scientist of the National Key Basic Research and Development Program (973), "The Mechanism of High-efficiency Light Energy Conversion in Photosynthesis and Its Application in Agriculture".

Xuchu Wang got his doctor's degree and majored in Plant Development Science from the Institute of Botany in Chinese Academy of Sciences. He is one of the five Country Representatives for China in International Plant Proteomics Organization (INPPO) (http://www.inppo.com/country.jsp). Now, he is a professor in College of Life Sciences, Hainan Normal University. As a group leader of Proteomics of Tropical Biology, he guides his group mainly focused on the uncovering the molecular mechanism of latex biosynthesis in the rubber tree Hevea brasiliensis and the regulation network of salt tolerance in halophytes. In the coming years, Dr. Xuchu Wang's group will focus on the quantitative proteomics, and then to determine the detail biological functions of several key factors for the specific modified proteins.

Xiaochun Qin obtained her Ph.D. in Botany in 2007 from Photosynthesis Research Center, Institute of Botany, Chinese Academy of Sciences (IBCAS). She had been an Assistant professor from 2007 and an Associate professor from 2016 in IBCAS anda Professor in School of Biological Science and Technology, University of Jinan from 2016. Dr. Qin focuses on structural study of photosynthetic protein-pigments super-complexes to gain insights into their structure-function relationship.

Shaojun Dai obtained his Ph.D degree in Northeast Forestry University, China, in 2002. He accomplished the postdoctoral training in the Institute of Botany, Chinese Academy of Sciences in 2006, and in University of Florida in 2009. Now, he is a professor in Shanghai Normal University. He is the secretary general of Asia Ocean Agricultural Proteomics Organization (AOAPO). He is also a proteomics committee member in China Human Proteomics Organization (CNHUPO), and the Leader of Plant Proteomics Committee in CNHUPO. Dr. Dai's laboratory mainly focuses on the functional proteomics of plant in response to stress.

Pingfang Yang got his bachelor's degree and majored in Economic Forestry from Northwest Sci-Tech University of Agriculture and Forestry in 1997, then his master's degree from Huazhong Agriculture University for Plant Embryology in 2002, and finally obtained his Ph.D degree in Institute of Botany, Chinese Academy of Sciences, China, in 2006. From 2005 to 2007, he worked as a visiting scholar in Hong Kong University of Science and Technology. Following this, he accomplished the postdoctoral training in the Plant Research Laboratory in Michigan State University in USA. Dr. Yang's lab is focused on studying the molecular mechanism of seed dormancy and proteomics of plant sexual reproduction.

Ling Li obtained her B.S. in biology and M.S. in Botany from Peking University in 1997 and 2000. She got her Ph.D. in Genetics (minor in Statistics) from Iowa State University in 2006. She had been an Adjunct Assistant Professor in the Department of Genetics, Development, and Cell Biology at Iowa

State University from 2011. She has been an Assistant Professor in the Department of Biological Sciences at Mississippi State University from 2017. Dr. Li has been developing an integrated experimental/biocomputational approach to identify the factors that regulate plant metabolism and plant adaptation to environmental changes, bridging basic research from Arabidopsis and application in crops.

Preface to "Plant Protein and Proteome Altlas–Integrated Omics Analyses of Plants under Abiotic Stresses"

With the annotation of genomes for thousands of plant species, plant biology study is dawning upon the post-genomic era. Biochemical, physiological and molecular studies have paved the way toward a comprehensive understanding of the complex biological processes for plants in response to stress conditions. In this new post-genomic time, atlas analysis of plants under different abiotic stresses has become more important for uncovering the potential key genes and proteins in different plant tissues. High-quality genomic data and integrated analyses of transcriptomic, proteomic, metabolomic and phenomic patterns have provided deeper understanding of how plants grow and survive under environmental stresses. Therefore, we edited this book titled, "Plant Protein and Proteome Atlas–Integrated Omics Analyses of Plants under Abiotic Stresses", based on the publication of 27 papers in the Special Issue in International Journal of Molecular Sciences in the year of 2019. This book is focused on concluding the recent progress in the Protein and Proteome Atlas in plants under different stresses. Based on the Integrated Omics Analyses of plants, it covers various aspects of plant protein ranging from agricultural proteomics, structure and function of proteins, novel techniques and approaches for protein identification and quantification, novel techniques for PTMs, and new insights into proteomics.

In this book, we included crop plants as well as model plant species for fundamental research of stress physiology and biochemistry. A total of 27 papers including two timely reviews were included into five parts in this special book. Part one (Page 8-123) begins with the topic of "Comparative Proteomics of Different Plants" and was edited by Dr. Xuchu Wang, from the College of Life Sciences in Hainan Normal University. In this part, six papers were included to describe the phenotypic changes and proteomic analyses of different plants under different conditions. These findings will allow to gain deeper insights into the stress-mitigating mechanisms in plants for translation into high productivity and stress tolerance ability. The second part (Page 124-248), focused on the topic of "Proteomics of Plants under Osmotic Stress" was edited by Dr. Shaojun Dai from the College of Life and Environmental Sciences in Shanghai Normal University, and includes six papers to introduce the recent comparative proteomics analyses of plants under osmotic stress, particularly the drought and salinity stresses in leaves of certain plant species. The third part (Page 249-362), on the topic of "Quantitative Proteomics and Phosphoproteomics of Crops", was edited by Dr. Pingfang Yang from the College of Life Sciences in Hubei University. In this part, five papers were included to study the proteomics of several plants including an energy plant for biodiesel Jatropha curcas, a biofuel tree Pongamia, and three economic crops.: radish, cotton and maize. The fourth part (Page 363-455), entitled "Integrative Omics of Plants during Development", was edited by Dr. Ling Li from the Department of Biological Sciences in Mississippi State University in the USA. Five papers using omics tools were included to demonstrate the recent omics studies on different plants during their development processes. The fifth and final section (Page 456-535), edited by Dr. Xiaochun Qin from the College of Life Science and Technology, in the University of Jinan, was set to identify more stress responsive genes and proteins in plants. In this section, five papers were included to determine the genes and proteins for regulation of the growth and development of plants.

1. Comparative Proteomics of Different Plants (Page 8-123)

Various plants provide a source of the main necessities of life for human beings. However, because of the environmental constraints of the nature, these plant species must cope with different abiotic and biotic stressed conditions during their whole life time; these stresses include high salinity, water logging, cold, drought, heat, UV radiation, heavy metals, anaerobic and toxic conditions, etc. It has become increasingly important for uncovering the potential key genes and proteins in different plant tissues under different stressed environments. In the first section of this book, we focused on the phenotypic changes and proteomics analyses of the six plant species including spinach, Brassica rapa, wheat, diatom, lily and rubber grass under different stressed conditions.

Investigating heat-responsive molecular mechanisms is important for breeding heat-tolerant crops. In this Section, a comparative proteomics of the leaves under both low-temperature and high-temperature stressed conditions in a heat-sensitive spinach (Spinacia oleracea L.) variety Sp73 has been performed by using 2DE and iTRAQ approaches, and a total of 257 heat-responsive proteins were identified by Li and coworkers. Their proteomics data demonstrated that both the photosynthesis process and reactive oxygen species (ROS) scavenging pathways were inhibited, whereas protein synthesis and turnover, carbohydrate and amino acid metabolism were promoted in the spinach Sp73 in response to high temperature. Comparison of the proteomics data in this study with that in a previous published study in a heat-tolerant spinach variety Sp75 revealed that the heat-decreased biosynthesis of chlorophyll, carotenoid and soluble sugar contents in the variety Sp73 was quite different from that in the variety Sp75. A similar proteomics study performed by Yuan and colleagues in this Section was focused on determining the heat-responsive proteins in a tolerant variety Wucai (Brassica campestris L.) to both low and high temperature stressed conditions. Wucai is a species of non-heading Chinese cabbage and becoming one of the most important leafy vegetables in China, and is also now extensively cultivated worldwide. Comparative proteomics resulted in 1022 differentially expressed proteins in Wucai under temperature stressed conditions, and most of these proteins were identified to be associated with redox homeostasis, photosynthesis, carbohydrate metabolism, and heat-shock response. These proteomics data also demonstrated that maintaining redox homeostasis is an important common regulatory pathway for tolerance to temperature stress in novel Wucai germplasm.

In wheat plants, pollen, as a highly specialized organ, develops in the anther. In the third paper, comparative cytological and proteomic analyses were conducted to better understand the mechanism on chemical hybridizing agent CHA-SQ-1 induced pollen abortion in wheat, and noticed pollen grains underwent serious structural injury in the CHA-SQ-1-treated plants. Finally, 60 proteins showed statistically significant differences in abundance were successfully identified by mass spectrometry. Gene ontology and pathway analyses of these proteins indicated that most proteins were involved in carbohydrate and energy metabolism. Furthermore, an iTRAQ-based proteomics was performed to explore the distinct cellular responses associated with oxidative stress in the diatom Skeletonema dohrnii to the silicate limitation. Diatoms are a successful group of marine phytoplankton that often thrives under adverse environmental stress conditions. In this proteomic study, 594 differentially expressed proteins were determined from 1768 proteins, and most down-regulated proteins were related to photosynthesis metabolism, light-harvesting complex, oxidative phosphorylation, inducing oxidative stress, and ROS accumulation. The proteomic data revealed that ATP-limited diatoms are unable to rely on photosynthesis, duo to break down of carbon metabolism to compensate for photosynthetic carbon fixation losses. The lily is a popular ornamental flower from ancient times to today and is often used in flower arrangement. Both hydrogen gas (H2) and nitric oxide (NO) could enhance the postharvest freshness of Cut Lilies. However, the deep mechanism of H2 in delaying the senescence and shelf life of perishable horticultural products needs to be further investigated. Comparative proteomics were used to investigate the relationship between hydrogen gas and NO and identified 50 differentially accumulated proteins during postharvest freshness in Cut Lilies leaves, which were classified into seven functional categories. These proteomics data suggested that NO and a chloroplastic alpha subunit of ATP synthase might play important roles during the process of hydrogen gas-improved freshness of cut lilies. At the end of this section, a proteomic landscape in the mature roots of a rubber-producing grass Taraxacum Kok-saghyz (TKS) was provided by Xie and coworkers. The rubber grass TKS contains large amounts of natural rubber (cis-1,4-polyisoprene) in its enlarged roots and it is an alternative crop source of natural rubber, its genome has just been annotated and many NRB-related genes have been determined. However, there is limited knowledge about the protein regulation mechanism for natural rubber biosynthesis (NRB) in TKS roots. The authors identified 371 protein species from the mature roots of TKS by combining 2-DE and MS. Meanwhile, a large-scale shotgun analysis of proteins in TKS roots at the enlargement stage was performed, and 3545 individual proteins were determined. Fifty-eight NRB-related proteins, including eight small rubber particle protein (SRPP) and two rubber elongation factor (REF) members, were identified from the TKS roots, and these proteins were involved in both mevalonate acid (MVA) and methylerythritol phosphate (MEP) pathways, which proved MVA and MEP pathways are important for NRB in TKS roots. It is the first high-resolution draft proteome map of the mature TKS roots, which provides new evidence on the roles of proteins in NRB.

Altogether, the six papers in this section focus on the phenotypic changes and proteomic analyses of different plants under different conditions. These findings will allow to gain deeper insights into the stress-mitigating mechanisms in plants for translation into high productivity and stress tolerance ability.

2. Proteomics of Plants under Osmotic Stress (Page 124-248)

Drought and salinity are two serious kinds of osmotic stresses that inhibit plant growth and crop yields. Recent comparative proteomics analyses have provided more information for understanding the drought- and salinity- responsive mechanisms in certain plant species. In this section, we included six papers and focus on proteomics analyses of plants under osmotic stress.

Drought, as an important abiotic stress, can seriously limits crop yields. Understanding of the drought tolerance mechanisms offers guidance for the genetic improvement of drought tolerance in this crop. To understand how a well-known wheat genotype Jinmai 47 responds to drought, Wang and colleagues adopted the iTRAQ and LC/MS approaches to determine the different proteins in wheat leaves after exposure to PEG, and finally identified 176 differentially expressed proteins (DEPs). Functional analysis of these DEPs indicated that wheat genotype Jinmai 47 can increase TCA cycle, plasma dehydration protection, and protein structure protection, as well as the biosynthesis of sucrose and proline to cope with polyethylene PEG-induced drought stress. In another drought-tolerant wheat variety, 335 drought-responsive proteins were found to mainly take part in photosynthesis, carbon fixation, glyoxylate and dicarboxylate metabolism. Among these, two proteins named TaDrSR1 and TaDrSR2 were proven to be important for drought tolerance by using virus-induced gene silencing technology. In addition to this, comparative analysis of filling-kernel proteomes from two maize inbred lines with different drought-tolerant ability identified 5175 drought-responsive proteins, which indicated that drought tolerant variety was attributable to its enhanced redox modification, epigenetic regulation, energy metabolism, secondary metabolites biosynthesis, and seed storage proteins. This proteomics study performed by Wang and coworkers presents an elaborate understanding of drought-responsive proteins and metabolic pathways mediating maize filling-kernel drought tolerance, and provides important candidate genes for subsequent functional validation. Proteomics results also implied that wood vinegar treatment can enhance the drought tolerance of wheat root through promoting stress response, carbohydrate metabolism, protein metabolism, and secondary metabolism.

Salt stress is one of the most important abiotic stresses and limiting factors for plant growth and agricultural production in the world. Interestingly, the salinity stress to plant can be alleviated by arbuscular mycorrhizal fungi. To reveal the mechanism of salinity stress alleviation by arbuscular mycorrhizal fungi, a label-free quantitative proteomics was performed to find the stress-responsive proteins in the leaves of E. angustifolia. The proteomics results revealed that E. angustifolia seedlings inoculated with arbuscular mycorrhizal fungi increased the phenylpropane metabolism, enhanced the signal transduction of Ca2+ and ROS scavenging, promoted the protein biosynthesis and folding, and inhibited the protein degradation under salt stress. This proteomics study implied that symbiosis of halophytes and arbuscular mycorrhizal fungi has potential as an application for the improvement of saline-alkali soils. An iTRAQ-based proteomics study was conducted on D. salina during early response to salt stress, and identified 141 differentially abundant proteins (DAPs). Functional analysis of these DAPs revealed photosynthesis and ATP synthesis were crucial for the modulation of early salinity-responsive pathways. These proteomics results presented an overview of the systematic molecular response to salt stress in D. salina and and potentially contributes to developing strategies to improve stress resilience.

Therefore, improving osmotic tolerance of crops such as rice, wheat, tomato, potato and other plants to salinity and drought stress has as yet not been realized by molecular engineering, and the need to obtain such highly salt and drought tolerant crops remains in a world with a rapidly growing population and a decreasing availability of fresh water for agriculture. Proteomics of plants under salinity and drought stress present an elaborate understanding of stress-responsive proteins and metabolic pathways and these osmotic-response proteins and genes are important candidates for further functional validation.

3. Quantitative Proteomics and Phosphoproteomics of Crops (Page 249-362)

Plant propagation and development are two important aspects that could largely determine the life cycle and economic values of different crops. Meanwhile, crops are always exposed to different biotic and abiotic stresses during their life span because of the sessile feature. In addition to the traditional crops, some plants showing special economic usage or stress resistance are also worth of studying, which might help to obtain not only in-depth understanding on some special traits in plant kingdom, but also very good plant germplasm. The plant materials used include energy plant for biodiesel (Jatropha curcas), a biofuel tree Pongamia (Millettia pinnata syn. Pongamia pinnata), and three economic crops: radish, cotton and maize. They were studied under stress conditions such as chilling stress and heat stress, or for research in development for leaf sheath maturation, somatic embryogenesis, and developing seeds. Liu et al. used combined analyses of the phosphoproteomics, physiological characteristics and ultrastructure studies to identify the responses of J. curcas seedling under chilling. They revealed significantly changed phosphoproteins, several protein kinases' phosphorylation in signaling pathways, and possible phosphorylation in chilling stress.

It is known that salinity is a serious challenge that constrains the production of maize, especially at the seedling stage. Comparative study between the salt resistance and sensitive cultivars with very close genetic background will undoubtedly help to explore the key regulator. Based on this idea, Zhao et al. conducted phosphorproteomic analysis between two maize inbred lines showing different resistance to salt stress. It seems that enhancement of potassium and sodium transportation, carbon and redox related metabolism could increase the salt resistance in maize. Maize is one of the three most important cereal crops with highest yield in the world. It provides the stable energy for the world population, which is largely determined by its kernel starch content. The pathway of starch biosynthesis has been constructed, with the characterization of 4 enzymes, AGPase, starch synthase, starch branch enzyme and starch debranch enzyme. Among them, AGPase catalyzes the rate-limiting reaction. However, regulation on AGPase activity has not yet been fully elucidated. In this section, Yu et al. applied a newly developed technique: Phos-tagTM technology. This helped the identification of 21 phosphorylated peptides of AGPase, which was further proven to be positively related to the enzyme activity in the study. This indicates a novel regulatory mechanism on AGPase.

Cotton is the most important natural fiber resource in the world, which makes it an important crop. Studies have been widely conducted focusing on many aspects of this crop. Although genetic modified cotton cultivars have been generated and widely used in its production, transformation of cotton could only be succeeded in very limited germplasm with low efficiency. During the induction of cotton somatic embryo, most of the embryos either redifferentiate in to calli or become necrotic, with very few being able to mature and regenerate into plant. It is very important to identify the key regulators that determine the fate of somatic embryos. Although several proteins have been identified in Arabidopsis, it is obviously that there is big difference between cotton and this model plant. Comparative proteomics seems to be a powerful method, which was successfully applied among the embryos destined into three fates.

Additionally, it also contains some special features, such as seed longevity, self-cleaning (lotus-effect) and floral thermogenesis. Since the releasing of its genome information in the year of 2013, more and more studies were conducted on this economic important plant. To obtain a comprehensive knowledge, it is necessary to summarize the latest studies and advancements, and perspective the future study and breeding on different aspects of this plant. Lotus is a perennial aquatic basal eudicot, an important horticultural plant, utilizing from ornamental, nutritional to medicinal values. Due to the important values, it has obtained a lot of attention from the scientific community recently, and might be an important model plant in horticulture. Lin and colleagues reviewed the latest advancement studies of lotus, including phylogeny, genomics and the molecular mechanisms underlying its unique properties, its economic important traits, and so on, providing the information of basic use of omics.

We believe that the proteomics-based results in this section will help the readers to gain novel insights for the understanding of complicated physiological processes in crops and other important plants. The identification of target genes (or proteins) may decipher the complex relationship between genes, proteins, metabolites, and their biological functions.

4. Integrative Omics of Plants during Development (Page 363-455)

Integrative Omics tools, including genomics, transcriptomics, proteomics and metabolomics, are very powerful to study the molecular basis of biological activity between biomolecules (DNA, RNA, proteins and metabolites). Integrative Omics methods has been widely used to study biomolecules for their interactions, biosynthesis, and the regulation of these interactions in the various systems of plants, plant development and their interaction with various environments. In this section, five papers using omics tools are included to demonstrate the recent omics studies on different plants during their development process.

Some uncanonical plants may have special features showing either high resistance to external stresses or economic values. Alligator weed is a dicotyledonous perennial herb that grows worldwide due to its high adaptability to harsh environments. Although, it is listed as an invasive species in many areas, study on the mechanism underlying its high adaptability will be very helpful in the breeding of other crops. Its high tolerance to potassium deficiency seems very important since potassium is essential for plant growth and development. In this section, Li and coworkers investigated the physiological and proteomic changes in alligator weed stems under low potassium stress, and identified 296 differentially abundant proteins, which were mainly involved in oxidative phosphorylation, plant-pathogen interactions, glycolysis, sugar metabolism, and transport in stems. These proteomics data provide valuable information on the adaptive mechanisms in alligator weed stems under low potassium stress and facilitate the development of efficient strategies for genetically engineering potassium-tolerant crops.

Combined metabolomic and transcriptomic analyses revealed dynamic regulation of purine metabolism and flavonoid synthesis in transdifferentiation during somatic embryogenesis is crucial for cotton regeneration via somatic embryogenesis. A total of 581 metabolites were present in embryogenic calli, and metabolites related to purine metabolism were significantly enriched. An association analysis of the metabolome and transcriptome data indicated that purine metabolism and flavonoid biosynthesis were co-mapped, and purine metabolism-related genes associated with signal recognition, transcription, stress, and lipid binding were significantly upregulated. These omics data provide a valuable foundation for a deeper understanding of the regulatory mechanisms underlying cell totipotency at the molecular and biochemical levels.

Furthermore, a transcriptomic analysis of leaf sheath maturation in maize was performed by Dong and coworkers, and they discovered leaf sheath transcriptome has dynamic perturbation and the processes and genes that are involved in sheath maturation and organ specialization. Although extensive transcriptional profiling of the tissues along the longitudinal axis of the developing maize leaf blade has been conducted, little is known about the transcriptional dynamics in sheath tissues. Using a comprehensive transcriptome dataset, the authors identified 15 genes expressed at significantly higher levels in the leaf sheath compared with their expression in the leaf blade. These genes may important for sheath maturation and organ specialization. By an integrative analysis of transcriptome and small RNA sequencing data from the radish young leaves under short-term heat stress, Yang and coworkers discovered differentially expressed mRNAs and potential pathways involved in heat stress in radish leaves. They detected 1802 differentially expressed mRNAs and 169 differentially expressed lncRNAs as well as three differentially expressed circRNAs through strand-specific RNA sequencing technology. They also identified one lncRNA–miRNA–mRNAs combination responsive to heat stress. These results will be helpful for further illustration of molecular regulation networks of how radish responds to heat stress. Jin et al. unraveled 1327 microRNA-mediated genes in regulation of Pongamia seeds by high-throughput small RNA profiling at three developmental stages and identified 236 conserved miRNAs within 49 families. They also detected 143 novel miRNAs within the families by deep sequencing of Pongamia seeds sampled at three developmental phases. These results provide valuable miRNA candidates for further functional characterization and breeding practice in Pongamia and other oilseed plants.

All these findings in the five papers may broaden basic knowledge of our understanding of molecular regulatory networks and may provide valuable information for further study in the molecular breeding in economic plants and crops.

5. Identification of Stress Responsive Genes and Proteins in Plants (Page 456-535)

With the development of sequencing technology and molecular biology, studies on plant growth and development, environmental adaptation and other mechanisms are no longer restricted to physiological phenomenon; they are more widely to be used to identify the stress responsive genes and proteins in different plants to decipher the intrinsic regulating mechanisms from transcriptome and proteome level. In this section, five papers were included to determine the genes and proteins for regulation of the growth and development of plants.

Nitrogen is an essential nutrient for plants and a key limiting factor of crop production. Tang and colleagues explored proteomics analysis the regulatory mechanism of nitrogen fertilization in cereal crops, and determined 511 differentially expressed proteins among the identified 6093 proteins in the two rice cultivars after nitrogen fertilizer treatment. These differentially expressed proteins were mainly involved in ammonium assimilation, amino acid metabolism, carbohydrate metabolism, etc. These proteomics results provided new insights in identifying potential molecular protein markers to assist the breeding for high nitrogen use efficiency cultivars. Micro-exons, a set of small exons with lengths no more than 51 nucleotides, distributed widely in various species, have special splicing properties thus be good objects investigating mechanism of splicing. They are of vital importance, functioning in multiple ways including alternative splicing, such as degradation of the transcripts via nonsense-mediated decay, altering protein domain architecture, and introducing novel post-translational modification sites. Therefore, micro-exons have been widely concerned and studied in recent years. Song and coworkers investigated the distribution, evolutionary conservation, and potential functions of micro-exons in the whole genome of two indica rice varieties of Zhenshan 97 and Minghui 63, improving the understanding of micro-exons and may contribute to gene expression regulation.

Polyamines play an important role in plant growth and development, and response to abiotic stresses. As deterioration of the ecological environment, such as soil salinization, drought and heat stresses, photosynthetic capacity of plants and crop yields are being severely restricted, leading it the hot spot issue to be widely concerned. Ji and coworkers applied proteomics of Sugar Beet M14 under salt stress, and found S-adenosylmethionine decarboxylase (SAMDC) protein differentially expressed. Further analysis of the function of SAMDC in Arabidopsis thaliana showed that this protein enhancing salt tolerance via mediating the biosynthesis of spermidine and spermin. Together, these proteomics results suggest that the BvM14-SAMDC mediated biosynthesis of Spm and Spd contributes to plant salt stress tolerance through enhancing antioxidant enzymes and decreasing ROS generation.

It is widely known that Myo-inositol-1-phosphate synthase plays important roles in plant growth and development, stress responses, and cellular signal transduction. Ma and coworkers focused on a cotton Myo-inositol-1-phosphate synthase gene, and find it can rescue the abnormal phenotype and promote the root cell elongation as a positive regulator. Inositol signaling is believed to play a crucial role in various aspects of plant growth and adaptation. As an important component in biosynthesis and degradation of myo-inositol and its derivatives, inositol phosphatases could hydrolyze the phosphate of the inositol ring, thus affecting inositol signaling. At the end of this section, a review on the function of inositol phosphatases in plant tolerance to abiotic stress was included. In this timely review, the authors concluded the functions of more than 30 members of inositol phosphatases in plants, and revised some current knowledge in relation to their substrates and function in response to abiotic stress. The potential mechanisms were also concluded with the focus on their activities of inositol phosphatases.

All these findings in the five parts may broaden basic knowledge of our understanding of molecular regulatory networks and may provide valuable information for further study in the molecular breeding in economic plants and crops. We hope the readers will enjoy these articles and believe that the proteomics-based results in this book will help the readers to gain novel insights for the understanding of complicated physiological processes in crops and other important plants. The identification of target genes and proteins may decipher the complex relationship between genes, proteins, metabolites, and their biological functions.

Tingyun Kuang, Xuchu Wang, Xiaochun Qin, Shaojun Dai, Pingfang Yang, Ling Li
Special Issue Editors

Editorial

Protein and Proteome Atlas for Plants under Stresses: New Highlights and Ways for Integrated Omics in Post-Genomics Era

Xuchu Wang

Key Laboratory for Ecology of Tropical Islands, College of Life Sciences, Ministry of Education, Hainan Normal University, Haikou 571158, China; xchwang@hainnu.edu.cn; Tel.: +86-898-6589-1065

Received: 15 October 2019; Accepted: 16 October 2019; Published: 21 October 2019

Abstract: In the post-genomics era, integrative omics studies for biochemical, physiological, and molecular changes of plants in response to stress conditions play more crucial roles. Among them, atlas analysis of plants under different abiotic stresses, including salinity, drought, and toxic conditions, has become more important for uncovering the potential key genes and proteins in different plant tissues. High-quality genomic data and integrated analyses of transcriptomic, proteomic, metabolomics, and phenomic patterns provide a deeper understanding of how plants grow and survive under environmental stresses. This editorial mini-review aims to synthesize the 27 papers including two timely reviews that have contributed to this Special Issue, which focuses on concluding the recent progress in the Protein and Proteome Atlas in plants under different stresses. It covers various aspects of plant proteins ranging from agricultural proteomics, structure and function of proteins, novel techniques and approaches for gene and protein identification, protein quantification, proteomics for post-translational modifications (PTMs), and new insights into proteomics. The proteomics-based results in this issue will help the readers to gain novel insights for the understanding of complicated physiological processes in crops and other important plants in response to stressed conditions. Furthermore, these target genes and proteins that are important candidates for further functional validation in economic plants and crops can be studied.

Keywords: integrated omics; plants under stress; post-genomics era; proteome atlas; quantitative proteomics

---

With the annotation of genomes for human [1] and hundreds of plant species, plant biology study is dawning upon the post-genomics era [2]. Biochemical, physiological, and molecular studies have paved the way toward a comprehensive understanding of the complex biological processes for plants in response to stress conditions. Among them, abiotic stresses are the foremost limiting factors for plant survival and development [3]. Different from animals, plants cannot move away from the stressed conditions and they must cope with all kinds of adverse external pressures via their intrinsic biological mechanisms [3]. In this new post-genomics time, atlas analysis of plants under different abiotic stresses, including salinity, water logging, cold, drought, heat, UV radiation, heavy metals, anaerobic, and toxic conditions in the root zone, among others, has become more important for uncovering the potential key genes and proteins in different plant tissues [4]. High-quality genomic data and integrated analyses of transcriptomic, proteomic, metabolomics, and phenomic patterns (Figure 1) provide a deeper understanding of how plants grow and survive under environmental stresses [5].

We organized sixteen researchers in the omics study area to edit this Special Issue titled "Plant Protein and Proteome Atlas: Integrated Omics Analyses of Plants under Abiotic Stresses", and this Special Issue book is edited based on these published papers. Therefore, this special book is focused on concluding the recent progress in the Protein and Proteome Atlas in plants under different stresses.

Integrated omics analysis covers various aspects of plant protein ranging from agricultural proteomics, structure and function of proteins, novel techniques and approaches for gene and protein identification, protein quantification, proteomics for post-translational modifications (PTMs), and new insights into proteomics. Proteome atlas aims to compare the quantified relative abundances of the genome-wide genes and proteins across different plant tissues or subcellular compartments. Large-scale analyses of post-translational modifications in proteins, such as phosphoproteomics, glycoproteomics, and ubiquiproteomics, have become more imperative to define and interpret the plant–environment relationships in terms of protection against abiotic stresses in multi-layers [5]. It helps to gain novel insights into the identification of target genes and proteins, which may decipher the complex relationship between genes, proteins, metabolites, and their biological functions. At the same time, combining the big-data-based multi-omics approaches and traditional molecular biology technologies gains deeper insights into the stress-mitigating mechanisms in plants for translation into higher productivity [5].

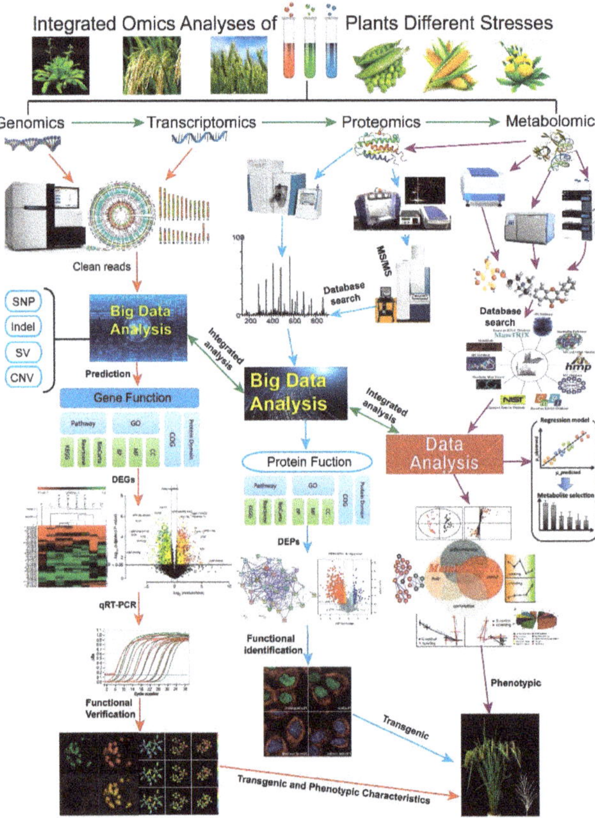

Figure 1. Integrated omics analyses of plants under different stresses. On the basis of the high-quality genomic data, integrated analyses using transcriptomic, proteomic, metabolomics, and phenomic methods have recently been performed in different plant species under different biotic and abiotic stressed conditions to determine their stress responsive genes and proteins, after which functional analyses of these target genes and proteins are conducted by molecular and biochemical methods. These integrated data have provided a deeper understanding of how plants grow and survive under different environments.

In this issue, we included crop plants as well as model plant species for fundamental research of stress physiology and biochemistry. The topics focus on integrative analyses of quantitative protein change in plants under abiotic stresses; transcriptomic, proteomic, metabolomics, and phenomic analyses of plant species and tissues under abiotic stresses; plant proteome atlas of different tissues and cell compartments; post-translational modifications in plant proteins upon stressed conditions; bioinformatics and computational tools for analyzing big data via various omics approaches; genetic and phenomic studies of plant species in different environments; and identification and functional validation of key genes and proteins obtained from integrative omics approaches in response to stresses in plants.

Finally, a total of 27 papers including two timely reviews were divided into five parts in this special book. Part 1 was edited by Dr. Xuchu Wang from Hainan Normal University. This part contains six papers describing the comparative proteomics of different plants under different conditions. Proteomics analyses were performed on six plant species, including spinach [6], *Brassica rapa* [7], wheat [8], diatom [9], lily [10], and a rubber grass *Taraxacum Kok-saghyz* [11], in order to gain deeper insights into the stress-mitigating mechanisms in plants under different conditions. Li et al. (2019) performed a comparative proteomics analysis of the leaves under both low-temperature and high-temperature stressed conditions in heat-sensitive spinach and identified 257 heat-responsive proteins [6]. Their proteomics data revealed that both the photosynthesis process and reactive oxygen species scavenging pathways are inhibited in response to high temperature stress. A similar proteomics study performed by Yuan and colleagues resulted in 1022 differentially expressed proteins in Wucai under temperature stressed conditions, and most of these proteins were identified to be associated with redox homeostasis, photosynthesis, carbohydrate metabolism, and heat-shock response [7]. In wheat plants, pollen, as a highly specialized organ, develops in the anther. Comparative cytological and proteomic analyses were conducted by Wang et al. (2019) to better understand the mechanism on the chemical hybridizing agent induced pollen abortion in wheat, and determined 60 significant different proteins [8]. Furthermore, Thangaraj et al. (2019) performed an iTRAQ-based proteomics analysis to explore the distinct cellular responses associated with oxidative stress in the diatom *Skeletonema dohrnii*, and determined 594 differentially expressed proteins from 1768 proteins. Their proteomics data also revealed that ATP-limited diatoms are unable to rely on photosynthesis, owing to break down of carbon metabolism to compensate for photosynthetic carbon fixation losses [9]. Comparative proteomics were used to investigate the relationship between hydrogen gas and NO, and identified 50 differentially accumulated proteins during postharvest freshness in the Cut Lilies leaves [10]. Xie and coworkers [11] provided a proteomic landscape of the mature roots of a rubber-producing grass *Taraxacum Kok-saghyz* and identified 371 protein species from the mature roots by combining 2-DE and MS. Meanwhile, 3545 individual proteins were determined by a large-scale shotgun proteomics analysis of the enlargement roots, and fifty-eight natural rubber biosynthesis-related proteins were identified; these proteins were involved in both mevalonate acid and methylerythritol phosphate pathways [11]. This is the first high-resolution draft proteome map of the mature roots of rubber grass.

Furthermore, proteomics analyses of different plants under both salinity and drought stresses were conducted to find the osmotic response proteins [12–17]. Drought and salinity are two serious kinds of osmotic stresses that inhibit plant growth and crop yields. Recent comparative proteomics analyses have provided more information for understanding the drought- and salinity-responsive mechanisms in certain plant species. In this part, we included six papers and focused on proteomics analyses of plants under osmotic stress. Drought, as an important abiotic stress, can seriously limits crop yields. Wang and colleagues adopted iTRAQ and LC/MS approaches to determine the different proteins in wheat leaves after exposure to PEG, and finally identified 176 differentially expressed proteins [12]. In another drought-tolerant wheat variety, 335 drought-responsive proteins were found to mainly take part in photosynthesis, carbon fixation, and glyoxylate and dicarboxylate metabolism [13]. In addition, a comparative analysis of filling-kernel proteomes from two maize inbred lines with different drought-tolerant ability identified 5175 drought-responsive proteins [14].

Proteomics study implied that wood vinegar treatment can enhance the drought tolerance of wheat root through promoting stress response, carbohydrate metabolism, protein metabolism, and secondary metabolism [15]. Two proteomics studies have been performed to find the stress-responsive proteins in the leaves of *E. angustifolia* seedlings [16] and *D. salina* during early response to salt stress [17], which implied that the symbiosis of halophytes and arbuscular mycorrhizal fungi has potential as an application for the improvement of saline-alkali soils. An iTRAQ-based proteomics analysis of *D. salina* revealed that photosynthesis and ATP synthesis are crucial for the modulation of early salinity-responsive pathways [17]. Improving osmotic tolerance of crops to salinity and drought stress has not yet been realized by molecular engineering, and proteomics of plants under salinity and drought stress present an elaborate understanding of stress-responsive proteins and metabolic pathways. These osmotic-response proteins and genes are important candidates for further functional validation.

Plant propagation and development are two important aspects that could largely determine the life cycle and economic values of different crops. The special issue studied different plants species, including an energy plant *Jatropha curcas* [18], a biofuel tree Pongamia [19], and three economic crops [20–22]. Liu et al. used combined analyses of the phosphoproteomics, physiological characteristics, and ultrastructure studies to identify the responses of *J. curcas* seedlings under chilling, and revealed significantly changed phosphoproteins under chilling stress [18]. A comparative study between the salt resistance and sensitive cultivars with very close genetic background will undoubtedly help to explore the key regulator. On the basis of this idea, Zhao et al. conducted a phosphorproteomic analysis between two maize inbred lines showing different resistance to salt stress, and found that the enhancement of potassium and sodium transportation, carbon, and redox-related metabolism could increase the salt resistance in maize [19]. In this issue, Yu et al. applied a newly developed Phos-tagTM technology to identify 21 phosphorylated peptides of AGPase [20].

Cotton is the most important natural fiber resource in the world, which makes it an important crop. Studies have been widely conducted focusing on many aspects of this crop. Although genetic modified cotton cultivars have been generated and widely used in its production, transformation of cotton could only be succeeded in very limited germplasm with low efficiency. It is very important to identify the key regulators that determine the fate of somatic embryos. Comparative proteomics were successfully applied by Guo et al. to determine 6730 proteins in the embryogenic calli of cotton [21]. Additionally, Lin and colleagues reviewed the latest advancement studies of lotus and provided more omics information on this plant [22].

Integrative omics tools, including genomics, transcriptomics, proteomics, and metabolomics, are very powerful to study the molecular basis of biological activity between biomolecules (DNA, RNA, proteins, and metabolites). Integrative omics methods have been widely used to study biomolecules for their interactions; biosynthesis; and the regulation of these interactions in the various systems of plants, plant development, and their interaction with various environments. Li and coworkers combined physiological and proteomic methods to study the changes in alligator weed stems under low potassium stress, and provided valuable information on the adaptive mechanisms in alligator [23]. Using combined metabolomic and transcriptomic analyses, Guo et al. revealed that dynamic regulation of purine metabolism and flavonoid synthesis in transdifferentiation during somatic embryogenesis is crucial for cotton regeneration [24]. A total of 581 metabolites were present in the embryogenic calli, and metabolites related to purine metabolism were significantly enriched. These omics data provide a valuable foundation for a deeper understanding of the regulatory mechanisms underlying cell totipotency at the molecular and biochemical levels.

Furthermore, many stress responsive genes were identified by transcriptomics analyses [25–27]. Dong and coworkers discovered that the leaf sheath transcriptome has dynamic perturbation, and the processes and genes involved in sheath maturation are important for organ specialization [25]. Yang and coworkers discovered differentially expressed mRNAs and potential pathways involved in heat stress in radish leaves. They detected 1802 differentially expressed mRNAs and 169 differentially expressed lncRNAs, as well as three differentially expressed circRNAs, through strand-specific RNA sequencing technology [26]. Jin et al. unraveled 1327 microRNA-mediated genes in the regulation of Pongamia seeds by high-throughput small RNA profiling and identified 236 conserved miRNAs and 143 novel miRNAs within the families by deep sequencing of Pongamia seeds [27]. These results provide valuable miRNA candidates for further functional characterization and breeding practice in Pongamia and other oilseed plants.

More stress responsive genes and proteins in different plants have been identified, and their functions are studied in the following part of this Special Issue. With the development of sequencing technology and molecular biology, studies on plant growth and development, environmental adaptation, and other mechanisms will no longer be restricted to physiological phenomenon, but will be more popular to use the omics-based methods to identify the stress responsive genes and proteins in different plants to decipher the intrinsic regulating mechanisms from the transcriptome and proteome level. Tang and colleagues explored a proteomics analysis of the regulatory mechanism of nitrogen fertilization in cereal crops, and determined 511 differentially expressed proteins among the identified 6093 proteins in the two rice cultivars [28]. Micro-exons, a set of small exons with lengths no more than 51 nucleotides, have been widely studied in recent years. Song and coworkers investigated the potential functions of micro-exons in the whole genome of two *indica* rice varieties [29].

Polyamines play an important role in plant growth and development, as well as response to abiotic stresses. Ji and coworkers applied proteomics of Sugar Beet M14 under salt stress, and found that S-adenosylmethionine decarboxylase can enhance salt tolerance via mediating the biosynthesis of spermidine and spermin [30]. Myo-inositol-1-phosphate synthase plays important roles in plant growth and development, stress responses, and cellular signal transduction. Ma and coworkers proved that Myo-inositol-1-phosphate synthase can rescue the abnormal phenotype and promote the root cell elongation [31]. Inositol signaling is believed to play a crucial role in various aspects of plant growth and adaptation. A mini-review on the function of inositol phosphatases in plant tolerance to abiotic stress was included. In this timely review, the authors concluded the functions of more than 30 members of inositol phosphatases in plants, and revised some current knowledge in relation to their substrates and function in response to abiotic stress [32].

We believe that the proteomics-based results in this issue will help the readers to gain novel insights for the understanding of complicated physiological processes in crops and other important plants in response to stressed conditions. The identification of target genes and proteins will decipher the complex relationship between genes, proteins, metabolites, and their biological functions in plants, and these genes and proteins are important candidates for further functional validation and may provide valuable information for further study in molecular breeding in economic plants and crops.

Funding: Thanks for the funding support provided by the National Key Research and Development Program of China (No. 2018YFD1000502) and the National Natural Science Foundation of China (No. 31570301, 31860224).

Acknowledgments: The author wants to thank in particular Quanliang Xie for his kind help in collecting the data and polishing this manuscript.

Conflicts of Interest: The author declares no conflict of interests.

References

1. Aldhous, P. Genomics: Beyond the book of life. *Nature* **2000**, *405*, 894–896. [CrossRef]
2. Wang, W.S.; Mauleon, R.; Hu, Z.Q.; Chebotarov, D.; Tai, S.; Wu, Z.C.; Li, M.; Zheng, T.Q.; Fuentes, R.R.; Zhang, F.; et al. Genomic variation in 3010 diverse accessions of Asian cultivated rice. *Nature* **2018**, *557*, 43–49. [CrossRef] [PubMed]
3. Zhu, J.K. Abiotic stress signaling and responses in plants. *Cell* **2016**, *167*, 313–324. [CrossRef] [PubMed]
4. Walley, J.W.; Sartor, R.C.; Shen, Z.X.; Schmitz, R.J.; Wu, K.J.; Urich, M.A.; Nery, J.R.; Smith, L.G.; Schnable, J.C.; Ecker, J.R.; et al. Integration of omic networks in a developmental atlas of maize. *Science* **2016**, *353*, 814–818. [CrossRef] [PubMed]
5. Zhu, G.; Wang, S.; Huang, A.; Zhang, S.; Liao, Q.; Zhang, C.; Lin, T.; Peng, M.; Yang, C.; Cao, X.; et al. Rewiring of the fruit metabolome in tomato breeding. *Cell* **2018**, *172*, 249–261. [CrossRef] [PubMed]

6. Li, S.S.; Yu, J.J.; Li, Y.; Zhang, H.; Bao, X.S.; Bian, J.Y.; Xu, C.X.; Wang, X.L.; Cai, X.F.; Wang, Q.H.; et al. Heat-responsive proteomics of a heat-sensitive spinach variety. *Int. J. Mol. Sci.* **2019**, *20*, 3872. [CrossRef] [PubMed]

7. Yuan, L.Y.; Wang, J.; Xie, S.L.; Zhao, M.R.; Nie, L.B.; Zheng, L.B.; Zhu, S.D.; Hou, J.F.; Chen, G.H.; Wang, C.G. Comparative proteomics indicates that redox homeostasis is involved in high- and low-temperature stress tolerance in a novel Wucai (*Brassica campestris* L.) genotype. *Int. J. Mol. Sci.* **2019**, *20*, 3760. [CrossRef]

8. Wang, S.P.; Zhang, Y.X.; Fang, Z.W.; Zhang, Y.M.; Song, Q.L.; Hou, Z.H.; Sun, K.K.; Song, Y.L.; Li, Y.; Ma, D.F.; et al. Cytological and proteomic analysis of wheat pollen abortion induced by chemical hybridization agent. *Int. J. Mol. Sci.* **2019**, *20*, 1615. [CrossRef]

9. Satheeswaran, T.; Shang, X.M.; Sun, J.; Liu, H.J. Quantitative proteomic analysis reveals novel insights into intracellular silicate stress-responsive mechanisms in the diatom *Skeletonema dohrnii*. *Int. J. Mol. Sci.* **2019**, *20*, 2540.

10. Huo, J.Q.; Huang, D.J.; Zhang, J.; Fang, H.; Wang, B.; Wang, C.L.; Ma, Z.J.; Liao, W.B. Comparative proteomic analysis during the involvement of nitric oxide in hydrogen gas-improved postharvest freshness in cut lilies. *Int. J. Mol. Sci.* **2018**, *19*, 3955. [CrossRef]

11. Xie, Q.L.; Ding, G.H.; Zhu, L.P.; Yu, L.; Yuan, B.X.; Gao, X.Q.; Wang, D.; Sun, Y.; Liu, Y.; Li, H.B.; et al. Proteomic landscape of the mature roots in a rubber-producing grass *Taraxacum Kok-saghyz*. *Int. J. Mol. Sci.* **2019**, *20*, 2596. [CrossRef] [PubMed]

12. Wang, Y.J.; Zhang, X.Y.; Huang, G.R.; Feng, F.; Liu, X.Y.; Guo, R.; Gu, F.X.; Zhong, X.L.; Mei, X.R. iTRAQ-Based quantitative analysis of responsive proteins under PEG-induced drought stress in wheat leaves. *Int. J. Mol. Sci.* **2019**, *20*, 2621. [CrossRef] [PubMed]

13. Wang, X.B.; Xu, Y.H.; Li, J.J.; Ren, Y.Z.; Wang, Z.Q.; Xin, Z.Y.; Lin, T.B. Identification of two novel wheat drought tolerance-related proteins by comparative proteomic analysis combined with virus-induced gene silencing. *Int. J. Mol. Sci.* **2018**, *19*, 4020. [CrossRef] [PubMed]

14. Wang, Y.Y.; Qiu, L.; Song, Q.L.; Wang, S.P.; Wang, Y.J.; Ge, Y.H. Root proteomics reveals the effects of wood vinegar on wheat growth and subsequent tolerance to drought stress. *Int. J. Mol. Sci.* **2019**, *20*, 943. [CrossRef]

15. Wang, X.; Tinashe, Z.; Liu, S.T.; Liu, G.; Jin, H.Y.; Dai, L.; Dong, A.Y.; Yang, Y.T.; Duan, H.J. Comparative proteomics and physiological analyses reveal important maize filling-kernel drought-responsive genes and metabolic pathways. *Int. J. Mol. Sci.* **2019**, *20*, 3743. [CrossRef]

16. Jia, T.T.; Wang, J.; Chang, W.; Fan, X.X.; Sui, X.; Song, F.Q. Proteomics analysis of *E. angustifolia* seedlings inoculated with *Arbuscular Mycorrhizal* fungi under salt stress. *Int. J. Mol. Sci.* **2019**, *20*, 788. [CrossRef]

17. Wang, Y.; Cong, Y.T.; Wang, Y.H.; Guo, Z.H.; Yue, J.R.; Xing, Z.Y.; Gao, X.N.; Chai, X.J. Identification of early salinity stress-responsive proteins in *Dunaliella salina* by isobaric tags for relative and absolute quantitation (iTRAQ)-based quantitative proteomic analysis. *Int. J. Mol. Sci.* **2019**, *20*, 599. [CrossRef]

18. Liu, H.; Wang, F.F.; Peng, X.J.; Huang, J.H.; Shen, S.H. Global phosphoproteomic analysis reveals the defense and response mechanisms of *Jatropha Curcas* seedling under chilling stress. *Int. J. Mol. Sci.* **2019**, *20*, 208. [CrossRef]

19. Zhao, X.Y.; Bai, X.; Jiang, C.F.; Li, Z. Phosphoproteomic analysis of two contrasting maize inbred lines provides insights into the mechanism of salt-stress tolerance. *Int. J. Mol. Sci.* **2019**, *20*, 1886. [CrossRef]

20. Yu, G.W.; Lv, Y.N.; Shen, L.Y.; Wang, Y.B.; Qing, Y.; Wu, N.; Li, Y.P.; Huang, H.H.; Zhang, N.; Liu, Y.H.; et al. The proteomic analysis of maize endosperm protein enriched by phos-tagtm reveals the phosphorylation of brittle-2 subunit of ADP-Glc pyrophosphorylase in starch biosynthesis process. *Int. J. Mol. Sci.* **2019**, *20*, 986. [CrossRef]

21. Guo, H.X.; Guo, H.H.; Zhang, L.; Fan, Y.P.; Fan, Y.P.; Tang, Z.M.; Zeng, F.C. Dynamic TMT-based quantitative proteomics analysis of critical initiation process of totipotency during cotton somatic embryogenesis transdifferentiation. *Int. J. Mol. Sci.* **2019**, *20*, 1691.

22. Lin, Z.Y.; Zhang, C.; Cao, D.D.; Damaris, R.N.; Yang, P.F. The latest studies on lotus (*Nelumbo nucifera*), an emerging horticultural model plant. *Int. J. Mol. Sci.* **2019**, *20*, 3680. [CrossRef] [PubMed]

23. Li, L.Q.; Lyu, C.C.; Li, J.H.; Tong, Z.; Lu, Y.F.; Wang, X.Y.; Ni, S.; Yang, S.M.; Zeng, F.H.; Lu, L.M. Physiological analysis and proteome quantification of alligator weed stems in response to potassium deficiency stress. *Int. J. Mol. Sci.* **2019**, *20*, 221. [CrossRef] [PubMed]

24. Guo, H.H.; Guo, H.X.; Zhang, L.; Tang, Z.M.; Yu, X.M.; Wu, J.F.; Zeng, F.C. Metabolome and transcriptome association analysis reveals dynamic regulation of purine metabolism and flavonoid synthesis in transdifferentiation during somatic embryogenesis in cotton. *Int. J. Mol. Sci.* **2019**, *20*, 2070. [CrossRef] [PubMed]

25. Dong, L.; Qin, L.; Dai, X.R.; Ding, Z.D.; Bi, R.; Liu, P.; Chen, Y.H.; Thomas, P.B.; Wang, X.L.; Li, P.H. Transcriptomic analysis of leaf sheath maturation in maize. *Int. J. Mol. Sci.* **2019**, *20*, 2472. [CrossRef] [PubMed]

26. Yang, Z.; Li, W.; Su, X.; Ge, P.F.; Zhou, Y.; Hao, Y.Y.; Shu, H.Y.; Gao, C.L.; Cheng, S.H.; Zhu, G.P.; et al. Early response of radish to heat stress by strand-specific transcriptome and miRNA analysis. *Int. J. Mol. Sci.* **2019**, *20*, 3321. [CrossRef]

27. Jin, Y.; Liu, L.; Hao, X.H.; David, E.H.; Zheng, Y.Z.; Huang, T.B.; Huang, J.Z. Unravelling the microRNA-mediated gene regulation in developing Pongamia seeds by high-throughput small RNA profiling. *Int. J. Mol. Sci.* **2019**, *20*, 3509. [CrossRef]

28. Tang, J.C.; Sun, Z.G.; Chen, Q.H.; Rebecca, N.D.; Lu, B.L.; Hu, Z.R. Nitrogen fertilizer induced alterations in the root proteome of two rice cultivars. *Int. J. Mol. Sci.* **2019**, *20*, 3674. [CrossRef]

29. Song, Q.; Lv, F.; Muhammad, T.Q.; Xing, F.; Zhou, R.; Li, H.; Chen, L.L. Identification and analysis of micro-exon genes in the rice genome. *Int. J. Mol. Sci.* **2019**, *20*, 2685. [CrossRef]

30. Ji, M.C.; Wang, K.; Wang, L.; Chen, S.X.; Li, H.Y.; Ma, C.Q.; Wang, Y.G. Overexpression of a S-adenosylmethionine decarboxylase from sugar beet M14 increased Araidopsis salt tolerance. *Int. J. Mol. Sci.* **2019**, *20*, 1990. [CrossRef]

31. Ma, R.D.; Song, W.Y.; Wang, F.; Cao, A.; Xie, S.Q.; Chen, X.F.; Jin, X.; Li, H.B. A cotton (*Gossypium hirsutum*) Myo-inositol-1-phosphate synthase (GhMIPS1D) gene promotes root cell elongation in Arabidopsis. *Int. J. Mol. Sci.* **2019**, *20*, 1224. [CrossRef] [PubMed]

32. Jia, Q.; Kong, D.F.; Li, Q.H.; Sun, S.; Song, J.L.; Zhu, Y.B.; Liang, K.J.; Ke, Q.M.; Lin, W.X.; Huang, J.W. The function of inositol phosphatases in plant tolerance to abiotic stress. *Int. J. Mol. Sci.* **2019**, *20*, 3999. [CrossRef] [PubMed]

International Journal of
Molecular Sciences

Article

Heat-Responsive Proteomics of a Heat-Sensitive Spinach Variety

Shanshan Li [1,2,3,†], Juanjuan Yu [2,4,†], Ying Li [2], Heng Zhang [1], Xuesong Bao [2], Jiayi Bian [1], Chenxi Xu [1], Xiaoli Wang [1], Xiaofeng Cai [1], Quanhua Wang [1], Pengcheng Wang [5], Siyi Guo [6], Yuchen Miao [6], Sixue Chen [1,7], Zhi Qin [1,*] and Shaojun Dai [1,2,*]

1   Development Center of Plant Germplasm Resources, College of Life Sciences, Shanghai Normal University, Shanghai 200234, China
2   Alkali Soil Natural Environmental Science Center, Northeast Forestry University, Key Laboratory of Saline-alkali Vegetation Ecology Restoration, Ministry of Education, Harbin 150040, China
3   College of Life Sciences and Agriculture and Forestry, Qiqihar University, Qiqihar 161006, China
4   College of Life Sciences, Henan Normal University, Xinxiang 453007, China
5   Shanghai Center for Plant Stress Biology, Chinese Academy of Sciences, Shanghai 201602, China
6   Institute of Plant Stress Biology, State Key Laboratory of Cotton Biology, Department of Biology, Henan University, Kaifeng 475004, China
7   Department of Biology, Genetics Institute, Plant Molecular and Cellular Biology Program, Interdisciplinary Center for Biotechnology Research, University of Florida, Gainesville, FL 32610, USA
*   Correspondence: qinzhi@shnu.edu.cn (Z.Q.); daishaojun@shnu.edu.cn (S.D.); Tel.: +86-21-64324576
†   These authors contributed equally to this work.

Received: 28 June 2019; Accepted: 6 August 2019; Published: 8 August 2019

Abstract: High temperatures seriously limit plant growth and productivity. Investigating heat-responsive molecular mechanisms is important for breeding heat-tolerant crops. In this study, heat-responsive mechanisms in leaves from a heat-sensitive spinach (*Spinacia oleracea* L.) variety Sp73 were investigated using two-dimensional gel electrophoresis (2DE)-based and isobaric tags for relative and absolute quantification (iTRAQ)-based proteomics approaches. In total, 257 heat-responsive proteins were identified in the spinach leaves. The abundance patterns of these proteins indicated that the photosynthesis process was inhibited, reactive oxygen species (ROS) scavenging pathways were initiated, and protein synthesis and turnover, carbohydrate and amino acid metabolism were promoted in the spinach Sp73 in response to high temperature. By comparing this with our previous results in the heat-tolerant spinach variety Sp75, we found that heat inhibited photosynthesis, as well as heat-enhanced ROS scavenging, stress defense pathways, carbohydrate and energy metabolism, and protein folding and turnover constituting a conservative strategy for spinach in response to heat stress. However, the heat-decreased biosynthesis of chlorophyll and carotenoid as well as soluble sugar content in the variety Sp73 was quite different from that in the variety Sp75, leading to a lower capability for photosynthetic adaptation and osmotic homeostasis in Sp73 under heat stress. Moreover, the heat-reduced activities of SOD and other heat-activated antioxidant enzymes in the heat-sensitive variety Sp73 were also different from the heat-tolerant variety Sp75, implying that the ROS scavenging strategy is critical for heat tolerance.

Keywords: heat response; heat-sensitive spinach variety; proteomics; ROS scavenging

---

1. Introduction

Global warming has adverse effects on crop yield [1,2]. Heat stress limits plant growth, development and reproduction [3,4] by affecting gene expression, protein synthesis and degradation, and membrane structure, as well as cytoskeleton dynamics [5,6]. In addition, heat can change the efficiency of intracellular enzymatic reactions, leading to internal metabolic imbalance, which in turn causes an excessive accumulation of toxic byproducts, such as reactive oxygen species (ROS) [7]. In order to shield the effects of heat stress on the internal metabolic processes, plants modulate the composition of corresponding transcripts, proteins, metabolites and lipids to establish a new metabolic homeostasis, as well as changing their growth and reproduction to cope with high-temperature environments [8,9].

High-throughput proteomics techniques have facilitated the large-scale identification of heat-responsive proteins (HRPs) in plants [9–12]. Proteomics data have revealed diverse expression patterns of HRPs in leaves of *Arabidopsis thaliana* [13], alfalfa (*Medicago sativa*) [10], *Oryza sativa* [14,15], *Oryza meridionalis* [16], wheat (*Triticum aestivum*) [17,18], maize (*Zea mays*) [19,20], soybean (*Glycine max*) [21,22], and *Apium graveolens* [23]. These HRPs are mainly involved in signal transduction, photosynthesis, ROS scavenging, transcription and post-transcriptional regulation, protein synthesis and degradation, as well as carbon and energy metabolism [13–23].

Spinach (*Spinacia oleracea* L.) is rich in vitamins, minerals and other nutrients, and is considered as one of the major green leafy vegetables in China. In general, spinach is a cold-tolerant but heat-sensitive species [24], and high temperatures cause a low germination rate of seeds and retarded growth, leading to a reduction of yield and nutrition. The investigation of heat-responsive molecular mechanisms in spinach will be instructive for breeding new varieties with heat tolerance capability. Previous studies have reported that heat stress (35 °C, 30 min) on whole spinach induced a significant decrease in the CO_2 assimilation rate [25]. Heat stress also induced a release of the extrinsic oxygen evolving complex (OEC) subunits (PsbO, PsbP and PsbQ) from Photosystem II (PSII), which results in significant D1 aggregation and degradation [26,27]. Our previous proteomics study has identified 911 heat-responsive proteins in the heat-tolerant spinach variety Sp75 [11]. The data showed that calcium-mediated signaling, ROS homeostasis, endomembrane trafficking, and cross-membrane transport pathways were enhanced under heat stress. Moreover, diverse primary and secondary metabolic pathways (e.g., glycolysis, pentose phosphate pathway, isoprenoid biosynthesis, as well as metabolisms of amino acid, fatty acid, nucleotide, and vitamins) were employed for heat tolerance [11].

To compare and contrast with our previous study on the heat-tolerant spinach variety Sp75, a heat-sensitive variety Sp73 was subjected to heat-responsive physiological and proteomic analyses. The abundance patterns of 257 heat-responsive proteins imply that photosynthesis was inhibited, but ROS scavenging pathways, protein turnover, carbohydrate metabolism were enhanced in the heat-sensitive spinach variety Sp73. This study provides important insights into the molecular mechanisms of the heat stress response of spinach.

2. Results

2.1. Morphology and Relative Water Content (RWC)

The morphology of leaves from spinach variety Sp73 were affected after 24 h of heat treatment (HHT), 48 HHT and 72 HHT, when compared with 0 HHT (Figure 1 A–D). The number of withered leaves increased, and the withering was more serious at 48 and 72 HHT (Figure 1C,D), although the length and width of leaves were not changed significantly. The RWC was decreased by 17.45%, 19.38% and 23.61% at 24 HHT, 48 HHT and 72 HHT, respectively. These results clearly show that spinach Sp73 is very sensitive to heat stress.

2.2. Photosynthesis and Chlorophyll Fluorescence Parameters

The photosynthesis rate (Pn), stomata conductance (Gs), intercellular CO_2 (Ci) and transpiration rate (Tr) in heat-treated spinach Sp73 were measured (Figure 2). Compared with 0 HHT, the Pn (Figure 2A), Gs (Figure 2B) and Tr (Figure 2D) were apparently decreased by 2.4-fold, 2.3-fold and 1.1-fold at 72 HHT, respectively, throughout the heat-stress process. In addition, Ci was increased under the heat treatment process (Figure 2C). Furthermore, chlorophyll fluorescence parameters were also monitored to evaluate the photosynthetic performance. The PSII maximum quantum yield (Fv/Fm) was decreased 1.1-fold at 24 HHT, 1.2-fold at 48 HHT and 1.4-fold at 72HHT (Figure 2E). The effective PSII quantum yield (Y(II)) was slightly increased at 24 HHT and 48 HHT, but decreased at 72 HHT (Figure 2F).

Figure 1. Morphology and relative water content (RWC) in spinach variety Sp73 under heat stress. Leaves were withered and the leaf RWC was decreased after 24 h of heat treatment (HHT) of 37 °C/32 °C (day/night). (**A**) 0 HHT; (**B**) 24 HHT; (**C**) 48 HHT; and (**D**) 72 HHT. Underneath each morphological image is the leaf RWC expressed as mean ± standard deviation (SD) (n = 3). Small letters (a, b) in the brackets indicate significant difference among different treatments ($p < 0.05$). Bar = 2.9 cm.

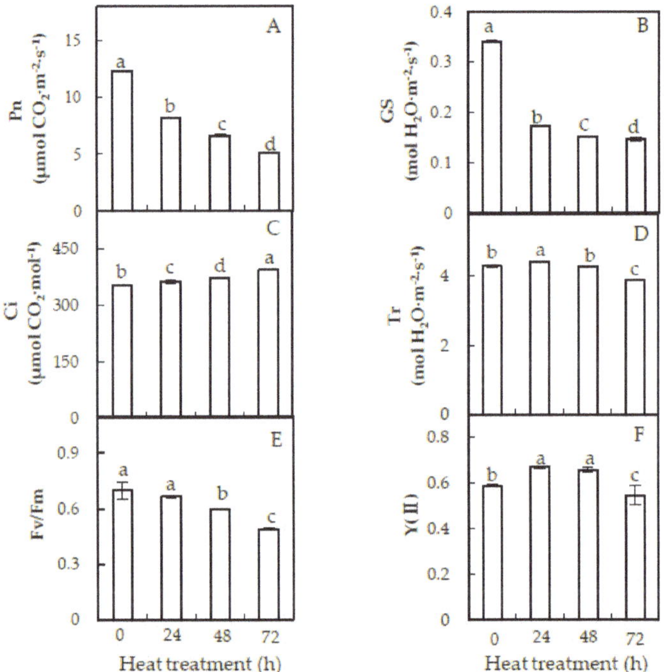

Figure 2. Photosynthetic characteristics and chlorophyll fluorescence parameters in leaves of spinach Sp73 under the heat-stress treatment. (**A**) Photosynthesis rate (Pn); (**B**) stomata conductance (Gs); (**C**) intercellular CO_2 (Ci); (**D**) transpiration rate (Tr); (**E**) Photosystem II (PSII) maximum quantum yield (Fv/Fm); (**F**) effective PSII quantum yield (Y(II)). The values were determined after plants were treated with heat stress at 37 °C/32 °C (day/night) for 0 h, 24 h, 48 h and 72 h and are presented as means ± SD (n = 3). The different small letters (a–d) indicate significant difference ($p < 0.05$) in different treatments.

2.3. Plasma Membrane Integrity and Osmolyte Accumulation in Leaves

To evaluate the effects of heat stress on membrane stability, the thiobarbituric acid reactive substance (TBARS) content and relative electrolyte leakage (REL) were detected in leaves under heat stress (Figure 3). TBARS contents increased from 8.6 ± 0.1 nmol·g⁻¹ fresh weight (FW) at 0 HHT to

10.6 ± 0.2 nmol·g^{-1} FW at 48 HHT (Figure 3A). RELs were increased 1.4-fold at 48 HHT and 1.8-fold at 72 HHT compared with 0 HHT (Figure 3B). These results indicate that long-term heat stress leads to severe oxidative damage to leaf cells in spinach Sp73.

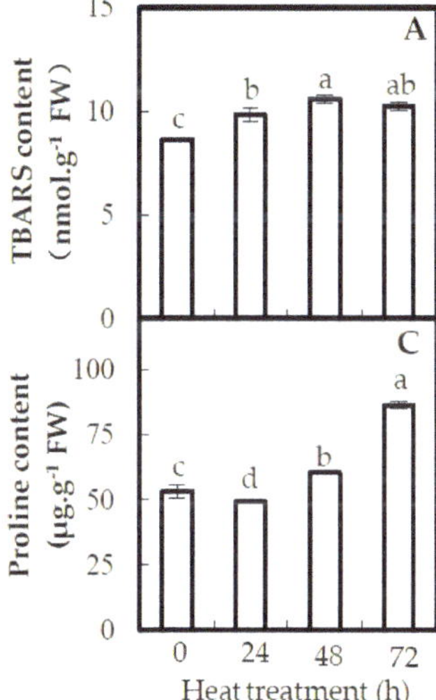
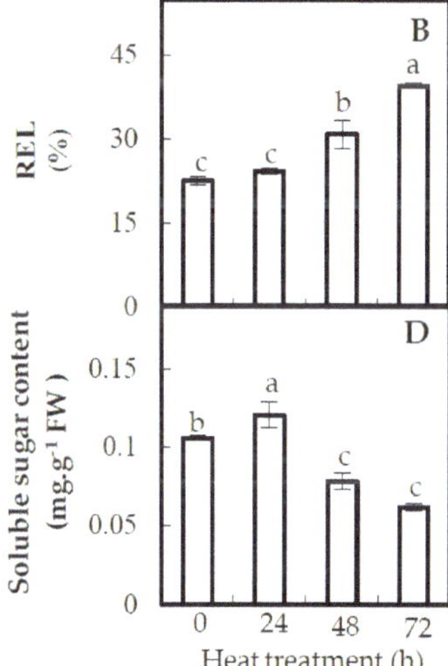

Figure 3. Membrane integrity and osmolyte accumulation in leaves of spinach Sp73 under heat stress. (**A**) Thiobarbituric acid reactive substance (TBARS) content in leaves; (**B**) relative electrolyte leakage (REL) of leaves; (**C**) proline in leaves; (**D**) soluble sugar content in leaves. The values were determined after plants were treated with heat stress at 37 °C /32 °C (day/night) for 0 h, 24 h, 48 h and 72 h and are presented as means ± SD (n = 3). The different small letters (a–d) indicate significant difference ($p < 0.05$) among different treatments.

Proline and soluble sugars function by maintaining osmotic balance and protein stabilization. Compared with 0 HHT, the proline content in leaves was increased 1.6-fold at 72 HHT (Figure 3C). Soluble sugar contents were increased at 24 HHT, but decreased at 48 HHT and 72 HHT (Figure 3D).

2.4. Activities of Antioxidant Enzymes and Antioxidant Contents in Leaves

We analyzed the activities of antioxidant enzymes and antioxidant contents to evaluate the ROS changes and dynamics of the ROS scavenging system in leaves under the heat-stress treatment. Compared with 0 HHT, the hydrogen peroxide (H_2O_2) content and superoxide anion radicals ($O_2^{\bullet-}$) generation rate were significantly increased in leaves at 48 HHT and 72 HHT (Figure 4A), implying that heat-increased ROS would cause oxidative damage to spinach leaves. The activities of superoxide dismutase (SOD) and catalase (CAT) were obviously decreased in the leaves throughout the heat-stress process (Figure 4B). However, the activities of ascorbate peroxidase (APX) and peroxidase (POD) were increased 1.9-fold and 1.7-fold at 48 HHT, respectively (Figure 4C). The activities of glutathione peroxidase (GPX) and glutathione S-transferase (GST) were increased 2.5-fold and 3.0-fold at 48 HHT, respectively (Figure 4D,F). Besides this, the activities of monodehydroascorbate reductase (MDHAR), dehydroascorbate reductase (DHAR) and glutathione reductase (GR) were all increased, except the MDHAR activity at 72 HHT, and all of them reached their highest level at 24 HHT and then decreased gradually at 48 HHT and 72 HHT (Figure 4E,F). For the antioxidants in the glutathione (GSH)-ascorbate (AsA) cycle, the AsA contents were decreased, while dehydroascorbate (DHA) contents were increased at 72 HHT (Figure 3G), and all the contents of reduced GSH and oxidized glutathione (GSSG) were increased in leaves under the heat-stress process (Figure 4H). The ratios of GSH/GSSG were decreased 1.3-fold and 1.2-fold at 24 HHT and 72 HHT, respectively, while they increased by 1.1-fold at 48 HHT

(Figure 4I). However, the ratios of AsA/DHA were increased 1.4-fold at both 24 HHT and 48 HHT, while they decreased by 2.0-fold at 72 HHT.

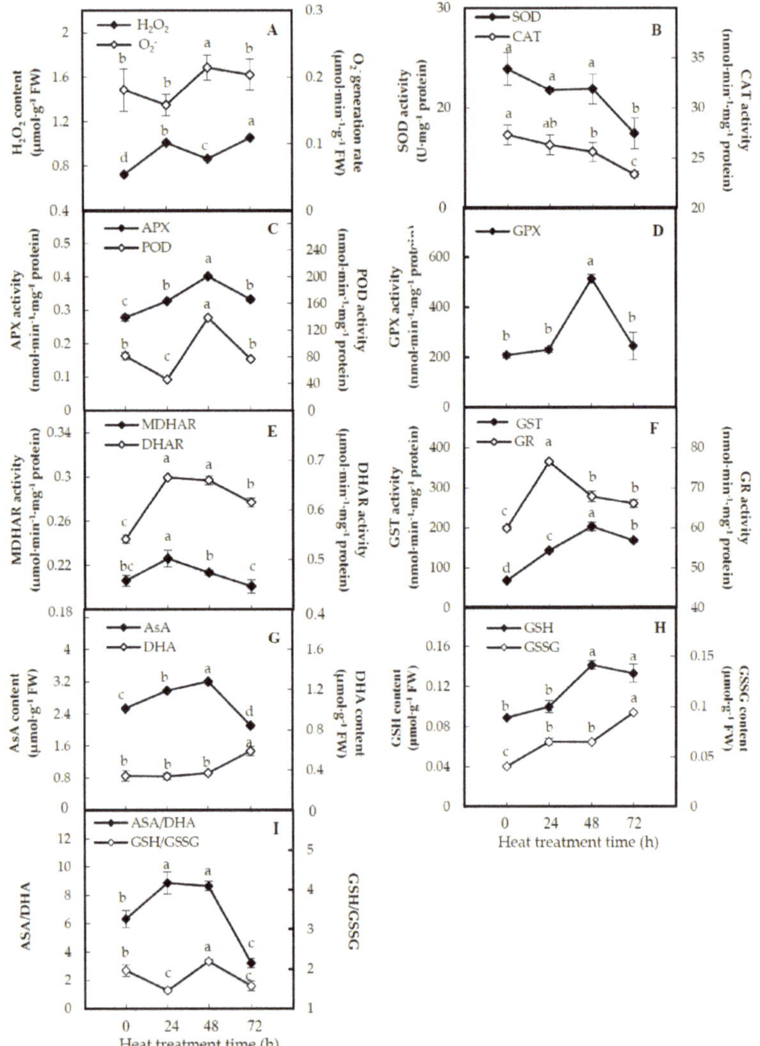

Figure 4. Activities of antioxidant enzymes and antioxidant contents in leaves of spinach Sp73 under heat-stress treatment. (**A**) Hydrogen peroxide (H_2O_2) content and superoxide anion radical ($O_2^{\bullet-}$) generation rate; (**B**) activities of superoxide dismutase (SOD) and catalase (CAT); (**C**) activities of ascorbate peroxidase (APX) and peroxidase (POD); (**D**) glutathione peroxidase (GPX) activity; (**E**) activities of monodehydroascorbate reductase (MDHAR) and dehydroascorbate reductase (DHAR); (**F**) activities of glutathione reductase (GR) and glutathione S-transferase (GST); (**G**) contents of ascorbate (AsA) and dehydroascorbate (DHA); (**H**) contents of reduced glutathione (GSH) and oxidized glutathione (GSSG); (**I**) ratios of AsA/DHA and GSH/GSSG. The values were determined after plants were treated with heat stress at 37 °C/32 °C (day/night) for 0 h, 24 h, 48 h and 72 h and are presented as means ± SD (*n* =3). The different small letters (a–d) indicate significant difference ($p < 0.05$) among different treatments.

2.5. Identification of Heat-Responsive Proteins by 2DE-Based and iTRAQ-Based Proteomics

Two complementary proteomics approaches, two-dimensional gel electrophoresis (2DE)-based and isobaric tags for relative and absolute quantification (iTRAQ)-based approaches, were applied to determine the heat-responsive protein abundances in leaves of spinach Sp73 under heat stress. From the results of 2DE-based proteomics, more than 1000 protein spots were detected on Coomassie Brilliant Blue (CBB)-stained gels (Figure 5, Supplementary Figure S1; Table S1). Among them, 93 reproducibly matched spots showed more than 1.5-fold changes in abundance in response to heat treatment ($p < 0.05$). Among them, 84 protein spots were identified by matrix-assisted laser desorption/ ionization tandem time of flight mass spectrometry (MALDI TOF-TOF MS) and Mascot searching with stringent criteria. The 84 protein spots all contained a single protein in each spot (Supplementary Table S1). Thus, the 84 proteins were taken as HRPs. Besides this, in the iTRAQ-based proteomics, 3526 proteins were identified and quantified in four independent biological replicates. Among them, 173 proteins showed differential abundances under heat stress (fold change > 1.2 and $p < 0.05$) (Supplementary Table S2). There were no overlaps between the results from 2DE-based and iTRAQ-based approaches. In total, 257 HRPs were identified in the leaves of spinach Sp73 (Figure 6, Supplementary Table S3). The HRPs were annotated against the National Center for Biotechnology Information non-redundant (NCBInr) protein database with Basic Local Alignment Search Tool (BLAST) analysis, and 51 proteins were reannotated according to the functional domain annotation from the NCBInr protein database (Supplementary Tables S4 and S5). Among these HRPs, 141 proteins were heat stress-increased and 112 heat-decreased under at least one heat stress condition compared with 0 HHT. The remaining four proteins exhibited different change patterns under heat stress.

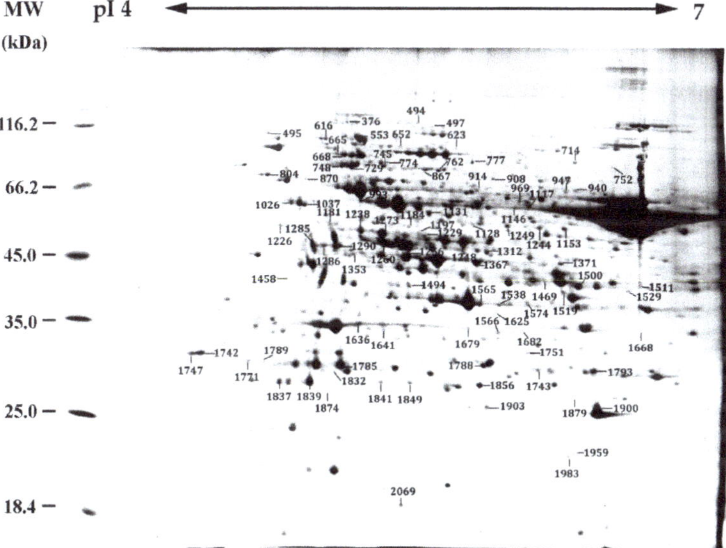

Figure 5. Representative two-dimensional gel electrophoresis (2DE) gel images of proteins in leaves of spinach Sp73. Proteins were separated on 24 cm linear gradient immobilized pH gradient (IPG) strips (pH 4–7) using isoelectric focusing (IEF) in the first dimension, followed by 12.5% sodium dodecyl sulfate polyacrylamide gel electrophoresis (SDS-PAGE) gels in the second dimension. The 2DE gel was stained with Coomassie Brilliant Blue. The molecular weight (MW) in kilodaltons (KDa) and isoelectric point (pI) of proteins are indicated on the left and top of the gel, respectively. A total of 84 heat-responsive proteins identified by matrix-assisted laser desorption/ionization tandem time of flight mass spectrometry (MALDI TOF-TOF MS) were marked with numbers on the gel. Detailed information can be found in Supplementary Figure S1 and Table S1.

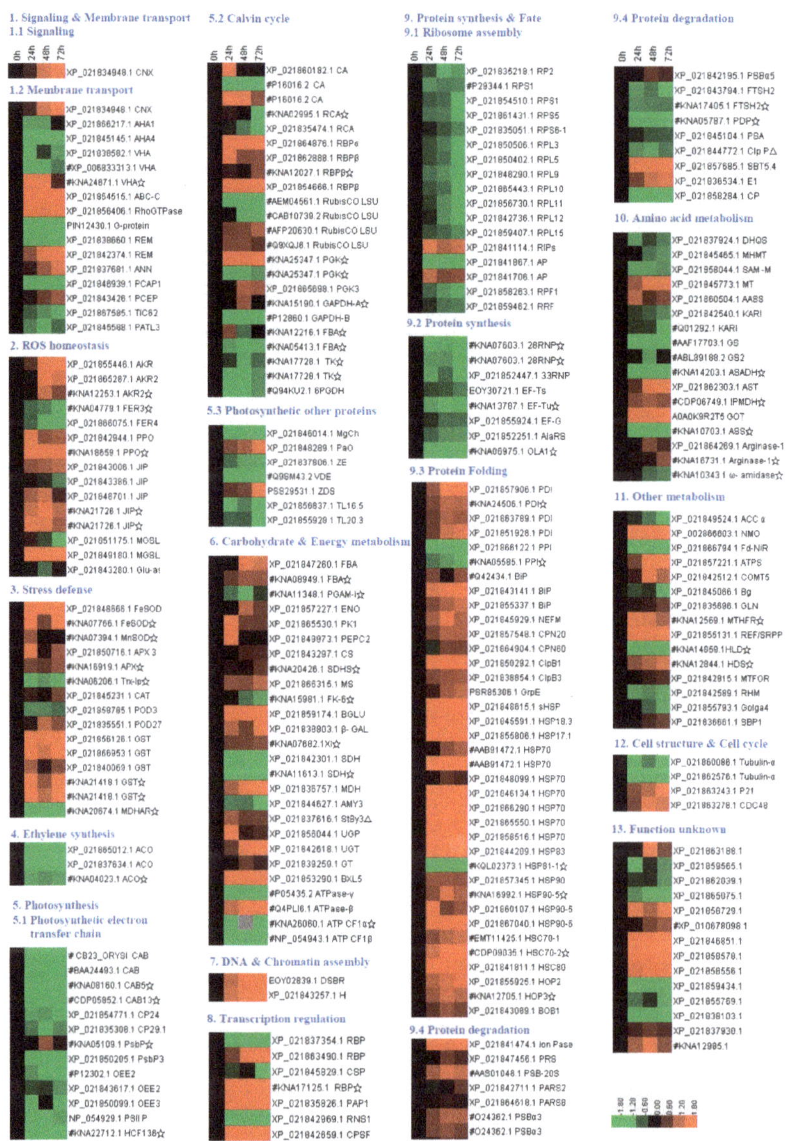

Figure 6. Abundance patterns of heat-responsive proteins in leaves of spinach Sp73 revealed from proteomic analysis. The four columns represent different heat treatments at 37 °C/32 °C (day/night) for 0 h, 24 h, 48 h, and 72 h. The rows represent individual proteins. The increased or decreased proteins are indicated in red or green, respectively. The color intensity increases with increasing abundance differences, as shown in the scale bar. The scale bar indicates the log (base2)-transformed relative protein abundance ratios ranging from −1.8 to 1.8. Database accession numbers and the abbreviations of protein names are listed on the right side (please refer to Supplementary Table S3 for the full protein names). The database accession numbers are from NCBInr. Those marked with the pound signs (#) indicate the proteins identified by the 2DE-based proteomics approach, and the rest were identified by the iTRAQ-based proteomics approach. The protein names marked with pentagrams (☆) were annotated according to the functional domain annotation from the NCBInr protein database.

2.6. Annotation and Functional Categorization of HRPs

Based on BLAST alignments, subcellular localization prediction, and literature information, the 257 proteins were classified into 13 functional categories, including signaling and membrane transport, ROS homeostasis, stress defense, ethylene synthesis, photosynthesis, carbohydrate and energy metabolism, DNA and chromatin assembly, transcription regulation, protein synthesis and fate, amino acid metabolism, other metabolisms, cell structure and cell cycle, and unknown function (Supplementary Table S3 and Figure 7A). Interestingly, protein synthesis and fate-related proteins accounted for the largest group (30.4%), followed by photosynthesis-related proteins (16.7%). A total of 54.1% of HRPs were heat-increased compared with 0 HHT (Figure 7C). For example, 12 out of 15 (80%) ROS scavenging proteins (e.g., SOD, APX, and GST) and 11 out of 17 (64.7%) stress defense-related proteins (e.g., aldo/keto reductase (AKR) and jasmonate-induced protein) were heat stress-increased. Besides this, carbohydrate and energy metabolism-related proteins (69.2%) and protein turnover-related proteins (57.7%) were increased by heat stress. Interestingly, 83.0% of protein folding/degradation-related proteins were increased in the heat-sensitive spinach Sp73. In contrast, most proteins involved in photosynthesis (65.1%), amino acid metabolism (58.8%), ribosome assembly (92.0%) and protein synthesis (100%) were decreased by heat stress. These results indicate that carbon assimilation, basic metabolism and protein synthesis were inhibited by heat stress in the heat-sensitive spinach Sp73.

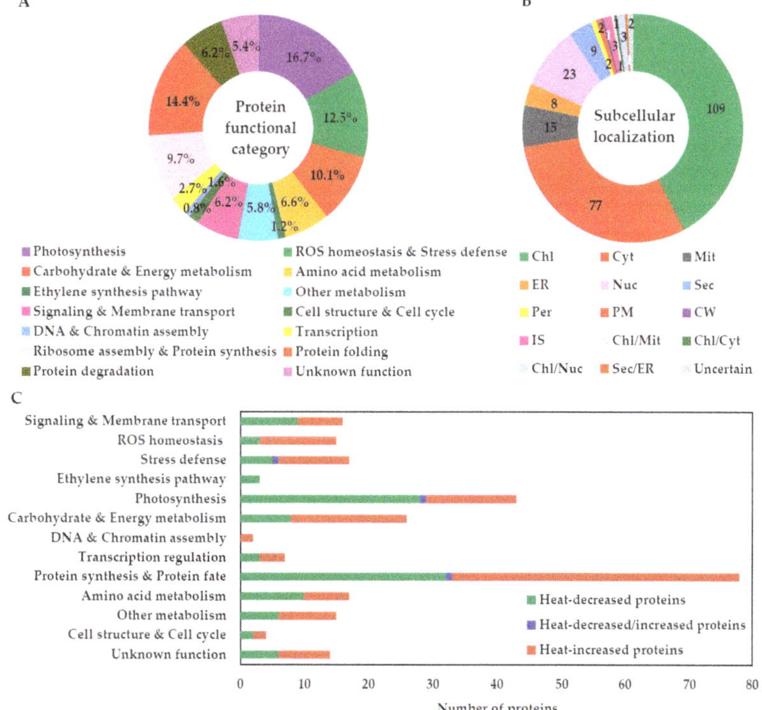

Figure 7. Functional categorization and subcellular localization of heat-responsive proteins. (**A**) Protein functional categories; (**B**) subcellular localization groups. The numbers of proteins with different locations are shown; (**C**) abundance patterns of heat-responsive proteins in each functional category. Chl, chloroplast; CW, cell wall; Cyt, cytoplasm; ER, endoplasmic reticulum; IS, intercellular space; Mit, mitochondrion; Nuc, nucleus; Per, peroxisome; PM, plasma membrane; Sec, secreted.

2.7. Subcellular Localization and Protein–Protein Interaction (PPI) Network of HRPs

The subcellular localization of HRPs was predicted using five different tools (*i.e.*, YLoc, LocTree3, Plant-mPLoc, ngLOC, and TargetP) (Figure 7B, and Supplementary Table S6). Among the 257 proteins, 109 proteins (42%) were predicted to be localized in chloroplasts, 77 (30%) in cytoplasm, 23 (9%) in nucleus, 15 (6%) in mitochondria, nine in the secreted pathway, and eight in the endoplasmic reticulum. The remaining proteins are localized in the plasma membrane (2), intercellular space (3), cell wall (1), peroxisomes (2) and uncertain locations (2). Clearly, most heat-affected proteins were localized in chloroplasts, indicating the heat-stress sensitivity of the chloroplasts in the spinach Sp73.

A total of 207 unique homologous proteins of HRPs were found in Arabidopsis (Supplementary Table S7), 123 of which were depicted in the PPI network. Six modules formed tightly connected clusters, and strong associations were represented by thick lines in the network (Figure 8). Module 1 contained 35 proteins mainly involved in photosynthesis. Module 2, Module 3 and Module 4 included those proteins mainly involved in protein synthesis, protein folding and protein degradation, respectively. Module 5 contained 19 proteins which are mainly involved in sugar metabolism, amino acid metabolism and other metabolisms. Module 6 contained 12 proteins important in ROS scavenging and stress defense.

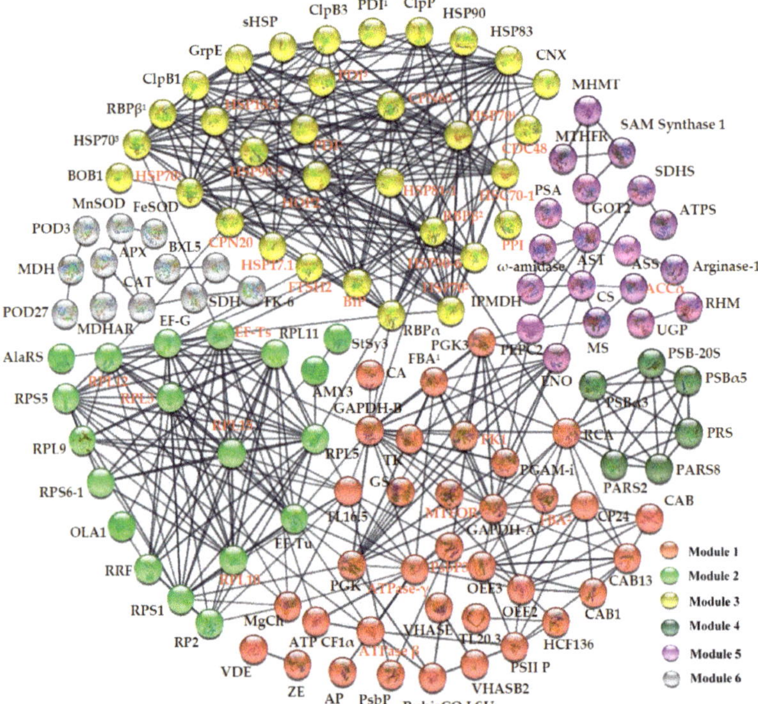

Figure 8. The protein–protein interaction (PPI) network of heat-responsive proteins in spinach Sp73 revealed by functional protein association networks (STRING) analysis. A total of 123 unique homologous proteins from *Arabidopsis thaliana* are shown in the PPI network. Six modules are indicated in different colors. Module 1: photosynthesis; Module 2: protein synthesis; Module 3: protein folding; Module 4: protein degradation; Module 5: sugar metabolism, amino acid metabolism, and other metabolisms; Module 6: ROS scavenging and stress defense. The PPI network is shown in the confidence view generated by STRING analysis. Stronger associations are represented by thicker lines. Please refer to Supplementary Table S7 for abbreviations.

3. Discussion

3.1. Heat-Inhibited Photosynthesis in Heat-Sensitive Spinach Variety Sp73

Heat stress has a negative impact on the photosynthetic capacity of plants [8], such as grape (*Vitis vinifera*) [28], wheat [17] and soybean [22]. Photosynthetic reactions which occur in thylakoid membranes (both stacked grana and lamellae) and carbon metabolism in the stroma of chloroplasts are primary sites of heat damage [29]. In general, spinach is a heat-sensitive vegetable species. Through genetic screening, we identified a heat-tolerant variety, Sp75 [11]. Our results in this study showed that photosynthesis was decreased in both the heat-tolerant variety Sp75 and heat-sensitive variety Sp73 under heat stress (Figure 2). Compared with the heat-tolerant Sp75, the heat-sensitive Sp73 exhibited obvious decreases in photosynthetic parameters, such as Pn and Gs, throughout the heat-stress process (Figure 2; Supplementary Figure S2).

The PSII is highly thermolabile and its activity is greatly diminished at high temperature [30], and this decrease may be due to PSII localization in the thylakoid membrane [31]. Our proteomics results showed that the abundances of most PSII proteins (e.g., chlorophyll a-b binding protein, oxygen-evolving enhancer protein, OEC subunit PsbP, and high chlorophyll fluorescence 136) were all decreased in the Sp73 under heat stress process. This indicates that OEC might be dissociated and the potential active center of PSII was damaged in the heat-sensitive spinach Sp73 [32]. This is similar to that found in heat-sensitive *A. stolonifera* [33], leading to the inhibition of light harvesting and electron transport under heat stress [34].

Moreover, we identified that 11 of the 19 HRPs in the Calvin cycle were decreased in the Sp73, resulting in inhibited carbon assimilation (Figure 6) [29]. This is consistent with previous studies in grape [35] and soybean [21]. Among the Calvin cycle enzymes, ribulose bisphosphate carboxylase/oxygenase (RuBisCO) and RuBisCO activase (RCA) were reported to be very sensitive to high temperature and showed diverse abundance patterns. For example, RCA in the wild downy grape leaves was decreased [35]. However, RCAs in *A. thaliana* [13] and rice [14] leaves were increased in response to heat stress. RCA is a molecular chaperone that plays an important role in maintaining the catalytic activity of RuBisCO [36] and is one of the limiting factors for photosynthesis under heat stress [37,38]. In addition, enzymes involved in chlorophyll synthesis (e.g., magnesium-chelatase subunit ChlI) and carotenoid biosynthesis (e.g., zeta-carotene desaturase) were decreased under heat stress. On the other hand, pheophorbide oxygenase involved in chlorophyll degradation was increased in the Sp73. This is quite different from the nine heat-increased chlorophyll biosynthetic enzymes in the heat-tolerant Sp75 (Supplementary Figure S3 and Table S8). This implies that the heat-sensitive Sp73 has a lower capability of chlorophyll maintenance than Sp75 for photosynthesis under heat stress [11].

In addition, the xanthophyll cycle is usually regarded as the most important photoprotection mechanism in higher plants. In this cycle, violaxanthin de-epoxidase and zeaxanthin epoxidase are critical enzymes. In Sp73, violaxanthin de-epoxidase and zeaxanthin epoxidase were decreased, indicating the photoprotection of the reaction center of PS II was reduced under heat stress [39].

3.2. Heat-Altered ROS Scavenging Pathways in Spinach Sp73

In spinach Sp73 leaves, the accumulation of H_2O_2 was obviously induced by heat stress, leading to membrane lipid peroxidation and damage (Figure 3) [40]. The accumulated ROS may be due to the low electron transport efficiency under heat stress [34]. Here, we found that various anti-oxidative enzymes and antioxidants were altered in intracellular H_2O_2 scavenging to cope with heat stress (Figure 4). Among them, the SOD activity was decreased by heat stress, in spite of the increased abundance of chloroplast-/mitochondrion-localized SODs under heat stress. This suggests that the first line of heat response to dismutate the intracellular $O_2^{\bullet-}$ to H_2O_2 was inhibited in the Sp73 [41], while the activities and abundances of SOD in the heat-tolerant Sp75 (Supplementary Figure S3) [11] and other heat-tolerant species (e.g., maize [20], poplar (*Populus yunnanensis*) [42] and wheat [43,44] were increased by heat stress. Interestingly, other antioxidant enzymes involved in converting intracellular

H_2O_2 to H_2O showed distinct activity patterns under heat stress. Among them, CAT activities were decreased, while APX and GPX activities were increased under the heat-stress process. The activity changes of CAT and APX in the Sp73 were opposite to those in the Sp75 (Supplementary Figure S2), indicating the activation of different antioxidant enzymes to scavenge ROS between the heat-sensitive Sp73 and heat-tolerant Sp75 [11]. Besides this, the accumulation of DHA, GSH and GSSG in the AsA-GSH pathway under the heat-stress process was observed in Sp73, similar to what happened in the heat-tolerant Sp75 (Supplementary Figure S3), maize [20], wheat [43] and poplar [42] under heat stress. It appears that the AsA-GSH pathway is conserved in plants to remove the heat-induced H_2O_2 [11]. Another similar strategy between the Sp73 and Sp75 is the increased abundances of GST and AKR (Supplementary Figure S3 and Table S8), which can contribute to detoxifying lipid peroxidation-derived reactive aldehydes and lead to enhanced salt-stress tolerance [45].

In addition, the accumulation of non-enzymatic antioxidants (e.g., proline and soluble sugars) can help to buffer redox changes and stabilize subcellular structures [46]. This is an effective strategy for dealing with heat stress-induced oxidative stress [47]. In the heat-tolerant Sp75, the contents of proline and soluble sugars were increased in response to heat stress (Supplementary Figure S2). Similar cases were also observed in tomato (*Lycopersicon esculentum*) [48], tobacco (*Nicotiana tabacum*) [49] and poplar [42], indicating that the heat stress-induced accumulation of soluble sugars and proline plays an important role in heat-stress tolerance. However, the soluble sugar content was decreased in the Sp73. This result may explain the heat sensitive phenotype of the Sp73.

3.3. Heat-Stress Signaling and Transport Pathways in Spinach Sp73

Plasma membrane fluidity and calcium ion channels are disturbed under high temperatures, leading to the entry of calcium ions into cells. Calcium ions can also be released from intracellular calcium stores, thereby activating calcium-dependent signal transduction pathways in response to high temperatures [4]. Our previous proteomics investigation revealed that several signaling proteins (e.g., calcium-dependent protein kinase 3, multiprotein bridging factor 1, dehydration-responsive element-binding, and 14–3-3 protein) were accumulated and the abundance of protein phosphatase 2C (a negative regulator in the mitogen-activated protein kinase (MAPK) pathway) was decreased by heat stress (Supplementary Figure S3 and Table S8), indicating the induction of a calcium-mediated MAPK cascade in the heat-tolerant Sp75 [11]. Besides this, several transporters involved in the transport of water, ions and metabolites were increased in the heat-tolerant Sp75 under high-temperature stress (Supplementary Figure S3 and Table S8) [11]. However, in the heat-sensitive Sp73, only a few calcium-related proteins and transporters were detected to be changed in levels. Among them, only annexin (ANN) and adenosine-triphosphate (ATP)-binding cassette (ABC) transporters were increased by heat stress. ANN is a calcium-dependent membrane-bound protein with ion channel activity. Its important role in the adaptation of plant cells to osmotic stress has been demonstrated [15]. AtANN1 is important in regulating the heat stress-induced $[Ca^{2+}]_{cyt}$ in Arabidopsis seedlings [18]. ATP-binding cassette transporters carry out the transmembrane transport of various biomolecules using the energy generated from hydrolyzing ATP. They play an important role in cellular detoxification, plant growth and development, and pathogen defense [28]. In addition, four H^+-ATPases were heat-decreased, indicating the decrease of trans-membrane proton movement, and the perturbed membrane potential and secondary solute transport in the heat-sensitive Sp73 [50,51].

3.4. Heat Stress-Perturbed Diverse Primary and Secondary Metabolisms

Carbohydrate metabolism plays an important role in plant growth, development and stress response [52]. Our proteomic results indicate that the carbohydrate metabolism tends to be induced in the heat-sensitive Sp73 under heat stress. For example, 18 out of 26 enzymes involved in glycolysis, the tricarboxylic acid (TCA) cycle, and other sugar metabolism were heat-increased. This result is similar to the heat-tolerant Sp75 (Supplementary Figure S3 and Table S8) [11]. It was also reported that fructose-bisphosphate aldolases (FBAs) in alfalfa seeds [10], wild downy grape leaves [35], and enolases in wheat spikelets and seeds were increased under heat stress [53,54]. In addition, we found here that a starch synthase was increased but an alpha amylase was decreased, which would favor starch accumulation and tolerance to heat stress [55].

Amino acid metabolism showed differential change patterns between the heat-sensitive Sp73 and the heat-tolerant Sp75 under heat stress (Supplementary Figure S3 and Table S8). In Sp73, 10 out of 17 enzymes involved in amino acid metabolism were heat-decreased. Among them, glutamine synthetase (GS) catalyzing the assimilation of ammonium to glutamine was significantly decreased, which would reduce the nitrogen metabolism in spinach under heat stress. This is consistent with the findings in *Agrostis* grass [56] and wheat [18]. In contrast, most of the amino acid metabolic enzymes (76%) in Sp75 were obviously increased under heat stress (Supplementary Figure S3 and Table S8). These results indicate that the enhancement of amino acid metabolism is critical for heat-stress tolerance [11].

3.5. Heat Stress-Induced Transcriptional Regulation and Protein Processing in Sp73

In the heat-stressed Sp73, the repair of chromosome/DNA and transcriptional regulation might be enhanced, because two crucial proteins (high-mobility group-Y-related protein A-like and DNA double-strand break repair rad50 ATPase) were significantly increased to facilitate nucleosome formation, transcriptional regulation [57], telomere maintenance, and DNA damage checkpoint control [58]. Besides this, we found that nucleus-localized RNA-binding protein family members (RBPs) and poly(A) polymerase were increased, but chloroplast-localized RBP was decreased in the Sp73 under stress. These results indicate that the RNA stability, maturation, and transport in nucleus were enhanced, but were decreased in chloroplasts when Sp73 experienced heat stress. In addition, our proteomics results indicated that protein synthesis was inhibited in the Sp73. A number of small ribosomal subunits (e.g., RP, S1, S5 and S6) and large ribosomal subunits (e.g., L3, L5, L9, L10, L11, L12 and L15), as well as translation-related factors (e.g., elongation factors Tu, elongation factors Ts and elongation factors G) were decreased. In contrast, ribosome-inactivating proteins and rRNA N-glycosidase proteins, which inactivate ribosomes, were increased in the Sp73 under heat stress.

Importantly, we found that 34 out of 37 (92%) proteins involved in protein folding and processing were significantly increased in Sp73 under heat stress, including heat shock protein 70 (HSP70), heat shock cognate 70 kDa protein 1 (HSC70–1), HSP70–2, HSP83, HSP90–5, HSP70-HSP90 organizing protein, chaperonin (Cpn), protein GrpE, and small HSPs. The heat-enhanced protein folding and processing in Sp73 were consistent with Sp75 (Supplementary Figure S3 and Table S8). These increased proteins may help to prevent proteins from improper folding, denaturation, and aggregation under heat stress [59]. It has been reported that the HSP70 family can interact with HSP90 to promote protein folding and maintain a stable structure under heat stress [60], and Cpn60 can protect RuBisCO activase from thermal denaturation and function in acclimating photosynthesis under heat stress [61]. A significant increase in the abundances of Cpn60, HSP70, HSP90, and small HSPs were also found in rice [14,15], soybean [21], purslane (*Portulaca oleracea*) [62], and *C. spinarum* [63]. For example, HSP70s, chaperonins and small HSPs were increased in rice seedlings at 45 °C for 48 h [14] and in rice leaves under 42 °C for 12 h and 24 h [15]. Besides this, HSP70, heat shock cognate (HSC) 70, and several low molecular weight HSPs (e.g., HSP22, HSP18.5, and HSP 17.5) were newly induced and/or highly increased in soybean leaves, stems, and roots under 40 °C [21]. Moreover, HSP70, HSP90, and the molecular chaperones DnaK and DnaJ were all increased in purslane under 35 °C [62]. Importantly, the HSPs and chaperonins accounted for the largest category (43.3%) of heat-responsive proteins in *C. spinarum* under 35 °C treatment. Among them, small HSPs were remarkably induced [63]. These results suggest the increased abundances of molecular chaperones are necessary for thermos-tolerance [64]. In addition, four protein disulfide-isomerases contributing to the formation of natural disulfide bonds were significantly increased under heat stress, allowing proteins to enter the normal folding pathway under heat stress.

In addition, there is an active protein degradation pathway to prevent the accumulation of non-functional or potentially toxic proteins in heat-stressed spinach leaves. In Sp73, ten proteins associated with protein degradation were increased under heat stress; e.g., ubiquitin-activating enzyme E1, ion protease, cysteine protease, 26S proteasome proteins, and 20S proteasome proteins. Interestingly, the abundance of ion protease was increased 23-fold at 48 h under heat stress, which can help to maintain the cellular protein turnover by mediating the abnormal or transient regulation of protein degradation. This is consistent with that shown in Sp75 (Supplementary Figure S3 and Table S8) [11].

4. Materials and Methods

4.1. Plant Cultivation and Treatment

Seeds of spinach (*Spinacia oleracea* L.) variety Sp73 were collected from the Germplasm Resources Center of Shanghai Normal University. Seeds were sown on the sterilized culture matrix and grown in a growth chamber with a 22 °C / 18 °C and 10 h / 14 h day/night cycle, and 60% relative humidity for 50 days. Plants were watered daily to avoid the occurrence of water deficit. On the 50th day, plants of the treatment groups were moved to another growth chamber with the same growth condition as the control, except for temperature (37 °C / 32°C day/night), and watered daily on a regular schedule as well. The spinach Sp73 were treated for 0 h, 24 h, 48 h and 72 h. After plant morphological changes were recorded, fully expanded true leaves were collected for both control and heat-treated plants, immediately frozen in liquid nitrogen and stored at −80 °C for future physiological and proteomics analyses. For each treatment, at least three biological replicates were performed [11].

4.2. RWC Measurement

To determine RWC, 0.2 g of fresh leaves were detached, weighed immediately and recorded as fresh weight (A). Subsequently, the leaves were then immersed in distilled water for 24 h, the turgid weight (B) was quickly measured, and they were dried at 80 °C for 2 h and then 60 °C to constant weight, and the dry weight (C) was recorded. The RWC as calculated as follows: RWC = [(A −C) / (B − C)] × 100% [11].

4.3. Photosynthesis and Chlorophyll Fuorescence Parameter Measurement

Photosynthetic parameters (Gs, Ci, Pn, and Tr) were measured on fully expanded leaves of each plant using a portable photosynthesis system LICOR 6400 (LI-COR Inc., Lincoln, NE, USA) [65]. Fv/Fm and Y (II) were determined using OS5p+ (Model OS5p+, OPTI-Sciences, Hudson, NH, USA). After the dark adaptation of spinach leaves for 0.5 h, Fv/Fm was measured 4 times. Y(II) was measured 4 times using a red-light source.

4.4. Determination of TBARS Content, REL, Total Soluble Sugar, and Proline Contents

Lipid peroxidation was measured as the amount of TBARS determined by the thiobarbituric acid reaction according to Lee et al. [15,66]. TBARS was extracted from fresh leaves in 10% trichloroacetic acid and 0.6% thiobarbituric acid solution under 100 °C for 5 min. The absorbance of the supernatant at 450 nm, 532 nm, and 600 nm was detected as OD450, OD532 and OD600, respectively. The TBARS concentration was calculated according to the following equations: C (nmol L^{-1}) = 6.45 × (OD532−OD600) − 0.56 × OD450; TBARS concentration (nmol·g^{-1} **FW**) =C × V/FW (V, volume of total extraction solution; FW, fresh weight of leaves).

The REL was determined as described by Wang et al. [66]. The electrical conductivity of deionized water (E0) was detected at room temperature, which was measured using a conductivity instrument (DDS-11A). The fresh leaves were cut and immersed in 20 mL deionized water, and then were incubated at 100 °C for 10 min. The electrical conductivity of samples before and after boiling were recorded as E1 and E2. The REL was calculated according to the equation REL (%) = (E1 − E0) / (E2 − E0) × 100%.

The contents of soluble sugars and proline were measured using ninhydrin reaction and sulfuric acid–anthrone reagents according to a previous report [66]. For soluble sugar assay, the fresh leaves were ground in deionized water and incubated at 100 °C for 30 min. The supernatant was collected and mixed with 2% (*w/v*) ethyl acetate solution of anthrone and concentrated sulfuric acid, and then was incubated at 100 °C for 1 min. After cooling down to room temperature, the absorbance of the solution was measured at 630 nm using a spectrophotometer. Soluble sugar concentration was calculated with the concentration (C) determined from glucose standard curves, volume of total extracion solution (V), and fresh weight of leaves (FW) according to the following equation: soluble sugar concentration (mg·g^{-1} FW) = C × V/FW. For proline assay, the fresh leaves were ground in 3% sulfosalicylic acid and incubated at 100 °C for an hour. The supernatant was collected after being centrifuged at 15,000 rpm for 5 min. Then, 1 mL supernatant, 2 mL glacial acetic acid, 2 mL ninhydrin were mixed and boiled for 30 min. After cooling down, 4 mL methylbenzene was added into the mixture and stood for two hours. The upper layer was collected for absorbance reading at 520 nm using a spectrophotometer. The proline concentrations in samples were calculated with concentration (C), volume of total extraction solution (V), and fresh weight of leaves (FW): proline concentration (μg·g^{-1} FW) = C × V/FW.

4.5. Determination of ROS and Antioxidant Substance Contents, and Antioxidant Enzyme Activity Assay

The H$_2$O$_2$ content in leaves was measured according to the method of Ibrahim et al. [67]. The leaves were ground with 0.1% trichloroacetic acid and then centrifuged at 15,000× *g* for 15 min at 4 °C. The supernatant was collected and determined spectrophotometrically at 390 nm after reacting with potassium iodide. To determine the O$_2$$^-$ generation rate and antioxidant enzyme activities, 0.2 g frozen leaves were ground in an extraction buffer containing 50 mM phosphate buffer solution (pH 7.8), 2% polyvinylpyrrolidone-40, and 2 mM reduced ascorbate (AsA) (for ascorbate peroxidase (APX) activity assay) at 4 °C. After centrifugation at 20,000× *g* for 15 min at 4 °C, the supernatant was collected for analysis. The O$_2$$^{\bullet-}$ generation rate was measured using a hydroxylamine oxidization method [68].

The activities of six antioxidant enzymes (SOD, CAT, POD, APX, GPX and GST) were determined according to the method of Yin et al. [68]. SOD activity was assayed on the basis of its ability to inhibit the photochemical reduction of nitro blue tetrazolium (NBT) at 560 nm. One unit of SOD activity was defined as the amount of enzyme that inhibited 50% of NBT photoreduction [68]. CAT activity was assayed by measuring H$_2$O$_2$ consumption at 240 nm [68]. POD activity was determined by a guaiacol method at 470 nm [65]. APX activity was measured by monitoring the absorbance decrease at 290 nm as the ascorbate was oxidized [20]. GPX activity was measured by recording the absorbance changes at 340 nm because of the oxidation of NADPH [69]. GST activity was measured by the product of CDNB conjugated with GSH absorbed at 340 nm [65]. The activities of MDHAR, DHAR and GR were measured by recording the absorbance changes at 340 nm due to the oxidation of NADH, at 265 nm due to the production of oxidized glutathione (GSSG) (ε = 14 mM^{-1}·cm^{-1}), and at 340 nm due to the oxidation of NADPH (ε = 6.22 mM^{-1}·cm^{-1}), respectively. Their activities were subsequently expressed as the amount of NADH oxidized, GSSG produced, and NADPH oxidized per minute per milligram protein, respectively [68]. For the contents of AsA and DHA, total AsA and reduced AsA were determined by recording the absorbance changes at 525 nm [68]. DHA content was estimated from the difference between assays with and without dithiothreitol (DTT) [68].

4.6. Quantitative Proteomics Analyses

The proteins were extracted using a phenol extraction method according to Yu et al. [65]. The protein pellet was dissolved in a lysis solution (7 M urea, 2 M thiourea, 4% 3-[(3-Cholamidopropyl) dimethyl-ammonio] propanesulfonic acid (CHAPS), 0.04 M DTT, and 4% protease inhibitor cocktail). Protein concentration was determined using a 2D Quant Kit according to the manufacturer's instructions (GE Healthcare, Uppsala, Sweden).

For 2DE-based proteomics analysis, about 1.6 mg protein extract was loaded into per gel, separated on linear gradient IPG strips (24 cm, pH 4–7) through isoelectric focusing (IEF) in the first dimension, then transferred into 12.5% SDS-PAGE for two-dimensional electrophoresis, and stained by Coomassie Brilliant Blue staining. Gel image acquisition and analysis were conducted as described in detail by Wang et al. [66]. The volume of each spot was normalized to the total volume of all the detected spots.

Protein spots considered to be differential abundant proteins needed to show consistent abundance changes from three biological replicates with greater than 1.5-fold changes and a p value smaller than 0.05 [65]. In-gel digestion was performed on the protein spots with abundance differences. MS was calibrated using a quality standard kit (AB Sciex Inc., Frammingham, MA, USA) and a bovine serum albumin standard (Sigma-Aldrich, St. Louis, MO, USA). MS/MS spectra were obtained using an ABI 5800 MALDI TOF/TOF MS (AB Sciex, Foster City, CA, USA). The mass spectrum error of MS and MS/MS was less than 30 ppm, and the resolution was 10,000. To determine the confidence of the protein search results, the following criteria were applied: (1) top-ranked search results (top five results); (2) the probability score obtained by molecular weight searching (MOWSE) should be greater than 50 ($p < 0.01$); (3) at least two peptide matches, all Y-ion series and partially complementary B-ion series should correspond to the high-intensity peaks.

For iTRAQ-based proteomics analysis, the protein samples of spinach leaves treated for 0 h, 24 h, 48 h and 72 h were digested using trypsin (1:50, *w/w*, trypsin: sample). The digested samples were labeled with iTRAQ reagents 113 & 117 (0 h), 114 & 118 (24 h), 115 & 119 (48 h), and 116 & 121 (72 h). Then, different samples were mixed together. The peptide mixture was fractionated on XBridge C18 column (150 mm × 4.6 mm, 5 μm, Waters, Milford, MA, USA) using a Shimadzu LC-20A high-performance liquid chromatography (HPLC) system (Shimadzu, Kyoto, Japan). Each of the fractionated components was desalted with 3 M Empore C18 solid-phase extraction disks (3 M Bioanalytical Technologies, St.Paul, MN, USA). Peptide samples were identified by an online nanoacquity ultraperformance LC (Waters, Milford, MA, USA) coupled with an Orbitrap Fusion Tribrid mass spectrometer (Thermo Fisher Scientific, San Jose, CA, USA) using a method by Zhao et al. [11]. The MS2 spectra obtained from the MS analysis were searched against a protein database [70] via ProteinPilot Software 4.5 (AB Sciex, Frammingham, MA, USA). The credible protein identification and quantitative results needed to meet the following criteria: unused protein score > 1.3 and number of unique peptides ≥ 2. In at least three biological replicates, the proteins with fold changes >1.2 and $p < 0.05$ in the treatment group compared to the control group were defined as HRPs.

4.7. Protein Function Classification and Cluster Analysis

Protein functional domains were analyzed by BLAST alignment against the NCBInr protein database [71], and the proteins were classified into different functional groups by combining with the knowledge from the Kyoto Encyclopedia of Genes and Genomes (KEGG) pathway database [72], UniProt database [73], Gene Ontology database [74], and literature information.

4.8. Protein Subcellular Localization Prediction

Protein subcellular localization was predicted with five different online tools: YLoc [75], LocTree3 [76], ngLOC [77], Plant-mPLoc [78], TargetP [79]. Only the consistent predictions from at least two tools were accepted as confident results listed in the column of "confirmed localization".

4.9. Protein-Protein Interaction Prediction

The homologs of spinach heat-stress response proteins in Arabidopsis were obtained by sequence BLAST alignment in the Arabidopsis Information Resource (TAIR) database. The homologs were used in the Web tool of STRING 10.5 [80] to create the PPI network.

4.10. Statistical Analysis

Each physiological/proteomics experiment result was obtained from at least three biological replicates. The experimental results were listed as mean ± standard deviation. The differences between different treatment samples in respect of physiological indexes were compared using one-way analysis of variance (one-way ANOVA) in SPSS 17.0. The differences between individual treatment group and the control group in respect of the protein abundance were compared using the Student t-test, and $p < 0.05$ was considered to be statistically significant.

5. Conclusions

In this study, the physiological and proteomic changes in the heat-sensitive spinach Sp73 were reported and compared with those in the heat-tolerant spinach Sp75. Interestingly, the two spinach varieties possess some common thermal response processes, but also show specificity in response to heat stress. Our results revealed that photosynthesis was heat stress-inhibited, but ROS scavenging pathways and stress defense, carbohydrate and energy metabolism, and protein folding and degradation were heat stress-enhanced in both Sp73 and Sp75, implying the conservation of these processes in the response of spinach to high-temperature stress. Notably, calcium signaling, endomembrane trafficking, as well as the regulation of the cell cycle and differentiation were specifically enhanced in heat-treated spinach Sp75. Moreover, signal transduction, protein synthesis, and amino acid metabolism were heat stress-suppressed in spinach Sp73 but enhanced in Sp75. All these data provide important insights into the molecular mechanisms underlying the heat-stress response/tolerance of the two contrasting spinach varieties Sp75 and Sp73.

Supplementary Materials: Supplementary materials can be found at http://www.mdpi.com/1422-0067/20/16/3872/s1.

Author Contributions: Formal analysis, H.Z.; Funding acquisition, S.D., Q.W., S.L., H.Z., C.X., X.C. and X.W.; Investigation, S.L., J.Y., Y.L., J.B., and X.B.; Methodology, P.W., S.G. and Y.M.; Project administration, Z.Q. and S.D.; Resources, C.X., X.W., X.C., Q.W. and Z.Q.; Supervision, Z.Q. and S.D.; Writing—original draft, S.L. and J.Y.; Writing—review & editing, S.C. and S.D. All authors have read and approved the final manuscript.

Funding: The project was supported by grants from the Foundation of Shanghai Science and Technology Committee, China (No. 17391900600 and No. 16391901000) to Shaojun Dai and Quanhua Wang, the China Postdoctoral Science Foundation Grant (No. 2019M651541 and No. 2019M651542) to Shanshan Li and Heng Zhang, the Found of Shanghai Engineering Research Center of Plant Germplasm Resources (No. 17DZ2252700), the National Natural Science Foundation of China (No. 31601769, No. 31501754, and No. 31601744) to Chenxi Xu, Xiaofeng Cai and Xiaoli Wang, and the Foundation of Shanghai Municipal Agricultural Commission (No. 201509) to Quanhua Wang.

Conflicts of Interest: The authors declare no conflict of interest.

Abbreviations

2DE	Two-dimensional gel electrophoresis
AKR	Aldo/keto reductase
ANN	Annexin
APX	Ascorbate peroxidase
AsA	Ascorbate
ATP	Adenosine-triphosphate
BLAST	Basic Local Alignment Search Tool
CAT	Catalase
Ci	Intercellular CO_2
DHA	Dehydroascorbate
DHAR	Dehydroascorbate reductase
Fv/Fm	PSII maximum quantum yield
FW	Fresh weight
GPX	Glutathione peroxidase
GR	Glutathione reductase
Gs	Stomata conductance
GSH	Reduced glutathione
GSSG	Oxidized glutathione
GST	Glutathione S-transferase
H_2O_2	Hydrogen peroxide
HHT	Hour of heat treatment
HRP	Heat stress-responsive protein
IEF	Isoelectric focusing
IPG	Immobilized pH gradient

iTRAQ	Isobaric tags for relative and absolute quantification
MALDI	Matrix-assisted laser desorption/ ionization
MAPK	Mitogen-activated protein kinase
MDHAR	Monodehydroascorbate reductase
MS	Mass spectrometry
NCBInr	National Center for Biotechnology Information non-redundant
$O_2^{\bullet-}$	Superoxide anion radicals
OEC	Oxygen evolving complex
pI	Isoelectric point
Pn	Photosynthesis rate
POD	Peroxidase
PPI	Protein-protein interactions
PSII	Photosystem II
RBP	RNA-binding protein family member
RCA	Ribulose bisphosphate carboxylase/oxygenase activase
REL	Relative electrolyte leakage
ROS	Reactive oxygen species
RubisCO	Ribulose-1,5-bisphosphate carboxylase/oxygenase
RWC	Relative water content
SDS-PAGE	Sodium dodecyl sulfate polyacrylamide gel electrophoresis
SOD	Superoxide dismutase
Sp73	Heat-sensitive spinach
Sp75	Heat-tolerant spinach
STRING	Search tool for recurring instances of neighbouring genes
TBARS	Thiobarbituric acid reactive substance
TOF	Time of flight
Tr	Transpiration rate
Y(II)	Effective PSII quantum yield

References

1. Fahad, S.; Bajwa, A.A.; Nazir, U.; Anjum, S.A.; Farooq, A.; Zohaib, A.; Sadia, S.; Nasim, W.; Adkins, S.; Saud, S.; et al. Crop production under drought and heat stress: Plant responses and management options. *Front Plant Sci.* **2017**, *8*, 1147. [CrossRef] [PubMed]

2. Lamaoui, M.; Jemo, M.; Datla, R.; Bekkaoui, F. Heat and drought stresses in crops and approaches for their mitigation. *Front. Chem.* **2018**, *6*, 26. [CrossRef] [PubMed]

3. Bita, C.E.; Zenoni, S.; Vriezen, W.H.; Mariani, C.; Pezzotti, M.; Gerats, T. Temperature stress differentially modulates transcription in meiotic anthers of heat-tolerant and heat-sensitive tomato plants. *BMC Genom.* **2011**, *12*, 384. [CrossRef] [PubMed]

4. Mittler, R.; Finka, A.; Goloubinoff, P. How do plants feel the heat? *Trends Biochem. Sci.* **2012**, *37*, 118–125. [CrossRef] [PubMed]

5. Bokszczanin, K.L.; Fragkostefanakis, S. Perspectives on deciphering mechanisms underlying plant heat stress response and thermotolerance. *Front Plant Sci.* **2013**, *4*, 315. [CrossRef] [PubMed]

6. Hasanuzzaman, M.; Nahar, K.; Alam, M.M.; Roychowdhury, R.; Fujita, M. Physiological, biochemical, and molecular mechanisms of heat stress tolerance in plants. *Int. J. Mol. Sci.* **2013**, *14*, 9643–9684. [CrossRef] [PubMed]

7. McClung, C.R.; Davis, S.J. Ambient thermometers in plants: From physiological outputs towards mechanisms of thermal sensing. *Curr. Biol.* **2010**, *20*, R1086–R1092. [CrossRef] [PubMed]

8. Mathur, S.; Agrawal, D.; Jajoo, A. Photosynthesis: Response to high temperature stress. *J. Photochem. Photobiol. B.* **2014**, *137*, 116–126. [CrossRef]

9. Wang, X.; Xu, C.; Cai, X.; Wang, Q.; Dai, S. Heat-responsive photosynthetic and signaling pathways in plants: Insight from proteomics. *Int. J. Mol. Sci.* **2017**, *18*, 2191. [CrossRef] [PubMed]

10. Li, W.; Wei, Z.; Qiao, Z.; Wu, Z.; Cheng, L.; Wang, Y. Proteomics analysis of alfalfa response to heat stress. *PLoS ONE* **2013**, *8*, e82725. [CrossRef]

11. Zhao, Q.; Chen, W.; Bian, J.; Xie, H.; Li, Y.; Xu, C.; Ma, J.; Guo, S.; Chen, J.; Cai, X.; et al. Proteomics and phosphoproteomics of heat stress-responsive mechanisms in Spinach. *Front Plant Sci.* **2018**, *9*, 800. [CrossRef] [PubMed]

12. Zhu, Y.; Zhu, G.; Guo, Q.; Zhu, Z.; Wang, C.; Liu, Z. A comparative proteomic analysis of *Pinellia ternata* leaves exposed to heat stress. *Int. J. Mol. Sci.* **2013**, *10*, 20614–20634. [CrossRef] [PubMed]

13. Rocco, M.; Arena, S.; Renzone, G.; Scippa, G.S.; Lomaglio, T.; Verrillo, F.; Scaloni, A.; Marra, M. Proteomic analysis of temperature stress-responsive proteins in *Arabidopsis thaliana* rosette leaves. *Mol. Biosyst.* **2013**, *9*, 1257–1267. [CrossRef] [PubMed]

14. Han, F.; Chen, H.; Li, X.J.; Yang, M.F.; Liu, G.S.; Shen, S.H. A comparative proteomic analysis of rice seedlings under various high-temperature stresses. *Biochim. Biophys. Acta.* **2009**, *1794*, 1625–1634. [CrossRef] [PubMed]

15. Lee, D.G.; Ahsan, N.; Lee, S.H.; Young Kang, K.; Dong Bahk, J.; Lee, I.J.; Lee, B.H. A proteomic approach in analyzing heat-responsive proteins in rice leaves. *Proteomics* **2007**, *7*, 3369–3383. [CrossRef] [PubMed]

16. Scafaro, A.P.; Haynes, P.A.; Atwell, B.J. Physiological and molecular changes in *Oryza meridionalis* Ng., a heat-tolerant species of wild rice. *J. Exp. Bot.* **2010**, *61*, 191–202. [CrossRef] [PubMed]

17. Lu, Y.; Li, R.; Wang, R.; Wang, X.; Zheng, W.; Sun, Q.; Tong, S.; Dai, S.; Xu, S. Comparative proteomic analysis of flag leaves reveals new insight into wheat heat adaptation. *Front Plant Sci.* **2017**, *8*, 1086. [CrossRef]

18. Wang, X.; Dinler, B.S.; Vignjevic, M.; Jacobsen, S.; Wollenweber, B. Physiological and proteome studies of responses to heat stress during grain filling in contrasting wheat cultivars. *Plant Sci.* **2015**, *230*, 33–50. [CrossRef]

19. Hu, X.; Wu, L.; Zhao, F.; Zhang, D.; Li, N.; Zhu, G.; Li, C.; Wang, W. Phosphoproteomic analysis of the response of maize leaves to drought, heat and their combination stress. *Front Plant Sci.* **2015**, *6*, 298. [CrossRef]

20. Zhao, F.; Zhang, D.; Zhao, Y.; Wang, W.; Yang, H.; Tai, F.; Li, C.; Hu, X. The difference of physiological and proteomic changes in maize leaves adaptation to drought, heat, and combined both stresses. *Front Plant Sci.* **2016**, *7*, 1471. [CrossRef]

21. Ahsan, N.; Donnart, T.; Nouri, M.Z.; Komatsu, S. Tissue-specific defense and thermo-adaptive mechanisms of soybean seedlings under heat stress revealed by proteomic approach. *J. Proteome Res.* **2010**, *9*, 4189–4204. [CrossRef] [PubMed]

22. Das, A.; Eldakak, M.; Paudel, B.; Kim, D.W.; Hemmati, H.; Basu, C.; Rohila, J.S. Leaf proteome analysis reveals prospective drought and heat stress response mechanisms in soybean. *Biomed Res. Int.* **2016**, *2016*, 6021047. [CrossRef] [PubMed]

23. Huang, W.; Ma, H.Y.; Huang, Y.; Li, Y.; Wang, G.L.; Jiang, Q.; Wang, F.; Xiong, A.S. Comparative proteomic analysis provides novel insights into chlorophyll biosynthesis in celery under temperature stress. *Physiol. Plant* **2017**, *161*, 468–485. [CrossRef] [PubMed]

24. Yan, J.; Yu, L.; Xuan, J.; Lu, Y.; Lu, S.; Zhu, W. De novo transcriptome sequencing and gene expression profiling of spinach (*Spinacia oleracea* L.) leaves under heat stress. *Sci. Rep.* **2016**, *6*, 19473. [CrossRef] [PubMed]

25. Tang, Y.; Wen, X.; Lu, Q.; Yang, Z.; Cheng, Z.; Lu, C. Heat stress induces an aggregation of the light-harvesting complex of photosystem II in spinach plants. *Plant Physiol.* **2007**, *143*, 629–638. [CrossRef] [PubMed]

26. Komayama, K.; Khatoon, M.; Takenaka, D.; Horie, J.; Yamashita, A.; Yoshioka, M.; Nakayama, Y.; Yoshida, M.; Ohira, S.; Morita, N.; et al. Quality control of photosystem II: Cleavage and aggregation of heat-damaged D1 protein in spinach thylakoids. *Biochim. Biophys. Acta.* **2007**, *1767*, 838–846. [CrossRef]

27. Yamashita, A.; Nijo, N.; Pospisil, P.; Morita, N.; Takenaka, D.; Aminaka, R.; Yamamoto, Y.; Yamamoto, Y. Quality control of photosystem II: Reactive oxygen species are responsible for the damage to photosystem II under moderate heat stress. *J. Biol. Chem.* **2008**, *283*, 28380–28391. [CrossRef]

28. Jianfu, J.; Liu, X.; Liu, G.-T.; Liu, C.; Li, S.; Wang, L. Integrating omics and alternative splicing reveals insights into grape response to high temperature. *Plant Physiol.* **2017**, *173*, 1502–1518.

29. Wise, R.; Olson, A.J.; Schrader, S.M.; Sharkey, T. Electron transport is the functional limitation of photosynthesis in field-grown Pima cotton plants at high temperature. *Plant Cell Environ.* **2004**, *27*, 717–724. [CrossRef]

30. Camejo, D.; Rodriguez, P.; Morales, M.A.; Dell'Amico, J.M.; Torrecillas, A.; Alarcon, J.J. High temperature effects on photosynthetic activity of two tomato cultivars with different heat susceptibility. *J. Plant Physiol.* **2005**, *162*, 281–289. [CrossRef]

31. McDonald, G.; Paulsen, G.M. High temperature effects on photosynthesis and water relations of grain legumes. *Plant Soil* **1997**, *196*, 47–58. [CrossRef]
32. As, W.; Galani, S.; Ashraf, M.; Foolad, M. Heat tolerance in plants: An overview. *Environ. Exp. Bot.* **2007**, *61*, 199–223.
33. Xu, C.; Huang, B. Differential proteomic response to heat stress in thermal *Agrostis scabra* and heat-sensitive *Agrostis stolonifera*. *Physiol. Plant.* **2010**, *139*, 192–204. [CrossRef] [PubMed]
34. Wang, X.; Cai, J.; Jiang, D.; Liu, F.; Dai, T.; Cao, W. Pre-anthesis high-temperature acclimation alleviates damage to the flag leaf caused by post-anthesis heat stress in wheat. *J. Plant Physiol.* **2011**, *168*, 585–593. [CrossRef] [PubMed]
35. Liu, G.T.; Ma, L.; Duan, W.; Wang, B.C.; Li, J.H.; Xu, H.G.; Yan, X.Q.; Yan, B.F.; Li, S.H.; Wang, J. Differential proteomic analysis of grapevine leaves by iTRAQ reveals responses to heat stress and subsequent recovery. *BMC Plant Biol.* **2014**, *14*, 110. [CrossRef] [PubMed]
36. Portis, A.R., Jr. Rubisco activase-Rubisco's catalytic chaperone. *Photosynth. Res.* **2003**, *75*, 11–27. [CrossRef] [PubMed]
37. Kurek, I.; Chang, T.K.; Bertain, S.M.; Madrigal, A.; Liu, L.; Lassner, M.W.; Zhu, G. Enhanced thermostability of *Arabidopsis* Rubisco activase improves photosynthesis and growth rates under moderate heat stress. *Plant Cell* **2007**, *19*, 3230–3241. [CrossRef]
38. Ristic, Z.; Momcilovic, I.; Bukovnik, U.; Prasad, P.V.; Fu, J.; Deridder, B.P.; Elthon, T.E.; Mladenov, N. Rubisco activase and wheat productivity under heat-stress conditions. *J. Exp. Bot.* **2009**, *60*, 4003–4014. [CrossRef]
39. Horton, P. *Crop Improvement through Alteration in the Photosynthetic Membrane*; ISB News Report; Virginia Tech: Blacksburg, VA, USA, 2002.
40. de Pinto, M.C.; Paradiso, A.; Leonetti, P.; De Gara, L. Hydrogen peroxide, nitric oxide and cytosolic ascorbate peroxidase at the crossroad between defence and cell death. *Plant J.* **2006**, *48*, 784–795. [CrossRef]
41. Mittler, R.; Vanderauwera, S.; Gollery, M.; Van Breusegem, F. Reactive oxygen gene network of plants. *Trends Plant Sci.* **2004**, *9*, 490–498. [CrossRef] [PubMed]
42. Li, X.; Yang, Y.; Sun, X.; Lin, H.; Chen, J.; Ren, J.; Hu, X.; Yang, Y. Comparative physiological and proteomic analyses of poplar (*Populus yunnanensis*) plantlets exposed to high temperature and drought. *PLoS ONE* **2014**, *9*, e107605. [CrossRef] [PubMed]
43. Kumar, R. Protection against heat stress in wheat involves change in cell membrane stability, antioxidant enzymes, osmolyte, H_2O_2 and transcript of heat shock protein. *Plant Physiol. Bioch.* **2012**, *4*, 83–91.
44. Sairam, R.K.; Srivastava, G.; Saxena, D. Increased antioxidant activity under elevated temperatures: A mechanism of heat stress tolerance in wheat genotypes. *Biol. Plant.* **2000**, *43*, 245–251. [CrossRef]
45. Sengupta, D.; Naik, D.; Reddy, A.R. Plant aldo-keto reductases (AKRs) as multi-tasking soldiers involved in diverse plant metabolic processes and stress defense: A structure-function update. *J. Plant Physiol.* **2015**, *179*, 40–55. [CrossRef] [PubMed]
46. Hayat, S.; Hayat, Q.; Alyemeni, M.N.; Wani, A.S.; Pichtel, J.; Ahmad, A. Role of proline under changing environments: A review. *Plant Signal Behav.* **2012**, *7*, 1456–1466. [CrossRef]
47. Wahid, A.; Close, T. Expression of dehydrins under heat stress and their relationship with water relations of sugarcane leaves. *Biol. Plant.* **2007**, *51*, 104–109. [CrossRef]
48. Rivero, R.; Jm, R.; Romero, L. Oxidative metabolism in tomato plants subjected to heat stress. *J. Hortic. Sci. Biotech.* **2004**, *79*, 560–564. [CrossRef]
49. Cvikrova, M.; Gemperlova, L.; Dobra, J.; Martincova, O.; Prasil, I.T.; Gubis, J.; Vankova, R. Effect of heat stress on polyamine metabolism in proline-over-producing tobacco plants. *Plant Sci.* **2012**, *182*, 49–58. [CrossRef]
50. Gaxiola, R.A.; Palmgren, M.G.; Schumacher, K. Plant proton pumps. *FEBS Lett.* **2007**, *581*, 2204–2214. [CrossRef]
51. Savchenko, A.; Vieille, C.; Kang, S.; Zeikus, J.G. Pyrococcus furiosus alpha-amylase is stabilized by calcium and zinc. *Biochemistry* **2002**, *41*, 6193–6201. [CrossRef] [PubMed]
52. Ruan, J.; Haerdter, R.; Gerendas, J. Impact of nitrogen supply on carbon/nitrogen allocation: A case study on amino acids and catechins in green tea [*Camellia sinensis* (L.) O. Kuntze] plants. *Plant Biol. (Stuttg.)* **2010**, *12*, 724–734. [CrossRef] [PubMed]
53. Laino, P.; Shelton, D.; Finnie, C.; De Leonardis, A.; Mastrangelo, A.; Svensson, B.; Lafiandra, D.; Masci, S. Comparative proteome analysis of metabolic proteins from seeds of durum wheat (cv. Svevo) subjected to heat stress. *Proteomics* **2010**, *10*, 2359–2368. [CrossRef] [PubMed]

54. Yang, F.; Jorgensen, A.D.; Li, H.; Sondergaard, I.; Finnie, C.; Svensson, B.; Jiang, D.; Wollenweber, B.; Jacobsen, S. Implications of high-temperature events and water deficits on protein profiles in wheat (*Triticum aestivum* L. cv. Vinjett) grain. *Proteomics* **2011**, *11*, 1684–1695. [CrossRef] [PubMed]

55. Majoul, T.; Bancel, E.; Triboi, E.; Ben Hamida, J.; Branlard, G. Proteomic analysis of the effect of heat stress on hexaploid wheat grain: Characterization of heat-responsive proteins from non-prolamins fraction. *Proteomics* **2004**, *4*, 505–513. [CrossRef] [PubMed]

56. Xu, C.; Huang, B. Root proteomic responses to heat stress in two *Agrostis* grass species contrasting in heat tolerance. *J. Exp. Bot.* **2008**, *59*, 4183–4194. [CrossRef]

57. Cai, X.; Gao, C.; Su, B.; Tan, F.; Yang, N.; Wang, G. Expression profiling and microbial ligand binding analysis of high-mobility group box-1 (HMGB1) in turbot (*Scophthalmus mavximus* L.). *Fish Shellfish Immunol.* **2018**, *78*, 100–108. [CrossRef] [PubMed]

58. Hopfner, K.P.; Karcher, A.; Shin, D.S.; Craig, L.; Arthur, L.M.; Carney, J.P.; Tainer, J.A. Structural biology of Rad50 ATPase: ATP-driven conformational control in DNA double-strand break repair and the ABC-ATPase superfamily. *Cell* **2000**, *101*, 789–800. [CrossRef]

59. Johnova, P.; Skalak, J.; Saiz-Fernandez, I.; Brzobohaty, B. Plant responses to ambient temperature fluctuations and water-limiting conditions: A proteome-wide perspective. *Biochim. Biophys. Acta.* **2016**, *1864*, 916–931. [CrossRef]

60. Larkindale, J.; Hall, J.D.; Knight, M.R.; Vierling, E. Heat stress phenotypes of *Arabidopsis* mutants implicate multiple signaling pathwayes in the acquisition of thermotolerance. *Plant Physiol.* **2005**, *138*, 882–897. [CrossRef]

61. Salvucci, M.E. Association of Rubisco activase with chaperonin-60beta: A possible mechanism for protecting photosynthesis during heat stress. *J. Exp. Bot.* **2008**, *59*, 1923–1933. [CrossRef] [PubMed]

62. Yang, Y.; Chen, J.; Liu, Q.; Ben, C.; Todd, C.D.; Shi, J.; Yang, Y.; Hu, X. Comparative proteomic analysis of the thermotolerant plant *Portulaca oleracea* acclimation to combined high temperature and humidity stress. *J. Proteome Res.* **2012**, *11*, 3605–3623. [CrossRef] [PubMed]

63. Zhang, M.; Li, G.; Huang, W.; Bi, T.; Chen, G.; Tang, Z.; Su, W.; Sun, W. Proteomic study of *Carissa spinarum* in response to combined heat and drought stress. *Proteomics* **2010**, *10*, 3117–3129. [CrossRef] [PubMed]

64. Perez, D.E.; Hoyer, J.S.; Johnson, A.I.; Moody, Z.R.; Lopez, J.; Kaplinsky, N.J. BOBBER1 is a noncanonical *Arabidopsis* small heat shock protein required for both development and thermotolerance. *Plant Physiol.* **2009**, *151*, 241–252. [CrossRef]

65. Yu, J.; Chen, S.; Zhao, Q.; Wang, T.; Yang, C.; Diaz, C.; Sun, G.; Dai, S. Physiological and proteomic analysis of salinity tolerance in *Puccinellia tenuiflora*. *J. Proteome Res.* **2011**, *10*, 3852–3870. [CrossRef] [PubMed]

66. Wang, X.; Chen, S.; Zhang, H.; Shi, L.; Cao, F.; Guo, L.; Xie, Y.; Wang, T.; Yan, X.; Dai, S. Desiccation tolerance mechanism in resurrection fern-ally *Selaginella tamariscina* revealed by physiological and proteomic analysis. *J. Proteome Res.* **2010**, *9*, 6561–6577. [CrossRef]

67. Ibrahim, M.H.; Jaafar, H.Z. Primary, secondary metabolites, H_2O_2, malondialdehyde and photosynthetic responses of *Orthosiphon stimaneus* Benth. to different irradiance levels. *Molecules* **2012**, *17*, 1159–1176. [CrossRef]

68. Yin, Z.; Balmant, K.; Geng, S.; Zhu, N.; Zhang, T.; Dufresne, C.; Dai, S.; Chen, S. Bicarbonate induced redox proteome changes in Arabidopsis suspension cells. *Front Plant Sci.* **2017**, *8*, 58. [CrossRef]

69. Suo, J.; Zhao, Q.; Zhang, Z.; Chen, S.; Cao, J.; Liu, G.; Wei, X.; Wang, T.; Yang, C.; Dai, S. Cytological and proteomic analyses of osmunda cinnamomea germinating spores reveal characteristics of fern spore germination and rhizoid tip growth. *Mol. Cell. Proteom.* **2015**, *14*, 2510–2534. [CrossRef]

70. Protein Database. Available online: http://www.spinachbase.org/ (accessed on 20 September 2018).

71. PSI-BLAST and PHI-BLAST Program. Available online: http://www.ncbi.nlm.nih.gov/BLAST/ (accessed on 15 April 2018).

72. KEGG Pathway Database. Available online: http://www.kegg.jp/kegg (accessed on 15 April 2018).

73. UniProt Database. Available online: http://www.ebi.uniprot.org/ (accessed on 15 May 2018).

74. Gene Ontology Database. Available online: http://geneontology.org (accessed on 15 May 2018).

75. YLoc: Interpretable Subcellular Localization Prediction. Available online: http://abi.inf.uni-tuebingen.de/Services/YLoc/webloc.cgi (accessed on 15 August 2018).

76. LocTree3: Protein Subcellular Localization Prediction System. Available online: https://rostlab.org/services/loctree3/ (accessed on 15 August 2018).

77. ngLOC: A Bayesian Method for Predicting Protein Subcellular Localization. Available online: http: //genome.unmc.edu/ngLOC/index.html (accessed on 15 August 2018).
78. Plant-mPLoc: Predicting Subcellular Localization of Plant Proteins Including Those with Multiple Sites. Available online: http://www.csbio.sjtu.edu.cn/bioinf/plant-multi/# (accessed on 15 August 2018).
79. TargetP Server. Available online: http://www.cbs.dtu.dk/services/TargetP/ (accessed on 15 August 2018).
80. Web Tool of STRING 10.5. Available online: http://string-db.org (accessed on 20 September 2018).

International Journal of
Molecular Sciences

Article

Comparative Proteomics Indicates That Redox Homeostasis Is Involved in High- and Low-Temperature Stress Tolerance in a Novel Wucai (*Brassica campestris* L.) Genotype

Lingyun Yuan [1,2,3,†], Jie Wang [1,2,†], Shilei Xie [1,2], Mengru Zhao [1,2], Libing Nie [1,2], Yushan Zheng [1,2], Shidong Zhu [1,2,3], Jinfeng Hou [1,2,3], Guohu Chen [1,2] and Chenggang Wang [1,2,3,*]

[1] College of Horticulture, Vegetable Genetics and Breeding Laboratory, Anhui Agricultural University, 130 West Changjiang Road, Hefei 230036, China

[2] Provincial Engineering Laboratory for Horticultural Crop Breeding of Anhui, 130 West of Changjiang Road, Hefei 230036, China

[3] Department of vegetable culture and breeding, Wanjiang Vegetable Industrial Technology Institute, Maanshan 238200, China

* Correspondence: cgwang@ahau.edu.cn; Tel./Fax: +86-0551-65786212

† These authors contributed equally to this work.

Received: 26 June 2019; Accepted: 30 July 2019; Published: 1 August 2019

Abstract: The genotype WS-1, previously identified from novel wucai germplasm, is tolerant to both low-temperature (LT) and high-temperature (HT) stress. However, it is unclear which signal transduction pathway or acclimation mechanisms are involved in the temperature-stress response. In this study, we used the proteomic method of tandem mass tag (TMT) coupled with liquid chromatography-mass spectrometry (LC-MS/MS) to identify 1022 differentially expressed proteins (DEPs) common to WS-1, treated with either LT or HT. Among these 1022 DEPs, 172 were upregulated in response to both LT and HT, 324 were downregulated in response to both LT and HT, and 526 were upregulated in response to one temperature stress and downregulated in response to the other. To illustrate the common regulatory pathway in WS-1, 172 upregulated DEPs were further analyzed. The redox homeostasis, photosynthesis, carbohydrate metabolism, heat-shockprotein, and chaperones and signal transduction pathways were identified to be associated with temperature stress tolerance in wucai. In addition, *35S:BcccrGLU1* overexpressed in *Arabidopsis*, exhibited higher reduced glutathione (GSH) content and reduced glutathione/oxidized glutathione (GSH/GSSG) ratio and less oxidative damage under temperature stress. This result is consistent with the dynamic regulation of the relevant proteins involved in redox homeostasis. These data demonstrate that maintaining redox homeostasis is an important common regulatory pathway for tolerance to temperature stress in novel wucai germplasm.

Keywords: wucai; low-temperature stress; high-temperature stress; proteomics; redox homeostasis; GLU1; glutathione

1. Introduction

As a consequence of climate change, temperature stress is becoming a major concern in plant research. Changes in low temperature (LT) and high temperature (HT) are expected to greatly affect enzyme activities, photosynthesis, carbon assimilation, DNA/RNA stability, membrane fluidity, and transcription and translation rates [1]. Exposure to extreme temperatures usually reduces photosynthesis efficiency, and the reduced photosynthesis disturbs cellular homeostasis and promotes lipid peroxidation, either by increasing the production of reactive oxygen species (ROS) or by decreasing the scavenging of superoxide anion ($O_2^{\bullet-}$) in the cell [2]. The accumulation of ROS depends on changes in the redox state of cells, and thus it could function to reset the redox state and maintain redox homeostasis [3]. Because the sulfur-containing amino acids cysteine and methionine are sensitive to redox potential, changes in the redox state can affect protein structure and folding [4]. Changes in redox potential may also alter enzyme activity, biochemical reactions, and plant physiological processes, which can negatively affect plant survival [5].

Wucai (*Brassica campestris* L. ssp. *chinensis* var. *rosularis* Tsen et Lee.), a species of non-heading Chinese cabbage (NHCC) and one of the most important leafy vegetables in China, is now extensively cultivated worldwide. Injury from temperature stress can decrease wucai yield and edible quality [6]. The novel genotype WS-1, which was previously identified from a series of wucai germplasms, exhibited higher tolerance to LT and HT than other genotypes used in previous experiments. Under HT, WS-1 had a greater net photosynthesis rate, antioxidative capacity, and carbon–nitrogen assimilation efficiency than other genotypes [6–8]. Under LT, the malondialdehyde (MDA) content and relative electric conductivity were lower in WS-1 than in other genotypes, while levels of soluble sugar, free proline, and chlorophyll were higher than in others [9]. However, it is unclear which regulatory pathway plays a dominant role in responding to thermal stress, and whether LT and HT tolerance could be improved by the same pathway.

In recent years, proteomics analysis has helped researchers understand responses to various environmental stresses such as cold [10], drought [11], heat [12,13], flooding [12], and salinity [14]. HT has been found to significantly induce heat-shockprotein expression and inhibit enzyme activities related to redox homeostasis and protein synthesis and degradation in rice, radish, and chickpea [15–18]. Furthermore, several studies have indicated that LT can downregulate photosynthesis-related proteins and upregulate proteins involved in carbohydrate metabolism, detoxification, ROS scavenging, and cell wall remodeling in wild wheat, *Arabidopsis*, and barley [19–21]. Response mechanisms that protect against the potentially harmful effects of HT or LT have been extensively studied in NHCC, but most studies have focused on biochemical responses and specific genes [6,7,22]. Song et al. [23] conducted a comprehensive analysis of responses to LT and HT treatments in NHCC using RNA-sequencing. Among 14,329 differentially expressed genes (DEGs), 33 and 25 genes were enriched by heat and cold stress, respectively. Among the identified DEGs, only 10 were found in response to heat and cold stress.

In this study, we used tandem mass tag (TMT) to evaluate the molecular changes that occur in the novel NHCC genotype WS-1 in response to LT and HT. TMT is a highly sensitive technique that improves the throughput and dynamic range of protein analysis [24,25]. This study aimed to identify and compare the regulatory mechanisms responsible for tolerance to LT and HT stress in WS-1.

2. Results

2.1. Identification of Differentially Expressed Proteins (DEPs) by Quantitative Proteomic Analysis

Using liquid chromatography-mass spectrometry (LC-MS/MS), we detected 207,427 total spectra, 105,424 spectra, 50,800 peptides, 23,059 unique peptides, and 6831 proteins (score sequence HT > 0 and unique peptides ≥ 1) (Figure 1A and Supplementary Table S1). The distribution of peptide numbers is shown in Figure 1B, and >81.3% of the proteins had at least two peptides (Figure 1B and Supplementary Table S2). Approximately 99.2% of the proteins had mass >10 kDa, which indicates very good coverage (Figure 1C and Supplementary Table S2). Sequence coverage distribution greater than 10% and 20% were 61.2% and 41.3% respectively, indicating that the data were of high quality (Figure 1D and Supplementary Table S2).

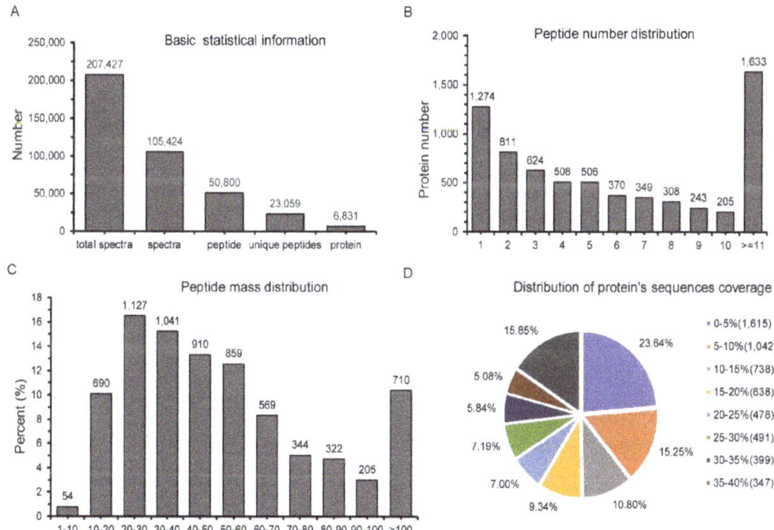

Figure 1. Identification and quantitative evaluation of identified proteins. (**A**) Spectra, peptides, and proteins identified by ProteomeDiscoverer. (**B**) Distribution of peptide numbers as determined by ProteomeDiscoverer. (**C**) Distribution of identified proteins according to molecular mass. (**D**) Distribution of identified protein sequences.

2.2. Functional Cataloging of DEPs Common to Both Low-Temperature (LT) and High-Temperature (HT) Groups

We regarded proteins in the LT or HT treatment with abundance >1.2-fold and $p < 0.05$ relative to the control treatment as upregulated, and those with abundance <0.83-fold and $p < 0.05$ as downregulated. A total of 1732 differentially expressed proteins (DEPs) were identified in the LT treatment, and 2806 DEPs were identified in the HT treatment (Figure 2A). These upregulated and downregulated DEPs were assigned to three groups: (1) 710 DEPs were specific to the LT group, (2) 1022 DEPs were common to both the LT and HT groups, including 172 common upregulated proteins, 324 common downregulated proteins, and 526 differently regulated proteins (Figure 2B), and (3) 1784 DEPs were specific to the HT group. To determine the mechanism underlying the temperature stress tolerance of WS-1, we focused on the functions of the 1022 DEPs that were regulated under both LT and HT conditions.

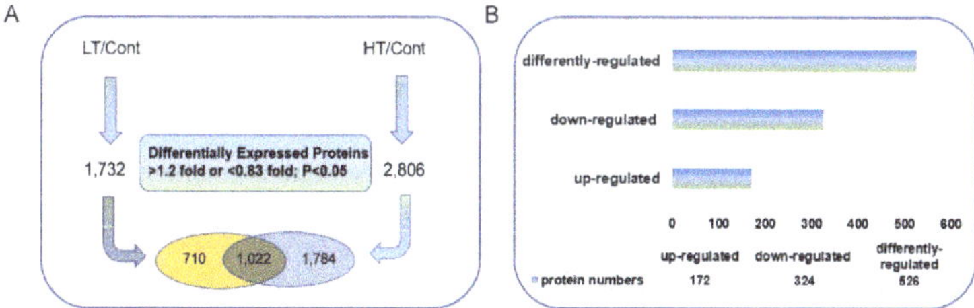

Figure 2. Summary of temperature stress-responsive proteins under low temperature (LT) and high temperature (HT) treatments. (**A**) Number of differentially expressed proteins in wucai leaves under LT and HT treatments compared to control. The value shared by the two ovals indicates the number of commonly regulated proteins, i.e., proteins that were differentially expressed relative to the control under both LT and HT treatments. (**B**) Distribution of commonly regulated proteins (1022 proteins) by LT and HT.

The functions of these "commonly regulated" proteins were assigned to several groups based on their Gene Ontology (GO) annotations (Figure 3A, Supplementary Tables S1 and S2), including biological process, cell component, and molecular function. As expected, several DEPs were assigned to multiple groups. In the LT/Cont treatment, single-organism metabolic process, small-molecule metabolic process, and single-organism biosynthetic process were the main GO enrichment pathways (Supplementary Table S1). Single-organism cellular process, response to stimulus, and single-organism metabolic process were the predominant pathways in the HT/Cont treatment (Supplementary Table S2). Biological process analysis of 1022 DEPs, commonly regulated proteins in LT/Cont and HT/Cont treatments, suggested that the predominant pathways were small-molecule metabolic process and response to abiotic stimulus. Most of the commonly regulated DEPs participate in various molecular metabolic processes (small-molecule, oxoacid, organic acid, single-organism, carboxylic acid, sulfur compound, and monocarboxylic acid), while the remainder are involved in response to stimulus. As shown in Figure 3A, and Supplementary Tables S1 and S2, most of the commonly regulated DEPs were predicted to localize in the cell part, cell, intracellular part, and cytoplasm, the same as in the LT/Cont and HT/Cont treatments, indicating that DEPs mainly functioned on these cell components. The main molecular functions of commonly regulated DEPs were RNA binding and protein binding in the HT/Cont treatment, whereas they were catalytic activity and RNA binding in LT/Cont (Figure 3A, Supplementary Tables S1 and S2).

To identify the metabolic pathways relevant to temperature stress tolerance, 1022 DEPs were further analyzed according to the Kyoto Encyclopedia of Genes and Genomes (KEGG) database (Figure 3B). Among the 1022 commonly regulated DEPs, 172 DEPs were upregulated in both LT and HT treatments, 324 were downregulated in both treatments, and 526 were differently regulated (Figure 2B). Most DEPs were enriched in porphyrin and chlorophyll metabolism, ribosome (4% of the downregulated DEPs), biosynthesis of secondary metabolites, glutathione metabolism, metabolic pathways, carbon metabolism, carbon fixation in photosynthetic organisms, 2-oxocarboxylic acid metabolism, biosynthesis of amino acids, and pyruvate metabolism (Figure 3C). Upregulated DEPs were enriched in 2-oxocarboxylic acid, carbon fixation in photosynthetic organisms, biosynthesis of amino acids, and carbohydrate metabolism.

2.3. Functional Cataloging of DEPs That Were Upregulated under both LT and HT

A total of 172 DEPs that were upregulated under both LT and HT were assessed using GO annotation and KEGG pathway analysis. GO analysis was used to assign these DEPs to three categories: biological processes, cell components, and molecular functions (Figure 4A). For biological processes, the analysis suggested that the 172 DEPs, which were predicted to be located in the cytoplasm and in chloroplasts and plastids, were involved in response to abiotic stimulus and stress, and in the metabolic process of carboxylic acid, oxoacid, organic acid, and monocarboxylic acid. In molecular functions, the most prevalent categories were protein binding and oxidoreductase activity. The KEGG pathway analysis indicated that most of the proteins were enriched in the secondary metabolite biosynthesis and metabolic pathway (Figure 4B). A more detailed ontological analysis of the response to temperature stress was then obtained for the 172 DEPs that were upregulated in both LT and HT treatments. According to GO annotation, these upregulated DEPs were divided into several metabolic groups: redox homeostasis, photosynthesis, carbohydrate metabolism, heat-shock proteins, signal transduction, and metabolic process (Supplementary Table S1). Because not all proteins have been identified in wucai, some unidentified proteins were mapped to the *Arabidopsis* genome.

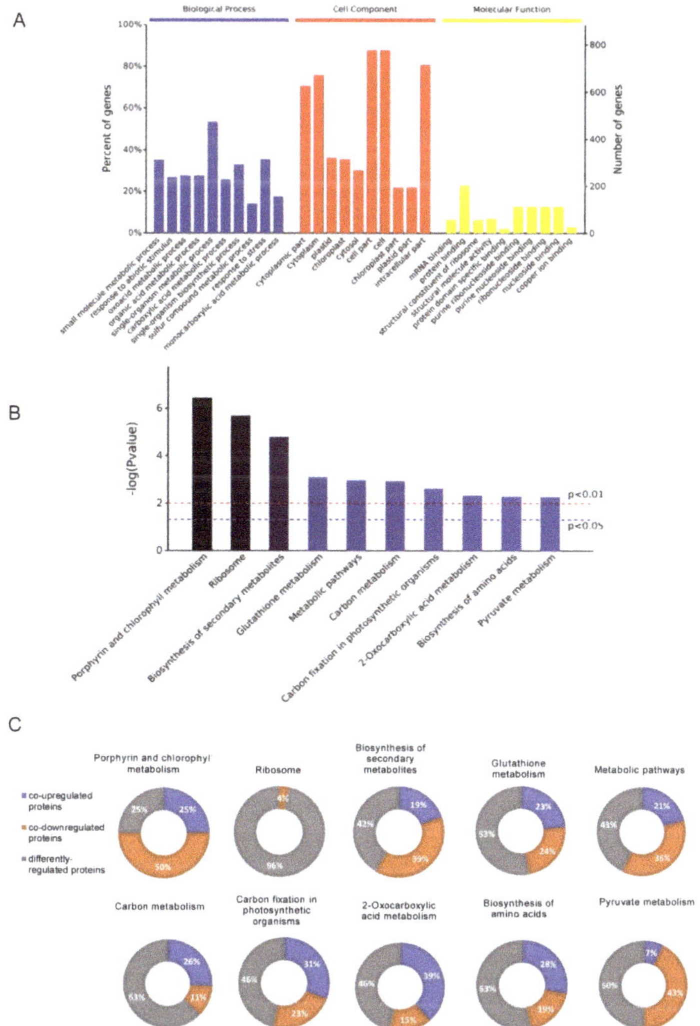

Figure 3. Summary of temperature stress-responsive proteins under LT and HT treatments. (**A**) Histogram presentation of Gene Ontology (GO) classification of 1022 commonly regulated proteins under LT and HT treatments. (**B**) Kyoto Encyclopedia of Genes and Genomes (KEGG) pathways of 1022 commonly regulated proteins. (**C**) Percentage of the 1022 differentially regulated proteins that were co-upregulated (i.e., upregulated under both LT and HT), co-downregulated (i.e., downregulated under both LT and HT), and differently regulated (upregulated in response to one temperature treatment but downregulated in response to the other) in each KEGG pathway.

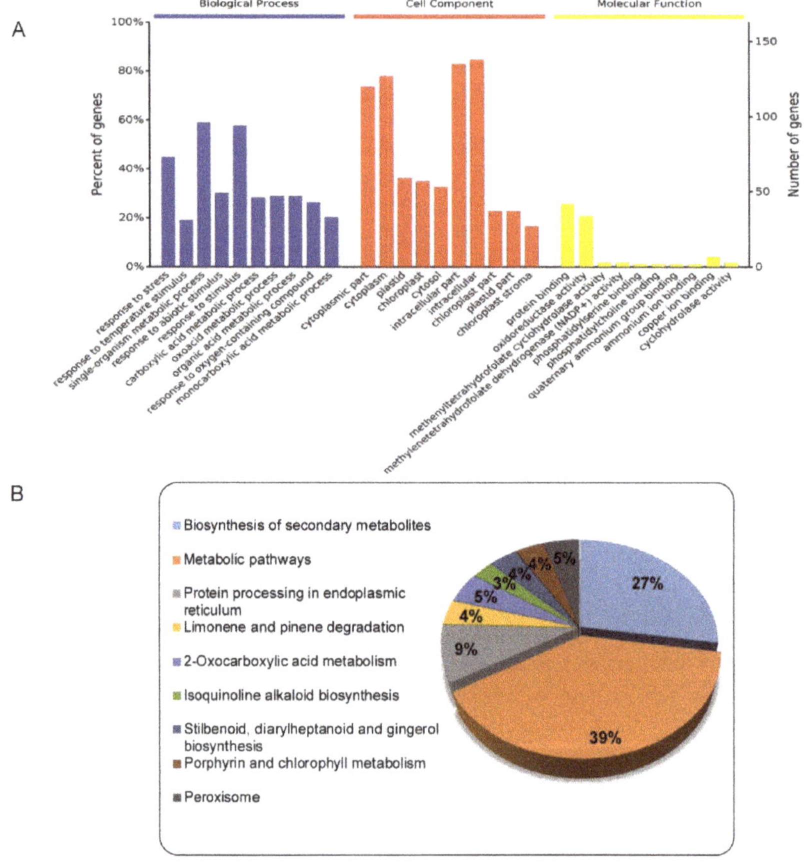

Figure 4. Summary of 172 proteins that were upregulated in LT and HT treatments. (**A**) Histogram of GO classification of upregulated proteins under both treatments. (**B**) Percentage of the 172 upregulated proteins in each KEGG pathway.

2.3.1. Proteins Involved in Redox Homeostasis

Of the 172 DEPs that were upregulated in both LT and HT treatments, those that were characterized were disulfide-isomerases, amine oxidase, isocitrate dehydrogenase (NADP) (ICDH) (A0A078FVI0), delta-1-pyrroline-5-carboxylate synthase (P5CS) (M4DE13), and ubiquinol oxidase (AOX) (J7GUS8) (Supplementary Table S3). The uncharacterized proteins—ferredoxin-dependent glutamate synthase (Fd-GOGAT) (A0A0D3EHZ6), glutathione hydrolase (A0A0D3BW42), betaine aldehyde dehydrogenase (ALDH) (A0A0D3CWL2), and ubiquinol oxidase (AOX) (J7GUS8)—were mapped to respective genes: *GLU1*, *GGT1*, *ALDH10A8*, and *AOX1A*. Among the proteins in the redox homeostasis category, upregulation was highest for GLU1: 3.46-fold under LT and 2.51-fold under HT.

2.3.2. Proteins Involved in Photosynthesis

Several crucial DEPs were categorized in the photosynthesis pathway (Supplementary Table S3). Two proteins related to the oxygen-evolving complex, PSBO1 (M4F7V3) and PSBP1 (A0A078FRX3), were mapped, and among the 172 commonly upregulated proteins, the fold increase was highest for *PsbO1*: upregulation was 18.87-fold under LT and 21.88-fold under HT. In addition,

NADPH-protochlorophyllide oxidoreductase (PORC) (M4EX79) was upregulated and is involved in pigment metabolism.

2.3.3. Proteins Involved in Carbohydrate Metabolism

Several crucial enzymes involved in starch and sucrose metabolism, including beta-amylase (BAM) (M4DEV3) and sucrose synthase (SUS) (M4CQT7), were also enhanced under LT and HT (Supplementary Table S3).

2.3.4. Heat-Shock Proteins and Chaperones

Putative small heat-shock proteins and chaperones were upregulated in response to LT and HT treatments: M4DG78, M4CEA8, M4CQW7, A0A0D3EDG5, M4F2A3, A0A0D3CBS0, and M4DH02 (Supplementary Table S3). These proteins were mapped to genes *Hsp17.7*, *Hsp18.1*, *ATJ2*, *ATJ3*, *DJA7*, *ERD10*, and *ERD14* respectively, in the *Arabidopsis* genome.

2.3.5. Proteins Relevant to Stimulus and Signaling Transduction

The signal transduction pathway from the initial perception of LT or HT to the final adaptive stress-responsive change in protein expression is very complex (Supplementary Table S3). Several characterized proteins thiamine thiazole synthase (THI) (A0A078I0E1) and UDP-glycosyltransferases (UGTs) (A0A0D3B6L4) and several uncharacterized proteins mapped to abscisic acid (ABA) and salicylic acid (SA) metabolism BAX inhibitor 1 (BI-1) (A0A0D3AUW9), BURP domain protein (RD22) (A0A078JCL4), and glycine-rich protein (GRP3) (A0A078INC4) were upregulated in response to both LT and HT.

2.3.6. Metabolic-Related Proteins

CytochromeP450s (CYPs) participate in the oxidation of lipophilic substrates as heme-thiolate proteins. Proteins mapped to the CYP family—A0A078IGH2, M4C991, M4E8T8, A0A078HAQ8, M4DYH4, A0A078HUG7, and A0A078GN44—accounted for almost half of this category. Their corresponding genes were *CYP38*, *CYP706A6*, *CYP71B3*, *CYP71B35*, *CYP72A15*, *CYP71B4*, and *CYP71B28*, respectively (Supplementary Table S3). The abundance of these proteins significantly increased in response to both LT and HT treatments.

2.3.7. Protein–Protein Interaction (PPI) Analysis

The present study used the Search Tool for the Retrieval of Interacting Genes/Proteins (STRING), a biological database and web resource of known and predicted protein–protein interactions (PPI). STRING PPI network analysis revealed the functional partnership and interaction networks. It was found that glutathione metabolism occupied a dominant position among all metabolic pathways. In addition, GLU1 had direct or indirect correlations with other metabolic processes, of which the major ones were SRK2G (P43292), FOLD4 (A0A0D3DZN5), RH14 (Q8H136), HSP17.7 (M4DG78), HSP18.1 (M4CEA8), GGT1 (A0A0D3BW42), BCAT3 (Q9M401), PDIL1-2 (Q9SRG3), ALDH10A8 (A0A0D3CWL2), ICDH (A0A078FVI0), and P5CSB (M4DE13) (Figure 5).

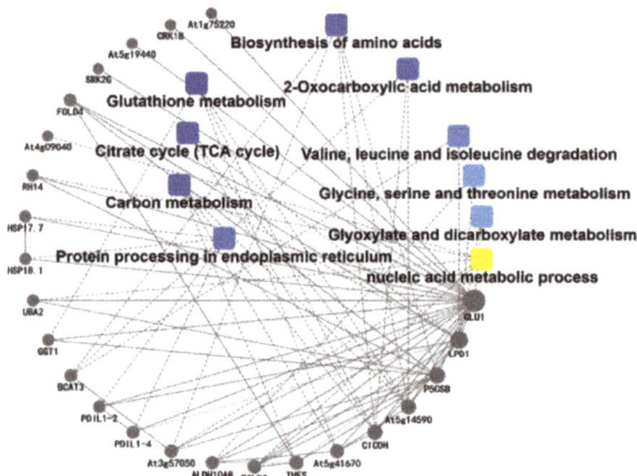

Figure 5. Protein–protein interaction (PPI) network analysis by the Search Tool for the Retrieval of Interacting Genes/Proteins (STRING) PPI network. ▪, GO/KEGG term; ●, protein. Symbol color corresponds to degree of interaction.

2.4. Expression Profiles of Genes Involved in DEPs That Were Upregulated under Both LT and HT

To evaluate the correlation between mRNA and protein expression levels, we selected 12 genes related to DEPs that were upregulated under both LT and HT for real-time polymerase chain reaction (RT-PCR) analysis. The results show that the expression of most of these genes was greater in the LT or HT group than in the control group. The expression of the following genes was higher in the LT group: *RD22*, *SUS1*, *P5CSB*, and *CYP38* (Figure 6). Compared to the control, expression levels of *sHSP17.7*, *sHSP18.1*, *ICDH*, *PSBP1*, and *ERD10* were significantly increased in response to both LT and HT treatments. These results indicate that the protein expression patterns were not always consistent with their transcriptional expression patterns, perhaps because stress-related genes dominated the response to LT and HT.

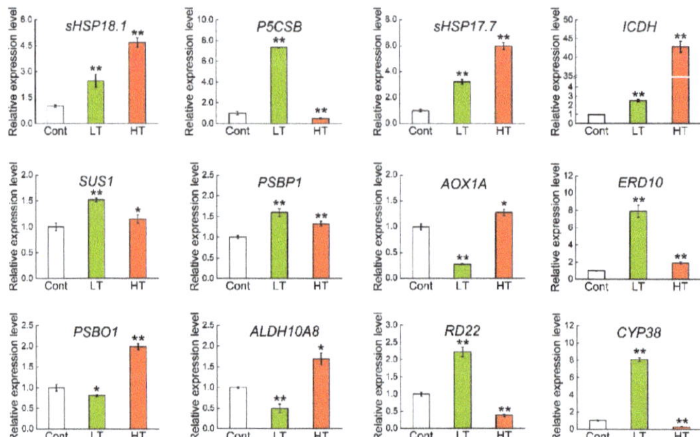

Figure 6. Gene expression related to upregulated differentially expressed proteins (DEPs) in leaves of plants subjected to LT and HT. *Actin* was used as the internal control to calculate the relative expression level. Data shown here are mean ± standard deviation (SD) of 3 biological replicates. ** Significant difference *at p* < 0.01.

2.5. Overexpression of BccrGLU1 in Arabidopsis Leads to Temperature Stress Tolerance

To investigate the effect of redox homeostasis on tolerance to temperature stress in plants, we generated transgenic *Arabidopsis* plants overexpressing *BccrGLU1* under the control of the CaMV 35S promoter (Figure 7A). Quantitative RT-PCR results show that the transcript level of *BccrGLU1* was highly induced in selected transgenic plants, confirming that the plants had been successfully transformed with *35S-BccrGLU1:GFP* (Figure 7B). No visible phenotype differences between wild-type (WT) and overexpressed (OE) transgenic lines were detected under normal growth conditions, indicating that overexpression of *BccrGLU1* did not cause phenotypic defects in transgenic *Arabidopsis* seedlings. In OE lines, ferredoxin (Fd)-GOGAT activity was significantly higher than in WT under normal conditions (Figure 7C).

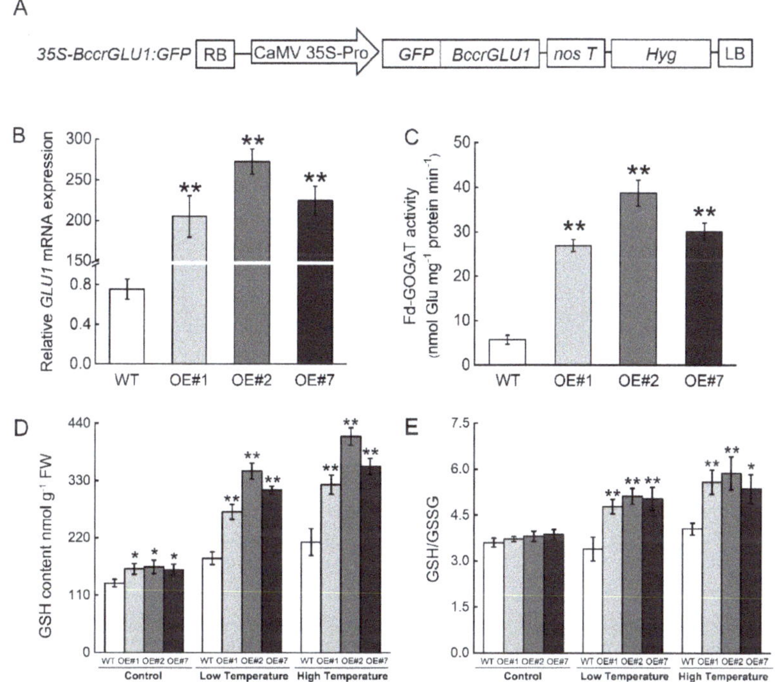

Figure 7. Schematic diagram of *35S-BccrGLU1:GFP* fusion protein construct and physiological changes associated with temperature stress response in wild-type (WT) and overexpressed (OE) lines of *Arabidopsis* plants. (**A**) Schematic diagram of *35S-ccrGLU1:GFP* fusion protein construct. (**B**) Relative expression of *BccrGLU1* in T3 transgenic plants. WT: Col-0; OE#1, OE#2, OE#7. T3 plants with *BccrGLU1* on the *AtCol-0* background. (**C**) Fd-GOGAT activity in WT and OE lines. (**D**) Glutathione (GSH) content in leaves. (**E**) GSH/oxidized glutathione (GSSG) ratio in leaves. Values represent mean ± SE (n = 3). ** and * indicate significant differences from wild-type plants at the level of $p < 0.01$ and $p < 0.05$ respectively, using Student's t-test.

To evaluate the role of *BccrGLU1*, physiological assays were performed with three OE lines. Glutathione (GSH) levels were higher in OE lines than in WT under normal conditions. Under low and high temperature, GSH levels in three overexpression plants (OE#1, OE#2, and OE#7) were significantly higher than in wild-type plants (Figure 7D). This increased degree in OE lines under temperature stress was greater than under normal conditions. The GSH/oxidized glutathione (GSSG) ratio in OE lines was not different from WT under normal conditions, while the ratio in OE lines was significantly higher than in WT plants under temperature stress (Figure 7E).

In addition, we determined oxidative damage parameters under temperature stresses to evaluate the redox homeostasis in OE lines. Under normal conditions, MDA content and electrolyte leakage in OE lines were not different than those in WT. When exposed to LT and HT, OE lines had lower MDA content and electrolyte leakage than WT (Figure 8A,B). Proline content in three OE lines was slightly higher than in WT under normal conditions, and proline content of OE lines had greater increases than WT responding to LT and HT (Figure 8C).

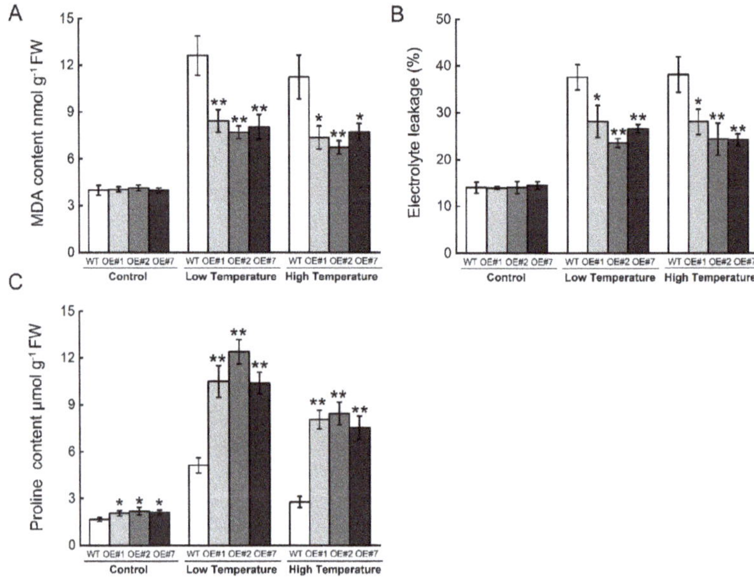

Figure 8. Oxidative damage and proline content in WT and OE lines of *Arabidopsis* plants. (**A**) Malondialdehyde (MDA) content in *Arabidopsis* seedlings. (**B**) Electrolyte leakage in *Arabidopsis* seedlings. (**C**) Proline content in *Arabidopsis* seedlings. Values represent mean ± SE ($n = 3$). ** and * indicate significant differences in wild-type plants at the level of $p < 0.01$ and $p < 0.05$ respectively, using Student's t-test.

The staining patterns indicated lower ROS levels in OE lines compared to WT under stress (Figure 9A). The hydrogen peroxide (H_2O_2) content and O_2^- generation rates were also decreased in OE lines compared with wild-type, which was consistent with the staining pattern (Figure 9B,C). These results indicate that *BccrGLU1* was associated with reactive oxygen species scavenging to counteract oxidative stress.

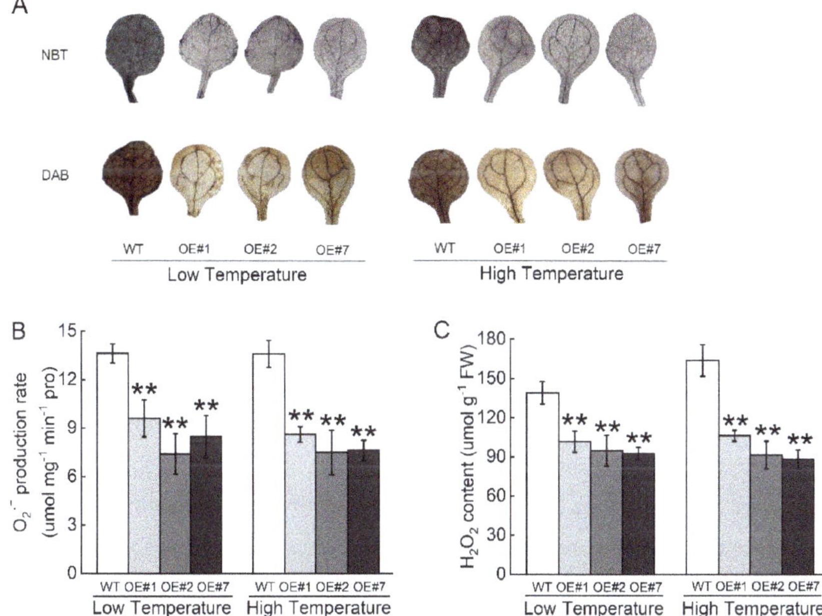

Figure 9. Oxidative damage of WT and OE lines of *Arabidopsis* plants. (**A**) Histochemical staining assay for superoxide anion ($O_2^{\bullet-}$) by nitro blue tetrazolium (NBT) and hydrogen peroxide (H_2O_2) in the upper row and by 3,3-diaminobenzidine (DAB) staining in the lower row. (**B**) Quantification of $O_2^{\bullet-}$ content. (**C**) Quantification of H_2O_2 content. Values represent mean ± SE (n = 3). ** indicate significant differences from wild-type plants at the level of $p < 0.01$, using Student's t-test.

3. Discussion

The mechanisms that enable plants to tolerate LT and HT are morphological, physiological, and molecular in nature and often vary among genotypes. The novel genotype WS-1, which is tolerant to both LT and HT, could facilitate the breeding of temperature tolerance in wucai [8,9]. Because protein metabolic processes are very susceptible to environmental temperature, proteomic research could greatly increase our understanding of the protein alteration response to temperature stress. In the present study, we found that redox homeostasis, photosynthesis, carbohydrate metabolism, heat-shock protein, and signal transduction pathways were associated with temperature stress tolerance in WS-1.

In the current research, 172 "commonly upregulated" DEPs (i.e., DEPs that are upregulated in response to both LT and HT) in WS-1 were identified to respond to LT and HT (Supplementary Table S1). There were significantly more DEPs than in previous transcriptome analysis of another NHCC variety according to Reference [23], which found seven upregulated and three downregulated in response to heat and cold stress, and most of the DEGs were involved in stress-related protein, such as *KIN2* and *LEA14*. In our study, more than 20 DEPs were assigned to the redox homeostasis pathway. The putative Fd-GOGAT protein, mapped to *GLU1*, was upregulated under both LT and HT. In a previous report, a *GLU1*-deficient mutant of *Arabidopsis* displayed a lethal phenotype, indicating that *GLU1* is a major gene that encodes Fd-GOGAT [26]. As indicated in another report [27], Fd-GOGAT might contribute significantly to photosynthesis, and its upregulation in response to LT and HT in the current study may enhance the production of glutathione. The abundance of putative gamma-glutamyl transferase (GGT) was enhanced in the current study and was mapped to *GGT1*. GGT was also found to be involved in glutathione metabolism, which can synthesize and degrade the glutathione in the gamma-glutamyl cycle [28]. In addition, ICDH, which plays a crucial role in the oxidative pentose phosphate pathway, was upregulated in response to LT and HT in our study. Redox-proteomics analysis also verified that ICDH is a key enzyme in the tricarboxylic acid cycle as a potential redox-regulated protein in *Arabidopsis* [29]. The upregulation of these DEPs in response to LT and HT indicates that glutathione metabolism has a dominant role in the redox balance of WS-1 plants exposed to low- or high-temperature stress.

Glycine betaine (GB) and proline, commonly referred to as osmoprotectants, protect plants from abiotic stress by contributing to cellular osmotic adjustment, detoxification of ROS, protection of membrane integrity, and stabilization of enzymes/proteins [30,31]. In the present study, two upregulated genes in WS-1, *ALDH10A8* and *P5CSB*, were identified to code for BADH and P5CS under LT and HT. Consistent with our results, *ALDH10A8* mutant was more sensitive to dehydration and salt stress than wild-type plants, indicating that the *ALDH10A8* gene can respond to abiotic stress in *Arabidopsis* [32]. Under cold stress, the *BADH* transgenic lines with higher GB content maintained better membrane integrity and had lower ROS production, which was attributed to enhanced cold stress [33]. In yet another study, *Arabidopsis* with an antisense proline dehydrogenase cDNA exhibited increased accumulation of proline and a constitutive tolerance to freezing temperatures (as low as −7 °C) [34]. Proline has been verified to function in multiple ways as a molecular chaperone, which can protect protein integrity and increase many enzyme activities [31]. *P5CS2* in *Arabidopsis* can be activated by bacteria, salicylic acid, and reactive oxygen species signals [35]. In the present study, we that found P5CS protein, encoded by *P5CSB*, which is homologous to *P5CS2*, was enhanced in response to temperature stress. Overexpressing *BccrGLU1 Arabidopsis* plants with a higher proline content in our study also exhibited stronger tolerance to temperature stress compared with WT (Figure 8C). Therefore, an increased abundance of BADH and P5CSB may modify hormone signal transduction in order to achieve redox balance in response to temperature stress. In agreement with this, an enhanced abundance of putative AOX in response to LT and HT mapped to *AOX1a*, also contributed to the intercellular redox homeostasis [36].

Among the upregulated proteins related to photosynthesis under both LT and HT, the abundance of the putative oxygen-evolving complex proteins PsbO and PsbP showed a large increase relative to the control. A *PsbO*-deficient mutant of *Arabidopsis* had retarded development, pale leaves, and increased sensitivity to light stress [37]. The PsbO protein, however, is downregulated under LT stress in *Spirulina*, which causes a decreased electron transfer rate and photosynthetic capacity [38]. This supports that PsbO plays a dominant role in protecting and stabilizing the catalytic center as an important protein for Mn_4Ca cluster stabilization [39]. An additional protein mapped to *PsbP1*, which is involved in electron transfer within PSII [40], was also increased in response to HT and LT stress in the current study. Researchers have speculated that the nitratable tyrosine residues of PsbO1 and PsbP1 are redox-active and sensitive to ROS [41]. Thus, the relatively stable redox state might facilitate the expression of PsbO and PsbP.

Enhanced expression of SUS and BAM proteins in the wucai genotype WS-1 was observed in the current study, confirming that LT and HT enhance the expression of these proteins. Expression of the gene encoding SUS is induced by cold in sugar beet [42], and an increased accumulation of soluble sugars has been documented in chickpea subjected to cold treatment [17]. Increased BAM abundance in WS-1 was favorable to catalyze the breakdown of starch to generate maltose and participate in the metabolism of sugar [43]. Our findings are consistent with previous reports that attributed the enhanced thermal tolerance of *Agrostis* grass species to the upregulation of enzymes involved in sucrose metabolism [44].

According to GO annotation, several uncharacterized DEPs are mapped to the following heat-shock proteins, heat-shock-related proteins, and chaperones: HSP17.7, sHSP18.1, ERD10, ERD14, DJA7, ATJ2, and ATJ3. These proteins were upregulated in both LT and HT treatment. Researchers have also reported that sHSPs can refold proteins that were denatured by freezing so as to prevent the aggregation of denatured proteins under cold stress [45,46]. Supporting this, transgenic carrot cell lines with *sHsp17.7* have an enhanced survival rate at high temperatures [47]. In peas, heat-denaturing proteins that initially interact with *sHSP18*.1 have remarkably improved subsequent refolding by *HSP70* [48]. Although the function of molecular chaperones *sHSP18.1* and *sHSP17.7* is not clear, the results from our study strongly support the notion that both sHSP18.1 and sHSP17.7 enhance the thermotolerance of WS-1. The identification of several unique small HSPs in our study suggests that these could now be used as biomarkers for evaluation of thermal-stress tolerance in wucai [15,49].

Several uncharacterized proteins that were mapped to *BI-1*, *RD22*, *GPR3*, and *UGTs* were related to plant hormone signal transduction under both LT and HT treatment. *BI-1* expression in *Arabidopsis thaliana* is induced by exogenous SA, and its overexpression increases the viability of rice cells after application of SA [50,51]. Wang et al. [52] demonstrated that overexpression of the soybean protein *GmRD22* in rice could be involved in the ABA regulation pathway [53]. *UGT73B5* belongs to the early SA-induced genes, whose pathogen-induced expression is co-regulated with genes related to cellular redox homeostasis [54]. The involvement of THI, which is required for plant development, has been implicated in ABA signaling in *Arabidopsis* guard cells [55] and salt stress [56]. These studies suggest that plant hormone signaling contributes to temperature-stress tolerance in WS-1.

Based on the above analysis and PPI network annotation, we considered that GLU1 has strong interactions with the proteins that contribute to several protein metabolic processes when encountering temperature stresses. Then we generated *35S:BccrGLU1* overexpression transgenic lines in *Arabidopsis*. In OE lines and WT, the glutathione redox status was assayed by measuring GSH and GSSG levels in two-week-old *Arabidopsis* seedlings. GSH is not only a key metabolite in the removal of ROS but also can control the expression of redox-sensitive proteins, which could alter the intercellular redox state and induce tolerance of ROS-dependent signaling pathways [57]. Higher GSH has been shown to positively modulate tolerance to HT stress in mung beans [58] and to LT stress in maize [59]. In the present study, GSH content in three OE lines was strongly induced, whereas GSSG had slight increases compared with WT under LT and HT. The higher ratio of GSH/GSSG in OE lines was beneficial to antioxidative protection, which could in turn improve thermotolerance [60]. In the *Arabidopsis cat2* mutant, in which the GSH/GSSG ratio is altered, glutathione helps regulate the oxidant-dependent salicylic acid and jasmonic acid signaling pathways [61,62]. In addition, compared to WT, OE lines exhibited lower MDA content and electronic leakage, supported by the lower H_2O_2 content and $O_2^{\cdot-}$ generation rate in response to temperature stresses. Higher proline content in OE lines reduced oxidative injury by lowering free radical generation as important osmolytes [63]. This provides evidence that *BccrGLU1* contributes to balance redox homeostasis in WS-1 to confer temperature stress response. Moreover, it is consistent with our proteomic data, where most of the identified DEGs were associated with or modulated by redox homeostasis.

4. Materials and Methods

4.1. Plant Materials and Growth Conditions

WS-1, which has stable genetic traits, was obtained by multigenerational self-inbreeding via manual pollination at Anhui Agricultural University. WS-1 seeds were sown in a peat/vermiculite (2:1) medium, and the seedlings were kept in a growth chamber at 25 ± 1 °C (day) and 18 ± 1 °C (night), with relative humidity of 70% and a 14/10 h light/dark photoperiod with 300 $\mu mol \cdot m^{-2} \cdot s^{-1}$ photosynthetically active radiation. After 20 days, the seedlings had developed 4–5 true leaves and were randomly divided into 3 groups (50 seedlings per group). Each group was subjected to 1 of 3 treatments: control (25/18 °C day/night), LT (3/8 °C), or HT (40/30 °C). The experiment, which used 3 growth chambers, had a completely randomized design and was repeated 3 times, such that each treatment was assigned to the same growth chamber only once. Thus, each repetition of the experiment was considered a block, for a total of 3 blocks.

After the seedlings were exposed to the treatments for 3 days, the leaves from each of 6 randomly selected plants per treatment group were collected for proteomics analysis and assessment of physiological properties.

4.2. Protein Extraction

For extraction of proteins, the leaf samples were crushed in 500 μL of extraction buffer before Tris-phenol buffer was added, and the preparation was mixed at 4 °C for 30 min. After the mixture was centrifuged at 7100× g and 4 °C for 10 min, the supernatant was collected. Five volumes of 0.1 M cold ammonium acetate–methanol buffer was added to the supernatant, which was precipitated at –20 °C for 12 h. The precipitated sample was centrifuged at 12,000× g and 4 °C for 10 min, and the supernatant was removed to obtain the precipitate. The precipitate was washed with cold methanol, gently mixed, and centrifuged again to collect additional precipitate. The methanol was then replaced with acetone to remove the methanol contaminants, and the steps used to obtain precipitate were repeated. The precipitate was dried at room temperature, mixed with lysis buffer, and dissolved for 3 h. Finally, the sample was centrifuged at 12,000× g and 4 °C for 5 min, and the supernatant was retained. The supernatant was subjected to a second centrifugation to completely remove the precipitate. Protein concentration was determined using the Bradford assay [64], and aliquots were stored at −80 °C.

4.3. Protein Reduction, Alkylation, Digestion, and TMT Labeling

The filter-aided sample preparation (FASP) method was used for protein reduction [65]. The steps of trypsin digestion were carried out as described by Cen et al. with partial modification [66]. First, 100 μg of protein extract was added to 120 μL of reduction buffer (10 mM DTT, 8 M urea, 100 mM triethylammonium bicarbonate (TEAB), pH 8.0). The solution was incubated at 60 °C for 1 h and iodoacetamide (IAA) was added to the solution with the final concentration of 50 mM in the dark for 40 min at room temperature. The protein was then digested and centrifuged at 10,000× g for 20 min at 4 °C and the flow-through was discarded from the collection tube. Then, 100 μL of 100 mM TEAB was added to the solutions and centrifuged at 10,000× g for 20 min, and this step was repeated twice. The filter units were transferred into new collection tubes and 100 μL of 100 mM TEAB was added, followed by 2 μL of sequencing-grade trypsin (1 μg·μL^{-1}) in each tube. Then the solutions were incubated for digestion at 37 °C for 12 h. Finally, the collections of digested peptides were centrifuged at 10,000× g for 20 min. After 50 μL of TEAB (100 mM) was added, the preparation was centrifuged again, and the solution was lyophilized and stored at −80 °C.

For TMT labeling, 100 μL of TEAB (50 mM) was added to the sample, and 40 μL of the sample was used for labeling. A 60 mL volume of TEAB was added to the 40 μL sample and mixed well, before 40 μL of anhydrous acetonitrile was added to the TMT reagent vial. The solution was thoroughly dissolved for 5 min and centrifuged at 10,000× g and 4 °C for 30 min. Next, 41 μL of TMT labeling reagent was added to 100 μL of sample and mixed well. The tubes were incubated at room temperature for 1 h. Finally, 8 μL of 5% hydroxylamine was added to each sample and incubated for 15 min to terminate the reaction. The solution was incubated for 1 h at room temperature. Finally, the reaction was terminated by adding 8 μL of 5% hydroxylamine to the sample. The labeling peptide solutions were lyophilized and stored at −80 °C. The samples were labeled with TMT tags as 126 (control), 127N (control), 127C (control), 128N (HT), 128C (HT), 129N (HT), 129C (LT), 130N (LT), and 130C (LT).

4.4. Reversed-phase liquid chromatography (RPLC) Analysis

Reversed-phase (RP) separation was performed on an Agilent 1100 HPLC System (Agilent Technologies Inc., CA, USA) using an Agilent Zorbax Extend RP column (5 μm, 150 mm × 2.1 mm). The RP separation was conducted with mobile phases A (2% acetonitrile in high-performance liquid chromatography (HPLC) water) and B (98% acetonitrile in HPLC water) as follows: 0−8 min, 98% A; 8.00−8.01 min, 98−95% A; 8.01−38 min, 95−75% A; 38−50 min, 75−60% A; 50−50.01 min, 60−10% A; 50.01−60 min, 10% A; 60−60.01 min, 10−98% A; and 60.01−65 min, 98% A. Tryptic peptides were separated at a fluent flow rate of 300 μL·min^{-1} and were monitored at 210 and 280 nm. Dried samples were harvested for 8−50 min, and the elution buffer was collected once per minute and numbered from 1 to 10 with pipeline. The separated peptides were lyophilized for MS detection.

4.5. Liquid Chromatography-Mass Spectrometry Analyses

LC-MS/MS was carried on an Agilent 1100 HPLC purifier system (Agilent Technologies Inc., CA, USA) according to the method of Cao et al. with slight modification [24]. The samples were loaded on a trap column (100 μm × 20 mm; RP-C18, Thermo Scientific Inc., MA, USA) with an autosampler and separated with an analysis column (75 μm × 150 mm; RP-C18, Thermo Scientific Inc.) at 300 nL·min^{-1}. The mobile phase consisted of buffer A (0.1% formic acid in water) and buffer B (0.1% formic acid in acetonitrile). The linear washing gradient was set as follows: 0–7 min, 4% buffer B; 8–51 min, 4–25% B; 52–60 min, 25–40% B; 61–70 min, 40–85% B; 71–74 min, 85% B. Full MS scans were acquired in the mass range of 375–1800 *m/z* with a mass resolution of 120,000 (at *m/z* 200), the automatic gain control (AGC) target value was set at 4×10^5, and the maximum injection time was 50 ms. The dynamic exclusion was set to 60.0 s and run under positive mode. The 10 most intense peaks in MS were fragmented with higher-energy collisional dissociation with collision energy of 38 eV. MS/MS spectra were obtained with a resolution of 50,000 (at *m/z* 200) and max injection time of 100 ms.

4.6. Protein Identification and Quantification

Proteome Discoverer (v.2.2, Thermo Fisher Scientific, Bremen, Germany) was used to thoroughly search all of the fusion raw data against the sample protein database. Database searches were carried out with trypsin digestion in *Brassica fasta*. Alkylation on cysteine was considered to be a fixed modification in the database search. The analysis results were filtered as follows: significance threshold $p < 0.05$ (with 95% confidence) and ion score or expected cutoff <0.05 (with 95% confidence).

The main parameters were set as per the method of Xie et al. [67]. TMT 10-plex was selected for the protein quantification method. A global false discovery rate was set at 0.01, and protein groups required at least 2 peptides to be considered for quantification. For protein quantification, the protein ratios were calculated as the median of only the unique peptides of the protein. All peptide ratios were normalized by the median protein ratio. Cutoff values of >1.20- and <0.83-fold were used to identify upregulated and downregulated proteins at $p < 0.05$.

4.7. Bioinformatics Analysis of Proteins

Bioinformatics analysis of proteins was performed as previously reported [68]. The proteins were described by searching against the UniProt database (28 January 2018; 172,630 sequences). GO (http://www.geneontology.org/) function entries for all aligned protein sequences were extracted using the mapping function of BLAST2GO (version 3.0). The DEPs were further analyzed using KEGG (http://www.genome.jp/kegg/) to determine the active biological pathways [69]. The chi-square test with a defining cutoff of 0.01 was performed to evaluate the functional category protein enrichment. A false discovery rate significance threshold of 0.05 was used as false-positive control. The STRING database (version 9.1; http://string-db.org/) was used to analyze the protein–protein interaction (PPI) networks among these DEPs.

4.8. Transcriptional Expression Analysis by Quantitative RT-PCR

Based on the functional category and the fold change in expression in the LT and HT groups relative to the control group, 15 genes were chosen for qRT-PCR analysis. Primer software version 5.0 (Premier Biosoft International, CA, USA) was used to design gene-specific primers, and primer sequences are indicated in Table 1. qRT-PCR analysis was performed using AceQ qPCR SYBR GREEN Master Mix (Vazyme Biotechnology Co., Nanjing, China), and relative expression levels were calculated using the $2^{-\triangle\triangle Ct}$ method [70]. The *BnaActin* gene was used as the control. Each experiment was replicated 3 times.

Table 1. Primers used for fragment amplification of differentially expressed genes.

Accession	Name	Forward Primer (5′→3′)	Reverse Primer (5′→3′)
XM_009106490.1	ALDH10A8	TCGTCAATCCAGCAACCCAA	TCAGTCACCTTAGCGGCAAT
XM_013806620.2	AOX1A	AGCCATCTCTTGAAACCTGC	AGCGATTCCTTTGTTACCTCC
XM_009151277.2	ERD10	ACTGTTTGACTTCTTGG	GAGGAGAGTAGGCTTATG
XM_022705499.1	RD22	CAAACACTCCCATACCA	TACACCTCCCTTTCCAA
XM_013841006.2	PSBP1	TTTCACTCTCCAAACCCGTCCA	AGCTTCACCATAGGCGGCATC
XM_013830831.2	PSBO1	CAACCTCTGCTCTCGTCGTC	CTTGCTAACACTCTCGGCCT
XM_009129212.2	ICDH	AGTGAGGGAGGCTATGTGTG	CTATGCTGTTTGTGCTGGT
XM_009122518.1	SUS1	ATTTCATCATCACCAG	AGTCAATCTACGCTTC
XM_009118014.2	P5CSB	CCTTTTCCACCAAGATGCAC	CCCAGGCTTCATAACTAAACGA
XR_002653900.1	CYP38	CGGGAACTTTGTGGACTTGG	GCGTTTTCTTCCCAGTCACC
XM_002873500.2	sHSP17.7	AAGACCCGTAACAACCCT	CTTTTTCCACTCACCACA
XM_013858088.2	sHSP18.1	CAGCATTCACAAACGC	CCTCCTCATAAACTTC
XM_009127097.2	Actin	TGGGTTTGCTGGTGACGAT	TGCCTAGGACGACCAACAATACT

4.9. Tolerance Assay in Transgenic Arabidopsis under Temperature Stress

The putative Fd-GOGAT protein, which was upregulated under HT and LT stress, was selected to confirm gene function. Fd-GOGAT, one type of GOGAT, is specifically distributed in photosynthetic organisms and has a major role in photosynthetic tissues [71]. There are 2 genes in *Arabidopsis* that encode Fd-GOGAT, GLU1 and GLU2. It has been proven that GLU1 is highly expressed, primarily in leaves [72]. To generate *35S:GFP-BccrGLU1*/Col-0 lines, the Co-Ding Sequence (CDS) encoding the full-length sequence of *BccrGLU1* were amplified from cDNA using gene-specific primers (LP, 5′-gggacaagtttgtacaaaaaagcaggcttcATGATTATTAATTTAATATACATATATGTAGCTGCGC-3′; RP, 5′-ggggaccactttgtacaagaaagctgggtcGTCGAAATTGACACCATACTTGGG-3′). Then, the fragments were introduced into the pDONR207 entry vector with BP clonase (Invitrogen, Carlsbad, CA, USA). LR clonase (Invitrogen) reaction was performed to transfer the inserted fragments to destination vector pMDC43 for GFP fusion. *35S-BccrGLU1:GFP* was transformed into WT *Arabidopsis* plants Col-0 by Agrobacterium (strain GV3101)-mediated transformation via floral dip [73]. OE were screened on 1/2 MS agar plates containing 50 μg/mL hygromycin B (Invitrogen). Quantitative RT-PCR was carried out to determine the expression level of transgene using gene-specific primers (LP, 5′-TCCAAGAGAACCAGACGCAGA-3′ and RP, 5′-TTTGCTATAAACCCGACACCAC-3′).

To determine temperature stress tolerance of the transgenic *Arabidopsis* seedlings, temperature stress treatments were carried out as follows: WT and OE seedlings were planted in 1/2 MS plate at 24 °C under 16/8 h (light/dark) conditions with 75% relative humidity and 200 mmol·m^{-2}·s^{-1} light intensity. For low-temperature stress, 2-week-old seedlings were transferred to 4 °C for 48 h. For high-temperature stress, the plants were transferred to 40 °C for 48 h. Two-week-old seedlings were sampled to analyze transcript level and physiological parameters. The 2 temperature stresses were performed in triplicate. T3 of *Arabidopsis* transgenic plants was analyzed.

Glutathione content and Fd-GOGAT enzyme activity were estimated according to previous studies [74,75]. MDA content and electrolyte leakage were measured with the methods of References [76,77]. Proline content was estimated by using the acid-ninhydrin method of Reference [78]. Hydrogen peroxide (H_2O_2) content and superoxide anion ($O_2^{\bullet-}$) production rate were assayed with methods according to References [79,80]. Histochemical staining of H_2O_2 and $O_2^{\bullet-}$ in leaves was carried out with 3,3-diaminobenzidine (DAB) and nitro blue tetrazolium (NBT) respectively, by the method of Reference [81].

4.10. Statistical Analysis

Values of physiology and biochemistry are expressed as means ± SE of 3 replications. Statistical significance ($p < 0.05$) was determined by ANOVA with SAS software (SAS Institute, Cary, NC, USA), and means were separated using Duncan's multiple range test.

5. Conclusions

In this study, we identified 172 proteins that were upregulated in response to both LT and HT stress. Proteomic analysis showed that the temperature tolerance of WS-1 may be related to several factors, including steady redox homeostasis, an intact oxygen-evolving complex, the accumulation of osmolytes, the induction of sHSPs and chaperones, and enhanced metabolic processes. We infer that tolerance to LT and HT stress involves a common regulatory pathway, and that increased expression of DEPs, including Fe-GOGAT, BADH, P5CS, GGT, and AOX, helps regulate redox homeostasis (Figure 10). To further confirm the effect of redox homeostasis, we overexpressed *BccrGLU1* in *Arabidopsis*. Under LT and HT stress, *35S:BccrGLU1* plants exhibited higher GSH content and GSH/GSSG ratio and less oxidative damage compared to WT. These results are in agreement with the data of our comparative proteomics in WS-1. Combined with these data, redox homeostasis apparently played an important role in enhancing LT and HT stress tolerance in WS-1.

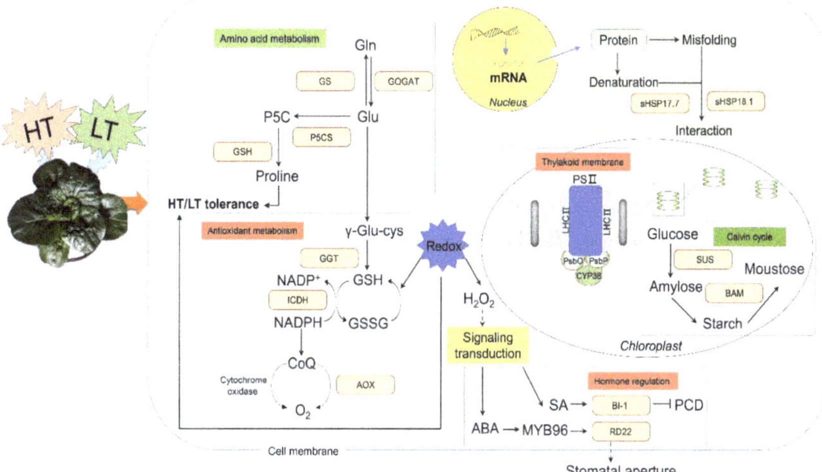

Figure 10. Schematic diagram of WS-1 proteins that were upregulated in response to both LT and HT. Red border represents upregulated proteins, arrows indicate positive regulation, and T-bars indicate negative regulation.

Supplementary Materials: Supplementary materials can be found at http://www.mdpi.com/1422-0067/20/15/3760/s1.

Author Contributions: Funding acquisition, Project Administration, Writing–Review & Editing, L.Y.; Data Curation, Writing–Original Draft Preparation, J.W.; Data Curation, Methodology, S.X.; Formal Analysis, M.Z. and G.C.; Investigation, L.N. and Y.Z.; Resources, S.Z.; Software, J.H.; Funding acquisition, Supervision, C.W.

Funding: This research was funded by the National Natural Science Foundation of China (No. 31701910) and the National Key R & D Program of China (2017YFD0101803). And the APC was funded by the National Key R & D Program of China.

Acknowledgments: This work was funded by the National Natural Science Foundation of China (No. 31701910), the National Key R & D Program of China (2017YFD0101803), and the Major Science and Technology Projects of Anhui Province, China (17030701013).

Conflicts of Interest: The authors declare no conflict of interest. The sponsors had no role in the design, execution, interpretation, or writing of the study.

Abbreviations

DAB	3,3-diaminobenzidine
H_2O_2	Hydrogen peroxide
HT	High temperature
LC-MS/MS	Liquid chromatography–mass spectrometry
LT	Low temperature
MDA	Malondialdehyde
NBT	Nitro blue tetrazolium
NHCC	Non-heading Chinese cabbage
$O_2{}^{\bullet-}$	Superoxide anion
ROS	Reactive oxygen species
TMT	Tandem mass tag

References

1. Niu, Y.; Xiang, Y. An Overview of Biomembrane Functions in Plant Responses to High-Temperature Stress. *Front. Plant Sci.* **2018**, *9*, 915. [CrossRef] [PubMed]
2. Tripathy, B.C.; Oelmüller, R. Reactive oxygen species generation and signaling in plants. *Plant Signal. Behav.* **2012**, *7*, 1621–1633. [CrossRef] [PubMed]
3. Awasthi, R.; Bhandari, K.; Nayyar, H. Temperature stress and redox homeostasis in agricultural crops. *Front. Env. Sci.* **2015**, *3*, 11. [CrossRef]
4. Feleciano, D.R.; Arnsburg, K.; Kirstein, J. Interplay between redox and protein homeostasis. *Worm* **2016**, *5*, e1170273. [CrossRef] [PubMed]
5. Paciolla, C.; Paradiso, A.; de Pinto, M.C. Cellular Redox Homeostasis as Central Modulator in Plant Stress Response. In *Redox State as a Central Regulator of Plant-Cell Stress Responses*; Springer International Publishing: Berlin, Germany, 2016; pp. 1–23.
6. Yuan, L.; Liu, S.; Zhu, S.; Chen, G.; Liu, F.; Zou, M.; Wang, C. Comparative response of two wucai (*Brassica campestris* L.) genotypes to heat stress on antioxidative system and cell ultrastructure in root. *Acta Physiol. Plant.* **2016**, *38*, 223. [CrossRef]
7. Yuan, L.; Tang, L.; Zhu, S.; Hou, J.; Chen, G.; Liu, F.; Wang, C. Influence of heat stress on leaf morphology and nitrogen-carbohydrate metabolisms in two wucai (*Brassica campestris* L.) genotypes. *Acta Soc. Bot. Pol.* **2017**, *86*, 16. [CrossRef]
8. Zou, M.; Yuan, L.; Zhu, S.; Liu, S.; Ge, J.; Wang, C. Response of osmotic adjustment and ascorbate-glutathione cycle to heat stress in a heat-sensitive and a heat-tolerant genotype of wucai (*Brassica campestris* L.). *Sci. Hortic.* **2016**, *211*, 87–94. [CrossRef]
9. Shao, L.; Wang, C.; Song, J.; Zhang, H.; Zou, M. Effects of low temperature stress on physiological characteristics of savoy cultivars and assessment of their cold tolerance at seedling stage. *J. Anhui Agric. Univ.* **2014**, *41*, 265–269.
10. Gao, F.; Ma, P.; Wu, Y.; Zhou, Y.; Zhang, G. Quantitative Proteomic Analysis of the Response to Cold Stress in Jojoba, a Tropical Woody Crop. *Int. J. Mol. Sci.* **2019**, *20*, 243. [CrossRef]
11. Benešová, M.; Holá, D.; Fischer, L.; Jedelský, P.L.; Hnilička, F.; Wilhelmová, N.; Rothová, O.; Kočová, M.; Procházková, D.; Honnerová, J.; et al. The physiology and proteomics of drought tolerance in maize: Early stomatal closure as a cause of lower tolerance to short-term dehydration? *PLoS ONE* **2012**, *7*, e38017. [CrossRef]
12. Mustafa, G.; Komatsu, S. Quantitative proteomics reveals the effect of protein glycosylation in soybean root under flooding stress. *Front. Plant Sci.* **2014**, *5*, 627. [CrossRef] [PubMed]
13. Echevarría-Zomeño, S.; Fernández-Calvino, L.; Castro-Sanz, A.B.; López, J.A.; Vázquez, J.; Castellano, M.M. Dissecting the proteome dynamics of the early heat stress response leading to plant survival or death in *Arabidopsis*. *Plant Cell Environ.* **2016**, *39*, 1264–1278. [CrossRef] [PubMed]
14. Hu, Y.; Guo, S.; Li, X.; Ren, X. Comparative analysis of salt-responsive phosphoproteins in maize leaves using Ti(4+)–IMAC enrichment and ESI-Q-TOF MS. *Electrophoresis* **2012**, *34*, 485–492. [CrossRef] [PubMed]
15. Lee, D.G.; Ahsan, N.; Lee, S.H.; Kang, K.Y.; Bahk, J.D.; Lee, I.J.; Lee, B.H. A proteomic approach in analyzing heat-responsive proteins in rice leaves. *Proteomics* **2007**, *7*, 3369–3383. [CrossRef] [PubMed]

16. Zhu, J.K. Salt and drought stress signal transduction in plants. *Annu. Rev. Plant Biol.* **2002**, *53*, 247–273. [CrossRef]

17. Parankusam, S.; Bhatnagar-Mathur, P.; Sharma, K.K. Heat responsive proteome changes reveal molecular mechanisms underlying heat tolerance in chickpea. *Environ. Exp. Bot.* **2017**, *141*, 132–144. [CrossRef]

18. Zhang, Y.; Xu, L.; Zhu, X.; Gong, Y.; Xiang, F.; Sun, X.; Liu, L. Proteomic Analysis of Heat Stress Response in Leaves of Radish (*Raphanus sativus* L.). *Plant Mol. Biol. Rep.* **2013**, *31*, 195–203. [CrossRef]

19. Gharechahi, J.; Alizadeh, H.; Naghavi, M.R.; Sharifi, G. A proteomic analysis to identify cold acclimation associated proteins in wild wheat (*Triticum urartu* L.). *Mol. Biol. Rep.* **2014**, *41*, 3897–3905. [CrossRef]

20. Rocco, M.; Arena, S.; Renzone, G.; Scippa, G.S.; Lomaglio, T.; Verrillo, F.; Scaloni, A.; Marra, M. Proteomic analysis of temperature stress-responsive proteins in *Arabidopsis thaliana* rosette leaves. *Mol. Biosyst.* **2013**, *9*, 1257–1267. [CrossRef]

21. Crosatti, C.; Pagani, D.; Cattivelli, L.; Stanca, A.M.; Rizza, F. Effects of growth stage and hardening conditions on the association between frost resistance and the expression of the cold-induced protein COR14b in barley. *Environ. Exp. Bot.* **2008**, *62*, 93–100. [CrossRef]

22. Yuan, L.; Yuan, Y.; Liu, S.; Wang, J.; Zhu, S.; Chen, G.; Hou, J.; Wang, C. Influence of High Temperature on Photosynthesis, Antioxidative Capacity of Chloroplast, and Carbon Assimilation among Heat-tolerant and Heat-susceptible Genotypes of Non-heading Chinese Cabbage. *HortScience* **2017**, *52*, 1464–1470. [CrossRef]

23. Song, X.; Liu, G.; Huang, Z.; Duan, W.; Tan, H.; Li, Y.; Hou, X. Temperature expression patterns of genes and their coexpression with LncRNAs revealed by RNA-Seq in non-heading Chinese cabbage. *BMC Genom.* **2016**, *17*, 297. [CrossRef] [PubMed]

24. Cao, J.Y.; Xu, Y.P.; Cai, X.Z. TMT-based quantitative proteomics analyses reveal novel defense mechanisms of *Brassica napus* against the devastating necrotrophic pathogen *Sclerotinia sclerotiorum*. *J. Proteom.* **2016**, *143*, 265–277. [CrossRef] [PubMed]

25. Liu, Y.; Cao, D.; Ma, L.; Jin, X.; Yang, P.; Ye, F.; Liu, P.; Gong, Z.; Wei, C. TMT-based quantitative proteomics analysis reveals the response of tea plant (*Camellia sinensis*) to fluoride. *J. Proteom.* **2018**, *176*, 71–81. [CrossRef] [PubMed]

26. Takabayashi, A.; Niwata, A.; Tanaka, A. Direct interaction with ACR11 is necessary for post-transcriptional control of GLU1-encoded ferredoxin-dependent glutamate synthase in leaves. *Sci. Rep.* **2016**, *6*, 29668. [CrossRef] [PubMed]

27. Suzuki, A.; Knaff, D.B. Glutamate synthase: Structural, mechanistic and regulatory properties, and role in the amino acid metabolism. *Photosynth. Res.* **2005**, *83*, 191–217. [CrossRef] [PubMed]

28. Masi, A.; Trentin, A.R.; Agrawal, G.K.; Randeep, R. Gamma-glutamyl cycle in plants: A bridge connecting the environment to the plant cell? *Front. Plant Sci.* **2015**, *6*, 16–252. [CrossRef] [PubMed]

29. Mhamdi, A.; Noctor, G. Analysis of the roles of the *Arabidopsis* peroxisomal isocitrate dehydrogenase in leaf metabolism and oxidative stress. *Environ. Exp. Bot.* **2015**, *114*, 22–29. [CrossRef]

30. Lee, M.R.; Kim, C.S.; Park, T.; Choi, Y.; Lee, K. Optimization of the ninhydrin reaction and development of a multiwell plate-based high-throughput proline detection assay. *Anal. Biochem.* **2018**, *556*, 57–62. [CrossRef]

31. Szabados, L.; Savoure, A. Proline: A multifunctional amino acid. *Trends Plant Sci.* **2010**, *15*, 89–97. [CrossRef]

32. Missihoun, T.D.; Schmitz, J.; Klug, R.; Kirch, H.H.; Bartels, D. Betaine aldehyde dehydrogenase genes from *Arabidopsis* with different sub-cellular localization affect stress responses. *Planta* **2011**, *233*, 369–382. [CrossRef] [PubMed]

33. Zhang, X.Y.; Liang, C.; Wang, G.P.; Luo, Y.; Wang, W. The protection of wheat plasma membrane under cold stress by glycine betaine overproduction. *Biol. Plant.* **2010**, *54*, 83–88. [CrossRef]

34. Nanjo, T.; Fujita, M.; Seki, M.; Kato, T.; Tabata, S.; Shinozaki, K. Toxicity of free proline revealed in an *Arabidopsis* T-DNA-tagged mutant deficient in proline dehydrogenase. *Plant Cell Physiol.* **2003**, *44*, 541–548. [CrossRef] [PubMed]

35. Fabro, G.; Kovacs, I.; Pavet, V.; Szabados, L.; Alvarez, M.E. Proline accumulation and *AtP5CS2* gene activation are induced by plant-pathogen incompatible interactions in *Arabidopsis*. *Mol. Plant Microbe Interact.* **2004**, *17*, 343–350. [CrossRef] [PubMed]

36. Vanlerberghe, G.C.; Ordog, S.H. Alternative Oxidase: Integrating Carbon Metabolism and Electron Transport in Plant Respiration. In *Photosynthetic Nitrogen Assimilation and Associated Carbon and Respiratory Metabolism*; Kluwer Academic Publishers: Dordrecht, The Netherlands, 2002; Volume 12, pp. 173–191.

37. Murakami, R.; Ifuku, K.; Takabayashi, A.; Shikanai, T.; Endo, T.; Sato, F. Characterization of an *Arabidopsis thaliana* mutant with impaired *psbO*, one of two genes encoding extrinsic 33-kDa proteins in photosystem II. *FEBS Lett.* **2002**, *523*, 138–142. [CrossRef]

38. Li, Q.; Chang, R.; Sun, Y.; Li, B. iTRAQ-Based Quantitative Proteomic Analysis of Spirulina platensis in Response to Low Temperature Stress. *PLoS ONE* **2016**, *11*, e0166876. [CrossRef] [PubMed]

39. Allahverdiyeva, Y.; Mamedov, F.; Holmström, M.; Nurmi, M.; Lundin, B.; Styring, S.; Spetea, C.; Aro, E.M. Comparison of the electron transport properties of the *psbo1* and *psbo2* mutants of *Arabidopsis thaliana*. *Biochim. Biophys. Acta Bioenerg.* **2009**, *1787*, 1230–1237. [CrossRef]

40. Nishimura, T.; Nagao, R.; Noguchi, T.; Nield, J.; Sato, F.; Ifuku, K. The N-terminal sequence of the extrinsic PsbP protein modulates the redox potential of Cyt b559 in photosystem II. *Sci. Rep.* **2016**, *6*, 21490. [CrossRef]

41. Takahashi, M.; Shigeto, J.; Sakamoto, A.; Morikawa, H. Selective nitration of PsbO1, PsbO2, and PsbP1 decreases PSII oxygen evolution and photochemical efficiency in intact leaves of *Arabidopsis*. *Plant Signal. Behav.* **2017**, *12*, e1376157. [CrossRef]

42. Hesse, H.; Willmitzer, L. Expression analysis of a sucrose synthase gene from sugar beet (*Beta vulgaris* L.). *Plant Mol. Biol.* **1996**, *30*, 863–872. [CrossRef]

43. Peng, T.; Zhu, X.; Duan, N.; Liu, J.H. PtrBAM1, a β-amylase-coding gene of Poncirus trifoliata, is a CBF regulon member with function in cold tolerance by modulating soluble sugar levels. *Plant Cell Environ.* **2015**, *37*, 2754–2767. [CrossRef]

44. Xu, C.; Huang, B. Root proteomic responses to heat stress in two *Agrostis* grass species contrasting in heat tolerance. *J. Exp. Bot.* **2008**, *59*, 4183–4194. [CrossRef]

45. Wang, W.; Vinocur, B.; Shoseyov, O.; Altman, A. Role of plant heat-shock proteins and molecular chaperones in the abiotic stress response. *Trends Plant Sci.* **2004**, *9*, 244–252. [CrossRef]

46. Mamedov, T.G.; Shono, M. Molecular chaperone activity of tomato (*Lycopersicon esculentum*) endoplasmic reticulum-located small heat shock protein. *J. Plant Res.* **2008**, *121*, 235–243. [CrossRef]

47. Sun, W.; Montagu, M.V.; Verbruggen, N. Small heat shock proteins and stress tolerance in plants. *Biochim. Biophys. Acta Gene Struct. Expr.* **2002**, *1577*, 1–9. [CrossRef]

48. Lee, G.J.; Vierling, E. A small heat shock protein cooperates with heat shock protein 70 systems to reactivate a heat-denatured protein. *Plant Physiol.* **2000**, *122*, 189–198. [CrossRef]

49. Gammulla, C.G.; Pascovici, D.; Atwell, B.J.; Haynes, P.A. Differential metabolic response of cultured rice (*Oryza sativa*) cells exposed to high- and low-temperature stress. *Proteomics* **2010**, *10*, 3001–3019. [CrossRef]

50. Matsumura, H.; Nirasawa, S.; Kiba, A.; Urasaki, N.; Saitoh, H.; Ito, M.; Kawai-Yamada, M.; Uchimiya, H.; Terauchi, R. Overexpression of Bax inhibitor suppresses the fungal elicitor-induced cell death in rice (*Oryza sativa* L.) cells. *Plant J.* **2010**, *33*, 425–434. [CrossRef]

51. Kawai-Yamada, M.; Ohori, Y.; Uchimiya, H. Dissection of *Arabidopsis* Bax inhibitor-1 suppressing Bax-, hydrogen peroxide-, and salicylic acid-induced cell death. *Plant Cell* **2004**, *16*, 21–32. [CrossRef]

52. Wang, H.; Zhou, L.; Fu, Y.; Cheung, M.Y.; Wong, F.L.; Phang, T.H.; Sun, Z.; Lam, H.M. Expression of an apoplast-localized BURP-domain protein from soybean (*GmRD22*) enhances tolerance towards abiotic stress. *Plant Cell Environ.* **2012**, *35*, 1932–1947. [CrossRef]

53. Matus, J.T.; Aquea, F.; Espinoza, C.; Vega, A.; Cavallini, E.; Santo, S.D.; Cañón, P.; de la Guardia, A.R.; Serrano, J.; Tornielli, G.B.; et al. Inspection of the Grapevine BURP Superfamily Highlights an Expansion of *RD22* Genes with Distinctive Expression Features in Berry Development and ABA-Mediated Stress Responses. *PLoS ONE* **2014**, *9*, e110372. [CrossRef]

54. Simon, C.; Langlois-Meurinne, M.; Didierlaurent, L.; Chaouch, S.; Bellvert, F.; Massoud, K.; Garmier, M.; Thareau, V.; Comte, G.; Noctor, G.; et al. The secondary metabolism glycosyltransferases UGT73B3 and UGT73B5 are components of redox status in resistance of Arabidopsis to *Pseudomonas syringae* pv. *tomato*. *Plant Cell Environ.* **2014**, *37*, 1114–1129. [CrossRef]

55. Machado, C.R.; de Oliveira, R.L.; Boiteux, S.; Praekelt, U.M.; Meacock, P.A.; Menck, C.F. Thi1, a thiamine biosynthetic gene in *Arabidopsis thaliana*, complements bacterial defects in DNA repair. *Plant Mol. Biol.* **1996**, *31*, 585–593. [CrossRef]

56. Rapala-Kozik, M.; Wolak, N.; Kujda, M.; Banas, A.K. The upregulation of thiamine (vitamin B$_1$) biosynthesis in *Arabidopsis thaliana* seedlings under salt and osmotic stress conditions is mediated by abscisic acid at the early stages of this stress response. *BMC Plant Biol.* **2012**, *12*, 2. [CrossRef]

57. Foyer, C.H.; Noctor, G. Stress-triggered redox signalling: What's in pROSpect? *Plant Cell Environ.* **2016**, *39*, 951–964. [CrossRef]

58. Kamrun, N.; Mirza, H.; Md, M.A.; Masayuki, F. Exogenous Spermidine Alleviates Low Temperature Injury in Mung Bean (*Vigna radiata* L.) Seedlings by Modulating Ascorbate-Glutathione and Glyoxalase Pathway. *Int. J. Mol. Sci.* **2015**, *16*, 30117–30132.

59. Kocsy, G.; Kobrehel, K.; Szalai, G.; Duviau, M.P.; Buzás, Z.; Galiba, G. Abiotic stress-induced changes in glutathione and thioredoxin *h* levels in maize. *Environ. Exp. Bot.* **2004**, *52*, 101–112. [CrossRef]

60. Turan, Ö.; Ekmekci, Y. Activities of photosymtem II and antioxidant enzymes in chickpea (*Cicer arietinum* L.) cultivars exposed to chilling temperatures. *Acta Physiol. Plant.* **2011**, *33*, 67–78. [CrossRef]

61. Han, Y.; Chaouch, S.; Mhamdi, A.; Queval, G.; Zechmann, B.; Noctor, G. Functional analysis of Arabidopsis mutants points to novel roles for glutathione in coupling H_2O_2 to activation of salicylic acid accumulation and signaling. *Antioxid. Redox Signal.* **2013**, *18*, 2106–2121. [CrossRef]

62. Han, Y.; Mhamdi, A.; Chaouch, S.; Noctor, G. Regulation of basal and oxidative stress-triggered jasmonic acid-related gene expression by glutathione. *Plant Cell Environ.* **2013**, *36*, 1135–1146. [CrossRef]

63. Verbruggen, N.; Hermans, C. Proline accumulation in plants: A review. *Amino Acids* **2008**, *35*, 753–759. [CrossRef]

64. Smith, P.K.; Krohn, R.I.; Hermanson, G.T.; Mallia, A.K.; Gartner, F.H.; Provenzano, M.D.; Fujimoto, E.K.; Goeke, N.M.; Olson, B.J.; Klenk, D.C. Measurement of protein using bicinchoninic acid. *Anal. Biochem.* **1985**, *150*, 76–85. [CrossRef]

65. Wisniewski, J.R.; Zougman, A.; Nagaraj, N.; Mann, M. Universal sample preparation method for proteome analysis. *Nat. Methods* **2009**, *6*, 359–362. [CrossRef]

66. Cen, W.; Liu, J.; Lu, S.; Jia, P.; Yu, K.; Han, Y.; Li, R.; Luo, J. Comparative proteomic analysis of QTL *CTS-12* derived from wild rice (*Oryza rufipogon* Griff.), in the regulation of cold acclimation and de-acclimation of rice (*Oryza sativa* L.) in response to severe chilling stress. *BMC Plant Biol.* **2018**, *18*, 163. [CrossRef]

67. Xie, S.; Nie, L.; Zheng, Y.; Wang, J.; Zhao, M.; Zhu, S.; Hou, J.; Chen, G.; Wang, C.; Yuan, L. Comparative proteomic analysis reveals that chlorophyll metabolism contributes to leaf color changes in wucai (*Brassica campestris* L.) responding to cold acclimation. *J. Proteome Res.* **2019**, *18*, 2478–2492. [CrossRef]

68. Yang, L.T.; Qi, Y.P.; Lu, Y.B.; Guo, P.; Sang, W.; Feng, H.; Zhang, H.X.; Chen, L.S. iTRAQ protein profile analysis of *Citrus sinensis* roots in response to long-term boron-deficiency. *J. Proteomics* **2013**, *93*, 179–206. [CrossRef]

69. Zhang, N.; Zhang, L.; Zhao, L.; Ren, Y.; Cui, D.; Chen, J.; Wang, Y.; Yu, P.; Chen, F. iTRAQ and virus-induced gene silencing revealed three proteins involved in cold response in bread wheat. *Sci. Rep.* **2017**, *7*, 7524. [CrossRef]

70. Schmittgen, T.D.; Livak, K.J. Analyzing real-time PCR data by the comparative CT method. *Nat. Protoc.* **2008**, *3*, 1101–1108. [CrossRef]

71. Oliveira, I.C.; Lam, H.; Coschigano, K.; Melooliveira, R.; Coruzzi, G.M. Molecular-genetic dissection of ammonium assimilation in *Arabidopsis thaliana*. *Plant Physiol. Biochem.* **1997**, *35*, 185–198.

72. Coschigano, K.T.; Melo-Oliveira, R.; Lim, J.; Coruzzi, G.M. *Arabidopsis gls* mutants and distinct Fd-GOGAT genes: Implications for photorespiration and primary nitrogen assimilation. *Plant Cell* **1998**, *10*, 741–752. [CrossRef]

73. Clough, S.J.; Bent, A.F. Floral dip: A simplified method for Agrobacterium-mediated transformation of *Arabidopsis thaliana*. *Plant J.* **1998**, *16*, 735–743. [CrossRef]

74. Baker, M.A.; Cerniglia, G.J.; Zaman, A. Microtiter plate assay for the measurement of glutathione and glutathione disulfide in large numbers of biological samples. *Anal. Biochem.* **1990**, *190*, 360–365. [CrossRef]

75. Lin, C.C.; Kao, C.H. Disturbed ammonium assimilation is associated with growth inhibition of roots in rice seedlings caused by NaCl. *Plant Growth Regul.* **1996**, *18*, 233–238. [CrossRef]

76. Heath, R.L.; Packer, L. Photoperoxidation in isolated chloroplasts: I. Kinetics and stoichiometry of fatty acid peroxidation. *Arch. Biochem. Biophys.* **1968**, *125*, 189–198. [CrossRef]

77. Lutts, S.; Kinet, J.M.; Bouharmont, J. NaCl-induced Senescence in Leaves of Rice (*Oryza sativa* L.) Cultivars Differing in Salinity Resistance. *Ann. Bot.* **1996**, *78*, 389–398. [CrossRef]

78. Bates, L.S.; Waldren, R.P.; Teare, I.D. Proline content Ref: Rapid determination of free proline for water-stress studies. *Plant Soil* **1973**, *39*, 205–207. [CrossRef]

79. Patterson, B.D.; Macrae, E.A.; Ferguson, I.B. Estimation of hydrogen peroxide in plant extracts using titanium(IV). *Anal. Biochem.* **1984**, *139*, 487–492. [CrossRef]

80. Elstner, E.F.; Heupel, A. Inhibition of nitrite formation from hydroxylammoniumchloride: A simple assay for superoxide dismutase. *Anal. Biochem.* **1976**, *70*, 616–620. [CrossRef]

81. Juszczak, I.; Cvetkovic, J.; Zuther, E.; Hincha, D.K.; Baier, M. Natural Variation of Cold Deacclimation Correlates with Variation of Cold-Acclimation of the Plastid Antioxidant System in *Arabidopsis thaliana* Accessions. *Front. Plant Sci.* **2016**, *7*, 305. [CrossRef]

International Journal of
Molecular Sciences

Article

Cytological and Proteomic Analysis of Wheat Pollen Abortion Induced by Chemical Hybridization Agent

Shuping Wang [1,2,*], Yingxin Zhang [1,3], Zhengwu Fang [1], Yamin Zhang [2], Qilu Song [2], Zehao Hou [1], Kunkun Sun [1], Yulong Song [2], Ying Li [2], Dongfang Ma [1], Yike Liu [4], Zhanwang Zhu [4], Na Niu [2], Junwei Wang [2], Shoucai Ma [2] and Gaisheng Zhang [2,*]

[1] Hubei Collaborative Innovation Center for Grain Industry/Hubei Key Laboratory of Waterlogging Disaster and Agricultural Use of Wetland/College of Agriculture, Yangtze University, Jingzhou 434000, China; zhangyingxin1985@126.com (Y.Z.); fangzhengwu88@163.com (Z.F.); houzehao1994@126.com (Z.H.); 201772390@yangtzeu.edu.cn (K.S.); madf@yangtzeu.edu.cn (D.M.)

[2] College of Agriculture, Northwest A&F University, Yangling 712100, China; ymzhang2017@163.com (Y.Z.); songqilu1234@163.com (Q.S.); sylbl1986@163.com (Y.S.); qiuxuewuying@163.com (Y.L.); niuna@nwsuaf.edu.cn (N.N.); wjw@nwsuaf.edu.cn (J.W.); mashoucai@nwsuaf.edu.cn (S.M.)

[3] Institute of Genetics and Developmental Biology, Chinese Academy of Sciences, Beijing 100101, China

[4] Food Crops Institute, Hubei Academy of Agricultural Sciences, Wuhan 450064, China; liuyike@webmail.hzau.edu.cn (Y.L.); zhuzhanwang@163.com (Z.Z.)

* Correspondence: wangshuping2003@126.com (S.W.); zhanggsh@public.xa.sn.cn (G.Z.)

Received: 21 February 2019; Accepted: 27 March 2019; Published: 1 April 2019

Abstract: In plants, pollen grain transfers the haploid male genetic material from anther to stigma, both between flowers (cross-pollination) and within the same flower (self-pollination). In order to better understand chemical hybridizing agent (CHA) SQ-1-induced pollen abortion in wheat, comparative cytological and proteomic analyses were conducted. Results indicated that pollen grains underwent serious structural injury, including cell division abnormality, nutritional deficiencies, pollen wall defect and pollen grain malformations in the CHA-SQ-1-treated plants, resulting in pollen abortion and male sterility. A total of 61 proteins showed statistically significant differences in abundance, among which 18 proteins were highly abundant and 43 proteins were less abundant in CHA-SQ-1 treated plants. 60 proteins were successfully identified using MALDI-TOF/TOF mass spectrometry. These proteins were found to be involved in pollen maturation and showed a change in the abundance of a battery of proteins involved in multiple biological processes, including pollen development, carbohydrate and energy metabolism, stress response, protein metabolism. Interactions between these proteins were predicted using bioinformatics analysis. Gene ontology and pathway analyses revealed that the majority of the identified proteins were involved in carbohydrate and energy metabolism. Accordingly, a protein-protein interaction network involving in pollen abortion was proposed. These results provide information for the molecular events underlying CHA-SQ-1-induced pollen abortion and may serve as an additional guide for practical hybrid breeding.

Keywords: CHA-SQ-1; cytomorphology; pollen abortion; proteomics; wheat

---

1. Introduction

In wheat plants, pollen develops in the anther, a highly specialized organ. Sporogenous cells (center of anther locule) give rise to microsporocyte. The microsporocyte undergoes two meiotic divisions, developing into a tetrad of haploid microspores (tetrad stage). Then the microspores are released, and each consists of a central nucleus (early-uninucleate stage). These microspores grow and undergo cell polarization until the nucleus is adjacent to the wall, and a single vacuole dominates the intracellular space (later-uninucleate stage). The polarized cell then divides to form one large vegetative cell and one small generative cell (binucleate stage). Later, this bicellular system produces tricellular pollen (a vegetative cell and two sperm cells; trinucleate stage), forming mature pollen grain [1–4]. Therefore, the development of mature pollen grain follows a tightly controlled sequence of events within the anther. Once this sequence is broken, pollen abortion occurs. In previous study, SQ-1 is an effective CHA for wheat and can impair the production and release of viable pollen [2]. Previous research studies on CHA-SQ-1-induced male sterility were mainly concentrated on reactive oxygen metabolism, aliphatic metabolism, and DNA methylation [5–7], the genetic and molecular mechanisms of CHA induced wheat pollen sterility still needs to be further elucidated.

In plant, pollen grains are highly useful model for investigating the molecular mechanisms underlying cell differentiation, polar cell growth, cell-to-cell communication, and cell fate determination. Additionally, the development of mature pollen grain was followed by a tightly controlled sequential process within anthers. Once this process is disturbed, pollen abortion occurs [2]. In the past two decades, pollen function specialization has been studied using biochemistry, functional genomics, and molecular genetics. More than 150 genes have been found to regulate pollen development, as indicated by reverse genetics, and some of these genes have been functionally characterized and expressed at specific stages of pollen maturation [8]. Additionally, a large number of transcripts expressed during pollen development have been identified during pollen development, which encode proteins involved in heat shock, cytoskeleton, pollen cell wall, allergen, cell division, signal transduction, pectin, carbohydrate, and energy metabolism [9]. Understanding the exact mechanisms underlying pollen development and fertilization may facilitate more advanced crop breeding and engineering. These issues have been investigated in higher plants in studies of male-sterile mutants [10–12], factors affecting pollen germination and tube growth [13,14], and the molecular mechanisms underlying self-incompatibility [15]. More importantly, pollen development is known to be very sensitive to abiotic stress [1]. In particular, high temperature [16,17], low temperature [18,19], drought [20], and water-stress [21] can result in male sterility in crops, including *Hordeum vulgare* [22], *Oryza sativa* [23], *Triticum aestivum* [18]. Meanwhile, proteomics is a powerful approach to study the molecular processes of pollen development, and also a powerful complement to the whole genome sequencing [24]. Based on the proteins identified in this work, a detailed pathway was successfully constructed and protein-protein interactions were more clearly understood [25]. To date, there has been an increased application of proteomics approaches to study the pollen reproduction or pollen responses to abiotic stress, for example, *Lycopersicon esculentum* (Response to heat-stress and ethylene) [18], *Oryza sativa* (Response to high temperature-stress) [23], *Triticum aestivum* (Response to drought-stress) [26]. Thus, proteomics has been extensively used to investigate the protein expression pattern in pollen development under several abiotic stresses [16,27].

Recently, there have been some progress in studies of CHA-SQ-1-induced male sterility of wheat, for example, metabolism [5,6,28], DNA methylation [7], cell morphology [2,3], transcriptome [29], and proteomic [30,31] of flag leaf [30], floret [32], and anther [2,4,31]. These results provide necessary theoretical basis for studying the mechanism of CHA-SQ-1 induced male sterility. The objective of this study was to uncover the cytological and biochemical mechanisms of CHA-SQ-1 induced pollen abortion in wheat. Towards this objective, a comprehensive analysis of pollen grain cytomorphology and proteome was performed. CHA-SQ-1 impaired pollen maturation and resulted in complete pollen sterile. As expected, 60 identified differential abundant proteins (DAPs) in this study did play important roles in cell growth and division, stress response, carbohydrate and energy metabolism, and protein metabolism.

2. Results

2.1. Cytological Changes in Pollen Abortion Induced by CHA-SQ-1 in Wheat

Morphological differences of pollen grains between control and CHA-SQ-1-treated plants were revealed by microscopic observation (Figure 1). The pollen grains of controls showed two sperm nuclei, detectable nuclear nutrients (Figure 1A), and a full complement of storage materials (Figure 1B and C) fostering pollen viability and facilitating function, which include proteins (Figure 1D) and starch granules (Figure 1E). However, the pollen grains of CHA-SQ-1 treated plants showed abnormal development (Figure 1G) and accumulated less nutritional material in the cytoplasm (Figure 1H–K). Unlike control pollen grains, which showed strong staining, none of the treated pollen grains were deeply stained with iodine-potassium iodide (2% I_2-KI; Figure 1E,K), which indicated that the plants were 100% pollen sterile.

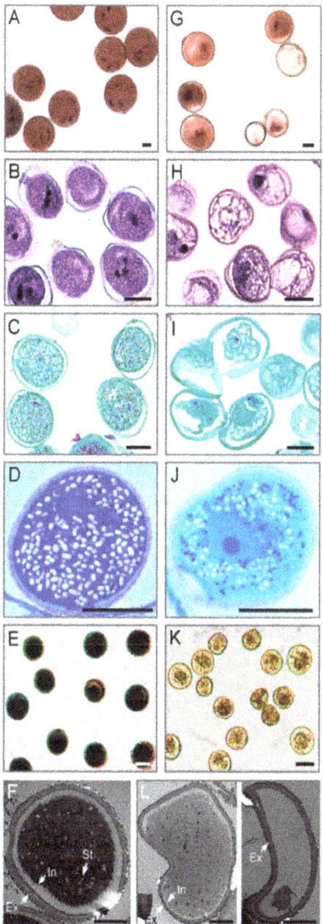

Figure 1. Comparison of pollen grain from the control (**A–F**) and CHA-SQ-1-treated wheat plants (**G–L**) during pollen maturation. (**A,G**) 1% acetocarmine staining. (**B,H**) Ehrlich's hematoxylin staining. (**C,I**) safranin O/fast green staining. (**D,J**) CBB-R250 staining. (**E,K**) I_2-KI staining. (**F,L**) Transmission electron micrograph. Ap, germination aperture; Ex, exine; In, intine; St, starch granule; Bars: 20 μm (**A–E,G–K**), 10 μm (**F,L**).

Detailed examination of treated and control pollen grains was performed to identify defects in cytological structural using transmission electron microscopy (TEM, Figure 1F,L). TEM analysis showed that the control pollen grains had fairly dense cytoplasm and normal pollen walls with distinguishable exine and intine layers (Figure 1F). Although the walls of pollen grains from CHA-SQ-1-treated plants seemed to have normal exine, the pollen grains had less dense cytoplasm and often empty chambers, and the intine was thin or undetectable (Figure 1L).

These results indicated that pollen grains were impaired by CHA-SQ-1, resulting in abnormal pollen development and shape, reduced storage materials, defective pollen intine, and collapsed pollen grains.

2.2. Pollen Grain Proteomic Analysis of CHA-SQ-1-Treated Wheat Plants

To determine which pollen grain proteins changed in abundance in response to CHA-SQ-1-treatment, a proteomic study was performed using two-dimensional gel electrophoresis (2-DE) and MALDI-TOF/TOF MS. Three independent biological replicates were performed in this 2-DE experiment. Figure 2 shows a representative gel image of proteins extracted from control and CHA-SQ-1 treated plants, respectively. From a spot-to-spot comparisons and statistical analysis, a total of 61 protein spots exhibited at least 1.5-fold ($P \leq 0.01$) difference in abundance between the control and CHA-SQ-1 treated plants (Figure 2 and Table S1). Of these, 19 spots were high-abundant and 42 spots were low-abundant in pollen grains from CHA-SQ-1-treated plants (Table S1).

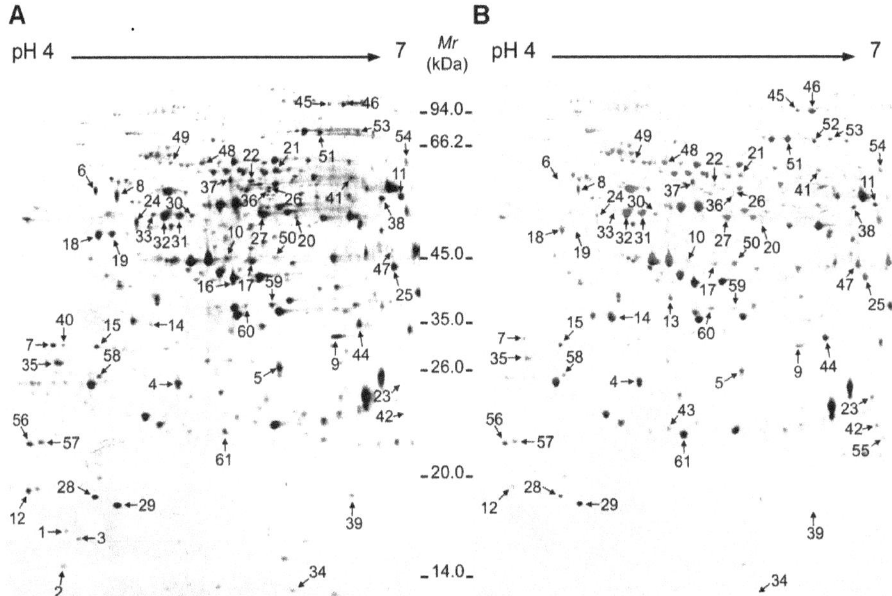

Figure 2. 2-DE electrophoresis gels of pollen grain proteomes in the control (**A**) and CHA-SQ-1-treated wheat plants (**B**). Proteins were stained with CBB-G250. About 550 μg of protein was loaded onto IPG strips at pH 4–7 (17 cm, linear). SDS-PAGE was performed using 12% gels. Identified proteins are numbered on the gels.

2.3. Identification and Classification of DAPs

All 61 DAPs were further analyzed using MALDI-TOF/TOF MS, and 60 of them were successfully identified (Table S2). Of these, 57 were functionally annotated in the current database, but the remaining three identities were unnamed proteins (spot 58, spot 59, and spot 60; Tables S2 and S3). To annotate these proteins, their sequences were used as queries and NCBI was searched for homologues using BLASTP. The result was listed in Table S4. Homologues were defined as proteins sharing at least 80% positive identity with the target at the amino acid level, which was here considered indicative of similar function. In this way, protein homologues were divided into appropriate categories. However, about one third of the identified proteins were detected in multiple spots with different pIs or molecular masses (Table S2). This suggested the existence of isoforms and posttranslational modification. Similar results were also found in others [24,33]. Taken together, the 60 identities represented 49 unique proteins (Tables S2 and S4).

Furthermore, based on the metabolic and functional features of wheat pollen, all of these identities were classified into nine functional groups (Figure 3A,C; Table S2), including cell growth and division (18%, eleven low-abundant proteins in CHA-SQ-1 treatment), glycolysis (18%, nine low-abundant and two high-abundant proteins in treatment), protein synthesis and destination (15%, nine high-abundant proteins in treatment), redox homeostasis and defense (13%, eight low-abundant proteins in treatment), energy metabolism (12%, seven low-abundant proteins in treatment), one carbon metabolism (9%, five high-abundant proteins in treatment), storage protein (5%, two low-abundant and one high-abundant proteins in treatment), TCA cycle (5%, three low-abundant proteins in treatment), and signal transduction (5%, two low-abundant and one high-abundant proteins in treatment). An impressive 85% of these identified proteins were implicated in the first six functional groups, whereas the largest functional group consisted of proteins involving cell growth and division (18%) and glycolysis (18%), which were greatly affected by CHA-SQ-1 treatment. Further analysis of the abundance changes of each group revealed that proteins involved in protein synthesis and destination (15%), redox homeostasis and defense (13%), energy metabolism (12%), and one carbon metabolism (9%) were overrepresented, either in number or in expression level, suggesting that these processes were susceptible to CHA-SQ-1 treatment during pollen maturation.

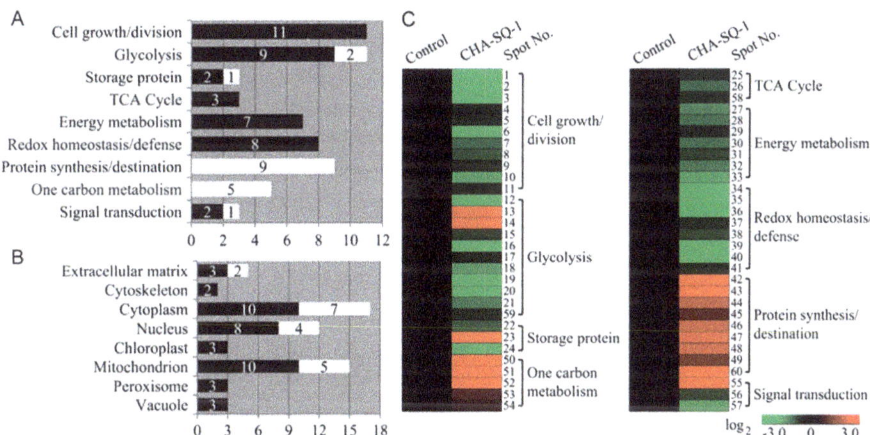

Figure 3. Functional classification, subcellular localization, and hierarchical clustering of the 60 DAPs in pollen grain of the control and CHA-SQ-1-treated wheat plants. The DAPs were divided into nine functional groups (**A**) and classified by predicted subcellular localization (**B**) using Euk-mPLoc 2.0 database (http://www.csbio.sjtu.edu.cn/bioinf/euk-multi-2/), BaCelLo database (http://gpcr.biocomp.unibo.it/bacello/pred.htm) and ESLpred database (http://www.imtech.res.in/raghava/eslpred2/). The black bars indicate high-abundant proteins and the white bars low-abundant proteins. (**C**) The hierarchical cluster analysis was conducted using the Genesis 1.7.6 procedure (Graz University of Technology, Austria, http://genome.tugraz.at/genesisclient/genesisclient_download.shtml) and the log$_2$-transformed values of -fold change ratios listed in Table S1.

The subcellular location of a protein can indicate its physiological function. Here, prediction showed that these differentially accumulated proteins were localized at the extracellular matrix (9%, three low-abundant and two high-abundant proteins in treatment), cytoskeleton (3%, two low-abundant proteins in treatment), cytoplasm (28%, ten low-abundant and seven high-abundant proteins in treatment), nucleus (20%, eight low-abundant and four high-abundant proteins in treatment), chloroplast (5%, three low-abundant proteins in treatment), mitochondrion (25%, ten low-abundant and five high-abundant proteins in treatment), peroxisome (5%, three low-abundant proteins in treatment), and vacuole (5%, three low-abundant proteins in treatment, Figure 3B). In order to visually portray the patterns of protein expression in all nine functional categories, hierarchical clustering of proteins was analyzed and a graphic was produced (Figure 3C).

2.4. Protein-Protein Interactions Network in CHA-SQ-1-Induced Pollen Abortion

In living cells, proteins do not act as single entities. Rather, they form a network of functional interconnections that underlie the cellular processes. To determine how CHA-SQ-1 interacts with the wheat pollen grains and affect cell functions, identified proteins were annotated using Arabidopsis thaliana TAIR10 protein database (Table S5), the corresponding AGI codes were then used to generate an interactome on STRING (Figure 4) and BiNGO (Figure 5).

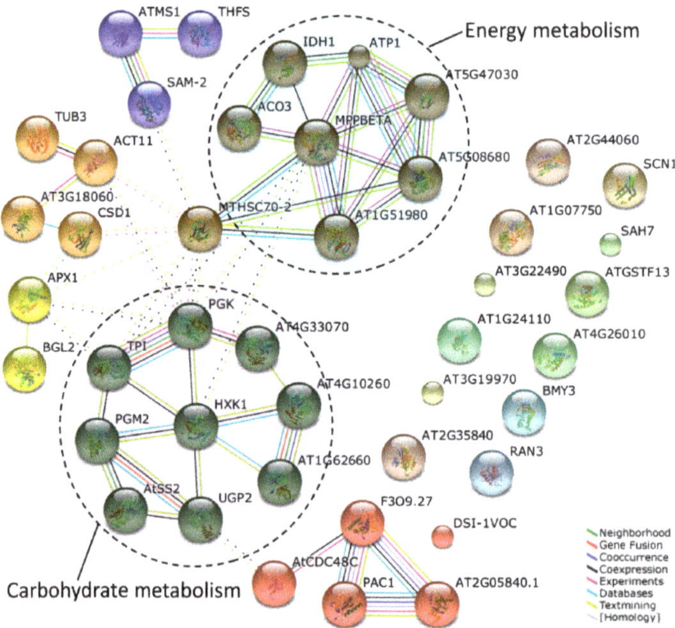

Figure 4. Analysis of protein interaction networks using the STRING system. TAIR homologous proteins among those identified here were mapped using STRING 10 software at a confidence of 0.4. All the homologous proteins are listed in Table S5. Abbreviations of the specific protein names in STRING network were presented in Table S6. Colored lines between the proteins indicate the various types of interaction evidence. The two clusters of protein nodes that interacted closely and frequently are indicated by circles. They include proteins involved in carbohydrate and energy metabolism.

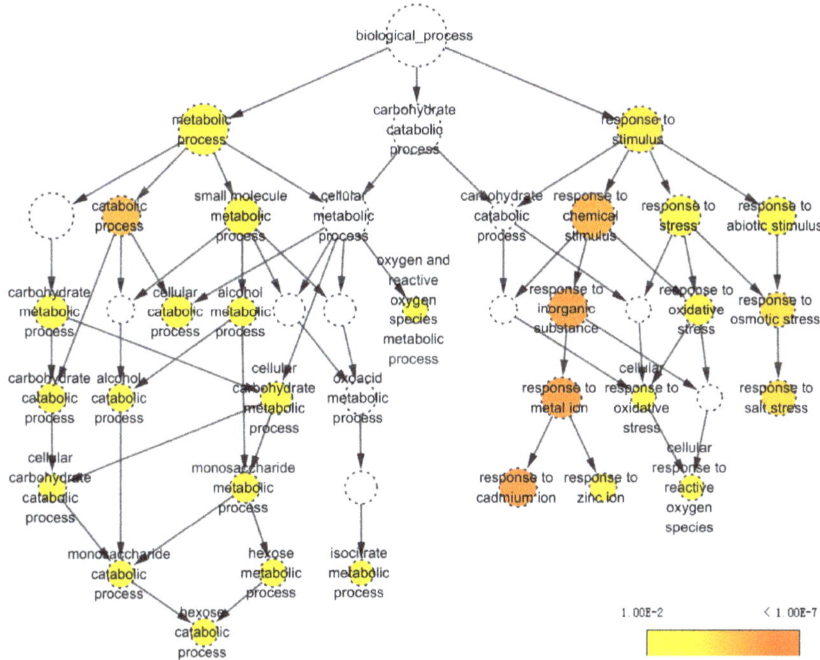

Figure 5. Biological pathway networks generated using the BiNGO plugin from Cytoscape tool. Homologous proteins were used here for GO analysis. Node size is shown as proportional to the number of proteins placed in the GO category. Color denotes the *P*-value of each enriched GO term (color scale, right bottom). White nodes are not enriched.

The STRING analysis revealed a protein association network (Figure 4). Specific protein names and abbreviations are given in Table S6. The two major clusters identified included proteins involved in carbohydrate and energy metabolism. CHA-SQ-1 treatment was found to inhibit the accumulation of a protein essential to carbohydrate metabolism, hexokinase (HXK1, gi I 475536774, spot15). Heat shock 70 kDa protein (MTHSC70-2; gi I 473970552, spot 48, and gi I 379645201, spot 49) is the core protein of this network, and it interacts with many other clusters (carbohydrate metabolism, energy metabolism, cell growth/division, one carbon metabolism).

BiNGO indicated statistically over- and under-represented biological pathways related to pollen grains of plants with CHA-SQ-1 treatment (Figure 5; Table S7), which provide a complete list of enriched Gene Ontology (GO) biological pathways of these proteins. Of them, two major biological categories were significantly overrepresented in pollen grains of plants with CHA-SQ-1 treatment, including: metabolic process ($P = 8.25 \times 10^{-5}$) and response to stimulus ($P = 8.51 \times 10^{-6}$). This indicates that some of biological pathways in pollen grains responded to resistance and susceptibility conditions under the CHA-SQ-1 treatment, especially oxygen and reactive oxygen species metabolic process ($P = 9.46 \times 10^{-4}$) and responded to oxidative stress ($P = 3.96 \times 10^{-5}$), which implied that the pollen grains of CHA-SQ-1 treated plants suffered from oxidative stress (Figure 6E,F). Of them, several of the most highly enriched DAPs were found to participate in carbohydrate metabolism, such as carbohydrate catabolic process ($P = 2.70 \times 10^{-5}$), hexose metabolic process ($P = 3.91 \times 10^{-4}$), alcohol metabolic process ($P = 2.64 \times 10^{-4}$), etc. These results suggested that CHA-SQ-1 disturbed pollen development by several biological pathways, particularly carbohydrate metabolism, oxidative/antioxidative system. More importantly, these abnormal changes in pollen grain of CHA-SQ-1 treatment results in blocking of the process of glucose metabolism and accumulation of starch grains maintaining pollen grain development (Figure 1E,K).

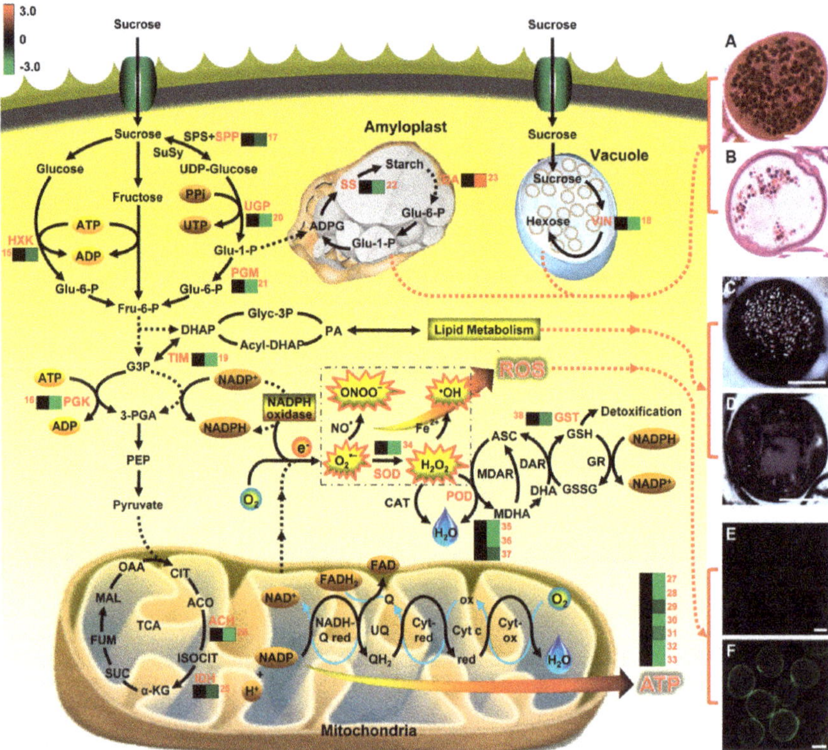

Figure 6. Schematic overview of the metabolic pathways associated with the differentially expressed proteins identified in pollen abortion of CHA-SQ-1-treated wheat plants. Hierarchical clustering and numbers represent protein identification and abundance listed in Tables S1 and S2. SPS, sucrose phosphate synthase; SPP, sucrose phosphate phosphatase; SuSy, sucrose synthase; UGP, UTP-glucose-1-phosphate uridylyltransferase; HXK, hexokinase; Glu-1-P; glucose-1-phosphate; PGM, phosphoglucomutase, Glu-6-P, glucose 6-phosphate; Fru-6-P, fructose 6-phosphate; G3P, glycerate 3-phosphate; DHAP, dihydroxyacetone phosphate; PGK, 3-phosphoglycerate kinase; 3-PGA, 3-phosphoglyceric acid; PEP, phosphoenolpyruvate; TIM, triosephosphat-isomerase; CIT, citrate; ACO, aconitate; ACH, aconitate hydratase; ISOCIT, isocitrate; IDH, isocitrate dehydrogenase; α-KG, α-ketoglutarate; SUC, succinate; FUM, fumarate; MAL, malate; OAA, oxaloacetate; red, reductase; ox, oxidase; UQ, ubiquinone; Q, quinone; QH₂, hydroquinone; Cyt, cytochrome; ADPG, ADP-glucose; SS, starch synthase; BA, beta-amylase; VIN, vacuolar invertase; Glyc-3P, glyceraldehyde 3-phosphate; PA, phosphatidic acid; SOD, superoxide dismutase; CAT, catalase; POD, peroxidase; MDHA, Monodehydroascorbate; MDHR, Monodehydroascorbate reductase; ASC, ascorbate; DHA, dehydroascorbate; DAR, dehydroascorbate reductase; GSH, reduced glutathione; GSSG, oxidized glutathione; GR, glutathione reductase; GST, glutathione S-transferase. (**A**–**F**) Comparison of insoluble polysaccharides, lipids, and ROS in pollen grains from control (**A**,**C**,**E**) and CHA-SQ-1-treated wheat plants (**B**,**D**,**F**) during pollen maturation. (**A**,**B**) PAS staining. (**C**,**D**) Sudan black B staining. (**E**,**F**) H₂DCF-DA staining. Bars: 20 μm.

Observations of the connectivity of proteins in this biological network collectively suggest that functional regulation of the cellular mechanisms of pollen abortion in response to CHA-SQ-1 treatment may be involved in significant physiological changes.

3. Discussion

3.1. CHA-SQ-1 Induced Complete Pollen Abortion

Commercial wheat hybrids have been produced using CHAs, which are growth regulators selectively interfering with the development of pollen or natural systems of male fertility [3,4]. In the present study, the cytological observations indicated that various pollen constituents occurred in CHA-SQ-1-treated plants were markedly different from those in control plants, including abnormal chromosome behavior (Figure S2A,B), low viability (Figure S2C,D), reduced insoluble polysaccharides (Figure 6A,B) and lipid particles (Figure 6C,D), and other storage materials (proteoplasts, starch; Figure 1D–K). These results confirmed that CHA-SQ-1 could induce complete (100%) pollen abortion. Indeed, CHA-SQ-1-induced male sterility is constantly used in China to produce heterosis in wheat [2,4,28]. More importantly, CHA-induced male sterility, with exact the same nuclear background, may circumvent the confounding factors of genotype in cytoplasmic male sterility and genetic male sterility [4], this provides a shortcut for revealing the mechanism of male sterility.

3.2. Proteins Involved in Cell Growth and Division

During normal development, the production of functional pollen grain is heavily dependent on timely cell growth and division. In the present study, 11 low-abundant proteins were found to be associated with cell growth and division under CHA-SQ-1 treatment (Table S2).

Mature pollen grains have an outer coat over the underlying wall. The pollen coat composition is critical for protection against environmental damage, such as lipids, phenolic compounds and several other proteins. In addition, some of the components are allergenic to humans, especially pollen allergens. Recent studies showed that pollen allergens serve the plant by fostering vegetative growth and play a role in reproductive development, which were clustered in more than 10 categories [34]. It is important to note that most allergenic pollen proteins are located inside the pollen. Previous studies indicated that CHA-SQ-1 impacts anther development through preceding programmed cell death, misshaping and shrinking extine pattern, and disturbing microspore development [2,31]. In this study, the amount of five pollen allergen proteins (spots 4–8) were significantly decreased in response to CHA-SQ-1 treatment, which resulted in defective pollen intine (Figure 1F,L), leading to membrane and pollen damage, cellular content spilling in pollen grain of CHA-SQ-1 treated plants.

In addition, three DAPs (spots 1–3) were identified as pollen-specific protein in the present study, which are only expressed in a temporally and regionally specific manner and synthesized after microspore mitosis, and then accumulated in the cytoplasm during pollen maturation [35]. Previous studies indicated that pollen-specific *F8-1* is a positive regulator of CHA-SQ-1-induced male sterility, and knockdown of this gene in wheat resulted in 43.56 percent of pollen sterility [34]. Therefore, changes of the structure or pattern of pollen-specific protein can render pollen development abnormal. The data presented here showed that low-abundant of pollen-specific proteins are observed in pollen grains of CHA-SQ-1 treated plants.

3.3. Proteins Involved in Carbohydrate and Energy Metabolism

A large proportion (37%) of the proteins was found to be related to carbohydrate and energy metabolism. Of these, 10 proteins were found to be associated with glycolysis (spots 12–21), three with the storage protein (spots 22–24), two with the TCA cycle (spots 25 and 26), and seven with the electron transport chain (spots 27–33; Figure 6 and Table S2).

Mature pollen grains store polysaccharides, lipids, proteins, hormones, and other substances that play important roles in pollen germination and early tube growth. Therefore, it is critical that pollen grains contain sufficient supplies of carbon and energy reserves to utilize at appropriate time [36]. Recent studies indicate that the starch content in CHA-SQ-1 treated anthers was approximately 45% of those in the fertile line; and the activities of vacuolar invertase (VIN) were significantly reduced [28,31]. Meanwhile, the expression of one sucrose transporter gene (TaSUT1) was decreased in CHA-SQ-1 treated anther [28]. Additionally, abscisic acid (ABA)- or cold-induced male sterility in rice involved a disruption of sugar transport in anthers [37]. In this case, sucrose synthesis and degradation play important roles in pollen maturation, germination and tube growth. In both photosynthetic and storage cells, sucrose synthesis involves two enzymes, sucrose phosphate synthase (SPS) and sucrose phosphate phosphatase (SPP). SPP catalyzes the final step in the pathway of sucrose biosynthesis [38], which decreased in abundance and might slow down the process of sucrose biosynthesis of CHA-SQ-1 treated plants (spot 17, Figure 6). Meanwhile, two glycoside hydrolyses (spots 13 and 14) were high-abundant in CHA-SQ-1 treatment, which accelerated the degradation of sucrose. Sucrose is consumed via glycolysis to provide energy necessary for cell expansion, division, differentiation, nutrient uptake, and maintenance during plant development [4]. This requires a series of enzymes to catalyze this process, including Hexokinase (HXK, spot 15), 3-phosphoglycerate kinase (PGK, spot 16), vacuolar invertase (VIN, spots 18), triosephosphat-isomerase (TIM, spots 19), UTP-glucose-1-phosphate uridylyltransferase (UGP, spots 21), and phosphoglucomutase (PGM, spots 21). Of those, as shown in Figure 6, the first step of glycolysis is the phosphorylation of glucose by HXK. PGK, as a major enzyme of glycolysis, catalyzes the first ATP-generating step. TIM is an enzyme that interconverts dihydroxyacetone phosphate and D-glyceraldehyde-3-phosphate very quickly. Its catalytic site located at the dimer interface [39]. UGP1 is an enzyme associated with glycogenesis, and its product, UDP-glucose, is involved in multiple pathways and is a precursor for other sugar nucleotides [36]. PGM is an enzyme that transfers a phosphate group and facilitates the interconversion of glucose 1-phosphate and glucose 6-phosphate. Additionally, glucose and fructose can be produced by VIN in vacuole; in situ hybridization revealed that VIN were highly expressed in the pollen grains [40]. Previous studies have demonstrated that decreased expression of VIN altered the hexose-to-sucrose ratio and VIN has long been believed to be a major player in cell expansion [28]. In the present study, in CHA-SQ-1 treatment, the decreased expression of these proteins (spot 15, spot 16, and spots 18–21) might repress the glycolysis process and further slowed down the pyruvate production rate of glycolysis (Figure 6). Pyruvate is a key intersection of metabolic pathway network, which can be converted to acetyl CoA by the mitochondria pyruvate dehydrogenase complex (mtPDC). The acetyl CoA could enter the TCA cycle and mitochondrial electron transport chain (mtETC), and went through a series of enzyme-catalyzed reaction (such as aconitase, isocitrate dehydrogenase, ATPase, etc.) to supply energy. However, the decreased co-expression of isocitrate dehydrogenase (NAD) regulatory subunit 1 (spot 25), aconitate hydratase (spot 26), and ATP synthase (spots 27–33) further limited the energy production and storage (Figure 6). Therefore, these cause a serious imbalance, and followed by a decrease rate of respiration and ATP production in anthers at trinucleate stage (Figure S3). Taken together, these physiological alterations occur upon disruption of the mitochondrion.

Pollen maturation requires accumulation of starch, which acts as an energy reserve to facilitate pollen germination. For this reason, starch content can serve as a marker of pollen maturity [35]. Recent studies of carbohydrate metabolism indicate that disruption of sugar balance in the pollen grain can impair pollen development significantly and cause male sterility [41]. Here, two key enzymes (spot 22, low-abundant; spot 23, high-abundant) involved in sucrose-to-starch showed different expression patterns in CHA-SQ-1 treated plants (Figure 6), which caused reduced or invisible starch grains (Figures 1K and 6), and resulted in non-functional pollens.

Taken together, the data collected here show that the different expression patterns of 22 proteins inhibited the carbohydrate and energy metabolism, which reduced the level of ATP and sucrose/starch. This indicates that pollens of CHA-SQ-1-treated plants are in a state short of nutrient and energy. With respect to high ATP- and sucrose-requiring processes during pollen maturation, there is not sufficient storage carbohydrates and energy to support cell metabolism. The effects of pollen abortion are likely to be more severe during the period of intense growth.

3.4. Stress Response Related Proteins

Proteins included in this group are associated with stress response related proteins, which included eight redox homeostasis- and defense-related proteins (spots 34–41, low-abundant), three signal transduction-related proteins (spot 55, high-abundant; spots 56 and 57, low-abundant). Pollen grains are free-floating and subject to various abiotic stresses, including drought and harsh temperatures. These stresses are often inextricably linked with reactive oxygen species (ROS). To some extent, ROS act as signal molecules in the regulation of biological processes such as growth, development, and responses to both biotic and abiotic stimuli [16,42]. However, excessive ROS production could cause oxidative damage [3,4]. Therefore, pollen grains have evolved a strategy to combat the ROS by inducing various protective enzymes and developing a balance between ROS production and clearance. In our results, many enzymes involved in the ROS detoxification were identified in wheat pollen, including superoxide dismutase (spot 34), peroxidase (spots 35 and 36), L-ascorbate peroxidase (spot 37), and glutathione-S-transferase (spot 38). These antioxidative proteins were low-abundant in CHA-SQ-1-treated pollen grains (Figure 6; Tables S1 and S2), promoting the accumulation of ROS. These conditions led to the occurrence of severe oxidative stress during pollen maturation (Figure 6E,F).

3.5. Proteins Related to Synthesis, Folding, and Proteolysis

In the present study, a total of eight DAPs were found to be involved in protein metabolism. Of these, three spots (spots 42–44) corresponding to proteasome involved in protein degradation were identified (Table S2). The other five proteins involved in protein synthesis such as three mitochondrial-processing peptidase subunits (spots 45–47) and two HSPs (spots 48–49) were also identified (Table S2). Other proteomic studies have shown that the same groups of proteins were also identified suffer from abiotic stress during pollen development [43,44]. Therefore, protein synthesis and degradation are important for pollen development. However, in CHA-SQ-1 treated pollen grains, although the process of protein synthesis was enhanced by the increased expression of mitochondrial-processing peptidase subunits (spots 45–47) and HSPs (spots 45–47), it was clearly not sufficient for the accelerated protein degradation caused by the increased expression of proteasome (spots 42–44). Ultimately, CHA-SQ-1-treated pollen grains exhibited an increase protein catabolism (Figure 1J).

3.6. Other Proteins

It was documented that DNA methylation reactions happened in CHA-SQ-1-treated anthers [7], and the similar phenomenon was observed in our previous work [32]. In this study, five high-abundant proteins (spots 50–54) associated with one carbon metabolism were identified in CHA-SQ-1-treated pollen grains, which might promote the DNA methylation reactions. Although 2-DE combined with mass spectrometry is likely to provide an extremely useful tool for identifying abundant proteins, there is also some limitation for identifying alkaline and hydrophobic proteins, and some identified proteins might be grouped as unknown proteins due to the limited information on their putative functions [45]. In this study, three spots (spots 58–60) were attributed to proteins with no clearly predicted, well-defined function. This is consistent with the findings of other pollen proteomic studies [14,46].

4. Materials and Methods

4.1. Plant Material and Treatment

Wheat plants were treated as described previously [4,32]. CHA-SQ-1 was administered to the wheat cultivar 'Xinong 1376' at the rate of 5.0 kg/ha starting when the plants averaged 8.5 on the Feekes' scale [4]. Control plants received an equal volume of water as a negative control. Pollen grains at trinucleate stage were collected from at least 1500 anthers from 100 plants by tapping the anthers on glass slides. This stage was checked as described previously [2,4]. The samples were examined under a dissecting microscope. Debris was removed with a fine needle. Sampled pollen grains were either used immediately or stored at -80 °C for later use. Three biological replicates were performed. Pollen development was assessed using 1% acetocarmine and fertility with 2% iodine-potassium iodide (2% I_2-KI) [2]. The activities of respiratory and ATPase were determined according to Wang et al. [4].

4.2. Sample Fixation and Infiltration

Pollen grains were fixed immediately in FAA (70% alcohol: 37% formaldehyde: acetic acid; 18:1:1, $v/v/v$) and performed for light microscopic observation. Pollen grains treated with 2.5% (v/v) glutaraldehyde were ready for ultrastructural microscopy observation, respectively.

4.3. Light Microscopic Observation

Pollen samples were first embedded in Epon-812 resin; semithin (1 μm) sections were cut using an Ultracut E ultramicrotome (Leica Microsystems, Wetzlar, Germany). Then they were stained with Ehrlich's hematoxylin and safranin O/fast green. The staining method for proteins, lipids, and insoluble polysaccharides was performed according to Konyar and Dane [47] and Tian et al. [48]. The procedure was described briefly as follows.

For Coomassie Brilliant Blue (CBB) staining, semi-thin sections of pollen gains were stained with CBB-R250 solution containing 45% (v/v) methanol, 10% (v/v) glacial acetic acid, 45% (v/v) water and 0.3 g CBB-R250, and then rinsed in distilled water. Proteins were stained blue.

For Sudan black B reaction, semi-thin sections were stained by 70% (v/v) alcohol saturated with Sudan black B, then rinsed in a mixture of 70% (v/v) alcohol and distilled water. This caused the lipids in solution to turn black.

A periodic acid-Schiff (PAS) reaction was performed, in which transverse sections were oxidized in 1% (w/v) periodic acid and stained for 1 h in Schiff's reagent. They were then destained in 0.5% (w/v) sodium bisulfite. This caused the insoluble polysaccharides to turn pink and purple.

Samples were imaged with a DS-U2 high-resolution camera and Nikon ECLIPSE E600 microscope equipped with the NIS-Elements software (Nikon, Tokyo, Japan).

4.4. Electron Microscopic Observation

For transmission electron microscopy (TEM), ultrathin sections (50 to 70 nm) were produced using a UC6 ultramicrotome (Leica, Wetzlar, Germany) and collected on copper grids. They were then double-stained in 2% (w/v) aqueous uranyl acetate and 2.6% (w/v) aqueous lead citrate. Finally, they were examined under a HT7700 transmission electron microscope (Hitachi, Tokyo, Japan).

4.5. Fluorescence Microscopic Observation

Then 4′,6-diamidino-2-phenylindole (DAPI; Sigma-Aldrich, Oakville, Ontario, Canad) was used to stain nuclei, and a fluorescein diacetate (FDA; Sigma-Aldrich, Oakville, Ontario, Canad) assay was performed to assess the vitality of fresh pollen grains. Samples were washed, embedded, and stained as described previously [2].

For ROS detection, fresh pollen grains were washed with PBS, then incubated in 5 μM H_2DCF-DA (Sigma-Aldrich) dissolved in anhydrous dimethyl sulfoxide (DMSO; Sigma-Aldrich) for 60 min in the dark at 25 °C. Excess probe was removed using PBS before detection.

Fluorescent signals were captured using a fluorescence microscope (Olympus BX 51, Olympus, Japan). Filter sets for blue (DAPI, 4′,6-diamidino-2-phenylindole) and green fluorescence (FDA, H_2DCF-DA) were Olympus part numbers.

4.6. Protein Extraction

The pollen protein was extracted with TCA/acetone methods described by Song et al. [30] and Zhang et al. [49] with modifications. Briefly, pollen gains were ground to fine powder in liquid nitrogen. This powder was then suspended in −20 °C pre-cooled solution of 10% TCA, 0.07% β-ME and 1 mM PMFS and kept at −20 °C overnight. The samples were centrifuged at 20,000× g for 20 min at 4 °C, and the pellets were rinsed with a pre-cooled acetone solution containing 0.07% (v/v) β-ME and 1 mM PMSF. They were then centrifuged at 25,000× g for 30 min. Three rounds of rinsing and centrifugation were performed. Then the vacuum-dried pellets were dissolved in lysis solution containing 2 M thiourea, 7 M urea, 65 mM DTT, 4% (w/v) CHAPS, and 0.5% (v/v) Bio-Lyte (Bio-Rad, Hercules, CA, USA) and 0.001% (w/v) bromophenol blue. Insoluble materials were centrifuged out, and protein concentration was determined using a Bio-Rad Protein Assay Kit II (Bio-Rad, Hercules, CA, USA) according to the manufacturer's instructions.

4.7. Two-Dimensional Gel Electrophoresis

2-DE was performed as described by Song et al. [30] and Wang et al. [4]. Approximately 550 μg of protein was separated by loading the sample onto a 17 cm pH 4–7 linear pH gradient IPG strip (Bio-Rad, Hercules, CA, USA). The sample was then subjected to electrophoresis on the IPGphor apparatus (Protean IEF Cell; Bio-Rad, Hercules, CA, USA) at 80 kV-h, and the conditions were as follows: constant power (50 μA/IPG strip) at 250 V for 1 h, 500 V for 1h, 1000 V for 1 h, 8000 V for 4 h and 8000 V for a total of 80,000 V-h (17 cm, pH 4–7). The second electrophoretic dimension was performed using 12% SDS-PAGE. The gels were stained with CBB-G250. There were three independent biological replicates per sample.

4.8. Image Analysis

The stained gels were imaged using a PowerLook 2100XL scanner (UMAX, Taiwan, China) and analyzed using PDQuest 2-DE 8.0.1 (Bio-Rad, Hercules, CA, USA). Briefly, the images were initially processed through transformation, filtering, automated spot detection, normalization, and matching. All protein spots detected in gel were matched to the corresponding spots of master gel (the control), and then each spot density was normalized against the whole gel densities. Analysis was based on total densities of gels as percent volume, and spots were considered to have significant differences in expression if the mean abundance changed more than 1.5-fold ($P < 0.05$) as indicated by the t test.

4.9. MS Analysis and Database Search

DAPs detected in stained gels were selected manually and excised for protein identification. In-gel digestion of DAPs was described by Song et al. [30] and Wang et al. [4]. Specifically, it was performed after destaining, reduction, and alkylation, and executed by incubation at 37 °C for 16 h in trypsin solution. All samples were subjected to MALDI-TOF/TOF mass spectroscopy on a 5800 MALDI Time of Flight (TOF)/TOF™ analyzer (AB Sciex, Foster City, CA, USA). Mass spectra were acquired by TOF/TOF™ Series Explorer™ Software (version 4.1, AB Sciex) that recorded across a range of mass from 700 to 4000 Da with a focus on 1700 Da. For one main MS spectrum 15 sub-spectra with 200 shots per sub-spectrum were accumulated, and for the MS/MS spectrum up to 25 sub-spectra with 250 shots per sub-spectrum were accumulated.

All of the MS/MS data were checked against the NCBInr database using the Mascot search engine (www.matrixscience.com) with the taxonomy parameter set to green plants. Other parameters were as follows: enzyme of trypsin, one missed cleavage site, fixed modification of carbamidomethyl (Cys), variable modification of Gln->pyro-Glu (N-term Q) and oxidation (Met), peptide tolerance of 100 ppm, MS/MS tolerance of 0.3 Da, and peptide charge of 1+.

For positive identification, the peptides were considered to be identified when the scoring value exceeded the identity or extensive homology threshold score of identity value, as calculated by Mascot, based on the MOWSE score. Sequences of proteins identified as unknown, hypothetical, or of uncharacterized function served as queries in a search for their homologues using the BLASTP algorithm. The mass spectrometry proteomics data have been deposited to the ProteomeXchange Consortium via the PRIDE [50] partner repository with the dataset identifier PXD012519.

4.10. Bioinformatic Analysis of Identified Proteins

The details of bioinformatics analysis have been described previously [30,32]. The DAPs defined by MS analysis were classified using the gene index and Uniprot accession number, which were entered into the Uniprot database (http://www.uniprot.org). Hierarchical clustering was performed on log 2-transformed data using Genesis 1.7.6 software. All identified proteins were blasted against *Arabidopsis thaliana* TAIR10 protein databases (http://www.arabidopsis.org/). The protein–protein interaction network was analyzed using STRING 10, which is publicly available (http://string-db. org/). Biological processes were predicted using the Cytoscape plugin BiNGO 3.02. Subcellular localization of the identified proteins was determined using Euk-mPLoc 2.0 database (http://www. csbio.sjtu.edu.cn/bioinf/euk-multi-2/), BaCelLo database (http://gpcr.biocomp.unibo.it/bacello/ pred.htm) and ESLpred database (http://www.imtech.res.in/raghava/eslpred2/).

5. Conclusions

In seed plants, pollen grains transport sperm cells to the female gametophyte. In this study, cytological and proteomic changes of wheat pollen abortion induced by CHA-SQ-1 were investigated. The cytological study indicated that pollen grain was impaired by CHA-SQ-1 during its maturation, which appeared as abnormal pollen development and shape, reduced storage materials, and defective pollen wall. A total of 60 identified DAPs with various functions were identified in mature pollen grains; some of these are central actors in biological processes (carbohydrate metabolism and energy metabolism) that regulate pollen maturation, especially some proteins related to sucrose and starch metabolism and ROS metabolism (Figures 5 and 6). Ultimately, these results induce complete (100%) pollen abortion in CHA-SQ-1-treated plants. Hence, this study has investigated the cytological, physiological and biochemical changes during pollen maturation in CHA-SQ-1-treated plants, which could provide a valuable resource for plant biology research, particularly for sexual reproduction in plants.

Supplementary Materials: Supplementary materials can be found at http://www.mdpi.com/1422-0067/20/ 7/1615/s1. Figure S1. The replicate 2-DE electrophoresis gels of pollen grain proteomes in the control (A, C) and CHA-SQ-1-treated wheat plants (B, D). Figure S2. Comparison of nuclei and vitality in pollen grains from control (A, C) and CHA-SQ-1-treated wheat plants (B, D) during pollen maturation. (A, B) DAPI staining. (C, D) FDA staining. Bars: 20 μm. Figure S3. Analysis of respiratory activity and ATPase activity. (A) Total respiration (V_t), and activities of the cytochrome pathway (V_{cyt}) and the alternative pathway activity (V_{alt}) in control and CHA-SQ-1-treated wheat plants. (B) Analysis of ATPase activity. Data are means ± SD of three independent experiments (biological replicates). The significant of differences was assessed by Student's *t*-test (*$P < 0.05$, **$P < 0.01$). Table S1. Folds change of differential abundance proteins (DAPs) in pollen grain between the control and CHA-SQ-1-treated wheat plants. Table S2. Identification of differentially expressed proteins. Table S3. Matched peptide sequences of identified proteins. Table S4. Corresponding homologues of the three unknown proteins. Table S5. Differentially expressed proteins blasted against the TAIR database. Table S6. Abbreviations of the specific protein names in the STRING network (Figure 6). Table S7. Biological pathways of differentially expressed proteins.

Author Contributions: S.W. and G.Z. designed the study and wrote the manuscript. S.W., Y.Z. (Yingxin Zhang), Z.F., Y.Z. (Yamin Zhang), Q.S. and Y.S. participated in experiments. Z.H., K.S., Y.L. (Ying Li), Z.Z., D.M., Y.L. (Yike

Liu), N.N., J.W. and S.M. discussed the results and revised the manuscript. All authors have read and approved the final manuscript.

Funding: This research was funded by the Nature Science Foundation of Hubei Province (2017CFB234), the National Transgenic Key Project of the Ministry of Agriculture of China (2018ZX0800909B), the Opening Fund of Engineering Research Center of Ecology and Agricultural Use of Wetland, Ministry of Education (KF201708), the National Support Program of China (2015BAD27B01), the National Natural Science Foundation of China (31171611 and 31371697) and Yangtze Fund for Youth Teams of Science and Technology Innovation (7011802111).

Conflicts of Interest: The authors declare no conflict of interest.

Abbreviations

2-DE	Two-dimensional gel electrophoresis
CBB	Coomassie Brilliant Blue
CHA	Chemical hybridizing agent
DAPI	4′, 6-diamidino-2-phenylindole
DAPs	Differential abundant proteins
DMSO	Dimethyl sulfoxide
FDA	Fluorescein diacetate
HXK	Hexokinase
mtETC	Mitochondrial electron transport chain
mtPDC	Mitochondria pyruvate dehydrogenase complex
PAS	Periodic acid-Schiff
PGK	3-phosphoglycerate kinase
PGM	Phosphoglucomutase
pI	Isoelectric point
ROS	Reactive oxygen species
SPP	Sucrose phosphate phosphatase
SPS	Sucrose phosphate synthase
TEM	Transmission electron microscopy
TIM	Triosephosphat-isomerase
UGP	UTP-glucose-1-phosphate uridylyltransferase
VIN	Vacuolar invertase

References

1. Shi, J.-X.; Cui, M.-H.; Yang, L.; Kim, Y.J.; Zhang, D.-B. Genetic and biochemical mechanisms of pollen wall development. *Trends Plant Sci.* **2015**, *20*, 741–753. [CrossRef] [PubMed]
2. Wang, S.-P.; Zhang, G.-S.; Song, Q.-L.; Zhang, Y.-X.; Li, Z.; Guo, J.-L.; Niu, N.; Ma, S.-C.; Wang, J.-W. Abnormal development of tapetum and microspores induced by chemical hybridization agent SQ-1 in wheat. *PLoS ONE* **2015**, *10*, e0119557. [CrossRef]
3. Wang, S.-P.; Zhang, G.-S.; Song, Q.-L.; Zhang, Y.-X.; Li, Y.; Chen, Z.; Niu, N.; Ma, S.-C.; Wang, J.-W. Programmed cell death, antioxidant response and oxidative stress in wheat flag leaves induced by chemical hybridization agent SQ-1. *J. Integr. Agr.* **2016**, *15*, 76–86. [CrossRef]
4. Wang, S.-P.; Zhang, Y.-X.; Song, Q.-L.; Fang, Z.-W.; Chen, Z.; Zhang, Y.-M.; Zhang, L.-L.; Zhang, L.; Niu, N.; Ma, S.-C.; et al. Mitochondrial dysfunction causes oxidative stress and tapetal apoptosis in chemical hybridization reagent-induced male sterility in wheat. *Front. Plant Sci.* **2018**, *8*, 2217. [CrossRef] [PubMed]
5. Ba, Q.-S.; Zhang, G.-S.; Wang, J.-S.; Che, H.-X.; Liu, H.-Z.; Niu, N.; Ma, S.-C.; Wang, J.-W. Relationship between metabolism of reactive oxygen species and chemically induced male sterility in wheat (*Triticum aestivum* L.). *Can. J. Plant Sci.* **2013**, *93*, 675–681. [CrossRef]
6. Ba, Q.-S.; Zhang, G.-S.; Che, H.X.; Liu, H.-Z.; Ng, T.B.; Zhang, L.; Wang, J.-S.; Sheng, Y.; Niu, N.; Ma, S.-C.; et al. Aliphatic metabolism during anther development interfered by chemical hybridizing agent in wheat. *Crop Sci.* **2014**, *54*, 1458–1467. [CrossRef]
7. Ba, Q.-S.; Zhang, G.-S.; Wang, J.-S.; Niu, N.; Ma, S.-C.; Wang, J.-W. Gene expression and DNA methylation alterations in chemically induced male sterility anthers in wheat (*Triticum aestivum* L.). *Acta Physiol. Plant* **2014**, *36*, 503–512. [CrossRef]

8. Nakamura, Y.; Teo, N.Z.; Shui, G.; Chua, C.H.; Cheong, W.F.; Parameswaran, S.; Koizumi, R.; Ohta, H.; Wenk, M.R.; Ito, T. Transcriptomic and lipidomic profiles of glycerolipids during *Arabidopsis* flower development. *New Phytol.* **2014**, *203*, 310–322. [CrossRef]

9. Rutley, N.; Twell, D. A decade of pollen transcriptomics. *Plant Reprod.* **2015**, *28*, 73–89. [CrossRef]

10. Zhang, D.-D.; Liu, D.; Lv, X.-M.; Wang, Y.; Xun, Z.-L.; Liu, Z.-X.; Li, F.-L.; Lu, H. The cysteine protease CEP1, a key executor involved in tapetal programmed cell death, regulates pollen development in *Arabidopsis*. *Plant Cell* **2014**, *26*, 2939–2961. [CrossRef]

11. Sumiyoshi, M.; Inamura, T.; Nakamura, A.; Aohara, T.; Ishii, T.; Satoh, S.; Iwai, H. UDP-arabinopyranose mutase 3 is required for pollen wall morphogenesis in rice (*Oryza sativa*). *Plant Cell Physiol.* **2015**, *56*, 232–241. [CrossRef]

12. Tan, C.; Liu, Z.Y.; Huang, S.N.; Feng, H. Mapping of the male sterile mutant gene ftms in *Brassica rapa* L. ssp. pekinensis via BSR-Seq combined with whole-genome resequencing. *Theor. Appl. Genet.* **2019**, *132*, 355–370. [CrossRef] [PubMed]

13. Dai, S.; Li, L.; Chen, T.; Chong, K.; Xue, Y.; Wang, T. Proteomic analyses of *Oryza sativa* mature pollen reveal novel proteins associated with pollen germination and tube growth. *Proteomics* **2006**, *6*, 2504–2529. [CrossRef]

14. Sheoran, I.S.; Pedersen, E.J.; Ross, A.R.; Sawhney, V.K. Dynamics of protein expression during pollen germination in canola (*Brassica napus*). *Planta* **2009**, *230*, 779–793. [CrossRef]

15. Kim, M.H.; Kim, Y.S.; Park, S.K.; Shin, D.I.; Park, H.S.; Chung, I.K. A genotype-specific pollen gene associated with self-incompatibility in *Lycopersicon peruvianum*. *Mol. Cells* **2003**, *16*, 260–265. [PubMed]

16. Keller, M.; Hu, Y.-J.; Mesihovic, A.; Fragkostefanakis, S.; Schleiff, E.; Simm, S. Alternative splicing in tomato pollen in response to heat stress. *DNA Res.* **2017**, *24*, 205–217. [CrossRef] [PubMed]

17. Jegadeesan, S.; Chaturvedi, P.; Ghatak, A.; Pressman, E.; Meir, S.; Faigenboim, A.; Rutley, N.; Beery, A.; Harel, A.; Weckwerth, W.; et al. Proteomics of heat-stress and ethylene-mediated thermotolerance mechanisms in tomato pollen grains. *Front. Plant Sci.* **2018**, *9*, 1558. [CrossRef]

18. Barton, D.A.; Cantrill, L.C.; Law, A.M.; Phillips, C.G.; Sutton, B.G.; Overall, R.L. Chilling to zero degrees disrupts pollen formation but not meiotic microtubule arrays in *Triticum aestivum* L. *Plant Cell Environ.* **2014**, *37*, 2781–2794. [CrossRef] [PubMed]

19. Ishiguro, S.; Ogasawara, K.; Fujino, K.; Sato, Y.; Kishima, Y. Low temperature-responsive changes in the anther transcriptome's repeat sequences are indicative of stress sensitivity and pollen sterility in rice strains. *Plant Physiol.* **2014**, *164*, 671–682. [CrossRef] [PubMed]

20. Jin, Y.; Yang, H.-X.; Wei, Z.; Ma, H.; Ge, X.C. Rice male development under drought stress: Phenotypic changes and stage-dependent transcriptomic reprogramming. *Mol. Plant* **2013**, *6*, 1630–1645. [CrossRef]

21. Koonjul, P.K.; Minhas, J.S.; Nunes, C.; Sheoran, I.S.; Saini, H.S. Selective transcriptional down-regulation of anther invertases precedes the failure of pollen development in water-stressed wheat. *J. Exp. Bot.* **2005**, *56*, 179–190. [CrossRef] [PubMed]

22. Abiko, M.; Akibayashi, K.; Sakata, T.; Kimura, M.; Kihara, M.; Itoh, K.; Asamizu, E.; Sato, S.; Takahashi, H.; Higashitani, A. High-temperature induction of male sterility during barley (*Hordeum vulgare* L.) anther development is mediated by transcriptional inhibition. *Sex. Plant Reprod.* **2005**, *18*, 91–100. [CrossRef]

23. Liao, J.-L.; Zhou, H.-W.; Zhang, H.-Y.; Zhong, P.-A.; Huang, Y.-J. Comparative proteomic analysis of differentially expressed proteins in the early milky stage of rice grains during high temperature stress. *J. Exp. Bot.* **2014**, *65*, 655–671. [CrossRef] [PubMed]

24. Wang, Y.-Y.; Qiu, L.; Song, Q.-L.; Wang, S.-P.; Wang, Y.-J.; Ge, Y.-H. Root proteomics reveals the effects of wood vinegar on wheat growth and subsequent tolerance to drought stress. *Int. J. Mol. Sci.* **2019**, *20*, 942. [CrossRef] [PubMed]

25. Dong, S.; Lau, V.; Song, R.; Ierullo, M.; Esteban, E.; Wu, Y.; Sivieng, T.; Nahal-Bose, H.K.; Gaudinier, A.; Pasha, A.; et al. Proteome-wide, structure-based prediction of protein-protein interactions/new molecular interactions viewer. *Plant Physiol.* **2019**. [CrossRef] [PubMed]

26. Fotovat, R.; Alikhani, M.; Valizadeh, M.; Mirzaei, M.; Salekdeh, G.H. A proteomics approach to discover drought tolerance proteins in wheat pollen grain at meiosis stage. *Protein Peptide Lett.* **2017**, *24*, 26–36. [CrossRef] [PubMed]

27. Das, S.; Krishnan, P.; Nayak, M.; Ramakrishnan, B. High temperature stress effects on pollens of rice (*Oryza sativa* L.) genotypes. *Environ. Exp. Bot.* **2014**, *101*, 36–46. [CrossRef]

28. Zhu, W.; Ma, S.; Zhang, G.; Liu, H.; Ba, Q.; Li, Z.; Song, Y.; Zhang, P.; Niu, N.; Wang, J. Carbohydrate metabolism and gene regulation during anther development disturbed by chemical hybridizing agent in wheat. *Crop Sci.* **2015**, *55*, 868–876. [CrossRef]

29. Zhu, Q.-D.; Song, Y.-L.; Zhang, G.-S.; Ju, L.; Zhang, J.-G.; Yu, Y.-G.; Niu, N.; Wang, J.-W.; Ma, S.-C. *De Novo* assembly and transcriptome analysis of wheat with male sterility induced by the chemical hybridizing agent SQ-1. *PLoS ONE* **2015**, *10*, e0123556. [CrossRef]

30. Song, Q.-L.; Wang, S.-P.; Zhang, G.-S.; Li, Y.; Li, Z.; Guo, J.-L.; Niu, N.; Wang, J.-W.; Ma, S.-C. Comparative proteomic analysis of a membrane-enriched fraction from flag leaves reveals responses to chemical hybridization agent SQ-1 in wheat. *Front. Plant Sci.* **2015**, *6*, 669. [CrossRef]

31. Liu, H.-Z.; Zhang, G.-S.; Wang, J.-S.; Li, J.-J.; Song, Y.-L.; Qiao, L.; Niu, N.; Wang, J.-W.; Ma, S.-C.; Li, L.-L. Chemical hybridizing agent SQ-1-induced male sterility in *Triticum aestivum* L.: A comparative analysis of the anther proteome. *BMC Plant Biol.* **2018**, *18*, 7. [CrossRef]

32. Wang, S.-P.; Zhang, G.-S.; Zhang, Y.-X.; Song, Q.-L.; Chen, Z.; Wang, J.-S.; Guo, J.-L.; Niu, N.; Wang, J.-W.; Ma, S.-C. Comparative studies of mitochondrial proteomics reveal an intimate protein network of male sterility in wheat (*Triticum aestivum* L.). *J. Exp. Bot.* **2015**, *66*, 6191–6203. [CrossRef]

33. Peng, Z.; Wang, M.; Li, F.; Lv, H.; Li, C.; Xia, G. A proteomic study of the response to salinity and drought stress in an introgression strain of bread wheat. *Mol. Cell Proteom.* **2009**, *8*, 2676–2686. [CrossRef] [PubMed]

34. Song, Y.-L.; Wang, J.-W.; Zhang, G.-S.; Zhao, X.-L.; Zhang, P.-F.; Niu, N.; Ma, S.-C. Isolation and characterization of a wheat F8-1 homolog required for physiological male sterility induced by a chemical hybridizing agent (CHA) SQ-1. *Euphytica* **2015**, *205*, 707–720. [CrossRef]

35. Dai, S.; Wang, T.; Yan, X.; Chen, S. Proteomics of pollen development and germination. *J. Proteome Res.* **2007**, *6*, 4556–4563. [CrossRef] [PubMed]

36. Wang, J.-P.; Nayak, S.; Koch, K.; Ming, R. Carbon partitioning in sugarcane (*Saccharum species*). *Front. Plant Sci.* **2013**, *4*, 201. [CrossRef] [PubMed]

37. Oliver, S.N.; Dennis, E.S.; Rudy, D. ABA regulates apoplastic sugar transport and is a potential signal for cold-induced pollen sterility in rice. *Plant Cell Physiol.* **2007**, *48*, 1319–1330. [CrossRef]

38. Tiessen, A.; Padilla-Chacon, D. Subcellular compartmentation of sugar signaling: Links among carbon cellular status, route of sucrolysis, sink-source allocation, and metabolic partitioning. *Front. Plant Sci.* **2013**, *3*, 306. [CrossRef]

39. Wierenga, R.-K.; Kapetaniou, E.-G.; Venkatesan, R. Triosephosphate isomerase: A highly evolved biocatalyst. *Cell. Mol. Life Sci.* **2010**, *67*, 3961–3982. [CrossRef]

40. Oliver, S.N.; Dongen, J.T.V.; Alfred, S.C.; Mamun, E.A.; Zhao, X.C.; Saini, H.S.; Fernandes, S.F.; Blanchard, C.L.; Sutton, B.G.; Geigenberger, P.; et al. Cold-induced repression of the rice anther-specific cell wall invertase gene *OSINV4* is correlated with sucrose accumulation and pollen sterility. *Plant Cell Environ.* **2010**, *28*, 1534–1551. [CrossRef]

41. Zhu, X.-L.; Liang, W.-Q.; Cui, X.; Chen, M.-J.; Yin, C.-S.; Luo, Z.-J.; Zhu, J.-Y.; Lucas, W.J.; Wang, Z.-Y.; Zhang, D.-B. Brassinosteroids promote development of rice pollen grains and seeds by triggering expression of *Carbon Starved Anther*, a MYB domain protein. *Plant J.* **2015**, *82*, 570–581. [CrossRef]

42. Signorelli, S.; Tarkowski, L.P.; Van den Ende, W.; Bassham, D.C. Linking autophagy to abiotic and biotic stress responses. *Trends Plant Sci.* **2019**. [CrossRef] [PubMed]

43. Chaturvedi, P.; Doerfler, H.; Jegadeesan, S.; Ghatak, A.; Pressman, E.; Castillejo, M.A.; Wienkoop, S.; Egelhofer, V.; Firon, N.; Weckwerth, W. Heat-Treatment-Responsive proteins in different developmental stages of tomato pollen detected by targeted mass accuracy precursor alignment (tMAPA). *J. Proteome Res.* **2015**, *14*, 4463–4471. [CrossRef]

44. Chaturvedi, P.; Ghatak, A.; Weckwerth, W. Pollen proteomics: From stress physiology to developmental priming. *Plant Reprod.* **2016**, *29*, 119–132. [CrossRef]

45. Kuntumalla, S.; Braisted, J.C.; Huang, S.T.; Parmar, P.P.; Clark, D.J.; Alami, H.; Zhang, Q.-S.; Donohue-Rolfe, A.; Tzipori, S.; Fleischmann, R.D.; et al. Comparison of two label-free global quantitation methods, APEX and 2D gel electrophoresis, applied to the *Shigella dysenteriae* proteome. *Proteome Sci.* **2009**, *7*, 22. [CrossRef] [PubMed]

46. Wang, L.-Q.; Zhang, X.-L.; Zhang, J.; Fan, W.; Lu, M.-Z.; Hu, J.-J. Proteomic analysis and identification of possible allergenic proteins in mature pollen of *Populus tomentosa*. *Int. J. Mol. Sci.* **2018**, *19*, 250. [CrossRef]

47. Konyar, S.T.; Dane, F. Cytochemistry of pollen development in *Campsis radicans* (L.) Seem. (Bignoniaceae). *Plant Syst. Evol.* **2013**, *299*, 87–95. [CrossRef]

48. Tian, Q.-Q.; Lu, C.M.; Li, X.; Fang, X.-W. Low temperature treatments of rice (*Oryza sativa* L.) anthers changes polysaccharide and protein composition of the anther walls and increases pollen fertility and callus induction. *Plant Cell Tissue Organ Culture* **2015**, *120*, 89–98. [CrossRef]

49. Zhang, J.; Wu, L.-S.; Fan, W.; Zhang, X.-L.; Jia, H.-X.; Li, Y.; Yin, Y.-F.; Hu, J.-J.; Lu, M.-Z. Proteomic analysis and candidate allergenic proteins in *Populus deltoides* CL. "2KEN8" mature pollen. *Front. Plant Sci.* **2015**, *6*, 548. [CrossRef] [PubMed]

50. Perez-Riverol, Y.; Csordas, A.; Bai, J.; Bernal-Llinares, M.; Hewapathirana, S.; Kundu, D.J.; Inuganti, A.; Griss, J.; Mayer, G.; Eisenacher, M.; et al. The PRIDE database and related tools and resources in 2019: Improving support for quantification data. *Nucleic Acids Res.* **2019**, *47*, D442–D450. [CrossRef]

International Journal of
Molecular Sciences

Article

Quantitative Proteomic Analysis Reveals Novel Insights into Intracellular Silicate Stress-Responsive Mechanisms in the Diatom *Skeletonema dohrnii*

Satheeswaran Thangaraj [1,2,3], Xiaomei Shang [1,2], Jun Sun [1,2,*] and Haijiao Liu [1,2,4]

1   Tianjin Key Laboratory of Marine Resources and Chemistry, Tianjin University of Science and Technology, No 29, 13th Avenue, TEDA, Tianjin 300457, China; satheeswaran1990@gmail.com (S.T.); shangxiaomei1987@126.com (X.S.); coccolith@126.com (H.L.)
2   Research Center for Indian Ocean Ecosystem, Tianjin University of Science and Technology, No 29, 13th Avenue, TEDA, Tianjin 300457, China
3   Faculty of Food Engineering and Biotechnology, Tianjin University of Science and Technology, No 29, 13th Avenue, TEDA, Tianjin 300457, China
4   Institute of Marine Science and Technology, Shandong University, No 27, Shanda Nan Road, Jinan 250110, China
*   Correspondence: phytoplankton@163.com; Tel.: +86-22-6060-1116

Received: 17 April 2019; Accepted: 20 May 2019; Published: 23 May 2019

Abstract: Diatoms are a successful group of marine phytoplankton that often thrives under adverse environmental stress conditions. Members of the *Skeletonema* genus are ecologically important which may subsist during silicate stress and form a dense bloom following higher silicate concentration. However, our understanding of diatoms' underlying molecular mechanism involved in these intracellular silicate stress-responses are limited. Here an iTRAQ-based proteomic method was coupled with multiple physiological techniques to explore distinct cellular responses associated with oxidative stress in the diatom *Skeletonema dohrnii* to the silicate limitation. In total, 1768 proteins were detected; 594 proteins were identified as differentially expressed (greater than a two-fold change; $p < 0.05$). In Si-limited cells, downregulated proteins were mainly related to photosynthesis metabolism, light-harvesting complex, and oxidative phosphorylation, corresponding to inducing oxidative stress, and ROS accumulation. None of these responses were identified in Si-limited cells; in comparing with other literature, Si-stress cells showed that ATP-limited diatoms are unable to rely on photosynthesis, which will break down and reshuffle carbon metabolism to compensate for photosynthetic carbon fixation losses. Our findings have a good correlation with earlier reports and provides a new molecular level insight into the systematic intracellular responses employed by diatoms in response to silicate stress in the marine environment.

Keywords: abiotic stress; silicate limitation; diatom; iTRAQ; proteomics; photosynthesis; carbon fixation

---

1. Introduction

Diatoms are a major group of phytoplankton, which play a significant role in the global carbon cycle [1], often thriving under adverse marine environmental conditions. It is known that diatoms can subsist during long periods of nutrient limitation [2] and form blooms following higher nutrient concentration [3]. Diatoms require a large quantity of silicon for their unique siliceous cell wall and cellular process [4]. It was reported that silicate deficiency led to diatom cell cycle arrest, which was reinitiated once silicate was provided [5,6].

In general, the response of nutrient limitations on phytoplankton photosynthetic mechanism is poorly understood [7], due to their cellular response variance to different nutrient stress [8]. A recent proteomic investigation on the diatom *Thalassiosira pseudonana* in response to iron and nitrate limitation showed PSII and PSI proteins were downregulated and, therefore, reduced its oxidoreductase activity [9–11]. Despite this, the diatom photosynthetic mechanism in response to silicate deprivation remains largely unknown. The response of nutrient limitations on diatom carbon metabolism has been widely discussed, i.e., nitrogen starvation [9], phosphorus limitation [12], silicon limitation [13], and iron limitation [11], however, intracellular mechanisms interlinked with electron transport, carbon metabolism, and ROS accumulation remains to be explored.

There have been few studies on diatoms' reactions with the oxidative response during nitrogen and phosphorus limitation [14] and iron starvation [15], while silicate limitation in these intracellular processes unknown. To date studies have mostly focused silicate limitation on diatom fundamental molecular mechanisms: cell cycle arrest [6], cell wall formation [4], carbon adjustment [13], and morphological changes [16]. Therefore, the mechanism underlying intracellular silicate stress-responses associated with oxidative stress, and ROS production are largely unknown.

Nutrient limitation on diatoms can produce reactive oxygen species (ROS) and induce programmed cell death (PCD) [17,18]. The preliminary characteristics of PCD have been described on diatoms *Thalassiosira weissflogii* under N stress or exposure to exogenous aldehyde [17,19]; *Ditylum brighwellii* under N and P stress [20]; *T. pseudonana* under Fe limitation [15]; and *Phaeodactylum tricornutum* under N stress [3], however, the mechanism underlying the association of PCD in diatoms intracellular silicate limitation remains elusive. In general, the cellular response of diatoms has some common characteristics to a variant stress condition, however, the ROS and PCD have been the exception of these due to their differential attribution to variant nutrient limitation [21]. In this manner, it is essential to investigate the possible activation of ROS and PCD instigated by Si-stress to understand the intracellular silicate stress-response on diatom.

New introduction methods of isobaric tags for relative and absolute quantitation (iTRAQ) is a well-suited method to examine the distinct proteomic changes in cells acclimated to a different environmental condition. Here an iTRAQ-LC-MS/MS proteomic approach was used to examine the comparative proteomic investigation in *Skeletonema dohrnii* under silicate deplete and replete condition, respectively. The purpose of this comparative study is to understand the different metabolic regulation involved in the intracellular silicate stress-response especially those associated mechanism controlling cell growth, replication, carbon fixation, and inducing oxidative stress and possible ROS production. The findings revealed underlying molecular mechanisms controlling cell growth, replication, and photosynthetic carbon fixation under depleted silicate conditions in the marine environment.

2. Results

2.1. Physiological Responses of the Cell

Physiological changes of *S. dohrnii* under Si-deplete and Si-replete conditions are shown in (Figure 1). Cells subject to Si restriction grew slower all through the experiment than the Si-included sample (Figure 1A). Furthermore, the cells were under Si-constrained declined quickly once they achieve a higher density, while Si-included cells declined gradually. The net primary productivity NPP showed a significant variation between the treatments, particularly in the expositional stage (day 5) where cells exposed to silicate supplementation NPP expanded radically, whereas, in the Si-limited group it was 50% slower than the Si-included example (Figure 1B). The photochemical effectiveness of PSII, *Fv/Fm* in the expositional stage notably increased in Si-added condition, while in Si-restricted cells it increased gradually (Figure 1C). Together these cellular performance measurements indicate the strong dependence of the *S. dohrnii* integrated physiology on Si-bioavailability.

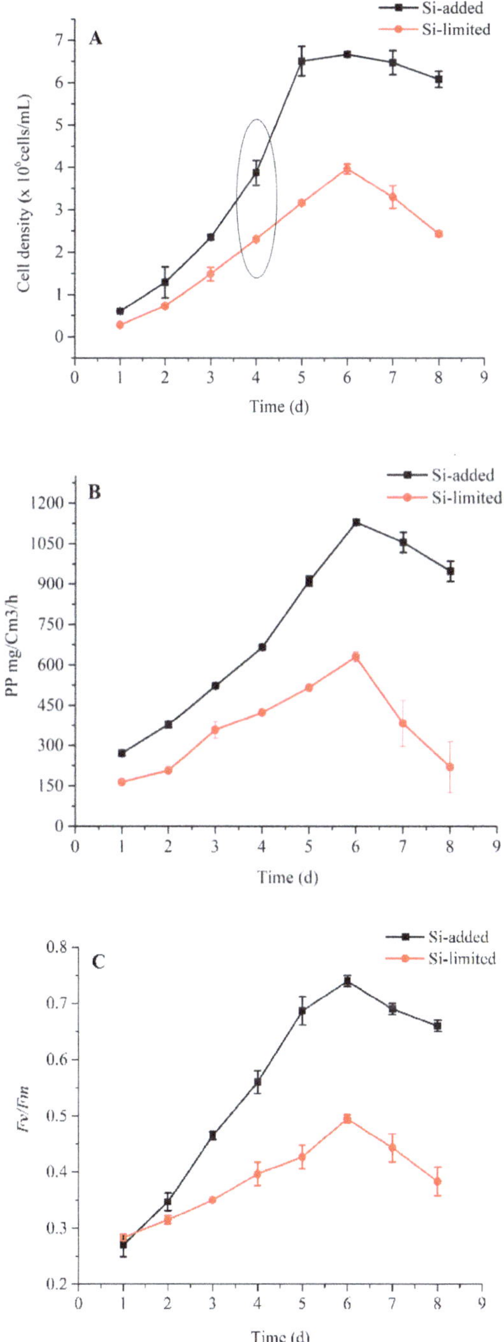

Figure 1. Cell density, net primary productivity, and photosynthetic efficiency in Si-limited and Si-added samples of *S. dohrnii*. Cell density (**A**), net primary productivity (**B**) and photochemical quantum yield of PS II (*FV/FM*) (**C**). The circle in Figure A shows the duration of sample collection for iTRAQ proteomics. The error bars represent the standard errors from triplicate measurements.

2.2. iTRAQ Results

To unravel *S. dohrnii* intracellular responses to Si-stress, we harvested cells for iTRAQ based proteomic analysis on day 4. Four biological replicates of the samples have been labeled and mixed for proteomic profiling. In whole 364,183 spectra, 3892 peptides, and 1768 proteins were identified with 1% of FDR. Detailed information including accession numbers, peptide sequences, protein description, number of spectra, *p*-value and *q*-value and repeatability test between the samples are given in (Supplementary Table S1). Of these identified proteins using a significance cut-off of a two-fold change and a *p*-value less than 0.05, we have identified 594 differentially expressed proteins between the Si-restricted and Si-added samples, of which 345 proteins were downregulated (Supplementary Table S2), and 249 were up-regulated (Supplementary Table S3), respectively.

2.3. Gene Ontology and COG Analysis

To additionally explore the potential role of overall identified proteins we used GO terms, which comprised three sets of ontologies: molecular functions (MF), cellular components (CC), and biological process (BP) (Figure 2A). As shown in (Figure 2A) the most represented GO terms in the MF category was a catalytic activity, and protein binding consisting of electron transport-transferring electrons within the cycling electron transport of photosynthesis activity. In the CC category, the cellular process and cell cyclic, including photosystem II, photosystem I, chloroplast thylakoid membrane, and proton transport proteins were significantly enriched. For the BP category biological regulation, including protein-chromophore linkage, regulation in ATP synthesis coupled proton transport, regulation in photosynthesis, and glycolysis, were the most important enriched term. Furthermore, the overall classified proteins were subjected to COG enrichment analysis to understand the cellular metabolism influenced by the Silicate deprivation. COG functional classification result shows that proteins were involved in almost every aspect of *S. dohrnii* metabolism (Figure 2B). However, among all groups, translations, ribosome structure, and biogenesis were the largest group (219) influenced by lower silicate, followed by post-translational modifications, protein turnover, chaperones (169), and general function proteins (149).

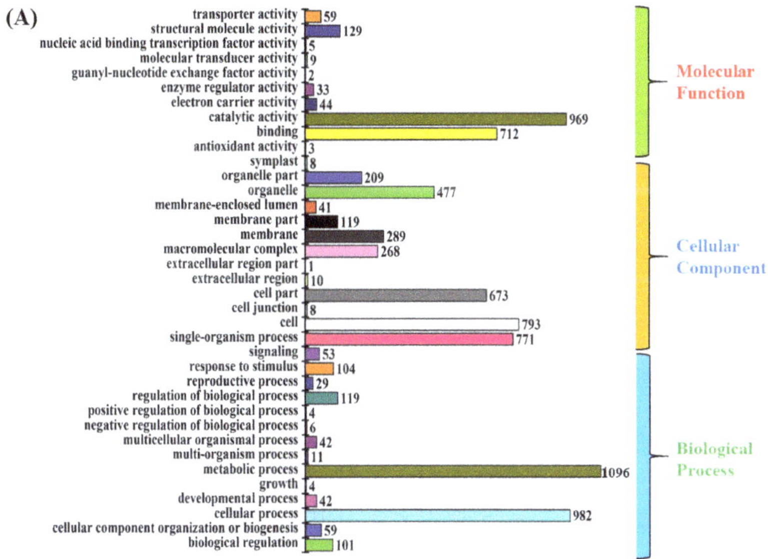

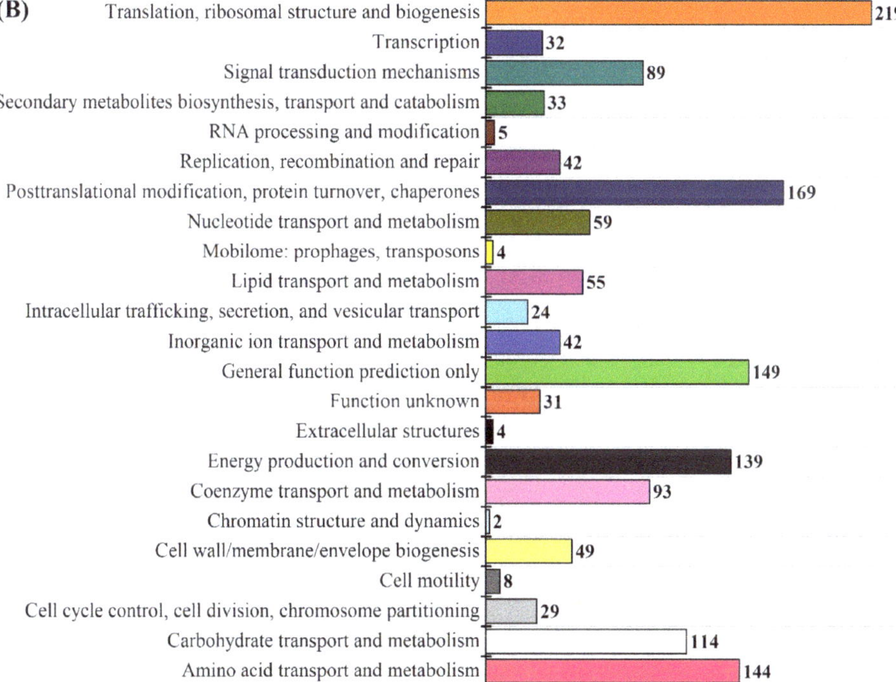

Figure 2. (**A**) Bar plot of Gene Ontology analysis; the different color shows different GO category. (**B**) The diagram shows the bar plot of COG analysis *x*-axis displays the COG term; *y*-axis displays the corresponding protein count illustrating the protein number of different functions.

2.4. KEGG Analysis of the DEPs in Skeleteonema dohrnii

The Kyoto Encyclopedia of Genes and Genomes (KEGG) enrichment analysis was performed to determine potential clustering of the differentially expressed protein in a specific metabolic pathway. Using the KEGG database 594 DEPs were annotated and classified into 91 metabolic pathways. The 'main KEGG pathway classification of DEPs are photosynthesis ($p < 0.01$), photosynthesis antenna ($p < 0.01$), and oxidative phosphorylation ($p < 0.01$), glycolysis/gluconeogenesis ($p < 0.03$), carbon fixation in photosynthesis organism ($p < 0.01$) and pentose phosphate ($p < 0.02$) (Figure 3).

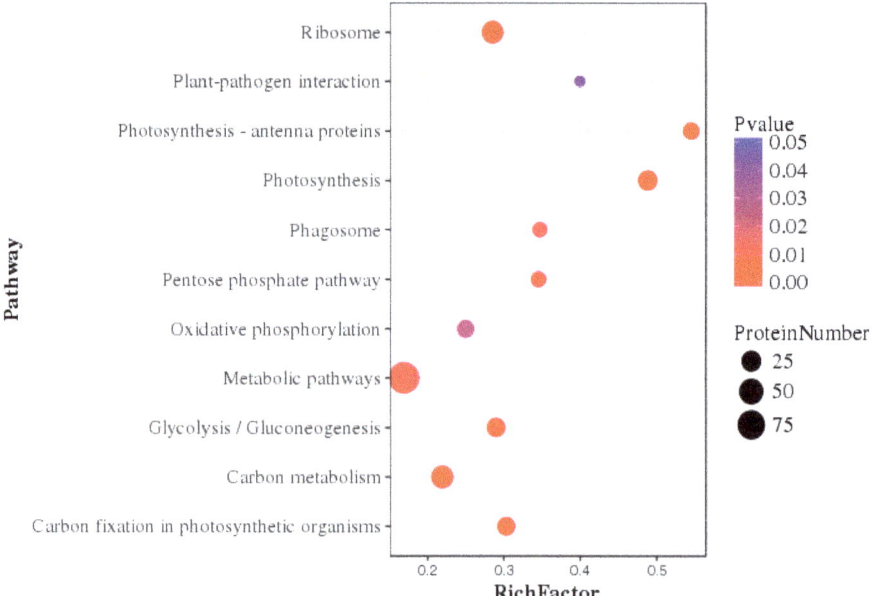

Figure 3. Statistics of pathway enrichment factor analysis of differentially expressed proteins comparing of Si-limited vs. Si-added conditions.

2.5. Photosynthesis Metabolism

The multi-subunit protein complex of PSII is embedded in the thylakoid membrane of algae that drives a water-plastoquinone oxido-reductase. The overall differentially abundant proteins in photosynthesis metabolism of this study are given in (Supplementary Table S4). Result shows that Si-limitation caused decreased abundance of seven primary proteins in the PSII complex, i.e., D1 (PabA, −1.52-fold), D2 (PsbD, −1.44-fold), CP43 (PsbC, −1.51-fold), CP47 (PsbB, −1.26-fold), cytochrome b559 subunits (PsbE, −1.64-fold; PsbF, −1.43-fold; and PsbQ, −0.12-fold). Decreased abundance of these proteins influenced in cytochrome b6/*f* complex; resulting in two primary proteins being decreased in abundance (PetD, −1.12-fold; PetA, −1.36-fold). The impact in PSII and cytochrome b6/*f* further downregulated six proteins in PSI, i.e., (PsaA, −1.35-fold; PsaB, −1.68-fold; PsaC, −0.63-fold; PsaF, −0.83-fold; PsaJ, −0.92-fold, and PsaL, −2.21-fold) which could damage the electron transfer from PSI to NADP [22].

Fucoxanthin and chlorophyll are known as major proteins in the LHCs of diatoms. In the present study 13 pigment proteins were downregulated, consisting of five FCPs and five Chl proteins and three accessory proteins (Supplementary Table S4), which may influence in the process of exciting energy and chlorophyll biosynthesis. In addition, 26 functional proteins associated with ATP metabolism were differentially expressed, of which 16 proteins were downregulated and 10 proteins were upregulated (Supplementary Table S4). Functional proteins associated with quinone binding,

ATP synthase, proton transport, and rotational mechanism were notable down-regulated proteins, whereas mitochondria, vascular, and sulfur proteins were noteworthy up-regulated proteins.

2.6. Carbon Metabolism

The state transition of photosynthesis metabolism due to silicate deprivation was expected to impact on carbon metabolism. Nevertheless, results showed an increased abundance of carbon metabolic proteins rather than decreasing. Nine proteins associated with glycolysis were differentially expressed and, among them, two were downregulated and seven were up-regulated (Supplementary Table S4). The up-regulated proteins of (fructose1,6-biphosphate; triose phosphate isomerase; glyceraldehyde-3-phosphate dehydrogenase; nicotinate-nucleotide pyrophosphorylase; phosphoglycerate kinase; and enolase) involved in the glycolytic process of step 1 (hexokinase), step 2 (phosphoglucose Isomerase) and step 3 (phosphofructokinase). As these upregulated proteins could enhance the production of pyruvate, six functional proteins were upregulated in pyruvate metabolism in this study. Upregulated proteins of (phosphoenolpyruvate; D-2 hydroxy-acid dehydrogenase; pyruvate dehydrogenase; pyruvate kinase; pyridine nucleotide transhydrogenase; acetyl-coenzyme) would enter mitochondria to produce GTP and FADH2. In addition, many functional proteins in TCA metabolism were upregulated, i.e., (isocitrate; oxoglutarate, succinyl; succinate; malate; dihydrolipoamide acetyltransferase; dihydrolipoyl dehydrogenase; pyruvate carboxylase; fumarate hydratase; and pyridine nucleotide), showing the interlinking pathway of diatoms' central carbon metabolism under silicate stress.

2.7. Response of Cellular Respiration

The respiratory chain has protein complexes of cytochrome I, II, III, IV, that transfer electrons from NADH to O2 via redox. The present result shows downregulation of two NADH-ubiquinone reductases (XP_002292597.1, −1.37-fold; AAZ99426.1, −1.32-fold), and cytochrome complex I (YP_316586.1, −1.49-fold), cytochrome *c* oxidase complex II (AOE43471.1), and cytochrome *c* oxidase subunit 6B1, complex IV (EED96133.1, −1.74-fold) (Supplementary Table S2), shows an influence in electron deliver. Further downregulation of mitochondrial alternative oxidase (AOX) (XP_002294653.1) and many FoF1-type ATP synthase proteins (Supplementary Table S4) indicates the possible impact on ATP production.

2.8. Identification PCD Related Proteins

It is known that superoxide dismutase and peroxiredoxin enzymes replace ROS scavenging function, however, in the present study, both proteins were down-regulated (superoxide dismutase, EJK75419.1; peroxiredoxin EED94096.1) which may impact the defense mechanism against ROS. In general, heat shock proteins protect the cellular mechanism when cells exposed to a stressful condition. Nevertheless, in this investigation downregulation of chaperonin GroES, EJK46404.1; molecular chaperone, XP_002286821.1; and some co-expression proteins like FKBP-peptidyl-prolyl cis-trans isomerase, EJK59733.1; and trypsin-serine proteases, EJK75980.1 may increase the impact of the molecular mechanism by Si stress. Additionally, downregulation of an anti-apoptotic protein of chaperone GroEL (HSP60 family, ACI64273.1) was observed in Si-limited cells. Furthermore, upregulation of RuBisCO large subunit (ABF60361.1) and downregulation of RuBisCO small subunit (BAA75795.1) may lead cells to oxidative stress under Si-limited conditions.

Moreover, in comparing with silicate-deplete and -replete cells, four nutrient transporters were differentially abundant. Among them, nitrate/nitrite transporter (NRT2) (EJK46860.1) was downregulated, whereas 5′-nucleotidase, (XP_002295180.1), sulfate adenylyltransferase, (XP_002286746.1), and sulfite reductase, (EED93364.1) were upregulated.

Additionally, photosynthesis and carbon metabolic regulation, results showed that silicate stress further regulated numerous cellular metabolism of *S. dohrnii*. To unravel the whole impact, an iPath global metabolic pathway was annotated (Supplementary Figure S1). The iPath biochemistry map shows Si-stress also regulated amino acid, nucleotide, and lipid metabolisms. Many proteins involved in the biosynthesis of purine and pyrimidine were down-regulated (Supplementary Table S4), which were associated with amino acid metabolisms, i.e., DNA binding (EJK63460.1), RNA polymerase (YP_009093166.1), adeno succinate synthase (XP_002292359.1), and de novo IMP biosynthetic process (EJK50540.1).

3. Discussion

The goal of this study was to better understand the marine diatom *S. dohrnii* metabolic responses when exposed to a silicate-limited environment by examining the expressed proteome at the exponential growth stage. However, why proteomic? Why not genomic or transcriptomic? Few investigators have already reported gene expression of diatoms to Si-limitation [13,23]. Furthermore, recent functional RNA study showed that although much of the genome was transcribed, not all the transcriptions were translated into functional proteins [24,25]. This suggests that genomic and transcriptomic studies may be limited to understand the cellular metabolomic changes to the changing environment deeply. Similarly, diatoms change their transcript expression into a small range of protein expression under nutrient limitation suggesting diatoms, like other organisms, may rapidly increase the transcript abundances, but not all translate into proteins [26]. Therefore, an iTRAQ quantitative proteomic approach was applied to *S. dohrnii* to understand accurately the underlying intracellular mechanisms in response to silicate stress.

3.1. Physiological Changes

The growth rate of *S. dohrnii* in the present study varied based on the silicate concentrations of which could be the reason for silicate demand to their cellular processes. Previous reports noted that Si-stress induced the aging early of diatoms, therefore, regulated metabolism inside and reduced cell growth rate and replication [6,13,27]. These reports also support the observed variation of NPP between Si-limited and Si-added samples. A further difference in *Fv/Fm* shows the inefficient activity of PSII complex, could be the attribution of LHCs responses to the Si-concentration [28,29]. The present study GO and COG results (Figure 2B) showed the similar detection of proteins on diatom *T. pseudonana* under silicon limitation, [4] indicates diatom's common metabolomic regulation under silicate stress.

3.2. Downregulation of Photosynthesis

PSII in plants referred to as an engine of earth and this complex affected first when plants expose to any abiotic stress [30]. Decreased abundance of PSII proteins in this study showed a reduction in catalyzing electron production, and transport, therefore, only limited electrons were pumped across a chloroplast. The schematic diagram (Figure 4) describes pathways of electron flow between chloroplast and mitochondria during oxygenic photosynthesis; also it shows five compensating mechanisms that rectifying process during the reduction in electron flow. Fascinatingly, results showed that, besides the reduction of electron production and transfer, Si-stress also influenced in the four compensating mechanisms of electron transfer, i.e., down-regulated malate carrier protein in mitochondria (XP_002295632.1, −1.4-fold) shows a reduction in first compensating mechanism [23], while serine/threonine-protein kinase (STT7kinase) downregulation (XP_002288169.1, −1.17-fold) displays a reduction in the third step and alteration in LHCs [31]. Furthermore, the decreased abundance of ferredoxin (EJK54785.1, −0.9-fold), and AOX protein (XP_002294653.1, −1.74-fold) indicates the failure of the fourth and fifth compensating mechanisms [32]. These results show that during silicate limitation of *S. dohrnii* cells the electron flow between chloroplasts and mitochondria were altered.

Regulation in PSII electron flow would damage the PSI and LHCs [33]; therefore, it is not surprising 13 proteins in the pigment metabolism were decreased in abundance due to Si-stress. Similarly, in diatom *T. pseudonana* six chlorophyll biosynthesis genes were downregulated due to silicon stress [13]. Chlorophyll is a nitrogenous compound; any impact on these proteins would decrease the nitrogen demand of the cell and diminishes their light-capturing efficiency, therefore, inducing oxidative stress and ROS production [9]. Downregulation of these proteins also affects the initial process in energy production, and carbon fixation [34], shows silicate is a vital nutrient for diatom carbon fixation to facilitate the proper electron transport and catalyze the reaction. Differential proteins abundance in this study because of Si-limitation reduced electron flow and increasing the chances of oxidative stress and ROS (Figure 4). Reduced silicate assimilation in diatoms caused the changes in protein abundance, resulting in metabolic imbalances to the oxidative stress [26]. Up-regulation of metabolomic sulfur protein in this study could also be linked to oxidative stress because it has a role as a compatible solute, is thought to be a part of the cellular antioxidant [35], and its synthesis and excretion might dissipate the amount of excess energy, carbon, and reducing the equivalents under nutrient starvation [36]. Of which indicates reducing demand for carbon skeletons to the central carbon metabolism under silicate stress conditions.

In total, our findings show *S. dohrnii* under Si-limitation retain PSII, cytochrome, and PSI complexes, however, major functional proteins of these subunits were down-regulated. Hence, failure to produce enough electrons and protons to pump across the thylakoid membrane (Figure 4); therefore, a regulation in antenna complex that limits the initial step of carbon storage. Furthermore, the inadequate flow of electrons and protons also affected ATP metabolism to drive ATP synthase, therefore, resulting in less ATP synthase complex during the silicate stress that controls *S. dohrnii* cell growth, replication, and energy storage for other metabolic functions depending on these energies.

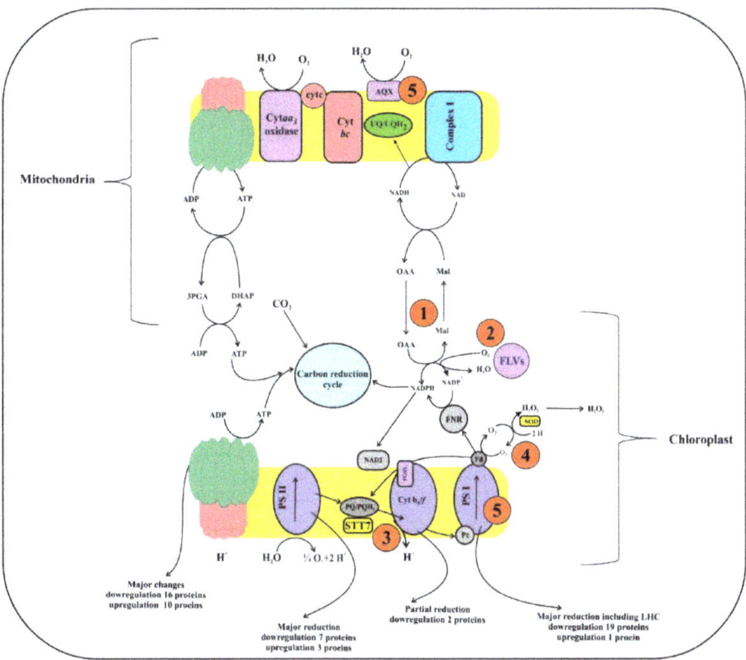

Figure 4. A schematic diagram of the electron transfer pathway in oxygenic photosynthesis. During the reaction, the reducing equivalents generated at PSII and transferred to plastoquinone then cytochrome, plastocyanin, PSI, Fd, and ferredoxin NADP+. Electron transfer/flow around PSI is mediated by PGRL1 and NAD2. Reduction in the electron flow around PSII, cytochrome, and PSI resulting in a deficiency in ATP, which is compensated by different mechanisms depending on the demand for ATP. For example, (1) A mitochondrial respiration-based first compensating mechanism likely involving malate/oxaloacetate shuttle, that reduced energy from the chloroplasts to cytosol, and eventually mitochondria.

After that, the reducing energy in the mitochondria are converted as ATP by electron transport, which would return to the chloroplast by the ATP translocator (2), the probable occurrence of oxygen photoreduction at the level of NADPH with the help of FLV proteins, which allows ATP biosynthesis via pseudo-cyclic phosphorylation (3). When mechanism (1) and (3) are highly activated, the plastoquinone becomes more reduced, resulting in STT7 kinase activation, therefore, altering LHC phosphorylation and state transition (4). The higher decrease in PSI acceptors would induce the Mehler reaction by direct O_2 photoreduction through reduced Fd, therefore making H_2O_2 (5), and ROS accumulation then induces the AOX and affects the PSI protein complex and decreases the PSII/PSI ratio. Furthermore, downregulation of PSII, cytochrome, and PSI proteins show it makes fewer reducing equivalents flow across a chloroplast and mitochondria, thus resulting in lower ATP production and impact on PSII and PSI, and the LHCs of *S. dohrnii*.

3.3. Reduction in the Photosynthetic Carbon Fixation

The upregulation of glycolic activity lead-carbon from intracellular stores to the central carbon metabolism. However, the upregulation of proteins catalyzed in the reverse reaction of gluconeogenesis in the present study showed an enhancement in the pyruvate metabolism. Similar results were observed in the transcript level of diatom *T. pseudonana* [13]. Observed upregulated proteins in the carbon metabolism show the direction of carbon flow in *S. dohrnii* (Figure 5). Pyruvate is the product of glycolysis, which can be converted into acetyl-CoA, also shows a unidirectional supporting of increasing glycolytic activity. Acetyl-CoA not only used for the carbon input to the TCA as a source of energy and reducing equivalents, but it also can be used for the fatty acid biosynthesis process, which provides a carbon skeleton for the silicate assimilation and biosynthesis compounds of fatty acids (Figure 5). Up-regulated proteins in the TCA cycle of this study shows carbon from glycolysis flowing through acetyl-CoA and then into the TCA cycle. All these findings reveal *S. dohrnii* cells in Si-stress reduced silicate assimilation; the demand of energy reducing equivalent, and carbon skeletons, in response to this photosynthetic carbon fixation was decreased.

If carbon fixation was decreased because of the less demanding carbon skeleton and reducing equivalent then why were the glycolysis, pyruvate, and TCA metabolic proteins upregulated? Interestingly, similar results were seen on cyanobacterium under nitrate limitation [22], diatoms under the iron [11], and nitrogen limitation [9]. These authors propose that, during nutrient limitation, cells' adaptation to the subsistence and the breakdown of intracellular stores is a more efficient source of carbon for the reassimilation of the nutrient than photosynthesis. These findings also show that the impact of Si-limitation in diatom carbon metabolism has a similar role as iron and nitrate.

Diatoms have a complete urea cycle [1], therefore, demanding higher carbon skeletons and nitrogen molecules than available in the organisms [37]. In the present study, reduction in the nitrogen molecule from chlorophyll pigments and carbon skeletons by Si-limitation regulated the *S. dohrnii* urea cycle (Supplementary Figure S1). Moreover, observed *S. dohrnii* carbon metabolism suggested up-regulation of acetyl-CoA was a forerunner for fatty acid biosynthesis (Figure 5). This statement provides a question to explain if acetyl-CoA is diverted into fatty acid biosynthesis what would be the origin for the TCA cycle? Notably, most of the TCA cycle processes are not only unidirectional, but it catalyzes the reverse reactions in a specific circumstance [9]. The TCA cycle processes are adjustable and expected to reflect the metabolomic and physiological processes as per the cell's demands [38]. This finding shows the carbon metabolic process of *S. dohrnii* under Si-limitation react to various TCA intermediates that decide the carbon flow to the entire central carbon metabolism than to be directed and altered to fatty acid metabolism. In total, reduction in carbon skeletons and nitrogen compound molecules by Si-limitation regulated central carbon and fatty acid metabolisms of *S. dohrnii* under silicate stress.

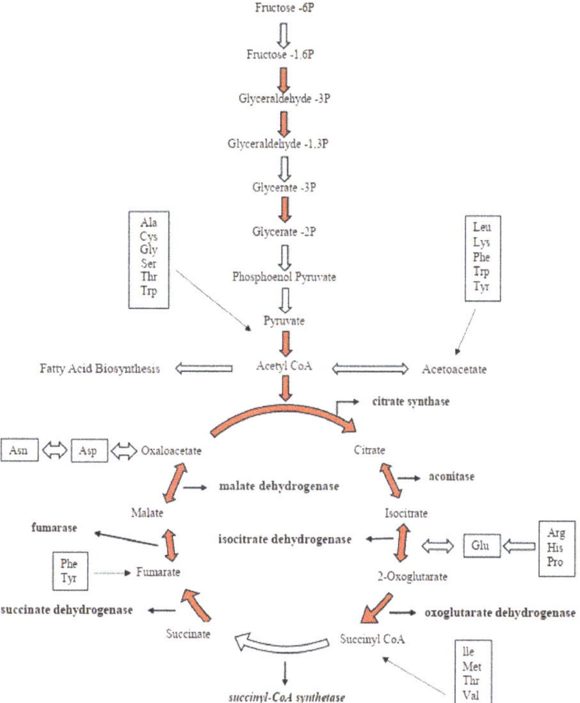

Figure 5. Representation of the changes in proteins abundance associated with carbon metabolism in *S. dohrnii* at the onset of silicate limitation. Increases are shown in red arrows. Boxes show where carbon skeletons from amino acid feed into the pathway.

3.4. The Response of Cellular Respiration to Possible ROS Accumulation

A schematic diagram (Figure 6) shows the identified energy and carbon metabolic pathways in this study and their metabolic regulation by Si-limitation to possible ROS production. Complexes I and III are the primary sites for electron transport and ROS generation [39]. In the present study downregulation of NADH-ubiquinone reductases and complex I suggested a possible enhancement of ROS from cytochrome complex I. Earlier investigations described that oxidation of either complex I or II substrates when complex III is altered and increased ROS accumulation [39,40]. Though cytochrome *c* oxidase is not a direct source of ROS [41], cytochrome *c* oxidase subunit 6B1 downregulation would facilitate the chances of ROS production from complexes I and III [42]. Therefore, it is proposed that the downregulation of cytochrome *c* oxidase subunit 6B1 in this study may inhibit complexes I and III and block the electron delivery to induce the ROS. Similar results were observed in transcriptome and proteome level in *T. pseudonana* at the onset of nitrate, phosphate, and iron limitation [14,15].

The mitochondrial alternative oxidase (AOX) protein used for removing an excess electron in the nutrient-limited diatom [43]; upregulation of this protein was involved in the mitigation of mitochondrial ROS production [14]. However, downregulation of this AOX protein and many essential enzymes of the F1 region of FoF1 type proteins (Supplementary Table S2) in this study suggests the reduction of ATP production with a coincided blockage of the respiratory chain and more possible chances of ROS production. Similar results were observed in the proteome level on diatom *T. pseudonana* in response to Fe limitation [15]. Although in the present study, direct observation of ROS production has not been identified, from the above findings in the proteome level, and comparing with similar findings on diatom it is predicted that due to inefficient electron transport in the photosynthetic and respiratory chain (Figure 6) leads to a possible ROS production under Si-limitation. These statements are consistent with the previously published reports [14,15] on diatoms in response to nutrient limitations.

3.5. Metabolic Regulation of PCD Related Proteins

The previous investigation noted that ROS caused by environmental stress can induce the PCD [44]; therefore, ROS cells must have a defense mechanism. However, in this study, some typical ROS defense proteins were decreased in abundance because of the Si constraint, such as superoxide dismutase, which has been earlier cloned and characterized in diatoms [42]; peroxidase I, which allows the enzymes to react with hydrogen peroxide [18]; peroxiredoxin, an antioxidant enzyme to detoxify peroxidase substrate [45]; and thiol-disulfide, an essential antioxidant protein that can combat oxidative stress by detoxifying [46]. Furthermore, the decreased abundance of chaperonin GroEL and other heat shock proteins shows the opportunity of increasing the impact on cytoplasm and apoptosis [47]. Photorespiration plays a vital role in excess energy dissipation under stress condition [48]. Along these lines, regulation of RuBisCO in this study shows an imbalance in photorespiration to alleviate oxidative stress. Interestingly, a serine protease was downregulated in this study, whereas adenine nucleotide translocator was upregulated, which would induce the caspase-specific activity and apoptosis [49,50]. Though the expression of these cell death-related domains may involve in the PCD of Si-limited *S. dohrnii* cells, the detailed mechanisms, i.e., biochemical role of this domain, detailed regulatory pathways in the *S. dohrnii* PCD require further investigation.

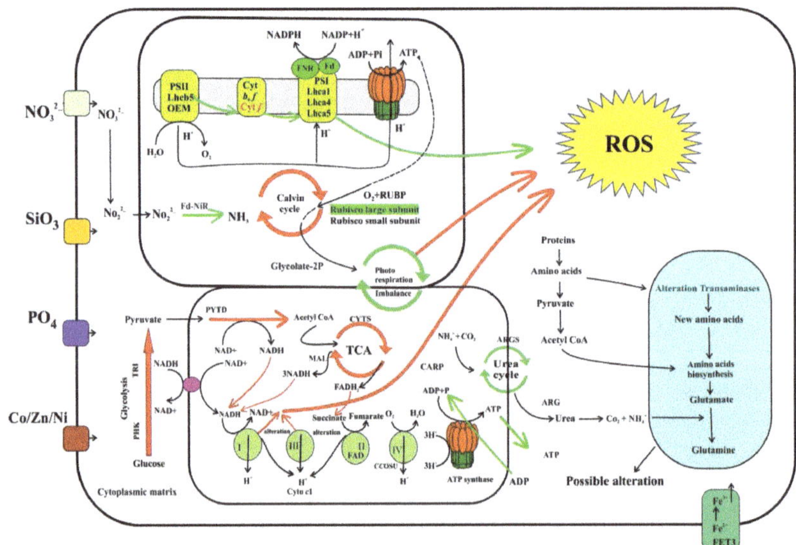

Figure 6. Identified energy and carbon metabolomic pathways and processes of *S. dohrnii* under Si-limited-Si-added condition. All red arrows indicate up-regulated proteins and processes and green shows downregulation proteins and processes. PS: photosystem, Lhc: light-harvesting complex, OEM: oxygen-evolving complex, Cyt: cytochrome, Fd: ferredoxin; FNR: ferredoxin-NADP+, Fd-NiR: ferredoxin nitrite reductase; TCA: tricarboxylic acid, RUBO: RuBisCO; ROS: reactive oxygen species, PHK: phosphoglycerate kinase; TRI: triosephosphate isomerase; PYTD: pyruvate dehydrogenase; CYTS: citrate synthase; MAL: malate/lactate dehydrogenase; CARP: carbamoyl-phosphate synthase; ARGS: argininosuccinate synthase; ARG: arginase, complex I; NADH ubiquinone reductase, complex III; cytochrome *c*1.

Nitrate transporters (NRT2) in plants are responsible for nitrate uptake and deliver to various cellular parts; regulation of this transporter by abiotic stress affects the cellular process [51].

Recently, Hippler et al. [52] reported that Cu stress leads to downregulation of NRT2 and affect the cellular process of *Arabidopsis*. Likewise, in this study downregulation of NRT2 shows alteration in the nitrate transport leads to lower nitrate assimilation for diatom cellular process. A similar finding was reported on the *T. pseudonana* response to nitrate limitation [53].

Generally, purine and pyrimidine compounds are nitrogen originates from amino acids [54]. Changes in purine and pyrimidine could regulate amino acid metabolites and cell growth [55]. Therefore, it is assumed that the downregulation of 14 proteins caused the impact of amino acid binding and further associated metabolic activity of *S. dohrnii*. These included multiple aminotransferases that can yield many fates for amino acids: intracellular recycling to alpha-keto acids, pyruvate, or ammonia, and re-organization into new amino acids and completing the intracellular activity of the cell [56]. The whole impact of Si constraints on *S. dohrnii* metabolism are shown in Supplementary Figure S1, indicating the dynamic role of silicate stress on intracellular diatom metabolism. Our findings are in line with previous transcriptomic and proteomic studies of diatoms and other phytoplankton in response to the nutrient limitation [4,9,11,12,14,15].

4. Materials and Methods

4.1. Species Description

The cylindrical type of marine centric diatom *Skeletonema dohrnii* was first discovered by Sarno et al. [57]. In general, members of *Skeletonema* genus has been considered as a key species in diatoms research due to their global distribution, tropic importance to grazers and their similar physiology with other diatoms [58], therefore, we used *S. dohrnii* as a model species in this experiment.

4.2. Culture Condition

The marine diatom *Skeletonema dohrnii* was isolated from the Yellow Sea and cultured in the f/2 medium in the laboratory (25 °C 100–120 µmol photons $m^{-2} \cdot s^{-1}$, with 14:10 light-dark cycle). Artificial seawater media (ASW) [59] was used with silicate stock solution (28.40 $g \cdot L^{-1}$) in this experiment, however, for the silicate-deplete and -replete conditions modified silicate concentrations (silicate limited 0.5 mL/L) and (silicate added 1.5 mL/L) were used in a controlled incubator (25 °C 150 µmol photons $m^{-2} \cdot s^{-1}$) with a 14/10 h light/dark ratio using cool white fluorescent light. Both the cultures (Si-limited and Si-added) were grown separately in 1000 mL Nalgene and, during the experiment, pH was in a range of 7.5–8.0. The duration of the experimental phase was eight days.

4.3. Analysis of Physiological Measurements

The cell numbers were counted by using an AE 2000 inverted microscope (Motic Group Co., Ltd., Xiamen, China) and Qiujing hemocytometer at the same time every day. The cell density was calculated as follows: CD = $(N/80) \times 400 \times 104$, where CD is the cell density and N is the cell abundance counted in 80 grids on the slide [60]. The photochemical quantum yield of photosystem (*Fv/Fm*) was measured by a Fast Ocean FRRF3 sensor with an Act2 system by the following procedure: Samples were initially kept in the low light to allow the oxidation of the electron transport chain (ETC) and relaxation of NPQ. The single-turnover protocol consisted of 100 flash lets (Fet, a single 1 µs excitation pulse from the LEDs within an FRRF3 sensor) with 2.0 µs Fet pitch (interval between the start of one Fet and the next) were set in the instrument. Then each sample was exposed sequentially to 12 actinic background irradiances spanning from 0–1200 µmol quanta $m^{-2} \cdot s^{-1}$ to retrieve fluorescence light response curves. In the dark-adapted state, the maximum quantum efficiency of PSII was calculated using the ratio of *Fv/Fm* as per Wei et al. [61].

The ^{14}C method was used for the net primary productivity (NPP) as follows: The inoculum was transferred to 30 mL incubation flasks in triplicates for the 24 h incubation. Before the incubation, 100 µL aliquots were transferred to a scintillation bottle from one sample and soaked in 10 mL scintillation cocktail. Three dark bottles were set up for control. The culture bottles were covered with different pore size meshes to create light gradients (0–800 PAR) after adding 20 $\mu Ci^{14}C$ sodium bicarbonate. The incubation samples analyzed by Wallac System 1400 liquid scintillation counter (PerkinElmer Life and Analytical Sciences Inc., Wellesley, MA, USA) according to the manufacturer's protocol.

4.4. Protein Extraction and Preparation

According to the cell cycle progression of diatoms, day 4 (mid-exponential growth) was chosen as the sampling time point for iTRAQ proteomic analysis. Protein extraction was carried out according to the description by Du et al. [4]. Briefly, one liter of culture from each sample (4 Si-deplete + 4 Si-replete = 8 samples = 8 L), collected through 2 μm pore-size filter membrane. Then the cells were resuspended with 10 mL medium into a 15 mL centrifuge tube. Once the pellets were collected by the centrifugation at 3000× *g* for 5 min, immediately 10 mL TRIzol Reagent added (Invitrogen, Life Technology, Carlsbad, CA, USA), and then the protein was extracted according to the manufacturer's protocols.

For the protein preparation, the protein pellets were resuspended in a lysis buffer (8 M Urea, 40 mM Tris-HCl, 2 mM EDTA, pH 8.5), then incubated with additional of 10 mM dithiothreitol (DTT) and placed at 56 °C for 1 h to reduce the disulfide bonds. After that 55 mM iodoacetamide (IAA) was added and samples were incubated with 45 min to block the cysteine residues proteins. The reduced protein mixtures were then precipitated by adding acetone at −20 °C overnight. After centrifugation at 4 °C, 25,000× *g*, for 20 min the pellet was dissolved in 0.5 M TEAB (triethylammonium bicarbonate, Applied Biosystems, Milan, Italy) and sonicated in ice. After centrifugation at 25,000× *g* at 4 °C, an aliquot was taken for the protein quantification by Bradford Assay according to the manufacturer's protocol. The supernatants were kept −80 °C until further analysis.

4.5. iTRAQ Labeling and Fraction

A total of 100 μg of proteins from each sample was digested with Trypsin Gold (Promega, Madison, WI, USA) with the ratio of protein/trypsin = 40:1, at 37 °C overnight. After trypsin digestion, peptides were dried with vacuum centrifugation then reconstituted in 0.5 M TEAB and proceeded based on the manufacturer's protocol. In the present study, an experiment setting of 4:4 (8-plex, four biological replicates) was selected and labeled with different iTRAQ tags. The silicate replete samples were labeled with iTRAQ tags 113, 115, 117, 119, and silicate deplete samples were labeled with iTRAQ tags 114, 116, 118, and 121, respectively. The labeled peptides with isobaric tags were then incubated at room temperature for 2 h, and labeled peptides were then pooled and dried by vacuum centrifugation.

The strong cationic exchange chromatography was performed using a Shimadzu LC-20AB HPLC equipped with a UV–VIS detector (Shimadzu, Kyoto, Japan). The fractionated peptides were first reconstituted with buffer A (5% ACN,95% H_2O, adjust pH to 9.8 with ammonia) then loaded onto Gemini C18 5-μM, 4.6 × 250 mm reverse phase column containing 5-μm particles (Phenomen-ex). The peptides were separated at a flow rate of 1 mL/min with a gradient of 5% buffer B (5% H_2O, 95% ACN, adjust pH to 9.8 with ammonia) for 10 min, 5–35% buffer B for 40 min, 35–95% buffer B for 1 min. The system is then maintained in 95% buffer B for 3 min before equilibrating buffer B for 10 min before the next injection. The elution was monitored by measuring absorbance at 214 nm, and fractions were collected at every 1 min. The eluted peptides were then pooled as 20 fractions, desalted with Strata X C18 column (Phenomenex) and vacuum-dried according to the manufacturer's protocol.

4.6. LC-MS/MS Analysis

Every fraction was then resuspended in buffer A (2% ACN, 0.1% FA in water) and centrifuged at 20,000× *g* for 10 min. The final concertation of the peptide was 0.5 μg/μL on average. The supernatant was then loaded onto LC-20AD nano-HPLC (Shimadzu, Kyoto, Japan) by the autosampler onto C18 trap column and the peptides were eluted to analytical C18 column (inner diameter 75 μm) packed in-house. Samples were then collected at a rate of 8 µL/min for 5 min. A 35 min gradient was run at 300 nL/min ranging from 8% to 35% of buffer B (2% H_2O and 0.1% FA in ACN), followed by a 5 min linear gradient to 60% buffer B, maintenance at 80% of buffer B for 5 min and return to 5% on 0.1 min and equilibrated for 10 min.

The peptide samples were then subjected to Nanospray III ionization of TripleTOF 5600 System, has a high mass accuracy, less than 2 ppm (SCIEX, Framingham, MA, USA). Data were acquired with the following MS conditions: ion spray voltage 2300 V, curtain gas of 30, nebulizer gas of 15, and interface heater temperature of 150 °C. The accumulation time for MS1 is 250 ms, and the mass ranges were from 350 to 1500 Da. Based on the intensity in MS1 survey as many as 30 product ion scans were collected if exceeding a threshold of 120 counts per second (counts/s) and with charge-state 2+ to 5+, dynamic exclusion was set for 1/2 of peak width (12 s). For iTRAQ data acquisition, the collision energy was adjusted to all precursor ions for collision-induced dissociation, and the Q2 transmission window for 100 Da was 100%.

4.7. Proteomic Data Analysis

Acquired raw MS/MS data were converted into MGF format by Proteo Wizard tool msConvert, and the exported MGF files were searched using Mascot version 2.3.02 against the selected database. The data attainment was performed with Analyst QS 2.0 software (Applied Biosystems/MDS SCIEX). Furthermore, the peptide and protein detection was performed through Mascot 2.3.02 (Matrix Science, London, UK), [62] against the NCBI for the *T. pseudonana* CCMP1335. For the protein identification mass tolerance of (0.05 Da) was permitted for intact peptide masses and (0.1 Da) for fragmented mass with an allowance of missed cleavage in the trypsin digests. Gln-pyro-Glu (N-term Q), oxidation (M), iTRAQ 8plex (Y) as variable modification, and carbamidomethyl (C), iTRAQ 8 plex (N-term), iTRAQ 8plex (K) as fixed modifications.

The charge-states of peptides were set +2 to +5. Furthermore, an automated software IQuant, [63] used for analyzing labeled peptides with isobaric tags, which integrates Mascot Percolator, a well-performing machine learning method for rescoring database search results. To assess, the confidence level of peptide the PSMs was pre-filtered at a PSM-level FDR of 1%. Then based on the "simple principle" (the parsimony principle), identified peptide sequences are assembled into a set of confident proteins. For the reducing the probability of false identification a protein FDR at 1%, which is based on the picked-up protein FDR strategy, will also be estimated after protein inference (protein-level FDR $\leq$ 0.01). For the protein quantification, it was required that protein should contain a minimum of two unique peptides, and the protein ratios were weighted and normalized by the median ration in Mascot. Additionally, Mascot (2.3.02) was used to perform Student's *t*-test. We only used ratio *p*-values < 0.05, and fold changes of >2 were considered as significant.

4.8. Functional Annotation

The COG (Cluster of Orthologous Groups of proteins) and then GO (Gene Ontology) analyses (http://www.geneontology.org) were performed according to the method reported in the earlier literature [64]. The differentially regulated proteins in GO terms was carried out using the following formula:

$$P = 1 - \sum_{i=0}^{m-1} \frac{\binom{M}{i}\binom{N-M}{n-i}}{\binom{N}{n}}$$

where N is the number of all proteins with GO annotation information, n is the number of the differentially regulated proteins with GO annotation information, M is the number of proteins with a given GO term annotation, and m is the number of the differentially regulated proteins with a given GO term annotation. Furthermore, proteins with the two-fold change between each sample and a p-value less than 0.05 were determined as differentially expressed proteins to regulate cellular metabolism. The metabolic pathway analysis of differentially identified proteins was conducted according to the KEGG Pathway Database (http://www.genome.jp/kegg/), which represents knowledge in molecular interactions and reaction networks.

5. Conclusions

In this study, iTRAQ proteomic profile and physiological characteristics represent not only comprehensive systematic analysis of the diatoms intracellular silicate stress responses but also provides many previously uncharacterized cellular metabolism processes and proteins involved in the energy production, carbon metabolism, and ROS accumulation as shown in Figures 4 and 6. The present finding demonstrates new molecular views of diatom response to silicate stress and their cell controlling mechanisms: (1) reduced electron flow in photosystem reduced the ATP production and controls the cell growth and replication, (2) reduction of reduced equivalents and carbon skeletons regulated central carbon metabolism and controlled carbon fixation in Si-limited cells, and (3) alteration of electron flow in the cellular process may lead to possible ROS accumulation. This proteomic study significantly facilitates our understanding of the intracellular processes of diatoms in response to silicate stress in the marine environments which could be valuable for future physiological research in diatoms.

Supplementary Materials: Supplementary materials can be found at http://www.mdpi.com/1422-0067/20/10/2540/s1.

Author Contributions: S.T. and J.S. conceived and designed the experiments; S.T. performed the laboratory experiments, analyzed the data, and drafted the paper; J.S. provided the facilities and funding to perform research; X.S. performed the monitoring analysis; and H.L. performed primary productivity analysis.

Funding: This work was supported by the National Nature Science Foundation of China (41876134, 41676112 and 41176136), the Key Project of Natural Science Foundation of Tianjin (17JCZDJC40000), the University Innovation Team Training Program for Tianjin (TD12-5003), the Tianjin 131 Innovation Team Program (20180314), and the Changjiang Scholar Program of Chinese Ministry of Education to Jun Sun.

Conflicts of Interest: The authors declare no conflict of interest

Abbreviations

iTRAQ	Isobaric tag for relative and absolute quantification
ROS	Reactive Oxygen Species
ATP	Adenosine triphosphate
PSII	Photosystem II
PSI	Photosystem I
PCD	Programmed Cell Death
LC-MS/MS	liquid chromatography-tandem mass spectrometry
NPP	Net Primary Productivity
GO	Gene Ontology
COG	Clusters of Orthologous Groups
KEGG	The Kyoto Encyclopedia of Genes and Genomes
DEPs	Diffrenccilaly Expressed Proteins
FCPs	Fucoxanthin-chlorophyll proteins
GTP	Guanosine triphosphate
FADH2	flavin adenine dinucleotide
TCA	The citric acid cycle
NADH	Nicotinamide adenine dinucleotide
RuBisCO	Ribulose-1,5-bisphosphate carboxylase/oxygenase

References

1. Armbrust, E.V. The Life of Diatoms in the World's Oceans. *Nature* **2009**, *459*, 185–192. [CrossRef]
2. Bowler, C.; Vardi, A.; Allen, A.E. Oceanographic and Biogeochemical Insights from Diatom Genomes. *Annu. Rev. Mar. Sci.* **2010**, *2*, 333–365. [CrossRef] [PubMed]
3. Rosenwasser, S.; Van Creveld, S.G.; Schatz, D.; Malitsky, S.; Tzfadia, O.; Aharoni, A.; Levin, Y.; Gabashvili, A.; Feldmesser, E.; Vardi, A. Mapping the Diatom Redox-Sensitive Proteome Provides Insight into Response to Nitrogen Stress in the Marine Environment. *Proc. Natl. Acad. Sci. USA* **2014**, *111*, 2740–2745. [CrossRef]
4. Du, C.; Liang, J.-R.; Chen, D.-D.; Xu, B.; Zhuo, W.-H.; Gao, Y.-H.; Chen, C.-P.; Bowler, C.; Zhang, W. iTRAQ-Based Proteomic Analysis of the Metabolism Mechanism Associated with Silicon Response in the Marine Diatom Thalassiosira Pseudonana. *J. Proteome Res.* **2014**, *13*, 720–734. [CrossRef] [PubMed]
5. Brzezinski, M.A.; Olson, R.J.; Chisholm, S.W. Silicon Availability and Cell-Cycle Progression in Marine Diatoms. *Mar. Ecol. Prog. Ser.* **1990**, *67*, 83–96. [CrossRef]
6. Shrestha, R.P.; Tesson, B.; Norden-Krichmar, T.; Federowicz, S.; Hildebrand, M.; Allen, A.E. Whole Transcriptome Analysis of the Silicon Response of the Diatom Thalassiosira Pseudonana. *BMC Genom.* **2012**, *13*, 499. [CrossRef] [PubMed]
7. Tirichine, L.; Bowler, C. Decoding Algal Genomes: Tracing Back the History of Photosynthetic Life on Earth. *Plant J.* **2011**, *66*, 45–57. [CrossRef] [PubMed]
8. Parkhill, J.P.; Maillet, G.; Cullen, J.J. Fluorescence-Based Maximal Quantum Yield for PSII as a Diagnostic of Nutrient Stress. *J. Phycol.* **2001**, *37*, 517–529. [CrossRef]
9. Hockin, N.L.; Mock, T.; Mulholland, F.; Kopriva, S.; Malin, G. The Response of Diatom Central Carbon Metabolism to Nitrogen Starvation is Different from that of Green Algae and Higher Plants. *Plant Physiol.* **2012**, *158*, 299–312. [CrossRef]
10. Jian, J.; Zeng, D.; Wei, W.; Lin, H.; Li, P.; Liu, W. The Combination of RNA and Protein Profiling Reveals the Response to Nitrogen Depletion in Thalassiosira Pseudonana. *Sci. Rep.* **2017**, *7*, 8989. [CrossRef]
11. Nunn, B.L.; Faux, J.F.; Hippmann, A.A.; Maldonado, M.T.; Harvey, H.R.; Goodlett, D.R.; Boyd, P.W.; Strzepek, R.F. Diatom Proteomics Reveals Unique Acclimation Strategies to Mitigate Fe Limitation. *PLoS ONE* **2013**, *8*, e75653. [CrossRef]
12. Brembu, T.; Muhlroth, A.; Alipanah, L.; Bones, A.M. The Effects of Phosphorus Limitation on Carbon Metabolism in Diatoms. *Philos. Trans. R. Soc. Lond. B Biol. Sci.* **2017**, *372*, 20160406. [CrossRef]
13. Mock, T.; Samanta, M.P.; Iverson, V.; Berthiaume, C.; Robison, M.; Holtermann, K.; Durkin, C.; BonDurant, S.S.; Richmond, K.; Rodesch, M. Whole-Genome Expression Profiling of the Marine Diatom Thalassiosira Pseudonana Identifies Genes Involved in Silicon Bioprocesses. *Proc. Natl. Acad. Sci. USA* **2008**, *105*, 1579–1584. [CrossRef]
14. Lin, Q.; Liang, J.-R.; Huang, Q.-Q.; Luo, C.-S.; Anderson, D.M.; Bowler, C.; Chen, C.-P.; Li, X.-S.; Gao, Y.-H. Differential Cellular Responses Associated with Oxidative Stress and Cell Fate Decision under Nitrate and Phosphate Limitations in Thalassiosira Pseudonana: Comparative Proteomics. *PLoS ONE* **2017**, *12*, e0184849. [CrossRef] [PubMed]
15. Luo, C.-S.; Liang, J.-R.; Lin, Q.; Li, C.; Bowler, C.; Anderson, D.M.; Wang, P.; Wang, X.-W.; Gao, Y.-H. Cellular Responses Associated with ROS Production and Cell Fate Decision in Early Stress Response to Iron Limitation in the Diatom Thalassiosira Pseudonana. *J. Proteome Res.* **2014**, *13*, 5510–5523. [CrossRef]
16. Javaheri, N.; Dries, R.; Burson, A.; Stal, L.; Sloot, P.; Kaandorp, J. Temperature Affects the Silicate Morphology in a Diatom. *Sci. Rep.* **2015**, *5*, 11652. [CrossRef]
17. Casotti, R.; Mazza, S.; Brunet, C.; Vantrepotte, V.; Ianora, A.; Miralto, A. Growth Inhibition and Toxicity of the Diatom Aldehyde 2-Trans, 4-Trans-Decadienal Onthalassiosira Weissflogii (Bacillariophyceae). *J. Phycol.* **2005**, *41*, 7–20. [CrossRef]
18. Chung, C.C.; Hwang, S.P.; Chang, J. Cooccurrence of ScDSP Gene Expression, Cell Death, and DNA Fragmentation in a Marine Diatom, Skeletonema Costatum. *Appl. Environ. Microbiol.* **2005**, *71*, 8744–8751. [CrossRef]
19. Berges, J.A.; Falkowski, P.G. Physiological Stress and Cell Death in Marine Phytoplankton: Induction of Proteases in Response to Nitrogen or Light Limitation. *Limnol. Oceanogr.* **1998**, *43*, 129–135. [CrossRef]
20. Brussaard, C.P.; Noordeloos, A.A.; Riegman, R. Autolysis Kinetics of the Marine Diatom Ditylum Brightwellii (Bacillariophyceae) under Nitrogen and Phosphorus Limitation and Starvation 1. *J. Phycol.* **1997**, *33*, 980–987. [CrossRef]

21. Mackey, K.; Morris, J.J.; Morel, F.; Kranz, S. Response of Photosynthesis to Ocean Acidification. *Oceanography* **2015**, *25*, 74–91. [CrossRef]

22. Tolonen, A.C.; Aach, J.; Lindell, D.; Johnson, Z.I.; Rector, T.; Steen, R.; Church, G.M.; Chisholm, S.W. Global Gene Expression of Prochlorococcus Ecotypes in Response to Changes in Nitrogen Availability. *Mol. Syst. Biol.* **2006**, *2*, 53. [CrossRef]

23. Shen, W.; Wei, Y.; Dauk, M.; Tan, Y.; Taylor, D.C.; Selvaraj, G.; Zou, J. Involvement of a Glycerol-3-Phosphate Dehydrogenase in Modulating the NADH/NAD+ Ratio Provides Evidence of a Mitochondrial Glycerol-3-Phosphate Shuttle in Arabidopsis. *Plant Cell* **2006**, *18*, 422–441. [CrossRef]

24. Fernie, A.R.; Stitt, M. On the Discordance of Metabolomics with Proteomics and Transcriptomics: Coping with Increasing Complexity in Logic, Chemistry, and Network Interactions Scientific Correspondence. *Plant Physiol.* **2012**, *158*, 1139–1145. [CrossRef] [PubMed]

25. Dyhrman, S.T.; Jenkins, B.D.; Rynearson, T.A.; Saito, M.A.; Mercier, M.L.; Alexander, H.; Whitney, L.P.; Drzewianowski, A.; Bulygin, V.V.; Bertrand, E.M.; et al. The Transcriptome and Proteome of the Diatom Thalassiosira Pseudonana Reveal a Diverse Phosphorus Stress Response. *PLoS ONE* **2012**, *7*, e33768. [CrossRef] [PubMed]

26. Lommer, M.; Specht, M.; Roy, A.-S.; Kraemer, L.; Andreson, R.; Gutowska, M.A.; Wolf, J.; Bergner, S.V.; Schilhabel, M.B.; Klostermeier, U.C.; et al. Genome and Low-Iron Response of an Oceanic Diatom Adapted to Chronic Iron Limitation. *Genome Biol.* **2012**, *13*, R66. [CrossRef]

27. Takabayashi, M.; Lew, K.; Johnson, A.; Marchi, A.; Dugdale, R.; Wilkerson, F.P. The Effect of Nutrient Availability and Temperature on Chain Length of the Diatom, Skeletonema Costatum. *J. Plankton Res.* **2006**, *28*, 831–840. [CrossRef]

28. Anning, T.; Harris, G.; Geider, R. Thermal Acclimation in the Marine Diatom Chaetoceros Calcitrans (Bacillariophyceae). *Eur. J. Phycol.* **2001**, *36*, 233–241. [CrossRef]

29. Falkowski, P.G. Light-Shade Adaptation in Marine Phytoplankton. In *Primary Productivity in the Sea*; Springer: Berlin, Germany, 1980; pp. 99–119.

30. Gururani, M.A.; Venkatesh, J.; Tran, L.S. Regulation of Photosynthesis during Abiotic Stress-Induced Photoinhibition. *Mol. Plant* **2015**, *8*, 1304–1320. [CrossRef] [PubMed]

31. Dang, K.V.; Plet, J.; Tolleter, D.; Jokel, M.; Cuine, S.; Carrier, P.; Auroy, P.; Richaud, P.; Johnson, X.; Alric, J.; et al. Combined Increases in Mitochondrial Cooperation and Oxygen Photoreduction Compensate for Deficiency in Cyclic Electron Flow in Chlamydomonas Reinhardtii. *Plant Cell* **2014**, *26*, 3036–3050. [CrossRef]

32. Munekage, Y.; Hojo, M.; Meurer, J.; Endo, T.; Tasaka, M.; Shikanai, T. PGR5 is Involved in Cyclic Electron Flow Around Photosystem I and is Essential for Photoprotection in Arabidopsis. *Cell* **2002**, *110*, 361–371. [CrossRef]

33. Tikkanen, M.; Mekala, N.R.; Aro, E.-M. Photosystem II Photoinhibition-Repair Cycle Protects Photosystem I from Irreversible Damage. *Biochim. Biophys. Acta* **2014**, *1837*, 210–215. [CrossRef] [PubMed]

34. Jordan, P.; Fromme, P.; Witt, H.T.; Klukas, O.; Saenger, W.; Krauß, N. Three-Dimensional Structure of Cyanobacterial Photosystem I at 2.5 Å Resolution. *Nature* **2001**, *411*, 909. [CrossRef]

35. Sunda, W.; Kieber, D.; Kiene, R.; Huntsman, S. An Antioxidant Function for DMSP and DMS in Marine Algae. *Nature* **2002**, *418*, 317. [CrossRef] [PubMed]

36. Stefels, J. Physiological Aspects of the Production and Conversion of DMSP in Marine Algae and Higher Plants. *J. Sea Res.* **2000**, *43*, 183–197. [CrossRef]

37. Allen, A.E.; Vardi, A.; Bowler, C. An Ecological and Evolutionary Context for Integrated Nitrogen Metabolism and Related Signaling Pathways in Marine Diatoms. *Curr. Opin. Plant Biol.* **2006**, *9*, 264–273. [CrossRef] [PubMed]

38. Sweetlove, L.J.; Beard, K.F.; Nunes-Nesi, A.; Fernie, A.R.; Ratcliffe, R.G. Not Just a Circle: Flux Modes in the Plant TCA Cycle. *Trends Plant Sci.* **2010**, *15*, 462–470. [CrossRef]

39. St-Pierre, J.; Buckingham, J.A.; Roebuck, S.J.; Brand, M.D. Topology of Superoxide Production from Different Sites in the Mitochondrial Electron Transport Chain. *J. Biol. Chem.* **2002**, *277*, 44784–44790. [CrossRef] [PubMed]

40. Herrero, A.; Barja, G. Sites and Mechanisms Responsible for the Low Rate of Free Radical Production of Heart Mitochondria in the Long-Lived Pigeon. *Mech. Ageing Dev.* **1997**, *98*, 95–111. [CrossRef]

41. Babcock, G.T.; Wikström, M. Oxygen Activation and the Conservation of Energy in Cell Respiration. *Nature* **1992**, *356*, 301–309. [CrossRef]

42. Ken, C.-F.; Hsiung, T.-M.; Huang, Z.-X.; Juang, R.-H.; Lin, C.-T. Characterization of Fe/Mn—Superoxide Dismutase from Diatom Thallassiosira Weissflogii: Cloning, Expression, and Property. *J. Agric. Food Chem.* **2005**, *53*, 1470–1474. [CrossRef] [PubMed]

43. Allen, A.E.; Laroche, J.; Maheswari, U.; Lommer, M.; Schauer, N.; Lopez, P.J.; Finazzi, G.; Fernie, A.R.; Bowler, C. Whole-Cell Response of the Pennate Diatom Phaeodactylum Tricornutum to Iron Starvation. *Proc. Natl. Acad. Sci. USA* **2008**, *105*, 10438–10443. [CrossRef]

44. Levine, R.L.; Mosoni, L.; Berlett, B.S.; Stadtman, E.R. Methionine Residues as Endogenous Antioxidants in Proteins. *Proc. Natl. Acad. Sci. USA* **1996**, *93*, 15036–15040. [CrossRef] [PubMed]

45. Dietz, K.J. Plant Peroxiredoxins. *Annu. Rev. Plant Biol.* **2003**, *54*, 93–107. [CrossRef]

46. Murata, H.; Ihara, Y.; Nakamura, H.; Yodoi, J.; Sumikawa, K.; Kondo, T. Glutaredoxin Exerts an Antiapoptotic Effect by Regulating the Redox State of Akt. *J. Biol. Chem.* **2003**, *278*, 50226–50233. [CrossRef]

47. Kaufman, B.A.; Kolesar, J.E.; Perlman, P.S.; Butow, R.A. A Function for the Mitochondrial Chaperonin Hsp60 in the Structure and Transmission of Mitochondrial DNA Nucleoids in Saccharomyces Cerevisiae. *J. Cell Biol.* **2003**, *163*, 457–461. [CrossRef]

48. Parker, M.S.; Armbrust, E.V. Synergistic Effects of Light, Temperature, and Nitrogen Source on Transcription of Genes for Carbon and Nitrogen Metabolism in the Centric Diatom Thalassiosira Pseudonana (Bacillariophyceae)1. *J. Phycol.* **2005**, *41*, 1142–1153. [CrossRef]

49. Hedstrom, L. Serine Protease Mechanism and Specificity. *Chem. Rev.* **2002**, *102*, 4501–4524. [CrossRef]

50. Coffeen, W.C.; Wolpert, T.J. Purification and Characterization of Serine Proteases that Exhibit Caspase-Like Activity and are Associated with Programmed Cell Death in Avena Sativa. *Plant Cell* **2004**, *16*, 857–873. [CrossRef]

51. Zhang, G.B.; Meng, S.; Gong, J.M. The Expected and Unexpected Roles of Nitrate Transporters in Plant Abiotic Stress Resistance and Their Regulation. *Int. J. Mol. Sci.* **2018**, *19*, 3535. [CrossRef] [PubMed]

52. Hippler, F.W.R.; Mattos-Jr, D.; Boaretto, R.M.; Williams, L.E. Copper Excess Reduces Nitrate Uptake by Arabidopsis Roots with Specific Effects on Gene Expression. *J. Plant Physiol.* **2018**, *228*, 158–165. [CrossRef]

53. Chen, X.H.; Li, Y.Y.; Zhang, H.; Liu, J.L.; Xie, Z.X.; Lin, L.; Wang, D.Z. Quantitative Proteomics Reveals Common and Specific Responses of a Marine Diatom Thalassiosira Pseudonana to Different Macronutrient Deficiencies. *Front. Microbiol.* **2018**, *9*, 2761. [CrossRef]

54. Zrenner, R.; Stitt, M.; Sonnewald, U.; Boldt, R. Pyrimidine and Purine Biosynthesis and Degradation in Plants. *Annu. Rev. Plant Biol.* **2006**, *57*, 805–836. [CrossRef] [PubMed]

55. Alipanah, L.; Rohloff, J.; Winge, P.; Bones, A.M.; Brembu, T. Whole-Cell Response to Nitrogen Deprivation in the Diatom Phaeodactylum Tricornutum. *J. Exp. Bot.* **2015**, *66*, 6281–6296. [CrossRef] [PubMed]

56. Berg, J.M.; Tymoczko, J.L.; Stryer, L. *Biochemistry*; WH Freeman: New York, NY, USA, 2002.

57. Sarno, D.; Kooistra, W.H.; Medlin, L.K.; Percopo, I.; Zingone, A. Diversity in the Genus Skeletonema (Bacillariophyceae). ii. An Assessment of the Taxonomy of S. Costatum-Like Species with the Description of Four New Species 1. *J. Phycol.* **2005**, *41*, 151–176. [CrossRef]

58. Zingone, A.; Percopo, I.; Sims, P.A.; Sarno, D. Diversity in the Genus Skeletonema (Bacillariophyceae). I. A Reexamination of the Type Material of S. Costatum with the Description of S. Grevillei Sp. NOV. 1. *J. Phycol.* **2005**, *41*, 140–150. [CrossRef]

59. Sunda, W.G.; Price, N.M.; Morel, F.M. Trace Metal Ion Buffers and Their Use in Culture Studies. *Algal Cult. Tech.* **2005**, *4*, 35–63.

60. Sun, J.; Ning, X. Marine Phytoplankton Specific Growth Rate. *Adv. Earth Sci.* **2005**, *20*, 939–945.

61. Wei, Y.; Zhao, X.; Sun, J. Fast Repetition Rate Fluorometry (FRRF) Derived Phytoplankton Primary Productivity in the Bay of Bengal. *Front. Microbiol.* **2019**. [CrossRef]

62. Charbonneau, M.-È.; Girard, V.; Nikolakakis, A.; Campos, M.; Berthiaume, F.; Dumas, F.; Lépine, F.; Mourez, M. O-Linked Glycosylation Ensures the Normal Conformation of the Autotransporter Adhesin Involved in Diffuse Adherence. *J. Bacteriol.* **2007**, *189*, 8880–8889. [CrossRef]

63. Van Domselaar, G.H.; Stothard, P.; Shrivastava, S.; Cruz, J.A.; Guo, A.; Dong, X.; Lu, P.; Szafron, D.; Greiner, R.; Wishart, D.S. BASys: A Web Server for Automated Bacterial Genome Annotation. *Nucleic Acids Res.* **2005**, *33*, W455–W459. [CrossRef] [PubMed]

64. Unwin, R.D. Quantification of Proteins by iTRAQ. *Methods Mol. Biol.* **2010**, *658*, 205–215. [PubMed]

International Journal of
Molecular Sciences

Article

Identification of Two Novel Wheat Drought Tolerance-Related Proteins by Comparative Proteomic Analysis Combined with Virus-Induced Gene Silencing

Xinbo Wang [1,2,3,†], Yanhua Xu [1,2,3,4,†], Jingjing Li [1,2,3], Yongzhe Ren [1,2,3,*], Zhiqiang Wang [1,2,3], Zeyu Xin [1,2,3] and Tongbao Lin [1,2,3,*]

1   College of Agronomy, Henan Agricultural University, Zhengzhou 450002, China; boxinw@163.com (X.W.); yanhuaxu66@163.com (Y.X.); 13592589578@163.com (J.L.); wzq78@163.com (Z.W.); mrxxtz@163.com (Z.X.)
2   State Key Laboratory of Wheat and Maize Crop Science, Henan Agricultural University, Zhengzhou 450002, China
3   Collaborative Innovation Center of Henan Grain Crops, Henan Agricultural University, Zhengzhou 450002, China
4   Shangqiu Normal University, Shangqiu 476000, China
*   Correspondence: yongzheren66@163.com (Y.R.); linlab@163.com (T.L.); Tel.: +86-371-569-90186 (Y.R. & T.L.)
†   These authors contributed equally to this work.

Received: 17 November 2018; Accepted: 10 December 2018; Published: 12 December 2018

Abstract: Drought is a major adversity that limits crop yields. Further exploration of wheat drought tolerance-related genes is critical for the genetic improvement of drought tolerance in this crop. Here, comparative proteomic analysis of two wheat varieties, XN979 and LA379, with contrasting drought tolerance was conducted to screen for drought tolerance-related proteins/genes. Virus-induced gene silencing (VIGS) technology was used to verify the functions of candidate proteins. A total of 335 differentially abundant proteins (DAPs) were exclusively identified in the drought-tolerant variety XN979. Most DAPs were mainly involved in photosynthesis, carbon fixation, glyoxylate and dicarboxylate metabolism, and several other pathways. Two DAPs (W5DYH0 and W5ERN8), dubbed *TaDrSR1* and *TaDrSR2*, respectively, were selected for further functional analysis using VIGS. The relative electrolyte leakage rate and malonaldehyde content increased significantly, while the relative water content and proline content significantly decreased in the *TaDrSR1*- and *TaDrSR2*-knock-down plants compared to that in non-knocked-down plants under drought stress conditions. *TaDrSR1*- and *TaDrSR2*-knock-down plants exhibited more severe drooping and wilting phenotypes than non-knocked-down plants under drought stress conditions, suggesting that the former were more sensitive to drought stress. These results indicate that *TaDrSR1* and *TaDrSR2* potentially play vital roles in conferring drought tolerance in common wheat.

Keywords: drought stress; *Triticum aestivum* L.; comparative proteomic analysis; iTRAQ; VIGS

---

1. Introduction

As a result of the continuous growth of the global population and the adverse effects of environmental change on food production, food security has become a major problem worldwide [1]. Wheat (*Triticum aestivum* L.) is one of the most important food crops globally. At present, wheat production in many parts of the world is still dependent on excessive use of water, which is ecologically and economically unsustainable [2–4]. The shortage of water resources in many regions of the world greatly hinders crop productivity and sustainability. It is estimated that wheat production is likely to decline by 23% to 27% in 2050 due to global drought conditions [4,5]. Further exploration of the mechanisms underlying drought tolerance in wheat is therefore of vital importance for both the genetic improvement of drought tolerance in this crop and the reduction of the impact of drought stress on wheat production.

The physiological changes of wheat plants under drought stress, and the molecular mechanisms underpinning the response to drought stress have been well studied [6–9]. Relative water content (RWC), relative conductivity, malonaldehyde (MDA), and proline content are important physiological indicators that reflect plant drought tolerance. When subjected to water stress for extended durations, drought-tolerant varieties have been found to maintain a relatively higher RWC than drought-sensitive varieties [10,11]. Under severe drought stress conditions, membrane proteins are damaged, the membrane is easily broken, and even the cytoplasm is extravasated, resulting in an increase in relative conductivity [12,13]. The content of the membrane lipid peroxide MDA is an important marker of structural damage to the membrane [14]. Moreover, the accumulation of free proline under drought stress is positively correlated with drought tolerance in plants. Drought-tolerant varieties accumulate larger amounts of proline than drought-sensitive varieties [15]. A recent study in pepper reported a drought-inducible bZIP family gene, *CaDILZ1*, which controls plant drought tolerance by altering ABA content, stomatal closure, and the expression of ABA responses and drought response marker genes, while a RING-type E3 ligase, CaDSR1, can interact with CaDILZ1 and negatively regulate ABA signaling via E3 ligase activity, through influencing CaDILZ1 stability [16]. Studies have also shown that the AP2/ERF, MYB/MYC, and NAC transcription factors are associated with the drought stress response in wheat [17–21]. The expression of the AP2/ERF transcription factor-encoding gene *TaERF1* has been shown to be induced by drought stress. *TaERF1*-overexpressing transgenic *Arabidopsis*, wheat, and tobacco plants exhibit up-regulated expression of multiple stress response-related genes, which significantly enhances the tolerance of these transgenic plants to osmotic stress [22]. Overexpression of *TaMYB30* in *Arabidopsis* conferred significant resistance to drought during the germination and seedling stage [17,23]. The expression level of NAC transcription factor-encoding gene *TaNAC69* has also been demonstrated to be significantly up-regulated in response to drought stress. The overexpression of *TaNAC69* in wheat improves the drought tolerance of transgenic plants [24,25]. Peng et al. identified 93 (root) and 65 (leaf) drought/salinity stress-responsive proteins in a drought-tolerant somatic hybrid wheat cv. SR3 and its parent JN177. Most drought-responsive differentially abundant proteins (DAPs) also responded to salinity. The enhanced drought tolerance of SR3 appears to depend on the superior capacity for osmotic and ionic homeostasis, more efficient removal of toxic by-products, and better potential for growth recovery [26]. However, although extensive work has been conducted and significant progress in the elucidation of the physiological and molecular mechanisms underlying crop drought stress tolerance has been made [27,28], the mechanisms have not been fully explored, especially in hexaploid wheat.

In recent years, transcriptomics, proteomics, and other high-throughput research approaches have also been applied in the study of wheat tolerance to varying biotic and abiotic stresses [29,30]. In most cases, proteins are the ultimate functional molecules; therefore, proteomics has become a powerful and promising tool for the study of plant stress response [31]. However, although many drought-responsive proteins have been characterized [32–37], most of them have not been functionally verified. One important reason is that functional verification by transgenic research in some polyploid crops, especially in hexaploid wheat, is time-consuming and not suitable for high-throughput studies. Virus-induced gene silencing (VIGS) is an alternative strategy for gene functional analysis by simultaneously knocking down the expression of multiple gene copies, which can overcome the inherent problems of polyploidy and limited transformation potential that hamper functional validation studies in hexaploid wheat [38,39]. Currently, it has been used for the analysis of gene function in resistance against wheat pathogen and wheat aphid, as well as cold and drought stress [39–43]. In this study, the proteome profile of a drought stress-tolerant variety XN979 and a drought stress-sensitive variety LA379 were obtained under control and drought stress conditions, using the iTRAQ-based proteome approach. A total of 335 DAPs specifically identified in drought tolerance variety XN979 were obtained through a comparative proteomic analysis. Two DAPs

(W5DYH0 and W5ERN8) were selected and dubbed *TaDrSR1* (drought stress response 1) and *TaDrSR2* (drought stress response 2), respectively, for further functional analysis using VIGS.

2. Results

2.1. Comparison of Drought Tolerance Between XN979 and LA379

As shown in Figure 1, 16 days after sowing, there were no significant phenotypic changes in drought-tolerant wheat varieties XN979 under drought stress (DS) conditions compared to no stress (NS) conditions. However, leaf drooping appeared in drought-sensitive wheat variety LA379 under drought stress conditions but did not under NS (Figure 1A). The RWC of LA379 dropped significantly by 40.39% but only dropped by 8.42% in XN979 (Figure 1B). As proline is an important osmotic regulator, we measured the content of proline in the leaves of both varieties under NS and DS conditions. Results showed that the content of proline significantly increased in both varieties: By 241.59% in XN979 under DS conditions relative to that under NS and 103.81% in LA379 (Figure 1C). MDA content and electrolyte leakage are important indices of membrane injury. The content of MDA in LA379 significantly increased by 169.64% under DS conditions relative to that under NS but only by 74.02% in XN979 (Figure 1D). Similarly, the electrolyte leakage was 73.05% higher than under NS in LA379 but only 20.05% higher than under NS in XN979 (Figure 1E).

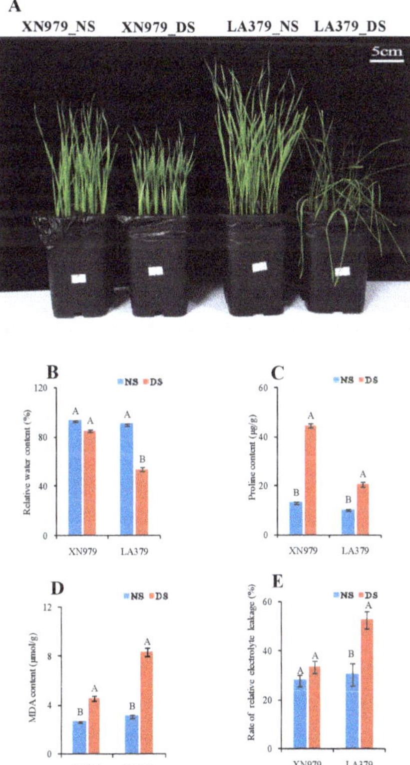

Figure 1. Phenotypic changes (**A**) and physiological responses (**B–E**) of XN979 and LA379 under drought stress (DS) and control (NS) conditions. (**A**) Phenotypic changes of XN979 and LA379 under drought stress (DS) and control (NS) conditions. (**B**) Relative water content (RWC); (**C**) Proline content; (**D**) malonaldehyde (MDA) content; (**E**) Rate of relative electrolyte leakage; NS, no stress; DS, drought stress. Three biological replicates were analyzed. Different letters indicate significant differences at $P \leq 0.01$ levels.

2.2. Protein Identification and DAPs Analysis

An iTRAQ-based comparative proteomic analysis was performed to identify wheat drought stress tolerance-related proteins. A total of 227076 spectra were generated and 38070 were matched to known spectra. These identified spectra were assigned to 17416 peptides, with 12126 unique peptides. Ultimately, 5369 proteins were identified (Figure 2A, Table S1). In the XN979_NS-XN979_DS comparison, a total of 482 proteins showed more than a 1.2-fold change ($P < 0.05$) in their respective abundances and were classified as DAPs (Table S2). Among the DAPs, 199 proteins were up-regulated and 283 proteins were down-regulated in XN979_DS compared to those in XN979_NS (Figure 2B,C). A comparison of LA379_NS and LA379_DS enabled the identification of 600 DAPs; these comprised 301 up-regulated and 299 down-regulated proteins (Figure 2B,C; Table S3). Among the DAPs identified above, 335 DAPs were exclusively identified in drought-tolerant variety XN979, but not in drought-sensitive variety XN979 (Figure 2B,C; Table S4). These 335 DAPs may be involved in the conferring drought stress tolerance in wheat.

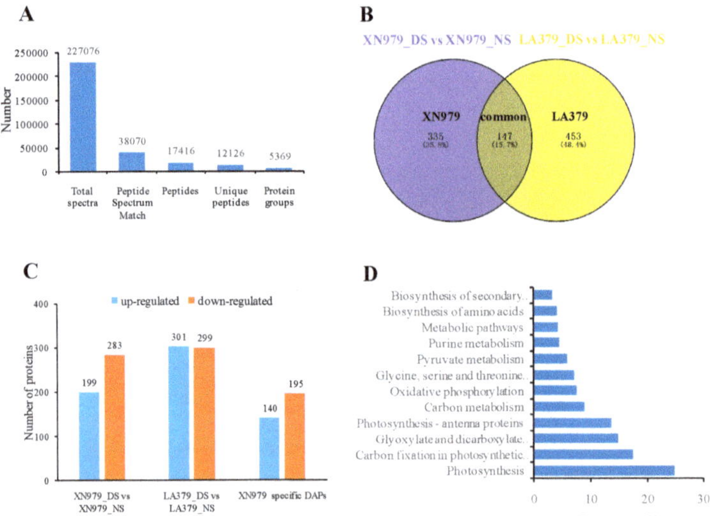

Figure 2. Statistical analysis of the proteome results and differentially abundant proteins (DAPs) under drought stress (DS) and control (NS) conditions. (**A**) Statistics for total spectra, matched spectra, matched peptides, unique peptides, and identified proteins. (**B**) Venn diagram analysis of DAPs in the one-to-one comparisons between NS and DS. (**C**) Number of up- and down-regulated DAPs in the XN979_DS-XN979_NS comparison, LA379_DS-LA379_NS, and drought-tolerant variety XN979 specific DAPs. (**D**) Enriched pathways of the DAPs specifically identified in XN979. The values on the abscissa indicate the percentage of the input number of DAPs among the number of the background proteins in the pathway; NS, no stress; DS, drought stress.

2.3. Pathway Analysis of the DAPs Involved in Drought Stress Tolerance

To explore the metabolic pathways that the present DAPs were involved in, the 335 proteins were further investigated using the web-based tool KOBAS 3.0. In total, 12 pathways were significantly enriched in the 335 DAPs (Table S5, Figure 2D). The top four significantly enriched pathways were photosynthesis (24.7%), carbon fixation (17.4%), glyoxylate and dicarboxylate metabolism (14.9%), and photosynthesis-antenna proteins (13.6%). Other pathways, such as carbon metabolism, oxidative phosphorylation, and oxidative phosphorylation, may also be involved in drought stress tolerance (Table S5, Figure 2D).

2.4. Real-Time PCR Verification

To verify the iTRAQ data and investigate the correlation of the abundance of proteins with their corresponding mRNA level, we randomly selected 4 XN979-specific DAPs (B1P766, W5DYH0, W5ERN8, and W5EI90), 5 LA379-specific DAPs (A0MA43, M8BCN0, M8BCR3, M8BTH4, and T1N5G8), and 3 common responsive proteins in both varieties (W5H6J0, V9QGR5, and Q8VYX1) for analysis of their RNA levels. *TaActin* and *TaGAPDH* were used as reference genes to normalize the expression level of target genes, respectively. Results showed that the expression levels of most DAPs were consistent with their protein expression levels (Figure 3). In most cases, the results of verification of expression using two different reference genes (*TaGAPDH* and *TaActin*) were similar. Only a few DAPs exhibited poor correlation between mRNA and protein expression levels (Figure 3).

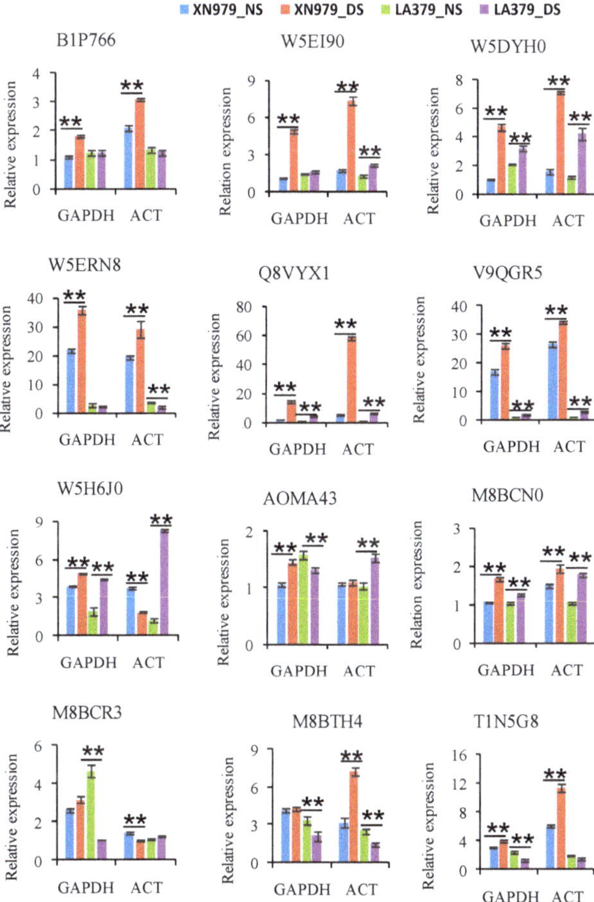

Figure 3. Relative mRNA expression analysis of twelve differentially abundant proteins (DAPs) using real-time PCR under drought stress (DS) and no stress (NS) conditions. Each bar shows the mean ± standard errors (SE) of three biological replicates. Two independent trials were conducted using *TaGAPDH* (*GAPDH*) and *TaActin* (*ACT*) as reference genes, respectively. The relative expression levels of each gene were calculated using the formula $2^{-\Delta\Delta Ct}$ (* $P \leq 0.05$; ** $P \leq 0.01$, Duncan's multiple range test).

2.5. VIGS of TaDrSR1 and TaDrSR2

As mentioned above, the DAPs that were exclusively identified in drought tolerance variety XN979 are most likely involved in conferring drought stress tolerance. Therefore, we selected two functional uncharacterized proteins (W5DYH0 and W5ERN8) from the four qRT-PCR verified, XN979-specific DAPs (B1P766, W5DYH0, W5ERN8, and W5EI90) for further functional analysis using VIGS technology. Here, the drought stress-tolerant variety XN979 and another Chinese wheat variety ZM9023 were used for viral infection in the VIGS experiment independently. Firstly, we investigated the expression levels of *TaDrSR1* and *TaDrSR2* in four independent BSMV$_{TaDrSR1}$- and BSMV$_{TaDrSR2}$-infected plants; results showed that the transcription levels of the *TaDrSR1* and *TaDrSR2* genes were significantly knocked down compared to those of the negative control (BSMV$_0$) and DS plants (Figure 4).

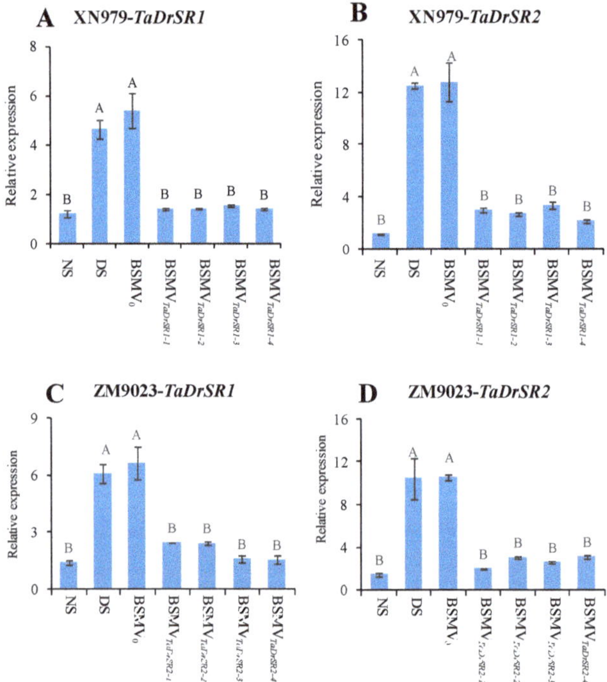

Figure 4. Detection of the expression levels of *TaDrSR1* and *TaDrSR2* in the corresponding knock-down plants. (**A**) The relative expression levels of the *TaDrSR1* gene in the *TaDrSR1*-knock-down plants of variety XN979. (**B**) The relative expression levels of the *TaDrSR2* gene in the *TaDrSR2*-knock-down plants of variety XN979. (**C**) The relative expression levels of the *TaDrSR1* gene in the *TaDrSR1*-knock-down plants of variety ZM9023. (**D**) The relative expression levels of the *TaDrSR2* gene in the *TaDrSR2*-knoc-down plants of variety ZM9023; NS, no stress; DS, drought stress. BSMV$_0$, negative control of the virus-induced gene silencing (VIGS) system; BSMV$_{TaDrSR1}$, *TaDrSR1*-knock-down plants; BSMV$_{TaDrSR2}$, *TaDrSR2*-knock-down plants. Four independent *TaDrSR1*- and *TaDrSR2*-knock-down plants were analyzed, respectively. Different letters indicate significant differences at $P \leq 0.01$ levels.

In XN979, ten days after infection, slight chlorosis was observed in all the BSMV construct-infected plants; this was attributed to the plant's response to viral infection. The BSMV$_{PDS}$-infected plants (positive control) exhibited bleached leaves (Figure 5A), indicating that the infection was successful [44]. Twenty days after infection, a substantial level of leaf drooping and wilting was observed in both BSMV$_{TaDrSR1}$- and BSMV$_{TaDrSR2}$-infected plants. However, no significant sagging and withering of the leaves was observed in plants infected with $BSMV_0$ (negative control) and DS plants (Figure 5B–F). The leaf RWC in both $BSMV_{TaDrSR1}$ and $BSMV_{TaDrSR2}$-infected plants reduced by 59.35% and 50.93% compared to that under NS but only reduced by 16.43% and 12.68% in the DS and $BSMV_0$-infected plants (Figure 6A). The relative electrolyte leakage rate in both $BSMV_{TaDrSR1}$- and $BSMV_{TaDrSR2}$-infected plants increased by 371.75% and 322.62% compared to that under NS, respectively, which was much higher than that in the DS and $BSMV_0$-infected plants (Figure 6B). MDA content in both $BSMV_{TaDrSR1}$- and $BSMV_{TaDrSR2}$-infected plants increased by 314.08% and 277.34% compared to that under NS respectively, which was significantly higher than that in the DS and $BSMV_0$-infected plants (Figure 6C). Proline content in both $BSMV_{TaDrSR1}$- and $BSMV_{TaDrSR2}$-infected plants increased by 114.89% and 98.43% compared to that under NS, respectively, which was significantly lower than that in the DS and $BSMV_0$-infected plants (Figure 6D). Similar results were obtained for ZM9023 (Figure 5G–K and Figure 6E–H).

Figure 5. The phenotypes of *TaDrSR1*- and *TaDrSR2*-knock-down plants. (**A**) Leaf; (**B–K**) Whole plants; XN979 was selected as the receptor for viral infection in the VIGS experiment; (**G–K**) ZM9023 was selected as the receptor for viral infection in the VIGS experiment. The pot on the left side of each picture represents the no stress (NS) treatment and the pot on the right side represents the drought stress treatment (DS). $BSMV_0$ represents the negative control of VIGS system; $BSMV_{PDS}$ represents the positive control monitoring time course of VIGS; $BSMV_{TaDrSR1}$ and $BSMV_{TaDrSR2}$ represent *TaDrSR1*- and *TaDrSR2*-knock-down plants, respectively.

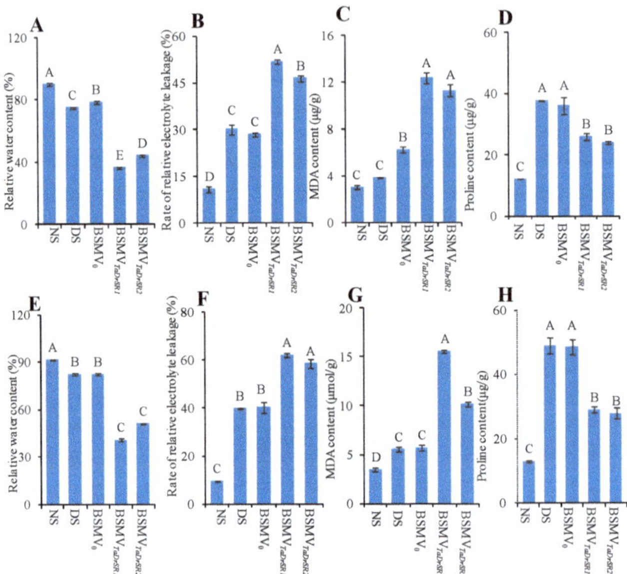

Figure 6. The changes in the physiological indices of the *TaDrSR1*- and *TaDrSR2*-knock-down plants. (**A–D**) XN979 was selected as the receptor for viral infection in the VIGS experiment; (**A**) relative water content; (**B**) rate of relative electrolyte leakage; (**C**) MDA content; (**D**) proline content. (**E–H**) ZM9023 was selected as receptor for viral infection in the VIGS experiment; (**E**) relative water content; (**F**) rate of relative electrolyte leakage; (**G**) MDA content; (**H**) proline content; NS, non-stressed plants; DS, drought-stressed plants; $BSMV_0$, negative control of the VIGS system; $BSMV_{TaDrSR1}$ and $BSMV_{TaDrSR2}$, *TaDrSR1*- and *TaDrSR2*-knock-down plants, respectively. Each bar shows the mean $\pm$ standard errors (SE) for three biological replicates. Different letters indicate significant differences at $P \leq 0.01$ levels.

3. Discussion

Drought stress occurs frequently worldwide and is a major limiting factor for plant growth and productivity. It has been reported that wheat can adapt in response to drought stress and exhibit significant differences in drought tolerance among different genotypes [45]. The elucidation of the molecular mechanisms and investigation of drought tolerance-related genes/proteins is crucial for the genetic improvement of plant drought tolerance.

Comparative proteomic analysis is a powerful tool for the study of plant stress response [31,46]. The changes in protein expression profiles under drought conditions have been investigated in several plants, including wheat, maize, rice, peanut, and soybean, and many drought-responsive proteins have been characterized [32–37]. However, despite this progress, most of these proteins have not been functionally verified. The molecular mechanism underlying plant drought tolerance is still far from fully elucidated, especially in hexaploid wheat. Here, a comparative proteomic analysis of two wheat varieties with contrasting drought tolerance enabled the identification of 335 DAPs specific to the drought-tolerant variety XN979, and 147 common responsive DAPs in both varieties (Table S4, Figure 2B). The XN979-specific drought-responsive proteins may participate in pathways mediating drought stress tolerance in wheat. Pathway enrichment analysis indicated that most of these proteins were involved in photosynthesis, carbon fixation, glyoxylate and dicarboxylate metabolism, and some other pathways (Figure 2D). In our study, 19 DAPs involved in the photosynthesis pathway were significantly enriched in drought-tolerant variety XN979. Previous studies have shown that the photosynthesis pathway is indeed affected by drought stress. The photosynthetic rate and the assimilation product are reduced, thereby further reducing the material basis of leaf growth [26,47–49]. The photosynthetic rate and the chlorophyll a content in a drought-tolerant variety SR3 were higher than in the leaves of its parent JN177. This higher capacity to maintain photosynthesis under stress may be achieved through a combination of a more robust cellular homeostasis and a more effective means of removing ROS and other toxic by-products [26]. Therefore, the 19 DAPs involved in the photosynthesis pathway may play important roles in drought tolerance in wheat. Other pathways, such as carbon fixation and photosynthesis-antenna proteins, are also associated with plant photosynthesis [50]. The other enriched pathways, such as carbon metabolism, pyruvate metabolism, amino acids metabolism, and oxidative phosphorylation, are mainly involved in the metabolism and allocation of carbohydrates. These results indicate that the regulation of the photosystem and carbohydrates metabolism play important roles in wheat drought tolerance.

Two DAPs (W5DYH0 and W5ERN8) were selected and dubbed *TaDrSR1* and *TaDrSR2*, respectively, for further functional analysis using VIGS technology. The ortholog of *TaDrSR1* in rice, r40c1, is up-accumulated in the roots of *DREB1A* transgenic plants, which may play an important role in the generation of drought-tolerant plants [51,52]. *TaDrSR2* encodes a functional unknown protein with 71 amino acid residues. Therefore, we know very little about the functions of both genes. We analyzed the chromosomal locations of *TaDrSR1* and *TaDrSR2* according to Yang et al. [53] and found that *TaDrSR1* and *TaDrSR2* were anchored on chromosome 4A and 2D, respectively. Interestingly, the chromosomal regions also have a number of QTLs for agronomic traits under drought stress conditions [54–56]. For example, a QTL for the root number under water-limited environments, *qRN.qgw-4A*, is linked with the *TaDrSR1* gene [54]. This chromosomal region also harbors QTLs for the drought stress susceptibility index (SSI) and drought stress tolerance (TOL) [56]. *TaDrSR2* is also linked with QTLs for plant height, spike length, spikelets per spike, and kernels per spike under drought stress conditions [55]. This evidence further confirms that the chromosomal regions in which *TaDrST1* and *TaDrST2* are located are important for wheat drought stress tolerance. As RWC, relative conductivity, MDA, and proline content are important physiological indicators reflecting plant drought tolerance [14,57–59], we measured these indices in the *TaDrSR1*- and *TaDrSR2*-knock-down plants and negative controls. Our results showed that the relative electrolyte leakage rate and MDA content significantly increased, while the RWC and proline content significantly decreased in the VIGS-mediated *TaDrSR1*- and *TaDrSR2*-knock-down plants compared to that in the controls (BSMV$_0$ and DS) under drought stress conditions (Figure 6). The *TaDrSR1*- and *TaDrSR2*-knock-down plants exhibited severe drooping and wilting phenotypes relative to the negative control (BSMV$_0$-infected plants) and non-infected plants under drought stress conditions, indicating that they were more sensitive to drought stress (Figure 5B–F). We performed the VIGS trial using another variety (ZM9023) as receptors for viral infection and obtained similar results (Figure 5G–K and Figure 6E–H). Previous studies have shown that the higher the MDA content of plants, the greater the level of membranous peroxidation and permeability. Further, the cytoplasm undergoes extravasation under drought stress, resulting in an increase in relative conductivity [14]. Under drought stress, proline reduces the osmotic potential, plays a role in osmotic regulation, and protects cell membrane structure; these changes represent an adaptive response of plants to adverse environments [60–62]. Our results, together with those of previous studies, strongly suggest that *TaDrSR1* and *TaDrSR2* potentially play vital roles in conferring drought tolerance in common wheat.

A comparative proteomic analysis approach can identify hundreds of wheat drought-tolerant protein candidates. For example, in this study, we identified 335 drought tolerance-related proteins, which were involved in 12 pathways (Figure 2D; Table S5). Some of them may play important roles, while others may only serve as an aid. The verification the function of these proteins is a big challenge in hexaploid wheat. VIGS has the advantages of simple operation and short test period [39,40]. It is suitable for preliminary functional verification on a large number of candidates. Therefore, comparative proteomic analysis combined with VIGS is a good choice for identifying novel genes for drought tolerance in wheat, although subsequent transgenic validation is necessary.

In summary, we identified 335 DAPs involved in wheat drought tolerance. Most of them were involved in the pathways of photosynthesis, carbon fixation, glyoxylate and dicarboxylate metabolism, carbon metabolism, and oxidative phosphorylation. Two DAPs *TaDrSR1* and *TaDrSR2* were selected for further functional analysis using VIGS. The *TaDrSR1*- and *TaDrSR2*-knock-down plants exhibited severe drooping and wilting phenotypes under drought stress conditions. Further physiological and molecular analyses indicated that these knock-down plants were more sensitive to drought stress, suggesting that *TaDrSR1* and *TaDrSR2* potentially play vital roles in conferring drought tolerance in common wheat. Our results also showed that comparative proteomic analysis combined with VIGS is an efficient way for identification of novel drought tolerance-related proteins in hexaploid wheat.

4. Materials and Methods

4.1. Plant Materials

XN979 and LA379, two wheat varieties with contrasting drought tolerance, were used as materials to conduct comparative proteomic analysis. XN979 exhibits far higher tolerance to drought stress than LA379. The drought stress-tolerant variety XN979 was used for viral infection in the VIGS experiment. A previous study reported that ZM9023 is susceptible to viral infection in VIGS trials [39]. To further confirm the results of VIGS in XN979, ZM9023 was also selected as a receptor for viral infection in this study.

4.2. Plant Growth Conditions and Sampling

Pot culture was used in this trial. Firstly, seeds were disinfected with 1% H_2O_2 for 10 min and then rinsed with distilled water for three times. Before sowing, substrate (Pindstrup Substrate Peat) and vermiculite were well mixed (volume ratio 1:2). Seeds were germinated for 16 h at 22 °C; then, 24 germinated seeds were sown in each pot with soil water content at 90% field capacity (FC). The pots were divided into two groups for two watering treatments, respectively: (i) A group of 20 pots (10 pots per cultivar) was grown under well-watered conditions (maintained at 80–90% FC, NS), and (ii) another 20 pots (10 pots per cultivar) were subjected to drought stress (DS, no watering after sowing). The temperature ranged from 22–23 °C in the daytime and 18–20 °C at night during the pot culture trial. Wheat plants were thinned to 20 plants per pot after emergence. Sixteen days after sowing (about 44% FC under DS conditions), fresh leaves of XN979 and LA379 under NS and DS conditions were sampled, respectively, and used for the measurement of physiological indices (proline, MDA content, and electrolyte leakage), with at least three biological replicates. The leaves of fifteen individual plants of XN979 and LA379 under NS and DS conditions were mixed respectively, fast-frozen in liquid nitrogen, and stored in a −80 °C freezer for protein extraction. This collection was repeated for the second biological replicate. The samples of two biological replicates of XN979 and LA379 under NS and DS conditions were named as XN979_NS-1, XN979_NS-2, LA379_NS-1, LA379_NS-2, XN979_DS-1, XN979_DS-2, LA379_DS-1, and LA379_DS-2, respectively. For RNA extraction and real-time PCR verification, three biological replicates were analyzed. Twenty days after sowing (about 40% FC under DS conditions), the leaves were collected for measurement of RWC.

4.3. The Measurement of Physiological Indices

The RWC, which was measured as described by Flexas et al. [63], was calculated based on the following formula: RWC (%) = [(FW − DW)/ (TW − DW)] × 100; FW represents fresh weight, TW refers to turgid weight, and DW represents dry weight. Electrolyte leakage was assayed according to Yan et al. [43]. Proline was extracted and determined according to the method of Bates et al. [64]. MDA content was measured following the methods of Hodgeset al. [65].

4.4. Protein Extraction, Digestion, and iTRAQ Labeling

Protein extraction was performed according to Thiellementet al. [66]. Protein digestion was performed using the FASP procedure [67]. Finally, a peptide mixture (100 µg) of each sample was labeled with iTRAQ reagents according to the manufacturer's instructions (Applied Biosystems,

USA). XN979_NS-1, XN979_NS-2, LA379_NS-1, LA379_NS-2, XN979_DS-1, XN979_DS-2, LA379_DS-1, and LA379_DS-2 were labeled with 113, 114, 115, 116, 117, 118, 119 and 121, respectively.

4.5. Strong Cationic Exchange (SCX) Fractionation and LC–ESI–MS/MS Analysis

iTRAQ labeled peptides were combined and dried under vacuum. Strong cationic exchange (SCX) chromatography was performed to fractionate the labeled peptides using the AKTA Purifier system (GE Healthcare) as previously described, with minor modifications [68,69]. In brief, the labeled peptide mixture was acidified using 10 mM KH_2PO_4 in 25% of ACN (buffer A, pH 3.0) and eluted at a flow rate of 1 mL/min with a gradient of 0–8% 500 mM KCl, 10 mM KH_2PO_4 in 25% of ACN (buffer B, pH 3.0) for 22 min, 8–52% buffer B for 22–47 min, 52%–100% buffer B for 47–50 min, 100% buffer B for 50–58 min, and buffer B was reset to 0% after 58 min. The elution was monitored by absorbance at 214 nm, and fractions were collected every 1 min. Samples were reconstituted with trifluoroacetic acid and stored at −80 ∘C for LC–MS/MS analysis.

Liquid chromatography–electrospray ionization tandem MS analysis was performed using a Q Exactive mass spectrometer coupled to an Easy nLC (Proxeon Biosystems) according to previously reported literature with minor modifications [68,69]. The peptide mixture was loaded onto a reverse phase trap column connected to the C18-reversed phase analytical column in 0.1% formic acid (buffer A) and separated with a linear gradient of 0.1% formic acid and 84% acetonitrile (buffer B) with a 300 nL/min flow rate. The linear gradient was 0–35% buffer B for 0–50 min, 35–100% buffer B for 5 min, and hold in 100% buffer B for another 5 min. The most abundant precursor ions were dynamically selected from the survey scan (300–1800 m/z) to acquire MS data. The instrument was run with the peptide recognition mode enabled [68,69].

4.6. Data Analysis

MS/MS spectra were searched using MASCOT engine (Matrix Science, version 2.2) embedded into Proteome Discoverer 1.4 and run against the UniProt_Poaceae database and the decoy database (http://www.uniprot.org). The Mascot search parameters were set according to previous studies [68,69]. Mascot search parameters were set as follows: Enzyme: Trypsin; max missed cleavage: 2; fixed modification: Carbamidomethyl (C), iTRAQ8plex(N-term), iTRAQ8plex(K); variable modification: Oxidation (M), iTRAQ8plex (Y); peptide mass tolerance: ± 20 ppm; fragment mass tolerance: 0.1 Da; peptide false discovery rate (FDR) $\leq$ 1%. All peptide ratios were normalized by the median protein ratio. The abundance of each protein was calculated as the median of unique peptides of the protein, and the fold change was defined based on the abundance of the protein under drought relative to their respective level under control. Protein species with an abundance ratio fold change of at least 1.2 and a P-value < 0.05 were defined as DAPs [70,71].

4.7. Pathway Enrichment Analysis

Pathway enrichment analysis was performed using a web-based tool KOBAS 3.0 (http://koba s.cbi.pku.edu.cn/anno_iden.php) [72]. The adjusted P-values with Benjamini–Hochberg correction under a threshold of 0.05 were considered to represent statistically significant differences.

4.8. Quantitative Real-Time PCR

Total RNA from the leaves was extracted using the total RNA kit (TaKaRa, Dalian, China). The Two-Step Prime-Script TM RT Reagent Kit (Perfect Real Time; TaKaRa) with gDNA Eraser was used for the reverse transcription reactions following the manufacturer's instructions. Twelve DAPs were randomly selected for RNA-level examination. The specific primers were designed using Primer 5.0 (Premier Biosoft, Palo Alto, CA, USA). The cDNA samples were used as templates and mixed with primers and SYBR Premix ExTaq (TaKaRa) for real-time PCR analysis. Real-time PCR was conducted using a BioRad IQ5 Real-time PCR Detection System (Bio-Rad, Hercules, CA, USA). The temperature settings were 95 °C for 5 min followed by 40 cycles of 95 °C for 15 s, 60 °C for 15 s, and 72 °C for 15 s.

TaActin and *TaGAPDH* were used as reference genes to normalize the expression level of target genes, respectively. Relative gene expression was computed using the $2^{-\Delta\Delta Ct}$ method [73]. All primers used in this study are listed in Table S6.

4.9. Vector Construction for VIGS

To further study the functions of DAPs, we constructed VIGS vectors of *TaDrSR1* and *TaDrSR2*. Firstly, 118 bp- and 107 bp-fragments of the coding regions of *TaDrSR1* and *TaDrSR2* were cloned, respectively. Vector construction was performed as previously described [74]. Linearized plasmids were used as templates to synthesize α, β, and γ RNAs of the BSMV genome with the RiboMAX TM Large Scale RNA Production System-T7 (Promega, Madison, WI, USA) [75]. In vitro transcripts of each RNA fragment were mixed in an equimolar ratio and added to triturated FES buffer [40]. Then, the mixture was diluted and treated with an equal volume of DEPC. Each of the silencing constructs consisted of BSMV α, β, and γ with the target gene insertion. The original $BSMV_0$ was constructed from α, β, and γ derived from the original empty pSL038-1 vector and used as the negative control. $BSMV_{PDS}$ (in which the wheat gene encoding phytoene desaturase, GenBank: FJ517553.1, was silenced) was used as a positive control of VIGS [41]. $BSMV_{TaDrSR1}$ was constructed from α, β, and γ with the insertion of the target gene *TaDrSR1*. $BSMV_{TaDrSR2}$ was constructed from α, β, and γ with the target gene *TaDrSR2*'s insertion.

4.10. Infection with VIGS Vectors

The drought stress-tolerant variety XN979 and virus-susceptible variety ZM9023 were selected as receptors for infection with VIGS vectors. Two-leaf-stage plants of XN979 and ZM9023 were used for infection, which was performed according to previously described procedures [40]. After the infection, the incubator temperature was set at 23 ± 2 °C, with darkness for 24 h, followed by a 16 h light/8 h dark photoperiod. Twenty days after the infection, the third and fourth leaves were collected and stored at -80 °C for measurement of physiological indices (proline and MDA content) and real-time PCR analysis. Twenty-four days after the infection, the leaves of the remaining plants were collected for the measurement of RWC and rate of relative electrolyte leakage.

4.11. Data Analysis

The means, standard errors (SE), and ranges of each measured morphological trait were analyzed using IBM SPSS statistics 21 software. For each morphological trait and gene expression level, $P = 0.05$ and $P = 0.01$ were used as thresholds to identify significant and extremely significant differences, respectively (Duncan's multiple range test). Venny 2.1 (a web-based tool) was used to generate Venn diagrams (http://bioinfogp.cnb.csic.es/tools/venny/index.html).

Supplementary Materials: Supplementary materials can be found at http://www.mdpi.com/1422-0067/19/12/4020/s1. All data related to this study have been deposited into the publicly accessible database iProX (ww.iprox.org) with identifier IPX0001326000/PXD011115.

Author Contributions: Conceptualization, Y.R. and T.L.; Data curation, X.W. and Y.R.; Formal analysis, X.W. and Y.R.; Investigation, X.W., Y.X., and J.L.; Methodology, X.W., Y.X., Z.W., and Z.X.; Project administration, Z.W. and Z.X.; Software, Y.X. and Y.R.; Supervision, T.L.; Validation, X.W.; Writing – original draft, X.W., Y.X., and Y.R.; Writing – review and editing, Y.R. and T.L.

Funding: This research was funded by the National Key Research and Development Program of China (2016YFD0300205), the Natural Science Foundation of Henan province (162300410133), and the State Key Laboratory Program (PCCE-KF-2017-04).

Acknowledgments: We thank Fen Chen (Henan Agricultural University) and Ning Zhang (Henan Agricultural University) for giving us the vectors for the VIGS experiment in this study.

Conflicts of Interest: The authors declare no conflict of interest.

Abbreviations

VIGS	Virus-induced gene silencing
DAPs	Differentially abundant proteins
MDA	Malonaldehyde
RWC	Relative water content
TaDrSR1	Drought stress response 1
TaDrSR2	Drought stress response 2
DS	Drought stress
NS	No stress
FC	Field capacity
SCX	Strong cationic exchange
BSMV	Barley stripe mosaic virus
FASP	Filter aided sample preparation
ACN	Acetonitrile

References

1. Ingram, J. A food systems approach to researching food security and its interactions with global environmental change. *Food Secur.* **2011**, *3*, 417–431. [CrossRef]
2. Rosegrant, M.W.; Ringler, C.; Zhu, T. Water for agriculture: Maintaining food security under growing scarcity. *Annu. Rev. Environ. Res.* **2009**, *34*, 205–222. [CrossRef]
3. Sinclair, T.R. Challenges in breeding for yield increase for drought. *Trends Plant Sci.* **2011**, *16*, 289–293. [CrossRef] [PubMed]
4. Jiao, X.; Lyu, Y.; Wu, X.; Li, H.; Cheng, L.; Zhang, C.; Yuan, L.; Jiang, B.; Rengel, Z.; Zhang, W.J.; et al. Grain production versus resource and environmental costs: Towards increasing sustainability of nutrient use in China. *J. Exp. Bot.* **2016**, *67*, 4935–4949. [CrossRef] [PubMed]
5. Thomas, L. Why Wheat Matters. International Maize and Wheat Improvement Center (CIMMYT). 2014. Available online: http://www.cimmyt.org (accessed on 14 August 2014).
6. Hu, R.; Xiao, J.; Gu, T.; Yu, X.; Zhang, Y.; Chang, J.; Yang, G.; He, G. Genome-wide identification and analysis of WD40 proteins in wheat (*Triticum aestivum* L.). *BMC Genom.* **2018**, *19*. [CrossRef]
7. Stallmann, J.; Schweiger, R.; Müller, C. Effects of continuous versus pulsed drought stress on physiology and growth of wheat. *Plant Biol.* **2018**, *20*, 1005–1013. [CrossRef] [PubMed]
8. Liu, Y.; Li, L.; Zhang, L.; Lv, Q.; Zhao, Y.; Li, X. Isolation and identification of wheat gene TaDIS1 encoding a RING finger domain protein, which negatively regulates drought stress tolerance in transgenic *Arabidopsis*. *Plant Sci.* **2018**, *275*, 49–59. [CrossRef] [PubMed]
9. Mutwali, N.I.A.; Mustafa, A.I.; Gorafi, Y.S.A.; Mohamed Ahmed, I.A. Effect of environment and genotypes on the physicochemical quality of the grains of newly developed wheat inbred lines. *Food Sci. Nutr.* **2016**, *4*, 508–520. [CrossRef] [PubMed]
10. Patanè, C.; Scordia, D.; Testa, G.; Cosentino, S.L. Physiological screening for drought tolerance in Mediterranean long-storage tomato. *Plant Sci.* **2016**, *249*, 25–34. [CrossRef]
11. Das, B.; Sahoo, R.N.; Pargal, S.; Krishna, G.; Verma, R.; Chinnusamy, V.; Sehgal, V.K.; Gupta, V.K. Comparison of different uni-and multi-variate techniques for monitoring leaf water status as an indicator of water-deficit stress in wheat through spectroscopy. *Biosyst. Eng.* **2017**, *160*, 69–83. [CrossRef]
12. Demirevska, K.; Simova-Stoliiva, L.; Vassileva, V.; Feller, U. Rubisco and some chaperone protein responses to water stress and rewatering at early seedling growth of drought sensitive and tolerant wheat varieties. *Plant Growth Regul.* **2008**, *56*, 97–106. [CrossRef]
13. Kocheva, K.V.; Landjeva, S.P.; Georgiev, G.I. Variation in ion leakage parameters of two wheat genotypes with different Rht-B1 alleles in response to drought. *J. Biosci.* **2014**, *39*, 753–759. [CrossRef] [PubMed]
14. Mirzaee, M.; Moieni, A.; Ghanati, F. Effects of drought stress on the lipid peroxidation and antioxidant enzyme activities in two canola (*Brassica napus* L.) cultivars. *J. Agric. Sci. Technol.* **2013**, *15*, 593–602.
15. Yang, C.W.; Lin, C.C.; Kao, C.H. Proline, ornithine, arginine and glutamic acid contents in detached rice leaves. *Biol. Plantarum* **2000**, *43*, 305–307. [CrossRef]

16. Lim, C.W.; Baek, W.; Lee, S.C. Roles of pepper bZIP protein CaDILZ1 and its interacting partner RING-type E3 ligase CaDSR1 in modulation of drought tolerance. *Plant J.* **2018**, *96*, 452–467. [CrossRef]

17. Zhang, L.C.; Zhao, G.Y.; Xia, C.; Jia, J.Z.; Liu, X.; Kong, X.Y. A wheat *R2R3-MYB* gene, *TaMYB30-B*, improved drought tolerance in transgenic *Arabidopsis*. *J. Exp. Bot.* **2012**, *63*, 5873–5885. [CrossRef] [PubMed]

18. Mao, X.; Chen, S.; Li, A.; Zhai, C.; Jing, R. Novel NAC transcription factor TaNAC67 confers enhanced multi-abiotic stress tolerances in *Arabidopsis*. *PLoS ONE* **2014**, *9*, e84359. [CrossRef] [PubMed]

19. Huang, Q.; Wang, Y.; Li, B.; Chang, J.; Chen, M.; Li, K.; Yang, G.; He, G. TaNAC29, a NAC transcription factor from wheat, enhances salt and drought tolerance in transgenic *Arabidopsis*. *BMC Plant Biol.* **2015**, *15*. [CrossRef]

20. Gahlaut, V.; Jaiswal, V.; Kumar, A.; Gupta, P.K. Transcription factors involved in drought tolerance and their possible role in developing drought tolerant cultivars with emphasis on wheat (*Triticum aestivum* L.). *Theor. Appl. Genet.* **2016**, *129*, 2019–2042. [CrossRef]

21. Xing, L.; Di, Z.; Yang, W.; Liu, J.; Li, M.; Wang, X.; Cui, C.; Wang, X.; Zhang, R.; Xiao, J. Overexpression of *ERF1-V* from *Haynaldia villosa* can enhance the resistance of wheat to powdery mildew and increase the tolerance to salt and drought stresses. *Front. Plant Sci.* **2017**, *8*. [CrossRef]

22. Xu, Z.S.; Xia, L.Q.; Chen, M.; Cheng, X.G.; Zhang, R.Y.; Li, L.C.; Zhao, Y.X.; Lu, Y.; Ni, Z.Y.; Liu, L. Isolation and molecular characterization of the *Triticum aestivum* L. Ethylene-responsive factor 1(*TaERF1*) that increases multiple stress tolerance. *Plant Mol. Biol.* **2007**, *65*, 719–732. [CrossRef] [PubMed]

23. Zhang, L.C.; Zhao, G.Y.; Jia, J.Z.; Liu, X.; Kong, X.Y. Molecular characterization of 60 isolated wheat *MYB* genes and analysis of their expression during abiotic stress. *J. Exp. Bot.* **2012**, *63*, 203–214. [CrossRef] [PubMed]

24. Xue, G.P.; Way, H.M.; Richardson, T.; Drenth, J.; Joyce, P.A.; Mclntyre, C.L. Overexpression of *TaNAC69* leads to enhanced transcript levels of stress up-regulated genes and dehydration tolerance in bread wheat. *Mol. Plant* **2011**, *4*, 697–712. [CrossRef] [PubMed]

25. Baloglu, M.C.; Oz, M.T.; Oktem, H.A.; Yucel, M. Expression analysis of *TaNAC69-1* and *TtNAMB-2*, wheat NAC family transcription factor genes under abiotic stress conditions in durum wheat (*Triticum turgidum*). *Plant Mol. Biol. Rep.* **2012**, *30*, 1246–1252. [CrossRef]

26. Peng, Z.; Wang, M.; Li, F.; Lv, H.; Li, C.; Xia, G. A proteomic study of the response to salinity and drought stress in an introgression strain of bread wheat. *Mol. Cell Proteom.* **2009**, *8*, 2676–2686. [CrossRef]

27. Singh, S.; Gupta, A.K.; Kaur, N. Differential responses of antixoxidation defence system to long-term filed drought in wheat (*Triticum aestivum* L.) genotypes differing in drought tolerance. *J. Agron. Crop Sci.* **2012**, *198*, 185–195. [CrossRef]

28. Ouyang, W.J.; Struik, P.C.; Yin, X.Y.; Yang, J.C. Stomatal conductance, mesophyll conductance, and transpiration efficiency in relation to leaf anatomy in rice and wheat genotypes under drought. *J. Exp. Bot.* **2017**, *68*, 5191–5205. [CrossRef]

29. Giuliani, M.M.; Palermo, C.; De Santis, M.A.; Mentana, A.; Pompa, M.; Giuzio, L.; Masci, S.; Centonze, D.; Flagella, Z. Differential Expression of Durum Wheat Gluten Proteome under Water Stress during Grain Filling. *J. Agric. Food Chem.* **2015**, *63*, 6501–6512. [CrossRef]

30. Liu, Z.S.; Xin, M.M.; Qin, J.X.; Peng, H.R.; Ni, Z.F.; Yao, Y.Y.; Sun, Q.X. Temporal transcriptome profiling reveals expression partitioning of homeologous genes contributing to heat and drought acclimation in wheat (*Triticum aestivum* L.). *BMC Plant Biol.* **2015**, *15*. [CrossRef]

31. Kosová, K.; Vítámvás, P.; Urban, M.O.; Prášil, I.T.; Renaut, J. Plant abiotic stress proteomics: The major factors determining alterations in cellular proteome. *Front. Plant Sci.* **2018**, *9*. [CrossRef]

32. Hajheidari, M.; Eivazi, A.; Buchanan, B.B.; Wong, J.H.; Majidi, I.; Salekdeh, G.H. Proteomics uncovers a role for redoxin drought tolerance in wheat. *J. Proteome Res.* **2007**, *6*, 1451–1460. [CrossRef] [PubMed]

33. Govind, G.; ThammeGowda, H.V.; Kalaiarasi, P.J.; Iyer, D.R.; Muthappa, S.K.; Nese, S.; Makarla, U.K. Identification and functional validation of a unique set of drought induced genes preferentially expressed in response to gradual water stress in peanut. *Mol. Genet. Genom.* **2009**, *281*, 591–605. [CrossRef] [PubMed]

34. Krishnan, H.B.; Nelson, R.L. Proteomic Analysis of High Protein Soybean (*Glycine max*) Accessions Demonstrates the Contribution of Novel Glycinin Subunits. *J. Agric. Food Chem.* **2011**, *59*, 2432–2439. [CrossRef] [PubMed]

35. Zhao, F.Y.; Zhang, D.Y.; Zhao, Y.L.; Wang, W.; Yang, H.; Tai, F.J. The difference of physiological and proteomic changes in maize leaves adaptation to drought, heat, and Combined Both Stresses. *Front. Plant Sci.* **2016**, *7*. [CrossRef] [PubMed]

36. Nutwadee, C.; Maiporn, N.; Narumon, P.; Michael, V.M.; Sittiruk, R.; Supachitra, C. Proteomic analysis of drought-responsive proteins in rice reveals photosynthesis-related adaptations to drought stress. *Acta Physiol. Plant* **2017**, *39*. [CrossRef]

37. Wang, X.; Komatsu, S. Proteomic approaches to uncover the flooding and drought stress response mechanisms in soybean. *J. Proteom.* **2018**, *172*, 201–215. [CrossRef] [PubMed]

38. Manmathan, H.; Shaner, D.; Snelling, J.; Tisserat, N.; Lapitan, N. Virus-induced gene silencing of *Arabidopsis thaliana* gene homologues in wheat identifies genes conferring improved drought tolerance. *J. Exp. Bot.* **2013**, *64*, 1381–1392. [CrossRef] [PubMed]

39. Zhang, N.; Huo, W.; Zhang, L.R.; Chen, F.; Cui, D.Q. Identification of winter-responsive proteins in bread wheat using proteomics analysis and virus-Induced gene silencing (VIGS). *Mol. Cell. Proteom.* **2016**, *15*, 2954–2969. [CrossRef] [PubMed]

40. Scofield, S.R.; Huang, L.; Brandt, A.S.; Gill, B.S. Development of a virus-induced gene-silencing system for hexaploid wheat and its use in functional analysis of the Lr21-mediated leaf rust resistance pathway. *Plant Physiol.* **2005**, *138*, 2165–2173. [CrossRef]

41. Zhou, H.; Li, S.; Deng, Z.; Wang, X.; Chen, T.; Zhang, J.; Chen, S.; Ling, A.; Wang, D.; Zhang, X. Molecular analysis of three new receptor-like kinase genes from hexaploid wheat and evidence for their participation in the wheat hypersensitive response to stripe rust fungus infection. *Plant J.* **2007**, *52*, 420–434. [CrossRef]

42. Cloutier, S.; McCallum, B.; Loutre, C.; Banks, T.; Wicker, T.; Feuillet, C.; Keller, B.; Jordan, M. Leaf rust resistance gene Lr1, isolated from bread wheat (*Triticum aestivum* L.) is a member of the large psr567 gene family. *Plant Mol. Biol.* **2007**, *65*, 93–106. [CrossRef] [PubMed]

43. Van Eck, L.; Schultz, T.; Leach, J.E.; Scofield, S.R.; Peairs, F.B.; Botha, A.M.; Lapitan, N.L. Virus-induced gene silencing of *WRKY53* and an inducible phenylalanine ammonia-lyase in wheat reduces aphid resistance. *Plant Biotechnol. J.* **2010**, *8*, 1023–1032. [CrossRef]

44. Yan, SP.; Zhang, Q.Y.; Tang, Z.C.; Sun, W.N. Comparative proteomic analysis provides new insights into chilling stress responses in rice. *Mol. Cell. Proteom.* **2006**, *5*, 484–496. [CrossRef] [PubMed]

45. Aprile, A.; Havlickova, L.; Panna, R.; Marè, C.; Borrelli, G.M.; Marone, D.; Perrotta, C.; Rampino, P.; Bellis, L.D.; Curn, V.; et al. Different stress responsive strategies to drought and heat in two durum wheat cultivars with contrasting water use efficiency. *BMC Genom.* **2013**, *14*, 14. [CrossRef] [PubMed]

46. Zadražnik, T.; Egge-Jacobsen, W.; Meglič, V.; Šuštar-Vozlič, J. Proteomic analysis of common bean stem under drought stress using in-gel stable isotope labeling. *J. Plant Physiol.* **2017**, *209*, 42–45. [CrossRef] [PubMed]

47. Ergen, N.Z.; Budak, H. Sequencing over 13000 expressed sequence tags from six subtractive cDNA libraries of wild and modern wheats following slow drought stress. *Plant Cell Environ.* **2009**, *32*, 220–236. [CrossRef]

48. Ergen, N.Z.; Thimmapuram, J.; Bohnert, H.J.; Budak, H. Transcriptome pathways unique to dehydration tolerant relatives of modern wheat. *Funct. Integr. Genom.* **2009**, *9*, 377–396. [CrossRef]

49. Kantar, M.; Lucas, S.J.; Budak, H. miRNA expression patterns of *Triticum dicoccoides* in response to shock drought stress. *Planta* **2011**, *233*, 471–484. [CrossRef]

50. Kuthanová Trsková, E.; Belgio, E.; Yeates, A.M.; Sobotka, R.; Ruban, A.V.; Kana, R. Antenna proton sensitivity determines photosynthetic light harvesting strategy. *J. Exp. Bot.* **2018**, *69*, 4483–4493. [CrossRef]

51. Paul, S.; Gayen, D.; Datta, S.K.; Datta, K. Dissecting root proteome of transgenic rice cultivars unravels metabolic alterations and accumulation of novel stress responsive proteins under drought stress. *Plant Sci.* **2015**, *234*, 133–143. [CrossRef]

52. Kumar, A.; Bimolata, W.; Kannan, M.; Kirti, P.B.; Qureshi, I.A.; Ghazi, I.A. Comparative proteomics reveals differential induction of both biotic and abiotic stress response associated proteins in rice during *Xanthomonas oryzae* pv. *oryzae* infection. *Funt. Integr. Genom.* **2015**, *15*, 425–437. [CrossRef] [PubMed]

53. Yang, M.; Gao, X.; Dong, J.; Gandhi, N.; Cai, H.; von Wettstein, D.H.; Rustgi, S.; Wen, S. Pattern of protein expression in developing wheat grains identified through proteomic analysis. *Front. Plant Sci.* **2017**, *8*. [CrossRef] [PubMed]

54. Christopher, J.; Christopher, M.; Jennings, R.; Jones, S.; Fletcher, S.; Borrell, A.; Manschadi, A.M.; Jordan, D.; Mace, E.; Hammer, G. QTL for root angle and number in population developed from bread wheats (*Triticum aestivum*) with contrasting adaptation to water-limited environments. *Theor. Appl. Genet.* **2013**, *126*, 1563–1574. [CrossRef] [PubMed]

55. Mwadzingeni, L.; Shimelis, H.; Rees, D.J.; Tsilo, T.J. Genome-wide association analysis of agronomic traits in wheat under drought-stressed and non-stressed conditions. *PLoS ONE* **2017**, *12*. [CrossRef] [PubMed]

56. Sukumaran, S.; Reynolds, M.P.; Sansaloni, C. Genome-wide association analyses identify QTL hotspots for yield and component traits in durum wheat grown under yield potential, drought, and heat stress environments. *Front. Plant Sci.* **2018**, *9*. [CrossRef] [PubMed]

57. Ardestani, A.; Yazdanparast, R. Antioxidant and free radical scavenging potential of Achillea santolina extracts. *Food Chem.* **2007**, *104*, 21–29. [CrossRef]

58. Isaakcara, K.; Petkaujay, C.; Karmin, O.; Ominski, Kim.; Carlos, R.L.; Siowyaw, L. Seasonal variations in phenolic compounds and antioxidant capacity of Cornus stolonifera plant material: Applications in agriculture. *Can. J. Plant Sci.* **2013**, *93*, 725–734.

59. Rezayian, M.; Niknam, V.; Ebrahimzadeh, H. Improving tolerance against drought in canola by penconazole and calcium. *Pestic. Biochem. Physiol.* **2018**, *149*, 123–136. [CrossRef]

60. Iqbal, M.J.; Maqsood, Y.; Abdin, Z.U.; Manzoor, A.; Hassan, M.; Jamil, A. SSR Markers associated with proline in drought tolerant wheat germplasm. *Appl. Biochem. Biotechnol.* **2016**, *178*, 1042–1052.

61. Mwadzingeni, L.; Shimelis, H.; Tesfay, S.; Tsilo, T.J. Screening of bread wheat genotypes for drought tolerance using phenotypic and proline analyses. *Front. Plant Sci.* **2016**, *7*. [CrossRef]

62. Rai, A.C.; Singh, M.; Shah, K. Effect of water withdrawal on formation of free radical, proline accumulation and activities of antioxidant enzymes in ZAT12-transformed transgenic tomato plants. *Plant Physiol. Biochem.* **2012**, *61*, 108–114.

63. Flexas, J.; Ribas-Carbó, M.; Bota, J.; Galmés, J.; Henkle, M.; Martínez-Cañellas, S.; Medrano, H. Decreased Rubisco activity during water stress is not induced by decreased relative water content but related to conditions of low stomatal conductance and chloroplast CO2 concentration. *New Phytol.* **2006**, *172*, 73–82. [CrossRef]

64. Bates, L.S.; Waldren, R.P.; Teare, I.D. Rapid determination of free proline for water stress studies. *Plant Soil* **1973**, *39*, 205–207. [CrossRef]

65. Hodges, D.M.; DeLong, J.M.; Forney, C.F.; Prange, R.K. Improving the thiobarbituric acid-reactive-substances assay for estimating lipid peroxidation in plant tissues containing anthocyanin and other interfering compounds. *Planta* **1999**, *207*, 604–611. [CrossRef]

66. Thiellement, H.; Zivy, M.; Damerval, C.; Mechin, V. *Plant Proteomics: Methods and Protocols*; Springer: Secaucus, NJ, USA, 2007; pp. 1–8. ISBN 978-1-59745-227-4.

67. Wisniewski, J.R.; Zougman, A.; Nagaraj, N.; Mann, M. Universal sample preparation method for proteome analysis. *Nat. Methods* **2009**, *6*, 359–362. [CrossRef]

68. Hu, X.L.; Li, N.N.; Wu, L.J.; Li, C.Q.; Li, C.H.; Zhang, L.; Liu, T.X.; Wang, W. Quantitative iTRAQ-based proteomic analysis of phosphoproteins and ABA regulated phosphoproteins in maize leaves under osmotic stress. *Sci. Rep.* **2015**, *5*. [CrossRef] [PubMed]

69. Zhao, Y.L.; Wang, Y.K.; Yang, H.; Wang, W.; Wu, J.Y.; Hu, X.L. Quantitative proteomic analyses identify ABA-related proteins and signal pathways in maize leaves under drought conditions. *Front. Plant Sci.* **2016**, *7*. [CrossRef]

70. Chu, P.; Yan, G.X.; Yang, Q.; Zhai, L.N.; Zhang, C.; Zhang, F.Q.; Guan, R.Z. iTRAQ-based quantitative proteomics analysis of Brassica napus leaves reveals pathways associated with chlorophyll deficiency. *J. Proteom.* **2015**, *113*, 244–259. [CrossRef]

71. Chen, Y.Y.; Fu, X.M.; Mei, X.; Zhou, Y.; Cheng, S.H.; Zeng, L.T.; Dong, F. Proteolysis of chloroplast proteins is responsible for accumulation of free amino acids in dark-treated tea (*Camellia sinensis*) leaves. *J. Proteom.* **2017**, *157*, 10–17. [CrossRef]

72. Xie, C.; Mao, X.; Huang, J.; Ding, Y.; Wu, J.; Dong, S.; Kong, L.; Gao, G.; Li, C.; Wei, L. KOBAS 2.0: A web server for annotation and identification of enriched pathways and diseases. *Nucleic Acids Res.* **2011**, *39*, 316–322. [CrossRef]

73. Livak, K.J.; Schmittgen, T.D. Analysis of relative gene expression data using real-time quantitative PCR and the $2^{-\Delta\Delta Ct}$ Method. *Methods* **2001**, *25*, 402–408. [CrossRef] [PubMed]

74. Cakir, C.; Scofield, S. Evaluating the ability of the barley stripe mosaic virus-induced gene silencing system to simultaneously silence two wheat genes. *Cereal Res. Commun.* **2008**, *36*, 217–222. [CrossRef]

75. Petty, I.T.; Hunter, B.G.; Wei, N.; Jackson, A.O. Infectious barley stripe mosaic virus RNA transcribed in vitro from full-length genomic cDNA clones. *Virology* **1989**, *171*, 342–349. [CrossRef]

International Journal of
Molecular Sciences

Article

Proteomic Landscape of the Mature Roots in a Rubber-Producing Grass *Taraxacum Kok-saghyz*

Quanliang Xie [1,2], Guohua Ding [2], Liping Zhu [1,2], Li Yu [1,2], Boxuan Yuan [1,2], Xuan Gao [1], Dan Wang [3], Yong Sun [3], Yang Liu [2], Hongbin Li [1,*] and Xuchu Wang [1,2,*]

[1] Key Laboratory of Xinjiang Phytomedicine Resource and Utilization of Ministry of Education, College of Life Sciences, Shihezi University, Shihezi 832003, China; xiequanliang001@163.com (Q.X.); zhuliping0903@163.com (L.Z.); yulixjnu@163.com (L.Y.); yuanboxuan111@163.com (B.Y.); gaoxuan850419@163.com (X.G.)

[2] Key Laboratory for Ecology of Tropical Islands, Ministry of Education, College of Life Sciences, Hainan Normal University, Haikou 571158, Hainan, China; dingguohuasw@163.com (G.D.); y1yang1y@163.com (Y.L.)

[3] Institute of Tropical Biosciences and Biotechnology, Chinese Academy of Tropical Agricultural Sciences, Haikou 571101, Hainan, China; wangdanqz2009@126.com (D.W.); sunyong_03119308@126.com (Y.S.)

* Correspondence: lihb@shzu.edu.cn (H.L.); xchwang@hainnu.edu.cn (X.W.); Tel.: +86-898-65891065 (H.L.); Fax: +86-898-65891065 (X.W.)

Received: 8 May 2019; Accepted: 24 May 2019; Published: 27 May 2019

Abstract: The rubber grass *Taraxacum kok-saghyz* (TKS) contains large amounts of natural rubber (cis-1,4-polyisoprene) in its enlarged roots and it is an alternative crop source of natural rubber. Natural rubber biosynthesis (NRB) and storage in the mature roots of TKS is a cascade process involving many genes, proteins and their cofactors. The TKS genome has just been annotated and many NRB-related genes have been determined. However, there is limited knowledge about the protein regulation mechanism for NRB in TKS roots. We identified 371 protein species from the mature roots of TKS by combining two-dimensional gel electrophoresis (2-DE) and mass spectrometry (MS). Meanwhile, a large-scale shotgun analysis of proteins in TKS roots at the enlargement stage was performed, and 3545 individual proteins were determined. Subsequently, all identified proteins from 2-DE gel and shotgun MS in TKS roots were subject to gene ontology and Kyoto Encyclopedia of Genes and Genomes (KEGG) enrichment analyses and most proteins were involved in carbon metabolic process with catalytic activity in membrane-bounded organelles, followed by proteins with binding ability, transportation and phenylpropanoid biosynthesis activities. Fifty-eight NRB-related proteins, including eight small rubber particle protein (SRPP) and two rubber elongation factor(REF) members, were identified from the TKS roots, and these proteins were involved in both mevalonate acid (MVA) and methylerythritol phosphate (MEP) pathways. To our best knowledge, it is the first high-resolution draft proteome map of the mature TKS roots. Our proteomics of TKS roots revealed both MVA and MEP pathways are important for NRB, and SRPP might be more important than REF for NRB in TKS roots. These findings would not only deepen our understanding of the TKS root proteome, but also provide new evidence on the roles of these NRB-related proteins in the mature TKS roots.

Keywords: natural rubber biosynthesis; mass spectrometry; rubber grass; rubber latex; shotgun proteomics; *Taraxacum kok-saghyz*; two-dimensional gel electrophoresis; visual proteome map

1. Introduction

Natural rubber (NR, cis-1,4-polyisoprene) is a biopolymer with high economic value and it is widely used as a strategic raw material in more than 40,000 products [1,2]. More than 2500 plant species can biosynthesize NR [3,4]. However, high-quality natural rubber in viable quantities is only observed in a few plant species, such as the *Para* rubber tree *Hevea brasiliensis* [5,6], Russian dandelion *Taraxacum kok-saghyz* Rodin (TKS, also called the rubber grass) or its close species *Taraxacum brevicorniculatum* [7], *Eucommia ulmoides Oliver* and a guayule shrub *Parthenium argentatum* [8,9]. Currently, the *Para* rubber tree is nearly the only commercial plant to cultivate the exclusive source of NR [5,6]. However, rubber production of the *Para* rubber tree has reached the limit due to the strict climatic requirements for its planting areas [10], little genetic variability, labor cost, and threats of fatal fungal plant diseases [1]. Therefore, it is critical to find an alternative source and a model plant for NR production.

TKS has drawn special attention since the 1940s. It belongs to the Compositae family and originates from the Tekes River Basin near Tianshan Mountain border between Kazakhstan and China [11]. TKS root contains NR ranging from 3% to 28% of total dry weight [1] and its rubber property and molecular characteristics are similar with NR in the *Para* rubber tree [12,13]. TKS grows widely in temperate and cold areas, and its root also contains about 28% of inulin, which is an important material for bioethanol and the food industry [14,15]. With its advantages of high rubber content and quality, wide planting area, a relatively short life cycle, ease of transformation and harvesting method [14,15], TKS is becoming a promising crop for natural rubber production. In addition, TKS is a perennial herb with a relatively simple genome (1.29 Gb), containing 46,731 protein-coding genes [16]. It is easily manipulated for genetic transformation and can be used as an ideal model plant for studying the rubber biosynthesis mechanism [17].

NR is biosynthesized by two pathways: the mevalonate acid (MVA) pathway [18,19] and the methylerythritol phosphate (MEP) pathway [20]. Recently, to determine the protein-based regulation mechanism of NRB, several proteomics have been conducted on the total latex [21–25] and rubber particles [26–31] of the rubber tree *H. brasiliensis*. But no proteomic analysis has been performed on TKS, and only one study on determining proteins from the rubber phase and pellet phase of the latex from *T. brevicorniculatum* (TBR) was reported and 278 unique proteins from the one- and two-dimensional gel electrophoresis (1-DE and 2-DE) gels were identified [32]. In order to gain further insight into the protein-based regulation mechanism of rubber biosynthesis, we performed a comprehensive proteomics analysis of the mature roots of TKS at six months (6M) by 2-DE and mass spectrometry (MS), and then identified thousands of proteins by a large-scale shotgun method. These results may deepen our understanding of the root proteome and provide valuable gene candidates for genetic improvement of TKS breeding as a commercial alternative crop.

2. Results

2.1. Morphological Observation of Rubber Particles in the Main Roots of TKS

Natural rubber in TKS is produced from the latex system of its enlarged roots, and the biomass of the main roots is important for rubber yield. Therefore, the mature 6-month TKS plants with flowers were selected to perform morphological analysis, and the enlarged main roots are highlighted (Figure 1A). The light microscope results revealed that the concentric rings of laticifers as specialized tubular vessels in roots can be detected, and the total number of laticifer vessels is about 137 ± 32 ($n = 20$) in the main roots of 6M TKS. The brown rubber latex granules, known as the specific cytoplasm of laticifer cells, can be clearly detected in the areas of endodermis and cortex in phloem tissues from both the crosscutting and slitting slices of the main roots. All the laticifer cells were considered to originate from the initial cells of vascular cambium to secondary laticifers in paralleling rings (Figure 1B), and most of these laticifer cells seem to connect into tubular vessels in the main roots (Figure 1C). Ultrastructural investigation of rubber particles in TKS root at 6 months was conducted under a transmission electron microscope (TEM). Many spherical or ovoid-ellipsoid rubber particles, ranging from 50 nm to 2000 nm, were examined in the laticifer cells (Figure 1D). We noticed that most rubber particles have a diameter less than 200 nm; they are termed small rubber particles (SRPs). However, rubber particles with diameter larger than 400 nm are fewer in number; they are traditionally called large rubber particles (LRPs). These particles, including SRPs and LRPs, are surrounded by a monolayer membrane (Figure 1D), in which many enzymes related to NRB are anchored or combined with each other [33].

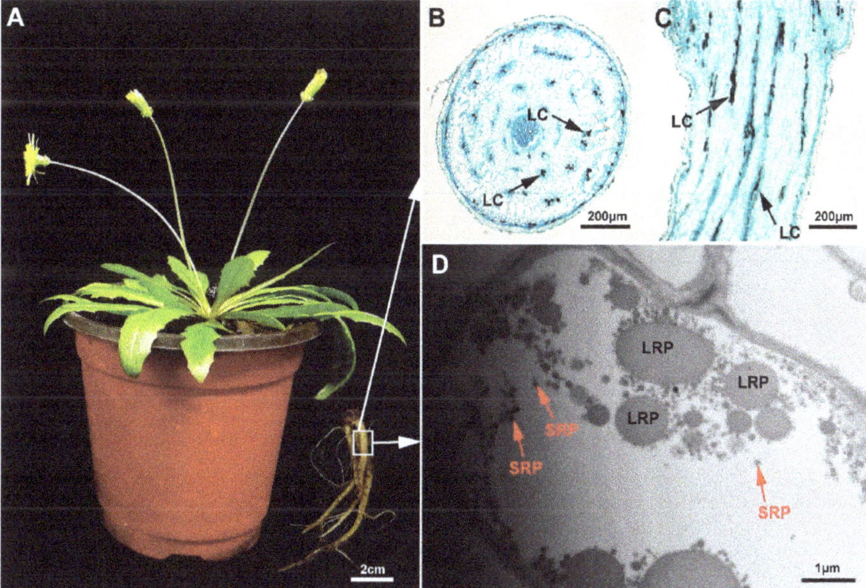

Figure 1. Morphological observation of the laticifer cells and rubber particles in the main roots of *Taraxacum kok-saghyz* (TKS). A six months old TKS plant and its main roots are highlighted (**A**). The laticifer cells containing brown rubber can be examined under the crosscutting (**B**) and slitting (**C**) slices of the 6M TKS roots. Black arrows indicate the positions of typical laticifer cells. Both small rubber particles (SRP) and large rubber particles (LRP) in the main roots of 6M TKS can be observed under a transmission electron microscope (TEM) (**D**).

2.2. Establishment of a Visual Two-Dimensional Gel Electrophoresis (2-DE) Proteome Map and Mass Spectrometry (MS) Identification of High Abundance Proteins in Taraxacum Kok-saghyz (TKS) Roots

To further obtain the protein accumulation profiling, the total proteins from the 6M TKS roots were extracted and performed 2-DE to separate the total proteins. Finally, a high-resolution 2-DE gel was obtained and a visual proteome map was established (Figure 2). In this reference 2-DE gel, 428 ± 15 protein spots, ranged from pH 4 to pH 7, were detected from three biological replicates (Figure S1). We manually excised all abundant protein spots from the 2-DE gel, and finally identified 371 protein species by MALDI TOF/TOF MS. These proteins occupied 84.38% volume of all the detected spots in the 2-DE gel and belonged to 231 gene products or named unique proteins (Table S1).

We compared the theoretical and experimental ratios of molecular weight (Mr) and isoelectric point (pI) of the 371 proteins identified from the 2-DE gel, and presented their ratios as radar axis labels (Mr for radial value; pI ratio for annular value) in radial chart (Figure 3A). Most proteins showed similar radial and annular values, and their ratios for theoretical and experimental Mr and pI are near line 1.0. However, some proteins demonstrated differentially experimental Mr and pI values in the 2-DE gel (Figure 2). In the proteomics study based on 2-DE gel, they are called different protein isoforms or protein species. In this study, 90 unique proteins were identified from at least two different spots, and these proteins contained 229 spots in the 2-DE gel. Among them, 56 proteins were identified from 112 spots, and they contained two protein isoforms in the 2-DE gel. Twenty-two proteins had three isoforms, and nine proteins were determined from 36 spots with four isoforms (Table S2). There are three unique proteins containing five isoforms; they are aldolase (spots 305, 313, 314, 403 and 405), enolase (spots 94, 100, 106, 150 and 158) and cytosolic glyceraldehyde-3-phosphate dehydrogenase (GAPC, spots 12, 163, 176, 272 and 275).

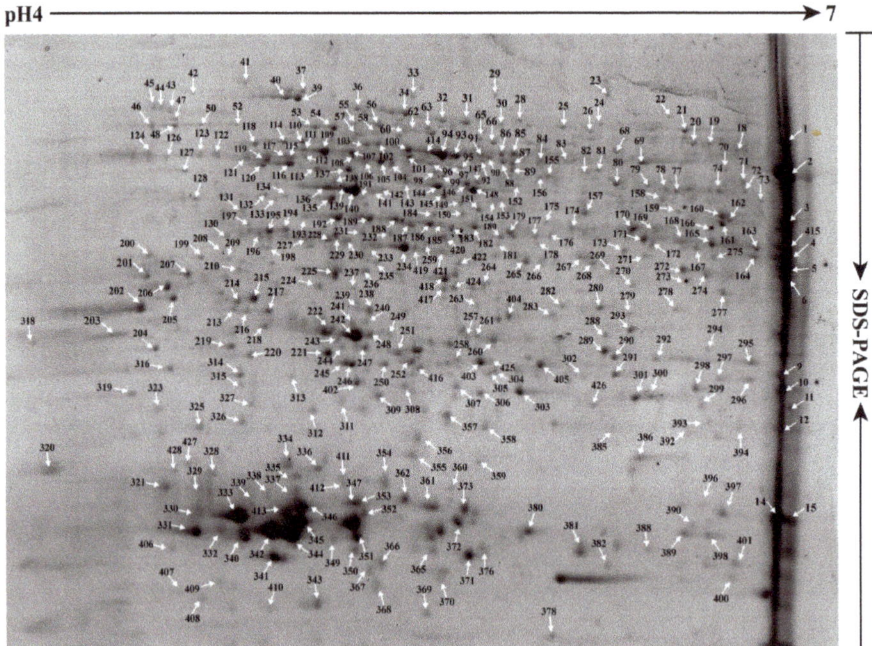

Figure 2. The reference proteome map and MS identification of abundant proteins in 2-DE gels of the 6M TKS roots. The proteins extracted from the 6M mature TKS roots were examined with two-dimensional gel electrophoresis (2-DE) to produce a high-resolution reference proteome map. The 371 major protein spots marked with numbers were positively identified by MALDI TOF/TOF MS and their positions are indicated by white arrows. The detailed identities of these proteins are listed in Table S1 and Figure S2.

It is noteworthy that these protein isoforms show the same theoretical M*r* and *p*I values. Most of their M*r* values are ranged from 20 kDa to 60 kDa, and their *p*I points are mainly distributed from pH 5.0 to 7.0 (Figure 3B). However, their experimental M*r* and *p*I values are different to each other, and most protein isoforms showed a different *p*I value in the 2-DE gel (Figure 1). These results indicated that these proteins could have post-modifications that resulted in protein forms/species in the 2-DE gel-based proteomics.

We further determined the protein abundance in the 2-DE gel and the most abundant 20 protein species are highlighted (Figure 3C). The results demonstrated that, based on the spot volume, the top-20 protein spots occupied almost a half of abundance volume (43.84%) for all proteins in the 2–DE gel of 6M TKS roots. The most abundant protein spots were identified as the protein species or isoforms of catalase (spot 355, 6.57%; spot 2, 3.38%), ferritin (spot 357, 3.77%; spot 359, 2.90%), actin (spots 142 and 354), actin (spots 142 and 354), GAPC2 (spots 4 and 118), and the subunit of ATPase (spots 11 and 344). Among them, the most abundant spot was identified as catalase-like isoform X1 (spot 355), followed by allene oxide cyclase (spot 346) and eukaryotic translation initiation factor 5A (eIF-5A, spot 361) (Table S1).

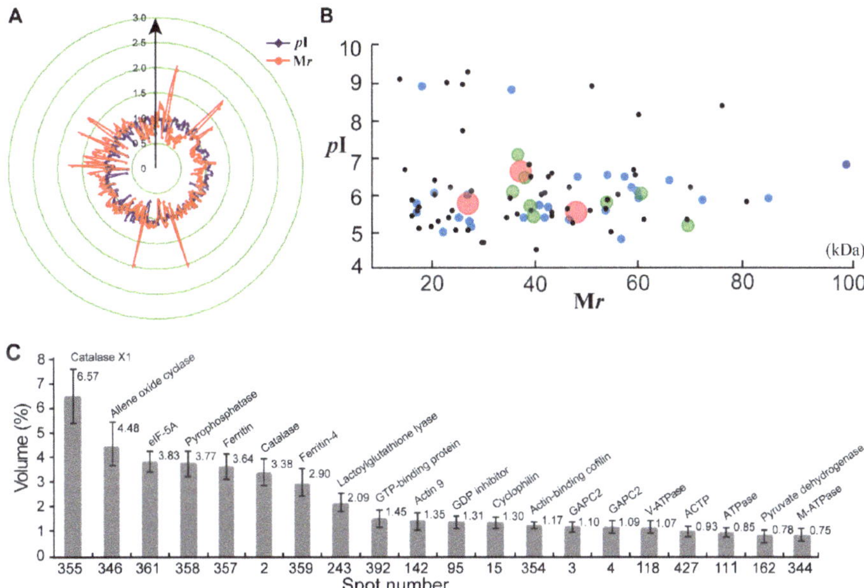

Figure 3. Distribution of the identified proteins on the reference 2-DE gel of 6M TKS roots. The theoretical and experimental ratios of M*r* and *p*I for the identified 371 are presented in radial charts with radial and annular radar axes (**A**). The theoretical distribution of M*r* and *p*I for the protein isoforms is highlighted (**B**). These protein isoforms were independently identified from two (black circle), three (blue circle), four (green circle), and five (red circle) spots on the reference 2-DE gels of the 6M TKS roots, respectively. The most abundant 20 proteins in the 2-DE gel of 6M roots are highlighted (**C**). The detailed identities for these protein isoforms are listed in Table S2.

2.3. High-Throughput Shotgun Proteomic Analysis of the 6M Roots of TKS

As proteins with very low abundance or out of the pH range of immobilized pH gradient (IPG) strips are difficult to separate and identify by 2-DE-based proteomic method, a shotgun analysis of the proteins was further performed. The proteins were extracted from the 6M TKS roots, and three biological repeats (R1, R2 and R3) were conducted to obtain a comprehensive proteomic profile. A total of 5205, 5323 and 5654 unique proteins were successfully identified from R1, R2 and R3, respectively (Table S3). Among them, 7481 proteins were identified from at least one independent shotgun experiment, and 5156 proteins were identified from at least two experiments. There were 3545 shared proteins in the three experiments (Figure 4A), and only these shared proteins from three experiments were considered as the 6M TKS root proteins produced by the shotgun proteomic method in the following study. We found 184 shared proteins in the 2-DE and shotgun proteomic methods. There were 3361 and 47 specific proteins in the shotgun and 2-DE methods, respectively. These results indicated that, although the shotgun method can produce large amounts of proteins, the traditional 2-DE gel-based proteomics also generates a few specific proteins from the 6M TKS roots (Figure 4B; Table S3). Our special interest is focusing on the potential biological functions of these shared proteins in TKS roots.

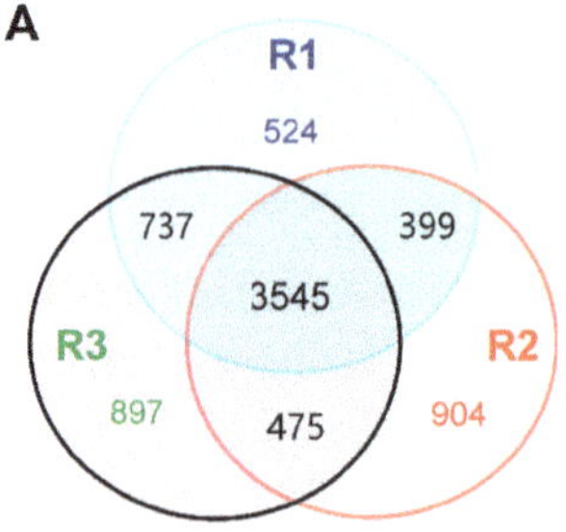
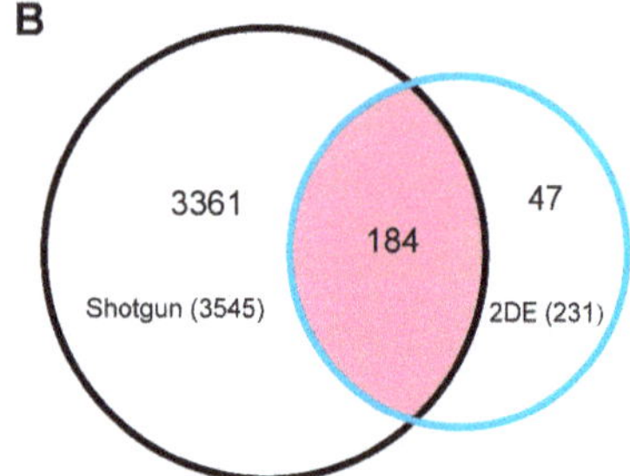

Figure 4. High-throughput proteomics analysis of the 6M TKS roots. Three shotgun proteomics were performed to determine the large-scale protein profiles in the 6M roots, and finally identify 3545 shared proteins in the three experiments (**A**) The blue, yellow, and pink circles represent for the first, second, and third biological replicates. Among these proteins, 184 proteins could also be detected from the 2-DE gel (**B**). The pink area represents for the 184 shared proteins that were identified by both the 2-DE and shotgun based proteomic methods.

2.4. Pathway Analysis of the Identified Proteins in 6M TKS Roots

To gain insight into the functional categories of the 3592 proteins (including the 231 and 3545 proteins identified by 2-DE and shotgun, respectively), we performed gene ontology (GO) classification and the enriched outputs of their biological process, cellular component and molecular function are presented (Figure 5). The GO enrichment revealed that many proteins were localized in membrane-bounded organelle, and they were mainly involved in metabolic process with catalytic activity or binding ability (Figure S3; Table S4). At the biological process level, 13 main processes were detected. Among them, the largest amount containing 1520 proteins were involved in the metabolic process, followed by 1033 proteins involved in the cellular process and 695 proteins in the single-organism process. A total of 302 proteins were considered to respond to external stimulus. At subcellular level, nine components were observed. Among them, almost half including 866 proteins were localized to the cell part. Many proteins were localized into organelle membrane and macromolecular complex. In molecular function classification, seven pathways were determined. Among them, the most portion including 1688 proteins were taken part into catalytic activity, followed by 1325 proteins with binding ability. There are 107 proteins with transporter activity and 101 proteins with structural molecule activity (Figure 5; Table S4). These different GO term distributions in TKS roots are related to their biological functions and indicate some important biological processes for secondary metabolite in the 6M TKS roots.

To further investigate the biological functions of the identified 3592 proteins, KEGG pathway analysis was performed using the BLAST2GO program. These proteins were clustered into 19 main pathways (Figure 6A), including translation process (616 proteins), carbohydrate metabolism (583 proteins), folding, sorting and degradation (440 proteins), global and overview metabolism (409 proteins), amino acid metabolism (380 proteins), transport and catabolism (342 proteins), lipid metabolism (308 proteins), etc. These results demonstrated that the main proteins in TKS roots were involved in translation and carbohydrate metabolism. Furthermore, 129 sub-pathways were enriched in KEGG annotation (Table S5), and the top 20 sub-pathways are highlighted (Figure 6B). Among them, the most abundant pathway containing 240 proteins is carbon metabolism. The second contains 224 proteins with ribosome activity. The third sub-pathway has 213 proteins for biosynthesis of amino acids, followed by 183 proteins for endocytosis, 182 proteins in the spliceosome, and 172 proteins in the endoplasmic reticulum. Proteins involved in starch and sucrose metabolism (164 proteins), glycolysis (121 proteins), phenylpropanoid biosynthesis (119 proteins), and amino sugar and nucleotide sugar metabolism (116 proteins), are also enriched in these KEGG sub-pathways (Table S5). These proteins are crucial for biosynthesis of isoprenoids in many plants, especially in natural rubber-producing plants [16].

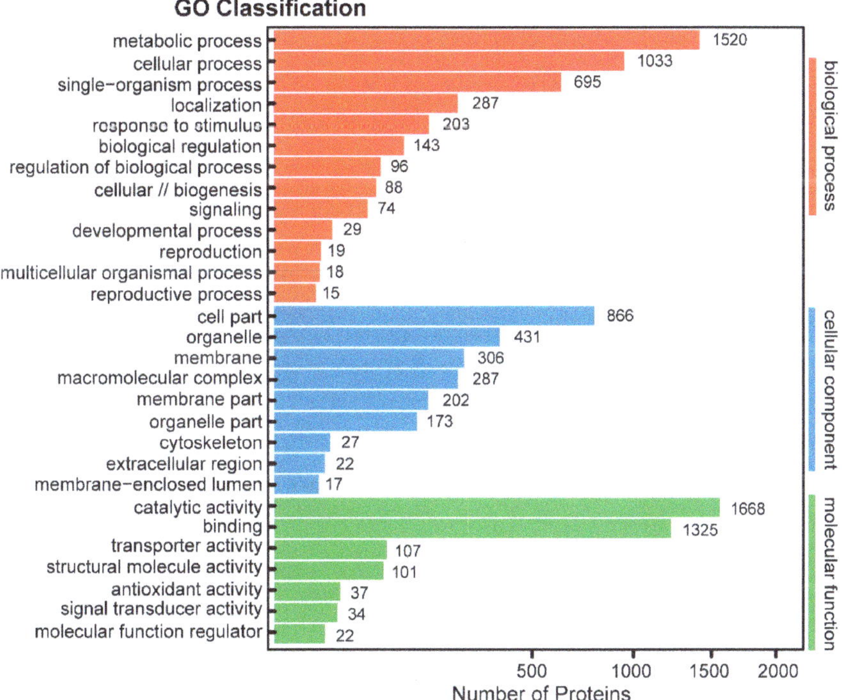

Figure 5. GO classification of the identified proteins. The 3592 proteins identified from the 6M roots by both shotgun and 2-DE methods were performed GO analysis and these proteins were classified into 3 main categories including biological process, cellular component and molecular function. The number of proteins denotes those with GO annotations.

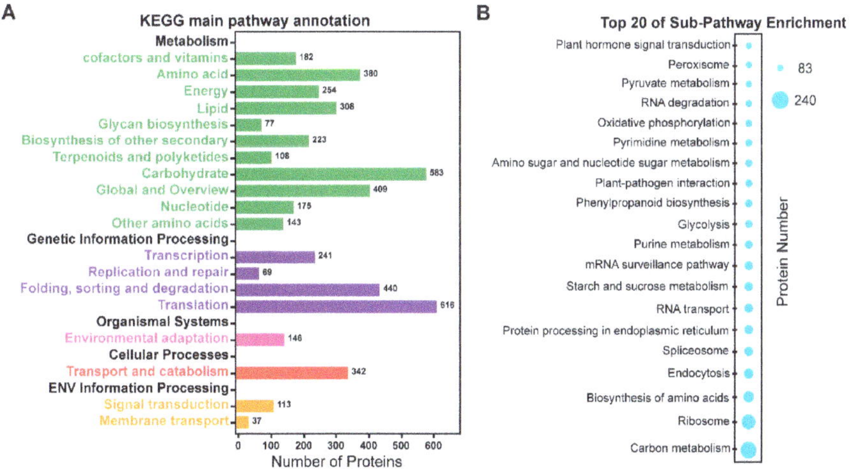

Figure 6. KEGG pathway analysis of all the identified proteins. KEGG pathways for the identified 3592 proteins are presented (**A**). The number of proteins enriched in the KEGG database is marked in the right side. Then, the most abundant 20 KEGG pathways are highlighted (**B**). The size of the circular ring, ranging from 83 to 240, stands for the number of proteins involved in these pathways.

2.5. The Proteins Involved in Natural Rubber Biosynthesis in TKS Roots

We checked the identified 3592 proteins from the 6M TKS roots and paid special attention to the proteins and enzymes involved in NRB. Our proteomic data demonstrated that a total of 58 unique proteins or gene products were identified at least one time by the three shotgun experiments (Table S6), and four of them, named ACAT7 (spots 157, 183 and 424), HMGR1 (spot 146), DXS3 (spot 257) and DXS8 (spots 199 and 204), were also identified from the 2-DE gel of 6M TKS roots (Figure 1; Table S1). Twenty kinds of proteins were determined from the 58 unique gene products (Table S6), and these proteins are involved in both MVA and MEP pathways for NBR in TKS roots (Figure 7). Genomic sequencing data demonstrated that both cytosolic MVA and plastidic MEP pathways are present for NBR, and a total of 102 NBR-related genes have been determined in the genome of TKS [16]. Our proteomic data demonstrated 22 unique proteins were involved in the MVA pathway and 13 proteins in MEP pathway in TKS roots (Figure 7; Table S6). In the MVA pathway, the identified 22 unique proteins belonged to six kinds of essential enzymes for rubber biosynthesis; they are ACAT, HMGS, HMGR, MVK, PMVK and MVD. Among them, seven members for ACAT (from ACAT2 to ACAT8), six members for mevalonate kinase (MVK 1, 2 and 7–11), and five members for phosphomevalonate kinase (PMVK 1 and 2–6), were positively identified from TKS roots by the shotgun method (Table S6). Our proteomics results also showed that, by contrast with rubber tree latex, TKS contains the MEP pathway for NRB in its roots. In the MEP pathway, seven kinds of proteins were identified from 13 unique proteins, including six family members for deoxyxylulose-5-phosphate synthase (DXS 1–3 and 8–10) and two members for D-xylulose 5-phosphate reductoisomerase (DXR 1 and 2).

In the following step for initiation of the synthesis of isopentenyl pyrophosphate (IPP) to form rubber molecule polymer, which is termed initiator synthesis, four crucial enzymes, including isopentenyl-diphosphate delta-isomerase (IPI 2–4), geranylgeranyl pyrophosphate synthase (GGPS 1–3), geranylgeranyl diphosphate synthase (GPS 3, 5 and 7) and farnesyl diphosphate synthase (FPS 1), were identified from 10 unique proteins.

In the final rubber elongation process, cis-isoprene transferase (CPT) can help GGPP to generate natural rubber hydrocarbons contain different length of carbons. Four kinds of proteins, named CPT, SRPP, REF, and HRBP, which is a Nogo-B receptor as a HRT1-REF bridging protein, are considered to play crucial roles for natural rubber elongation. In this proteomic study, we identified the above four members from 12 unique proteins. Among them, eight SRPP members (SRPP 1–7 and 9) and two REF members were determined from the 6M TKS roots.

We checked the annotated genomes of the rubber tree *H. brasiliensis* and the rubber grass *T. kok-saghyz*, and found 85 and 106 NRB-related proteins from *H. brasiliensis* and TKS genomes, respectively. It is noteworthy that 58 out of 106 NRB-related unique proteins were identified from the 6M TKS roots, and 13 out of the identified 58 unique proteins were crucial members in the MEP pathway (Table S6). These results revealed that both MVA and MEP pathways are important in controlling NRB and rubber production in the mature TKS roots.

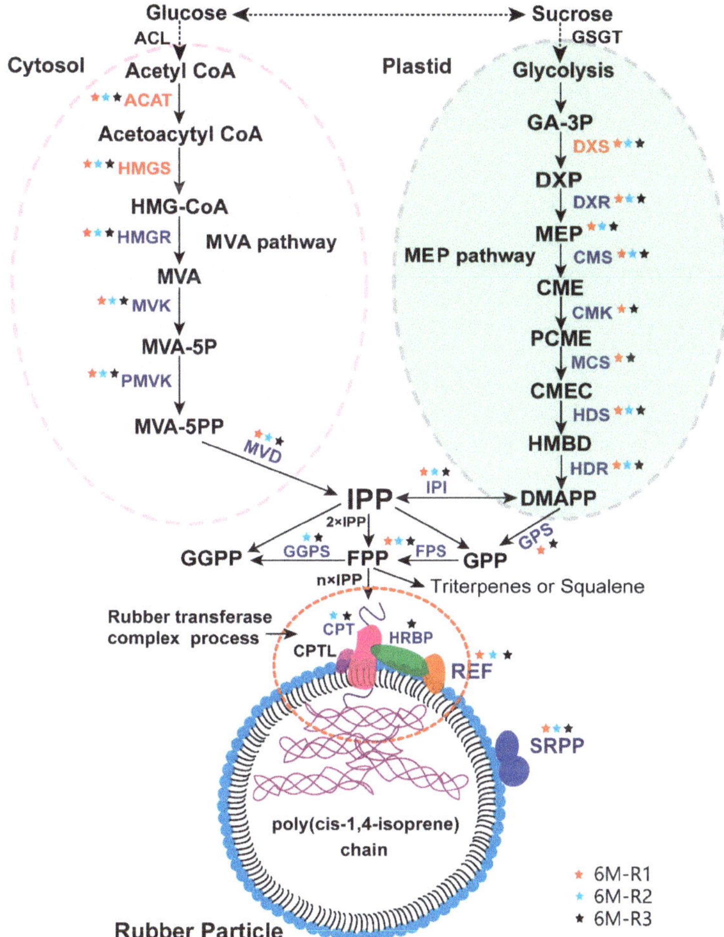

Figure 7. Schematic diagram of natural rubber biosynthesis (NRB)-related proteins in the 6M TKS roots. A schematic diagram of the identified proteins involved in the mevalonate acid (MVA, left) and methylerythritol phosphate (MEP, right) pathways of NRB is highlighted. In this schematic diagram, the identified proteins involved in NRB from the 6M TKS roots were marked with different color. The name of proteins identified from both 2-DE and shotgun methods was marked in red color. The proteins only determined by shotgun analysis are marked in blue color. The proteins identified from the first (6M-1), second (6M-2) and third (6M-3) shotgun experiments are marked with red, blue and black stars, respectively.

3. Discussion

3.1. The First Visual Proteome Based on 2-DE Gel Demonstrated Many Protein Isoforms in the Mature TKS Roots

The ease and availability of 2-DE combined MS is a classical standard for proteomics study, and it is still a valuable tool to provide a visual proteome for plants [34]. Large amounts of proteomics studies have been performed in many plant species, however, only several proteomic research works were reported on different latex components of the *Para* rubber tree [2,35], and only one proteomic analysis was performed to identify 278 unique proteins from the rubber particle phase and pellet phase of the *T. brevicorniculatum* latex [32]. Our 2-DE gel-based proteomics of TKS roots resulted in 371 abundant protein species (Figure 2; Table S1), but the protein profiles are very different to that in *T. brevicorniculatum* latex. Compared with the 2-DE gels for a clear aqueous phase and a yellowish pellet phase of the fractionated samples from *T. brevicorniculatum* latex [32], our 2-DE gel for the mature TKS roots is much clearer and contains more protein spots (Figure 2). Among the identified proteins, 90 unique proteins contained different protein isoforms, which have different experimental M*r* and *p*I values on the 2-DE gel (Table S1), probably due to the existence of modification protein variants.

In the past decade, only 1208 proteins were identified from all the rubber-producing plants, and only several proteomics studies were performed on the rubber particles [36]. In a previous high throughput proteomics study, a total of 186 proteins were positively identified from rubber particles by shotgun tandem MS [28]. After removal of the protein bands corresponding to the RP-abundant REF and SRPP, 137 protein species including 115 unique proteins were identified in a LC-MS/MS based proteome analysis [31]. By investigating the protein profile between LRPs and SRPs, 53 protein spots corresponding to 22 gene products were detected to have significant difference. Among them, most up-regulated proteins in SRPs were identified as SRPP, HMGS, phospholipase D, ethylene response factor, eukaryotic translation initiation factor, etc., but the most abundant proteins in LRPs were REF, glucanase and several hypothetical proteins [27]. Interaction network analysis of rubber particle proteomics revealed the formation of the protein complex consisting of HRT1, REF and HRBP might play crucial roles as a NRB machinery [31], and these proteins are associated with the endoplasmic reticulum [37].

Recently, more than 1600 proteins were identified from total latex of the *Para* rubber tree by using isobaric tags for relative and absolute quantitation (iTRAQ) method [2] and 1839 unique proteins were determined by LC-MS/MS from the whole translated draft genome of the rubber tree *H. brasiliensis* [24]. Our high throughput shotgun proteomics produced 7481 proteins from at least one independent shotgun experiment and 3545 shared proteins in the three experiments from the 6M TKS roots (Figure 4B; Table S3), which covered almost all the previously identified proteins from the latex-producing plants. To our best knowledge, it is the first visual proteome based on the combination of 2-DE gel and high throughput proteomics methods, and the results may help us to deepen our understanding of the roles of protein isoforms in the mature TKS roots.

3.2. Large-Scale Shotgun Proteomics Landscape Revealed both Mevalonate Acid (MVA) and Methylerythritol Phosphate (MEP) Pathways are Important for Natural Rubber Biosynthesis (NRB) in the Mature TKS Roots

Natural rubber is an isoprenoid polymer that is synthesized on the rubber particles in many latex-producing plants [16]. Although more than 2500 plants can biosynthesize natural rubber [31], the *Para* rubber tree *H. brasiliensis* is the only plant commercially cultivated to produce natural rubber for industry [2]. Rubber latex is a kind of specialized cytoplasm of laticifer cells in the bark phloem of the rubber tree, and natural rubber, as an elastomer with physical and chemical properties, cannot be fully matched by synthetic rubber [31]. Annotation of the rubber tree genome revealed that although 22 MEP genes can be identified, only two DXS genes (*DXS7* and *DXS10*) show substantial and preferential expression in the rubber latex; on the other hand, at least one gene for the identified 18 enzymes in MVA pathway shows latex-biased abundant expression [10]. Proteomics analysis also supported that MVA pathway is more important than MEP pathway for NRB in the latex of the *Para* rubber tree [2,24,25,29].

Similar results were observed in the recently assembled TKS genome, which contains 102 rubber biosynthesis-related genes. Among them, 40 genes in the MVA pathway and 23 genes in the MEP pathway were identified, respectively. Comparison of gene expression level proved that, for each process in the MVA pathway in TKS, at least one enzyme shows a predominant gene expression level in latex and roots, but most genes in the MEP pathway have a medium or low expression level in the TKS latex [16]. These results indicated that it is the MVA pathway rather than MEP pathway that is involved in the NBR in the rubber latex of both the rubber tree *H. brasiliensis* and the rubber grass TKS.

However, our comprehensive proteomics data showed that many proteins involved in both MVA (22 unique proteins) and MEP (unique 13 proteins) pathways have been positively identified from the mature TKS roots (Figure 7; Table S6). Among the identified 13 MEP-pathway proteins, two unique proteins named DXS 3 (spot 257) and DXS 8 (spots 199 and 204) are visible on the 2-DE gel (Figure 1), and their relative abundance is very high (Table S1). These proteomic data suggest that, by contrast with that in the inner bark of the *Para* rubber tree [10], both the MEP pathway and the MVA pathway might be the main source of IPP, and MVA pathway is also crucial for rubber biosynthesis in the mature TKS roots.

3.3. Almost All NRB-Related Proteins Can Be Identified From the Mature TKS Roots

In the *Para* rubber tree, NRB is known to begin with IPP synthesis in MVA pathway [38,39], which is a cytosolic pathway for rubber biosynthesis [40]. In the early steps of the MVA pathway, ACAT is important for generating acetoacytyl-CoA, then HMGS and HMGR activate the supply of mevalonate substrates [2,41]. ACAT can catalyze a Claisen-type condensation of two acetyl-CoA units to form acetoacetyl-CoA, which is the first step in the MVA pathway and is also important for providing the malonyl-CoA substrate for the biosynthesis of fatty acids [42]. This enzyme was found to mainly accumulate in the mature TKS roots (Table S1; Table S3), but it could not be detected in total latex of the *Hevea* rubber tree [2]. Through a series of enzymatic reactions, acetyl-CoA can be catalyzed by ACAT to form IPP [43]. Four ACAT genes were determined in the *Hevea* rubber tree [10] and eight ACAT genes were found in the TKS genome [16]. Overexpression of *AtACAT* results in the increased accumulation of ACAT in the latex of *T. brevicorniculatum*, and ultimately increased the pentacyclic triterpene and sterol levels [44]. Our proteomic results demonstrated that seven (ACAT 2-8) out of the eight ACAT members can be identified by the shotgun proteomic method, and one member (ACAT 7, spots 157, 183 and 424) can also be detected on the 2-DE gel (Figure 2), indicating that ACAT 7 is high-abundant in TKS roots.

In the following steps, HMGS and HMGR activate the supply of mevalonate substrates [2,41]. HMGS, the second enzyme in the MVA pathway, catalyzes aldol-type condensation of acetoacetyl-CoA and acetyl-CoA to produce HMG-CoA, and it is located as a cytosolic protein [45,46]. In our proteomic data, one HMGS member (HMGS 1, evm.model.utg 270.25; spot 146) was identified from both high-through and 2-DE gel-based proteomics methods (Figure 7; Table S6). Furthermore, HMG-CoA can be converted to MVA by HMGR [47]. HMGR controls the carbon flow and metabolic reaction rate in MVA [42], and it is also a critical player for the regulation of triterpenoid metabolism [38,48]. It was reported that ethylene cannot activate the enzyme activity of HMGR [49]. Five gene members have been characterized in the genome of the *Para* rubber tree [10], but only HMGR1 gene was found to be induced by ethylene stimulation [49]. In the TKS genome, 12 HMGR members were determined, and only two out of them (*TkHMGR*1 and *TkHMGR*2) were predominantly expressed in roots, with the highest expression level in latex [16]. In this proteomic study, two members of HMGR (HMGR1 and HMGR12) were detected from TKS roots by the shotgun method (Table S3). Based on these genomic and proteomic results, we consider that HMGR1 might be the most important player for NRB in TKS roots.

In the following steps, mevalonate kinase (MVK) converts mevalonate into isopentyl pyrophosphate and phosphomevalonate kinase (PMVK) catalyzes mevalonate-5-phosphoate to form mevalonate-5-pyrophosphate [50]. In *Hevea*, three MVK genes were determined [10]; and in TKS genome, 11 MVK genes were observed [16]. In our proteomic data, we identified eight MVK protein members from the predicted 11 MVK genes, and five PMVK members from the six PMVK genes (Table S6). Finally, MVD catalyzes mevalonate-5-pyrophosphate to form IPP [49]. Two MVD members were determined from both *Hevea* and TKS genomes [16], and in this proteomic study, they were identified from the mature TKS roots (Table S6). These results revealed that enzymes involved in the conversion of mevalonate into isopentyl pyrophosphate and diphosphate to form IPP are important for NRB in the mature TKS roots.

The MEP pathway is mainly occurred in the plastid and it is an alternative IPP synthesis pathway [51,52]. Recently, the plastidic MEP pathway has also been considered as a possible route for NRB [53–55]. The ^{13}C-labelling of *Hevea* seedlings revealed that MEP pathway mainly contributes IPP for carotenoid biosynthesis [38]. Genomic analysis demonstrated that all genes involved in the MEP pathway can be detected in both *Hevea* and TKS, but most of the MEP-related genes have a medium or low expression level in the latex of the *Hevea* rubber tree [10] and TKS [16]. However, our proteomics results demonstrated that 13 unique proteins involved in the MEP pathway were determined in the TKS roots; these proteins are DXS, DXR, CMS, CMK, HDS and HDR (Table S6). These results indicated that the MEP pathway may also contribute IPP for NRB in the mature TKS roots.

In the initiator synthesis process, GPS and IPI catalyze IPP and DMAPP to form GPP, and FPS to produce farnesyl diphosphate by adding IPP onto GPP to form polyisoprene [40,56], then GGPS and GPS catalyze farnesyl diphosphate to form geranylgeranyl diphosphate [31,42]. Our proteomics data showed that four crucial enzymes, named IPI, GGPS, GPS and FPS, can be identified from the TKS roots.

Then, in the natural rubber elongation process, cis-isoprene transferase (CPT), also named *Hevea* rubber transferase, catalyzes multiple isoprene units (C5H8) and polymerizes into IPP long chain molecules and determine the size of the rubber molecule [46,57]. It is closely associated with the rubber particle membrane and is widely known to be a key enzyme in rubber biosynthesis [28,31]. Eleven CPT members were determined in the *Para* rubber tree genome [10], but only CPT2 shows rubber transferase activity [40], and a rubber transferase activator (RTA) or CPT-like protein (CPTL) can interact with CPTs to activate NRB in the *Hevea* rubber latex [57]. In the TKS genome, eight CPT members and two CPT-like proteins (CPTL1 and CPTL2) have been characterized, and *TkCPT1*, *TkCPT2* and *TkCPTL1* genes play a critical role for the elongation of rubber polymers in the latex and roots of TKS [16]. However, our proteomics data only identified CPT1 in the TKS roots (Table S6), indicated CPT1 may be the crucial player for NRB in TKS roots.

In the final elongation process, three rubber particle membrane binding proteins, named REF, SRPP and a Nogo-B receptor or HRT1-REF bridging protein (HRBP), are widely known to play crucial roles [35], and they were positively identified from the TKS roots by our proteomics study (Figure 7). Among them, REF is anchored inside rubber particle membrane by its auto-assembly ability [42], whereas SRPP largely covers rubber particle surface in an oriented anisotropic manner [58]. REF is also suggested to be a conservative regulator of lipid droplets and it is related to form an intracellular structure [59,60]. However, proteomics of the detergent-washed rubber particles identified none lipid droplet proteins [28,31], indicated that the maturation of rubber particles might be different from that in lipid droplets. Histochemical localization indicated that rubber biosynthesis capability in *Hevea* laticifer is mostly concentrated in SRPs, and SRPP, a rubber biosynthesis-related protein that accumulated mainly in SRPs, is more important than REF for rubber biosynthesis in the *Hevea* rubber tree [48]. Five SRPP genes (*TbSRPP 1-5*) were found in TBR [7]. Among them, except for *TbSRPP 2* [61], four members can enhance NRB. But the average molecular weight of rubber hydrocarbons has not been significantly affected by these genes [62]. Overexpression of *TbSRPP3* only resulted in a slight change in the content of rubber in TBR root [63]. In this proteomics of TKS roots, eight out of 10 SRPP members were identified (Table S6). Different with that in *Hevea* genome, which contains eight REF members, only two REF members have been characterized in the TKS genome [16]. Our proteomics data also identified the two REF members in the mature TKS roots (Table S6). These published results, as well as our proteomics of the 6M TKS roots, revealed that SRPP might be more important than REF for rubber biosynthesis in the mature roots of TKS, and more attention should be paid on determination of the detail roles of SRPP in TKS roots in future.

4. Materials and Methods

4.1. Plant Growth Conditions and Root Collection

TKS seeds were collected from the Tekes River basin in Xinjiang, China. Following germination, seedlings were transplanted into the pots in a greenhouse of Shihezi University (Shihezi, China), containing nutritive soil and vermiculite with a ratio of 5/3, and a half strength of Hoagland liquid medium was irrigated with 55% relative humidity at 22 °C in the light and 18 °C in the dark. After growing for 6 months (termed as 6M roots), the main roots from plants were collected, rinsed with tapped water, blotted dry on a filter paper, and the middle parts were dissected into approximately 0.5 cm-thick slices, and then frozen in liquid nitrogen for further study.

4.2. Morphological Observation of Rubber Particles in TKS Roots

For histochemical staining, a 0.5 cm long section from the middle parts of the fresh main roots of TKS were fixed in 4% glutaraldehyde (0.1 M phosphate buffer, pH 7.5) at room temperature, then treated with bromine and iodine in glacial acetic acid as described [64]. Part of the sample was dehydrated through a graded series of ethanol, and embedded in paraffin. Sections (12 μm thickness and area of 5 mm × 5 mm) were cut with microtome (Leica Microsystems, Bannockburn IL, Germany) stained with mercury–bromophenol blue, which is effective in showing the latex protein [65]. Then, the slides were examined under a leica DMLN electron microscope (leica, Wetzlar, Germany). TKS laticifer cells in the root sections could be recognized due to iodine-bromine treatment of the rubber in the laticifers became deep brown or dark color. Another part of the sample was dehydrated in a range of acetone (30%, 50%, 70%, 90%, 1 h for once and pure acetone 2 h for twice). Then it was permeated with acetone and Epon 812 embedding agent and pure Epon 812 embedding agent permeates overnight as described [33]. After that, the samples were polymerized in the oven, flattened and cut into semi-thin slices. Observed under an optical microscope, the laticifers could be recognized by tracing rubber inclusions, which is brown with iodine-bromine staining [65]. For electron microscopy, samples were cut into a smaller size, immediately fixed in the glutaraldehyde solution at 4 °C for 24 h, and then fixed in phosphate buffer (pH 7.2, 2% OsO4 in 0.1 M) for 6 h at room temperature. Ultra-thin sections, double stained with uranium acetate and lead citrate again. A transmission electron microscope (TEM, Hitachi JEM-1230, JEOL, Japan)), country was used to examine these stained slices.

4.3. Protein Extraction and Two-Dimensional Gel Electrophoresis

Total proteins were extracted from TKS roots using a modified BPP phenol method as described [22]. The washed protein pellets were air-dried for 10 min and dissolved with the Lysis buffer (30 mM Tris-HCl, pH 8.5, 2 M thiourea, 7 M 17 urea, 4% 3-[(3-Cholamidopropyl) dimethylammonio] propanesulfonate (CHAPS)). Protein concentration was determined by the Bradford assaying, and BSA was used as the protein standard. For 2-DE assay, 1300 μg of proteins were diluted to 450 μL with lysis buffer and loaded onto a 24 cm, linear pH gradient 4–7 IPG strips (GE Healthcare, Uppsala, Sweden). Then, they were rehydrated for 24 h at 22 °C. Isoelectric focusing system and gel electrophoresis were performed as described [26]. The gels were stained with Coomassie Brilliant Blue G-250, and analyzed by the ImageMaster 2-D Platinum Software as described [2]. Biological variation analysis module was employed to identify the accumulated protein spots in roots samples with statistically significant (confidence above 95%, $p < 0.05$). Three biological replicates for each separation were conducted and the three typical 2-DE gels are presented in the supplementary Figure S1.

4.4. Identification of Protein Spots via MALDI-TOF/TOF-MS

The main protein spots on the 2-DE gel of 6M roots were manually excised from 2-DE gels and subjected to perform an in-gel digestion with the modified bovine trypsin (Cat. 11418025001, Roche, Basel, Switzerland) as described [64]. After digestion, mass spectra of trypsin-digested peptides map fingerprinting were obtained on an AB 5800 MALDI-TOF/TOF mass spectrometry (AB SCIEX,

Foster City, MA, USA) with a laser wavelength of 349 nm as described [2]. Ten main spectra were selected for MS/MS analysis and searched for a self-constructed database derived from the original TKS genome. The nucleotide and protein sequences were deposited in the Genome Warehouse (GWH; http://bigd.big.ac.cn/gwh/) under the accession number PRJCA000437 (https://academic.oup.com/nsr), which includes 46,731 gene sequences [16]. Protein matches were considered as a positive identification with a Mascot score higher than 75 and at least two matched peptides, meanwhile individual ion score > 31 ($p < 0.05$). Detailed information for the identified proteins is provided as the supplementary data (Table S1 and Figure S2).

4.5. Large-Scale Shotgun Proteomics of TKS Roots

Approximately 300 µg proteins from the main roots of TKS were reduced and alkylated by dithiothreitol and iodoacetamide, then digested by trypsin. Enzymic hydrolysates were separated using a Pepmap C18 column (Thermo Fisher, USA) following a 65 min 5%–35% organic gradient by an UltiMate 3000 instrument (Thermo Scientific, Rockford, USA). Fifty washed fractions were collected in 1.5 mL/min tubes. Fractions were dehydrated and merged into 15 fractions, and these samples were subjected to a high-resolution mass spectrometer Triple TOF 6600 system (AB SCIEX). Peptides were automatically selected by ProGroupTM algorithm and ProteinPilot™ software V5.0 (AB SCIEX) to calculate the error factor (EF), reporter peak area and p value as described [30]. A protein database was established by the target protein sequences (https://academic.oup.com/nsr) for the corresponding species. Protein with an unused score > 1.3 (confidence ≥ 95%) was considered as a positive identification. Finally, an in-house BlastP search of the UniProt database was performed for each protein to identify its homologues and potential functions.

4.6. Functional Classification, Hierarchical Clustering and Pathway Analysis of All the Identified Proteins From TKS Roots

The sequences of all the identified proteins were search against the UniProt database (http://www.uniprot.org/) to confirm their functional annotations. The identified proteins were then classified by Gene Ontology (GO, http://www.geneontology.org/) as described [66]. The GO terms were further determined by the WEGO software (http://wego.genomics.org.cn) on biological process, molecular functions and cellular components. Finally, these proteins were performed KEGG pathway analysis (http://www.genome.jp/kegg) to determine their molecular reaction networks.

5. Conclusions

In this study, we obtained the first visualization proteome profiles of mature TKS roots. Combination of 2-DE and shotgun proteomics leads to more than 3,000 unique proteins in TKS roots. Functional category of all TKS proteins revealed that most proteins were involved in carbon metabolic process with catalytic activity in membrane-bounded organelle, followed by proteins with binding ability, transportation activity, starch and sucrose metabolism, and phenylpropanoid biosynthesis. They are important for biosynthesis of isoprenoids in natural rubber-producing plants. Fifty-eight NRB-related proteins were identified, and they are involved in both MVA and MEP pathways. Our results indicated the MEP pathway is also important by contributing IPP for rubber biosynthesis in TKS roots. Almost all NRB-related proteins can be identified, and several members of many key proteins, including ACAT, HMGR, MVK, PMVK, DXS, DXR, GPS, FPS, IPI, REF and SRPP, were determined from the mature TKS roots. Future work should pay more attention to determination of the detailed roles of different members of these NRB-related proteins in the mature TKS roots.

Supplementary Materials: The following are available online at http://www.mdpi.com/1422-0067/20/10/2596/s1.

Author Contributions: X.W. and H.L. conceived and designed the experiments; Q.X., L.Z., G.D., D.W. and Y.S. performed the experiments; Q.X. wrote the original manuscript. X.W. revised this manuscript. Q.X., L.Z., B.Y., D.W., Y.S., X.G. and L.Y. analyzed the data. All authors have read and approved the final manuscript.

Funding: This research was funded by the Scientific Research Innovation Team Program of Hainan Province, grant number 2018CXTD341, the National Natural Science Foundation of China, grant numbers 31570301 and 31860224, and International scientific and technological cooperation to advance the project by the Shihezi University, grand number GJHZ201708.

Conflicts of Interest: The authors declare that they have no conflict of interests.

Abbreviations

ACL	ATP citrate lyase
ACAT	acetoacetyl-CoA thiolase
CHAPS	3-[(3-Cholamidopropyl) dimethylammonio] propanesulfonate
CPT	cis-isoprene transferase
CPTL	cis-isoprene transferase like
DMAPP	dimethylallyl diphosphate
DXS	1-deoxy-D-xylulose-5-phosphate synthase
eIF	eukaryotic translation initiation factor
FPS	farnesyl diphosphate synthase
GAPC	glyceraldehyde-3-phosphate dehydrogenase
GGPPS	geranylgeranyl diphosphate synthase
GPP	geranyl pyrophosphate
GPS	geranylgeranyl diphosphate synthase
GO	Gene Ontology
KEGG	Kyoto Encyclopedia of Genes and Genomes
HMGS	3-hydroxy-glutaryl-CoA reductase
HMGR	hydroxy-methyl-glutaryl-CoA reductase
IPI	isopentenyl-diphosphate delta-isomerase
IPG	immobilized pH gradient
iTRAQ	isobaric tags for relative and absolute quantification
LRP	large rubber particle protein
MEP	methyl-D-erythritol-4-phosphate
MVA	mevalonate acid
MVD	diphosphomevalonate decarboxylase
MVK	mevalonate kinase
MS	mass spectrometry
NR	natural rubber
NRB	natural rubber biosynthesis
PMVK	mevalonate disphosphate decarboxylase
REF	rubber elongation factor
RTA	rubber transferase activating factor
SRPP	small rubber particle protein
2-DE	two-dimensional gel electrophoresis

References

1. Mooibroek, H.; Cornish, K. Alternative sources of natural rubber. *Appl. Microbiol. Biotechnol.* **2000**, *53*, 355–365. [CrossRef]
2. Wang, X.C.; Wang, D.; Sun, Y.; Yang, Q.; Chang, L.L.; Wang, L.M.; Meng, X.R.; Huang, Q.X.; Jin, X.; Tong, Z. Comprehensive proteomics analysis of laticifer latex reveals new insights into ethylene stimulation of natural rubber production. *Sci. Rep.* **2015**, *5*, 13778. [CrossRef] [PubMed]
3. Schmidt, T.; Hillebrand, A.; Wurbs, D.; Wahler, D.; Lenders, M.; Gronover, C.; Prüfer, D. Molecular cloning and characterization of rubber biosynthetic genes from *Taraxacum kok-saghyz*. *Plant Mol. Biol. Rep.* **2010**, *28*, 277–284. [CrossRef]
4. Cornish, K.; Kopicky, S.L.; Mcnulty, S.K.; Amstutz, N.; Chanon, A.M.; Walker, S.; Kleinhenz, M.D.; Miller, A.R.; Streeter, J.G. Temporal diversity of *Taraxacum kok-saghyz* plants reveals high rubber yield phenotypes. *Biodiversitas* **2016**, *17*, 847–856. [CrossRef]

5.   Edathil, T.T. South American leaf blight: a potential threat to the natural rubber industry in Asia and Africa. *Trop. Pest Manag.* **1986**, *4*, 296–303. [CrossRef]

6.   Ramirez-Cadavid, D.A.; Cornish, K.; Michel, J. *Taraxacum kok-saghyz* (TK): compositional analysis of a feedstock for natural rubber and other bioproducts. *Ind. Crop. Prod.* **2017**, *107*, 624–640. [CrossRef]

7.   Collins-Silva, J.; Nural, A.T.; Skaggs, A.; Scott, D.; Hathwaik, U.; Woolsey, R.; Schegg, K.; McMahan, C.; Whalen, M.; Cornish, K.; et al. Altered levels of the *Taraxacum kok-saghyz* (Russian dandelion) small rubber particle protein, TKSRPP3, result in qualitative and quantitative changes in rubber metabolism. *Phytochemistry* **2012**, *79*, 46–56. [CrossRef] [PubMed]

8.   Kim, I.J.; Ryu, S.B.; Kwak, Y.S.; Kang, H. A novel cDNA from *Parthenium argentatum* Gray enhances the rubber biosynthetic activity in vitro. *J. Exp. Bot.* **2004**, *396*, 377–385. [CrossRef]

9.   Ikeda, Y.; Junkong, P.; Ohashi, T.; Phakkeeree, T.; Sakaki, Y.; Tohsan, A.; Kohjiya, S.; Cornish, K. Strain-induced crystallization behaviour of natural rubbers from guayule and rubber dandelion revealed by simultaneous time-resolved WAXD/tensile measurements: indispensable function for sustainable resources. *RSC Adv.* **2016**, *6*, 95601–95610. [CrossRef]

10.   Tang, C.R.; Yang, M.; Fang, Y.J.; Luo, Y.F.; Gao, S.H.; Xiao, X.H.; An, Z.W.; Zhou, B.H.; Zhang, B.; Tan, X.Y.; et al. The rubber tree genome reveals new insights into rubber production and species adaptation. *Nat. Plants* **2016**, *6*, 16073. [CrossRef] [PubMed]

11.   Whaley, W.G.; Bowen, J.S. Russian dandelion (*kok-saghyz*). An emergency source of natural rubber. *Misc. Publ. US Dept. Agric.* **1947**, *618*, 1–212.

12.   van Beilen, J.B.; Poirier, Y. Establishment of new crops for the production of natural rubber. *Trends Biotechnol.* **2007**, *25*, 522–529. [CrossRef] [PubMed]

13.   Kirschner, J.; Stepanek, J.; Cerny, T.; Heer, P.; Dijk, P.J. Available ex situ germplasm of the potential rubber crop *Taraxacum kok-saghyz* belongs to a poor rubber producer, *T. brevicorniculatum* (Compositae–Crepidinae). *Genet. Resour. Crop Evol.* **2013**, *2*, 455–471. [CrossRef]

14.   Kreuzberger, M.; Hahn, T.; Zibek, S.; Schiemann, J.; Thiele, K. Seasonal pattern of biomass and rubber and inulin of wild Russian dandelion (*Taraxacum kok-saghyz* L. Rodin) under experimental field conditions. *Eur. J. Agron.* **2016**, *80*, 66–77. [CrossRef]

15.   Stolze, A.; Wanke, A.; van Deenen, N.; Geyer, R.; Prüfer, D.; Gronover, C.S. Development of rubber-enriched dandelion varieties by metabolic engineering of the inulin pathway. *Plant Biotechnol. J.* **2017**, *6*, 740–753. [CrossRef]

16.   Lin, T.; Xu, X.; Ruan, J.; Liu, S.Z.; Wu, S.G.; Shao, X.J.; Wang, X.B.; Gan, L.; Qin, B.; Yang, Y.S.; et al. Genome analysis of *Taraxacum kok-saghyz* Rodin provides new insights into rubber biosynthesis. *Natl. Sci. Rev.* **2017**, *5*, 78–87. [CrossRef]

17.   Ramirez-Cadavid, D.A.; Valles-Ramirez, S.; Cornish, K.; Michel, F.C. Simultaneous quantification of rubber, inulin, and resins in *Taraxacum kok-saghyz* (TK) roots by sequential solvent extraction. *Plant Biotechnol. J.* **2018**, *122*, 647–656. [CrossRef]

18.   Bach, T.J. Some new aspects of isoprenoid biosynthesis in plants a review. *Lipids* **1995**, *3*, 191–202. [CrossRef]

19.   Walther, T.C.; Farese, R.V. The life of lipid droplets. *Biochim. Biophys. Acta* **2009**, *6*, 459–466. [CrossRef]

20.   Seetang-Nun, Y.; Sharkey, T.D.; Suvachittanont, W. Molecular cloning and characterization of two cDNAs encoding 1-deoxy-D-xylulose 5-phosphate reductoisomerase from *Hevea brasiliensis*. *J. Plant Physiol.* **2007**, *9*, 991–1002. [CrossRef]

21.   Yagami, T.; Haishima, Y.; Tsuchiya, T.; Tomitaka-Yagami, A.; Kano, H.; Matsunaga, K. Proteomic analysis of putative latex allergens. *Int. Arch. Allergy Immunol.* **2004**, *135*, 3–11. [CrossRef] [PubMed]

22.   Wang, X.C.; Shi, M.J.; Lu, X.L.; Ma, R.F.; Wu, C.G.; Guo, A.P.; Peng, M.; Tian, W.M. A method for protein extraction from different subcellular fractions of laticifer latex in *Hevea brasiliensis* compatible with 2-DE and MS. *Proteome Sci.* **2010**, *8*, 35. [CrossRef] [PubMed]

23.   D'Amato, A.; Bachi, A.; Fasoli, E.; Boschetti, E.; Peltre, G.; Sénéchal, H.; Sutrad, J.P.; Citteriob, A.; Righettib, P.G. In-depth exploration of *Hevea brasiliensis* latex proteome and hidden allergens via combinatorial peptide ligand libraries. *J. Proteom.* **2010**, *73*, 1368–1380. [CrossRef]

24.   Habib, M.H.; Yuen, G.C.; Othman, F.; Zainudin, N.N.; Latiff, A.A.; Ismail, M.N. Proteomics analysis of latex from *Hevea brasiliensis* (clone RRIM 600). *Biochem. Cell Biol.* **2017**, *95*, 232–242. [CrossRef]

25. Tong, Z.; Wang, D.; Sun, Y.; Yang, Q.; Meng, X.R.; Wang, L.M.; Feng, W.Q.; Li, L.; Wurtele, E.S.; Wang, X.C. Comparative proteomics of rubber latex revealed multiple protein species of REF/SRPP family respond diversely to ethylene stimulation among different rubber tree clones. *Int. J. Mol. Sci.* **2017**, *18*, 958. [CrossRef] [PubMed]

26. Rojruthai, P.; Sakdapipanich, J.T.; Takahashi, S.; Hyegin, L.; Noike, M.; Koyama, T.; Tanaka, Y. In vitro synthesis of high molecular weight rubber by *Hevea* small rubber particles. *J. Biosci. Bioeng.* **2010**, *109*, 107–114. [CrossRef] [PubMed]

27. Xiang, Q.L.; Xia, K.C.; Dai, L.J.; Kang, G.J.; Li, Y.; Nie, Z.Y.; Duan, C.F.; Zeng, R.Z. Proteome analysis of the large and the small rubber particles of *Hevea brasiliensis* using 2D-DIGE. *Plant Phys. Biochem.* **2012**, *60*, 207–213. [CrossRef] [PubMed]

28. Dai, L.J.; Kang, G.J.; Li, Y.; Nie, Z.Y.; Duan, C.F.; Zeng, R.Z. In-depth proteome analysis of the rubber particle of *Hevea brasiliensis* (para rubber tree). *Plant Mol. Biol.* **2013**, *82*, 155–168. [CrossRef]

29. Dai, L.J.; Kang, G.J.; Nie, Z.Y.; Li, Y.; Zeng, R.Z. Comparative proteomic analysis of latex from *Hevea brasiliensis* treated with ethrel and methyl jasmonate using iTRAQ-coupled two-dimensional LC-MS/MS. *J. Proteom.* **2016**, *132*, 167–175. [CrossRef]

30. Wang, D.; Sun, Y.; Tong, Z.; Yang, Q.; Chang, L.L.; Meng, X.R.; Wang, L.M.; Tian, W.M.; Wang, X.C. A protein extraction method for low protein concentration solutions compatible with the proteomic analysis of rubber particles. *Electrophoresis* **2016**, *37*, 2930–2939. [CrossRef]

31. Yamashita, S.; Yamaguchi, H.; Waki, T.; Aoki, Y.; Mizuno, M.; Yanbe, F.; Ishii, T.; Funaki, A.; Tozawa, Y.; Miyagi-Inoue, Y.; et al. Identification and reconstitution of the rubber biosynthetic machinery on rubber particles from *Hevea brasiliensis*. *eLife* **2016**, *5*, e19022. [CrossRef]

32. Wahler, D.; Colby, T.; Kowalski, N.A.; Harzen, A.; Wotzka, S.Y.; Hillebrand, A.; Fischer, R.; Helsper, J.; Schmidt, J.; Gronover, C.S.; et al. Proteomic analysis of latex from the rubber-producing plant *Taraxacum brevicorniculatum*. *Proteomics* **2012**, *12*, 901–905. [CrossRef]

33. Singh, A.P.; Wi, S.G.; Chung, G.C.; Kim, Y.S.; Kang, H. The micromorphology and protein characterization of rubber particles in *Ficus carica, Ficus benghalensis* and *Hevea brasiliensis*. *J. Exp. Bot.* **2003**, *54*, 985–992. [CrossRef]

34. Jorrin-Novo, V.J.; Pascual, J.; Sanchez-Lucas, R.; Romero-Rodriguez, M.C.; Rodriguez-Ortega, M.J.; Lenz, C.; Valledor, L. Fourteen years of plant proteomics reflected in proteomics: moving from model species and 2DE-based approaches to orphan species and gel-free platforms. *Proteomics* **2015**, *15*, 1089–1112. [CrossRef]

35. Wang, D.; Sun, Y.; Chang, L.L.; Tong, Z.; Xie, Q.L.; Jin, X.; Zhu, L.P.; He, P.; Li, H.B.; Wang, X.C. Subcellular proteome profiles of different latex fractions revealed washed solutions from rubber particles contain crucial enzymes for natural rubber biosynthesis. *J. Proteom.* **2018**, *182*, 53–64. [CrossRef]

36. Cho, W.K.; Jo, Y.; Chu, H.; Park, S.H.; Kim, K.H. Integration of latex protein sequence data provides comprehensive functional overview of latex proteins. *Mol. Biol. Rep.* **2014**, *41*, 1469–1481. [CrossRef] [PubMed]

37. Brown, D.; Feeney, M.; Ahmadi, M.; Lonoce, C.; Sajari, R.; Di Cola, A.; Frigerio, L. Subcellular localization and interactions among rubber particle proteins from *Hevea brasiliensis*. *J. Exp. Bot.* **2017**, *68*, 5045–5055. [CrossRef] [PubMed]

38. Sando, T.; Takeno, S.; Watanabe, N.; Okumoto, H.; Kuzuyama, T.; Yamashita, A.; Hattori, M.; Ogasawara, N.; Fukusaki, E.; Kobayashi, A. Cloning and characterization of the 2-C-methyl-D-erythritol 4-phosphate (MEP) pathway genes of a natural-rubber producing plant, *Hevea brasiliensis*. *Biosci. Biotechnol. Biochem.* **2008**, *11*, 2903–2917. [CrossRef]

39. Wu, C.T.; Li, Y.; Nie, Z.Y.; Dai, L.J.; Kang, G.J.; Zeng, R.Z. Molecular cloning and expression analysis of the mevalonate diphosphate decarboxylase gene from the latex of *Hevea brasiliensis*. *Tree Genet. Genomes* **2017**, *13*, 22. [CrossRef]

40. Lau, N.S.; Makita, Y.; Kawashima, M.; Taylor, T.D.; Kondo, S.; Othman, A.S.; Shu-Chien, A.C.; Matsui, M. The rubber tree genome shows expansion of gene family associated with rubber biosynthesis. *Sci. Rep.* **2016**, *6*, 28594. [CrossRef]

41. Puskas, J.E.; Gautriaud, E.; Deffieux, A.; Kennedy, J.P. Natural rubber biosynthesis-a living carbocationic polymerization. *Prog. Polym. Sci.* **2006**, *31*, 533–548. [CrossRef]

42. Cornish, K. Similarities and differences in rubber biochemistry among plant species. *Phytochemistry* **2001**, *57*, 1123–1134. [CrossRef]

43. Kirby, J.; Keasling, J.D. Biosynthesis of plant isoprenoids perspectives for microbial engineering. *Annu. Rev. Plant Biol.* **2009**, *60*, 335–355. [CrossRef]

44. Putter, K.M.; van Deenen, N.; Unland, K.; Prufer, D.; Gronover, C.S. Isoprenoid biosynthesis in dandelion latex is enhanced by the overexpression of three key enzymes involved in the mevalonate pathway. *BMC Plant Biol.* **2017**, *17*, 88. [CrossRef]

45. Sirinupong, N.; Suwanmanee, P.; Doolittle, R.F.; Suvachitanont, W. Molecular cloning of a new cDNA and expression of 3-hydroxy-3-methylglutaryl-CoA synthase gene from *Hevea brasiliensis*. *Planta* **2005**, *221*, 502–512. [CrossRef] [PubMed]

46. Grabinska, K.A.; Park, E.J.; Sessa, W.C. *cis*-Prenyltransferase: new insights into protein glycosylation, rubber synthesis, and human diseases. *J. Biol. Chem.* **2016**, *291*, 18582–18590. [CrossRef]

47. Schaller, H.; Crausem, B.; Benveniste, P.; Chye, M.L.; Tan, C.T.; Song, Y.H.; Chua, N.H. Expression of the *Hevea brasiliensis* Müll. Arg. 3-Hydroxy-3-Methylglutaryl-Coenzyme A Reductase 1 in tobacco results in sterol overproduction. *Plant Physiol.* **1995**, *109*, 761–770. [CrossRef]

48. Sando, T.; Hayashi, T.; Takeda, T.; Akiyama, Y.; Nakazawa, Y.; Fukusaki, E.; Kobayashi, A. Histochemical study of detailed laticifer structure and rubber biosynthesis-related protein localization in *Hevea brasiliensis* using spectral confocal laser scanning microscopy. *Planta* **2009**, *230*, 215–225. [CrossRef]

49. Chye, M.L.; Tan, C.T.; Chua, N.H. Three genes encode 3-hydroxy-3-methylglutaryl-coenzyme A reductase in *Hevea brasiliensis*: hmg1 and hmg3 are differentially expressed. *Plant Mol. Biol.* **1992**, *19*, 473–484. [CrossRef]

50. Williamson, B.Y.; Kekwick, R.G. The enzymes forming isopentenyl pyrophosphate from 5-phosphomevalonate (mevalonate 5-phosphate) in the latex of *Hevea brasiliensis*. *Biochem. J.* **1971**, *124*, 407–417.

51. Ko, J.H.; Chow, K.S.; Han, K.H. Transcriptome analysis reveals novel features of the molecular events occurring in the laticifers of *Hevea brasiliensis* (para rubber tree). *Plant Mol. Biol.* **2003**, *53*, 479–492. [CrossRef]

52. Swiezewska, E.; Danikiewicz, W. Polyisoprenoids: Structure, biosynthesis and function. *Prog. Lipid Res.* **2005**, *44*, 235–258. [CrossRef] [PubMed]

53. Chow, K.S.; Wan, K.L.; Isa, M.N.; Bahari, A.; Tan, S.H.; Harikrishna, K.; Yeang, H.Y. Insights into rubber biosynthesis from transcriptome analysis of *Hevea brasiliensis* latex. *J. Exp. Bot.* **2007**, *58*, 2429–2440. [CrossRef] [PubMed]

54. Chow, K.S.; Mat-Isa, M.N.; Bahari, A.; Ghazali, A.K.; Alias, H.; Mohd-Zainuddin, Z.; Hoh, C.C.; Wan, K.L. Metabolic routes affecting rubber biosynthesis in *Hevea brasiliensis* latex. *J. Exp. Bot.* **2012**, *63*, 1863–1871. [CrossRef] [PubMed]

55. Makita, Y.; Ng, K.K.; Veera Singham, G.; Kawashima, M.; Hirakawa, H.; Sato, S.; Othman, A.S.; Matsui, M. Large-scale collection of full-length cDNA and transcriptome analysis in *Hevea brasiliensis*. *DNA Res.* **2017**, *24*, 159–167.

56. Uthup, T.K.; Saha, T.; Ravindran, M.; Bini, K. Impact of an intragenic retrotransposon on the structural integrity and evolution of a major isoprenoid biosynthesis pathway gene in *Hevea brasiliensis*. *Plant Physiol. Biochem.* **2013**, *73*, 176–188. [CrossRef]

57. Epping, J.; Deenen, N.V.; Niephaus, E.; Stolze, A.; Fricke, J.; Huber, C.; Eisenreich, W.; Twyman, R.M.; Prufer, D.; Gronover, C.S. A rubber transferase activator is necessary for natural rubber biosynthesis in dandelion. *Nat. Plants* **2015**, *1*, 15048. [CrossRef]

58. Berthelot, K.; Lecomte, S.; Estevez, Y.; Peruch, F. *Hevea brasiliensis* REF (Hev b 1) and SRPP (Hev b 3): An overview on rubber particle proteins. *Biochimie* **2014**, *106*, 1–9. [CrossRef] [PubMed]

59. Gidda, S.K.; Park, S.; Pyc, M.; Yurchenko, O.; Cai, Y.; Wu, P.; Andrews, D.W.; Chapman, K.D.; Dyer, J.M.; Mullen, R.T. Lipid drop let associated proteins (LDAPs) are required for the dynamic regulation of neutral lipid compartmentation in plant cells. *Plant Physiol.* **2016**, *170*, 2052–2071. [CrossRef]

60. Kim, E.Y.; Park, K.Y.; Seo, Y.S.; Kim, W.T. *Arabidopsis* Small rubber particle protein homolog SRPs play dual roles as positive factors for tissue growth and development and in drought stress responses. *Plant Physiol.* **2016**, *4*, 2494–2510. [CrossRef]

61. Post, J.; van Deenen, N.; Fricke, J.; Kowalski, N.; Wurbs, D.; Schaller, H.; Eisenreich, W.; Huber, C.; Twyman, R.M.; Prüfer, D.; et al. Laticifer-specific cis-prenyltransferase silencing affects the rubber, triterpene, and inulin content of *Taraxacum brevicorniculatum*. *Plant Physiol.* **2012**, *158*, 1406–1417. [CrossRef]

62. Hillebrand, A.; Post, J.J.; Wurbs, D.; Wahler, D.; Lenders, M.; Krzyzanek, V.; Prufer, D.; Gronover, C.S. Down-regulation of small rubber particle protein expression affects integrity of rubber particles and rubber content in *Taraxacum brevicorniculatum*. *PLoS ONE* **2012**, *7*, e41874. [CrossRef]

63. Tata, S.K.; Choi, J.Y.; Jung, J.Y.; Lim, K.Y.; Shin, J.S.; Ryu, S.B. Laticifer tissue-specific activation of the *Hevea* SRPP promoter in *Taraxacum brevicorniculatum* and its regulation by light, tapping and cold stress. *Ind. Crop. Prod.* **2012**, *40*, 219–224. [CrossRef]

64. Wang, X.C.; Shi, M.; Wang, D.; Chen, Y.; Cai, F.; Zhang, S.; Wang, L.; Tong, Z.; Tian, W.M. Comparative proteomics of primary and secondary lutoids reveals that chitinase and glucanase play a crucial combined role in rubber particle aggregation in *Hevea brasiliensis*. *J. Proteome Res.* **2013**, *12*, 5146–5159. [CrossRef] [PubMed]

65. Hao, B.; Wu, J. Laticifer differentiation in *Hevea brasiliensis* induction by exogenous jasmonic acid and linolenic acid. *Ann. Bot.* **2000**, *85*, 37–43. [CrossRef]

66. Conesa, A.; Gotz, S.; García-Gomez, J.M.; Terol, J.; Talon, M.; Robles, M. Blast2GO: a universal tool for annotation, visualization and analysis in functional genomics research. *Bioinformatics* **2005**, *21*, 3674–3676. [CrossRef]

International Journal of
Molecular Sciences

Article

iTRAQ-Based Quantitative Analysis of Responsive Proteins Under PEG-Induced Drought Stress in Wheat Leaves

Yajing Wang [1,2], Xinying Zhang [1,2], Guirong Huang [1,2], Fu Feng [1,2], Xiaoying Liu [1,2], Rui Guo [1,2], Fengxue Gu [1,2], Xiuli Zhong [1,2,*] and Xurong Mei [1,2,*]

[1] Institute of Environment and Sustainable Development in Agriculture, Chinese Academy of Agricultural Sciences, 100081 Beijing, China; wyj8664@126.com (Y.W.); zhxyhb516@126.com (X.Z.); hguirong0920@126.com (G.H.); 82101176076@caas.cn (F.F.); liuxiaoying@caas.cn (X.L.); guorui01@caas.cn (R.G.); gufengxue@caas.cn (F.G.)

[2] State Engineering Laboratory of Efficient Water Use and Disaster Mitigation for Crops/Key Laboratory for Dryland Agriculture of Ministry of Agriculture, 100081 Beijing, China

* Correspondence: zhongxiuli@caas.cn (X.Z.); meixurong@caas.cn (X.M.);
   Tel.: +86-0108-210-6023 (X.Z.); +86-0108-210-6332 (X.M.)

Received: 2 May 2019; Accepted: 21 May 2019; Published: 28 May 2019

Abstract: Drought is an important abiotic stress that seriously restricts crop productivity. An understanding of drought tolerance mechanisms offers guidance for cultivar improvement. In order to understand how a well-known wheat genotype Jinmai 47 responds to drought, we adopted the iTRAQ and LC/MS approaches and conducted proteomics analysis of leaves after exposure to 20% of polyethylene glycol-6000 (PEG)-induced stress for 4 days. The study identified 176 differentially expressed proteins (DEPs), with 65 (36.5%) of them being up-regulated, and 111 (63.5%) down-regulated. DEPs, located in cellular membranes and cytosol mainly, were involved in stress and redox regulation (51), carbohydrate and energy metabolism (36), amino acid metabolism (24), and biosynthesis of other secondary metabolites (20) primarily. Under drought stress, TCA cycle related proteins were up-regulated. Antioxidant system, signaling system, and nucleic acid metabolism etc. were relatively weakened. In comparison, the metabolism pathways that function in plasma dehydration protection and protein structure protection were strongly enhanced, as indicated by the improved biosynthesis of 2 osmolytes, sucrose and Proline, and strongly up-regulated protective proteins, LEA proteins and chaperones. SUS4, P5CSs, OAT, Rab protein, and Lea14-A were considered to be important candidate proteins, which deserve to be further investigated.

Keywords: proteomics; wheat; drought; leaf; iTRAQ

1. Introduction

Wheat (*Triticum aestivum* L.), the second staple human food crop, is regarded as the leading source of vegetable protein in human nutrition [1] and has been subjected to intensive breeding and selection for almost a century [2]. Drought stress is one of the main factors restricting crop productivity and limiting the distribution of species worldwide. Thus, selection efforts have been made to improve drought tolerance to ensure good yield in drought-prone areas.

Studies on molecular and physiological mechanisms of plants in response to drought stress have been extensively conducted to guide cultivar improvement. Once subjected to drought stress, the response processes begin with perception and transduction of drought signal, which usually evokes other subsequent processes. Stomatal adjustment, namely rapid stomatal closure, is triggered by an ABA increase to decrease water loss from leaves [3]. Osmolyte, such as proline, glutamate, glycine betaine and sugars (mannitol, sorbitol and trehalose), accumulate to protect protoplasm from dehydration and enzyme inactivation [4]. Antioxidant systems enhance scavenging reactive oxygen species (ROS) strongly, which attack cellular membrane and organelle through peroxidation damage. These responses involve multiple biochemical pathways and significant changes in gene expression. A large number of studies on wheat plants have identified, cloned, and characterized new genes involved in drought response [5–8].

Accelerated by high throughput technology, genomics, transcriptomics, and proteomics have been rapidly advancing, which facilitate both the elucidation of underlying mechanisms of stress tolerance as well as advancement in breeding technology. A chromosome-based draft sequence of the bread wheat genome was published by the International Wheat Genome Sequencing Consortium in 2012 [9]. It is expected to enable a more effective and focused approach to the breeding of high-yield varieties with increased stress tolerance. Recently, with improvements in sequencing, an annotated reference genome with a detailed analysis of gene content among the structural organization for all the chromosomes and subgenomes was presented by the International Wheat Genome Sequencing Consortium. Quantitative trait mapping and CRISPR-based genome modification show crucial roles in applying this genome in agricultural research and breeding [10]. Moreover, high-throughput transcriptomic studies have provided substantial quantities of data to explore mRNA levels under stresses. However, since protein functions determine the final biological processes that are involved in adaption to drought stress, the changes in gene expression levels do not correspond directly to protein expression levels, let alone the growth phenotypes in wheat, due to the post-translation modification of the protein, which cannot be detected by transcriptomics analyze. Proteomics, as a study on gene products, namely protein, enables the observation of the products of gene expression that have a physiological effect on the plant. Thus, large-scale screening of drought-responsive proteins using comparative proteomic analysis is becoming one of the best strategies to investigate the stress responses of plants. Several recent studies have attempted to describe changes in proteome in response to drought stress [4,11–13]. The known drought-responsive proteins are mainly involved in various metabolic pathways, ranging from regulation of carbohydrate, nitrogen, energy and redox and amino acid metabolism to antioxidant capacity, cytoskeleton stability, signal transduction, as well as mRNA, and protein processing [4,11]. By linking the differentially expressed proteins (DEPs) back to the genes, candidate genes for agronomic traits can be selected, leading to the advances of functional molecular markers for expediting and assisting crop breeding practices [14].

In the North China Plain (NCP), the main wheat production region of China, only less than 30% of the rainfall occurs in the wheat-growing season, which meets only about 25–40% of the water requirements of wheat. As a result, more than 70% of the irrigation water used is for winter wheat [15]. Irrigation usage for wheat threatens the sustainability of the groundwater resource [16]. Breeding prominent rain-fed cultivars, which are capable of actively resisting drought stress, also fully utilizing rainfall, is an important pathway to ensure good harvest stably. Jinmai 47, grown in a large area in the NCP, where proper irrigation is unavailable, is a well-known rain-fed cultivar and is also employed as a parent material in cultivar breeding. Proteomics analysis will hopefully shed light on the key response mechanism to drought stress, and offer useful information for the breeding of new rain-fed cultivars.

Two-dimensional gel electrophoresis (2-DE) and Mass Spectrometry (MS) have been adopted for many years to identify proteins. Gel-based approaches were widely used because of their simplicity and reproducibility. However, the identification of 2-DE is limited by protein abundance (proteins with low abundance are unable to be detected), narrow pI range coverage [17]. On the other hand, isobaric tags for the relative and absolute quantitation (iTRAQ) technique is a high-throughput proteomic technique with higher sensitivity that allows simultaneous identification and quantification of proteins, including low-abundant proteins, in no more than eight samples with high coverage [18,19]. The technique has been applied to *Arabidopsis thaliana* [20], *Zea mays* L. [21], *Brassica napus* [22,23], *Triticum aestivum* L. [24] and *Nicotiana tabacum* [25] in recent years. In this study, iTRAQ-based proteomics analysis, a complement to transcript analysis, was implemented to elucidate the responses of wheat cultivar Jinmai 47 to PEG-induced drought stress.

2. Results

2.1. Physiological Changes in Wheat Seedlings under Drought Stress

Stress severity depends on both stress intensity and stress time. Seedlings of Jinmai 47 were constantly treated with a high concentration of 20% of PEG solution, which means that due to drought severity the seedlings' suffering increased over time. Ion leakage of cellular membrane closely relates to its integrity and stability, and is thus frequently used as a stress indicator. The membrane ion leakage demonstrated an increasing trend with stress time prolonged. It became significantly higher than controls (CK) after 4 days of stress and drastically increased after 5–6 days, reaching 125.6% higher than that of CK ($p < 0.01$) after 6 days of stress (Figure 1). The drastic permeability increase predicted severe irreversible damage to the cellular membrane, as well as plant tissues when stress continued for 5–6 days. However, 4 days of stress is considered to evoke a strong resistant response.

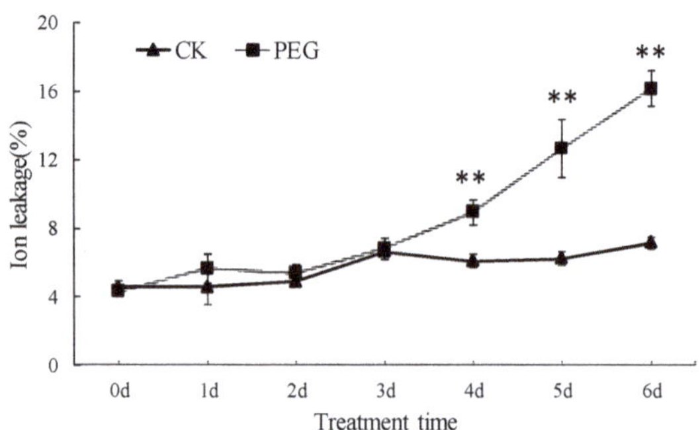

Figure 1. Ion leakage of wheat leaves under drought stress and well-watered condition. ** represents that there is significant differences in wheat in two treatments at $p < 0.01$. "d" in treatment time represents day here.

2.2. Identification of Proteins in Response to Drought Stress

Correlation analysis was conducted between the four parallel replicates (Figure 2). The correlation coefficient between every two replicates was more than 0.98, indicating the excellent biological repeatability of protein expression. PCA was also performed to characterize DEPs in the four replicates of the two treatments here. As Figure 3 showed, samples in two treatments were separated into two correctly, indicating that drought stress caused a significant difference in protein expression.

Comparative proteome analysis was carried out in the seedling leaves of cultivar Jinmai 47 under drought stress and well-watered conditions. 60,109 unique spectra were generated. Of them, 33,561 unique peptides can be matched to 5437 proteins. Among them, 3932 proteins were identified quantitatively (Table S1). 176 proteins presented significant differences in their accumulation levels, compared between the two water conditions, at a relative ratio of >1.2 or <0.83, $p < 0.05$ (Table S3). 65 (36.9%) proteins displayed declined accumulation, while 111 (63.1%) proteins showed increased levels under drought stress.

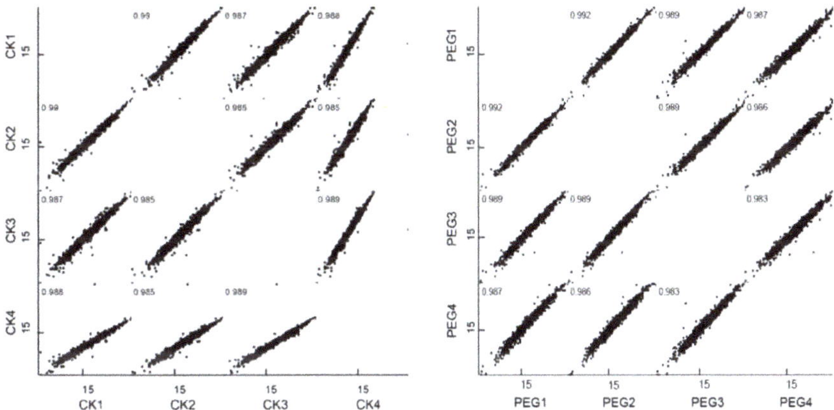

Figure 2. Correlations of replicates of wheat leaves under drought stress and well-watered condition.

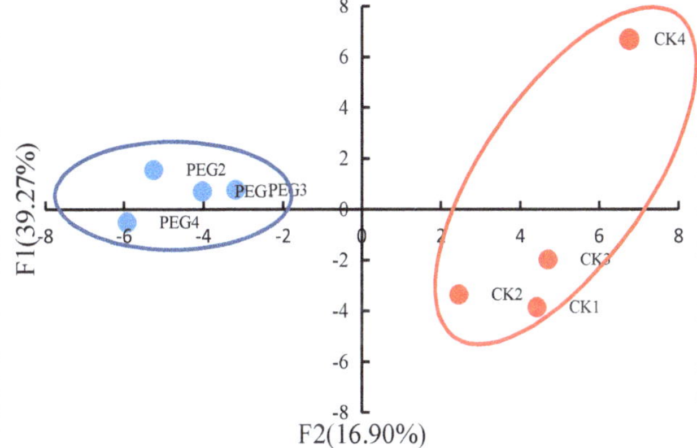

Figure 3. Principal component analysis (PCA) of wheat under drought for 4 days and controls.

2.3. Functional Annotation, Classification and Subcellular Localization of DEPs under Drought Stress

Functional annotation of 176 DEPs was conducted using the Blast2GO program against the non-redundant (nr) NCBI database (Viridiplantae database (txid: 33090, sequence: 5104920)). After that, the proteins were mapped to the pathways in the Kyoto Encyclopedia of Genes and Genomes (KEGG) database (Table S2). Of 176 DEPs, 70 were enriched in 62 pathways, mainly involved in metabolism pathways such as biosynthesis of antibiotics (20), phenylpropanoid biosynthesis (15), starch and sucrose metabolism (10), galactose metabolism (8), glutathione metabolism (7), glycolysis/gluconeogenesis (6), cyanoamino acid metabolism (6), drug metabolism - cytochrome P450 (6), metabolism of xenobiotics by cytochrome P450 (6), and thiamine metabolism (6).

To gain complete functional information of all 176 DEPs, we classified them on the basis of the NCBI database. All DEPs were classified into 13 categories (Figure 4). The largest category was stress and redox regulation (51), followed by carbohydrate metabolism, energy and photosynthesis metabolism (36), amino acid metabolism (24), biosynthesis of secondary metabolites (20) and unknown proteins (25).

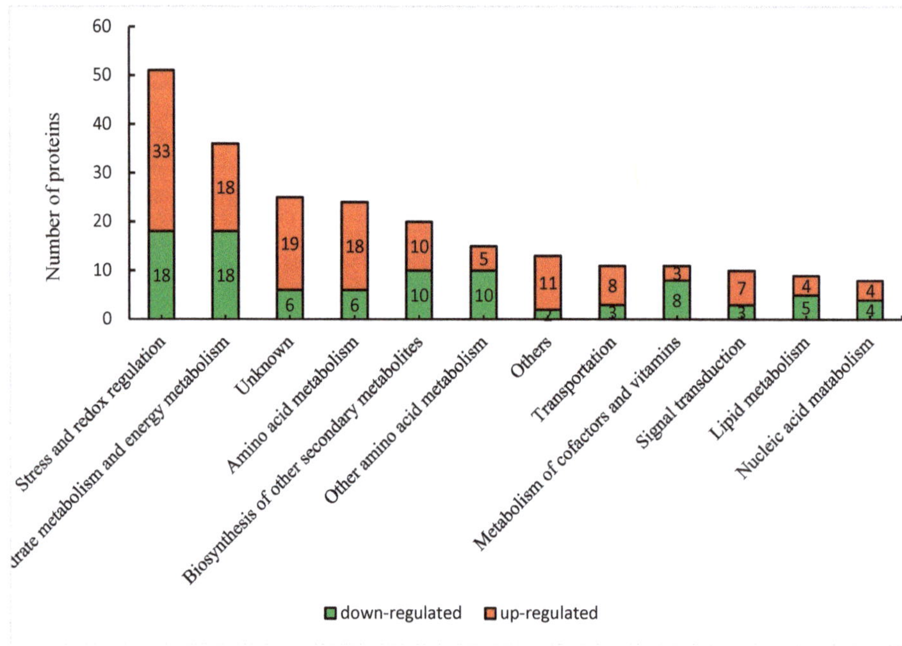

Figure 4. Functional classification of DEPs under drought stress.

The stress and redox regulation category also ranked first in the number of up-regulated DEPs (33). LEA protein family of this category increased most significantly, with nine family members, including 11 kDa LEA protein (A0A3B6TPI1), salt-induced YSK2 dehydrin 3 (A0A3B6QMR3), Group2 LEA protein (Q8LP43), LEA protein (A0A3B5Y269), rab protein (A0A3B6PTJ6), LEA14-A (A0A3B6GU21), LEA protein (A0A3B5Y545), dehydrin 5 (A0A3B6QLV9), LEA protein 31 (A0A3B6TU18) displayed 1.3 to 2.53 fold of that in CK. In addition, 2 chaperone protein ClpDs (A0A3B6MVC4, A0A3B6LR88) increased by 1.82 and 1.55 fold, respectively. For antioxidant proteins, however, 4 peroxidases (PODs) (A0A3B6B862, A0A3B5YTC1, A0A3B6SFH1, A0A3B6MX35), 2 peroxiredoxins (PRXs) (A0A3B6SJF8, A0A3B6RN44), 1 probable glutathione S-transferases (GST) (A0A3B6G148), and 1 ferritin (A0A3B6LKC3) increased slightly, far less than some LEA proteins. Furthermore, catalase (CAT, A0A3B6NJS8), 2 PODs (A0A3B6MKF9, A0A3B6ET60) decreased significantly. The number of up-regulated DEPs in amino acid metabolism category ranked second. This category was characterized by a significant increase in enzymes related to proline biosynthesis. Delta-1-pyrroline-5-carboxylate synthase (P5CS, W5ACM8), P5CS2, pyrroline-5-carboxylate reductase (P5CR, A0A3B6H2J0), and ornithine aminotransferase (OAT, A0A3B6MXE9) increased from 1.21 to 1.98 fold. Carbohydrate metabolism category, including 18 up-regulated proteins, ranked the third. Proteins related to sucrose biosynthesis, sucrose synthase 4 (SUS4, A0A3B6JH89) and galactinol-sucrose galactosyltransferase (RFS, A0A3B6EFU9, A0A3B6GR62), increased 1.6 fold, 1.22 and 1.46 fold, respectively. However, those involved in sucrose degradation, cell wall invertase (INV, A0A3B6JRS1) and fructokinase-2 (A0A3B6RKI2) decreased to 0.74 fold and 0.80 fold, respectively. Pyruvate dehydrogenase E1 component subunit alpha (PDHE1α-2, A0A3B6RHJ4), citrate synthase (CS), aconitate hydratase (ACO, A0A3B6NVD9) participated in TCA cycle, and fructose and mannose metabolism were increased to 1.25–1.53 fold. Surprisingly, except for the only down-regulated rubisco activase small subunit (RACS, AOA3B6JGN7), no other enzymes related to photosynthesis showed significant changes. Other categories, including lipid metabolism, signal transduction, nucleic acid metabolism, cofactor and vitamin metabolism (thiamine metabolism and riboflavin metabolism mainly), as well as metabolisms of other amino acid (glutathione metabolism and cyanoamino acid metabolism mainly), seemed to respond relatively weakly to drought stress (Table S3).

114 of 176 DEPs were successfully annotated subcellular location. Membrane, cytosol and chloroplast were predominant regions where these DEPs were located (Figure 5). Subcellular location of proteins further analyzed aimed at the pathways in which most proteins were up-regulated. Seventeen of 33 up-regulated DEPs involved in stress and redox regulation were located in chloroplast (6), extracellular region (plasmodesma and cell wall) (4), nucleus (3) mainly (Figure 6a). Lea14-A, which functions in dehydration protection of proteins and increased by 2.53-fold under stress, was distributed in the cytosol. Fourteen of 18 up-regulated proteins involved in amino acid metabolism were located in the cytosol (8), mitochondrion (2), and extracellular region (2) predominantly (Figure 6b). P5CS2, which is involved in proline biosynthesis and increased by 1.98-fold, was located in cytosol. Thirteen of 18 up-regulated DEPs related to carbohydrate and energy metabolism were located in the cytosol (6), mitochondrion (3), and chloroplast (2) primarily (Figure 6c). The subcellular localization of 15 in 25 unknown DEPs was also analyzed. Interestingly, most increased DEPs existed in the membrane (Figure 7a), while the majority of downregulated DEPs appeared in chloroplast and membrane (Figure 7b).

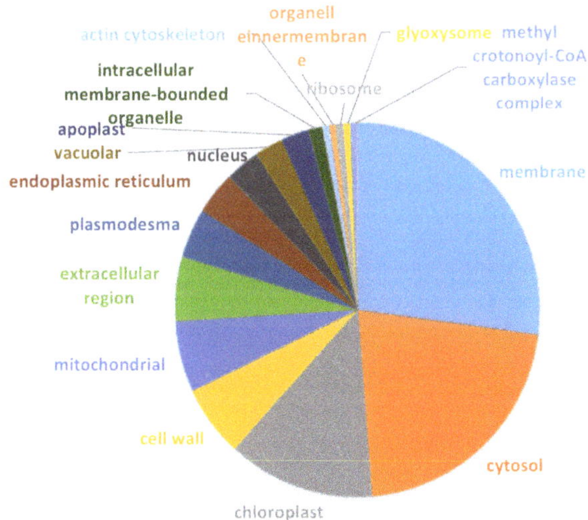

Figure 5. Subcellular localization of the DEPs under drought stress.

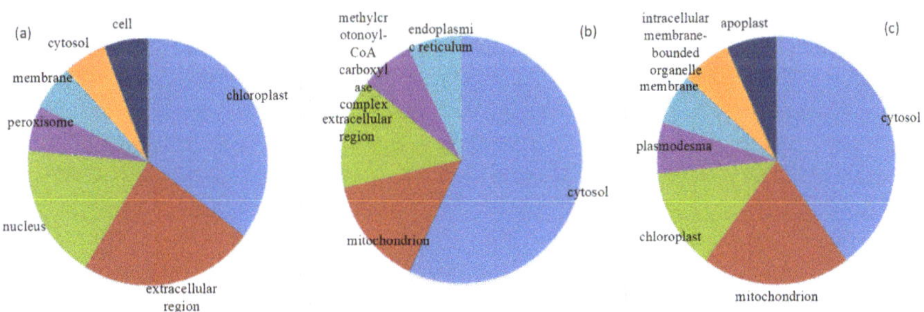

Figure 6. Subcellular location of up-regulated proteins involved in stress and redox regulation (**a**), amino acid metabolism (**b**) as well as carbohydrate and energy metabolism (**c**).

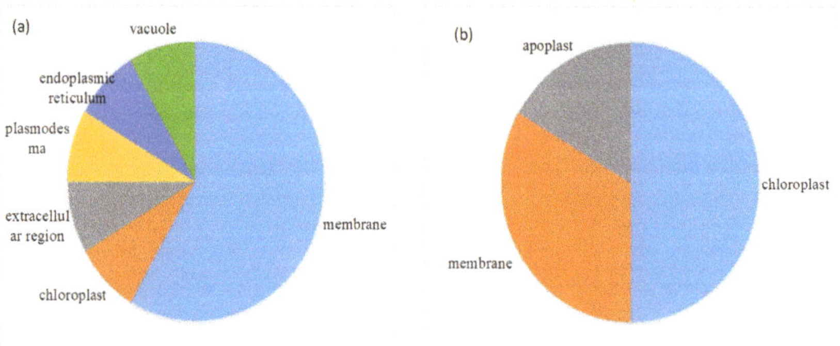

Figure 7. Subcellular location of function unknown up-regulated DEPs (**a**) and down-regulated DEPs (**b**).

2.4. Transcriptional Expression Analysis of Genes Encoding Several Selected DEPs

Several GSTs and a CAT, the important antioxidant enzymes, were down-regulated significantly under stress in Jinmai 47. LEA protein, a marked dehydration responsive protein; and P5CS, a crucial synthase of proline, was enhanced massively under stress. In order to reveal their expression in transcriptional levels, we measured mRNA levels of genes encoding the 4 proteins by quantitative real-time PCR (qRT-PCR) (Figure 8). GST2 and CAT were down-regulated in both transcript level and protein expression level; P5CS was up-regulated in both levels. LEA protein increased to 1.91-fold in protein level, while it showed no significant difference in mRNA level. The results showed that proline biosynthesis was enhanced, while the two antioxidant enzymes, GST and CAT decreased in response to drought. Regarding LEA protein, the poor agreement between mRNA level and protein expression level might be ascribed to the post-transcriptional regulation and/or post-translation modifications.

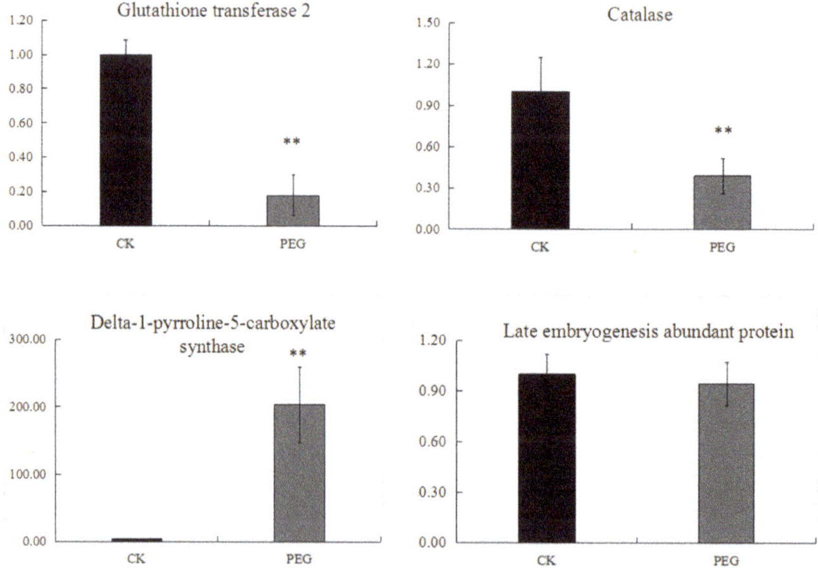

Figure 8. Transcript levels of selected DEPs by qRT-PCR. The data shown here are the mean ± SD of 3 biological replicates. ** represents significant difference at $p < 0.01$.

3. Discussion

The excellent wheat cultivar, which is suitable for growing in drought rain-fed areas, was subjected to a proteomics investigation after being exposed to drought stress for 4 days. The high throughput approach combined with bioinformatics analysis was hoped to comprehensively explain and deeply probe how the cultivar responds to drought stress. Here, 176 DEPs in 13 categories were obtained. 151 of the DEPs were successfully annotated function, and 114 of them were annotated subcellular localization. The highly expressed or suppressed DEPs as well as the metabolism pathways were paid more attention, in order to obtain useful information for breeding of new rain-fed cultivars.

3.1. Carbohydrate Metabolism and Photosynthesis

Carbon is the source of energy, and carbohydrate metabolism in organisms play an essential role in maintaining normal growth and development under stress conditions [26]. Our data showed that SUS4 in plants under drought stress increased to 1.6-fold of control under drought stress (Table S3). SUS catalyzes sucrose synthesis by transferring the glucosyl moiety of ADP glucose to the non-reducing end of an existing a-1,4-glucan chain [4]. It has been reported to be involved in the biosynthesis of sugar polymers, mainly including starch and cellulose, and generation of energy (ATP) [27,28]. Moreover, UDP-glucose-4-epimerase (A0A3B6A3S2) and RFS, which participate in the biosynthesis of raffinose and sucrose, were promoted under drought. On the other hand, INV and fructokinase-2, which associate with sucrose degradation, declined under stress. INV plays a key role in primary metabolism and plant development by hydrolyzing sucrose into glucose and fructose [29,30]. Responses of these DEPs involved in carbohydrate metabolism suggest that wheat genotype Jinmai 47 strongly enhanced sugar biosynthesis when subjected to drought stress. Elevating sucrose under stresses is a common strategy to relief damage, since sucrose can play its role as a compatible osmolyte and can have a protective effect on protein stabilization [31].

On the other hand, TCA was enhanced under drought in Jinmai 47. Three proteins involved in TCA cycle and fructose and mannose metabolism—CS, PDHE1α-2 and ACO—were up-regulated under drought stress, in agreement with previous research on wheat [4]. Enhancing TCA cycle activity may provide energy for diverse metabolisms towards tolerance of drought stress. Regarding photosynthesis, many photosynthesis-related proteins declined under stress, as reported by previous studies [4,11]. In the present study, however, only RACS was suppressed. The small subunit is key and speed-limit in photosynthesis, as proved by Andrew's study in transgenic tobacco (*Nicotiana tabacum* W38). They found that hemizygous leaves with a single antisense gene directed against rubisco's small subunit had 35% of rubisco content of control leaves, and the CO_2 assimilation rate was reduced to 40% of that in controls [32]. The fact that no other proteins about photosynthesis were down-regulated might demonstrate the strong drought resistance of Jinmai 47. Another possibility is that most down-regulated members in the unknown protein category were located in the chloroplast (Figure 7b); those suppressed proteins might be photosynthesis-related enzymes. These unknown proteins remain to be further studied. In terms of carbohydrate metabolism and photosynthesis, Jinmai 47, after exposure to drought stress for 4 days, was distinctly characterized by enhanced sucrose biosynthesis and TCA cycle.

3.2. Amino Acid Metabolism

Many amino acid pathways in plants participate in regulation of tolerance and adaption to stresses. In this study, proteins involved in metabolisms of glutamate, proline (Pro), methionine (Met), cysteine (Cys), and lysine (Lys) changed significantly. P5CR, P5CS2, P5CS and OAT, involved in important pathways for Pro biosynthesis, were significantly higher under drought stress. P5CS2, P5CS and glutamate decarboxylase (GAD, D8L9S2) were related to glutamate degradation. Up-regulation of these enzymes accelerated glutamate reduction. Methioninegamma-lyase (MGL, A0A3B5YXW9) here increased by 1.58-fold under stress treatment. Most cellular Met is converted to S-adenosyl-Met, which was used in essential plant processes such as synthesis of ethylene, cell walls, chlorophyll, DNA replication as well as secondary metabolites. However, *AtMGL*, methionine homeostasis gene *Methionine gamma-lyase* in Arabidopsis, catabolizes Met in an alternative pathway, resulting in the synthesis of Isoleucine (Ile) [33]. Ile, similar to Pro, accumulate as osmoprotectants under drought [34]. The increase of MGL here suggested the decrease of Met and accumulation of Ile (Figure 9), but no direct evidence of the elevation of Ile was found. Rebeille [35] reported that the up-regulation of *AtMGL* occurred in response to concurrent biotic and abiotic stresses. The decrease of Cysteine synthase (CYS, D6QX85) and the increase of MGL, which function in biosynthesis and degradation of Cys here indicate that Cys biosynthesis might be impaired under drought stress. Moreover, lysine-ketoglutarate reductase/saccharopine dehydrogenase1 (LKR/SDH, A0A3B6NTI8) and aldehyde dehydrogenase family 7 member (ALDH7A1, A0A3B6LKM3) related to lysine degradation increased under drought, while bifunctional aspartokinase/homoserine dehydrogenase 2 (AK/HseDH, A3B6LKX6) involved in lysine biosynthesis decreased, implying that lysine accumulation was suppressed under stress. In terms of amino acid metabolism, Jinmai 47, after being subjected to drought stress for 4 days, enhanced conversion and degradation process of some amino acids, including glutamate, Met, Cys and Lys, but strongly improved synthesis process of Pro, which can also be proved by the transcriptional expression increase of gene encoding of the key proline synthesis enzyme P5CS (Figure 8). Similar to sucrose, Pro, as a compatible osmolyte, plays an important role in protecting protoplasm from dehydration [36].

3.3. Stress and Redox Regulation

Plants enhance antioxidant systems to protect cells against the damage caused by high levels of ROS under stress conditions. However, in Jinmai 47 seedlings under drought stress, though most antioxidant enzymes were up-regulated, such as 4 PODs (A0A3B6B862, A0A3B5YTC1, A0A3B6SFH1, A0A3B6MX35), few of them were highly expressed. Moreover, 2 PODs (A0A3B6ET60, A0A3B6MKF9) and CAT were reduced. Even the common antioxidant enzyme, superoxide dismutase (SOD), did not significantly differ from the control. Blue copper proteins (BCPs), which were found to mediate response and tolerance to Aluminum stress and oxidative stress [37–39], decreased to 0.5 to-0.6- fold of control. GSTs, participating in the detoxification of xenobiotics and limiting oxidative damage, were down-regulated, also declining in transcript level (Figure 7). However, a probable GST (A0A3B6G148) was 1.47-fold up-regulated. Zhu et al. reported the contrast expression of GST8 and GST6 under ABA [22]. Ferritin, serving as a regulatory role in iron storage and homeostasis and contributing to plant resistance responses [40], was 1.36-fold up-regulated. The reaction between ferrous iron and H_2O_2 could result in the formation of ·OH, the most dangerous ROS under osmotic stress [41]. Thus, the increased ferritin here contributes to neutralization of ROS-induced damage. To conclude, in Jinmai 47 seedlings under drought stress, the antioxidant system did not respond as strongly as expected, except that some enzymes, such as 4 PODs, a probable GST, and ferritin were increased to play protective roles against ROS.

In comparison, LEA proteins in Jinmai 47 seedlings strongly increased in response to drought stress. Nine up-regulated proteins increased by 1.3 to 2.53 fold, with Group 2 LEA protein (Q8LP43), LEA protein(A0A3B5Y269), Rab protein (A0A3B6PTJ6), and Lea14-A (A0A3B6GU21) rising by 1.88, 1.91, 2.51, and 2.53 fold of that in CK. This result is in agreement with previous researches. Several wheat sequences encoding group 2 LEA proteins have been isolated mainly from cDNA libraries' investigative tissues of stressed seedlings. These genes were induced by cold [36,42], dehydration [43,44], salt [45], and ABA [46]. However, transcript analysis of LEA protein gene showed no significant difference in mRNA level between stress and CK, differing from the 1.91-fold increase of protein level. The poor agreement between mRNA level and protein expression level was ascribed to post-transcriptional regulation and/or post-translation modifications. Three mechanisms that help LEA proteins protect plant cells against abiotic stress have been revealed—binding of metal ions or lipid vesicles [47,48], hydration or ion sequestration [49], and, remarkably, stabilization of proteins and membranes in adaption to abiotic stress. The other function is the protection of proteins against heat-induced inactivation or aggregation under stress conditions [50,51]. Besides, Caseinolytic proteases (Clps) perform important roles in removing protein, which aggregates from cells and can otherwise prove to be highly toxic [52]. Over-expression of rice ClpD1 protein enhanced tolerance to salt and desiccation stresses in transgenic Arabidopsis plants [52]. In this study, ClpD1 increased by 1.82-fold (A0A3B6MVC4) and 1.55- fold (A0A3B6LR88), respectively, to play its protective role under drought stress.

Thus, it can be seen that under drought stress, although the wheat genotype Jinmai 47 enhanced antioxidant systems to some extent, it spent more energy on cell stabilization and protein protection.

3.4. Signal Transduction

Calreticulin-3 (CRT) is an important binding protein in the endoplasmic reticulum (ER). It plays its role in cellular processes, ranging from Ca^{2+} storage and release, protein synthesis, to molecular chaperone activity [51]. Our data showed that CRT protein was impaired under drought. However, in tobacco, CRT protein was overexpressed in response to drought [46]; in Hanxuan-10, a drought-tolerant cultivar, CRT protein abundance increased significantly by 6h after stress and then decreased to a great degree [46]. The 0.77-fold CRT after PEG treatment for 4 days might reach a low level after a great increase. Purple acid phosphatases (PAPs, A0A3B6KQ30, A0A3B6LHH5) also displayed significant decrease under stress. PAPs work in ROS generation and scavenge or stress-activated signal transduction. PAPs participate in thiamine and riboflavin metabolism. Oxidative and NaCl stress, but sufficient Phosphorus, promotes the expression of GmPAP3 gene in soybean. However, in alfalfa, inactive purple acid phosphatase-like protein decreased under osmotic stress [53]. Cytochrome P450 71A1 (CYP71A1, A0A3B6KH30) here were suppressed under PEG treatment. Cytochrome P450 (CYP) takes part in the synthesis of numerous secondary metabolites that play roles as growth and developmental signals or keep plants from various biotic and abiotic stresses [54]. Plant PP2Cs have been found as regulators of signal transduction pathways, such as ABA signaling [55]. Probable protein phosphatase 2C 59(A0A3B6RL15) and probable protein phosphatase 2C 70 isoform X1 (A0A3B6MX95) were higher under drought that in CK. Therefore, most proteins playing roles in signal transduction were impaired, except for probable protein phosphatase 2C 59 (A0A3B6RL15), probable protein phosphatase 2C 70 isoform X1 (A0A3B6MX95), and serine/threonine-protein kinase SAPK3 (A0A3B5Y0E9).

3.5. Transportation

Outer envelope pore protein 16-2 (OEP162, A0A3B6IT89) increased significantly to 1.82 fold of CK. *OEP162* expression was reported to mediate ABA signaling in seeds. The germination of *OEP162* knockout mutant seeds showed ABA hypersensitivity and amino acid metabolism imbalance [56]. OEP162 was located in chloroplast. Thus, its high expression suggests that this enzyme may modulate amino acid transport in chloroplast under stress. Plant non-specific lipid transfer proteins (LTPs) transfer phospholipids as well as galactolipids across membranes (https://www.uniprot.org/). In this study, 4 LTPs (A0A3B6GKQ2, W5D2I6, A0A341ZAA7, A0A3B6FF82) showed significant increase under drought stress, increasing by 1.34 to 1.91-fold of CK. LTPs are involved in the transfer of lipids by the extracellular matrix to form cuticular wax. Previous researches showed that LTP3 positively regulated plant response to abiotic stresses. The overexpression of *LTP3* had enhanced the freezing tolerance of Arabidopsis constitutively [57] *LTP3* knockout mutant was more sensitive to drought stress than wild type plants [57]. In tobacco, a 6-fold increase of *LTP* transcripts was observed after three drying events [58]. Thus, under serious drought stress, Jinmai 47 might begin to improve cuticular wax formation by enhancing LTPs in order to prevent water loss as well as high-temperature damage to leaves accompanying transpiration reduction.

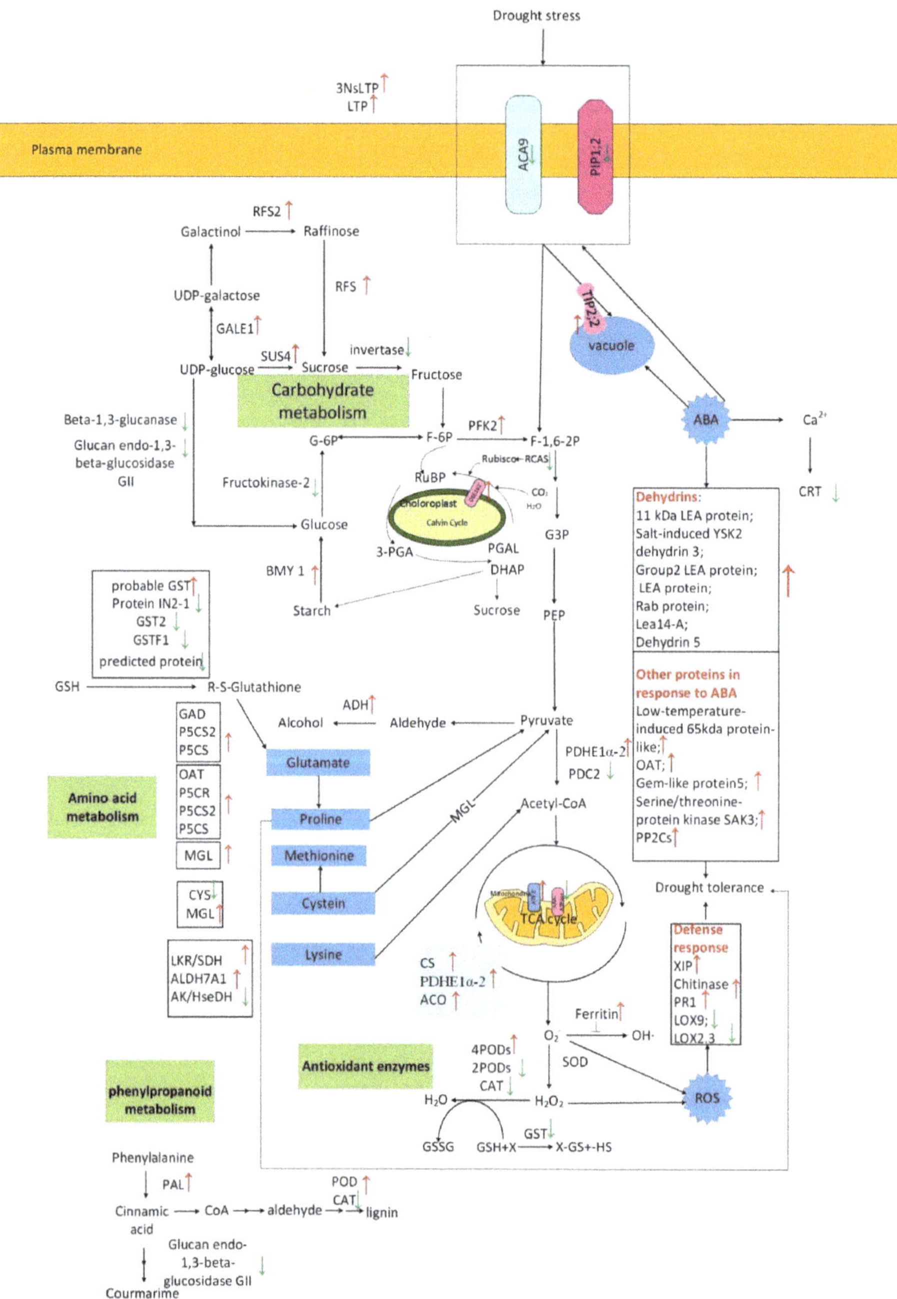

Figure 9. A pathway model of drought stress responses in Jinmai 47 seedlings. Abbreviations: NsLTP: non-specific lipid-transfer protein; LTP: lipid transfer protein; ACA9: probable calcium-transporting ATPase 9; GALE1: UDP-glucose 4-epimerase 1; RFS: probable galactinol-sucrose galactosyltransferase; PIP2-5: aquaporin PIP2-5; TIP2;2: probable aquaporin TIP2-2; SUS4: sucrose synthase 4; Invertase: cell wall invertase; PFK2: ATP-dependent 6-phosphofructokinase 2; PDHE1α-2: pyruvate dehydrogenase

E1 component subunit alpha-2; PDC2: pyruvate decarboxylase 2; BMY1:beta-amylase1; OEE162: outer envelope pore protein 16-2; RCAS: rubisco activase small subunit; probable GST: probable glutathione S-transferase; GST2: glutathione transferase 2; GSTF1: probable glutathione S-transferase GSTF1; GAD: Glutamate decarboxylase; ADH4: alcohol dehydrogenase-like 4; P5CS2: delta-1-pyrroline-5-carboxylate synthase 2; P5CS: deltal-pyrroline-5-carboxylate synthetase; OAT: ornithine aminotransferase; P5CR: pyrroline-5-carboxylate reductase; MGL: methionine gamma-lyase; CYS: beta-cyanoalanine synthase; LKR/SDH: lysine-ketoglutarate reductase/saccharopine dehydrogenase1; ALDH7A1: aldehyde dehydrogenase family 7 member; AK/HseDH: bifunctional aspartokinase/homoserine dehydrogenase 2; PAL: phenylalanine ammonia-lyase; ATP e: ATP synthase E chain; AAA-ATPase: AAA-ATPase ASD; CS: citrate synthase; PDHE1α-2: pyruvate dehydrogenase E1 component subunit alpha-2; ACO: aconitate hydratase; PODs: Peroxidases; CAT: catalase; SOD: superoxide dismutase; XIP: xylanase inhibitor protein 1; PR1: pathogenesis-related protein 1; PP2Cs: protein phosphatase 2C; LEA protein: late embryogenesis abundant protein; LOX9: linoleate 9S-lipoxygenase; LOX2.3: lipoxygenase 2.3.

3.6. Secondary Metabolites

20 DEPs are participating in the biosynthesis of secondary metabolites, 15 of which are involved in phenylpropanoid biosynthesis. The phenylpropanoid pathway leads to the biosynthesis of lignin, flavonoids, condensed tannins and so on, all of which possess antioxidant properties in protecting plants against stresses [59]. Phenylalanine ammonia-lyase (PAL) catalyzes the first reaction in the biosynthesis from l-phenylalanine to trans-cinnamate, a precursor of various metabolites produced in response to stresses [40], and also catalyzes the reaction from tyrosine to p-coumaric acid. Here, pal accumulation was enhanced to 1.31-fold under stress. Beta-1,3-glucanase (GLB3, Q9XEN5), glucan endo-1,3-beta-glucosidase gii (D8L9Q2), beta-glucosidase 5-like (A0A3B6ERB4), and 2 predicted proteins (A0A3B6H5C6, A0A3B6H5R7) belonging to gentiobiase, which also serve as the precursor of coumarine, were decreased. Suppression of these gentiobiases under drought might lead to coumarine biosynthesis reduction. PODs and CAT here play their roles as lactoperoxidase, which catalyze the reaction from p-Coumaryl alcohol to various lignins, such as p-Hydroxy-phenyl lignin and guaiacyl lignin. Most PODs here were promoted, while CAT was suppressed. The changes in phenylpropanoid biosynthesis metabolism mainly affected accumulation of cinnamic trans-cinnamate, coumarine and lignin.

3.7. Other Metabolism Categories

The majority of DEPs involved in signal transduction were impaired, except probable protein phosphatase 2C 59, probable protein phosphatase 2C 70 isoform X1, and serine/threonine-protein kinase SAPK3. Signaling system might be strongly initiated at an early stage of the stress response. Here, after 4 days of severe drought stress, membrane ion leakage began to increase. The plants are supposed to be concentrating on resistance, instead of signaling to evoke a response. In addition to the signaling system, lipid metabolism, nucleic acid metabolism, cofactor and vitamin metabolism, as well as metabolisms of other amino acid seemed to show a relatively weak response to drought stress.

4. Materials and Methods

4.1. Materials and Growth Conditions

Wheat seeds were surface-sterilized by 10% sodium hypochlorite for 10 min, rinsed with distilled water for several times, and soaked in distilled water for 8 h. After soaking, they were placed on water-wetted filter papers in culture dishes to germinate. The experiment was conducted in a growth chamber with 16 h/8 h photoperiod at day/night temperature of 23 °C/20 °C. Seedlings were grown in distilled water at the same temperature cycle. After 8 days, the seedlings were transplanted in rectangular pots with 1/2 Hoagland nutrient solution for 7 days in the same environment. Similar seedlings were selected and divided into two groups. A group was cultivated in 1/2 Hoagland nutrient solution continuously (CK); the other one was cultivated in 1/2 Hoagland nutrient solution with 20% Polyethylene glycol-6000 (PEG-6000).

4.2. Membrane Permeability Measurements

The method of measurement of membrane permeability referred to [60]. Some minor alterations were made here. The second fully expanded leaves were detached from plants under stress or well-watered conditions and washed briefly with deionized water. The leaves for each treatment were cut into 1 cm fragments and immersed in 10 mL deionized water. They were exposed to air by pump for 30 min, followed by agitating for 3 h. Afterwards, the first conductivity was measured and recorded. The leaf fragments in bathing solution were boiled for 15 min to obtain the second conductivity (total conductivity). Three replicates were measured per treatment.

$$\text{Relative Conductivity} = \frac{\text{1st Conductivity}}{\text{2nd Conductivity}} \times 100\% \tag{1}$$

4.3. Protein Extraction

Leaves of wheat treated by PEG for 4 days were harvested and frozen in liquid nitrogen and stored at −80°C. Each treatment contained 4 biological replicates. Acetone extraction method was adopted here to extract total protein from the leaves. A lysis buffer containing 7 M urea, 2 M thiourea, 0.1% CHAPS and 1% protease inhibitor cocktail was conducted in extracting proteins. After centrifugation (13000 g) for 15 min at 4 °C, centrifuge tubes with four volumes of precipitation solution (trichloroacetic acid:acetone, 1:9) were used to transfer the supernatants into, which were stored at −80 °C for 24 h. Bradford assay was used to perform quantification of proteins.

4.4. Protein Digestion and iTRAQ Labeling of Samples

For each sample, 100 μg protein was mixed with lysis buffer containing 25 mM DTT and 50 mM iodoacetamide. The filters were then washed three times using 300 μL dissolution buffer (20 mM triethylammonium bicarbonte), and then being centrifuged at 10000× *g* for 10 min. Subsequently, the proteins were digested at 37 °C overnight with 2 μg trypsin at a ratio of 1:50. The resulting peptides were labeled with 8-plex iTRAQ tags (ABsciex, Framingham, MA, USA), as shown as Table 1. Four replicates in 2 treatments were processed. The labeled mixture was incubated at room temperature for 2.5 h and vacuum-dried.

Table 1. Four replicates in 2 treatments were labelled by different iTRAQ regeants.

NO	113	114	115	116	117	118	119	121
1	CK1	CK2	PEG1	PEG2				POOL1
2			CK3		CK4	PEG3	PEG4	POOL2

4.5. High pH Reversed-Phase (HpRP) Chromatography

The samples were combined into one tube and dried in vacuo after the labeling. Dried peptides were resuspended in 100 μL of mobile phase A. Afterwards, peptides in phase A were centrifuged at 14,000 g for 20 min. The supernatants were loaded on the XBridge® peptide BEH C18 column (130 Å, 3.5 μm, 4.6 mm × 150 mm) (Waters, Milford, MA, USA.) and eluted stepwise by applicating mobile B in the RIGOL L-3000 system (RIGOL, Beijing, China). Mobile phase A was composed of 2% (*v/v*) acetonitrile, 98% (*v/v*) ddH$_2$O and pH 10, and Phase B consisted of 98% (*v/v*) acetonitrile, 2% (*v/v*) ddH$_2$O and pH 10. Peptides were eluted by 5% mobile B for 10 min, 5–20.5% mobile B for 21.7 min, followed by 20.5–95% mobile B for 15.3 min, and returned to 5% mobile B for 4 min finally at a 1 mL/min flow rate. The fractions were eluted at 1-min intervals and collected using step gradients of mobile B. 40 fractions were collected and pooled into 10 final samples, which were dried using a vacuum centrifuge.

4.6. Mass Spectrometer Analysis

LC-MS/MS analysis was performed on an Orbitrap Q-Exactive-plus mass spectrometer (Thermo Fisher Scientific, Waltham, MA, USA) combined with an EASY-nLC 1000 nanosystem (Thermo Fisher Scientific, Waltham, MA, USA). The dried iTRAQ-labelled peptides fractions were dissolved in 20 μL of liquid containing 0.1% formic acid. After centrifugation at 12,000 g for 10 min, the supernatant was collected. The peptide mixtures were then separated using a binary solvent system with 99.9% H_2O, 0.1% formic acid (phase A) and 80% acetonitrile, 0.1% formic acid (phase B). Linear gradients of 8–18% B for 20 min., 18–32% B for 85 min, 32–50% B for 28 min, 50–95% B for 17 min, with a flow rate of 600 nL/min, were employed. The eluent was input into a QE plus mass spectrometer through an EASY-Spray ion source. The mass spectrometer was adjusted automatically between MS and MS/MS mode. The full scan MS mode was operated with the following parameters: automatic gain control (AGC) target, 3e6; resolution, 70,000 FWHM; and scan range, 350–1800 m/z. The MS/MS mode was set as follows: activation type, HCD; collision energy, 32%; resolution, 17,500 FWHM; scan range, 300–1800 m/z AGC target, 1e5.

4.7. Database Searching and Protein Identification

All MS and MS/MS spectrum data were analyzed using ProteoWizard (version 3.0.8789). The MS/MS spectra were searched using the Mascot search engine against the UniProt-wheat_UP000019116 (130,673 entries, 8 January 2019). 2 missed cleavages, ion mass tolerance of 0.05 Da, and parent ion tolerance of 10 ppm for a peptide fragment were permitted. Carbamidomethylation at cysteine was set as a fixed modification, and oxidation at methionine was defined as a variable modification. The proteomic software Scaffold Q+ (version 4.6.2) was used for protein quality controlling and quantitation. The quantitative analysis was carried out on the log2-values of the measured intensities, and the data were normalized based on the median using Perseus software (version 1.5.5). DEPs were analyzed for significant down-regulation or up-regulation. Ratio of the abundance of the proteins identified in plants under PEG treatment to that of CK was used to assess their fold changes. Moreover, a two-sample t-test was used to identify significant ($p < 0.05$) differences in means between the two treatments. DEPs were defined on the basis of thresholds of >1.2- or <0.83-fold change ratios and p value < 0.05 in plants under PEG treatment compared to those of CK. The false discovery rate (FDR) of peptides was less than 1.0%. At least two specific peptides should be identified in each protein, and normalized by median data.

4.8. Functional Annotation and Analysis of DEPs

Functional annotation of proteins was conducted using the Blast2GO software against non-redundant (nr) NCBI databases (Viridiplantae database (txid: 33090, sequence: 5104920)). Kyoto Encyclopedia of Genes and Genomes (KEGG) pathway analyses were performed on DEPs.

4.9. qRT-PCR Analysis

Total RNA was extracted from leaves of wheat with RNeasy Plant Mini Kit (Qiagen, Inc., Hilden, Germany) under 20% PEG for four days and that of control plants. DNA contamination was removed with an on-column DNase (Qiagen, Inc.) treatment. IScript cDNA synthesis kit (BioRad Laboratories, Hercules, CA, USA) was used here for 1 mg of total RNA to reverse transcribe into first strand cDNA in a 20 μL reaction. cDNA was then diluted into 10 ng/μL, 2 ng/μL, 0.4 ng/μL and 0.08 ng/μL concentration series. An ABI 7500 Sequence Detection System from Applied Biosystems (Applied Biosystems, Singapore) was adopted here for three replicates of real-time PCR experiments for each concentration. GAPDH served as the reference gene here. The primers for target genes were designed by Primer Express software (Applied Biosystems) and the sequences are shown in Table 2. No primer dimmers or false amplicons interfered with the result since primer titration and dissociation experiments were performed. Ct number was extracted for both reference gene and target gene with auto baseline and manual threshold after the real-time PCR experiment.

Table 2. The primer sequences used in qRT-PCR.

Accession Number of Protein	Name of Protein	Accession Number of Related Gene	Primer Sequences (5′-3′)	Product Length (bp)
I7FHT3	Glutathione transferase 2	TraesCS1D02G190000	F: GCCCGTGCTCATCCACAA R: CAGCCCCTCCGCCTTCT	220
A0A3B6NJS8	Catalase	TraesCS6A02G041700	F: CCCAAACTACCTGATGCTCC R: TGATCCTCGTCTTCTCCCTTC	203
A0A077RXE4	Delta-1-pyrroline-5-carboxylate synthase	TraesCS3B02G395900	F: ACCCTGAAGGCTGGAAAGATA R: GCATCAGGACGAGACTCAAAA	176
A0A3B5Y545	late embryogenesis abundant protein	TraesCS1A02G364000	F: GGACCAGACCGCCAGCAC R: CCCATGCCCAGCGTGTT	261
W5GYX5	GAPDH	TraesCS6D02G196300	F: GTTTGGCATTGTTGAGGGTT R: ATCATAGGTTGCTGGCTTCG	268

4.10. Statistical Analysis

The data were subjected to T-test with SAS software. We use * and ** to present there are significant differences between plants under stress or well-watered at the levels $p < 0.05$ and $p < 0.01$, respectively. Microsoft Excel was used to plot the results.

5. Conclusions

The proteomics analysis of wheat genotype Jinmai 47 revealed some of its drought response characteristics. The antioxidant system was not highly expressed as expected. Expression of photosynthesis-related proteins was not affected significantly by drought treatment, except rubisco activase small subunit. TCA cycle related proteins were up-regulated, which may enhance energy supply for multiple biological processes to resist stress. The wheat genotype Jinmai 47, when subjected to severe drought stress, seems to be characterized by substantial enhancement in plasma and protein protection mechanism. Improving sucrose and Pro biosynthesis, the important compatible osmolytes, contributes to stabilizing cells by protecting plasma from dehydration under drought stress. Pro biosynthesis was proven in mRNA levels as well. The high expressions of LEA proteins and chaperone proteins play essential roles in safeguarding proteins when dehydrated. However, qRT-PCR analysis showed that the mRNA level of LEA protein had a poor agreement with its protein level, which might be due to post-transcriptional regulation or post-translational modification. An excellent rani-fed cultivar, which can stably gain higher yield in rain-fed areas, must be capable of tolerating more severe and longer drought, and promptly recovering growth after release from stress, or even keeping photosynthesis to some extent under stress conditions. The robust protective mechanism might not only improve drought resistance, but also ensure strong and prompt recovery. Based on this study, SUS4 (1.60), P5CSs (1.34, 1.98), OAT (1.22), Rab protein (2.51), and Lea14-A (2.53) were considered to be important candidate proteins, which deserve to be further investigated.

Supplementary Materials: Supplementary materials can be found at http://www.mdpi.com/1422-0067/20/11/2621/s1.

Author Contributions: Conceptualization, X.Z. (Xiuli Zhong), X.M. and Y.W.; Data curation, X.Z., G.H. and F.F.; Formal analysis, Y.W.; Methodology, Y.W.; Visualization, X.L., R.G. and F.G.; Writing—original draft, Y.W.; Writing— review & editing, Y.W., X.Z. (Xinying Zhang), X.Z. (Xiuli Zhong).

Funding: This research was funded by National Key Research and Development Program (2017YFD0201702).

Conflicts of Interest: The authors declare no conflict of interest.

Abbreviations

DEP	Differentially expressed protein
ROS	Reactive oxygen species
PEG	polyethylene glycol-6000
iTRAQ	Isobaric tags for relative and absolute quantitation
PCA	Principal component analysis
KEGG	KyotoEncyclopedia of Genes and Genomes
qRT-PCR	Quantitative real-time PCR

References

1. Raj, P.M.; Sendhil, R.; Tripathi, S.C.; Subhash, C.; Chhokar, R.S.; Sharma, R.K. Hydro-priming of seed improves the water use efficiency, grain yield and net economic return of wheat under different moisture regimes. *SAARC J. Agric.* **2013**, *11*, 149–159. [CrossRef]
2. Peng, Z.; Wang, M.; Li, F.; Lv, H.; Li, C.; Xia, G. A proteomic study of the response to salinity and drought stress in an introgression strain of bread wheat. *Mol. Cell Proteom.* **2009**, *8*, 2676–2686. [CrossRef]
3. Israelsson, M.; Siegel, R.S.; Young, J.; Hashimoto, M.; Iba, K.; Schroeder, J.I. Guard cell ABA and CO2 signaling network updates and Ca2+ sensor priming hypothesis. *Curr. Opin. Plant Biol.* **2006**, *9*, 654–663. [CrossRef]
4. Deng, X.; Liu, Y.; Xu, X.; Liu, D.; Zhu, G.; Yan, X.; Wang, Z.; Yan, Y. Comparative proteome analysis of wheat flag leaves and developing grains under water deficit. *Front. Plant Sci.* **2018**, *9*. [CrossRef]
5. Gao, H.; Wang, Y.; Xu, P.; Zhang, Z. Overexpression of a WRKY transcription factor TaWRKY2 enhances drought stress tolerance in transgenic wheat. *Front. Plant Sci.* **2018**, *9*. [CrossRef]
6. Rong, W.; Qi, L.; Wang, A.; Ye, X.; Du, L.; Liang, H.; Xin, Z.; Zhang, Z. The ERF transcription factor TaERF3 promotes tolerance to salt and drought stresses in wheat. *Plant Biotechnol. J.* **2014**, *12*, 468–479. [CrossRef] [PubMed]
7. Xue, G.; Way, H.M.; Richardson, T.; Drenth, J.; Joyce, P.A.; McIntyre, C.L. Overexpression of TaNAC69 leads to enhanced transcript levels of stress Up-Regulated genes and dehydration tolerance in bread wheat. *Mol. Plant.* **2011**, *4*, 697–712. [CrossRef]
8. Zhang, H.; Mao, X.; Zhang, J.; Chang, X.; Jing, R. Single-nucleotide polymorphisms and association analysis of drought-resistance gene TaSnRK2.8 in common wheat. *Plant Physiol. Biochem.* **2013**, *70*, 174–181. [CrossRef] [PubMed]
9. Brenchley, R.; Spannagl, M.; Pfeifer, M.; Barker, G.L.A.; D'Amore, R.; Allen, A.M.; McKenzie, N.; Kramer, M.; Kerhornou, A.; Bolser, D.; et al. Analysis of the breadwheat genome using whole-genome shotgun sequencing. *Nature* **2012**, *491*, 705–710. [CrossRef] [PubMed]
10. Appels, R.; Eversole, K.; Feuillet, C.; Keller, B.; Rogers, J.; Stein, N.; Pozniak, C.J.; Stein, N.; Choulet, F.; Distelfeld, A.; et al. Shifting the limits in wheat research and breeding using a fully annotated reference genome. *Science* **2018**, *361*, 661. [CrossRef]
11. Cheng, L.; Wang, Y.; He, Q.; Li, H.; Zhang, X.; Zhang, F. Comparative proteomics illustrates the complexity of drought resistance mechanisms in two wheat (*Triticum aestivum* L.) cultivars under dehydration and rehydration. *BMC Plant Biol.* **2016**, *16*. [CrossRef] [PubMed]
12. Ford, K.L.; Cassin, A.; Bacic, A. Quantitative proteomic analysis of wheat cultivars with differing drought stress tolerance. *Front. Plant Sci.* **2011**, *2*. [CrossRef]
13. Zhang, Y.; Huang, X.; Wang, L.; Wei, L.; Wu, Z.; You, M.; Li, B. Proteomic analysis of wheat seed in response to drought stress. *J. Integr. Agric.* **2014**, *13*, 919–925. [CrossRef]
14. Varshney, R.K.; Graner, A.; Sorrells, M.E. Genomics-assisted breeding for crop improvement. *Trends Plant Sci.* **2005**, *10*, 621–630. [CrossRef]
15. Mei, X.; Zhong, X.; Vincent, V.; Liu, X. Improving water use efficiency of wheat crop varieties in the north china plain: Review and analysis. *J. Integr Agr.* **2013**, *12*, 1243–1250. [CrossRef]
16. Zhang, X.Y.; Pei, D.; Hu, C.S. Conserving groundwater for irrigation in the North China Plain. *Irrig. Sci.* **2003**, *21*, 159–166. [CrossRef]
17. Gygi, S.P.; Corthals, G.L.; Zhang, Y.; Rochon, Y.; Aebersold, R. Evaluation of two-dimensional gel electrophoresis-based proteome analysis technology. *Proc. Natl. Acad. Sci. USA* **2000**, *97*, 9390–9395. [CrossRef]

18. Zhang, Y.; Lou, H.; Guo, D.; Zhang, R.; Su, M.; Hou, Z.; Zhou, H.; Liang, R.; Xie, C.; You, M.; et al. Identifying changes in the wheat kernel proteome under heat stress using iTRAQ. *CROP J.* **2018**, *6*, 600–610. [CrossRef]

19. Casado-Vela, J.; Jose Martinez-Esteso, M.; Rodriguez, E.; Borras, E.; Elortza, F.; Bru-Martinez, R. ITRAQ-based quantitative analysis of protein mixtures with large fold change and dynamic range. *Proteomics* **2010**, *10*, 343–347. [CrossRef] [PubMed]

20. Pang, Q.; Chen, S.; Dai, S.; Chen, Y.; Wang, Y.; Yan, X. Comparative Proteomics of Salt Tolerance in Arabidopsis thaliana and Thellungiella halophila. *J. Proteome Res.* **2010**, *9*, 2584–2599. [CrossRef]

21. Zhao, F.; Zhang, D.; Zhao, Y.; Wang, W.; Yang, H.; Tai, F.; Li, C.; Hu, X. The difference of physiological and proteomic changes in maize leaves adaptation to drought, heat, and combined both stresses. *Front. Plant Sci.* **2016**, 7. [CrossRef]

22. Zhu, M.; Simons, B.; Zhu, N.; Oppenheimer, D.G.; Chen, S. Analysis of abscisic acid responsive proteins in Brassica napus guard cells by multiplexed isobaric tagging. *J. Proteom.* **2010**, *73*, 790–805. [CrossRef]

23. Koh, J.; Chen, G.; Yoo, M.; Zhu, N.; Dufresne, D.; Erickson, J.E.; Shao, H.; Chen, S. Comparative proteomic analysis of brassica napus in response to drought stress. *J. Proteome Res.* **2015**, *14*, 3068–3081. [CrossRef] [PubMed]

24. Alvarez, S.; Choudhury, S.R.; Pandey, S. Comparative quantitative proteomics analysis of the ABA response of roots of Drought-Sensitive and Drought-Tolerant wheat varieties identifies proteomic signatures of drought adaptability. *J. Proteome Res.* **2014**, *13*, 1688–1701. [CrossRef]

25. Xie, H.; Yang, D.; Yao, H.; Bai, G.; Zhang, Y.; Xiao, B. ITRAQ-based quantitative proteomic analysis reveals proteomic changes in leaves of cultivated tobacco (*Nicotiana tabacum*) in response to drought stress. *Biochem. Biophys. Res. Commun.* **2016**, *469*, 768–775. [CrossRef]

26. Hossain, Z.; Komatsu, S. Contribution of proteomic studies towards understanding plant heavy metal stress response. *Front. Plant Sci.* **2012**, *3*, 310. [CrossRef]

27. Coleman, H.D.; Yan, J.; Mansfield, S.D. Sucrose synthase affects carbon partitioning to increase cellulose production and altered cell wall ultrastructure. *Proc. Natl. Acad. Sci. USA* **2009**, *106*, 13118–13123. [CrossRef]

28. Chourey, P.S.; Taliercio, E.W.; Carlson, S.J.; Ruan, Y.L. Genetic evidence that the two isozymes of sucrose synthase present in developing maize endosperm are critical, one for cell wall integrity and the other for starch biosynthesis. *Mol. Gen. Genet.* **1998**, *259*, 88–96.

29. Barratt, D.H.; Derbyshire, P.; Findlay, K.; Pike, M.; Wellner, N.; Lunn, J.; Feil, R.; Simpson, C.; Maule, A.J.; Smith, A.M. Normal growth of Arabidopsis requires cytosolic invertase but not sucrose synthase. *Proc. Natl. Acad. Sci. USA* **2009**, *106*, 13124–13129. [CrossRef] [PubMed]

30. Ruan, Y.L.; Jin, Y.; Huang, J. Capping invertase activity by its inhibitor: Roles and implications in sugar signaling, carbon allocation, senescence and evolution. *Plant Signal. Behav.* **2009**, *4*, 983–985. [CrossRef] [PubMed]

31. Lee, J.C.; Timasheff, S.N. The stabilization of proteins by sucrose. *J. Biol. Chem.* **1981**, *256*, 7193–7201. [PubMed]

32. Andrews, T.J.; Hudson, G.S.; Mate, C.J.; von Caemmerer, S.; Evans, J.R.; Arvidsson, Y.B.C. Rubisco: The consequences of altering its expression and activation in transgenic plants. *J. Exp. Bot.* **1995**, *46*, 1293–1300. [CrossRef]

33. Atkinson, N.J.; Lilley, C.J.; Urwin, P.E. Identification of genes involved in the response of arabidopsis to simultaneous biotic and abiotic stresses. *Plant Physiol.* **2013**, *162*, 2028–2041. [CrossRef] [PubMed]

34. Nambara, E.; Kawaide, H.; Kamiya, Y.; Naito, S. Characterization of an Arabidopsis thaliana mutant that has a defect in ABA accumulation: ABA-dependent and ABA-independent accumulation of free amino acids during dehydration. *Plant Cell Physiol.* **1998**, *39*, 853–858. [CrossRef]

35. Rebeille, F.; Jabrin, S.; Bligny, R.; Loizeau, K.; Gambonnet, B.; Van Wilder, V.; Douce, R.; Ravanel, S. Methionine catabolism in Arabidopsis cells is initiated by a gamma-cleavage process and leads to S-methylcysteine and isoleucine syntheses. *Proc. Natl. Acad. Sci. USA* **2006**, *103*, 15687–15692. [CrossRef]

36. Houde, M.; Danyluk, J.; Laliberte, J.F.; Rassart, E.; Dhindsa, R.S.; Sarhan, F. Cloning, characterization, and expression of a cDNA encoding a 50-kilodalton protein specifically induced by cold acclimation in wheat. *Plant Physiol.* **1992**, *99*, 1381–1387. [CrossRef]

37. Houde, M.; Diallo, A.O. Identification of genes and pathways associated with aluminum stress and tolerance using transcriptome profiling of wheat near-isogenic lines. *BMC Genom.* **2008**, *9*, 400. [CrossRef]

38. Ezaki, B.; Gardner, R.C.; Ezaki, Y.; Matsumoto, H. Expression of aluminum-induced genes in transgenic arabidopsis plants can ameliorate aluminum stress and/or oxidative stress. *Plant Physiol.* **2000**, *122*, 657–665. [CrossRef]

39. Ezaki, B.; Sivaguru, M.; Ezaki, Y.; Matsumoto, H.; Gardner, R.C. Acquisition of aluminum tolerance in Saccharomyces cerevisiae by expression of the BCB or NtGDI1 gene derived from plants. *FEMS Microbiol. Lett.* **1999**, *171*, 81–87. [CrossRef]

40. Glijin, A.; Mita, E.; Levitchi, A.; Acciu, A.; Calmis, A.; Duca, M. Phenylalanine ammonia-lyase in normal and biotic stress conditions. *Lucrari Stiintifice, Universitatea de Stiinte Agricole Si Medicina Veterinara "Ion Ionescu de la Brad" Iasi, Seria Horticultura* **2011**, *54*, 97–102.

41. Jiang, Q.; Li, X.; Niu, F.; Sun, X.; Hu, Z.; Zhang, H. ITRAQ-based quantitative proteomic analysis of wheat roots in response to salt stress. *Proteomics* **2017**, *17*, 1600265. [CrossRef] [PubMed]

42. Ouellet, F.; Houde, M.; Sarhan, F. Purification, characterization and cDNA cloning of the 200 kDa protein induced by cold acclimation in wheat. *Plant Cell Physiol.* **1993**, *34*, 59–65.

43. King, S.W.; Joshi, C.P.; Nguyen, H.T. DNA sequence of an ABA-responsive gene (rab 15) from water-stressed wheat roots. *Plant Mol. Biol.* **1992**, *18*, 119–121. [CrossRef]

44. Labhilili, M.; Joudrier, P.; Gautier, M.O. Characterization of cDNAs encoding Triticum durum dehydrins and their expression patterns in cultivars that differ in drought tolerance. *Plant Sci.* **1995**, *112*, 219–230. [CrossRef]

45. Masmoudi, K.; Brini, F.A.; Hassairi, A.; Ellouz, R. Isolation and characterization of a differentially expressed sequence tag from Triticum durum salt-stressed roots. *Plant Physiol. Biochem.* **2001**, *39*, 971–979. [CrossRef]

46. Jia, X.Y.; Xu, C.Y.; Jing, R.L.; Li, R.Z.; Mao, X.G.; Wang, J.P.; Chang, X.P. Molecular cloning and characterization of wheat calreticulin (CRT) gene involved in drought-stressed responses. *J. Exp. Bot.* **2008**, *59*, 739–751. [CrossRef]

47. Rahman, L.N.; Bamm, V.V.; Voyer, J.A.; Smith, G.S.; Chen, L.; Yaish, M.W.; Moffatt, B.A.; Dutcher, J.R.; Harauz, G. Zinc induces disorder-to-order transitions in free and membrane-associated Thellungiella salsuginea dehydrins TsDHN-1 and TsDHN-2: A solution CD and solid-state ATR-FTIR study. *Amino Acids* **2011**, *40*, 1485–1502. [CrossRef]

48. Kovacs, D.; Kalmar, E.; Torok, Z.; Tompa, P. Chaperone activity of ERD10 and ERD14, two disordered stress-related plant proteins. *Plant Physiol.* **2008**, *147*, 381–390. [CrossRef]

49. Tompa, P.; Banki, P.; Bokor, M.; Kamasa, P.; Kovacs, D.; Lasanda, G.; Tompa, K. Protein-water and protein-buffer interactions in the aqueous solution of an intrinsically unstructured plant dehydrin: NMR intensity and DSC aspects. *Biophys. J.* **2006**, *91*, 2243–2249. [CrossRef] [PubMed]

50. Sun, X.; Rikkerink, E.H.A.; Jones, W.T.; Uversky, V.N. Multifarious roles of intrinsic disorder in proteins illustrate its broad impact on plant biology. *Plant Cell.* **2013**, *25*, 38–55. [CrossRef]

51. Boucher, V.; Buitink, J.; Lin, X.; Boudet, J.; Hoekstra, F.A.; Hundertmark, M.; Renard, D.; Leprince, O. MtPM25 is an atypical hydrophobic late embryogenesis-abundant protein that dissociates cold and desiccation-aggregated proteins. *Plant Cell Environ.* **2010**, *33*, 418–430. [CrossRef]

52. Mishra, R.C.; Richa; Grover, A. Constitutive over-expression of rice ClpD1 protein enhances tolerance to salt and desiccation stresses in transgenic Arabidopsis plants. *Plant Sci.* **2016**, *250*, 69–78. [CrossRef] [PubMed]

53. Zhang, C.; Shi, S. Physiological and Proteomic Responses of Contrasting Alfalfa (*Medicago sativa* L.) Varieties to PEG-Induced Osmotic Stress. *Front. Plant Sci.* **2018**, *9*. [CrossRef]

54. Xu, J.; Wang, X.; Guo, W. The cytochrome P450 superfamily: Key players in plant development and defense. *J. Integr. Agric.* **2015**, *14*, 1673–1686. [CrossRef]

55. Schweighofer, A.; Hirt, H.; Meskiene, I. Plant PP2C phosphatases: Emerging functions in stress signaling. *Trends Plant Sci.* **2004**, *9*, 236–243. [CrossRef] [PubMed]

56. Pudelski, B.; Schock, A.; Hoth, S.; Radchuk, R.; Weber, H.; Hofmann, J.; Sonnewald, U.; Soll, J.; Philippar, K. The plastid outer envelope protein OEP16 affects metabolic fluxes during ABA-controlled seed development and germination. *J. Exp. Bot.* **2012**, *63*, 1919–1936. [CrossRef]

57. Guo, L.; Yang, H.; Zhang, X.; Yang, S. Lipid transfer protein 3 as a target of MYB96 mediates freezing and drought stress in Arabidopsis. *J. Exp. Bot.* **2013**, *64*, 1755–1767. [CrossRef]

58. Cameron, K.D.; Teece, M.A.; Smart, L.B. Increased accumulation of cuticular wax and expression of lipid transfer protein in response to periodic drying events in leaves of tree tobacco. *Plant Physiol.* **2006**, *140*, 176–183. [CrossRef]

59. Pallavi, S.; Jha, A.B.; Dubey, R.S.; Pessarakli, M. Reactive oxygen species, oxidative damage, and antioxidative defense mechanism in plants under stressful conditions. *J. Bot.* **2012**, *2012*, 217037.
60. Hong, Y.; Zheng, S.; Wang, X. Dual functions of phospholipase Dalpha1 in plant response to drought. *Mol. Plant* **2008**, *1*, 262–269. [CrossRef]

International Journal of
Molecular Sciences

Article

Comparative Proteomic Analysis during the Involvement of Nitric Oxide in Hydrogen Gas-Improved Postharvest Freshness in Cut Lilies

Jianqiang Huo, Dengjing Huang, Jing Zhang, Hua Fang, Bo Wang, Chunlei Wang, Zhanjun Ma and Weibiao Liao *

College of Horticulture, Gansu Agricultural University, Lanzhou 730070, China; huojq3061@163.com (J.H.); huangdj3032@163.com (D.H.); zhangj4517@163.com (J.Z.); fangh1610@163.com (H.F.); wangb0447@163.com (B.W.); wangchunlei@gsau.edu.cn (C.W.); mazhanjun@gsau.edu.cn (Z.M.)
* Correspondence: liaowb@gsau.edu.cn; Tel.: +86-138-9328-7942

Received: 29 October 2018; Accepted: 6 December 2018; Published: 9 December 2018

Abstract: Our previous studies suggested that both hydrogen gas (H_2) and nitric oxide (NO) could enhance the postharvest freshness of cut flowers. However, the crosstalk of H_2 and NO during that process is unknown. Here, cut lilies (*Lilium* "Manissa") were used to investigate the relationship between H_2 and NO and to identify differentially accumulated proteins during postharvest freshness. The results revealed that 1% hydrogen-rich water (HRW) and 150 μM sodium nitroprusside (SNP) significantly extended the vase life and quality, while NO inhibitors suppressed the positive effects of HRW. Proteomics analysis found 50 differentially accumulated proteins in lilies leaves which were classified into seven functional categories. Among them, ATP synthase CF1 alpha subunit (chloroplast) (AtpA) was up-regulated by HRW and down-regulated by NO inhibitor. The expression level of *LlatpA* gene was consistent with the result of proteomics analysis. The positive effect of HRW and SNP on ATP synthase activity was inhibited by NO inhibitor. Meanwhile, the physiological-level analysis of chlorophyll fluorescence and photosynthetic parameters also agreed with the expression of AtpA regulated by HRW and SNP. Altogether, our results suggested that NO might be involved in H_2-improved freshness of cut lilies, and AtpA protein may play important roles during that process.

Keywords: proteomic; postharvest freshness; ATP synthase; ATP synthase CF1 alpha subunit (chloroplast); chlorophyll fluorescence parameters; photosynthetic parameters

---

1. Introduction

Hydrogen gas (H_2), a colorless and odorless gas, is the lightest and structurally simplest gas in the world. As an important signaling molecule, H_2 has been shown to be involved in many plant developmental processes [1]. More recently, some researchers found that H_2 could alleviate aluminum (Al) toxicity [2], mercury (Hg) toxicity [3], and UV-A irradiation [4] by increasing the activity of antioxidant enzymes. Meanwhile, H_2 could promote lateral root formation through nitric oxide (NO) synthesis induced by auxin [5] or in a heme oxygenase-1/carbon monoxide-dependent manner [6]. Significantly, H_2 was also reported to play an important role in delaying senescence and maturity [7,8]. Hydrogen-rich water (HRW) treatments could prolong the shelf life of kiwifruit by regulating the antioxidant defense [7]. Our study has shown that H_2 enhanced the vase life and postharvest quality of cut lily (*Lilium* spp.) and cut rose (*Rosa hybrid* L.) flowers through maintaining water balance and membrane stability [8]. However, the deep mechanism of H_2 in delaying the senescence and shelf life of perishable horticultural products needs to be further investigated.

NO is a signaling molecule that interacts with other hormones and growth regulators. Recently, NO was reported to play vital roles in delaying senescence and improving the quality of horticultural products. NO inhibited the production of ethylene by modulating the expressions of some genes and proteins during postharvest of horticulture plants [9]. NO, as a preservative solution to cut flowers, can extend the vase life of cut gerbera flowers by increasing water uptake and promoting antioxidant activity [10]. Exogenous NO could also promote the vase life of cut gladiolus flowers by increasing the scavenging mechanism of reactive oxygen species (ROS) and down-regulating the expression of senescence-associated genes (SAGs) [11]. The vase life of cut carnation flowers was significantly prolonged by exogenous NO, which improved the activity of antioxidant enzymes including superoxide dismutase (SOD), peroxidase (POD), catalase (CAT), and ascorbate peroxidase (APX) [12]. Our previous study reported that NO could decrease ethylene production in cut roses by inhibiting the activity of 1-aminocyclopropane-1-carboxylate oxidase (ACO), thus promoting the vase life of cut roses [13].

As mentioned above, H_2 and NO as exogenous gaseous signaling molecules played exceedingly positive roles in the postharvest preservation of horticultural products. The relationship between H_2 and NO in plants has been reported in recent years. H_2 was reported to regulate stomatal movement, which is involved in the abscisic acid (ABA) signaling cascade by promoting the generation of NO [14]. Meanwhile, Zhu et al. [15] also found that the adventitious root formation in cucumber explants induced by H_2 was dependent on the NO pathway [15]. Furthermore, H_2 could alleviate the Al-induced inhibition of alfalfa root elongation by inhibiting the production of NO [16]. H_2 also was reported to be involved in auxin-induced lateral root formation, at least partially via a nitrate reductase (NR)-dependent NO synthesis [5]. Up to now, far too little attention has been paid to the crosstalk between H_2 and NO during the postharvest preservation of horticultural plants. In this study, pharmacological approaches and comparative proteomic analysis were applied to investigate the roles of H_2 and NO during the postharvest storage of cut lily (*Lilium* "Manissa") flowers and to identify the differentially accumulated proteins during that process. Thus, the study offers some important insights into the protein changes in the NO–H_2-regulated postharvest preservation of cut flowers.

2. Results

2.1. Effects of HRW, Sodium Nitroprusside (SNP), and NO Inhibitors on Vase Life

Compared with the control (distilled water), the vase life of cut lilies was extended by applying 150 µM SNP or 1% HRW (Figure 1). However, there was no significant difference between SNP and HRW. Compared with the HRW, 1% HRW together with 50 µM NaN_3 or 100 µM tungstate significantly decreased vase life, indicating the involvement of NO in the HRW-enhanced vase life of cut lilies (Figure 1).

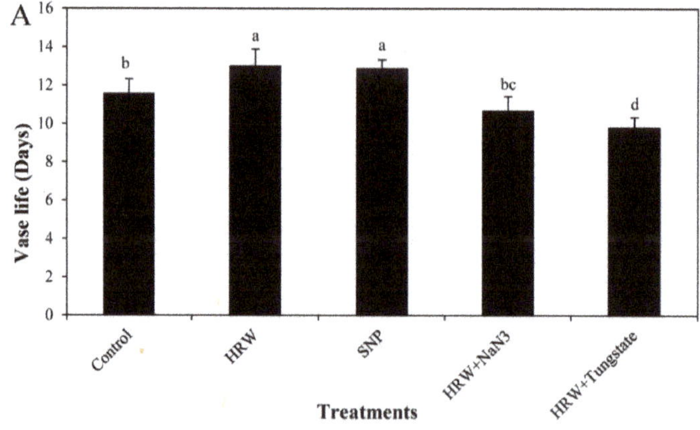

Figure 1. *Cont.*

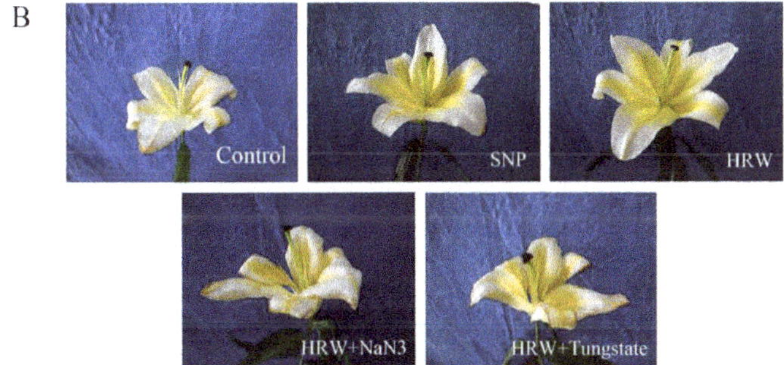

Figure 1. Effects of hydrogen-rich water (HRW), sodium nitroprusside (SNP), and HRW plus NaN_3 or tungstate on the vase life of cut lily flowers. The cut flowers were placed randomly in distilled water (control), 150 µM SNP, 1% HRW, 1% HRW + 50 µM NaN_3 and 1% HRW + 100 µM tungstate to investigate. The values of vase life (**A**) are the mean ± SE of three independent experiments. Bars with different letters illustrate significant differences ($p < 0.05$) according to Duncan's multiple range test. Photos (**B**) were taken after 8 days of treatments.

2.2. Effects of HRW, SNP, and NO Inhibitors on Maximum Flower Diameter and Rate of Fresh Weight Change

As shown in Figure 2A, the maximum value of the maximum flower diameter in the control and SNP treatment was obtained on the sixth day. The maximum value in HRW treatment appeared at the seventh day, while when HRW was applied with NaN_3 or tungstate, the maximum values were detected on the fifth day (Figure 2A). Interestingly, various vase solutions had no effects on the maximum flower diameter, suggesting that HRW merely delayed the flowering time rather than expanding the flower diameter.

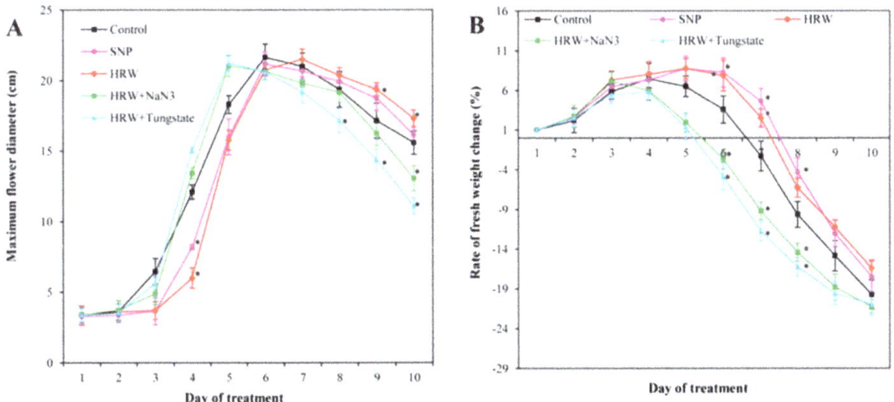

Figure 2. Effects of HRW, SNP, and HRW plus NaN_3 or tungstate on flower diameter and fresh weight of the cut lilies. Maximum flower diameter (**A**) and rate of fresh weight change (**B**) were expressed as mean ± SE of three independent experiments. Asterisks indicate significant difference ($p < 0.05$ by Duncan's multiple range test) compared to the control within the same day.

As time passed, the rate of fresh weight change initially increased and then decreased (Figure 2B). Compared with the control, the decrease of fresh weight in HRW or SNP treatment was postponed for one day after treatment, whereas the decrease of fresh weight in HRW plus tungstate treatment significantly appeared one day in advance. The decrease of fresh weight in HRW or SNP was significantly lower than in HRW with NaN$_3$ or tungstate (Figure 2B), indicating that the inhibition of endogenous NO could decrease the effect of HRW.

2.3. Two-Dimensional Electrophoresis Analysis and Identification of Proteins

In the study, the differentially accumulated proteins between control and treatments (SNP, HRW, HRW + NaN$_3$ or tungstate) were analyzed. In comparison of these two-dimensional electrophoresis (2-DE) gel images (Figure 3), 77 protein spots where the abundance was detected at ratios over 1.5-fold and false discovery rate (FDR) less than 5% were obtained on these images to analyze their basic information and function by 2-DE coupled to MALDI-TOF/TOF-MS. From these protein spots, 50 differentially accumulated proteins were successfully identified from the NCBI and Uniprot databases by Mascot analysis (Table 1 and Table S1). The molecular weights and isoelectric points (pIs) of identified proteins presented a different degree of variation, with molecular weights ranging from 16.70 kDa to 81.88 kDa and with pIs ranging from 4.83 to 9.35 (Table 1). Eleven protein spots were identified in the HRW treatment, while 10 spots were identified in the SNP treatment (Figure 3 and Figure S1). However, 20 or 9 protein spots were identified in HRW with NaN$_3$ or tungstate, respectively.

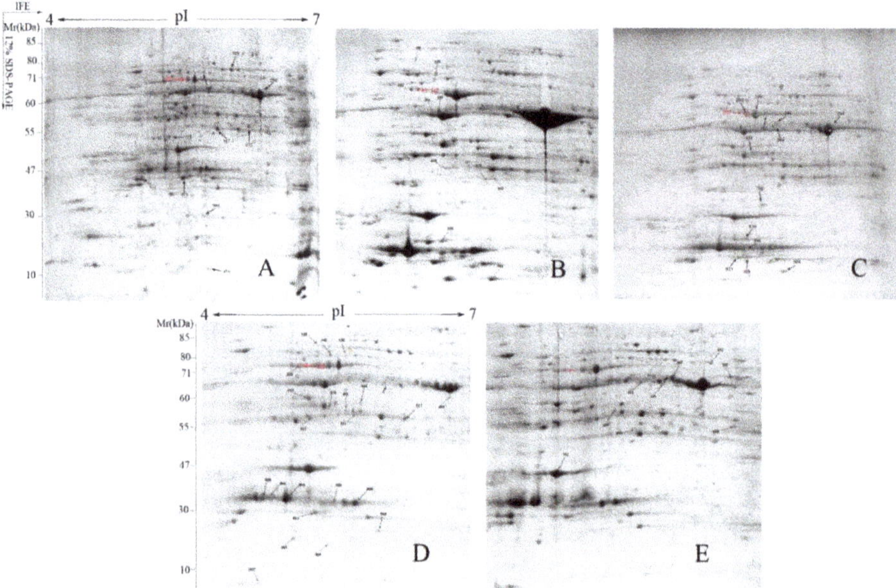

Figure 3. Two-dimensional electrophoresis (2-DE) image analysis of cut lily leaf proteins. Arrows indicate the 77 protein spots that were detected in control (**A**); 150 µM SNP (**B**); 1% HRW (**C**); 1% HRW + 50 µM NaN$_3$ (**D**); and 1% HRW + 100 µM tungstate (**E**). The red arrows show the spots of differentially accumulated ATP synthase CF1 alpha subunit (chloroplast) (AtpA) protein under different treatments.

Table 1. Identification and analysis of proteins in leaves of cut lily after HRW, SNP, and HRW plus NaN$_3$ or tungstate treatment.

Spot No.	Protein Name	Species	Accession No.	Expressed MW (Da)/pI	Theoretical MW (Da)/pI	Peptide Count	Score	Protein Score Confidence level (C.I.%)	Up/Down
201	ATP synthase CF1 alpha subunit (chloroplast)	*Lilium superbum*	YP_009130198.1	55,285.2/5.41	55,319.38/5.41	21	962	100	↑
259	ATP synthase alpha subunit, partial (mitochondrion)	*Erythronium dens-canis*	AFM91753.1	28,195.8/6.51	28,213.43/6.50	10	114	100	↑
202	ATP synthase CF1 alpha subunit (chloroplast)	*Lilium superbum*	YP_009130198.1	55,285.2/5.41	55,319.38/5.41	22	936	100	↑
908	Photosystem II oxygen evolving complex protein 2 precursor	*Fritillaria agrestis*	AAC04809.1	28,094.2/8.31	28,111.52/8.31	5	60	98.435	↓
284	ATP synthase CF1 beta subunit, partial (plastid)	*Lilium superbum*	AEZ48850.1	53,576.9/5.22	53,610.53/5.22	15	59	98.118	↑
294	Ribulose-1,5-bisphosphate carboxulase/oxygenase large subunit, partial (chloroplast)	*Heloniopsis kawanoi*	AIW53238.1	50,960.6/6.23	50,992.85/6.24	29	1180	100	↑
180	ATP synthase CF1 alpha subunit (chloroplast)	*Lilium superbum*	YP_009130198.1	55,285.2/5.41	5319.38/5.41	21	865	100	↑
896	Carbonic anhydrase	*Musa acuminata subsp.*	Tr I M0TL28	22,422.1/5.07	22,436.69/5.06	4	104	99.985	↑
716	PDZ domain-containing protein	*Cynara cardunculus var.*	Tr I A0A118JU51	36,198.3/6.18	36,220.78/6.18	6	113	99.998	↑
913	Chlorophyll a-b binding protein, chloroplastic	*Musa acuminata subsp.*	Tr I M0SBM9	29,718.2/8.96	29,737.07/8.96	4	171	100	↑
431	Actin	*Lilium davidii var. Davidii*	ALO18835.1	41,649.0/5.31	41,675.77/5.31	15	447	100	↑
511	Glutamine synthetase	*Tulipa pulchella*	BAM84282.1	38,673.4/5.64	38,697.60/5.64	5	90	99.999	↑
479	Actin	*Lilium regale*	AFU06383.1	41,619.0/5.31	41,645.75/5.31	16	421	100	↑
492	Monodehydroascorbate reductase	*Lilium longiflorum*	ADF43731.1	46,732.5/5.89	46,761.56/5.89	14	86	99.996	↑
220	ATP synthase CF1 alpha subunit (chloroplast)	*Lilium superbum*	YP_009130198.1	55,285.2/5.41	55,319.38/5.41	19	730	100	↓
988	Ribulose-1,5-bisphosphate carboxylase/oxygenase large subunit, partial (chloroplast)	*Gagea wilczekii*	AAM29162.1	50,739.4/5.96	50,771.69/5.96	12	494	100	↑
1060	Pathogenesis-related protein 10	*Lilium regale*	AHG94651.1	16,709.4/5.31	16,719.85/5.31	7	536	100	↓
136	Ribulose-1,5-bisphosphate carboxylase/oxygenase large subunit, partial (chloroplast)	*Trillium camschatcense*	AFP48691.1	44,763.6/6.52	44,792.01/6.53	9	63	99.14	↑
985	ATP synthase beta subunit, partial (chloroplast)	*Fritillaria acmopetala*	AKG96681.1	51,914.1/5.13	51,946.57/5.13	16	68	99.774	↓
142	ATP synthase CF1 alpha subunit (chloroplast)	*Ripogonum album*	ANO45506.1	55,341.1/5.26	55,375.27/5.26	10	72	99.91	↓
415	Glutamine synthetase	*Erythranthe guttata*	Tr I A0A022RZ30	39,028.6/5.40	39,053.05/5.39	7	393	100	↓
405	6-Phosphogluconate dehydrogenase, decarboxylating	*Citrus sinensis*	Tr I A0A067G3F9	53,519.6/6.38	53,553.34/6.38	13	437	100	↓
444	Elongation factor Tu	*Vigna angularis var. Angularis*	Tr I A0A0S3RGB1	52,659.2/6.34	52,692.26/6.34	14	671	100	↓
868	Chlorophyll a-b binding protein, chloroplastic	*Kalanchoe fedtschenkoi*	Tr I A0A089WZX0	28,226.3/5.15	28,244.20/5.15	7	195	100	↑
517	Glutamine synthetase	*Lolium perenne*	Tr I C5IW59	38,973.5/5.40	38,998.03/5.40	10	348	100	↓

Table 1. *Cont.*

Spot No.	Protein Name	Species	Accession No.	Expressed MW (Da)/pI	Theoretical MW (Da)/pI	Peptide Count	Score	Protein Score Confidence level (C.I.%)	Up/Down
866	Chlorophyll a-b binding protein, chloroplastic	*Carya cathayensis*	Tr I Q1KLZ3	28,296.3/5.15	28,314.25/5.15	5	101	99.969	↑
880	Beta carbonic anhydrase 3	*Arabidopsis thaliana*	Sp I Q9ZUC2	28,810.8/6.54	28,829.03/6.54	6	94	99.83	↑
864	Chlorophyll a-b binding protein, chloroplastic	*Kalanchoe fedtschenkoi*	Tr I A0A089WZX0	28,226.3/5.15	28,244.20/5.15	10	294	100	↓
860	Carbonic anhydrase	*Zea mays*	Tr I Q41729	71,291.9/8.93	71,337.55/8.93	10	101	99.969	↓
542	Ribulose bisphosphate carboxylase/oxygenase activase	*Medicago truncatula*	Tr I G7JTD2	52,135.9/5.42	52,169.06/5.42	15	358	100	↓
1037	Type II peroxiredoxin	*Medicago truncatula*	Tr I A0A072U4Q3	25,893.6/9.35	25,909.84/9.35	10	191	100	↓
304	ATP synthase CF1 beta subunit, partial (plastid)	*Lilium superbum*	AEZ48850.1	53,576.9/5.22	53,610.53/5.22	28	1190	100	↓
964	Ribulose-1,5-bisphosphate carboxylase/oxygenase large subunit, partial (chloroplast)	*Cardiocrinum giganteum var. Yunnanense*	AAM29161.1	50,201.1/6.04	50,233.05/6.04	12	88	99.998	↑
299	ATP synthase CF1 beta subunit, partial (plastid)	*Lilium superbum*	AEZ48850.1	53,576.9/5.22	53,610.53/5.22	21	112	100	↓
130	70 kDa heat shock protein	*Sandersonia aurantiaca*	AAL85887.1	36,768.6/4.83	36,791.57/4.82	3	260	100	↓
633	NADP-dependent alkenal double bond reductase P2	*Morus notabilis*	Tr I W9SE47	40,693.7/6.23	40,719.76/6.22	7	96	99.89	↓
90	Elongation factor G, mitochondrial	*Medicago truncatula*	Tr I A0A072UPP0	81,881.8/5.50	81,933.70/5.50	17	579	100	↓
671	Cysteine synthase	*Populus trichocarpa*	Tr I B9HJY5	34,176.2/7.64	34,197.70/7.64	11	132	100	↑
839	Putative L-ascorbate peroxidase 2, cytosolic-like	*Solanum chacoense*	Tr I A0A0V0HVQ3	28,638.6/5.75	28,656.76/5.75	9	324	100	↑
618	Trypsin-like serine protease	*Medicago truncatula*	Tr I G7KIR6	45,774.3/6.79	45,802.44/6.80	10	525	100	↑
182	FtsH-like protein Pftf	*Nicotiana tabacum*	Tr I Q9ZP50	74,335.8/6.00	74,382.14/6.00	24	658	100	↑
307	ATP synthase alpha subunit, partial (mitochondrion)	*Lilium lancifolium*	AAR28047.1	41,824.8/6.47	41,850.81/6.47	13	397	100	↑
313	Atpb (chloroplast)	*Lilium distichum*	AMT85217.1	53,546.9/5.22	53,580.50/5.22	17	113	100	↓
580	Glyceraldehyde-3-phosphate dehydrogenase, partial	*Lilium regale*	AHZ94971.1	36,779.2/7.11	36,802.04/7.11	12	255	100	↑
349	Ribulose-1,5-bisphosphate carboxylase/oxygenase large subunit, partial (chloroplast)	*Cardiocrinum giganteum var. Yunnanense*	AAM29161.1	50,201.1/6.04	50,233.05/6.04	19	159	100	↑
489	Monodehydroascorbate reductase	*Lilium longiflorum*	ADF43731.1	46,732.5/5.89	46,761.56/5.89	18	271	100	↓
752	Photosystem II oxygen evolving complex protein 1 precursor	*Fritillaria agrestis*	AAC04808.1	34,847.8/6.26	34,869.39/6.25	16	695	100	↓
263	Dihydrolipoyl dehydrogenase	*Salvia miltiorrhiza*	Tr I A0A0G2SJN7	53,520/6.96	53,553.72/6.96	6	259	100	↑
192	Malic enzyme	*Phaseolus angularis*	Tr I A0A0L9UG31	73,189.2/8.33	73,235.49/8.33	9	177	100	↓
604	Fructose-bisphosphate aldolase	*Oxytropis ochrocephala*	Tr I A0A0K1JSG5	42,894.2/6.39	42,920.76/6.39	9	477	100	↑

Note: Assigned spot number as indicated in Figure 3. Arrows indicate up- (↑) and down- (↓) regulation of the proteins.

2.4. Functional Classification and Analysis of Differentially Accumulated Proteins

These differentially accumulated proteins were analyzed in order to classify them in terms of their biological functions according to Gene Ontology and UniProt Protein Knowledgebase. The 50 differentially accumulated proteins identified in this study were classified into seven functional categories, as shown in below: photosynthesis (40%), energy metabolism (26%), defense-related protein (16%), amino acid metabolism (8%), transcription and translation (4%), cytoskeleton (4%), and signal transduction (2%) (Figure 4A). Subsequently, the differentially accumulated proteins among HRW, SNP, HRW + NaN$_3$, and HRW + tungstate treatments were analyzed. In HRW treatment, some differentially accumulated proteins that were associated with photosynthesis were detected, and many of them were increased when compared with the control. Among them, ATP synthase CF1 alpha subunit (chloroplast) (AtpA) was identified with three spots (180, 201, and 202), and all were up-regulated (Table 1). Under the SNP treatment, differentially accumulated proteins including FtsH-like protein Pftf (spot 182), trypsin-like serine protease (spot 618), and putative L-ascorbate peroxidase 2, cytosolic-like (spot 839) were observed. The expression of these three proteins related to defense was higher than that of the control. In HRW + NaN$_3$ treatment, two spots were identified as AtpA, and they were down-regulated when compared with the control. The expression of proteins related with defense was significantly decreased, such as Type II peroxiredoxin (spot 1037) and pathogenesis-related protein 10 (spot 1060). Under the treatment with HRW plus tungstate, photosystem II oxygen evolving complex protein 1 precursor (spot 752) belonged to photosynthesis, and malic enzyme (spot 192) associated with energy metabolism were all down-regulated in comparison with the control (Table 1). There were four differentially accumulated proteins overlapping between HRW and HRW + NaN$_3$ (Figure 4B). After statistical analysis, 28 proteins in total were up-regulated while 22 proteins were down-regulated (Table 1). Among them, the number of up-regulated proteins was significantly higher in HRW than in HRW + NaN$_3$ or tungstate (Figure 4C). However, the number of down-regulated proteins was lower in HRW than in HRW + NaN$_3$ or tungstate (Figure 4C). Thus, exogenous H$_2$ could up-regulated some proteins during postharvest storage of cut lilies.

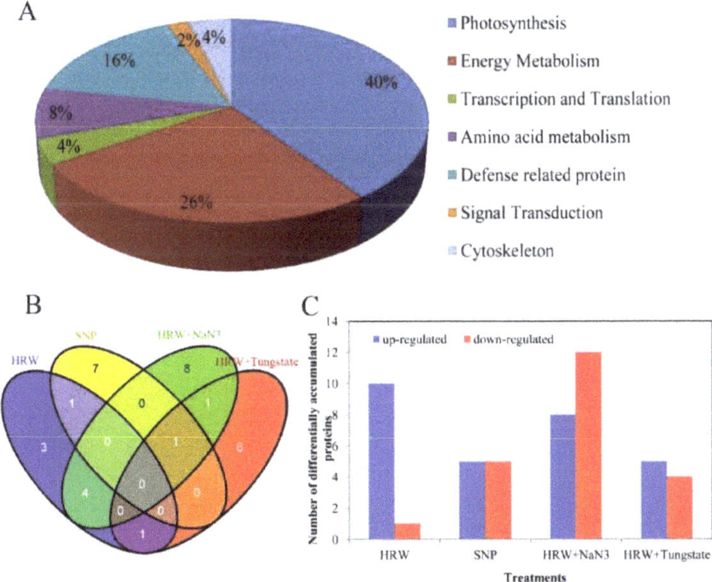

Figure 4. *Cont.*

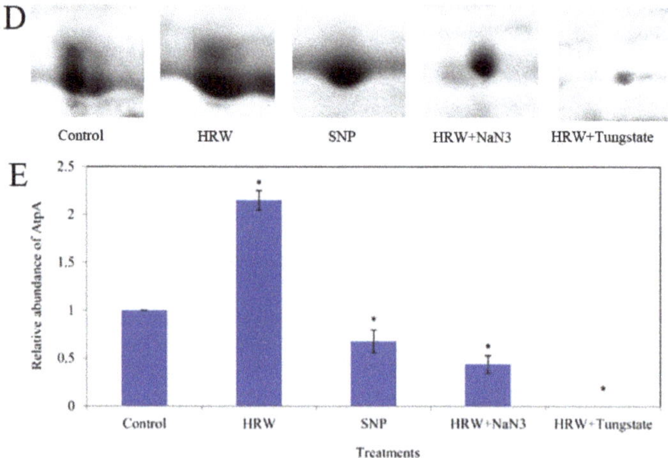

Figure 4. Functional classification and analysis of differentially accumulated proteins in cut lilies. Pie chart showed percentage of differentially accumulated proteins in different functional categories (**A**); Venn diagram showed the number of overlap proteins regulated by HRW, SNP, HRW + NaN$_3$ or tungstate compared with the control (**B**); Column chart showed the number of up- or down-regulated proteins in comparison with the control (**C**); The 2-DE gel sections showed the magnified views of differentially accumulated spots of AtpA protein in treatments. Spot positions corresponding to AtpA protein were shown with red arrows in Figure 2 (**D**); The column chart showed the differential relative abundance patterns among HRW, SNP, HRW + NaN$_3$ and HRW + tungstate treatment (**E**).

Interestingly, among them, only AtpA was up-regulated in HRW treatment and down-regulated in the HRW with NaN$_3$ treatment. Therefore, AtpA was selected to further investigate. The magnified views of 2-DE image showed the differential accumulation of AtpA protein in treatments of the control, HRW, HRW + NaN$_3$ and HRW + tungstate (Figure 4D). The relative abundance of AtpA protein has significant difference in 4 treatments (Figure 4E). The relative abundance of AtpA in HRW treatment was significantly higher than that in the control, while the relative abundance of AtpA was less in SNP than in the control. However, the relative abundance of AtpA protein in HRW + NaN$_3$ treatment was significantly less than that in HRW treatment, and the relative abundance of AtpA was not detected in HRW + tungstate (Figure 4E), which may be caused by too low differential accumulation of AtpA in HRW + tungstate. Taken together, H$_2$ could enhance the expression of AtpA protein, and the inhibitors of NO (NaN$_3$ and tungstate) may have inhibited the effect of H$_2$ on the expression of AtpA protein.

2.5. Relative Expression of LlatpA Gene and the Activity of ATP Synthase (ATPase)

The qRT-PCR analysis revealed that the relative expression of the *LlatpA* gene was significantly higher in HRW treatment than in the control (Figure 5). There was no significant difference in the gene expression between the control and SNP treatment. In comparison with HRW treatment, the relative expression of *LlatpA* gene was significantly inhabited by HRW plus NaN$_3$ or tungstate. As shown in Figure 5, compared with control, the activity of ATPase was remarkably enhanced by SNP or HRW. However, the activity of ATPase in HRW plus NaN$_3$ or tungstate treatment was decreased in comparison with HRW treatment.

2.6. Chlorophyll Fluorescence and Photosynthetic Parameters

The chlorophyll fluorescence parameters analysis result is shown in Figure 6. After 6 days of treatment, the maximum quantum yield of photosystem II complex (PSII) photochemistry (Fv/Fm) in HRW or SNP groups was higher than that in the control, whereas HRW in combination with NaN$_3$

or tungstate significantly inhibited the positive effects of HRW (Figure 6A,B). After treatment for 6 days, when compared with the control, the effective quantum yield of PSII (ΦPSII) and photochemical quenching (qP) were increased in HRW treatment. However, significantly decreased ΦPSII and qP appeared in the SNP treatment group. ΦPSII and qP in HRW plus NaN$_3$ or tungstate treatment were significantly decreased in comparison with those in the HRW treatment (Figure 6C,D).

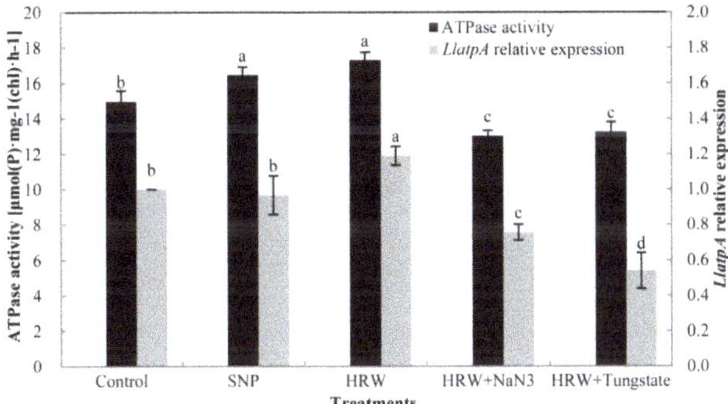

Figure 5. Effects of SNP, HRW, and HRW in combination with NaN$_3$ or tungstate on *LlatpA* gene expression and ATP synthase (ATPase) activity. Values of relative expression of *LlatpA* gene and activity of ATP synthase are the mean $\pm$ SE of three independent experiments with three repeats for each. Bars with different letters illustrate significant differences ($p < 0.05$) according to Duncan's multiple range test.

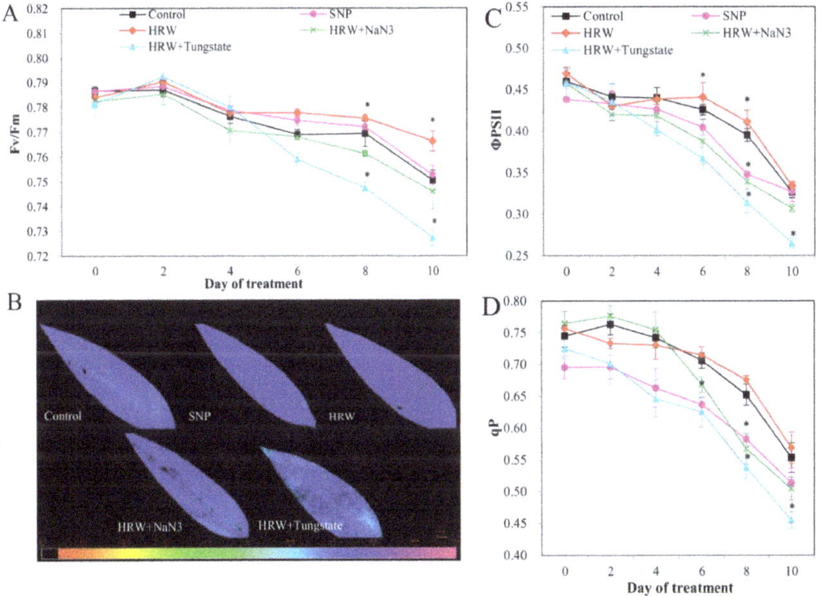

Figure 6. Effects of HRW, SNP, and HRW plus NaN$_3$ or tungstate on chlorophyll fluorescence parameters. Values of the Fv/Fm (maximum quantum yield of PSII photochemistry) (**A**); effective quantum yield of PSII (ΦPSII) (**C**); and photochemical quenching (qP) (**D**) are the mean $\pm$ SE of three independent experiments with three repeats for each. Fluorescent images (**B**) are given in colors that represent the absolute values of the ratio ranging from 0 (black) to 1.0 (purple) and were taken on the 8th day of treatment. Asterisks indicate significant difference ($p < 0.05$ by Duncan's multiple range test) compared to the control within the same day.

A downward trend was also observed in the net photosynthesis rate (Pn; Figure 7A) and stomatal conductance (Gs; Figure 7B) after 6 days of treatment. In contrast, intercellular CO_2 concentration (Ci; Figure 7C) was slightly increased after treatment for 6 days. Transpiration rate (Tr) showed a tendency to decrease during the experiment (Figure 7D). Pn, Gs, and Tr were significantly increased by HRW or SNP in comparison with those in the control group. HRW plus NaN_3 or tungstate treatment significantly decreased Pn, Gs, and Tr when compared with HRW treatment (Figure 7A,B,D). Interestingly, Ci in HRW treatment was lower than that in the control, whereas Ci in HRW together with NaN_3 or tungstate treatment was higher than that in HRW treatment (Figure 7C).

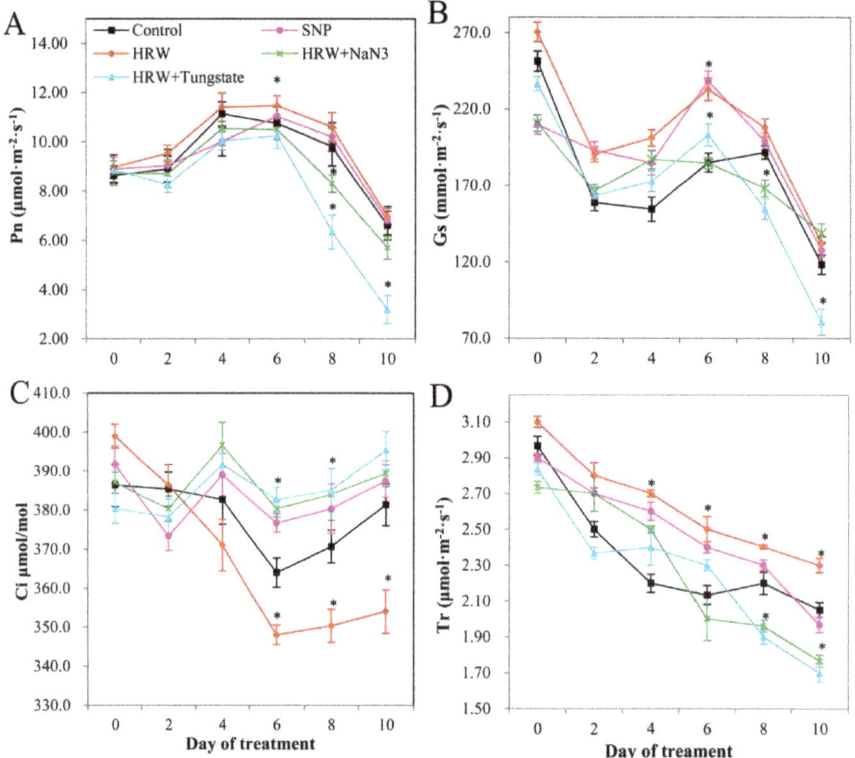

Figure 7. Effects of HRW, SNP, and HRW plus NaN_3 or tungstate on photosynthetic parameters. Values of net photosynthetic rate (Pn) (**A**), stomatal conductance (Gs) (**B**), the intercellular CO_2 concentration (Ci) (**C**), and transpiration rate (Tr) (**D**) are the mean ± SE of three independent experiments with three repeats for each. Asterisks indicate significant difference ($p < 0.05$ by Duncan's multiple range test) compared to the control within the same day.

3. Discussion

H_2 is considered as a novel signaling molecule involved in plant developmental and physiological processes [17]. Our previous studies showed that exogenous H_2 could enhance adventitious root development in marigold [18]. In addition, the shelf life of kiwifruit was prolonged by H_2 by decreasing ethylene biosynthesis [19] and reducing oxidative damage [7]. In this study, H_2 delayed the flowering time of cut lilies in the preservation process. H_2 improved the vase life of cut lilies by maintaining the fresh weight of cut lilies. The results were consistent with those of Ren et al. (2017) [8], who showed that exogenous H_2 enhanced the vase life of cut flowers by maintaining suitable water balance. In the current study, the vase life of cut lily was also enhanced by SNP treatment, suggesting that exogenous NO may play an important role in extending vase life. The vase life of cut gerbera was significantly extended by exogenous NO [10]. Exogenous NO also could delay petal wilting in cut carnation flowers by maintaining water metabolism and antioxidant enzyme activity [12]. In our study, NaH_3 and tungstate—inhibitors of nitrate reductase (NR) that can inhibit the reduction of nitrate to nitrite and further inhibit the production of NO—were used to investigate whether NO participates in H_2-regulated postharvest preservation. NO inhibitors NaN_3 or tungstate depressed the positive effects of H_2 on the vase life, the maximum flower diameter, and fresh weight of cut lilies, suggesting that NO played vital roles in H_2-induced cut flowers freshness. Our previous studies revealed that H_2 increased NO generation through regulated NR and NOS activity to induce the formation of adventitious root in cucumber [20]. H_2, as a mediator, activated cell cycle by NO pathway during adventitious root formation [15]. Additionally, H_2 was also involved in auxin-induced lateral root formation via an NR-dependent NO synthesis [5]. Here, for the first time, the involvement of NO in hydrogen gas-improved vase life in cut flowers was reported.

Two-dimensional electrophoresis (2-DE)-based proteomics analysis has been applied in plant proteomic research. Here, the results of the comparative proteomic analysis showed that 50 differentially accumulated proteins were successfully identified by Mascot analysis in cut lily leaves. Among them, 28 proteins were up-regulated while 22 proteins were down-regulated. Exogenous H_2 could increase the number of up-regulated proteins, while inhibitors of NO increased the number of down-regulated proteins. In chrysanthemum cuttings during adventitious root formation, 42 differentially accumulated protein spots were successfully matched to NCBI database entries [21]. In cut rose flowers, 103 proteins were obtained, and these proteins were involved in plant growth regulators, natural resistance, protein metabolism, and methionine synthesis [22]. In the current study, the 50 differentially accumulated proteins were involved in photosynthesis, energy metabolism, defense, amino acid metabolism, etc. In a *Medicago sativa* cadmium resistance study, the proteins related to photosynthesis were not detected in H_2 treatment [23]. However, we found that the proteins involved in photosynthesis showed a high expression level in H_2 treatment. This may be caused by different experimental materials and conditions. The proteins related to the stress response and defense changed significantly after NO treatment in the processes of peach fruit ripening, such as glutathione S-transferase (GST) and ascorbate peroxidase (APX) [24]. Simultaneously, we found that the expression of proteins associated with defense were up-regulated by exogenous NO in the cut lilies during preservation. This suggested that NO could promote the expression of proteins related to defense. The proteins related to energy metabolism were decreased during strawberry fruit ripening [25]. In litchi pulp, malate dehydrogenase (related to energy metabolism) was down-regulated in the later storage period [26]. In this study, the proteins related to energy metabolism were down-regulated in HRW plus tungstate, suggesting that NO played an important role in the proteins' expression, regulated by H_2. Thus, H_2 and NO could regulate the expression of proteins related to photosynthesis, defense, and energy metabolism while delaying the senescence of cut lilies.

ATP synthase CF1 alpha subunit (AtpA) protein is a key enzyme for the chloroplast thylakoid membranes, and plays a vital role in synthesizing ATP from ADP and phosphate [27]. ATP synthase CF1 α-subunit was obtained and showed an initial increase and then a decrease in *Kandelia candel* under salt stress [28]. The expression of ATP synthase CF1 α-subunit was decreased in the treatment of MAP kinase kinase (MEK) inhibitor in *Chlamydomonas reinhardtii* [29]. In this study, we revealed that the differential relative abundance of AtpA protein was significantly different between experimental treatments. H_2 could up-regulate the expression of AtpA protein during postharvest freshness of cut lilies, while the accumulation of AtpA protein was not significantly up-regulated by NO. Interestingly, AtpA protein was down-regulated in HRW + NaN_3, but no accumulation spots of AtpA protein were detected in HRW + tungstate treatment. This may be because the expression of AtpA was too low (abundance $\leq$ 1.5-fold) to detect in the HRW + tungstate group. This suggested that the positive effect of H_2 on the expression of AtpA protein was inhibited by the inhibitors of NO. The positive roles of H_2 on polyphenol oxidase activity were impaired by cPTIO (NO scavenger), L-NAME (NO synthase enzyme inhibitor), and NaN_3 [22]. H_2-promoted NO accumulation and stomata closure were greatly prevented by L-NAME or tungstate [14]. The above results suggest that the positive effects of H_2 were reversed when the generation of NO was blocked by an inhibitor or scavenger. Thus, it may be that H_2 at least partially played its positive roles through endogenous NO. In the study, the relative expression of *LlatpA* gene and the activity of ATPase were determined in order to further investigate the expression of AtpA protein at the transcriptional and biochemical levels. ATPase is embedded in the same coupling membrane, and is composed of several subunits, including an alpha subunit. The a-subunit is composed of five transmembrane helices (TMHs), including a four-helix bundle [30]. The prerequisite of ATPase exerting its proton-driven role is intersubunit mobility. Thus, the CF1 a-subunit plays an important role in ATPase. Additionally, ATPase is a key thylakoid membrane protein encoded by the *atpA* gene of the chloroplast genome [31]. In our study, the RT-qPCR results showed that exogenous H_2 could increase the expression of the *LlatpA* gene. The expression of the *atpA* gene of cucumber was increased by exogenous putrescine in salt stress [32]. H_2 could get into soluble spinach chloroplast to activate ATPase by exchanging into internal parts of the molecule on energized membranes [33]. Simultaneously, the change in the expression of the *atpA* gene was positively related to ATPase activity at the transcriptional level under low temperature conditions [34]. Meanwhile, we revealed that the activity of ATPase was also promoted by H_2, which was positively related to the relative expression of the *LlatpA* gene. In another study, NO was found to play a vital role in stimulating H^+-ATPase activity during the early stages of maize lateral root development [35]. Exogenous NO could alleviate the inhibition of H^+-ATPase in plasma membrane or tonoplast which was induced by $CuCl_2$ [36]. In the present study, exogenous NO could also enhance the activity of ATPase. However, the positive effects of H_2 on the *LlatpA* gene and ATPase were inhibited by inhibitors of NO. The transcription levels of the cyclin-dependent kinase B decreased when H_2 was used together with cPTIO, L-NAME, and NaN_3, respectively [15]. It was suggested that NO may act as a signaling molecule involved in H_2 to increase the expression of the *LlatpA* gene and the activity of ATPase. All of these results were consistent with the expression of AtpA protein. Taken together, H_2 may play its positive role in the expression of the *LlatpA* gene and the activity of ATPase by regulating endogenous NO.

The results of transcriptional level and biochemical level analysis were consistent with the results of AtpA protein expression, suggesting that the involvement of NO in the H_2-promoted vase life of cut lilies may be through regulation of the expression of AtpA protein. Since AtpA is a protein related to photosynthesis, and ATPase plays significant roles in photosynthesis-dependent membrane hyperpolarization and energy transfer [37], in the next study, the chlorophyll fluorescence parameters and photosynthetic parameters were determined to further validate the effects of the ATP protein on NO and H_2 co-regulated postharvest preservation at the physiological level. We found that exogenous H_2 could increase the value of Fv/Fm, ΦPSII, and qP. A previous study also found that exogenous H_2 could significantly alleviate high light induced-damage to PSII [38]. In this study, proteomics analysis suggested that the expression of AtpA was up-regulated by H_2. Therefore, H_2 played positive roles in enhancing the light energy conversion efficiency of PSII, possibly by regulating the expression of AtpA protein. Exogenous NO could alleviate paraquat-induced decline of Fv/Fm [39]. Exogenous NO also could significantly increase Fv/Fm, and thereby the toxic effects of arsenic (As) on photosynthesis were alleviated in *Luffa* seedlings [40]. The exogenous NO could decrease qP to inhibit the electron transport rate (ETR) [41]. In this study, the ratio of Fv/Fm was increased by NO. However, exogenous NO did not significantly change the values of ΦPSII and qP, suggesting that NO had no obvious role in the capture and distribution of light energy. Exogenous NO could remarkably alleviate the inhibition of Fv/Fm induced by chilling stress, while inhibitors of NO could reduce Fv/Fm, ΦPSII, and qP [42]. Our data also revealed that the positive effects of H_2 on Fv/Fm, ΦPSII, and qP were inhibited by inhibitors of NO. This result was consistent with the result that inhibitors of NO inhibited the positive effects of H_2 on the expression of the AtpA protein. Thus, the involvement of NO in the H_2-regulated electron transport of PSII may be by regulating AtpA protein. Photosynthesis leads to the storage of solar energy in organic compounds. As a key enzyme related to photosynthesis, AtpA protein may affect photosynthesis efficiency [43]. Here, the results of photosynthetic analysis showed that Pn and Gs were increased by exogenous H_2 or NO, but the value of Ci decreased. The above results were consistent with the proteomics analysis, indicating that H_2 enhanced photosynthesis by regulating the expression of AtpA protein. Exogenous H_2 could also increase the Pn in a concentration-dependent manner in maize seedlings [38]. Exogenous NO effectively inhibited the decrease in Pn as a result of non-stomatal factors under acid rain stress [44]. These results suggested that NO and H_2 played a positive role in improving the photosynthetic performance of cut lily leaves. Chen et al. (2014) [45] reported that the Pn of transgenic and wild-type rice plants was significantly increased by NO, while the effect of NO on the Pn was inhibited by an NO scavenger. The positive effect of H_2 on alleviating the Al-induced inhibition of alfalfa root growth was inhibited by cPTIO (a scavenger of NO) and tungstate [16]. In the present study, the effect of H_2 on Pn and Gs were decreased by inhibitors of NO. However, the Ci was increased when the roles of H_2 were inhibited by NO inhibitors. This may be due to the positive relationship between intercellular CO_2 concentration (Ci) and net photosynthetic rate (Pn) under stomatal opening. Proteomics analysis also showed that the positive effect of H_2 on improving the expression of AtpA protein was inhibited by NO inhibitors. Therefore, the involvement of NO in H_2-regulated photosynthesis may be through regulating the expression of AtpA protein.

In conclusion, exogenous H_2 or NO significantly promoted the vase life and quality of cut lilies, and NO might play an important role in the H_2-improved postharvest freshness of cut lilies. Additionally, H_2 also significantly regulated the expression of AtpA protein and the activity ATPase, as well as photosynthesis in the postharvest freshness of cut lilies (Figure 8). Interestingly, NO may be involved in this process. Collectively, our results also revealed that NO was involved in the H_2-enhanced shelf-life and quality of cut lilies, possibly through regulating the expression of the photosynthesis-related AtpA.

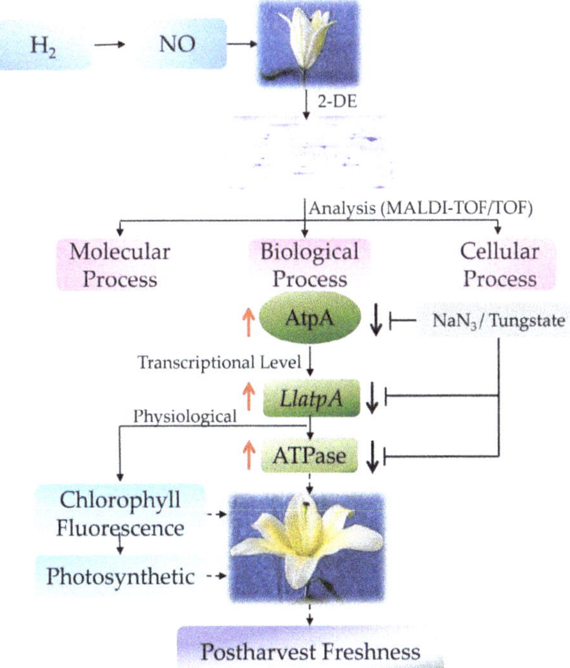

Figure 8. Schematic diagram of key proteins during nitric oxide–hydrogen gas-improved postharvest freshness in cut lily by comparative proteomic analysis.

4. Materials and Methods

4.1. Plant Material and Treatments

Cut lily (*Lilium* "Manissa") flowers with a single green bud and similar flowering degree were obtained from a commercial grower (Qianxi Florist, Lanzhou, China) and transferred rapidly to the laboratory. Flowers were held in water for 12 h and then cut under water to a length of 45 cm and every flower having five leaves on the top was kept to provide homogenous samples. Finally, the flowers were inserted into 1 L of treatment solution: distilled water (the control), 1% hydrogen-rich water (HRW, the preparation of HRW was based on the method of Zhu et al. [20]), 150 µM sodium nitroprusside (SNP, a donor of NO), 1% HRW plus NO inhibitors 50 µM sodium azide (NaN_3) or 100 µM tungstate. The above chemicals were purchased from Sigma-Aldrich (St Louis, MO, USA) except for tungstate (tungstate was provided from Shanghai Zhaoyun chemical Co., Ltd., Shanghai, China). The treatment solution was exchanged every day at regulating time. Furthermore, each treatment was conducted in three replicates, with each replication including five cut flowers. The laboratory was maintained at $25 \pm 3\,^{\circ}\text{C}$, $60\% \pm 5\%$ relative humidity, and 15 $\mu\text{mol}\cdot\text{m}^{-2}\cdot\text{s}^{-1}$ photons irradiance.

4.2. Determination of Vase Life and Maximum Flower Diameter

The vase life of cut lily flowers was determined according to time beginning on the first day when flowers were inserted in the vase solution. The vase life was considered to the termination when the flower was wilted. A Vernier caliper was used to measure the maximum flower diameter, which is the maximum distance between buds and petals. The cross method was conducted to measure the maximum flower diameter. After treatment, the maximum flower diameters were measured and recorded every day.

4.3. Determination of the Rate of Fresh Weight Change

The fresh weight of each flower before treatment was measured using an electronic balance and recorded as W_0 (W_0 is the fresh weight of the cut lilies at the first day). Then, the water at the base of the flower stem was blotted with filter paper and its fresh weight was measured. The value was recorded as W_d (W_d is the fresh weight of the cut lilies at $d = 1, 2, 3 \ldots$ days). The rate of fresh weight change was calculated according to the formula: $[(W_d - W_0)/W_0] \times 100$.

4.4. Protein Extraction

Three biological replicates were performed in the comparative proteomic analysis. The sample of 2 g cut lily leaves was used to extract protein in the eighth day of treatment. Each sample was ground to a fine powder with 0.04 g polyvingypyrrolidone (PVPP) in liquid nitrogen in a pre-cooled mortar. The powders were transferred into six 2 mL tubes. To each tube was added ice-cold trichloroacetic acid (TCA)/acetone (containing 10% (v/v) TCA and 0.07% (v/v) β-mercaptoethanol (β-ME)). Samples were mixed with a vortex and placed in a freezer at $-20\,^{\circ}\text{C}$ overnight. The next day, the pellet was centrifuged at $20,000 \times g$ for 30 min at $4\,^{\circ}\text{C}$. Two milliliters of 100% (v/v) pre-cooled acetone (containing 0.07% β-ME) was added, mixed with a vortex and placed at $-20\,^{\circ}\text{C}$ for 1 h, centrifuging at $20,000 \times g$ for 20 min at $4\,^{\circ}\text{C}$ and discarding the supernatant. Then, the step was repeated again. Two milliliters of 80% (v/v) pre-cooled acetone (containing 0.07% β-ME) was added, mixed by vortex and placed at $-20\,^{\circ}\text{C}$ for 30 min, centrifuged at $20,000 \times g$ for 15 min at $4\,^{\circ}\text{C}$, and the supernatant was discarded. Then, the step was repeated twice. The pellet was placed in a 2 mL centrifuge tube precooled with liquid nitrogen and placed in an in situ ordinary type freeze dryer (Scientz-10ND, Ningbo Xinzhi Biotechnology Co., Ltd., Ningbo, China) to dry to a white powder. A certain amount of 7 M protein lysate containing DL-dithiothreitol (DTT) was added and cracked at room temperature for 2 h, mixed by vortex one time per 30 min during cracking and centrifuged at $20,000 \times g$ for 30 min at $4\,^{\circ}\text{C}$. The supernatant was the total protein of lily leaves. The resulting protein was stored at $-80\,^{\circ}\text{C}$ for next use. The protein concentration was determined according to the method described by Bradford assay (Bio-Rad, Hercules, CA, USA).

4.5. Two-Dimensional Electrophoresis (2-DE) and Gel Image Analysis

For two-dimensional electrophoresis (2-DE), a total of 0.8 mg protein was first subjected to isoelectric focusing (IEF) and separated by two-dimensional SDS-PAGE. First-dimension IEF was done using pH 4–7 NL IPG strips (ReadyStrip, 17 cm, BioRad, USA). The strips were rehydrated in a rehydration solution (7 M urea, 2 M thiourea, 4% CHAPS, 65 mM DTT, 0.2% (w/v) Bio-Lyte, and 0.001% bromophenol blue) containing the protein sample for 12 h at room temperature. IEF was conducted in a Protean® IEF Cell (Bio-Rad) at 20 °C, and the voltage was set as 50 V for 14 h, 250 V for 3 h, 1000 V for 5 h, 9000 V for 5 h, and 9000 V until maximum 90,000 Vh, then under 500 V run for 24 h. After first-dimension IEF, the strips were equilibrated instantly for 15 min in equilibration buffer I (6 M urea, 2% (w/v) SDS, 0.375 M Tris-HCl (pH 8.8), 20% (v/v) glycerol, 2% (w/v) DTT). Then, 5 mL equilibration buffer II (6 M urea, 2% (w/v) SDS, 0.375 M Tris-HCl (pH 8.8), 20% (v/v) glycerol, 2.5% (w/v) iodoacetamide) was added and incubated for 15 min. Second dimension SDS-PAGE was conducted in 12% (v/v) polyacrylamide-SDS gel (Protean® Plus Dodeca Cell, Bio-Rad). After electrophoresis, the gels were stained with Coomassie Brilliant Blue (Bio-Rad). Stained 2-DE gels were scanned with a GS-800 Calibrated Densitometer (Bio-Rad), and data were analyzed by PDQuest software version 8.0 (Bio-Rad) as described by the manufacturer. The spots were automatically detected by the software and then subjected to careful manual editing and confirmation. Each spot of the standard gel accorded the following criteria: it was present in at least two of the three gels and was qualitatively consistent in size and shape in the replicate gels. The relative volume of each spot was assumed to represent the expression level of its protein. The volume of each well-separated spot was compared between control and different treatments to identify differentially accumulated protein spots. A spot abundance ratio of greater than 1.5 ($p < 0.05$) (a spot present uniquely or present in two-fold abundance in one sample relative to the other) was used as the threshold for a protein being differentially accumulated in subsequent studies. Every treatment was done with three biological replicates.

4.6. Protein Identification and Database Searching

Protein spots that appeared as differentially expressed between the control and treatment samples were excised from the gels and digested with trypsin (Promega, Madison, WI, USA), based upon the procedure described by Liu et al. [46]. MS and tandem mass spectrometry (MS/MS) data for protein identification were obtained by using a matrix-assisted laser desorption ionization time-of-flight (MALDI-TOF/TOF) instrument (4800 Plus MALDI TOF/TOFTM Analyzer; AB SCIEX, Framingham, MA, USA), as previously described by Sheffield et al. [47]. The MS/MS spectra searches were submitted to the NCBI database (http://www.ncbi.nlm.nih.gov) downloaded on 9 January 2017 (15,653 sequences) and Uniprot database (http://www.uniprot.org/) downloaded on 21 September 2016 (3,887,742 sequences) to identify proteins in the MASCOT (version 2.2, Matrix Science, London, UK) search engine using the following search parameters: MS tolerance of 50 ppm, MS/MS tolerance of 0.5 Da, peptide molecular mass ranging from 400 to 4000 Da, with one missing cleavage site, fixed modifications of carbamidomethyl (Cys) and variable modifications of oxidation (Met). The Percolator algorithm was used to estimate the false discovery rate (FDR) based on p-value, and only peptides at the 99% confidence level were counted as the identified protein. Proteins were considered as identified when one protein had to contain at least two peptides and when average of fold change was $\geq$1.5 in the experimentally treated groups compared to the control group.

4.7. Quantitative RT-PCR (qRT-PCR) Analyses

After 8 days of treatment, 1 g of the cut lily leaves was ground to a powder in liquid nitrogen and the total RNA was abstracted using TaKaRa MiniBEST plant RNA extraction kit (Takara Bio

Inc, Kusatsu, Shiga, Japan) according to the manufacturer's instructions. For these samples, 1 μg of total RNA was converted to cDNA using PrimeScript RT Master Mix kit (Takara Bio Inc, Kusatsu, Shiga, Japan) according to the manufacturer's instructions. Quantitative RT-PCR was conducted with SYBR® Premix Ex Taq™ II (Takara Bio Inc, Kusatsu, Shiga, Japan) and LightCycler®96 (Roche Applied Science, Mannheim, Germany) automated PCR system using two-step cycling conditions of 95 °C pre-degeneration for 30 s, followed by 40 cycles of 95 °C for 5 s, and then annealing at 60 °C for 20 s. The reaction mixture (20 μL) contained 1 μL of cDNA solution and primers at a concentration of 10 μM each. The lily gene *actin* (JX826390) was used as a reference for calculating relative transcript abundance. The primers of *actin* were: forward (5′-TGCTGGATTCTGGTGATGGT-3′) and reverse (5′-TCCCGTTCAGCTGTAGTTGT-3′). The CDS of ATP synthase CF1 alpha subunit (AtpA) was acquired according to accession no. by NCBI. The *atpA* gene *Lilium* "Manissa" was named the *LlatpA* gene. The *LlatpA* gene-specific primers were designed based on the cDNA sequences. The primers of *LlatpA* were: forward (5′-AAGCTTGTGCCTGTTTGGAG-3′) and reverse (5′-AACGGCAGATTCACCTGCTA-3′). The method of comparative Ct ($2^{-\Delta\Delta Ct}$) [48] was used to calibrate the relative quantification of RNA expression. Each sample was set three biological replicates.

4.8. Determination of ATP Synthase Activity

A 1 g sample of the cut lily leaves was ground with a crude enzyme extracting solution (2.75 mL β-mercaptoethanol and 0.688 g EDTA-Na$_2$, volume fixed to 1 L using 20 mmol/L Tris-HCl) to form a slurry. The slurry was filtered through four layers of gauze. Then, 1.5 mL of the filtrate was centrifuged at a speed of 12,000× g at 4 °C for 10 min. Subsequently, 100 μL supernatant was added in 200 μL of the reaction solution (0.5448 g ATP-Na and 0.6517 g MgCl$_2$·6H$_2$O were fixed to a volume of 300 mL using a 20 mmol/L maleic acid buffer) and incubated for 30 min at 38 °C. The reaction was terminated by 200 μL TCA. Finally, 1.5 mL phosphorus reagent (6 M H$_2$SO$_4$: distilled water: 2.5% ammonium molybdate aqueous solution: ascorbic acid = 1:2:1:1) was added to the reaction to enact a color reaction for 20 min at 45 °C. After finishing the color reaction, 3 mL distilled water was added to measure the OD value at 660 nm using a UV spectrophotometer (UV-2800A, Unico® (Shanghai) Instrument Co., Ltd., Shanghai, China). The inorganic phosphorus content was calculated according to the OD$_{660}$ and the standard curve that prepared with KH$_2$PO$_3$ at different concentrations was: $y = 8.235x - 0.0859$, $R^2 = 0.9984$. ATPase activity was calculated and the unit was μmol (Pi) × (mg (chl) h)$^{-1}$.

4.9. Determination of Chlorophyll Fluorescence and Photosynthetic Parameters

Chlorophyll fluorescence parameters were investigated using an Imaging-PAM Chlorophyll Fluorometer (Walz, Effeltrich, Germany) at 2, 4, 6, 8, and 10 days after treatment. Before measurement, the cut lily leaves were kept in darkness for 30 min to allow all reaction centers to open. The maximum quantum yield of PSII (Fv/Fm = (Fm − Fo)/Fm) and the effective quantum yield of PSII (ΦPSII) [ΦPSII = (Fm′ − Fs)/Fm′] was calculated according to Genty et al. [49]. Photochemical quenching (qP) [qP = (Fm′ − Fs)/(Fm′ − Fo′)] was calculated according to Van Kooten and Snel (1990) [50]. Photosynthetic parameters were measured using a CIRAS-2 Portable Photosynthesis and Chlorophyll Fluorescence System (PP Systems Ltd., Hitchin, Herts, UK) at a photon irradiance of 1500 Lmol m^{-2} s^{-1}. Three plants in each treatment were randomly selected for gas exchange measurement at 9:00 to 11:00 a.m. on a sunny morning at 2, 4, 6, 8, and 10 days after treatment. Net photosynthetic rate (Pn), transpiration rate (Tr), stomatal conductance (Gs), and intercellular CO$_2$ concentration (Ci) were recorded.

4.10. Statistical Analysis

Values are means ± SE of three various experiments with three replicated measurements. Multiple comparisons were performed using Duncan's multiple range test to determine the significance of the results between different treatments at the $p < 0.05$ level. The data analysis was conducted using the software SPSS 22.0 (SPSS Inc., Chicago, IL, USA).

Supplementary Materials: Supplementary materials can be found at http://www.mdpi.com/1422-0067/19/12/3955/s1.

Author Contributions: W.L. conceived the research idea and designed the experiment. J.H., J.Z., H.F., B.W., and Z.M. conducted the experiments. J.H., D.H., and C.W. finished the data analysis. J.H. prepared the manuscript. W.L. revised the manuscript. All the authors read and approved the submission of the manuscript.

Funding: This research was founded by Discipline Construction Funds for Horticulture, Gansu Agricultural University, China (GAU-XKJS-2018-228), the National Natural Science Foundation of China (31860568, 31560563 and 31160398), the Post Doctoral Foundation of China (20100470887 and 2012T50828), the Research Fund of Higher Education of Gansu, China (2018C-14), Scientific Research Foundation for the Yong Graduate Supervisor of Gansu Agricultural University in Lanzhou China (GAU-QNDS-201709) and Feitian and Fuxi Excellent Talents in Gansu Agricultural University.

Conflicts of Interest: The authors declare no conflict of interest.

Abbreviations

AtpA	ATP synthase CF1 alpha subunit (chloroplast)
Ci	intercellular CO_2 concentration
DTT	DL-dithiothreitol
Fv/Fm	the maximum quantum yield of PSII
Gs	stomatal conductance
HRW	hydrogen-rich water
Pn	net photosynthetic rate
qP	photochemical quenching
SNP	sodium nitroprusside
TCA	trichloroacetic acid
Tr	transpiration rate
2-DE	two-dimensional electrophoresis
β-ME	β-mercaptoethanol
ΦPSII	the effective quantum yield of PSII

References

1. Li, C.X.; Gong, T.Y.; Bian, B.T.; Liao, W.B. Roles of hydrogen gas in plants: A review. *Funct. Plant Biol.* **2018**, *45*, 783–792. [CrossRef]
2. Zhao, X.Q.; Chen, Q.H.; Wang, Y.M.; Shen, Z.G.; Shen, W.B.; Xu, X.M. Hydrogen-rich water induces aluminum tolerance in maize seedlings by enhancing antioxidant capacities and nutrient homeostasis. *Ecotoxicol. Environ. Saf.* **2017**, *144*, 369–379. [CrossRef] [PubMed]
3. Cui, W.T.; Fang, P.; Zhu, K.K.; Mao, Y.; Gao, C.Y.; Xie, Y.J.; Wang, J.; Shen, W.B. Hydrogen-rich water confers plant tolerance to mercury toxicity in alfalfa seedlings. *Ecotoxicol. Environ. Saf.* **2014**, *105*, 103–111. [CrossRef] [PubMed]
4. Su, N.N.; Wu, Q.; Liu, Y.Y.; Cai, J.T.; Shen, W.B.; Xia, K.; Cui, J. Hydrogen-rich water reestablishes ROS homeostasis but exerts differential effects on anthocyanin synthesis in two varieties of radish sprouts under UV-A irradiation. *J. Agr. Food Chem.* **2014**, *62*, 6454–6462. [CrossRef] [PubMed]
5. Cao, Z.Y.; Duan, X.L.; Yao, P.; Cui, W.T.; Cheng, D.; Zhang, J.; Jin, Q.J.; Chen, J.; Dai, T.S.; et al. Hydrogen gas is involved in auxin-induced lateral root formation by modulating nitric oxide synthesis. *Int. J. Mol. Sci.* **2017**, *18*, 2084. [CrossRef] [PubMed]
6. Lin, Y.T.; Zhang, W.; Qi, F.; Cui, W.T.; Xie, Y.J.; Shen, W.B. Hydrogen-rich water regulates cucumber adventitious root development in a heme oxygenase-1/carbon monoxide-dependent manner. *J. Plant Physiol.* **2014**, *171*, 1–8. [CrossRef] [PubMed]
7. Hu, H.L.; Li, P.X.; Wang, Y.N.; Gu, R.X. Hydrogen-rich water delays postharvest ripening and senescence of kiwifruit. *Food Chem.* **2014**, *156*, 100. [CrossRef] [PubMed]
8. Ren, P.J.; Jin, X.; Liao, W.B.; Wang, M.; Niu, L.J.; Li, X.P.; Zhu, Y.C. Effect of hydrogen-rich water on vase life and quality in cut lily and rose flowers. *Hortic. Environ. Biotechnol.* **2017**, *58*, 576–584. [CrossRef]
9. Manjunatha1, G.; Lokesh, V.; Neelwarne, B.; Singh, Z.; Gupta, K.J. Nitric oxide applications for quality enhancement of horticulture produce. Jules Janick. *Hortic. Rev.* **2014**, *42*, 121–156. [CrossRef]

10. Shabaniana, S.; Esfahania, M.N.; Karamian, R.; Tran, L.S.P. Physiological and biochemical modifications by postharvest treatment with sodium nitroprusside extend vase life of cut flowers of two gerbera cultivars. *Postharvest Biol. Technol.* **2018**, *137*, 1–8. [CrossRef]

11. Dwivedic, S.K.; Aroraa, A.; Singha, V.P.; Sairama, R.; Bhattacharya, R.C. Effect of sodium nitroprusside on differential activity of antioxidants and expression of SAGs in relation to vase life of gladiolus cut flowers. *Sci. Hortic.* **2016**, *210*, 158–165. [CrossRef]

12. Zeng, C.L.; Liu, L.; Xu, G.Q. The physiological responses of carnation cut flowers to exogenous nitric oxide. *Sci. Hortic.* **2011**, *127*, 424–430. [CrossRef]

13. Liao, W.B.; Zhang, M.L.; Yu, J.H. Role of nitric oxide in delaying senescence of cut rose flowers and its interaction with ethylene. *Sci. Hortic.* **2013**, *155*, 30–38. [CrossRef]

14. Xie, Y.J.; Mao, Y.; Zhang, W.; Lai, D.W.; Wang, Q.Y.; Shen, W.B. Reactive oxygen species-dependent nitric oxide production contributes to hydrogen-promoted stomatal closure in Arabidopsis. *Plant Physiol.* **2014**, *165*, 759–773. [CrossRef] [PubMed]

15. Zhu, Y.C.; Liao, W.B.; Niu, L.J.; Wang, M.; Ma, Z.J. Nitric oxide is involved in hydrogen gas-induced cell cycle activation during adventitious root formation in cucumber. *BMC Plant Biol.* **2016**, *16*, 146. [CrossRef] [PubMed]

16. Chen, M.; Cui, W.T.; Zhu, K.K.; Xie, Y.J.; Zhang, C.H.; Shen, W.B. Hydrogen-rich water alleviates aluminum-induced inhibition of root elongation in alfalfa via decreasing nitric oxide production. *J. Hazard. Mater.* **2014**, *267*, 40–47. [CrossRef]

17. Baudouin, E.; Hancock, J.T. Nitric oxide signaling in plants. *Front. Plant Sci.* **2014**, *4*, 553. [CrossRef] [PubMed]

18. Zhu, Y.C.; Liao, W.B. The metabolic constituent and rooting-related enzymes responses of marigold explants to hydrogen gas during adventitious root development. *Theor. Exp. Plant Physiol.* **2017**, *29*, 1–9. [CrossRef]

19. Hu, H.L.; Zhao, S.P.; Li, P.X.; Shen, W.B. Hydrogen gas prolongs the shelf life of kiwifruit by decreasing ethylene biosynthesis. *Postharvest Biol. Technol.* **2018**, *135*, 123–130. [CrossRef]

20. Zhu, Y.C.; Liao, W.B.; Wang, M.; Niu, L.J.; Xu, Q.Q.; Jin, X. Nitric oxide is required for hydrogen gas-induced adventitious root formation in cucumber. *J. Plant Physiol.* **2016**, *195*, 50. [CrossRef]

21. Liu, R.X.; Chen, S.M.; Jiang, J.F.; Zhu, L.; Zheng, C.; Han, S.; Gu, J.; Sun, j.; Wang, H.B.; Song, A.P.; Chen, F.D. Proteomic changes in the base of chrysanthemum cuttings during adventitious root formation. *BMC Genom.* **2013**, *14*, 919. [CrossRef] [PubMed]

22. Song, J.; Fan, L.; Hughes, T.; Palmer Campbell, L.; Li, L.; Li, X.H. Quantitative proteomic investigation on the effect of 1-methylcyclopropene treatments on postharvest quality of selected cut flowers. *Acta Hortic.* **2015**, *1104*, 311–318. [CrossRef]

23. Dai, C.; Cui, W.T.; Pan, J.C.; Xie, Y.J.; Wang, J.; Shen, W.B. Proteomic analysis provides insights into the molecular bases of hydrogen gas-induced cadmium resistance in Medicago sativa. *J. Proteom.* **2017**, *152*, 109–120. [CrossRef] [PubMed]

24. Kang, R.Y.; Zhang, L.; Jiang, L.; Yu, M.L.; Ma, R.J.; Yu, Z.F. Effect of postharvest nitric oxide treatment on the proteome of peach fruit during ripening. *Postharvest Biol. Technol.* **2016**, *112*, 277–289. [CrossRef]

25. Li, L.; Song, J.; Kalt, W.; Forney, C.; Tsao, R.; Pinto, D.; Pinto, D.; Chisholm, K.; Campbell, L.; Fillmore, S.; et al. Quantitative proteomic investigation employing stable isotope labeling by peptide dimethylation on proteins of strawberry fruit at different ripening stages. *J. Proteom.* **2013**, *94*, 219–239. [CrossRef] [PubMed]

26. Li, T.T.; Zhu, H.; Wu, Q.X.; Yang, C.W.; Duan, X.W.; Qu, H.X.; Yun, Z.; Jiang, Y.M. Comparative proteomic approaches to analysis of litchi pulp senescence after harvest. *Food Res. Int.* **2015**, *78*, 274–285. [CrossRef] [PubMed]

27. Hisabori, T.; Konno, H.; Ichimura, H.; Strotmann, H.; Bald, D. Molecular devices of chloroplast F1-ATP synthase for the regulation. *BBA Bioenerg.* **2002**, *1555*, 140–146. [CrossRef]

28. Wang, L.X.; Pan, D.Z.; Li, J.; Tan, F.L.; Hoffmann-Benning, S.; Liang, W.Y.; Chen, W. Proteomic analysis of changes in the *Kandelia candel* chloroplast proteins reveals pathways associated with salt tolerance. *Plant Sci.* **2015**, *231*, 159. [CrossRef]

29. Lee, C.; Rhee, J.K.; Kim, D.G.; Choi, Y.E. Proteomic study reveals photosynthesis as downstreams of both MAP kinase and cAMP signaling pathways in *Chlamydomonas reinhardtii*. *Photosynthetica* **2015**, *53*, 625–629. [CrossRef]

30. Wolfgang, J.; Nathan, N. ATP Synthase. *Annu. Rev. Biochem.* **2015**, *84*, 631–657. [CrossRef]

31. Shao, H.; Cao, Q.; Tao, X.; Gu, Y.; Chang, M.; Huang, C.; Zhang, Y.; Feng, H. Cloning and characterization of ATP synthase CF1 α gene from sweet potato. *Afr. J. Biotechnol.* **2011**, *10*, 19035–19042. [CrossRef]
32. Shu, S.; Yuan, Y.H.; Chen, J.; Sun, J.; Zhang, W.H.; Tang, Y.Y.; Zhong, M.; Guo, S. The role of putrescine in the regulation of proteins and fatty acids of thylakoid membranes under salt stress. *Sci. Rep.* **2015**, *5*, 14390. [CrossRef] [PubMed]
33. Viale, A.; Vallejos, R.; Jagendorf, A.T. Hydrogen exchange into soluble spinach chloroplast coupling factor during heat activation of its ATPase. *BBA–Bioenerg.* **1981**, *637*, 496–503. [CrossRef]
34. He, W.X.; Xu, Y.; Tang, L.; Wei, Q.; Li, J.; Chen, F. Molecular cloning regulation of chilling–repressed gene *atpA* in Elumus sibiricus. *Prog. Biochem. Biophys.* **2005**, *32*, 67–74.
35. Zandonadi, D.B.; Santos, M.P.; Dobbss, L.B.; Olivares, F.L.; Canellas, L.P.; Binzel, M.L.; Okorokova-Façanha, A.L.; Façanha, A.R. Nitric oxide mediates humic acids-induced root development and plasma membrane H+-ATPase activation. *Planta* **2010**, *231*, 1025–1036. [CrossRef] [PubMed]
36. Zhang, Y.K.; Han, X.X.; Chen, X.L.; Jin, H.; Cui, X.M. Exogenous nitric oxide on antioxidative system and ATPase activities from tomato seedlings under copper stress. *Sci. Hortic.* **2009**, *123*, 217–223. [CrossRef]
37. Okumura, M.; Inoue, S.I.; Kuwata, K.; Kinoshita, T. Photosynthesis activates plasma membrane H+-ATPase via sugar accumulation in *Arabidopsis* leaves. *Plant Physiol.* **2016**. [CrossRef] [PubMed]
38. Zhang, X.N.; Zhao, X.Q.; Wang, Z.Q.; Shen, W.B.; Xu, X.M. Protective effects of hydrogen-rich water on the photosynthetic apparatus of maize seedlings (*Zea mays* L.) as a result of an increase in antioxidant enzyme activities under high light stress. *Plant Growth Regul.* **2015**, *77*, 43–56. [CrossRef]
39. Cui, J.X.; Zhou, Y.H.; Ding, J.G.; Xia, X.J.; Shi, K.; Chen, S.C.; Asami, T.; Yu, J.Q. Role of nitric oxide in hydrogen peroxide-dependent induction of abiotic stress tolerance by brassinosteroids in cucumber. *Plant Cell Environ.* **2011**, *34*, 347. [CrossRef] [PubMed]
40. Singh, V.P.; Srivastava, P.K.; Prasad, S.M. Nitric oxide alleviates arsenic-induced toxic effects in ridged *Luffa* seedlings. *Plant Physiol. Biochem.* **2013**, *71*, 155–163. [CrossRef] [PubMed]
41. Wodala, B.; Deák, Z.; Vass, I.; Erdei, L. Nitric oxide modifies photosynthetic electron transport in pea leaves. *Acta Biol. Szegediensis* **2005**, *49*, 7–8.
42. Dong, N.G.; Li, Y.F.; Qi, J.X.; Chen, Y.H.; Hao, Y.B. Nitric oxide synthase-dependent nitric oxide production enhances chilling tolerance of walnut shoots in vitro via involvement chlorophyll fluorescence and other physiological parameter levels. *Sci. Hortic.* **2018**, *230*, 68–77. [CrossRef]
43. Spetea, C.; Schoefs Spetea, B. Solute transporters in plant thylakoid membranes: Key players during photosynthesis and light stress. *Commun. Integr. Biol.* **2010**, *3*, 122–129. [CrossRef]
44. Wang, T.; Yang, W.H.; Xie, Y.F.; Shi, D.W.; Ma, Y.L.; Sun, X. Effects of exogenous nitric oxide on the photosynthetic characteristics of bamboo (Indocalamus barbatus McClure) seedlings under acid rain stress. *Plant Growth Regul.* **2017**, *82*, 69–78. [CrossRef]
45. Chen, P.B.; Li, X.; Huo, K.; Wei, X.D.; Dai, C.C.; Lv, C.G. Promotion of photosynthesis in transgenic rice over-expressing of maize C_4 phosphoenolpyruvate carboxylase gene by nitric oxide donors. *J. Plant Physiol.* **2014**, *171*, 458–466. [CrossRef] [PubMed]
46. Liu, S.J.; Gao, J.D.; Chen, Z.J.; Qiao, X.Y.; Huang, H.L.; Cui, B.Y.; Zhu, Q.F.; Dai, Z.; Wu, H.; Pan, Y.; Yang, C. Comparative proteomics reveals the physiological differences between winter tender shoots and spring tender shoots of a novel tea (*Camellia sinensis* L.) cultivar evergrowing in winter. *BMC Plant Biol.* **2017**, *17*, 206. [CrossRef] [PubMed]
47. Sheffield, J.; Taylor, N.; Fauquet, C.; Chen, S. The cassava (Manihot esculenta Crantz) root proteome: Protein identification and differential expression. *Proteomics* **2006**, *6*, 1588–1598. [CrossRef]
48. Livak, K.J.; Schmitgen, T.D. Analysis of relative gene expression data using real-time quantification PCR and the 2–ΔΔCt method. *Methods* **2001**, *25*, 402–408. [CrossRef]
49. Genty, B.; Briantais, J.M.; Baker, N.R. The relationship between the quantum yield of photosynthetic electron transport and quenching of chlorophyll fluorescence. *BBA-Gen. Subj.* **1989**, *990*, 87–92. [CrossRef]
50. Van Kooten, O.; Snel, J.F.H. The use of chlorophyll fluorescence nomenclature in plant stress physiology. *Photosynth. Res.* **1990**, *25*, 147–150. [CrossRef]

International Journal of
Molecular Sciences

Article

Comparative Proteomics and Physiological Analyses Reveal Important Maize Filling-Kernel Drought-Responsive Genes and Metabolic Pathways

Xuan Wang [1,2], Tinashe Zenda [1,2], Songtao Liu [1,2], Guo Liu [1,2], Hongyu Jin [1,2], Liang Dai [1,2], Anyi Dong [1,2], Yatong Yang [1,2] and Huijun Duan [1,2,*]

[1]   Department of Crop Genetics and Breeding, College of Agronomy, Hebei Agricultural University, Baoding 071001, China
[2]   North China Key Laboratory for Crop Germplasm Resources of the Education Ministry, Hebei Agricultural University, Baoding 071001, China
*   Correspondence: hjduan@hebau.edu.cn; Tel.: +86-139-3127-9716

Received: 5 June 2019; Accepted: 29 July 2019; Published: 31 July 2019

Abstract: Despite recent scientific headway in deciphering maize (*Zea mays* L.) drought stress responses, the overall picture of key proteins and genes, pathways, and protein–protein interactions regulating maize filling-kernel drought tolerance is still fragmented. Yet, maize filling-kernel drought stress remains devastating and its study is critical for tolerance breeding. Here, through a comprehensive comparative proteomics analysis of filling-kernel proteomes of two contrasting (drought-tolerant YE8112 and drought-sensitive MO17) inbred lines, we report diverse but key molecular actors mediating drought tolerance in maize. Using isobaric tags for relative quantification approach, a total of 5175 differentially abundant proteins (DAPs) were identified from four experimental comparisons. By way of Venn diagram analysis, four critical sets of drought-responsive proteins were mined out and further analyzed by bioinformatics techniques. The YE8112-exclusive DAPs chiefly participated in pathways related to "protein processing in the endoplasmic reticulum" and "tryptophan metabolism", whereas MO17-exclusive DAPs were involved in "starch and sucrose metabolism" and "oxidative phosphorylation" pathways. Most notably, we report that YE8112 kernels were comparatively drought tolerant to MO17 kernels attributable to their redox post translational modifications and epigenetic regulation mechanisms, elevated expression of heat shock proteins, enriched energy metabolism and secondary metabolites biosynthesis, and up-regulated expression of seed storage proteins. Further, comparative physiological analysis and quantitative real time polymerase chain reaction results substantiated the proteomics findings. Our study presents an elaborate understanding of drought-responsive proteins and metabolic pathways mediating maize filling-kernel drought tolerance, and provides important candidate genes for subsequent functional validation.

Keywords: proteomes; iTRAQ; filling kernel; drought stress; heat shock proteins; *Zea mays* L.

---

1. Introduction

Field grown crops, similar to other sessile organisms, often endure numerous environmental instabilities throughout their life spans [1,2]. Such constant exposure hampers plant growth and development, consequently resulting in reduced crop yields [3]. Among the environmental stress factors (such as salinity, drought, freezing, heat, etc.), drought is the single factor that imposes the most severe limitations to agricultural production [4–6]. In the backdrop of a continually increasing global human population, heightened food demand, and continuing global climate change, drought is anticipated to increase in occurrence and intensity under future agricultural practice [7,8]. Therefore, understanding how crop plants respond to drought stress at the molecular level remains critical for guiding the genetic improvement in drought tolerance, so as to maintain sustainable higher productivity under such climate change conditions.

Maize (*Zea mays* L.) is the third most important cereal crop in the world after rice (*Orzya sativa* L.) and wheat (*Triticum aestivum* L.) [9]. Aslam et al. [10] have dubbed maize "a multidisciplinary crop" because of its multiple uses in the human food, animal feed, and fodder, as well as biofuel production [11,12]. However, maize production is under severe threat from drought stress. The crop is most susceptible to drought from the flowering stage to grain filling stage [13,14]. Kernel development in maize is well-characterized, with grain filling occurring from ~15 days post pollination (DPP) to ~45 DPP [15]. Maize kernels are particularly sensitive to the negative effects of drought stress during grain filling [16], a period also considered critical from an aflatoxin (*Aspergillus flavus* L.) resistance perspective [17]. Moisture stress at this stage could result in seed abortion and decreased productivity [18]. The numbers of kernels set and filled under drought stress account for most of the variation in maize grain yield under drought and directly affects the harvesting index. Additionally, kernels near the ear tip will often abort after several weeks of growth if they are drought-affected [7]. Therefore, minimizing moisture deficit and maintenance of an active supply of photo-assimilates at the grain filling stage are essential in reducing the effects of drought on kernel final weight.

To overcome environmental perturbations within their habitats, plants institute several adaptive strategies at different levels, ranging from physiological through metabolic to molecular [10,19]. Maintenance of water status and physiological activity of the plant cell is achieved through metabolic adjustment; plants synthesize and accumulate various protective molecules, such as antioxidant enzymes (peroxidase (POD), superoxide dismutase (SOD) etc.), polyamines, amino acids (predominantly proline and glycine), and some sugars [4,20]. To protect themselves against reactive oxygen species (ROS) and photo inhibition, plants activate osmoprotection via osmotic adjustment and antioxidant scavenging defense systems, aided by plant growth regulators [2,21]. At the molecular level, plants institute stress responsive proteins, transcription factors, and signalling pathways among other strategies to respond to drought stress [21]. These molecules confer drought tolerance through protection of cellular contents or via regulation of stress responsive genes [10].

Although there have been a lot of studies [22–26] on the physiological, biochemical, and molecular bases of maize dehydration tolerance, most of them attached importance to the seedling and vegetative stage responses, and mostly employed cDNA micro arrays and transcriptome analyses. Thus far, very few studies [3,18] have focused on the proteomic analysis of maize drought stress response at the grain filling stage, despite this being the most critical stage at which soil moisture deficit stress has a direct effect on the final grain yield in maize [7,16], and despite proteomic analysis approach being more expedient to transcriptomic approaches [27,28].

With recent advances in the technologies and approaches in abiotic stress response evaluation, large scale, high-throughput proteomics has become a very powerful tool for performing comprehensive analysis of crop proteins and identification of stress responsive proteins in comparative abiotic stress studies [29,30]. Proteomics describes the study and characterization of a complete set of proteins present in a cell, organ, or organism at a given time (known as a proteome) [31,32]. Generally, proteomic approaches are useful for proteome profiling, comparative expression analysis of two or more protein samples, and localization and identification of post translational modifications (PTMs). Proteomics involves the detection of protein diversity, abundance, isoforms, compartmentalization, and interaction with other proteins [33]. Additionally, it provides for both qualitative and quantitative measurements of proteomes in specific plant tissues at specific developmental and physiological stages [32,34]. Protein profiles reflect changes in the protein expression of a given tissue and its cellular compartments in response to endogenous or external perturbations [35]. Besides being complementary to genomics, proteomics provides information on the molecular mechanisms underlying plant growth and stress responses, and is a crucial link between transcriptomics and metabolomics [29]. Whereas the structural and expressional variation identified at genetic or transcriptional levels is not always translated into the predicted phenotype because of post-translational modifications [36], the proteome (unlike the static genome) is dynamic, and the evaluation of proteins caters for PTMs, thus providing more knowledge in understanding biological functions [33].

Gel-free methods involving digestion of intact proteins into peptides prior to separation have now become very popular as compared to gel-based methods (such as 2DE), which have a low rate of throughput. Application of gel-free protein separation techniques and the next generation of proteomic techniques, including quantitative proteomics approaches, such as isobaric tags for relative and absolute quantitation (iTRAQ) and isotope-coded affinity tags (ICAT), have become widely employed in descriptive and comparative plant abiotic stress adaptation proteomic studies [37,38]. Particularly, the iTRAQ-based method provides a gel-free shotgun quantitative analysis and uses isobaric reagents to label tryptic peptides and to monitor relative protein and peptide mass tolerance (PMT) abundance changes [39]. It also provides for multiplexing of up to eight samples in a single experiment, thereby allowing for the time-dependent analysis of plant stress responses or biological replicates in a single experiment [40]. The technique has become widely useful in plant abiotic stress response studies [36,41–43].

Previously [44], we employed an iTRAQ-based method to evaluate, through comparative analysis approach, the seedling-stage responses of two contrasting maize inbred lines (drought tolerant YE8112 and drought sensitive MO17) to drought stress. We realized that YE8112 was comparatively more tolerant to drought stress than MO17 owing to its enhanced activation of photosynthesis proteins involved in thermal dissipation of light energy, enhanced lipid metabolism, improved cellular detoxification capacity, improved chaperon activities, as well as reduced synthesis of redundant proteins to help save energy for enduring stress [44]. Additionally, our transcriptomic study [45] suggested improved stress sensing and signaling, enhanced carbohydrate synthesis and cell wall remodeling, enhanced amino acid biosynthesis, as well as YE8112's ability to sustainably accumulate greater amounts of proline and POD activity under water-limited conditions to be critical for drought tolerance.

In order to clarify the molecular mechanisms underlying maize response to drought stress at the kernel filling stage, herein we employed an iTRAQ-based approach to examine the protein expression profiles in developing kernels, and to compare the drought stress responses of the same (YE8112 and MO17) inbred lines after moisture deficit exposure for 14 days. Moreover, comparative phenotypic and physiological drought-stress response analyses buttress the proteomic analysis results. Our findings bring to light new insights into the principal responsive proteins and metabolic pathways and processes underpinning maize drought tolerance at the kernel filling stage. Additionally, building on our previous findings [44,45], we examine the similarities or differences between maize seedling and filling-kernel responses to drought in the context of gene and protein expression regulation and at the physiological level. Further, drought responsive proteins and genes identified herein could be harnessed for molecular breeding and biotechnological applications aimed at developing drought resilient cultivars.

2. Results

2.1. Inbred Lines Contrasting Phenotypic and Physiological Responses to Drought Stress

At the 26 DPP stage, we recorded some phenotypic and morphological measurements of the sample ears from different treatment groups from both MO17 and YE8112 inbred lines. Phenotypically, under water-sufficient (control) condition, both inbred lines exhibited decent grain-filling condition and no prominent visual variances (sensitive line under water-sufficient conditions, SC in Figure 1A; tolerant line under water-sufficient condition, TC in Figure 1B).

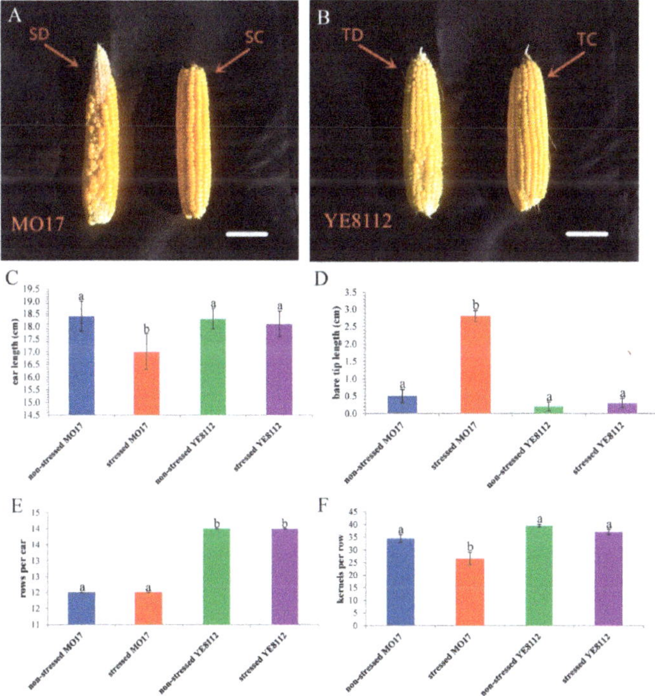

Figure 1. Phenotypic characterization of the two maize inbred lines (sensitive MO17, S; and tolerant YE8112, T) ears' responses to drought stress. Observations and measurements were made at 26 days post pollination (DPP) under both water-sufficient (control, C) and water-deficit (drought, D) conditions. (**A,B**) Ear phenotypes; (**C**) ear length; (**D**) ear bare tip length; (**E**) kernel rows per ear; (**F**) kernel number per row. Data are presented as mean ± standard errors ($n = 3$). Different letters on error bars mean significant difference at $p < 0.05$. (**C–F**) Each replication is an average for the measurement of 10 ears. Scale bars = 4 cm for both Figure 1A,B.

However, under moisture-deficit conditions, significant differences in ear appearances were observed, with sensitive line MO17 showing uneven kernel arrangement and an obvious phenomenon of tip barrenness (sensitive line under drought treatment; SD in Figure 1A), whilst tolerant line YE8112 had even kernel arrangement and minimal barren tips (tolerant line under drought treatment; TD in Figure 1B). Drought stress resulted in significant ($p < 0.05$) decrease in ear length (EL) only in sensitive line MO17 (Figure 1C). Additionally, drought stress significantly increased ear tip barrenness in MO17, but not in YE8112 (Figure 1D). Meanwhile, relative to the control conditions, kernel rows per ear (KRPE) was not significantly influenced by drought treatment in both lines (Figure 1E). However, under drought conditions, kernels per row (KPR) was significantly ($p < 0.05$) decreased in sensitive line MO17, but not in tolerant line YE8112 (Figure 1F). Our results reveal vivid contrasting phenotypic responses of the two lines to drought stress at the kernel filling stage, with MO17 being comparably more susceptible than YE8112.

To understand the physiological responses of the maize plants to drought, we determined some physiological indices in the kernels. Compared to the control groups, the kernel relative water content (RWC) significantly ($p < 0.05$) declined in both inbred lines under drought stress, but the rate of decline in MO17 was much sharper than that of YE8112 (Figure 2A).

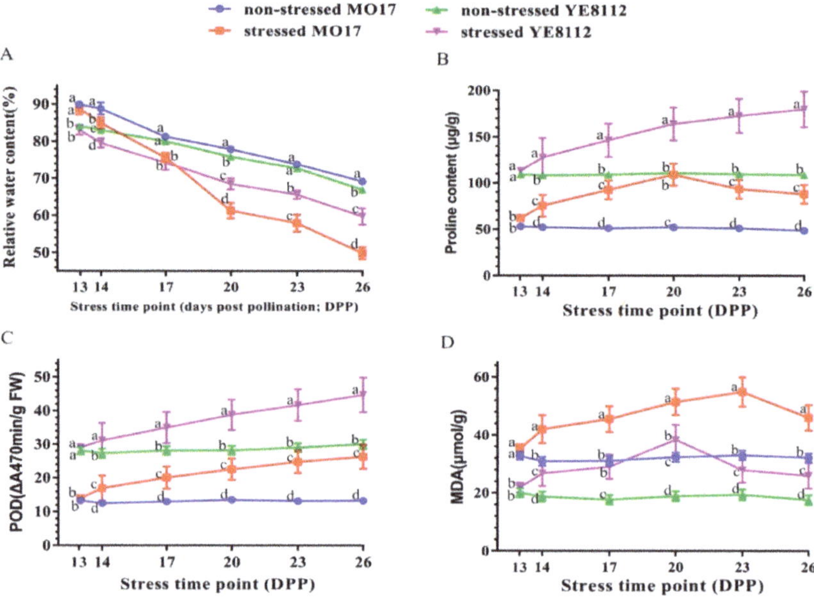

Figure 2. Physiological changes in the kernels of two contrasting maize inbred lines (sensitive MO17 and tolerant YE8112) in response to drought stress. Physiological changes were measured at different time points (13, 14, 17, 20, 23, and 26 days post pollination; DPP) under both water-sufficient (control) and water-deficit conditions. (**A**) Relative water content; (**B**) proline content; (**C**) peroxidase (POD) activity; (**D**) malonaldehyde (MDA) content. Data are presented as mean ± standard errors (*n* = 3). Different letters above line graphs show significant difference ($p \leq 0.05$) among treatments at a given stress time point.

Our results showed that compared to the control conditions, proline content increased in both lines from 14 DPP to 20 DPP under drought conditions (Figure 2B). Proline increase was mantained up to 26 DPP in the tolerant line YE8112, wheras in the susceptible line MO17, it decreased from 20 DPP onwards (Figure 2B). Compared to control conditions, POD activity significantly increased in both inbred lines under drought stress conditions, starting from 14 DPP (Figure 2C). The rate of POD increase was statistically similar between the two lines (Supplementary Figure S1). Furthermore, as compared to the corresponding control, MDA content was significantly elevated by drought stress in both inbred lines. However, there was a sharp decline in MDA content in YE8112 from 20 DPP, and moderate decrease in MO17 starting from 23 DPP (Figure 2D).

2.2. Summary Inventory of Maize Filling-Kernel Proteins Identified by iTRAQ Analysis

The proteomes of MO17 and YE8112 kernels were collected under water-sufficient (control) and water-deficit conditions to make four libraries of three replicates each. The samples were then used for iTRAQ analysis. With the aid of Mascot software (version 2.2), 303, 288 spectra were matched with theoretical spectra from the Uniprot *Zea mays* L. database (132,339-12 January 2018). A total 50,566 peptides (comprising 27,288 unique peptides) corresponding to 6640 proteins (Supplementary Table S1) were identified. A broad coverage was observed on protein molecular weight (MW) distribution (Supplementary Figure S2A), with MW of the identified proteins ranging from 3 to 500 kDa. Among them, 162 (2.44%) weighed < 10 kDa, 5406 (81.42%) weighed 10–70 kDa, 634 (9.55%) weighed 70–100 kDa, and 438 (6.59%) weighed > 100 kDa (Supplementary Figure S2A). The number of peptides defining each protein is distributed in Supplementary Figure S2B, and over 77% of the total (6640) proteins matched with at least two peptides. In addition, the protein sequence coverage was generally below 30% (Supplementary Figure S2C). Further, the distribution of the peptide lengths defining each protein showed that most (90%) of the peptides ranged between 5 to 20 amino acids, with 7–9 and 9–11 amino acids as modal lengths (Supplementary Figure S2D). Proteins possessing at least two unique peptides were used in subsequent analysis of differentially abundant proteins (DAPs).

2.3. Analysis of Drought-Responsive Differentially Abundant Proteins (DAPs)

We employed a comparative proteomic analysis approach to investigate the protein profile alterations in kernels of MO17 (drought-sensitive, S) and YE8112 (drought-tolerant, T) lines under water deficit conditions. A pairwise comparison of prior and post treatments (drought, D; control, C) was performed in MO17 (SC_SD) and YE8112 (TC_TD) individually. Additionally, we carried out a comparative analysis of the proteome between the sensitive and tolerant lines, under water-sufficient (control) (SC_TC) and drought (SD_TD) conditions, yielding four experimental comparisons (Figure 3A). A search for proteins with fold changes > 1.2 (up) or <0.83 (down) at $p < 0.05$ identified a total of 5175 DAPs among the four comparison groups. Prior to drought treatment, a total of 2172 DAPs were differentially expressed (SC_TC comparison), among which 1102 were up-regulated and 1070 down-regulated. Post drought exposure, 897 DAPs showed differential expression (SD_TD), with 447 being up-regulated and 450 down-regulated. Within the tolerant line YE8112, 155 proteins displayed differential abundance before and after drought treatment (TC_TD), with 81 DAPs being up-regulated and 74 down-regulated. Meanwhile, 1951 DAPs were observed in sensitive line MO17 before and after treatment (SC_SD), among which 950 were up-regulated and 1001 down-regulated (Figure 3A).

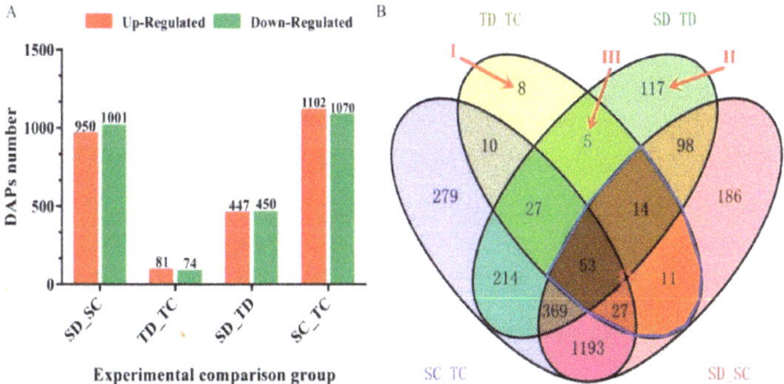

Figure 3. Analysis of differentially abundant proteins (DAPs) identified in four experimental comparisons. (**A**) Total number of DAPs identified in each experimental comparison group, by expression type. Up-regulated means DAPs with increased differential abundance. Down-regulated means DAPs with decreased differential abundance. (**B**) Venn diagram analysis of DAPs. Overlapping regions of the Venn diagrams indicate DAPs shared between or among corresponding groups. DAPs uniquely expressed in TC_TD (I), SD_TD (II), and TC_TD and SD_TD (III) are indicated with arrows. Area IV shows 105 overlapping DAPs within line.

The Venn diagram (Figure 3B) displays a comparative analysis of the DAPs described above. The combinations of the four experimental comparisons reflect the impact of treatment or lines. With regards to drought tolerance, some combinations are more important than others. The sets of DAPs (marked Areas I–IV in Figure 3B) are the most critical and assumed more relevant to drought stress response. Among the 8 DAPs specific to YE8112 under drought stress conditions (Area I), half were up-regulated and another half down-regulated (Table 1). Of the 117 DAPs (Supplementary Table S2) unique to SD_TD (Area II; DAPs shared between sensitive and tolerant lines after drought treatment), 59 were up-regulated and 58 down-regulated (Supplementary Table S2). Meanwhile, the five DAPs unique to Area III are shown in Table 2. The 105 DAPs shared between TC_TD and SC_SD (Area IV), that is, the overlapping (common) drought-responsive DAPS within line, are listed in Supplementary Table S3. Additionally, the 186 DAPs uniquely expressed in sensitive line MO17 after drought treatment are listed in Supplementary Table S4.

We further conducted an overview hierarchical clustering analysis of the DAPs in the two inbred lines under drought stress conditions (proteins unique to SD_TD; Area II in Figure 3). Our results revealed that the DAP in different replicates of the same line exhibited similar expression patterns. However, comparisons of the same DAP in YE8112 and MO17 revealed differences in expression patterns (Figure 4A). Meanwhile, analysis of the differential expression changes of the DAPs showed that down-regulated and up-regulated DAPs were symmetrical (Figure 4B).

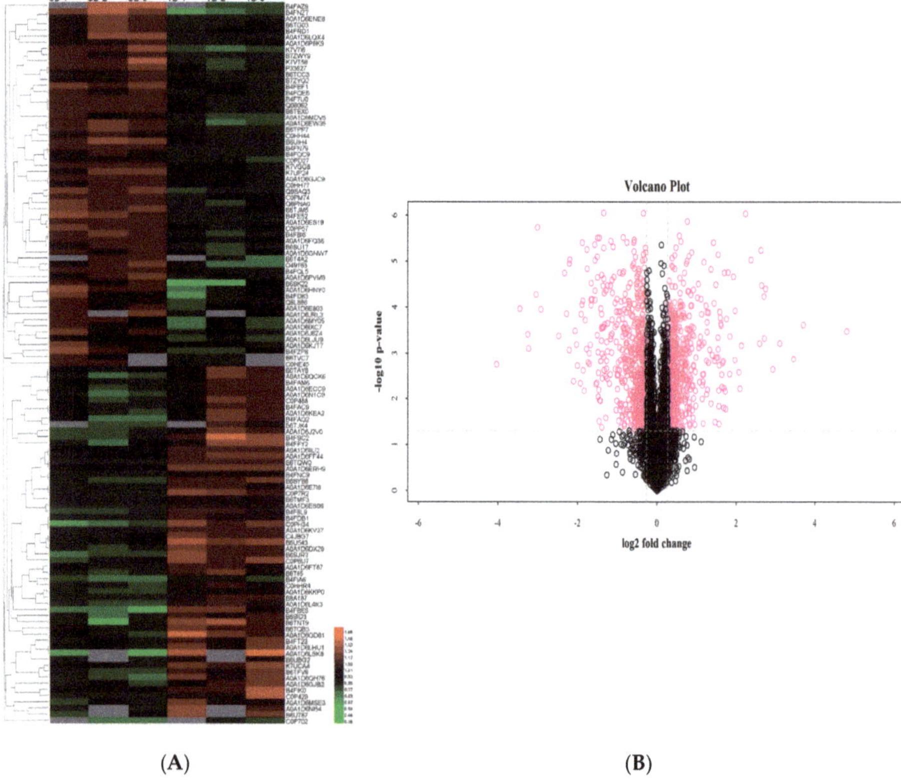

(A)   (B)

Figure 4. Clustering analysis of differentially abundant proteins (DAPs). (**A**) Heat map of DAPs overlapping in in SD_TD experimental comparison. Each row represents a significantly abundantly expressed protein. SD1-3 refers to the biological replicate number for MO17, whilst TD1-3 refers to the replicate number for YE8112. The DAPs were clustered based on the differentially expressed levels. The scale bar indicates the logarithmic value (log 2) expression of the DAPs, up-regulated (red) and down-regulated (green); (**B**) volcano plot showing the (log 2; −log 10 false discovery rate, FDR) expression of the DAPs in SD_TD comparison. Purple bubbles represent differentially expressed proteins and black bubbles represent proteins with non-differential expression.

Table 1. Drought-responsive maize kernel proteins identified specifically in tolerant line YE8112.

No	Accession [1]	Gene Name/ID [2]	Description [3]	Covrg. [4]	Pept. [5]	Log_2FC [6]	p Value [7]	Expr. [8]	Pathway [9]
1	A0A1D6N230	Zm00001d042191	Uncharacterized protein	11.16	6	1.234928	7.25×10^{-4}	Up	
2	B6TDF8	100282981	Glyceraldehyde-3-phosphate dehydrogenase	61.07	19	1.219132	1.36×10^{-4}	Up	Carbon fixation in photosynthetic organisms//Gluconeogenesis
3	A0A1D6L6C5	100284660	Cytosolic purine 5-nucleotidase	5.85	4	1.212466	6.72×10^{-4}	Up	
4	B4FWF5	100273587	Histone deacetylase 6	18.38	2	1.211052	1.39×10^{-3}	Up	
5	B6TPB9	100277111	Pentatricopeptide repeat-containing protein mitochondrial	16.47	6	0.824045	2.04×10^{-2}	Down	
6	B4FVQ0	100273462	Pentatricopeptide repeat-containing protein mitochondrial	4.07	2	0.798673	3.42×10^{-2}	Down	
7	B6U9Q8		mTERF family protein	5.52	2	0.773541	4.68×10^{-2}	Down	
8	A0A1D6HT77	100283204	Galactose-1-phosphate uridyl transferase-like protein	7.34	2	0.717849	1.57×10^{-3}	Down	Galactose metabolism//Amino sugar and nucleotide sugar metabolism

[1] Accession = unique protein identifying number in the UniProt database; [2] gene name/ID = name or ID number of the corresponding gene of the identified differentially abundant protein DAP as searched against the maize sequence database Gramene (http://ensemble.gramene.org/Zeamays); [3] description = annotated biological functions based on Gene Ontology (GO) analysis; [4] Covrg. (%) = sequence coverage is calculated as the number of amino acids in the peptide fragments observed divided by the protein amino acid length; [5] Pept. = peptide fragments, refer to the number of matched peptide fragments generated by trypsin digestion; [6] Log2FC = fold change (log 2), is expressed as the ratio of intensities of up-regulated or down-regulated proteins between drought stress treatments and control (well-watered conditions). All the fold change values below 1 represents that the proteins were down-regulated; [7] p value = statistical significant level (using a paired t-test) < 0.05; [8] Expr. = gene expression level; Up = up-regulated; Down = down-regulated; [9] pathways = metabolic pathways in which the identified protein was found to be significantly enriched.

Table 2. Drought-responsive DAPs of the tolerant line that were also differentially expressed between tolerant and sensitive lines after drought treatment.

No	Accession	Gene Name/ID	Description	Covrg.	Pept.	YE8112 Fold Change		SD_TD Fold Change		Pathway
						Log2FC	p Value	Log2FC	p Value	
1	B6SHX8		Uncharacterized protein	32.43	3	1.618419	1.51×10^{-3}	0.77178	2.65×10^{-2}	
2	B6SGF3	100280456	Glyoxalase family protein superfamily	38.13	3	0.514965	3.06×10^{-2}	1.912391	2.96×10^{-2}	
3	B4FQG0	100282825	Hydrogen peroxide-induced 1	31.67	2	1.266202	4.56×10^{-2}	0.807211	2.67×10^{-2}	
4	B4FPJ4		ADP, ATP carrier protein	13.3	8	0.668951	3.03×10^{-3}	1.308112	2.30×10^{-3}	
5	B4FGT5	P4H7	Prolyl 4-hydroxylase 7	13.76	4	1.219904	4.69×10^{-2}	0.825568	1.46×10^{-2}	Arginine and proline metabolism

For detailed description of the columns, please refer to Table 1 caption above.

2.4. Functional Annotation and Classification of Drought-Responsive DAPs

We used Blast2GO web-based application (https://www.blast2go.com) to perform gene ontology (GO) functional annotation, that is, to assign level 2 GO terms to the identified DAPs. GO functional analysis revealed that several GO terms were shared between MO17 and YE8112, including metabolic process (GO:0008152), cellular process (GO:0009987), regulation of biological process (GO:0050789), and response to stimulus (GO:0050896), within the biological process (BP) category. Within the molecular function (MF) category, binding (GO:0005488), catalytic activity (GO:0003824), nutrient reservoir activity (GO:0045735), and transporter activity (GO:0005488) were the most prominent shared GO terms between the two lines. With regards to the cellular components (CC) compartmentalization, most of the DAPs in both inbred lines were located in the cell (GO:0005623), cell part (GO:0044464), membrane (GO:0016020), membrane part (GO:0044425), and organelle (GO:0043226) (Supplementary Figure S3).

However, the analysis of the most significantly enriched top 20 GO terms in each line showed vivid differences in the GO terms between the two lines (Figure 5). In drought stressed YE8112, response to stress (GO:0006950), response to stimuli (GO:0050896), response to abiotic stimuli (GO:0009628), and hydrogen peroxide catabolic process (GO:0042744) were apparent under the biological process (BP) category. The GO terms nutrient reservoir activity (GO:0045735), oxidoreductase activity acting on a sulphur group (GO:0016667), and sulfite reductase (ferredoxin) activity (GO:0050311) were prominent in the molecular function (MF) category, whereas extracellular region (GO:0005576), monolyase-surrounded lipid storage body (GO:0012511), lipid droplet (GO:0005811), and apoplast (GO:0048046) were the apparent cellular component (CC) locations for the DAPs (Figure 5A). On the other hand, in drought stressed MO17, small molecule metabolic process (GO:0044281), cellular carbohydrate metabolic process (GO:0044262), and regulation of generation of precursor metabolites and energy (GO:0043467) were dominant terms in the BP category. In the MF category, catalytic activity (0003824), oxidoreductase activity (GO:0016491), and hydrolyase activity (GO:0016787) were most apparent. For the CC functions, photosystem II (GO:0009654) and thylakoid membrane (GO:0042651) were the most significantly enriched terms (Figure 5B). These differences in the most significantly enriched GO terms may be the main reason for the two inbred lines' divergent drought tolerance, and hence arouse our keen interest for further discussion.

To further analyze the functional fates of the identified DAPs, we mapped and assigned them to various metabolic pathways, based on the Kyoto Encyclopedia of Gene and Genomes (KEGG, available online: https://www.genome.jp/kegg/; accessed on 10 November 2018) database. We observed that the linoleic acid metabolism pathway responded to drought stress in both inbred lines (Figure 6), and that higher protein numbers were observed in MO17 enriched pathways than in YE8112 enriched pathways (Supplementary Figure S4). Using a hypergometric test, KEGG metabolic pathways with a p-value < 0.05 were labeled as significantly influenced by drought stress. Resultantly, protein processing in the endoplasmic reticulum and tryptophan metabolism pathways were the most significantly enriched in tolerant line YE8112 (Figure 6A). Contrastingly, starch and sucrose metabolism, oxidative phosphorylation, and phenylpropanoid biosynthesis pathways were the most significantly enriched in sensitive line MO17 (Figure 6B). For the full view of the top most significantly enriched pathways in YE8112 and MO17 lines, we refer you to Supplementary Figure S5. The diverse response exhibited by the two lines, in terms of the significantly enriched metabolic pathways, may be another clear contributing factor in their drought tolerance divergence.

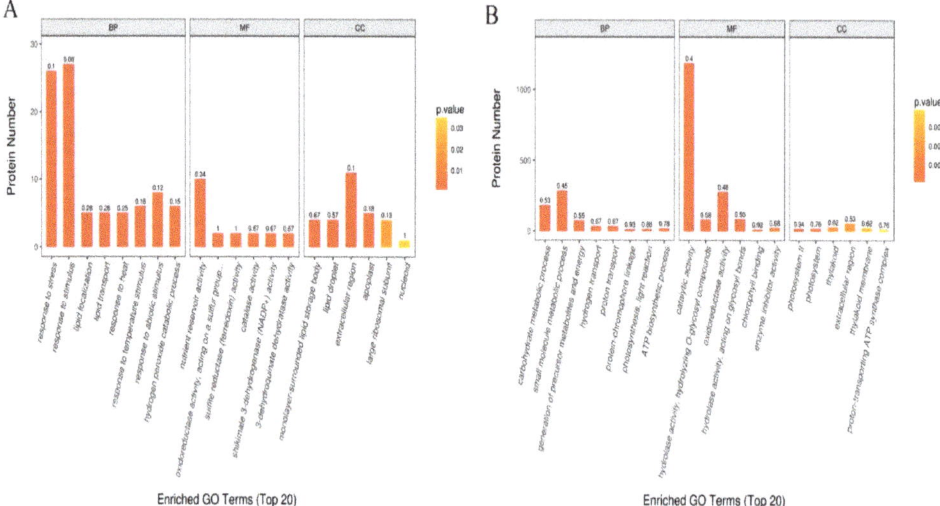

Figure 5. Gene ontology (GO) functional classification of differentially abundant proteins (DAPs). (**A**) Most significantly enriched GO terms (top 20) in tolerant line YE8112 under drought conditions; (**B**) most significantly enriched GO terms in sensitive line MO17 under drought conditions. The number above each bar graph shows the enrichment factor of each GO term (rich factor ≤ 1).

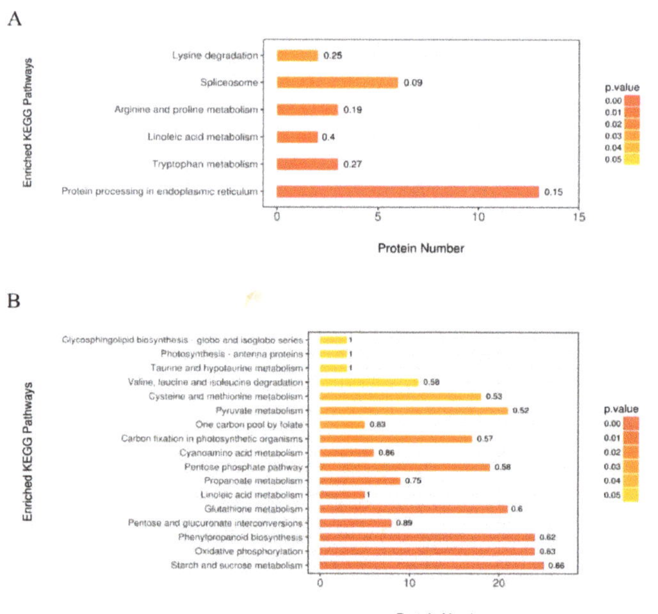

Figure 6. KEGG pathway enrichment analysis of the DAPs. (**A**) Most significantly enriched pathway in TD_TC; (**B**) most significantly enriched pathways in SD_SC, based on the hypergeometric test, $p < 0.05$. The color gradient represents the size of the p value; the color is from orange to red, and the nearer to red represents a smaller p value, and a higher significance level of enrichment of the corresponding KEGG pathway. The label above the bar graph shows the enrichment factor (rich factor ≤ 1).

2.5. Analysis of Protein-Protein Interactions

To understand how the maize kernel cells relay drought stress signals to influence certain cellular functions, the identified DAPs unique to YE8112 and MO17 inbred lines were further analyzed using the String database (Version 10.5, http://www.string-db.org/; accessed 10 March 2019). Selecting those DAPs with confidence scores higher than 0.7, two networks and two pairs of interacting proteins were observed in YE8112 (Figure 7A). The largest cluster comprises ten proteins, which are mostly heat shock proteins (Supplementary Table S5). The second group is made up of eight proteins; these participate in energy and carbohydrate (CHO) metabolism (Supplementary Table S5). Additionally, two protein pairs were identified to participate in drought stress response (Figure 7A), with the first pair being involved in ROS clearance and the second pair (of phosphogluconate dehydrogenases) involved in response to abscisic acid (Supplementary Table S5). Meanwhile, four distinct protein interaction networks were identified in MO17, comprising one sub-hub and three small clusters, as well as three protein pairs (Figure 7B).

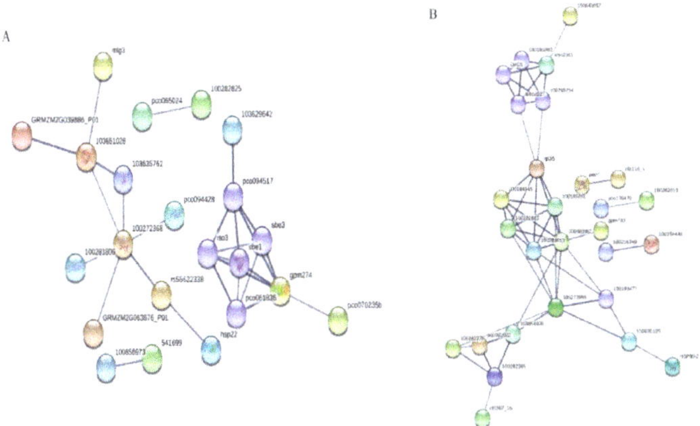

Figure 7. Protein–protein interaction analysis of the maize kernel drought-responsive differentially abundant proteins (DAPs). (**A**) DAPs differentially expressed in YE8112 after drought treatment (TC_TD). (**B**) MO17-specific DAPs. String database (version 10.5; http://www.string-db.org/) was used to construct the network. The nodes represent proteins, and the thickness of connectors between nodes represents the strength of the supporting data.

2.6. Quantitative Real-Time RT-PCR (qRT-PCR) Analysis

To confirm the accuracy of the iTRAQ results, the transcription of twenty-five representative genes (Supplementary Table S6) encoding proteins whose abundance was differentially expressed in response to drought stress was examined through a supporting quantitative real-time PCR (qRT-PCR) experiment. Resultantly, the transcription patterns and levels of all the twenty five sampled genes were consistent with our iTRAQ-based results (Figure 8; Supplementary Table S7). Thus, the qRT-PCR analysis results confirmed our proteomic analysis findings. Correlation coefficient analysis results (between qRT-PCR and iTRAQ data) are provided in Supplementary Figure S6.

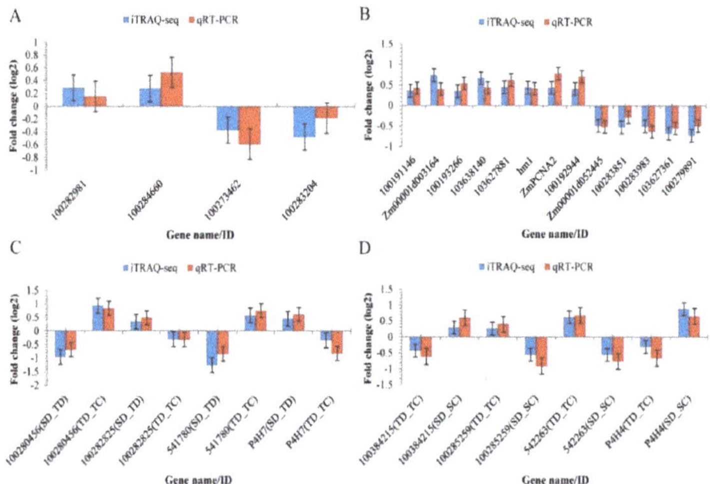

Figure 8. Quantitative real-time PCR (qRT-PCR) analysis results of the maize kernel drought-responsive genes encoding differentially abundant proteins (DAPs) from different experimental comparisons. (**A**) DAPs unique to TC_TD; (**B**) DAPs specific to SD_TD; (**C**) DAPs shared between SD_TD and TC_TD; (**D**) common or overlapping DAPs between TC_TD and SC_SD. All negative expression level values mean that the genes were down-regulated. *GAPDH* (accession No. X07156) was used as the house keeping gene. Error bars represent the SE (*n* = 3).

3. Discussion

Drought or soil moisture deficit stress, particularly at the flowering and grain-filling stages, has devastating effects on the final yield in maize [7]. Therefore, understanding how maize plants respond to drought stress at the molecular level remains critical for guiding the genetic improvement in drought tolerance so as to maintain sustainable higher productivity under such climate change conditions. In order to gain comprehensive understanding of the molecular mechanisms, and identify putative proteome signatures associated with maize drought stress response at the kernel filling stage, in this manuscript, we have performed comparative proteomic analysis (between two inbred lines with contrasting drought sensitivity) using iTRAQ approach. Additionally, comparative phenotypic and physiological analyses of the two maize lines buttressed the proteomic analysis results. Our findings offer better insights into the molecular mechanisms underpinning maize drought tolerance at the kernel filling stage.

3.1. Vivid Contrasting Phenotypic and Physiological Responses of the Two Inbred Lines to Drought Stress

Significant differences were shown in ear appearances under moisture-deficit conditions, with sensitive line MO17 exhibiting more pronounced drought symptoms than tolerant line YE8112 (Figure 1A,B). Growth inhibition of ears and kernels was evident, with more significant decrease in EL and increased ear tip barrenness in MO17 than in YE8112 (Figure 1C,D). Further, the significant decrease in EL in MO17 was accompanied with corresponding reduction in KPR (Figure 1F). Drought stress results in cell growth arrest [10], slow ear growth [11], and fewer kernels [46]. The kernel-filling period is tightly linked to maize yield potential [17], with drought at this stage resulting in kernel abortion and yield reduction (emanating from diminished source strength and sink capacity) [10]. Retarded ear growth may sometimes be reflected by a failure to set kernels even when pollinated with sufficient fresh pollen [7]. Here, we suggest that drought stress decreased the remobilization of photosynthetic assimilates [4,47], causing a reduction in ear growth, kernel filling, and size, with more pronounced occurrence in the susceptible line MO17 [17]. Further, reduction in sucrose and starch synthesizing enzyme activities is also suggested to have resulted in corresponding kernel filling decrease [48]. Overall, our results revealed vivid contrasting phenotypic responses of the two lines to drought stress at the kernel filling stage, with MO17 being comparably more sensitive and less productive than YE8112.

Under environmental stress, reactive oxygen species (ROS), such as singlet oxygen (O_2^-) and hydrogen peroxide (H_2O_2), are greatly increased within cells. Excessive ROS will destroy biological macromolecules, such as protein DNA and lipids, resulting in strong oxidative effect on cells [49,50]. Our results showed that under drought stress conditions, there was a sustained increase in POD activities in both lines (Figure 2C). Peroxidases act to eliminate ROS and harmful substances within the cells. This helped the cells to cope with drought stress [51]. Proline can protect the integrity of cell membranes and reduce ROS damage to cells through stable osmotic regulation under abiotic stress [52]. With the increase of proline content in the cells, the water potential of the cells is reduced, thereby promoting absorption of water into the cells to cope with hydropenia [53]. Our results revealed that increase in proline content was sustainably maintained up to 26 DPP in YE8112 but decreased in MO17 from 20 DPP under drought conditions (Figure 2B). This could possibly point to why tolerant line YE8112 could endure drought stress better that sensitive line MO17. MDA content can reflect the degree of lipid peroxidation in the cells or tissue [54]. It is generally considered that the higher the MDA content in cells, the greater the damage to the membrane system. When the two inbred lines were exposed to drought stress, we observed that MDA content initially increased (until 20 DPP) in both lines, and then declined sharply in YE8112 (from 20 DPP), and gradually in MO17 (from 23 DPP) (Figure 2D). Increase in MDA content in both lines at the initial stages may suggest the damage of the cell membranes by ROS generation in the cells emanating from drought stress [41]. However, the sustained increase of proline content (combined with sustained increase in POD activity) beyond 20 DPP in tolerant line YE8112 might have aided better ROS quenching in YE8112 than in sensitive line MO17. Although sustained increase in POD activity was also witnessed in MO17, ROS could not be effectively quenched, probably as a result of lack of additional contribution by proline content, which started to decline 20 DPP. Consequently, drought stress caused a greater deal of damage to MO17 than to YE8112 kernel cells. In summary, we hypothesize that tolerant line YE8112's better drought tolerance emanated from its enhanced ROS quenching ability, improved cell homeostasis regulation, better cell membrane integrity, and improved water retention capacity, as compared to sensitive line MO17 [44].

3.2. Differentially-Regulated Drought-Responsive Proteins in Tolerant Line YE8112

We performed comparative proteomic analysis using iTRAQ approach in order to identify putative proteome signatures in the tolerant maize inbred line YE8112. We observed that there was a pronounced change in the proteome of the stressed plants as compared to control conditions, as revealed by Venn diagram analysis (Figure 3B). We analyzed the iTRAQ data of YE8112 under drought conditions to screen out critical DAPs uniquely expressed in the tolerant line. Through bioinformatics analysis of these DAPs, proteins or genes and metabolic pathways responsive to drought stress were identified and clarified.

3.2.1. Redox Post Translational Modifications (PTMs) and Epigenetic Regulation Mechanisms

Among the eight drought-responsive proteins that were expressed specifically in tolerant line YE8112 under drought stress conditions, half were up-regulated. Among these proteins were glyceraldehyde-3-phosphate dehydrogenase (GAPDH/G3PDH), cytosolic purine 5-nucleotidase, and histone deacetylase (Table 1). We hypothesize that these differentially up-regulated proteins are involved in drought stress tolerance in tolerant line YE8112. Plant GAPDH is a ubiquitous enzyme involved in glycolysis and has been proven a moonlighting protein, that is, to perform alternative non-metabolic functions, including transpiration control in Arabidopsis [55], phytopathogenic virus replication regulation [56], gene expression regulation, and oxidative stress sensing in plant cells [57]. Thus, these non-glycolytic roles of GADPH, via oxidative modifications of the catalytic cysteine, may be essential in drought stress response in maize. Cytosolic purine 5-nucleotidase in *Zea mays* L. cells is involved in metal ion binding, although its exact function remains unknown (www.uniprot.org/uniprot/A0A1D6L6C5). There is, therefore, need for its further characterization. Histone deacetylase (HDAC) is a key epigenetic enzyme for the binding of regulatory proteins to chromatin. Thus, histone deacetylation may represent a unique regulatory mechanism in the early steps of gene activation [58]. Taken collectively, these up-regulated proteins involved in redox PTMs and epigenetic regulation, via sequence-specific histone modification of genes, may be critical in drought tolerance in maize kernels.

3.2.2. Response to Stress- and Response to Stimuli-Related Proteins under Drought Conditions

Our GO analysis of the DAPs expressed in YE8112 under drought (TC_TD) revealed that protein groups were enriched in "response to stress" and "response to stimuli" under the biological processes category (Figure 5). Among these proteins, eleven were up-regulated under drought stress (Supplementary Table S8, Sheet S1). These included several HSPs, chaperons, late embryogenesis abundant (LEA) protein group 3, defensin-like protein, and catalase, among others (Supplementary Table S8, Sheet S1). Under stress conditions, HSP70 plays a critical role in preventing aggregation and assisting in refolding of non-natural proteins [6,59]. The production of ROS and H_2O_2 under stress triggered the synthesis of HSP70, further enhanced the antioxidant enzyme activities, inhibited further accumulation of H_2O_2, and led to anti-stress and anti-apoptosis [60,61]. HSP90 is crucially involved in the management of protein folding [59,62], signal transduction networks, cell cycle control, protein degradation, and cellular protein transport [63]. Furthermore, as chaperones, small heat shock proteins (sHSPs) can protect proteins from aggregation under stress, thus promoting the cell's anti-stress ability [64].

LEA proteins' role in plant abiotic tolerance has been recognized due to their heat stability and hydrophilicity [6]; the group 3 LEA proteins accumulated in mature seeds and were regulated by abiotic stresses, such as dehydration, salinity, and low temperature [65]. LEA proteins particularly protect mitochondrial membranes against dehydration damage [66]. Zhang et al. [5] identified a group 3 LEA protein as a phosphorylated protein in two *Triticum aestivum* L. cultivars, with its phosphorylation level changing significantly under drought stress. Here, we suggest that the up-regulation of LEA group 3 proteins under drought stress helps protect YE8112 kernel cells and contributes immensely to maintaining cell homeostasis and normal energy metabolism. Taken collectively, we believe that when YE8112 is subjected to drought stress, it synthesizes more HSPs, enhances the chaperone functions of the protective proteins, maintains the cellular homeostasis, and improves the cell protection and tolerance to stress.

3.2.3. Energy Metabolism and Secondary Metabolite Biosynthesis-Related Proteins under Drought

As we have already opined [45], carbohydrate or energy metabolism is critical for meeting the energy needs of a cell enduring drought stress. A number of enzymes related to CHO metabolism and secondary metabolite biosynthesis responded to drought stress in YE8112 inbred line kernels. The GAPDH, which plays a major role in the glycolysis, was up-regulated under drought (Table 1). Additionally, malate dehydrogenase (MDH), which is involved in CHO synthesis [67], the chalcone-flavonone isomerase family protein, involved in flavonoid biosynthesis, a part of secondary metabolite biosynthesis, and putative RUB1 conjugating enzyme and putative ubiquitin conjugation factor E4, involved in ubiquitin mediated proteolysis, were also up-regulated in response to drought stress (Supplementary Table S2). Further, abundance of other metabolism-related proteins, such as glycosyltransferase, galactose oxidase, inositol-1-monophosphatase, and proline iminopeptidase, also increased under drought conditions (Supplementary Table S2). These enzymes coordinate in carbohydrate energy metabolism and secondary metabolite biosynthesis and may be considered an essential drought response strategy in the tolerant line YE8112 kernels.

Besides, NADPH-dependent HC-toxin reductase (HCTR), which detoxifies the HC-toxin produced by the fungus *Cochliobolus carbonun*, and encoded by the maize *Hm1* gene [68] was also up-regulated in response to drought stress (Supplementary Table S2). This may suggest that the biotic-abiotic stress crosstalk mechanism was activated to holistically resist combined stress from drought and fungus, probably aflatoxin—which is most prevalent particularly at this physiological growth stage. Taken together, several proteins related to energy metabolism and defense response showed up-regulation 14 days after cessation of watering, suggesting that a global metabolic change occurred in maize kernels in response to drought stress. These proteins are suggested to play vital roles in maize filling-kernel drought stress response. Further research is needed to determine each of these proteins' exact contribution in drought response.

3.2.4. Seed Storage Proteins under Drought Stress

As revealed by our GO analysis results, nutrient reservoir activity (NRA) (GO:0045735) was the most significantly enriched level 2 GO term under the molecular function category (Figure 5A). Of the eight NRA-related DAPs that were significantly enriched in this GO term, six were up-regulated in response to drought stress. Among these were globulin-1 S allele, legumin 1, vicilin-like seed storage protein, Z1D alpha zein protein, and 50kD γ-zein protein (Supplementary Table S8, Sheet S1). Globulin and legumin are the main storage proteins in most angiosperms and gymnosperms [69]. Zeins are insoluble storage proteins with high proline content. The 50 kDa γ-zein plays an important role in the formation and stability of the protein bodies [70]. The up-regulation of 50kD gamma zein might have helped in reducing the dependence of maize kernels on water. Further, larger zein protein bodies ensure well filled kernel endosperm with a rigid matrix and lesser starch proportion [71]. Taken together, the up-regulated expression of seed storage proteins is hypothesized to increase the sink capacity of filling kernels and maintenance of intracellular homeostasis in drought-stressed YE8112.

3.2.5. Most Significantly Enriched Metabolic Pathways of DAPs in Drought-Treated YE8112

The kernel proteome revealed changes in major metabolic pathways under drought conditions. Protein processing in endoplasmic reticulum (PPER), followed by 'tryptophan metabolism', were the most significantly enriched pathways in tolerant line YE8112 (Figure 6A). For the PPER pathway, eleven proteins were up-regulated, mostly HSPs (Supplementary Table S9, Sheet S1). These results resonate well with our GO analysis results where HSPs have been most significantly enriched in the prominent GO terms, suggesting their role as the major drought tolerance signature in YE8112 inbred line. Plants use the tryptophan pathway to provide precursors for the synthesis of secondary metabolites such as glucosinolates, and both indole- and anthranilate-derived alkaloids. Tryptophan metabolism, therefore, plays a direct role in regulating plant development and pathogen defense responses [72]. Here, we suggest that the tryptophan pathway may be involved in mediating biotic-abiotic cross talk in the kernels, thereby preventing cells from double trouble of drought stress and aflatoxin attack.

3.3. Differentially-Regulated Drought-Responsive Proteins in Sensitive Line MO17

For comparative analysis, in order to identify key drought stress-response differences between tolerant line YE8112 and sensitive line MO17, we also zeroed in on the most response strategies in MO17. Similar to our previous observation [44], tolerant inbred line YE8112 had fewer number of DAPs that were regulated in response to drought stress as compared to higher number in sensitive line MO17 (Figure 3). This may be probably because YE8112 perceived the drought treatment as moderate (as a result of its inherent better drought tolerance) and instituted limited proteomic response, whereas MO17, devoid of drought tolerance capabilities, perceived the same conditions as severe and hence deployed a heightened proteomic response.

3.3.1. Plant Hormone Signal Transduction, Chaperone Activities and Protein Ubiquitination Processes under Drought Stress

Analysis of the sensitive line MO17 specific DAPs showed that indole-3-acetic acid (IAA), amido synthetase (GH3), three serine/threonine-protein kinase SRK2A, six chaperones, several ubiquitin carboxyl-terminal hydrolases (UCHs), and other ubiquitin-associated (UBA)/TS-N domain containing proteins, as well as proteasomes, were the most up-regulated in response to drought stress (Supplementary Table S4). GH3 catalyzes the conjugation of IAA to amino acids, an essential step in auxin homeostasis [73]. This may be vital for stress signal transduction. Serine/threonine-protein kinase SRK2A mediates plant hormone signal transduction (polar auxin transport) via the mitogen activated protein kinase (MAPK) signalling pathway [74]. Molecular chaperones play an essential mechanistic role of re-establishing normal protein conformation, and eventually cellular homeostasis in response to stress [75]. The UCHs, UBA-domain-containing proteins, and proteasomes have been implicated in proteosomal degradation, thereby regulating the proper protein turnover under stress conditions [44,76,77]. Taken together, these results suggest that a coordinated interaction between plant hormone signaling, chaperones, and ubiquitination processes forms a significant portion of maize drought responses.

3.3.2. Decreased Mitochondrial Oxidative Phosphorylation Is Vital in Reducing Cellular ROS Generation in MO17

Analysis of the MO17 specific DAPs identified a total of 75 DAPs to be down-regulated in response to drought stress. Among these were NADH dehydrogenases, NADH ubiquinone oxidoreductases, and other enzymes involved in ATP synthesis (Supplementary Table S4). These enzymes play various roles in the oxidative phosphorylation process in the cells. NADH ubiquinone oxidoreductases can generate superoxide and H_2O_2 through multiple pathways, which are an important source of ROS [78]. In response, plant mitochondria prevent ROS generation within themselves via employing the alternative oxidase (AOX) pathway, in which complexes III and IV are bypassed and electrons are directly transferred to oxygen, resulting in thermal energy being generated instead of ATP [79]. A previous study has also shown that lack of NADH ubiquinone oxidoreductases will cause lower stomatal and hydraulic conductance so that the plants have better drought tolerance [80]. In the current study, the oxidative phosphorylation pathway of sensitive inbred line MO17 was greatly weakened as a result of drought stress, and was unable to supply sufficient energy. We believe that although the down-regulated expression of NADH ubiquinone oxidoreductases is conducive to reducing the accumulation of ROS, it was not enough to relieve the stress caused by drought stress on cells. Taken collectively, we herein infer that uncoupling of oxidative phosphorylation and electron transport is one of sensitive line MO17's main drought stress response strategies enhancing the reduction of cellular ROS generation, although in this case, the phenomenon was not effective enough as a result of greater mitochondrial damage imposed by the 14-day drought treatment exposure.

3.3.3. Most Significantly Enriched Metabolic Pathways of DAPs in Drought-Sensitive MO17

"Starch and sucrose metabolism" and "oxidative phosphorylation" were the most significantly enriched metabolic pathways in sensitive line MO17 (Figure 6B). Under drought conditions, the grain filling rate increased, but the duration shortened. However, the increased filling rate could not compensate for the shortened filling duration, resulting in decreased yield [81]. Nevertheless, plants under drought stress require a lot of energy to resist that stress and sustain their own metabolism. Studies have shown that speeding up carbon metabolism can provide more energy for stress resistance. However, enhancing energy production results in decreased carbohydrate synthesis, leading to a reduction in biomass [82].

Drought stress greatly limited the kernel filling process in MO17 (Figure 1A). At the initial stage of drought stress, the grout rate in MO17 was accelerated, but the loss of water in kernel cells led to the decomposition of starch and sucrose into soluble sugar to maintain the osmotic homeostasis and to supply a large amount of energy consumed by MO17, as well as enhancing signal transduction. We herein infer that the energy consumed by various pathways involved in MO17's resistance to drought stress could not be compensated by enhanced assimilate remobilization, consequently contributing to the significant decline of MO17 yield under drought conditions. Nonetheless, our results may suggest that a coordinated activation of proteins or genes, enzymes, and pathways associated with starch metabolism and uncoupling of mitochondrial oxidative phosphorylation may be critical in effecting drought stress tolerance in sensitive maize inbred line MO17 filling kernels.

3.4. Similarities and Differences in Drought Stress Responses between Maize Seedlings and Filling Kernels

Comparative analysis of our previous [44,45] and current study findings revealed notable similarities and differences between maize seedling and filling-kernel drought stress responses in the context of gene or protein expression regulation and at the physiological level. "Photosynthesis antenna proteins" [44] and "nitrogen metabolism", as well as secondary metabolites biosynthesis-related pathways (phenylalanine metabolism and phenylpropanoid biosynthesis) [45], were the most significantly enriched pathways in drought-stressed seedlings, whilst PPER and "tryptophan metabolism" were most significantly enriched in drought-stressed kernels. Meanwhile, according to the current study, filling kernel drought tolerance can be attributed to proteins involved in redox PTMs and epigenetic regulation mechanisms, elevated expression of HSPs, enriched CHO metabolism and secondary metabolites biosynthesis, up-regulated expression of seed storage proteins, and activated biotic–abiotic crosstalk mechanisms.

From the above findings, we deduced the following: (a) YE8112's sustained maintenance of proline content and POD increase alongside decreased MDA content under drought stress conditions is essential for drought tolerance in both seedling leaves and filling kernels; (b) activated expression of HSPs, secondary metabolites biosynthesis, and CHO metabolism are universal in both seedlings and kernels, with differences only being noticed in the exact individual specific proteins or genes involved; (c) biotic–abiotic cross tolerance mechanisms were activated at both maize seedling and filling kernel stages, with non-specific lipid transfer proteins involved in plant bacteria and fungus attack response and long distance stress signalling being apparent in seedlings, whereas HC-toxin reductase—probably as a response to fungus *Aspergillus flavus* L. attack—was prominent in kernels; (d) enhanced thermal dissipation of light energy in photosynthesis machinery is critical for seedling leaf cells' survival, whilst enhanced expression of seed storage proteins is essential in helping kernels endure drought stress; (e) even for a particular maize inbred line, a greater number of drought responsive proteins was observed in kernels than in the seedlings, highlighting the increased drought sensitivity of the kernel filling stage as compared to the seedling stage. Our combined results presented here will guide us in the selection of critical specific individual genes or proteins for targeted cloning and downstream analyses.

4. Materials and Methods

4.1. Plant Materials and Growth Conditions

Our laboratory (North China Key Laboratory for Crop Germplasm Resources of Education Ministry, Hebei Agricultural University, Baoding, China) provided the two drought tolerance divergent maize inbred lines used in this study. Previously, MO17 had shown high drought stress susceptibility, whilst YE8112 had shown high drought stress tolerance. For the detailed criterion on our selection of these inbred lines, we refer you to our recent study [44]. The field experiment was conducted at the Qing Yuan Experiment Station of Hebei Agricultural University, Baoding, China (115.560279' E, 38.795093' N, 118 m), under a fully automated rain-proof shelter. The experiment was set up in a randomized complete block design, with the control and drought stress treatment groups replicated three times. Each experimental unit or plot measured 5 m × 5 m, with plants spaced at 0.6 m × 0.3 m, giving 128 plants per plot. One week prior to anthesis, strict bagging was instituted in order to manage selfing of the plants during the anthesis stage. Both treatments received normal irrigation until the anthesis stage (at 12 DPP). At the grain filling stage (from 13 DPP to 26 DPP), drought stress treatment was then affected onto the treatment groups by withholding irrigation for 14 days, whereas the control group continued to receive adequate watering. Grain samples for physiological indices of assays were collected (from both control and drought treatment groups) at 14 DPP, and after every three days thereafter. Samples for proteomic analysis were collected at 26 DPP, from the middle part of each selected maize cob on the plant. At every sample collection, we ensured that the collected samples were immediately frozen in liquid nitrogen before storage at extremely low temperature ($-80\ ^{\circ}$C) while awaiting subsequent use.

4.2. Phenotypic and Physiological Characterizations

The phenotypes and physiological characteristics of the grains of the two inbred lines at the filling stage were determined under both moisture-abundant and moisture deficit stress conditions. Ten ears were randomly selected from each group for the measurement of KRPE and KPR and the mean values were calculated. Further, we used visual observation to assess some phenotypic characteristics of the harvested cobs. The relative RWC of the kernels was estimated using the method developed by Lee [83]. The guaiacol protocol of Han et al. [84] was used to determine the activity of peroxidase (POD) in the kernels. Meanwhile, the content of free proline in the samples was determined using the ninhydrin chromogenic method developed by Bates et al. [85]. Lipid peroxidation was determined as the amount of MDA produced by the thiobarbituric acid reaction, as per the method developed by Dhindsa et al. [86].

4.3. Protein Extraction

Total proteins were extracted from the non-stressed and stressed kernel tissues of two maize inbred lines with three biological replicates (each containing 500 mg maize kernels) using the cold acetone method, as fully described in our recent paper [44]. Total protein concentrations of the extracts were determined as per the manufacturer's instructions, using Coomassie Bradford Protein Assay Kit (23200, Thermo Fisher Scientific, Shanghai, China), with bovine serum albumin (BSA) as standard. The absorbance was determined at 562 nm using an xMark microplate absorbance spectrophotometer (Bio-Rad Laboratories Inc., Hercules, CA, USA) and protein extract quality was examined with tricine-sodium dodecyl sulfate polyacrylamide gel electrophoresis (SDS-PAGE) [87]. The protein samples were then stored at $-80\ ^{\circ}$C.

4.4. Protein Digestion and Isobaric Tags for Relative and Absolute Quantification (iTRAQ) Labeling

Before protein digestion, reduction of disulfide bonds and alkylation of free cysteine residues were performed as previously described [88]. Each sample was put into a new tube and adjusted to 100 µL using 8 M urea mixed with 11 µL 1 M DTT. The sample was then incubated at room temperature (37 $^{\circ}$C) for 1 h followed by vortexing at 4 $^{\circ}$C and 14,000× g for 10 min. For alkylation, 120 µL 55 mM iodoacetamide was added to the supernatant and then incubated in a dark room for 30 min. Then, washing of the supernatant was achieved by using 100 µL 100 mM TEAB and centrifugation at 14,000× g for 10 min at 4 $^{\circ}$C, followed by discarding the eluate. The washing step was repeated thrice.

Total proteins (100 µg samples) were digested using trypsin (Promega, Madison, WI, USA) at a ratio of protein:trypsin of 30:1 at 37 $^{\circ}$C overnight (16 h) [88]. Post digestion step, the peptides were dried in a centrifugal vacuum concentrator and reconstituted in 0.5M TEAB. Applied Protein Technology Co. Limited (Shanghai, China) conducted the protein iTRAQ labeling process using an iTRAQ Reagents 8-plex kit (AB Sciex, Foster City, CA, USA) as per the manufacturer's instructions. As described in our recent paper [44], one unit of iTRAQ reagent (defined as the amount of reagent required to label 100 µg of protein) was thawed and reconstituted in 70 µL isopropanol. The control replicates were labeled with iTRAQ tag 115 for drought-sensitive MO17 and tag 117 for drought-tolerant YE8112. The drought treated replicates were labeled with tags 114 and 116 for MO17 and YE8112 lines, respectively. Three technical replicates were performed.

4.5. Strong Cation Exchange (SCX) and LC-MS/MS Analysis

Sample fractionation was conducted before liquid chromatography-tandem mass spectrometry (LC-MS/MS) analysis using slightly modified procedures to those developed by Ross et al. [39]. Each SCX fraction was subjected to reverse phase nanoflow HPLC separation and quadruple time-of-flight (QSTAR XL) mass spectrometry analyses, as fully described in a previous report [43]. Peptides were subjected to nano electrospray ionization followed by tandem mass spectrometry (MS/MS). After fractionation, each 10 μL peptide was loaded onto an Eksigent nano LC System (AB SCIEX, CA, USA) with a trap C18 column (5 μm, 100 μm × 2 cm). Peptide elution was performed in an analytical C18 column (3 μm, 75 μm × 15 cm). The samples were loaded at 5 μL/min for 10 min before the 78 min LC gradient was run at 300 nL/min. The LC gradient started from 2 to 35% Buffer D (95% ACN, 0.1% FA) followed by the following schedule: 60% Buffer D at 5 min, 80% Buffer D at 2 min and maintained for 4 min, and finally return to 5% Buffer D in 60 s [88].

The mass spectrometry was analyzed by a Q-Exactive mass spectrometer (Thermo Fisher Scientific, Shanghai, China) after the sample had been analyzed by chromatography. The MS spectra with a mass range of 300–1800 m/z were acquired at a resolving power of 120 K, the primary mass spectrometry resolution of 70000 at 200 m/z, AGC (automatic gain control) target of 1×10^6, maximum IT of 50 ms, and dynamic exclusion time (active exclusion) of 60.0 s The mass charge ratio of polypeptides and polypeptide fragments were set according to the following parameters: 20 fragments (MS2 scan) were collected after each scan (full scan), MS2 activation type was HCD, isolation window 2 m/z, two-grade mass spectrometry resolution of 17,500 at 200 m/z, the normalized collision energy of 30 eV, underfill of 0.1%. The electrospray voltage applied was 1.5 kV. Maximum ion injection times for the MS and MS/MS were 50 and 100 ms, respectively [44].

4.6. Protein Identification and Quantification

Mass spectrometry data from the LC-MS/MS raw files were obtained using Mascot software version 2.2 (Matrix Science, London, UK) and converted into MGF files using Proteome Discovery 1.4 (Thermo Fisher Scientific Inc., Waltham, MA, USA). The MGF files were analyzed against the Uniprot *Zea mays* L. database (132 339 sequences, 12 January 2018) using Mascot search engine. The following search parameters were set as follows: trypsin as the cleavage enzyme; two maximum missed cleavages allowed; fragment mass tolerance was set at ±0.1 Da; peptide mass tolerance was set at ± 20 ppm; monoisotopic as the mass values; iTRAQ 8 plex (Y) and oxidation (M) as variable modifications; and Carbamidomethyl (C), iTRAQ 8 plex (N-term) and iTRAQ 8 plex (K) selected as fixed modifications. Only peptides with $p < 0.05$ and false discovery rate (FDR) ≤ 1% were counted as being successfully identified [44].

As detailed in a previous report [88], protein relative quantification was dependent on the reporter ion ratios, from which relative abundance of peptides was estimated. Only proteins observed in all the samples were considered for quantification, with shared peptides being omitted. Peak intensities of the reporter ions were used for determining reporter ion ratios, with control-treated YE8112 sample serving as reference. Final protein quantification ratios were analyzed using the median average of those ratios; the median of unique peptide ratios represented the protein ratio. Fold changes were calculated as the average ratios of TD to TC and SD to SC, and the ratios for each protein in each experimental comparison were normalized to 1. Then, a paired *t*-test was used to compare the differences between the groups [43,88], and a protein was considered statistically significant at $p < 0.05$ and a fold change of >1.2 (up) or <0.83 (down) [3].

4.7. Functional Classification, Pathway Enrichment and Hierarchal Clustering Analyses of DAPs

The successfully identified DAPs were used as queries to search the Interpro (https://www.ebi.ac.uk/interpro/), Pfam (http://pfam.xfam.org/), Gene Ontology (GO, http://www.geneontology.org/), and the Kyoto Encyclopedia of Genes and Genomics (KEGG, http://www.genome.jp/kegg/) databases.

Additionally, by searching the maize sequence database Gramene (http://ensemble.gramene.org/Zea_mays/), we obtained the corresponding gene sequences of the DAPs. The GO (protein) terms were assigned to each DAP based on BLASTX similarity (*E*-value $< 1.0 \times 10^5$) and known GO annotations using the Blast2GO tool (available online: https://www.blast2go.com; accessed on 15 January 2019). For the functional annotation and classification of the identified DAPs, GO analysis [89] was performed to categorize the DAPs into their BP, MF, and CC involvement in response to drought stress. Moreover, the DAPs were assigned to various metabolic pathways using the KEGG pathway analysis. Further, significant KEGG pathway enrichment analysis was performed using the hypergeometric test, with Q (Bonferroni-corrected *p*-value) < 0.05 set as the statistically significant threshold. Meanwhile, we used String web-based program (version 10.5) (http://www.string-db.org/) to construct a protein interaction network for the identified DAPs.

4.8. RNA Extraction, cDNA Synthesis, and qRT-PCR Analysis

Total RNA was extracted from non-stressed and stressed kernels of the two inbred lines (YE8112 and MO17) and prepared for quantitative real-time polymerase chain reaction (qRT-PCR) analysis using the Omini Plant RNA Kit (DNase I) (CWBIO, Beijing, China) as per the manufacturer's protocol. To generate cDNA template, 1 μg of total RNA was reverse-transcribed in a total volume of 20 μL, using HiFiscript cDNA Synthesis Kit (CWBIO, Beijing, China) according to the manufacturer's instructions. We randomly selected twenty-five DAPs and designed gene-specific primers (Supplementary Table S10) for qRT-PCR using Primer Premier 5 Designer software. Then, qRT-PCR was performed with a C1000 (CFX96 Real-Time System) Thermal Cycler (Bio-Rad), using 2 × Fast Super EvaGreen ® qPCR Mastermix (US Everbright Inc., Suzhou, China). Each total 20 μL qRT-PCR reaction mixture comprised 1 μL of template cDNA, 1 μL of each primer (50 pmol), and 10 μL of 2 × Fast Super EvaGreen ® qPCR Mastermix (US Everbright Inc., Suzhou, China). The amplification program was as follows: 95 °C for 2 min followed by 40 cycles of 95 °C for 10 s and 55 °C for 30 s [44]. A stable and constitutively expressed maize gene *GAPDH* (accession No. X07156), forward (GAPDH-F: 5′-ACTGTGGATGTCTCGGTTGTTG-3′), and reverse (GAPDH-R: 5′-CCTCGGAAGCAGCCTTAATAGC-3′) primers were used for housekeeping. The relative mRNA abundance was estimated using Livak and Schmittgen's $2^{-\Delta\Delta Ct}$ method [90]. Three biological replicates were performed for each sample.

4.9. Physiological Data Analysis

Student's *t*-test was used to compare the data means of well-watered (control) and drought-treated plants of MO17 and YE8112 inbred lines. The statistical analyses of physiological data were perfomed with SPSS statistical package (Version 19.0; SPSS Institute Ltd., Armonk, NY, USA) using One-Way ANOVA, followed by Duncan's multiple range test (DMRT) to evaluate the significant differences at $p \leq 0.05$.

5. Conclusions

We have performed a comprehensive comparative iTRAQ proteomics analysis of maize filling-kernel proteomes of two inbred lines contrasting in drought tolerance exposed to 14 days drought treatment. In total, 5175 DAPs were identified from the four experimental comparisons. By way of Venn diagram analysis, four critical sets of drought-responsive proteins were screened and further analyzed by bioinformatics techniques. The DAPs uniquely expressed in YE8112 chiefly participated in pathways related to "protein processing in the endoplasmic reticulum" and "tryptophan metabolism", whereas MO17 DAPs were involved in "starch and sucrose metabolism" and "oxidative phosphorylation" pathways. Most notably, our results revealed that YE8112 kernels were comparatively drought tolerant to MO17 kernels because of their enhanced redox PTMs and epigenetic regulation mechanisms, elevated expression of HSPs, enriched CHO metabolism, and up-regulated expression of seed storage proteins. Further, our comparative physiological analysis and qRT-PCR results substantiated the proteomics results. Our study presents an elaborate understanding of drought-responsive proteins and metabolic pathways mediating maize kernel drought tolerance, and provides important candidate genes for subsequent functional validation.

Supplementary Materials: Supplementary materials can be found at http://www.mdpi.com/1422-0067/20/15/3743/s1.

Author Contributions: Conceptualization, X.W., T.Z., S.L., and H.D.; data curation, X.W., T.Z., S.L., G.L., H.J., L.D., and H.D.; formal analysis, X.W., T.Z., H.J., L.D., and H.D.; funding acquisition, H.D.; investigation, X.W., T.Z., S.L., G.L., H.J., L.D., A.D., and Y.Y.; methodology, X.W., T.Z., S.L., and H.D.; project administration, H.D.; resources, H.D.; software, S.L.; supervision, H.D.; validation, X.W., T.Z., S.L., G.L., L.D., A.D., Y.Y., and H.D.; visualization, X.W. and T.Z.; writing—original draft, X.W. and T.Z.; writing—review and editing, X.W., T.Z., S.L., G.L., H.J., L.D., A.D., Y.Y., and H.D.

Acknowledgments: We are grateful for the funding support received from the National Key Research and Development Project of China (Selection and Efficient Combination Model of Wheat and Maize Water Saving, High Yield and High Quality Varieties) (Grant No. 2017YFD0300901).

Conflicts of Interest: The authors declare that they have no conflict of interest. Additionally, the funders had no role in the design of the study; in the collection, analyses, or interpretation of data; in the writing of the manuscript, or in the decision to publish the results.

Abbreviations

CFI	Chalcone flavonone isomerase
DAP	Differentially abundant protein
DPP	Days post pollination
GAPDH/G3PDH	Glyceraldehyde-3-phosphate dehydrogenase
GO	Gene ontology
HCTR	HC-toxin reductase
HDAC	Histone deacetylase
HSPs; sHSPs	Heat shock proteins; small HSPs
iTRAQ	Isobaric tags for relative and absolute quantification
KEGG	Kyoto Encyclopedia of Genes and Genomes
LEA	Late embryogenesis abundant (proteins)
LC-MS/MS	Liquid chromatography-tandem mass spectrometry
MDA	Malondialdehyde
MDH	Malate dehydrogenate
NRA	Nutrient reservoir activity
POD	Peroxidases
PTMs	Post translational modifications
PPER	Protein processing in the endoplasmic reticulum
qRT-PCR	Quantitative real-time polymerase chain reaction
ROS	Reactive oxygen species
UCHs	Ubiquitin carboxyl-terminal hydrolases

References

1. Zhu, J.K. Abiotic stress signaling and responses in plants. *Cell* **2016**, *167*, 313–324. [CrossRef] [PubMed]
2. Mohanta, T.K.; Bashir, T.; Hashem, A.; Abd_Allah, E.F. Systems biology approach in plant abiotic stresses. *Plant Physiol. Biochem.* **2017**, *121*, 58–73. [CrossRef] [PubMed]
3. Yang, L.; Jiang, T.; Fountain, J.C.; Scully, B.T.; Lee, R.D.; Kemerait, R.C.; Chen, S.; Guo, B. Protein profiles reveal diverse responsive signaling pathways in kernels of two maize inbred lines with contrasting drought sensitivity. *Int. J. Mol. Sci.* **2014**, *15*, 18892–18918. [CrossRef] [PubMed]
4. Farooq, M.; Wahid, A.; Kobayashi, N.; Fujita, D.; Basra, S.M.A. Plant drought stress: Effects, mechanisms and management. *Agron. Sustain. Dev.* **2009**, *29*, 185–212. [CrossRef]
5. Zhang, J.Y.; Cruz de Carvalho, M.H.; Torres-Jerez, I.; Kang, Y.; Allen, S.N.; Huhman, D.V.; Tang, Y.; Murray, J.; Sumner, L.W.; Udvardi, M.K. Global reprogramming of transcription and metabolism in *Medicago truncatula* during progressive drought and after rewatering. *Plant Cell Environ.* **2014**, *37*, 2553–2576. [CrossRef]

6. Ghatak, A.; Chaturvedi, P.; Nagler, M.; Roustan, V.; Lyon, D.; Bachmann, G.; Postl, W.; Schröfl, A.; Desai, N.; Varshney, R.K.; et al. Comprehensive tissue-specific proteome analysis of drought stress responses in *Pennisetum glaucum* (L.) R. Br. (Pearl millet). *J. Proteom.* **2016**, *143*, 122–135. [CrossRef] [PubMed]

7. Edmeades, G.O. *Progress in Achieving and Delivering Drought Tolerance in Maize-An Update*; ISAA: Ithaca, NY, USA, 2013; pp. 1–39.

8. Feller, U.; Vaseva, I.I. Extreme climatic events: Impacts of drought and high temperature on physiological processes in agronomically important plants. *Front. Environ. Sci.* **2014**, *2*, 39. [CrossRef]

9. Thirunavukkarasu, N.; Sharma, R.; Singh, N.; Shiriga, K.; Mohan, S.; Mittal, S.; Mittal, S.; Mallikarjuna, M.G.; Rao, A.R.; Dash, P.K.; et al. Genomewide expression and functional interactions of genes under drought stress in maize. *Int. J. Genet.* **2017**, *2017*, 2568706. [CrossRef]

10. Aslam, M.; Maqbool, M.A.; Cengiz, R. Drought stress in maize (*Zea mays* L.): Effects, resistance mechanisms, global achievements and biological strategies for improvement. In *SpringerBriefs in Agriculture*; Springer: Cham, Switzerland, 2015; ISBN 978-3-319-25440-1.

11. Campos, H.; Cooper, M.; Habben, J.E.; Edmeades, G.O.; Schussler, J.R. Improving drought tolerance in maize: A view from industry. *Field Crops Res.* **2004**, *90*, 19–34. [CrossRef]

12. Shiferaw, B.; Prasanna, B.M.; Hellin, J.; Bänziger, M. Crops that feed the world 6. Past successes and future challenges to the role played by maize in global food security. *Food Secur.* **2011**, *3*, 307–327. [CrossRef]

13. Grant, R.F.; Jakson, B.S.; Kiniry, J.R.; Arkin, G.F. Water deficit timing effects on yield components in maize. *Agron. J.* **1989**, *81*, 61–65. [CrossRef]

14. Chapman, S.C.; Edmeades, G.O.; Crossa, J. Pattern analysis of grains from selection for drought tolerance in tropical maize population. In *Plant Adaptation and Crop Improvement*; Cooper, M., Hammer, G.L., Eds.; CAB INTERNATIONAL: Walling Ford, UK, 1996; pp. 513–527.

15. Doll, N.M.; Depe'ge-Fargeix, N.; Rogowsky, P.M.; Widiez, T. Signaling in Early Maize Kernel Development. *Mol. Plant* **2017**, *10*, 375–388. [CrossRef] [PubMed]

16. Bruce, W.B.; Edmeades, G.O.; Barker, T.C. Molecular and physiological approaches to maize improvement for drought tolerance. *J. Exp. Bot.* **2002**, *53*, 13–25. [CrossRef] [PubMed]

17. Yang, L.; Fountain, J.C.; Ji, P.; Ni, X.; Chen, S.; Lee, R.D.; Kemerait, R.C.; Guo, B. Deciphering drought-induced metabolic responses and regulation in developing maize kernels. *Plant Biotechnol. J.* **2018**, *16*, 1616–1628. [CrossRef] [PubMed]

18. Luo, M.; Liu, J.; Lee, D.R.; Scully, B.T.; Guo, B.Z. Monitoring the expression of maize genes in developing kernels under drought stress using oligo-microarray. *J. Integr. Plant Biol.* **2010**, *52*, 1059–1074. [CrossRef] [PubMed]

19. Bhargava, S.; Sawant, K. Drought stress adaptation: Metabolic adjustment and regulation of gene expression. *Plant Breed.* **2013**, *132*, 21–32. [CrossRef]

20. Priya, M.; Siddique, K.H.M.; Dhankhar, O.P.; Prasad, P.V.V.; Rao, B.H.; Nair, R.M.; Nayyar, H. Molecular breeding approaches involving physiological and reproductive traits for heat tolerance in food crops. *Ind. J. Plant Physiol.* **2018**, *23*, 697–720. [CrossRef]

21. Osmolovskaya, N.; Shumilina, J.; Kim, A.; Didio, A.; Grishina, T.; Bilova, T.; Keltsieva, O.A.; Zhukov, V.; Tikhonovich, I.; Tarakhovskaya, E.; et al. Methodology of Drought Stress Research: Experimental Setup and Physiological Characterization. *Int. J. Mol. Sci.* **2018**, *19*, 4089. [CrossRef]

22. Maiti, R.K.; Maiti, L.E.; Sonia, M.; Maiti, A.M.; Maiti, M.; Maiti, H. Genotypic variability in maize (*Zea mays* L.) for resistance to drought and salinity at the seedling stage. *J. Plant Physiol.* **1996**, *148*, 741–744. [CrossRef]

23. Ahmadi, A.; Emam, Y.; Pessarakli, M. Biochemical changes in maize seedlings exposed to drought stress conditions at different nitrogen levels. *J. Plant Nutr.* **2010**, *33*, 541–556. [CrossRef]

24. Zheng, J.; Fu, J.; Gou, M.; Huai, J.; Liu, Y.; Jian, M.; Huang, Q.; Guo, X.; Dong, Z.; Wang, H.; et al. Genome-wide transcriptome analysis of two maize inbred lines under drought stress. *Plant Mol. Biol.* **2010**, *72*, 407–421. [CrossRef] [PubMed]

25. Min, H.; Chen, C.; Wei, S.; Shang, X.; Sun, M.; Xia, R.; Liu, X.; Hao, D.; Chen, H.; Xie, Q. Identification of drought tolerant mechanisms in maize seedlings based on transcriptome analysis of recombination inbred lines. *Front. Plant Sci.* **2016**, *7*, 1080. [CrossRef] [PubMed]

26. Zhang, X.; Lei, L.; Lai, J.; Zhao, H.; Song, W. Effects of drought stress and water recovery on physiological responses and gene expression in maize seedlings. *BMC Plant Biol.* **2018**, *18*, 68. [CrossRef] [PubMed]

27. Kosova, K.; Vitamvas, P.; Prasil, I.T.; Renaut, J. Plant proteome changes under abiotic stress-contribution of proteomics studies to understanding plant stress response. *J. Proteom.* **2011**, *74*, 1301–1322. [CrossRef] [PubMed]

28. Wu, S.; Ning, F.; Zhang, Q.; Wu, X.; Wang, W. Enhancing omics research of crop responses to drought under field conditions. *Front. Plant Sci.* **2017**, *8*, 174. [CrossRef] [PubMed]

29. Tan, C.T.; Lim, Y.S.; Lau, S.E. Proteomics in commercial crops: An overview. *J. Protozool.* **2017**, *169*, 176–188. [CrossRef]

30. Komatsu, S. Plant Proteomic Research 2.0: Trends and Perspectives. *Int. J. Mol. Sci.* **2019**, *20*, 2495. [CrossRef] [PubMed]

31. Wilkins, M.R.; Sanchez, J.-C.; Gooley, A.A.; Appel, R.D.; Smith, I.H.; Hochstrasser, D.F.; Williams, K.L. Progress with proteome projects: Why all proteins expressed by a genome should be identified and how to do it. *Biotechnol. Genet. Eng. Rev.* **1995**, *13*, 19–50. [CrossRef]

32. Ghatak, A.; Chaturvedi, P.; Weckwerth, W. Cereal crop proteomics: Systemic analysis of crop drought stress responses towards marker-assisted selection breeding. *Front. Plant Sci.* **2017**, *8*, 757. [CrossRef]

33. Labuschagne, M.T. A review of cereal grain proteomics and its potential for sorghum improvement. *J. Cereal Sci.* **2018**, *84*, 151–158. [CrossRef]

34. Kalisa, S.K.; Wu, H.F. Recent developments in nanoparticle-based MALDI mass spectrometric analysis of phosphoproteomes. *Microchim. Acta* **2014**, *181*, 853–864. [CrossRef]

35. Agrawal, G.K.; Sarkar, A.; Righetti, P.G.; Pedreschi, R.; Carpentier, S.; Wang, T.; Barkla, B.J.; Kohli, A.; Ndimba, B.K.; Bykova, N.V.; et al. A decade of plant proteomics and mass spectrometry: Translation of technical advancements to food security and safety issues. *Mass Spectrom. Rev.* **2013**, *32*, 335–365. [CrossRef] [PubMed]

36. Cui, D.; Wu, D.; Liu, J.; Li, D.; Xu, C.; Li, S.; Li, P.; Zhang, H.; Liu, X.; Jiang, C.; et al. Proteomic analysis of seedling roots of two maize inbred lines that differ significantly in the salt stress response. *PLoS ONE* **2015**, *10*, e0116697. [CrossRef] [PubMed]

37. Matros, A.; Kaspar, S.; Witzel, K.; Mock, H.P. Recent progress in liquid chromatography-based separation and label-free quantitative plant proteomics. *Phytochemistry* **2011**, *72*, 963–974. [CrossRef] [PubMed]

38. Hu, X.; Wu, L.; Zhao, F.; Zhang, D.; Li, N.; Zhu, G.; Li, C.; Wang, W. Phosphoproteomic analysis of the response of maize leaves to drought, heat and their combination stress. *Front. Plant Sci.* **2015**, *6*, 298. [CrossRef] [PubMed]

39. Ross, P.L.; Huang, Y.N.; Marchese, J.N.; Williamson, B.; Parker, K.; Hattan, S.; Khainovski, N.; Pillai, S.; Dey, S.; Daniels, S.; et al. Multiplexed protein quantitation in Saccharomyces cerevisiae using amine-reactive isobaric tagging reagents. *Mol. Cell. Proteom.* **2004**, *3*, 1154–1169. [CrossRef] [PubMed]

40. Wu, X.; Wang, W. Increasing confidence of proteomics data regarding the identification of stress-responsive proteins in crop plants. *Front. Plant Sci.* **2016**, *7*, 702. [CrossRef]

41. Zhao, Y.; Wang, Y.; Yang, H.; Wang, W.; Wu, J.; Hu, X. Quantitative proteomic analyses identify aba-related proteins and signal pathways in maize leaves under drought conditions. *Front. Plant Sci.* **2016**, *7*, 1827. [CrossRef]

42. Li, G.K.; Gao, J.; Peng, H.; Shen, Y.O.; Ding, H.P.; Zhang, Z.M.; Pan, G.T.; Lin, H.J. Proteomic changes in maize as a response to heavy metal (lead) stress revealed by iTRAQ quantitative proteomics. *Genet. Mol. Res.* **2016**, *15*, 1–14. [CrossRef]

43. Luo, M.; Zhao, Y.; Wang, Y.; Shi, Z.; Zhang, P.; Zhang, Y.; Song, W.; Zhao, J. Comparative proteomics of contrasting maize genotypes provides insights into salt-stress tolerance mechanisms. *J. Proteome Res.* **2018**, *17*, 141–153. [CrossRef]

44. Zenda, T.; Liu, S.; Wang, X.; Jin, H.; Liu, G.; Duan, H. Comparative proteomic and physiological analyses of two divergent maize inbred lines provide more insights into drought-stress tolerance mechanisms. *Int. J. Mol. Sci.* **2018**, *19*, 3225. [CrossRef] [PubMed]

45. Zenda, T.; Liu, S.T.; Wang, X.; Liu, G.; Jin, H.Y.; Dong, A.Y.; Yang, Y.T.; Duan, H.J. Key Maize Drought-Responsive Genes and Pathways Revealed by Comparative Transcriptome and Physiological Analyses of Contrasting Inbred Lines. *Int. J. Mol. Sci.* **2019**, *20*, 1268. [CrossRef] [PubMed]

46. Edmeades, G.O.; Bolaños, J.; Elings, A.; Ribaut, J.M.; Bänziger, M.; Westgate, M.E. The Role and Regulation of the Anthesis-Silking Interval in Maize. In *Physiology and Modeling Kernel Set in Maize*; Westgate, M.E., Boote, K.J., Eds.; CSSA Special Publication: WI, USA, 2000; Volume 29, pp. 43–73.

47. Yadav, R.S.; Hash, C.T.; Bidinger, F.R.; Devos, K.M.; Howarth, C.J. Genomic regions associated with grain yield and aspects of post flowering drought tolerance in pearl millet across environments and tester background. *Euphytica* **2004**, *136*, 265–277. [CrossRef]

48. Anjum, S.A.; Xie, X.Y.; Wang, L.C.; Saleem, M.F.; Man, C.; Lei, W. Morphological, physiological and biochemical responses of plants to drought stress. *Afr. J. Agric. Res.* **2011**, *6*, 2026–2032.

49. Das, K.; Roychoudhury, A. Reactive oxygen species (ROS) and response of antioxidants as ROS-scavengers during environmental stress in plants. *Front. Environ. Sci.* **2014**, *2*, 53. [CrossRef]

50. Mallick, N.; Mohn, F.H.; Soeder, C.J. Impact of physiological stresses on nitric oxide formation by green alga, scenedesmus obliquus. *J. Microbiol. Biotechnol.* **2000**, *110*, 300–307.

51. Sharma, P.; Jha, A.B.; Dubey, R.S.; Pessarakli, M. Reactive Oxygen Species, Oxidative Damage, and Antioxidative Defense Mechanism in Plants under Stressful Conditions. *J. Bot.* **2012**, *10*, 26. [CrossRef]

52. Reddy, P.S.; Jogeswar, G.; Rasineni, G.K.; Maheswari, M.; Reddy, A.R.; Varshney, R.K.; Kishor, P.B.K. Proline over-accumulation alleviates salt stress and protects photosynthetic and antioxidant enzyme activities in transgenic sorghum [*Sorghum bicolor* (L.) Moench]. *Plant Physiol. Biochem.* **2015**, *94*, 104–113. [CrossRef]

53. Kaur, G.; Asthir, B. Proline: A key player in plant abiotic stress tolerance. *Biol. Plant.* **2015**, *59*, 609–619. [CrossRef]

54. Davey, M.W.; Stals, E.; Panis, B.; Keulemans, J.; Swennen, R.L. High-throughput determination of malondialdehyde in plant tissues. *Anal. Biochem.* **2005**, *347*, 201–207. [CrossRef]

55. Guo, L.; Devaiah, S.P.; Narasimhan, R.; Pan, X.; Zhang, Y.; Zhang, W.; Wang, X. Cytosolic glyceraldehyde-3-phosphate dehydrogenases interact with phospholipase D delta to transduce hydrogen peroxide signals in the Arabidopsis response to stress. *Plant Cell* **2012**, *24*, 2200–2212. [CrossRef] [PubMed]

56. Prasanth, K.R.; Huang, Y.W.; Liou, M.R.; Wang, R.Y.; Hu, C.C.; Tsai, C.H.; Meng, M.; Lin, N.S.; Hsu, Y.H. Glyceraldehyde 3-phosphate dehydrogenase negatively regulates the replication of Bamboo mosaic virus and its associated satellite RNA. *J. Virol.* **2011**, *85*, 8829–8840. [CrossRef] [PubMed]

57. Zaffagnini, M.; Fermani, S.; Costa, A.; Lemaire, S.D.; Trost, P. Plant cytoplasmic GAPDH: Redox post-translational modifications and moonlighting properties. *Front. Plant Sci.* **2013**, *4*, 450. [CrossRef] [PubMed]

58. Yang, H.; Liu, X.Y.; Xin, M.M.; Du, J.K.; Hu, Z.R.; Peng, H.R.; Rossi, V.; Sun, Q.X.; Ni, Z.F.; Yao, Y.Y. Genome-Wide Mapping of Targets of Maize Histone Deacetylase HDA101 Reveals Its Function and Regulatory Mechanism during Seed Development. *Plant Cell* **2016**, *28*, 629–645. [CrossRef] [PubMed]

59. Frydman, J. Folding of newly translated proteins in vivo: The role of molecular chaperones. *Annu. Rev. Biochem.* **2001**, *70*, 603–647. [CrossRef]

60. Volkov, R.A.; Panchuk, I.; Mullineaux, P.M.; Schöffl, F. Heat stress-induced H_2O_2 is required for effective expression of heat shock genes in Arabidopsis. *Plant Mol. Biol.* **2006**, *61*, 733–746. [CrossRef]

61. Tripathy, B.C.; Oelmüller, R. Reactive oxygen species generation and signaling in plants. *Plant Signal. Behav.* **2016**, *7*, 1621–1633. [CrossRef]

62. Buchner, J. Hsp90 & Co.—A holding for folding. *Trends Biochem. Sci.* **1999**, *24*, 136–141. [CrossRef]

63. Young, J.C.; Moarefi, I.; Hartl, F.U. Hsp90: A specialized but essential protein-folding tool. *J. Cell Biol.* **2001**, *154*, 267–273. [CrossRef]

64. Bakthisaran, R.; Tangirala, R.; Rao, C.M. Small heat shock proteins: Role in cellular functions and pathology. *Biochim. Biophys. Acta* **2015**, *1854*, 291–319. [CrossRef]

65. Wang, X.S.; Zhu, H.B.; Jin, G.L.; Liu, H.L.; Wu, W.R.; Zhu, J. Genome-scale identification and analysis of LEA genes in rice (*Oryza sativa* L.). *Plant Sci.* **2006**, *172*, 414–420. [CrossRef]

66. Tolleter, D.; Hincha, D.K.; Macherel, D. A mitochondrial late embryogenesis abundant protein stabilizes model membranes in the dry state. *Biochim. Biophys. Acta* **2010**, *1798*, 1926–1933. [CrossRef] [PubMed]

67. Kim, S.G.; Lee, J.-S.; Kim, J.-T.; Kwon, Y.S.; Bae, D.-W.; Bae, H.H.; Son, B.-Y.; Baek, S.-B.; Kwon, Y.-U.; Woo, M.-O.; et al. Physiological and Proteomics Analysis of the response to drought stress in an inbred Korean maize line. *Plant Omics* **2015**, *8*, 159–168.

68. Hayashi, M.; Takahashi, H.; Tamura, K.; Huang, J.; Yu, L.H.; Kawai-Yamada, M.; Tezuka, T.; Uchimiya, H. Enhanced dihydroflavonol-4-reductase activity and NAD homeostasis leading to cell death tolerance in transgenic rice. *Proc. Natl. Acad. Sci. USA* **2005**, *102*, 7020–7025. [CrossRef] [PubMed]

69. Shewry, P.R.; Napier, J.A.; Tatham, A.S. Seed Storage Proteins: Structures' and Biosynthesis. *Plant Cell* **1995**, *7*, 945–956. [CrossRef] [PubMed]

70. Bicudo, T.C.; Forato, L.A.; Batista, L.A.R.; Colnago, L.A. Study of the conformation of γ-zeins in purified maize protein bodies by FTIR and NMR spectroscopy. *Anal. Bioanal. Chem.* **2005**, *383*, 291–296. [CrossRef] [PubMed]

71. Holding, D.R. Recent advances in the study of prolamin storage protein organization and function. *Front. Plant Sci.* **2014**, *5*, 276. [CrossRef]

72. Radwanski, E.R.; Last, R.L. Tryptophan Biosynthesis and Metabolism: Biochemical and Molecular Genetics. *Plant Cell* **1995**, *7*, 921–934. [CrossRef]

73. Ludwig-Müller, J. Auxin conjugates: Their role for plant development and in the evolution of land plants. *J. Exp. Bot.* **2011**, *62*, 1757–1773. [CrossRef]

74. Jagodzik, P.; Tajdel-Zielinska, M.; Ciesla, A.; Marczak, M.; Ludwikow, A. Mitogen-Activated Protein Kinase Cascades in Plant Hormone Signaling. *Front. Plant Sci.* **2018**, *9*, 1387. [CrossRef]

75. Wang, W.; Vinocur, B.; Shoseyov, O.; Altman, A. Role of plant heat-shock proteins and molecular chaperones in the abiotic stress response. *Trends Plant Sci.* **2004**, *9*, 244–252. [CrossRef] [PubMed]

76. Su, V.; Lau, A.F. Ubiquitin-like and ubiquitin-associated domain proteins: Significance in proteasomal degradation. *Cell. Mol. Life Sci.* **2009**, *66*, 2819–2833. [CrossRef] [PubMed]

77. Zhang, Z.; Li, J.; Liu, H.; Chong, K.; Xu, Y. Roles of ubiquitination-mediated protein degradation in plant responses to abiotic stresses. *Environ. Exp. Bot.* **2015**, *114*, 92–103. [CrossRef]

78. Murphy, M.P. How mitochondria produce reactive oxygen species. *Biochem. J.* **2013**, *417*, 1–13. [CrossRef] [PubMed]

79. Siedow, J.N.; Umbach, A.L. The mitochondrial cyanideresistant oxidase: Structural conservation amid regulatory diversity. *Biochim. Biophys. Acta* **2000**, *1459*, 432–439. [CrossRef]

80. Djebbar, R.; Rzigui, T.; Pétriacq, P.; Mauve, C.; Priault, P.; Fresneau, C.; De, P.M.; Florez-Sarasa, I.; Benhassaine-Kesri, G.; Streb, P.; et al. Respiratory complex I deficiency induces drought tolerance by impacting leaf stomatal and hydraulic conductances. *Planta* **2012**, *235*, 603–614. [CrossRef] [PubMed]

81. Wardlaw, I.F. Interaction between drought and chronic high temperature during kernel filling in wheat in a controlled environment. *Ann. Bot.* **2002**, *90*, 469–476. [CrossRef]

82. Li, W.; Wei, Z.; Qiao, Z.; Wu, Z.; Cheng, L.; Wang, Y. Proteomics analysis of alfalfa response to heat stress. *PLoS ONE* **2013**, *8*, e82725. [CrossRef]

83. Lee, H.S. *Principles and Experimental Techniques of Plant Physiology and Biochemistry*, 1st ed.; Higher Education Press: Beijing, China, 2000. (In Chinese)

84. Han, L.B.; Song, G.L.; Zhang, X. Preliminary observation of physiological responses of three turfgrass species to traffic stress. *HortTechnology* **2008**, *18*, 139–143. [CrossRef]

85. Bates, T.S.; Waldren, R.P.; Teare, I.D. Rapid determination of free proline for water-stress studies. *Plant Soil* **1973**, *39*, 205–207. [CrossRef]

86. Dhindsa, R.S.; Plumb-Dhindsa, P.; Thorpe, T.A. Leaf senescence: Correlated with increased leaves of membrane permeability and lipid peroxidation, and decreased levels of superoxide dismutase and catalase. *J. Exp. Bot.* **1981**, *32*, 93–101. [CrossRef]

87. Swägger, H. Tricine-SDS-PAGE. *Nat. Protoc.* **2006**, *1*, 16–22. [CrossRef] [PubMed]

88. Zhang, C.; Shi, S. Physiological and Proteomic Responses of Contrasting Alfalfa (*Medicago sativa* L.) Varieties to PEG-Induced Osmotic Stress. *Front. Plant Sci.* **2018**, *9*, 242. [CrossRef] [PubMed]

89. Ashburner, M.; Ball, C.A.; Blake, J.A.; Botstein, D.; Butler, H.; Cherry, J.M.; Davis, A.P.; Dolinski, K.; Dwight, S.S.; Eppig, J.T.; et al. Gene ontology: Tool for the unification of biology. *Nat. Genet.* **2000**, *25*, 25–29. [CrossRef] [PubMed]

90. Livak, K.J.; Schmittgen, T.D. Analysis of relative gene expression data using real-time quantitative PCR and the $2^{-\Delta\Delta CT}$ method. *Methods* **2001**, *25*, 402–408. [CrossRef] [PubMed]

International Journal of
Molecular Sciences

Article

Root Proteomics Reveals the Effects of Wood Vinegar on Wheat Growth and Subsequent Tolerance to Drought Stress

Yuying Wang [1,2], Ling Qiu [2,3,*], Qilu Song [1], Shuping Wang [4], Yajun Wang [2,3] and Yihong Ge [5]

1 College of Agronomy, Northwest A&F University, Yangling 712100, China; wyy0624@126.com (Y.W.); songqilu1234@163.com (Q.S.)

2 Northwest Research Center of Rural Renewable Energy, Exploitation and Utilization of Ministry of Agriculture, Northwest A&F University, Yangling 712100, China; wangyajunkaka@outlook.com

3 College of Mechanical and Electronic Engineering, Northwest A&F University, Yangling 712100, China

4 Hubei Key Laboratory of Waterlogging Disaster and Agricultural Use of Wetland, College of Agronomy, Yangtze University, Jingzhou 434025, China; wangshuping2003@126.com

5 Biogas Institute of Ministry of Agriculture and Rural Affairs, Chinese Academy of Agricultural Science, Chengdu 610041, China; geyihong@caas.cn

* Correspondence: xbgzzh@163.com; Tel.: +86-029-87092391

Received: 2 February 2019; Accepted: 19 February 2019; Published: 21 February 2019

Abstract: Wood vinegar (WV) or pyroligneous acid (PA) is a reddish-brown liquid created during the dry distillation of biomass, a process called pyrolysis. WV contains important biologically active components, which can enhance plant growth and tolerance to drought stress. However, its mechanism of action remains unknown. Our results after presoaking wheat seeds with various concentrations of WV indicate that a 1:900 WV concentration can significantly enhance growth. To investigate the response of wheat roots to drought stress, we compared quantitative proteomic profiles in the roots of wheat plants grown from seeds either presoaked (treatment) or non-presoaked (control) with WV. Our results indicated that the abscisic acid (ABA) content of wheat roots in the WV treatment was significantly increased. Reactive oxygen species (ROS) and malonaldehyde (MDA) levels roots were significantly lower than in the control treatment under drought stress, while the activity of major antioxidant enzymes was significantly increased. Two-dimensional electrophoresis (2D-PAGE) identified 138 differentially accumulated protein (DAP) spots representing 103 unique protein species responding to drought stress in wheat roots of the control and WV-treated groups. These DAPs are mostly involved in the stress response, carbohydrate metabolism, protein metabolism, and secondary metabolism. Proteome profiles showed the DAPs involved in carbohydrate metabolism, stress response, and secondary metabolism had increased accumulation in roots of the WV-treated groups. These findings suggest that the roots from wheat seeds presoaked with WV can initiate an early defense mechanism to mitigate drought stress. These results provide an explanation of how WV enhances the tolerance of wheat plants to drought stress.

Keywords: wheat; root; wood vinegar; drought stress; ROS; ABA; proteome

---

1. Introduction

Wood vinegar (WV) or pyroligneous acid (PA), a translucent reddish-brown aqueous liquid, is a by-product of the carbonization of tree branches, crop straw, bamboo, wood residue, and other biomaterials [1,2]. WV is a complex mixture, which contains various types of complicated chemical ingredients, namely organic acids, phenolic, alkane, furan derivatives, esters, alcohol, sugar derivatives, and nitrogen compounds [1,3]. The chemical composition of WV mainly depends on the heating rate, temperature, residence time, particle size, and the feedstock [4]. As a natural agricultural material, it contains important biologically active components, such as organic acid and phenolic compounds, and has been widely applied in the areas of medicine, food, and agriculture [5,6]. Most notably in agriculture, WV has been widely utilized as an insect repellent, soil ameliorant, and foliar fertilizer [1,7,8]. Studies indicate that WV improves seed germination rate and accelerates the growth of roots, stems, leaves, flowers, and fruits [2,9]. Further studies have investigated the antioxidant activities of the acids with regard to their radical-scavenging activity and reducing power [10,11]. There has also been an increasing interest in the antioxidant activities of WV and its use in food to replace synthetic antioxidants; however, previous studies based on typical chemical assays have valued the antioxidant for agricultural use only. There are no reports to date that focus on the regulation of the molecular mechanism of WV in plants under stress.

Wheat (*Triticum aestivum* L.) is one of the most important food sources. Demand for wheat is rising continually as a result of population growth and increasing consumption per capita for half a century [12]. However, wheat growth and yield are seriously influenced by drought stress, most notably at the seedling, stem elongation, booting, anthesis, and grain formation stages [13]. Young plants are susceptible to water deficit due to their low biomass, undeveloped protective structure, and water requirements for growth [14]. Roots are the initial receptors that signal a water deficit, followed by a series of responses at the morphological, physiological, and cellular levels. A well-developed root system can assist water uptake during drought conditions [15]. Hence, roots are important for maintaining crop yields, especially when plants are suffering drought stress.

Proteomics has become a powerful tool for creating a proteome profile of plants in response to drought stress [16]. In recent years, several comparative proteomics studies of wheat roots have been undertaken to assess response to drought stress [17–20]. Studies indicate that proteins related to defense and oxidative stress responses and involved in protein folding, such as heat shock proteins (HSPs), accumulate in greater abundance in wheat, soybean, and rice roots in response to drought stress [18,21,22]. Such increased abundance plays a vital role in scavenging accumulated ROS [23] and preventing aggregation and refolding of non-active proteins. There is evidence that proteins involved in bioenergy metabolism, such as acetyl CoA synthesis or the tricarboxylic acid cycle (TCA), are accumulated in rice roots during drought for increased demand for energy [21]. Meanwhile, proteins involved in cell wall biogenesis, amino acid metabolism, secondary metabolism, and signal transduction show abundant changes in response to drought in plants [18,21,22,24,25]. Exogenous WV pretreatment enhances root growth and tolerance in some plants to subsequent drought stress; however, the mechanisms have remained obscure. In this paper, we investigate the proteome pattern of wheat roots following a WV seed presoaking treatment under drought stress to explore further the molecular mechanisms underlying WV induced drought tolerance.

2. Results

2.1. Effects of WV Pretreatment on Phenotype and Growth Parameters of Wheat Seedlings

Wheat seeds were soaked in different concentrations of WV (primary WV:ddH$_2$O$_2$ (*V:V*) = 1:300–1:1500) for 3 days. We found that 1:900 was the optimal ratio for seedling growth and that high concentrations had adverse effects (1:300; Figure 1). Results from fresh weight (FW) and dry weight (DW) of shoots, and roots results were confirmed by quantitative analysis (Tables 1 and 2 and Table S2A). The FW of wheat shoots and roots that had been pretreated with 1:900 WV for 3 days, were significantly higher than those of the control, by 10.0% and 15.7% at 2 days, 15.5% and 38.6% at 3 days, 13.4% and 22.1% at 4 days, 14.9% and 20.8% at 5 days, and 16.5% and 15.9% at 6 days, respectively (Tables 1 and 2 and Table S2A). The DW of wheat shoots and roots that had been pretreated with 1:900 WV for 3 days, were significantly increased by 1.68 and 1.89-fold at 2 days, 1.66 and 2.95-fold at 3 days, 1.34 and 1.92-fold at 4 days, 1.39 and 1.94-fold at 5 days, and 1.47 and 1.94-fold at 6 days, respectively (Tables 1 and 2 and Table S2A). Meanwhile, the total length, surface area, total volume, and mean diameter of the roots from the root system scanning analysis were significantly higher than the control, and there was a maximum promotion at the 1:900 WV concentration (Table S2B, Figure S2C). The height of the shoots pretreated with 1:900 WV for 3 days were significantly greater than those of the control and other WV concentations (Figure 1 and Figure S2B, Table S2C).

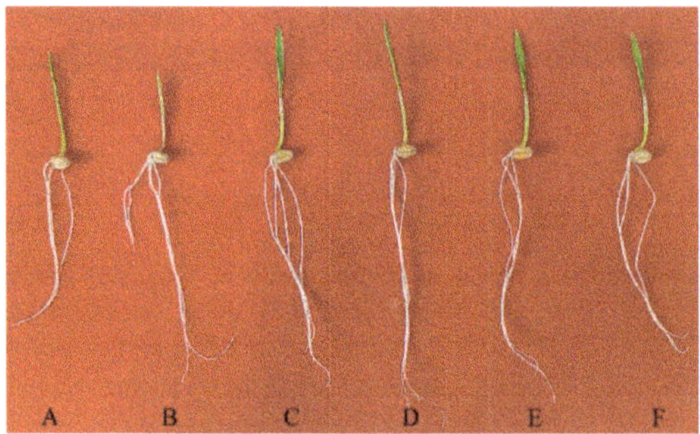

Figure 1. Phenotypic changes in wheat plants following seed soaking treatments with different concentrations of wood vinegar (WV). The various concentrations include 0 (**A**, CK group), 1:300, 1:600, 1:900, 1:1200, 1:1500 (primary WV:ddH$_2$O$_2$ (*V:V*); **B–F**, WV treated groups).

Table 1. Effect of different concentrations of wood vinegar on the fresh weight (FW) and dry weight (DW) of shoots of five wheat seedlings.

Treatments		CK	1:300	1:600	1:900	1:1200	1:1500
2d	FW(mg)[1]	71.1 ± 1.75b	60.8 ± 1.87a	76.9 ± 1.5d	78.2 ± 1.6d	74.1 ± 1.43c	73.1 ± 1.25bc
	DW(mg)[1]	7.7 ± 0.35b	6.9 ± 0.11a	10.9 ± 0.49d	12.9 ± 0.61e	9.9 ± 0.51c	9.6 ± 0.25c
3d	FW(mg)[1]	116.6 ± 6.26b	90 ± 2.2a	127.6 ± 4.05de	134.7 ± 5.29e	126.5 ± 3.65cd	119.7 ± 1.89bc
	DW(mg)[1]	14.1 ± 0.53a	12.5 ± 0.6a	19.5 ± 1.35c	23.5 ± 1.12d	17.7 ± 0.76b	16.8 ± 0.64b
4d	FW(mg)[1]	179 ± 3.63b	152.9 ± 3.12a	189.4 ± 3.72c	203 ± 4.54d	192.3 ± 2c	188.7 ± 1.46c
	DW(mg)[1]	21.8 ± 0.67b	17.9 ± 0.66a	26.4 ± 0.38d	29.2 ± 0.5e	26.4 ± 0.65d	23.7 ± 0.23c
5d	FW(mg)[1]	249.7 ± 1.45b	211.8 ± 1.71a	272.1 ± 1.96d	286.9 ± 2.11e	270.7 ± 1.45d	262.7 ± 1.59c
	DW(mg)[1]	29.9 ± 0.64b	24.2 ± 0.32a	37.9 ± 0.75d	42.2 ± 0.2e	37.6 ± 0.35d	32 ± 0.21c
6d	FW(mg)[1]	312.3 ± 2.07b	302.9 ± 1.53a	340.3 ± 1.8d	363.7 ± 2.1e	335.3 ± 1.8c	332.8 ± 2.03c
	DW(g)[1]	37.7 ± 0.47b	35.4 ± 0.67a	48.3 ± 0.72d	54 ± 0.31e	45.7 ± 0.47c	38 ± 0.71b

[1] Values are mean ± SD from three dependent experiments (biological replicates). Data with different letters in the same row indicate a significant difference at the $p < 0.05$ level.

Table 2. Effect of different concentrations of wood vinegar on the FW and DW of roots of five wheat seedlings.

Treatments		CK	1:300	1:600	1:900	1:1200	1:1500
2d	FW(mg)[1]	77.9 ± 1.4b	72 ± 1.96a	81.2 ± 1.31c	90.1 ± 1.42d	82.3 ± 1.25c	80.1 ± 1.18bc
	DW(mg)[1]	5.6 ± 0.35a	5.2 ± 0.3a	7.9 ± 0.35c	10.5 ± 0.2d	6.7 ± 0.25b	6.2 ± 0.21b
3d	FW(mg)[1]	91.5 ± 3.27ab	89.8 ± 2.39a	101.1 ± 3.72c	126.8 ± 3e	119.8 ± 4.58d	93.1 ± 2.69b
	DW(mg)[1]	6.3 ± 0.32a	6.8 ± 0.36a	10.1 ± 0.25c	18.7 ± 0.56d	10.4 ± 0.53c	7.7 ± 0.42b
4d	FW(mg)[1]	169.3 ± 1.65a	178 ± 1.61b	184.7 ± 1.55c	206.6 ± 1.67e	190.3 ± 1.05d	171 ± 1.15a
	DW(mg)[1]	14.3 ± 0.2a	15.5 ± 0.46b	19.9 ± 0.45c	27.4 ± 0.72e	23.9 ± 0.32d	15.7 ± 0.2b
5d	FW(mg)[1]	213.2 ± 2.05a	218.2 ± 2.11b	232.5 ± 1.56c	257.4 ± 1.22d	217.8 ± 1.76b	210.9 ± 2.27a
	DW(mg)[1]	17.8 ± 0.26a	21.7 ± 1.03b	26.5 ± 0.6c	33.7 ± 1.21d	26.7 ± 0.86c	18.1 ± 0.99a
6d	FW(mg)[1]	267.6 ± 2.12a	286.3 ± 2.03b	295.1 ± 4.41c	310.2 ± 1.65d	282.3 ± 2.04b	263.2 ± 1.78a
	DW(mg)[1]	21.4 ± 0.7a	27.9 ± 0.38b	35.2 ± 0.36d	41.5 ± 0.84e	33.7 ± 0.85c	21.6 ± 0.45a

[1] Values are mean ± SD from three dependent experiments (biological replicates). Data with different letters in the same row indicate a significant difference at the $p < 0.05$ level.

2.2. Physiological Changes in Wheat Seedlings Under Drought Conditions Following WV Pretreatment

To demonstrate the effect of the WV (1:900) seed soaking treatment on drought tolerance, wheat plants of both the control and WV treated groups were exposed to polyethylene glycol (PEG)-induced drought stress (PEG-6000, -1 MPa) for 2 days. Results showed that seedlings in the control group were stunted and wilted; in contrast, WV treated seedlings exhibited less wilting (Figure 2). The FW and DW of shoots were significantly higher than those of the control, by 33.0% and 45.7% at 5 days (the first day after drought stress), and 46.6% and 52.7% at 6 days (the second day after drought stress; Table S2D). The FW and DW of roots were significantly higher than those of the control, by 37.4% and 42.9% at 5 days (the first day after drought stress), and 40.0% and 58.9% at 6 days (the second day after drought stress; Table S2D). The total length, mean diameter, total area, and total volume from root system scanning analysis showed a lower impact of drought in wheat roots of the WV treated groups compared with those of the control (Figure S2D, Table S2E).

Figure 2. Phenotypic changes of wheat seedlings in both the control and WV treated groups under a drought stress treatment for 2 days. The concentration of WV was 0 and 1:900 in control and WV treated groups, respectively. Drought stress was simulated with PEG 6000(-1 MPa).

To explore the dynamic changes of the ABA content on drought tolerance, ABA levels in wheat shoots and roots of both the control and the WV treated groups (1:900) were measured before and after drought stress treatments. The ABA content of the roots was higher than in the shoots of both groups (Figure 3A). In the control group, there was no significant change in ABA content in shoots and roots at 3 and 4 days (Figure 3A); however, ABA content was significantly higher in shoots and roots at 5 and 6 days (an increase over day 3 of 4.62 and 6.70-fold in shoots and 4.70 and 5.40-fold in roots, respectively; Figure 3A). Meanwhile, in the WV treated group, ABA content increased from day 3 to day 6 (an increase over day 3 of 1.32, 3.02, and 4.41-fold in shoots, and 1.49, 2.30, and 2.61-fold in roots, respectively; Figure 3A); furthermore, there were significantly higher levels than in the control group (increased by 1.79 and 2.59-fold at 3 days; 2.10, and 3.35-fold at 4 days; 1.17 and 1.26-fold at 5 days; and 1.15 and 1.09-fold at 6 days in shoots and roots, respectively; Figure 3A). Real-time PCR results of 9-cis-epoxycarotenoid dioxygenase gene [*TaNECD*; National Coalition Building Institute (NCBI) accession: KX711890.1], the key gene of ABA biosynthesis, showed the same changing trend as ABA content in both groups (Figure 3B).

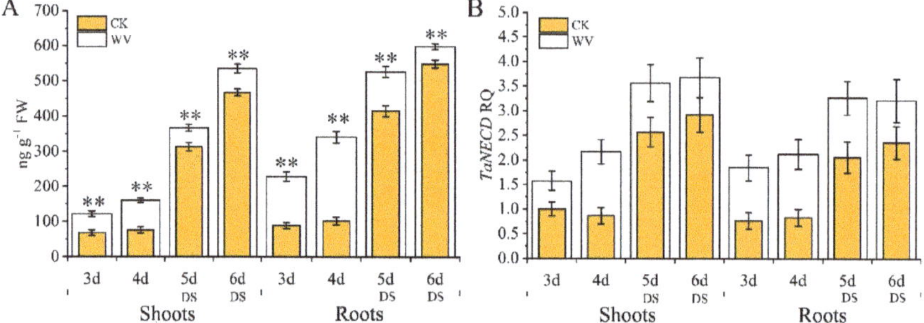

Figure 3. Dynamic changes in abscisic acid (ABA) content and expression pattern analysis of *TaNECD* of wheat shoots and roots in both the control and WV treated groups before and after the drought stress treatments. The concentration of WV was 0 and 1:900 in control and WV treated groups, respectively. The stages of non-drought stress and drought stress were from day 3 to day 4 and from day 5 to day 6, respectively, in the control and WV treated groups. (**A**) Changes of ABA content; (**B**) expression pattern of *TaNECD*. Data are means ± SD of three independent experiments (biological replicates). The significance of differences was assessed by Student's *t*-test (* $p < 0.05$, ** $p < 0.01$). RQ means relative quantification.

In order to explore whether the generation and accumulation of ROS in both groups of wheat roots before and after drought stress treatment, ROS content and antioxidant enzyme activity were measured. The content of O^{2-} and H_2O_2 increased from day 3 and day 6, respectively, in both the control and WV treated groups. The WV roots had higher levels of O^{2-}, H_2O_2, and MDA than the control group from 3 to 4 days, and content increased by 1.06 and 1.19-fold at 3 days, and 1.21 and 1.14-fold at 4 days, respectively (Figure 4A). Meanwhile, levels of O^{2-} and H_2O_2 in the WV treated group decreased by 0.89 and 0.89-fold at 5 days, and 0.85 and 0.82-fold at 6 days, respectively, as compared to the control group (Figure 4A). At the same time, the activities of superoxide dismutase (SOD, EC.1.15.1.1), guaiacol peroxidase (POD, EC1.11.1.7), and catalase (CAT, EC 1.11.1.6) were measured, and the results showed that the WV roots had higher activity of all 3 antioxidant enzymes than did the control group (Figure 4B). These enzymes increased significantly in the WV treated group from day 3 to day 6; there was no significant change in activity of all 3 antioxidant enzymes from day 3 to day 4 in the control group, and their activity increased significantly from day 5 to day 6 under subsequent drought stress (Figure 4B). Quantification of related antioxidant genes performed by real-time PCR, including peroxidase 1 gene (*TaPOX1*, NCBI accession: X85227.1; spot49, EMS54484.1), Cu/Zn superoxide dismutase gene (*TaSOD*, NCBI accession: AK457377), L-ascorbate peroxidase 1 gene (*TaAPX1*, NCBI accession: XM_020316778; spot 108, 110, 111, 112, EMS61931.1), glutathione transferase gene (*TaGST*, NCBI accession: AJ414697; spot 114, CAC94001.1), alcohol dehydrogenase gene (*TaADH1A*, NCBI accession: AK457420 ; spot 56, ABL74258.1), and monodehydroascorbate reductase gene (*TaMDAR*, NCBI accession: KC884831.1; spot 47, EMS50440.1), showed the same results as the antioxidant enzymes (Figure 4C,D). Excessive ROS can oxidize membrane lipids and generate MDA, which can aggravate damage to membrane structure. In order to explore whether the generation and accumulation of MDA caused membrane lipid peroxidation damage, MDA content was measured. No significant difference in MDA concentrations was found between the control and WV treated groups at 3 days and 4 days; however, MDA concentrations were notably higher in the control group than the WV treated group under drought stress (Figure 4A).

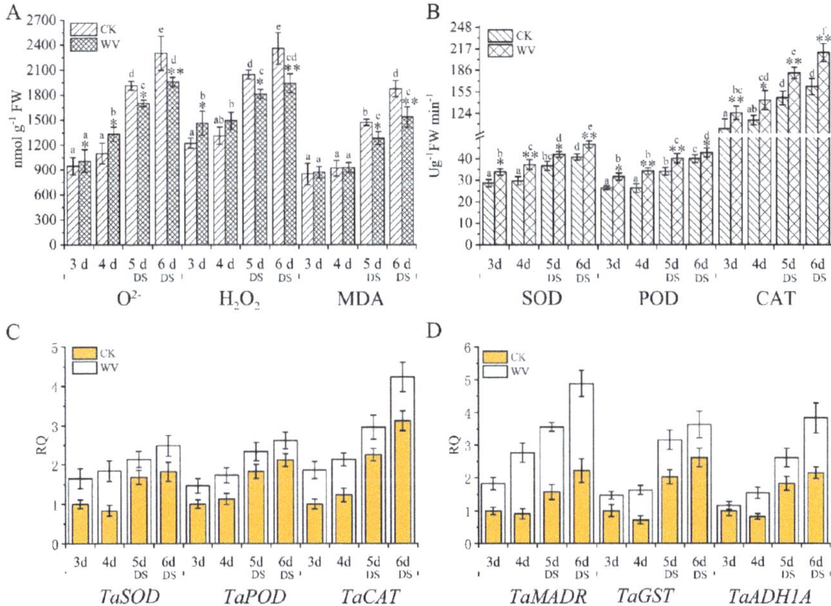

Figure 4. Assessment of ROS content, oxidative stress analysis, and expression pattern analysis of related antioxidative genes in wheat roots in both the control and WV treated groups before and after drought stress. The concentration of WV was 0 and 1:900 in the control and WV treated groups, respectively. The stages of non-drought stress and drought stress were from day 3 to day 4 and from day 5 to day 6, respectively, in both the control and WV treated groups. (**A**) The generation rate of O_2^- and content of H_2O_2 in control wheat roots and WV treated roots; the content of malonaldehyde (MDA) in the control wheat roots and WV treated shoots. (**B**) The activity of superoxide dismutase (SOD), guaiacol peroxidase (POD), and catalase (CAT) in the control wheat roots and WV treated roots. (**C**) Expression pattern analysis of *TaSOD*, *TaPOX1*, and *TaAPX1*. (**D**) Expression pattern analysis of *TaMDAR*, *TaGST*, and *TaADH1A*. Data are means ± SD of three independent experiments (biological replicates). The significance of differences was assessed by Student's t-test (* $p < 0.05$, ** $p < 0.01$). RQ, relative quantification.

2.3. Analysis of Differentially Accumulated Protein Spots (DAPs) in Control and WV Pretreated Roots Under Drought Tolerance

To understand the proteome response to short-term drought stress of wheat roots after WV pretreatment, and the changes in proteomes of wheat roots from the control, WV treated groups (1:900; drought stress treatment condition for 2 days) were analyzed by Two-dimensional gel electrophoresis (2-DE). The protein maps produced from three independent biological replicates showed a high reproducibility based on analysis using PDQuest software (Figure S3). PCA analysis indicated the homogeneity of biological replicates and difference of the treatments (Figure S4).

Figure 5 shows a representative gel image of proteins extracted from the control and WV treated groups. Protein spots [1799 (±97) and 1803 (±26)] were reproducibly detected using PDQuest software from the roots of the control and WV-treated groups, respectively (biological replicates, $n = 3$). From a spot-to-spot comparison and based on statistical analysis, a total of 138 spots (numbered from 1 to 138) exhibited at least a 1.5-fold (Student's t-Test, $p < 0.05$) difference in abundance between the control and WV treated groups (Figure 5, Table S4). In total, 77 spots had a >1.5-fold change in abundance ($p < 0.05$) and 61 spots showed a >2.0-fold change by comparing the two groups; meanwhile, in the roots of the WV treated group, 106 spots exhibited up-regulated expression (53 spots >1.5-fold and 53 spots showed a >2.0-fold change) and 32 spots down-regulated expression (24 spots >1.5-fold and 8 spots showed a >2.0-fold change; Table S4) compared with the control group. In all, 138 differential protein species showed quantitative changes (Figure 5, Table S4). Master gel and several typical examples of DAPs showing different profiles are exhibited in Supplementary Figures S5 and S6.

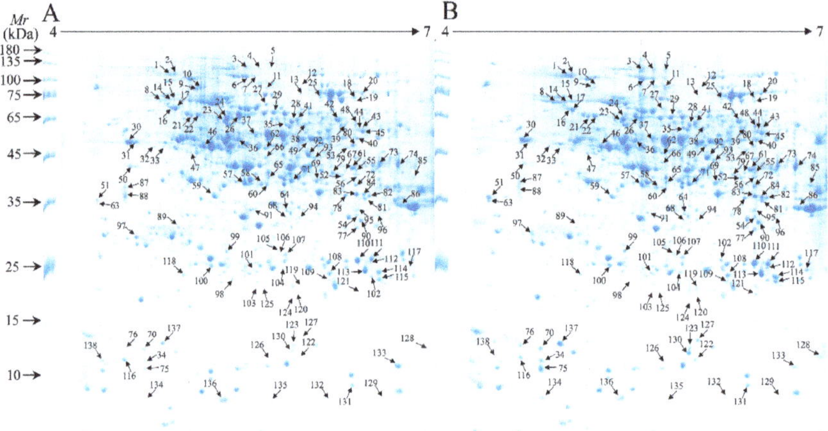

Figure 5. Two-dimensional gel electrophoresis (2-DE) image analysis of proteomes in control and WV treated wheat roots. (**A**) control group; (**B**) WV treated group (1:900).

2.4. Identification and Functional Classification of DAPs

A total of 138 DAPs were analyzed by MALDI-TOF/TOF MS and all of them were successfully identified by MS/MS (Table S5). All of the 138 identified protein species were functionally annotated in the current database (Table S5). In summary, 138 identities represented 103 unique protein species.

Based on the metabolic and functional features of the wheat roots, all of the 138 identified protein species were classified into 13 major categories, including carbohydrate metabolism, cell rescue and defense, protein folding and assembly, amino acid metabolism, protein metabolism, secondary metabolism, ATP synthesis and ion transport, signal transduction, cell motility and structural components, photosynthesis, cell division cycle process, transcription factor, and protein transport (Figure 6). Eighty-four percent of these identified protein species were implicated in the first six functional groups, whereas the largest functional groups that were greatly affected by drought stress were the protein species involved in carbohydrate metabolism and cell rescue and defense (25.4% and 24.6%). Further analysis of the change of abundance in each group revealed that proteins involved in protein folding and assembly (8.7%), amino acid metabolism (8.7%), protein metabolism (8.7%), and secondary metabolism (8.0%) were overrepresented, either in number or in expression level, suggesting that these processes were susceptible to drought stress. In order to visualize the protein expression patterns of all 13 categories, the hierarchical clustering of proteins was analyzed (Figure S7).

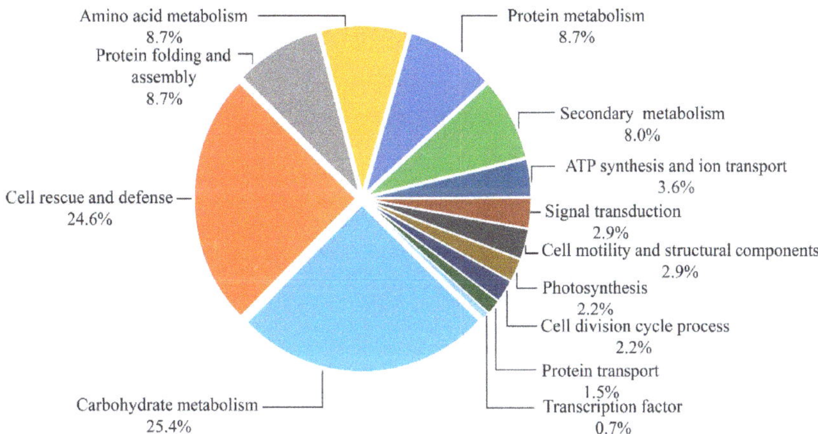

Figure 6. Functional classification of 138 identified proteins. Distribution of proteins according to their biological functions.

In general, the monoisotopic mass (*Mr*) calculated by SDS-PAGE with protein standard markers has about a ±10% error compared with the theoretical *Mr* value. In our work, 27 identities among all 138 identities had a smaller observed *Mr* value than theoretical *Mr* value (Table S5). This result indicated these protein species might be partially degraded. Besides, 13 identities among all 138 identities had larger experimental *pI* values than theoretical *pI* values (Table S5), suggesting that these identities may be modified. Some identified protein spots from different positions in the same gel with different observed *Mr* and *pI* were found to have the same name and NCBI accession number, whereas these proteins spots should be considered different protein species. For instance, spot 101, 104, and 105 was identified as triosephosphat-isomerase (TPI; CAC14917.1), and spot 110, 111 and 112 were identified as L-ascorbate peroxidase 1 (EMS61931.1). These protein spots were recognized to be different protein species due to gene polymorphisms, alternatively spliced transcripts, proteolytically processed protein species, and PTMs, which might have differential biological function [26].

2.5. PPI Analysis of Identified Membrane Proteins

The PPI network of all 138 DAPs was constructed using on line STRING 10.5 Software. All 138 DAPs were blasted against the *Arabidopsis thaliana* proteins database (TAIR 10; Table S7). DAPs were functional clusters according to the biological processes in which they are involved. STRING analysis revealed the functional links between DAPs in which the protein species involved in carbohydrate and energy metabolism, response to stress, protein metabolism, and protein folding processes were major clusters (Figure 7). Actually, these four clusters were not separated and together they formed a related-network in response to drought stress. The carbohydrate and energy metabolism, response to stress, and protein metabolism process groups contained more members, with interaction being concentrated on LOS2, ATARCA, and RPSAb, respectively. With respect to carbohydrate and energy metabolism, response to stress, and protein folding processes, CPN60A and HSP 70 were the most important nodes. Abbreviations of the specific protein species names in the network are shown in Table S7.

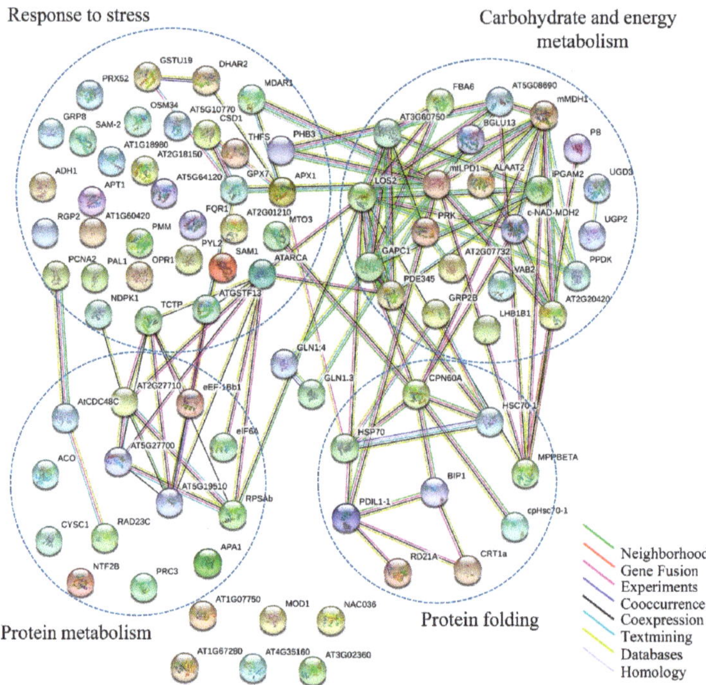

Figure 7. Analysis of protein interaction network by STRING 10.5. The Arabidopsis Information Resource (TAIR) homologous proteins from identified protein species were mapped by searching the STRING 10.5 software with a confidence level of 0.67. The colored lines between the proteins indicate the various types of interaction evidence.

To obtain statistically over- or under-represented categories of biological pathways and molecular functions related to drought treatment, BiNGO was used to analyze identified differential protein species (Figure 8, Tables S8 and S9). The results revealed that several overrepresented biological pathways were mostly significant (Figure 8A, Table S8), including response to metal ion ($p = 1.06 \times 10^{-28}$), response to inorganic substance ($p = 9.37 \times 10^{-28}$), response to cadmium ion ($p = 2.24 \times 10^{-26}$), response to stimulus ($p = 1.70 \times 10^{-22}$), response to stress ($p = 3.35 \times 10^{-21}$), and response to chemical stimulus ($p = 3.06 \times 10^{-20}$). More specifically, metabolic processes ($p = 2.99 \times 10^{-11}$), small molecule metabolic process ($p = 7.46 \times 10^{-8}$), cellular metabolic processes ($p = 7.60 \times 10^{-8}$), and the S-adenosylmethionine metabolic process ($p = 8.63 \times 10^{-6}$), were significantly overrepresented. Meanwhile, a complete list of the enriched Gene Ontology (GO) molecular functions for the proteins is presented in Figure 8B and Table S9. Of them, several of the most highly enriched molecular functions included copper ion binding ($p = 1.27 \times 10^{-16}$), catalytic activity ($p = 3.69 \times 10^{-11}$), oxidoreductase activity ($p = 1.14 \times 10^{-7}$), methionine adenosyltransferase activity ($p = 1.20 \times 10^{-7}$), antioxidant activity ($p = 1.81 \times 10^{-7}$), transition metal ion binding ($p = 8.68 \times 10^{-7}$), ion binding ($p = 1.91 \times 10^{-6}$), and cation binding ($p = 1.91 \times 10^{-6}$).

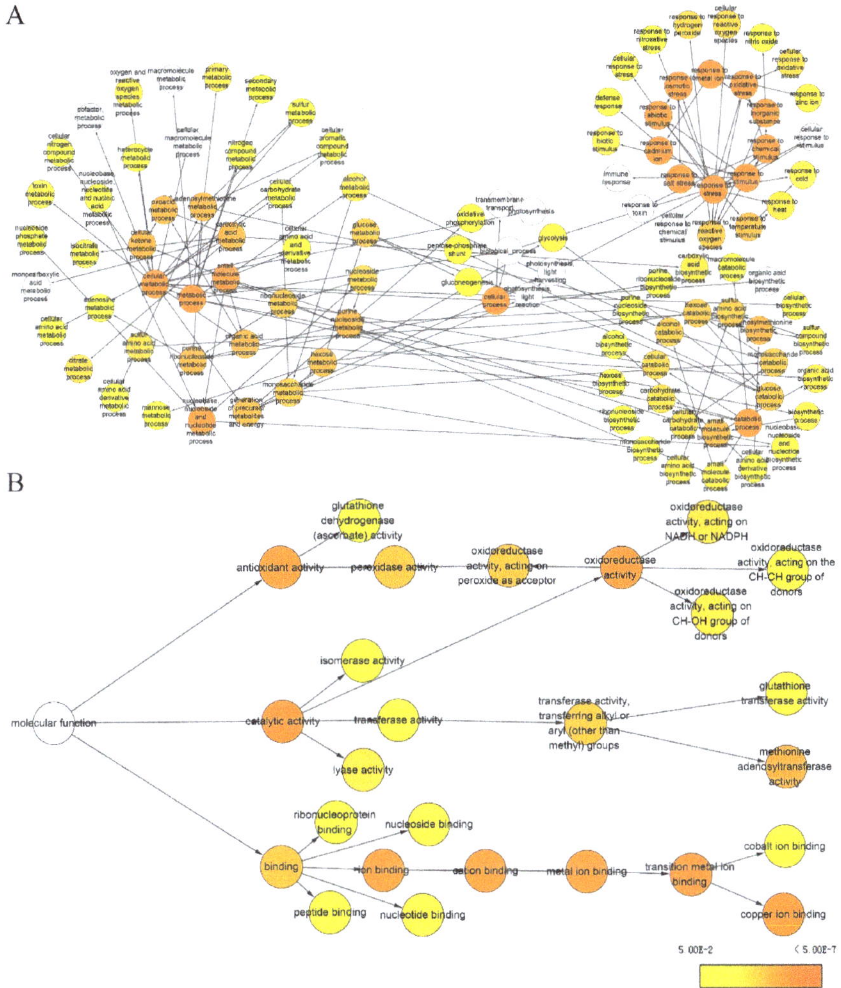

Figure 8. Biological pathway (**A**) and molecular function (**B**) networks generated by BiNGO. Homologous proteins were used for the gene ontology (GO) analysis. The size of the node is related to the number of proteins and the color represents the *p*-value for statistical significance of the overrepresented GO term (see the color scale on the right bottom).

3. Discussion

3.1. Morphological and Physiological Response of Wheat Seedlings to Exogenous WV Pretreatment

The effect of the exogenous WV seed soaking treatment on growth and stress tolerance of plants depended on the use of an optimum concentration because WV applied beyond a certain range might be detrimental. Our results indicated that soaking seeds with 1:900 WV gave optimal promotion to wheat seedlings.

ABA is a stress phytohormone that often accumulates in plants exposed to abiotic and biotic stress [27–31]. ABA is involved in defense priming in plants [28] and can activate antioxidative defense systems that contribute to the alleviation of stress [29,32]. Moreover, elevated ABA levels in plants can trigger signaling cascades downstream of phytohormones, such as salicylic acid (SA), which may also mitigate oxidative stress [30,31,33–36]. In our research, ABA content of shoots and roots were increased significantly in the WV treated group over those of the control group under both drought or non-drought stress. Our results indicated that WV can induce ABA biosynthesis in wheat seedlings. On the one hand, accumulation of ABA can regulate the stomatal apertures of leaves to prevent water loss; on the other hand, it can activate downstream antioxidative gene expression of shoots and roots to better resist subsequent drought stress.

Exogenous low concentrations of WV can cause slight oxidative stress in wheat roots but not oxidative damage. Our results showed that O^{2-} and H_2O_2 contents were higher in the WV treated group before drought stress than in the control group, whereas the content of MDA showed no significant difference between the two groups before drought stress. Moreover, the contents of O^{2-}, H_2O_2, and MDA were significantly lower in the WV treated group than the control group after drought stress. Meanwhile, the activities of antioxidant enzymes were higher and related antioxidant genes were upregulated in the WV treated group under both drought and non-drought stress, which enabled them to cope better with continuous ROS production. This result indicated that a low concentration of WV acts as a stressor, having a slight effect on the oxidative status of the plant similar to that of stress-acclimating processes.

3.2. Protein Species Involved in Carbohydrate Metabolism and Energy Production

Carbohydrate metabolism regulates sugar synthesis and transformation as well as carbon partitioning, while drought stress disrupts carbohydrate metabolism in plants. In this study, a large proportion of the protein species whose abundance changed significantly under drought stress are associated with carbohydrate metabolism and energy production processes. Triosephosphate isomerase (TPI; spot 101, 104, 105), glyceraldehyde-3-phosphate dehydrogenase (GAPDH; spot 83, 84, 85), enolase (spot 37), and fructose bisphosphate aldolase (FBA; spot 72, 73, 74), were present at higher levels in the WV treated group over the control group. In plants, TPI are located in cytosol and chloroplast and are involved in several metabolic pathways, including glycolysis, gluconeogenesis, and Calvin cycle [37]. GAPDH, as a moonlighting protein, is involved in glycolysis and the Calvin cycle, but also played a vital role in redox signal transduction in plants [38,39]. In higher plants, FBA is located in cytosol and plastids, functioning in the Calvin cycle, glycolysis and gluconeogenesis. Previous studies showed that FBA could be redox-modified by glutathione (GSH) and participated in redox regulatory of *Arabidopsis thaliana* [40]. In our study, 3, 2, and 3 DAPs were identified as TPI, NAPDH, and FBA, respectively. They were considered as different protein species and were involved in different metabolic pathways. During drought stress in plants, the increased abundance of TPI, GAPDH, FBA and enolase could be related to the cellular requirement for extra energy in order to deal with stress and repair damage [41]. Our study indicated that WV can promote the wheat glycolysis metabolic pathway to produce more energy under drought stress. The study showed that the TCA cycle may be fueled by products derived from the degradation of protein and other macromolecules, in order to produce sufficient ATP to meet energetic demands under stress [42]. Aconitate hydratase (spot 6, 7) and malate dehydrogenase (spot 86) are components of the TCA cycle, an important source of energy for cells, and were present at higher levels in the WV treated group under drought stress. Aconitase isoforms are located in the mitochondria and cytosol [43]. Another role of aconitase is a "circuit breaker" that reduced electron flow through the mitochondrial electron transport chain and to a subsequent decrease of ROS [44]. In our study, this result indicated that WV can enhance TCA cycle speed in wheat roots to enable them to cope with subsequent drought stress. ATP synthase is the universal enzyme that manufactures ATP from ADP and provides energy for a large number of fundamental biological processes [13]. In the present study, the proteins related to ATP production (spot 35, 36, 38, 127) were found to be increased in abundance in the WV treated group under drought stress. During biotic and abiotic stress in plants, energy costs are high during stress acclimation, for example, the increased relative abundance of components of ATP-synthase [45]. Here, as a whole, the abundance of different subunits of ATP synthesis was increased in the WV treated group compared to the control plants; this change protected multiple normal metabolic processes dependent on ATP under drought stress. Our study suggested that the WV pretreatment regulated carbohydrate metabolism and, under drought stress, further enhanced carbohydrate synthesis and ATP production in wheat roots.

3.3. Protein Species Involved in the Stress Response

In the present study, O^{2-}, H_2O_2 and MDA, which have the potential to cause peroxide damage and membrane lipid peroxidation. In general, within a certain threshold of abiotic stress, plants have a series of protective mechanisms to scavenge or reduce ROS and MDA levels and maintain the stability of cellular homeostasis [46]. These protective mechanisms include the activity of antioxidative proteins. In our study, protein species involved in the oxidative stress response were also identified; some anti-stress protein species, such as peroxidase 8 (spot 39, 40), peroxidase 1 (spot 49), pox1 (spot 78, 94), peroxidase 70 (spot 96), L-ascorbate peroxidase 1 (spot 99, 108, 110, 111, 112), glutathione transferase (spot 103, 113, 114), superoxide dismutase (spot 130), monodehydroascorbate reductase (spot 47), and dehydroascorbate reductase (spot 109) were more abundant in roots of the WV group than in control plants. Real-time PCR results of related antioxidative proteins (spot 47, 49, 56, 108, 114, 130), showed the same changing trend as the abundance of these protein species in both groups (Figure 4B). In plants, peroxidase was a protein superfamily and involved in countering effects of stress through signal transduction, strengthening of the cell wall [47], as well as the scavenging of toxic peroxides and ROS, accumulated under oxidative stresses [48]. Studies have indicated that peroxidase abundance and activity of peroxidase increased significantly in soybean roots under drought stress [49]. In plants, SOD is highly efficient at eliminating O^{2-}, which can convert O^{2-} to molecular oxygen and H_2O_2. Subsequently, H_2O_2 is reduced to H_2O by peroxidase [50]. Ascorbate peroxidase is one of the most important components for scavenging H_2O_2 [51]. GSH can combine with glutathione and a wide variety of hydrophobic and electrophilic compounds to eliminate cytotoxic compounds [52]. Studies have shown that GSTs were upregulated significantly in drought stressed wheat [53]. In our study, the contents of ROS and MDA were significantly increased in roots of the control and WV-treated groups under drought stress. However, the contents of ROS and MDA accumulated at a lower level in the WV-treated roots compared with the control group. There is no doubt that increased abundance of these anti-stress proteins restrained the accumulation of ROS and lowered damage induced by MDA in the WV treated group. These results suggest that WV pretreatment enhanced the antioxidant defense system to decrease oxidative damage under drought stress and provided a favorable environment for growth and development.

In addition to the above described DAPs involved in the stress response, *S*-adenosylmethionine synthetase is a member of the stress-induced family of genes [54]. Previous studies indicate that overexpression of S-adenosyl-L-methionine synthetase increase tomato plant tolerance to alkali stress through polyamine and hydrogen peroxide cross-linked networks [55]. In our study, four DAPs were identified as S-adenosyl-methionine synthase (spot 52, spot 53, spot 54, spot 106); we, therefore, suggest that its greater abundance in the WV treated group enhanced the capacity of plants to resist drought.

3.4. Protein Metabolism-Related Proteins

Protein synthesis, assembling, folding, and degradation are the main biologic process of protein metabolism [56]. In the present study, 24 DAP spots were involved in protein metabolism and were grouped into three functional subgroups: Proteins involved in protein synthesis, folding and degradation. In the first subgroup, elongation factor 1-delta (EF1D) and elongation factor 1-beta (EF1B; spot 87, spot 88) had a higher accumulation in the control group than the WV treated group. Elongation factors are proteins that play a central role in the elongation phase of protein synthesis in plants. Spots 133 and 138 were identified as 40S ribosomal protein and 60S acidic ribosomal protein, respectively, and their abundance was greater in the control group than in the WV treated group. 40S ribosomal protein and 60S acidic ribosomal protein are components of the ribosome machinery and are required for protein synthesis [57]. In our case, proteins suffered damage due to accumulated ROS in the control group root, increased abundance of the elongation factor and the related ribosomal protein could have caused the accumulated synthesis of proteins and replaced damaged proteins caused by peroxidation under drought stress in the control group plants. In contrast, ROS was eliminated over time by antioxidative proteins and this protected the stability of proteins in the roots of the WV treated group.

In the second subgroup, 3 DAPs (spot 8, spot 14, spot 15) were identified as 70 kDa heat shock proteins (HSPs), whose abundance was increased in roots of the WV treated group (Figure 3, Supplementary Tables S2 and S3). HSPs play crucial roles in protecting plants against stress and they are involved in a wide range of crucial cellular processes [58]. Previous studies indicate that HSP 70 can prevent the aggregation of denatured proteins and assist in the refolding of nonnative proteins caused by environmental stress [59]. Our results indicated that the accumulation of ROS caused instability of proteins under drought treatments. An increased abundance of HSPs in roots of the WV treated group provided a more effective protective mechanism in response to oxidative stress.

In the third subgroup, 3 DAPs (spot 79, spot 80, spot 89) were identified as aspartic proteinase proteins (Table S5). Aspartic proteinase is an endopeptidase and is active under acidic pH conditions [60]. Previous studies show that *SPAP1*, which encodes a typical aspartic protease protein, is responsible for leaf senescence in the sweet potato [60]. Many studies indicate that aspartic proteinase participates in the PCD process of many plant organs [61–64]. In our case, excessive ROS was not effectively removed from roots in the control group, which resulted in the accumulation of dysfunctional amounts of proteins. An increased abundance of aspartic proteinase in the control group could have effectively hydrolyzed dysfunctional proteins; moreover, aspartic proteinase will have promoted apoptosis of damaged cells by participating in the PCD process. Spot 102 was identified as a proteasome subunit, which controlled the protein degradation process. A previous study indicated that the ubiquitin-proteasome system (UPS) plays an important role in response to environmental stress such as drought, salinity, cold, and nutrient deprivation. Moreover, UPS has shown to be related to the production of ABA and participate in signal transduction pathway [65]. In our case, the abundance of the proteasome subunit was increased in roots of the control group. This indicated that excessive ROS induced oxidative damage to protein structure and function and these dysfunctional proteins needed to be degraded immediately in roots of the control group to maintain the stability of the normal mechanical processes of cellular homeostasis.

3.5. Proteins Involved in Secondary Metabolism

Jasmonic acid (JA) and salicylic acid (SA) are important secondary metabolites in plants, being involved in the various metabolic process, particularly in response to biotic and abiotic stress in plants [66,67]. Previous studies indicate that JA accumulates rapidly after biotic and abiotic stressors [67,68], that trigger the biosynthesis of JA from linolenic acid, suggesting that JA is an important stress-signaling molecule in plants. SA has also been identified as an endogenous regulatory signal in plants, particularly during plant defense against pathogens and drought stress [13,66,69]. In addition, pretreatment with low concentration SA significantly enhances the growth of wheat seedlings and the shoots and roots of soybean [13,70]. In our case, 4 DAPs (spot 64, spot 67, spot 68, spot 69) were identified as 12-oxophytodienoate reductase (OPR), which is a key enzyme in JA biosynthesis, and their abundance was higher in the roots of the WV treated group compared with the control. The expression pattern analysis of *TaOPR* showed the same results as for protein quantification (Figure S8). Previous studies indicate that wounded plants rapidly accumulate JA, and this signal activates the expression of early response genes [71]. In the WV treated group, increased accumulation of OPR may have accelerated the biosynthesis of JA; accumulated JA triggers expression of defense genes via the octadecanoid pathway or by acting directly on the genes. When wheat plants suffered drought stress, a faster and effective response triggered by accumulated JA was initiated in WV-treated roots. In our study, spot 12 and spot 13 were identified as phenylalanine ammonia-lyase (PAL), which plays a significant role in the biosynthesis of SA. A previous study indicates that PAL activity and the content of SA in pharbitis were both up-regulated under the stress treatment [72]. In our case, an increase of PAL in the WV treated group promoted the biosynthesis of SA. Moreover, the q-PCR result of *TaPAL* showed up-regulated expression in the WV treated group under drought stress and non-stress conditions (Figure S8). SA activates various genes that encode antioxidants, chaperones, and heat shock proteins to resist drought stress.

3.6. WV can Initiate An Early Defense Mechanism to Mitigate Subsequent Drought Stress

Our results showed ABA levels were significantly increased in the shoots and roots of the WV treated group. ABA accumulated in the shoots and roots of the WV treated group to rapidly regulate stomatal aperture and the expression of defense-related genes when the wheat plants underwent drought stress, thus conferring resistance to drought stress. Meanwhile, comparative proteomic analysis revealed that WV promoted the biosynthesis of JA and SA, which regulated downstream related anti-stress gene expression with ABA through signal transduction. In our present work, soaking with WV launched an early defense mechanism before drought stress began. Soaking with WV induced the production of ROS, which remained within a safe threshold because of increased activities of antioxidant enzymes and the effective opening of the defense system. During drought stress, ROS content was significantly higher in the roots of the control group and they suffered oxidative damage. Comparative proteomic analysis revealed that carbohydrate metabolism was inhibited, and this accelerated the degradation of damaged proteins in the control roots under drought stress condition. However, proteomic analysis and results of the determination of physiological indices indicated that ROS was effectively removed by the increased abundance of antioxidative and related stress proteins in WV pretreated roots after drought stress. An overview of the main metabolic pathways regulated by WV under drought stress is shown in Figure 9. These results indicated that WV can promote the growth of wheat shoots and roots, and also improve their tolerance to drought stress.

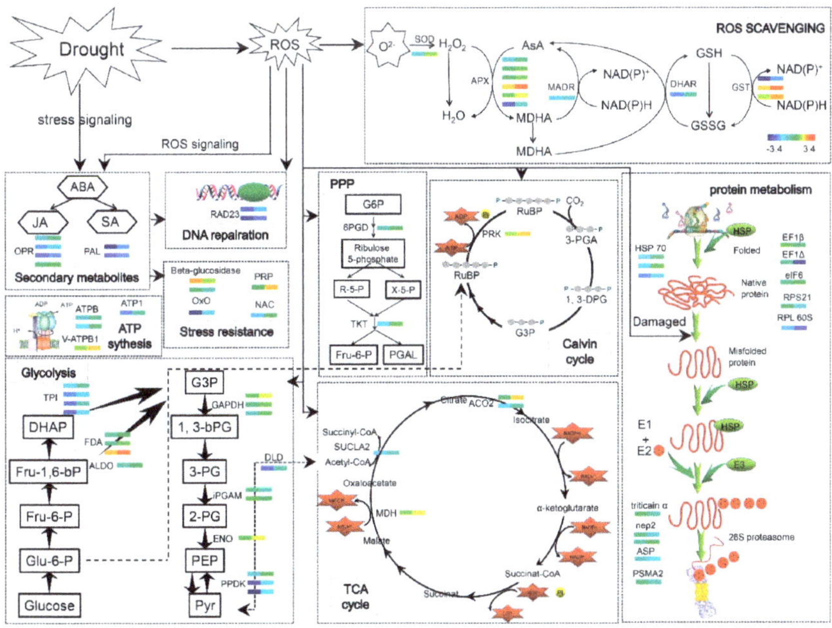

Figure 9. Diagram of the main metabolic pathways regulated by WV under drought stress. Each small colored square represents an individual protein under different treatments (from left to right, control and WV treated groups, respectively). Relative levels of expression are shown by a color gradient from low (blue) to high (red). SOD: Superoxide dismutase; APX: L-ascorbate peroxidase; MADR: Monodehydroascorbate reductase; DHAR: dehydroascorbate reductase; GST: glutathione transferase; OPR: 12-oxophytodienoate reductase; PAL: Phenylalanine ammonia-lyase; RAD23: DNA repair protein RAD23; ATPB: ATP synthase beta subunit; V-ATPB1: Vacuolar ATPase subunit B1; Oxo: Oxalate oxidase GF-2.8; PRP: Pathogenesis-related protein; NAC: NAC transcription factor; TPI: Triosephosphate isomerase; FDA: Fructose-bisphosphate aldolase cytoplasmic isozyme; ALDO: Aldolase; GAPDH: Lyceraldehyde-3-phosphate dehydrogenase; iPGAM: 2,3-bisphosphoglycerate-independent phosphoglycerate mutase; ENO: Enolase; PPDK: Phosphate dikinase 1; DLD: Dihydrolipoyl dehydrogenase 1; SUCLA2: Succinyl-CoA ligase [ADP-forming] subunit beta; MDH: Malate dehydrogenase 1; ACO2: Aconitate hydratase; 6PGD: 6-phosphogluconate dehydrogenase; TKT: Transketolase; PRK: Phosphoribulokinase; EF1β: Elongation factor 1-beta; EF1Δ: Elongation factor 1-delta; eIF6: Eukaryotic translation initiation factor 6; RPS21: 40S ribosomal protein S21; RPL 60S: 60s acidic ribosomal protein-like protein; nep2: Aspartic proteinase nepenthesin-2; ASP: Aspartic proteinase; PSMA2: Proteasome subunit alpha type-2.

4. Materials and Methods

4.1. Plant Materials

Wheat (*Triticum aestivum* L.) cultivar "Zhoumai 18" seeds were sterilized with 70% ethanol and 10% NaClO followed by a thorough washing with sterilized water. Seeds were then soaked in sterilized water supplemented with various volumes of primary WV (Yixin Bio-energy Technology Development Co. LTD, Yangling, China; composition of primary WV is listed in Table S1) for 3 days. The seeds were grown in a greenhouse under a day/night temperature regime of 25 °C, under 12 h d^{-1} illumination, light intensity of 300 µmol m^{-2}·s^{-1} and a relative humidity of 60%–70%. The experimental design was presented in Figure S1.

To explore the optimal various concentrations of WV on wheat seedling growth, 25 sterilized seeds were soaked in sterilized water supplemented with control 0 (control group), 1:300, 1:600, 1:900, 1:1200, and 1:1500 various volumes of WV [primary WV:ddH$_2$O$_2$ (V:V); WV treated groups], respectively, for 3 days. The experiments were laid out in a completely randomized design (CRD). Three biological replications (5 wheat plants) were set for each treatment. Then soaking seeds were distributed in a 115 × 115 mm sterile germination box with two layers of filter paper saturated with 10 mL of sterilized water. Seeds were dampened with 5 mL of water every day, for one week. The aerial parts (shoots) and roots of both the control and WV treated groups (1:300, 1:600, 1:900, 1:1200, 1:1500) from day 2 to day 6 after seed germination were collected for fresh weight (FW) and dry weight (DW) analysis.

In order to explore the effects of WV treatments on the drought tolerance of wheat, 50 sterilized seeds were soaked in sterilized water supplemented with control 0 (control group) and 1:900 primary WV [primary WV:ddH$_2$O$_2$ (V:V); WV treated groups], respectively, for 3 days. Three biological replications (50 wheat plants) were set for each treatment. Subsequently soaked seeds were distributed in a 115 × 115 mm sterile germination box with two layers of filter paper saturated with 10 mL of sterilized water, for 4 days. On the fifth day after germination, seedlings of both the control and WV treated groups (1:900) were transferred to a Hoagland nutrient solution with −1 MPa PEG 6000 (simulated drought stress), respectively, for 2 days. Thereafter, shoots and roots were collected from the control and WV treated groups (1:900) from day 3 to day 6 to determine the abscisic acid (ABA) content. Roots were then collected from the control and WV treated groups (1:900) from day 3 to day 6 to determine O^{2-}, H$_2$O$_2$, and malonaldehyde (MDA) content, and the generation rate and the activities of superoxide dismutase (SOD, EC.1.15.1.1), catalase (CAT, EC 1.11.1.6), and guaiacol peroxidase (POD, EC1.11.1.7). For proteomics, roots from both the control and WV treated groups (1:900) were collected on the sixth day (under drought stress treatment conditions for 2 days). For real-time PCR analysis, shoots and roots were collected from the control and WV treated groups (1:900) from day 3 to day 6.

4.2. Determination of O^{2-} Formation Rate and H_2O_2 Content

Determination of O^{2-} and the H_2O_2 content were performed according to Song et al. [73], with minor modifications: Roots were ground to powder in liquid nitrogen; centrifugal force was $7000 \times g$. Briefly, the reaction was initiated in assay solution (65 pH 7.8 mM phosphate buffer, 10 mM hydroxylamine chlorhydrate, 17 mM sulfanilamide and 7 mM α-naphthylamine). Absorbance at 530 nm was measured and the formation rate of O^{2-} was calculated from a standard curve of $NaNO_2$. Determination of H_2O_2 content was performed according to Song et al. [73].

4.3. Determination of Antioxidant Enzyme Activity

The activity of SOD, CAT, and POD was determined according to Song et al. [73]. To extract antioxidant enzymes, 0.5 g of fresh roots were ground in liquid nitrogen, then root powder was transferred to a 50 mM cool phosphate buffer [containing 1% (w/v) polyvinylpyrrolidone, pH 7.0] and centrifuged at 4 °C and $15,000 \times g$ for 20 min. The supernatant was used for enzyme activity assays.

For the estimation of SOD activity, the reaction was initiated in an activity assay solution (50 μM NBT, 1.3 μM riboflavin, 13 mM methionine, 75 nM EDTA, 50 mM pH 7.8 phosphate buffer, and enzyme extract). The absorbance at 560 nm was determined with a spectrophotometer. For the measurement of CAT activity, the reaction was initiated in the activity assay solution (50 mM pH 7.8 phosphate buffer, 15 mM H_2O_2, and enzyme extract). The decrease in absorbance of activity assay solution at 240 nm was read every 20 s. For the determination of POD activity, the reaction was initiated in the activity assay solution (50 mM pH 5.0 sodium acetate buffer, 20 mM guaiacol, 40 mM H_2O_2, and enzyme extract). The increase in absorbance of activity assay solution at 470 nm was recorded every 20 s.

4.4. Determination of MDA Content

MDA content was estimated according to Song et al. [73]. Briefly, roots (0.5 g) were homogenized in 20% (v/v) TCA and 0.5 (v/v) thiobarbituric acid (TBA). The supernatants after centrifugation were incubated at 95 °C for 10 min and cooled in ice immediately. The absorbance at 532 nm was read.

4.5. Quantitative Determination of ABA Content

The plant hormone ABA was extracted according to the method described by Shi et al. [74], with minor modifications. Fresh root samples (ca. 1 g) were ground in liquid nitrogen, then powder was suspended in 10 mL of 80% (v/v) methanol containing 200 mg·L^{-1} of butylated hydroxytoluene and 500 mg·L^{-1} of citric acid monohydrate on ice. The mixture was subsequently shaken overnight at 4 °C before centrifugation for 30 min at $8000 \times g$ and 4 °C. The supernatant was collected. The precipitate was extracted twice for 2 h, the supernatants were combined and subsequently dried under N_2 and resuspended in 500 μL of 80% methanol. The phytohormone concentrations in the extracts were analyzed using an LC-20AT high performance liquid chromatography system (Shimadzu, KinhDo, Japan) and an API 2000™ electrospray tandem mass spectrometer (AB Sciex, Foster City, CA, USA). Two microliter samples were separated within a Wondasil™ C18 column (5 μm, 4.6 × 150 mm; Shimadzu). ABA (($\pm$)-ABA, A1049; Sigma, St. Louis, MO, USA) was used to prepare standard curves for the determination of hormone concentrations in samples.

4.6. Protein Extraction

Protein extraction was performed as described by Valledor et al. [75], with minor modifications. Wheat roots were ground to a fine powder with liquid nitrogen. The ground root powder was homogenized with −20 °C ice-cold extraction buffer [10% (w/v) TCA, 0.07% β-mercaptoethanol (β-ME; v/v), and 1mM PMSF], then proteins were precipitated overnight. The following day, the mixture was centrifuged at 20,000× g for 30 min and the pellet was resuspended in 2 mL of −20 °C ice-cold acetone [0.07% β-mercaptoethanol (β-ME; v/v), and 1 mM PMSF]; this was repeated 3 times. Finally, the pellet was collected, lyophilized with vacuum freeze-drying equipment, and stored at −80 °C.

4.7. 2-DE and Gel Image Analysis

2D-PAGE was performed according to the method described by Valledor et al. [75], with minor modifications. The root proteins were solubilized in lysis solution and proteins concentration was determined using Bio-Rad Protein Assay Kit II (Bio-Rad, Shanghai, China), with Bovine serum albumin (BSA) as a standard protein. About 900 μg of protein was separated on a 17 cm pH 4–7 linear IPG strip (Bio-Rad) and actively rehydrated at 50 V for 14 h at 20 °C. Subsequently, focusing was performed under following conditions: 250 V for 1 h, 500 V for 1 h, 1000 V for 1 h, 8000 V for 4 h, and 8000 V to achieve 80,000 V-h. Strips were immediately equilibrated twices. The Second-dimension electrophoresis was performed on 12% polyacrylamide gels. Gels were stained with Coomassie Brilliant Blue (CBB) G-250. Each sample was run in 3 independent biological replicates.

Gels were visualized using a GS-900 Calibrated Densitometer (Bio-Rad, Taiwan, China) at a resolution of 600 dpi. Images were analyzed using the analytical software PDQuest 2-DE 8.0.1 (Bio-Rad, Hercules, CA, USA) for spot detection, gel matching, and statistical analysis of spots. The selection of protein spots of interest for analysis by MS was based on a fold change ≥ 1.5 ($p < 0.05$).

4.8. In-gel Digestion and MALDI-TOF/TOF MS Analysis

The DAP spots were excised, washed, de-stained, and dehydrated. Subsequently, protein spots were digested with trypsin. The supernatant was collected, and the resultant peptides were extracted twice with 0.1% trifluoroacetic acid (TFA) and 60% ACN. Then, the supernatants were combined. Mass spectra were collected using a 5800 MALDI Time of Flight (TOF)/TOF^{TM} analyzer (AB Sciex, Foster City, CA, USA) and analyzed using TOF/TOF^{TM} Series Explorer^{TM} Software V4.1.0 (AB Sciex, Redwood City, CA, USA).

MS/MS mass spectra data were searched against the NCBInr databases with a taxonomy parameter set to *Viridiplantae* using the Mascot search engine. The search parameters were set as follows: One missed cleavage, peptide tolerance set to 100 ppm, MS/MS tolerance of 0.5 Da, peptide charge of 1+, carbamidomethylation and oxidation of methionines allowed as fixed modification variable modification.

4.9. Total RNA Isolation and Real-Time PCR

Total RNA was extracted from wheat shoots and roots of the control and WV treated groups using an OMEGA plant RNA kit (R6827, Omega Bio-tek, Norcross, GA, USA), and cDNA was reverse transcribed from 1 μg of total RNA using the GoScript^{TM} Reverse Transcription System (A5001, Promega, Madison, WI, USA). Relative quantification of gene expression by qPCR was performed on a QuantStudio 3 Real-Time PCR System (Thermo Fisher Scientific, Singapore, Singapore). The primers used for qPCR were designed using the qPrimerDB database [76], Oligo 7 and Beacon Designer^{TM} 8.0 software. The sequence of the primers can be found in the Supplementary Table S3. Wheat actin gene was used as the endogenous control which remained stable throughout the drought treatment [77,78]. qPCR was performed in an optical 96-well plate, including 10 μL 2 × GoTaq® qPCR Master Mix (A6002, Promega, Madison, WI, USA), 2 μL 1:5-diluted template cDNA, and 0.2 μM of each gene-specific primer, in a final volume of 20 μL, using the following thermal cycles: 95 °C for 1 min, 40 cycles of 95 °C for 10 s, 60 °C for 1 min. Disassociation curve analysis was performed as follows: 95 °C for 15 s, 60 °C for 1min, and 95 °C for 15 min. Relative expression levels were calculated by the $2^{-\Delta\Delta Ct}$ method [79].

4.10. Bioinformatic Analysis

The prediction of transmembrane domains (TMDs) of the identified DAPs was carried out using TMpred (http://www.ch.embnet.org/software/TMPRED_form.html). The grand average of hydropathicity (GRAVY) value for each DAP was calculated using the Protein GRAVY tool (http://www.bioinformatics.org/sms2/protein_gravy.html). Cellular locations of DAPs were performed through WoLF PSORT (https://wolfpsort.hgc.jp/) and (http://www.csbio.sjtu.edu.cn/bioinf/plant-multi/). All identified DAPs were blasted against the *Arabidopsis thaliana* TAIR 10 (The Arabidopsis Information Resource) protein database (http://www.arabidopsis.org/) for obtaining the annotated protein information to conduct a PPI network using the online analysis tool STRING 10.5. Biological processes and cellular component were predicted by the BiNGO plugin of Cytoscape software (version 3.6.0, San Diego, CA, USA).

4.11. Statistical Analysis

Principal component analysis (PCA) was performed [80,81] by SPSS software (version 22.0, IBM Corporation, Armonk, NY, USA) to identify homogeneous biological replicates and the difference between the control group and WV-treated group. In our study, coefficient and KMO and Bartlett's test of sphericity were used for dimension reduction analysis. The volume of DAPs was estimated using the built-in statistical modules of PDQuest 8.01 by applying a log transformation and a *t*-test. The results were presented as mean $\pm$ standard deviation (SD) from three independent biological replicates. One-way analysis of variance (ANOVA) multiple comparisons was performed to calculate statistical significance; $p < 0.05$ was considered statistically significant. Graphical presentation of the data was performed using Originlab 2018b software (OriginLab Corporation, Northampton, MA, USA).

5. Conclusions

During wheat seedling growth, young seedlings are susceptible to water deficiency. However, a well-developed root system can improve wheat plants' ability to defend against drought stress. Pretreatment soaking in appropriate concentrations of wood vinegar significantly promoted root and seedling growth. Moreover, WV was able to initiate an early defense mechanism to mitigate subsequent drought stress. In this process, ROS was effectively removed through the increased abundance of antioxidative and other related stress proteins. A battery of protective mechanisms in the WV soaked seed treatment helped to maintain the stability of the normal mechanical processes of cellular homeostasis and metabolism.

Supplementary Materials: Supplementary materials can be found at http://www.mdpi.com/1422-0067/20/4/943/s1. Figure S1. Experimental design used in the study; Figure S2. Supplementary pictures in the study; Figure S3. 2-DE image of wheat root proteomes in the control and WV-treated groups; Figure S4. Principal component analysis of the control and WV-treated groups; Figure S5. Master gel; Figure S6. Typical examples of DAPs showing different profiles; Figure S7. Hierarchical clustering of protein species of all 13 categories in both the control and WV treated groups; Figure S8. Expression pattern analysis of *TaOPR* and *TaPAL* genes in wheat roots in both the control and WV treated groups before and after the drought stress; Table S1. Chemical compounds identified in the WV; Table S1. Supplementary data in the study; Table S3. Specific primers used in this study; Table S4. Fold change (Ratio) of DAPs between the WV-treated and the control groups under drought stress; Table S5. MS/MS data, GRAVY, numbers of TMDs and subcellular localization of DAPs in wheat roots under drought stress; Table S6 Log-transformed values of each DAP in both the control and WV-treated groups; Table S7. DAPs blasted against the TAIR database; Table S8. Biological pathways generated for wheat roots; Table S9. Molecular functions generated for wheat roots. The mass spectrometry proteomics data have been deposited to the ProteomeXchange (http://proteomecentral.proteomexchange.org) Consortium via the PRIDE partner repository with the dataset identifier PXD012150.

Author Contributions: Y.W. (Yuying Wang) and L.Q. conceived the study. Y.W. (Yuying Wang) and Q.S. performed the experiments. S.W. performed the MS/MS analysis. Y.W. (Yuying Wang) carried out the analysis of the data, made the identification of the proteins and drafted the manuscript. Y.W. (Yuying Wang) and Q.S. contributed in the preparation of the final draft of the manuscript. Y.W. (Yajun Wang), Q.S., S.W., L.Q. and Y.G. provided reagents, materials and analysis tools. All authors read and approved the final manuscript.

Funding: This research was funded by Agricultural Ecological Environment Special Fund of Ministry of Agriculture (No. 2110402-7), Agricultural Special Fund of Shaanxi Province (No. 2018-43).

Conflicts of Interest: The authors declare no conflicts of interest.

Abbreviations

2-DE	Two-dimensional gel electrophoresis
6PGD	6-phosphogluconate dehydrogenase
ABA	Abscisic acid
ACO2	Aconitate hydratase
ADH1A	Alcohol dehydrogenase
ALDO	Aldolase
APX1	L-ascorbate peroxidase 1
ASP	Aspartic proteinase
ATPB	ATP synthase beta subunit
CAT	Catalase
DEPs	Differential expression proteins
DHAR	Dehydroascorbate reductase
DLD	Dihydrolipoyl dehydrogenase 1
EF1β	Elongation factor 1-beta
EF1Δ	Elongation factor 1-delta
eIF6	Eukaryotic translation initiation factor 6
ENO	Enolase
FDA	Fructose-bisphosphate aldolase cytoplasmic isozyme
GAPDH	Glyceraldehyde-3-phosphate dehydrogenase
GRAVY	Grand average of hydropathicity value
GST	Glutathione transferase
iPGAM	2,3-bisphosphoglycerate-independent phosphoglycerate mutase
JA	Jasmonic acid
MDA	Malonaldehyde
MDAR	Monodehydroascorbate reductase
MDH	Malate dehydrogenase 1
Mr	Monoisotopic mass
NAC	NAC transcription factor
NECD	9-*cis*-epoxycarotenoid dioxygenase
nep2	Aspartic proteinase nepenthesin-2
OPR	12-oxophytodienoate reductase
Oxo	Oxalate oxidase
PA	Pyroligneous acid
PAL	Phenylalanine ammonia-lyase
pI	Isoelectric point
POD	Guaiacol peroxidase
POX1	Peroxidase 1
PPDK	Phosphate dikinase 1
PPI	Protein–protein interaction
PRK	Phosphoribulokinase
PRP	Pathogenesis-related protein
PSMA2	Proteasome subunit alpha type-2
q-PCR	Real-time PCR
RAD23	DNA repair protein RAD23
ROS	Reactive oxygen species
RPL 60S	60s acidic ribosomal protein-like protein
RPS21	40S ribosomal protein S21
RQ	Relative quantification
SA	Salicylic acid

SOD	Superoxide dismutase
SUCLA2	Succinyl-CoA ligase [ADP-forming] subunit beta
TKT	Transketolase
TMDs	Transmembrane domains
TPI	Triosephosphate isomerase
V-ATPB1	Vacuolar ATPase subunit B1
WV	Wood vinegar

References

1. Wei, Q.; Ma, X.; Dong, J. Preparation, chemical constituents and antimicrobial activity of pyroligneous acids from walnut tree branches. *J. Anal. Appl. Pyrolysis* **2010**, *87*, 24–28. [CrossRef]
2. Mungkunkamchao, T.; Kesmala, T.; Pimratch, S.; Toomsan, B.; Jothityangkoon, D. Wood vinegar and fermented bioextracts: Natural products to enhance growth and yield of tomato (*Solanum lycopersicum* L.). *Sci. Hortic.* **2013**, *154*, 66–72. [CrossRef]
3. Pimenta, A.S.; Fasciotti, M.; Monteiro, T.V.C.; Lima, K.M.G. Chemical composition of pyroligneous acid obtained from eucalyptus gg100 clone. *Molecules* **2018**, *23*, 426. [CrossRef]
4. Grewal, A.; Abbey, L.; Gunupuru, L.R. Production, prospects and potential application of pyroligneous acid in agriculture. *J. Anal. Appl. Pyrolysis* **2018**, *135*, 152–159. [CrossRef]
5. Cai, K.; Jiang, S.; Ren, C.; He, Y. Significant damage-rescuing effects of wood vinegar extract in living *Caenorhabditis elegans* under oxidative stress. *Sci Food* **2012**, *92*, 29–36. [CrossRef] [PubMed]
6. Dissatian, A.; Sanitchon, J.; Pongdontri, P.; Jongrungklang, N.; Jothityangkoon, D. Potential of wood vinegar for enhancing seed germination of three upland rice varieties by suppressing malondialdehyde production. *J. Agric. Sci.* **2018**, *40*, 371–380. [CrossRef]
7. Mohan, D.; Pittman, C.; Steele, P. Pyrolysis of wood/biomass for bio-oil: A critical review. *Energy Fuels* **2006**, *20*, 848–889. [CrossRef]
8. Jung, K. Growth inhibition effect of pyroligneous acid on pathogenic fungus, *Alternaria mali*, the agent of Alternaria blotch of apple. *Biotechnol. Bioprocess Eng.* **2007**, *12*, 318–322. [CrossRef]
9. Kulkarni, M.G.; Sparg, S.G.; Light, M.E.; van Staden, J. Stimulation of rice (*Oryza sativa* L.) seedling vigour by smoke-water and butenolide. *J. Agron. Crop Sci.* **2006**, *192*, 395–398. [CrossRef]
10. Siddhuraju, P.; Becker, K. Antioxidant properties of various solvent extracts of total phenolic constituents from three different agroclimatic origins of drumstick tree (*Moringa oleifera* Lam.) leaves. *J. Agric. Food Chem.* **2003**, *51*, 2144–2155. [CrossRef]
11. Loo, A.Y.; Jain, K.; Darah, I. Antioxidant and radical scavenging activities of the pyroligneous acid from a mangrove plant, *Rhizophora apiculata. Food Chem.* **2007**, *104*, 300–307. [CrossRef]
12. Curtis, T.; Halford, N.G. Food security: the challenge of increasing wheat yield and the importance of not compromising food safety. *Ann. Appl. Biol.* **2014**, *164*, 354–372. [CrossRef] [PubMed]
13. Kang, G.; Li, G.; Xu, W.; Peng, X.; Han, Q.; Zhu, Y.; Guo, T. Proteomics reveals the effects of salicylic acid on growth and tolerance to subsequent drought stress in wheat. *J. Proteome Res.* **2012**, *11*, 6066–6079. [CrossRef] [PubMed]
14. Loutfy, N.; El-Tayeb, M.A.; Hassanen, A.M.; Moustafa, M.F.M.; Sakuma, Y.; Inouhe, M. Changes in the water status and osmotic solute contents in response to drought and salicylic acid treatments in four different cultivars of wheat (*Triticum aestivum*). *J. Plant Res.* **2012**, *125*, 173–184. [CrossRef] [PubMed]
15. Bengough, A.G.; McKenzie, B.M.; Hallett, P.D.; Valentine, T.A. Root elongation, water stress, and mechanical impedance: A review of limiting stresses and beneficial root tip traits. *J. Exp. Bot.* **2011**, *62*, 59–68. [CrossRef] [PubMed]
16. Pandey, A.; Chakraborty, S.; Datta, A.; Chakraborty, N. Proteomics approach to identify dehydration responsive nuclear proteins from chickpea (*Cicer arietinum* L.). *Mol. Cell. Proteomics* **2008**, *7*, 88–107. [CrossRef] [PubMed]
17. Alvarez, S.; Choudhury, S.R.; Pandey, S. Comparative quantitative proteomics analysis of the ABA response of roots of drought-sensitive and drought-tolerant wheat varieties identifies proteomic signatures of drought adaptability. *J. Proteome Res.* **2014**, *13*, 1688–1701. [CrossRef] [PubMed]
18. Faghani, E.; Gharechahi, J.; Komatsu, S.; Mirzaei, M.; Khavarinejad, R.A.; Najafi, F.; Farsad, L.K.; Salekdeh, G.H. Comparative physiology and proteomic analysis of two wheat genotypes contrasting in drought tolerance. *J. Proteom.* **2015**, *114*, 1–15. [CrossRef]

19. Liu, H.; Sultan, M.A.R.F.; Liu, X.L.; Zhang, J.; Yu, F.; Zhao, H.X. Physiological and comparative proteomic analysis reveals different drought responses in roots and leaves of drought-tolerant wild wheat (*Triticum boeoticum*). *PLoS ONE* **2015**, *10*. [CrossRef]

20. Peng, Z.Y.; Wang, M.C.; Li, F.; Lv, H.J.; Li, C.L.; Xia, G.M. A Proteomic study of the response to salinity and drought stress in an introgression strain of bread wheat. *Mol. Cell. Proteomics* **2009**, *8*, 2676–2686. [CrossRef]

21. Agrawal, L.; Gupta, S.; Mishra, S.K.; Pandey, G.; Kumar, S.; Chauhan, P.S.; Chakrabarty, D.; Nautiyal, C.S. Elucidation of complex nature of peg induced drought-stress response in rice root using comparative proteomics approach. *Front. Plant Sci.* **2016**, *7*. [CrossRef] [PubMed]

22. Yu, X.W.; Yang, A.J.; James, A.T. Comparative proteomic analysis of drought response in roots of two soybean genotypes. *Crop Pasture Sci.* **2017**, *68*, 609–619. [CrossRef]

23. Abid, M.; Ali, S.; Qi, L.K.; Zahoor, R.; Tian, Z.; Jiang, D.; Snider, J.L.; Dai, T. Physiological and biochemical changes during drought and recovery periods at tillering and jointing stages in wheat (*Triticum aestivum* L.). *Sci. Rep.* **2018**, *8*, 4615. [CrossRef] [PubMed]

24. Mohammadi, P.P.; Moieni, A.; Komatsu, S. Comparative proteome analysis of drought-sensitive and drought-tolerant rapeseed roots and their hybrid F1 line under drought stress. *Amino Acids* **2012**, *43*, 2137–2152. [CrossRef] [PubMed]

25. Prinsi, B.; Negri, A.S.; Failla, O.; Scienza, A.; Espen, L. Root proteomic and metabolic analyses reveal specific responses to drought stress in differently tolerant grapevine rootstocks. *BMC Plant Biol.* **2018**, *18*, 126. [CrossRef] [PubMed]

26. Schlüter, H.; Apweiler, R.; Holzhütter, H.G.; Jungblut, P.R. Finding one's way in proteomics: A protein species nomenclature. *Chem. Cent. J.* **2009**, *3*, 11. [CrossRef] [PubMed]

27. Cao, X.; Jia, J.; Zhang, C.; Li, H.; Liu, T.; Jiang, X.; Polle, A.; Peng, C.; Luo, Z.B. Anatomical, physiological and transcriptional responses of two contrasting poplar genotypes to drought and re-watering. *Physiol. Plant.* **2014**, *151*, 480–494. [CrossRef] [PubMed]

28. Luo, Z.B.; Janz, D.; Jiang, X.; Gobel, C.; Wildhagen, H.; Tan, Y.; Rennenberg, H.; Feussner, I.; Polle, A. Upgrading root physiology for stress tolerance by ectomycorrhizas: Insights from metabolite and transcriptional profiling into reprogramming for stress anticipation. *Plant Physiol.* **2009**, *151*, 1902–1917. [CrossRef] [PubMed]

29. Wang, J.; Chen, J.; Pan, K. Effect of exogenous abscisic acid on the level of antioxidants in *Atractylodes macrocephala* Koidz under lead stress. *Environ. Sci. Pollut. Res.* **2013**, *20*, 1441–1449. [CrossRef] [PubMed]

30. Ma, Y.; Cao, J.; He, J.; Chen, Q.; Li, X.; Yang, Y. Molecular mechanism for the regulation of ABA homeostasis during plant development and stress responses. *Int. J. Mol. Sci.* **2018**, *19*, 3643. [CrossRef] [PubMed]

31. Cruz, T.M.; Carvalho, R.F.; Richardson, D.N.; Duque, P. Abscisic acid (ABA) regulation of Arabidopsis SR protein gene expression. *Int. J. Mol. Sci.* **2014**, *15*, 17541–17564. [CrossRef] [PubMed]

32. Sharma, S.S.; Kumar, V. Responses of wild type and abscisic acid mutants of *Arabidopsis thaliana* to cadmium. *J. Plant Physiol.* **2002**, *159*, 1323–1327. [CrossRef]

33. Disante, K.B.; Cortina, J.; Vilagrosa, A.; Fuentes, D.; Hernandez, E.I.; Ljung, K. Alleviation of Zn toxicity by low water availability. *Physiol. Plant.* **2014**, *150*, 412–424. [CrossRef] [PubMed]

34. Noriega, G.; Caggiano, E.; Lecube, M.L.; Cruz, D.S.; Batlle, A.; Tomaro, M.; Balestrasse, K.B. The role of salicylic acid in the prevention of oxidative stress elicited by cadmium in soybean plants. *Biometals* **2012**, *25*, 1155–1165. [CrossRef] [PubMed]

35. Stroiński, A.; Chadzinikolau, T.; Giżewska, K.; Zielezińska, M. ABA or cadmium induced phytochelatin synthesis in potato tubers. *Bio. Plant.* **2010**, *54*, 117–120. [CrossRef]

36. Trinh, N.N.; Huang, T.L.; Chi, W.C.; Fu, S.F.; Chen, C.C.; Huang, H.J. Chromium stress response effect on signal transduction and expression of signaling genes in rice. *Physiol. Plant.* **2014**, *150*, 205–224. [CrossRef] [PubMed]

37. López-Castillo, L.M.; Jiménez-Sandoval, P.; Baruch-Torres, N.; Trasviña-Arenas, C.H.; Diaz-Quezada, C.; Lara-González, S.; Winkler, R.; Brieba, L.G. Structural basis for redox regulation of cytoplasmic and chloroplastic triosephosphate isomerases from *Arabidopsis thaliana*. *Front. Plant Sci.* **2016**, *7*, 1817. [CrossRef] [PubMed]

38. Schneider, M.; Knuesting, J.; Birkholz, O.; Heinisch, J.J.; Scheibe, R. Cytosolic GAPDH as a redox-dependent regulator of energy metabolism. *BMC Plant Biol.* **2018**, *18*. [CrossRef] [PubMed]

39. Vescovi, M.; Zaffagnini, M.; Festa, M.; Trost, P.; Lo Schiavo, F.; Costa, A. Nuclear accumulation of cytosolic glyceraldehyde-3-phosphate dehydrogenase in cadmium-stressed Arabidopsis roots. *Plant Physiol.* **2013**, *162*, 333–346. [CrossRef] [PubMed]

40. Dixon, D.P.; Skipsey, M.; Grundy, N.M.; Edwards, R. Stress-induced protein *S*-glutathionylation in Arabidopsis. *Plant Physiol.* **2005**, *138*, 2233–2244. [CrossRef] [PubMed]

41. Žďarska, M.; Zatloukalová, P.; Benítez, M.; Šedo, O.; Potěšil, D.; Novák, O.; Svačinova, J.; Pešek, B.; Malbeck, J.; Vašíčkova, J.; et al. Proteome analysis in Arabidopsis reveals shoot- and root-specific targets of cytokinin action and differential regulation of hormonal homeostasis. *Plant Physiol.* **2013**, *161*, 918–930. [CrossRef]

42. Simova-Stoilova, L.P.; Romero-Rodriguez, M.C.; Sanchez-Lucas, R.; Navarro-Cerrillo, R.M.; Medina-Aunon, J.A.; Jorrin-Novo, J.V. 2-DE proteomics analysis of drought treated seedlings of *Quercus ilex* supports a root active strategy for metabolic adaptation in response to water shortage. *Front. Plant. Sci.* **2015**, *6*, 627. [CrossRef]

43. Igamberdiev, A.U.; Ratcliffe, R.G.; Gupta, K.J. Plant mitochondria: Source and target for nitric oxide. *Mitochondrion* **2014**, *19*, 329–333. [CrossRef] [PubMed]

44. Delledonne, M.; Xia, Y.J.; Dixon, R.A.; Lamb, C. Nitric oxide functions as a signal in plant disease resistance. *Nature* **1998**, *394*, 585–588. [CrossRef]

45. Kosova, K.; Vitamvas, P.; Prasil, I.T. Proteomics of stress responses in wheat and barley-search for potential protein markers of stress tolerance. *Front. Plant. Sci.* **2014**, *5*, 711. [CrossRef]

46. Gill, S.S.; Tuteja, N. Reactive oxygen species and antioxidant machinery in abiotic stress tolerance in crop plants. *Plant Physiol. Biochem.* **2010**, *48*, 909–930. [CrossRef]

47. Mittler, R.; Vanderauwera, S.; Suzuki, N.; Miller, G.; Tognetti, V.B.; Vandepoele, K.; Gollery, M.; Shulaev, V.; Van Breusegem, F. ROS signaling: the new wave? *Trends Plant Sci.* **2011**, *16*, 300–309. [CrossRef]

48. Herrero, J.; Esteban-Carrasco, A.; Zapata, J.M. Looking for *Arabidopsis thaliana* peroxidases involved in lignin biosynthesis. *Plant Physiol. Biochem.* **2013**, *67*, 77–86. [CrossRef] [PubMed]

49. Khan, M.N.; Komatsu, S. Proteomic analysis of soybean root including hypocotyl during recovery from drought stress. *J. Proteom.* **2016**, *144*, 39–50. [CrossRef]

50. Navrot, N.; Finnie, C.; Svensson, B.; Hägglund, P. Plant redox proteomics. *J. Proteom.* **2011**, *74*, 1450–1462. [CrossRef]

51. Bhatt, I.; Tripathi, B.N. Plant peroxiredoxins: catalytic mechanisms, functional significance and future perspectives. *Biotechnol. Adv.* **2011**, *29*, 850–859. [CrossRef]

52. Dixon, D.P.; Skipsey, M.; Edwards, R. Roles for glutathione transferases in plant secondary metabolism. *Phytochemistry* **2010**, *71*, 338–350. [CrossRef]

53. Bazargani, M.M.; Sarhadi, E.; Bushehri, A.A.; Matros, A.; Mock, H.P.; Naghavi, M.R.; Hajihoseini, V.; Mardi, M.; Hajirezaei, M.R.; Moradi, F.; et al. A proteomics view on the role of drought-induced senescence and oxidative stress defense in enhanced stem reserves remobilization in wheat. *J. Proteom.* **2011**, *74*, 1959–1973. [CrossRef]

54. Gong, B.; Wang, X.; Wei, M.; Yang, F.; Li, Y.; Shi, Q. Overexpression of *S-adenosylmethionine synthetase 1* enhances tomato callus tolerance to alkali stress through polyamine and hydrogen peroxide cross-linked networks. *Plant Cell Tissue Organ. Cult.* **2015**, *124*, 377–391. [CrossRef]

55. Gong, B.; Li, X.; VandenLangenberg, K.M.; Wen, D.; Sun, S.; Wei, M.; Li, Y.; Yang, F.; Shi, Q.; Wang, X. Overexpression of S-adenosyl-L-methionine synthetase increased tomato tolerance to alkali stress through polyamine metabolism. *Plant Biotechnol. J.* **2014**, *12*, 694–708. [CrossRef]

56. Saikawa, N.; Akiyama, Y.; Ito, K. FtsH exists as an exceptionally large complex containing HflKC in the plasma membrane of *Escherichia coli*. *J. Struct. Biol.* **2004**, *146*, 123–129. [CrossRef]

57. Nagaraj, S.; Senthil-Kumar, M.; Ramu, V.S.; Wang, K.; Mysore, K.S. Plant Ribosomal Proteins, RPL12 and RPL19, Play a Role in Nonhost Disease Resistance against Bacterial Pathogens. *Front. Plant. Sci.* **2015**, *6*, 1192. [CrossRef]

58. Wang, W.; Vinocur, B.; Shoseyov, O.; Altman, A. Role of plant heat-shock proteins and molecular chaperones in the abiotic stress response. *Trends Plant Sci.* **2004**, *9*, 244–252. [CrossRef]

59. Al-Whaibi, M.H. Plant heat-shock proteins: A mini review. *J. King Saud Univ. Sci.* **2011**, *23*, 139–150. [CrossRef]

60. Chen, H.J.; Huang, Y.H.; Huang, G.J.; Huang, S.S.; Chow, T.J.; Lin, Y.H. Sweet potato *SPAP1* is a typical aspartic protease and participates in ethephon-mediated leaf senescence. *J. Plant Physiol.* **2015**, *180*, 1–17. [CrossRef]

61. Fendrych, M.; van Hautegem, T.; van Durme, M.; Olvera-Carrillo, Y.; Huysmans, M.; Karimi, M.; Lippens, S.; Guérin, C.J.; Krebs, M.; Schumacher, K.; et al. Programmed cell death controlled by ANAC033/SOMBRERO determines root cap organ size in Arabidopsis. *Curr. Biol.* **2014**, *24*, 931–940. [CrossRef] [PubMed]

62. Niu, N.; Liang, W.; Yang, X.; Jin, W.; Wilson, Z.A.; Hu, J.; Zhang, D. EAT1 promotes tapetal cell death by regulating aspartic proteases during male reproductive development in rice. *Nat. Commun.* **2013**, *4*, 1445. [CrossRef] [PubMed]

63. Phan, H.A.; Iacuone, S.; Li, S.F.; Parish, R.W. The MYB80 transcription factor is required for pollen development and the regulation of tapetal programmed cell death in *Arabidopsis thaliana*. *Plant Cell* **2011**, *23*, 2209–2224. [CrossRef] [PubMed]

64. Van Durme, M.; Nowack, M.K. Mechanisms of developmentally controlled cell death in plants. *Curr. Opin. Plant Biol.* **2016**, *29*, 29–37. [CrossRef] [PubMed]

65. Stone, S.L. The role of ubiquitin and the 26S proteasome in plant abiotic stress signaling. *Front. Plant. Sci.* **2014**, *5*, 135. [CrossRef] [PubMed]

66. Clarke, J.D.; Volko, S.M.; Ledford, H.; Ausubel, F.M.; Dong, X. Roles of salicylic acid, jasmonic acid, and ethylene in cpr-induced resistance in Arabidopsis. *Plant Cell* **2000**, *12*, 2175–2190. [CrossRef]

67. Kamal, A.H.M.; Komatsu, S. Jasmonic acid induced protein response to biophoton emissions and flooding stress in soybean. *J. Proteom.* **2016**, *133*, 33–47. [CrossRef]

68. Rao, M.V.; Lee, H.; Creelman, R.A.; Mullet, J.E.; Davis, K.R. Jasmonic acid signaling modulates ozone-induced hypersensitive cell death. *Plant Cell* **2000**, *12*, 1633–1646. [CrossRef]

69. Sharma, M.; Gupta, S.K.; Majumder, B.; Maurya, V.K.; Deeba, F.; Alam, A.; Pandey, V. Salicylic acid mediated growth, physiological and proteomic responses in two wheat varieties under drought stress. *J. Proteom.* **2017**, *163*, 28–51. [CrossRef]

70. Gutiérrez-Coronado, M.A.; Trejo-López, C.; Larqué-Saavedra, A. Effects of salicylic acid on the growth of roots and shoots in soybean. *Plant Physiol. Biochem.* **1998**, *36*, 563–565. [CrossRef]

71. Koo, A.J.; Gao, X.; Jones, A.D.; Howe, G.A. A rapid wound signal activates the systemic synthesis of bioactive jasmonates in Arabidopsis. *Plant J.* **2009**, *59*, 974–986. [CrossRef] [PubMed]

72. Wada, K.C.; Mizuuchi, K.; Koshio, A.; Kaneko, K.; Mitsui, T.; Takeno, K. Stress enhances the gene expression and enzyme activity of phenylalanine ammonia-lyase and the endogenous content of salicylic acid to induce flowering in pharbitis. *J. Plant Physiol.* **2014**, *171*, 895–902. [CrossRef] [PubMed]

73. Song, Q.; Wang, S.; Zhang, G.; Li, Y.; Li, Z.; Guo, J.; Niu, N.; Wang, J.; Ma, S. Comparative proteomic analysis of a membrane-enriched fraction from flag leaves reveals responses to chemical hybridization agent SQ-1 in wheat. *Front. Plant. Sci.* **2015**, *6*, 669. [CrossRef] [PubMed]

74. Shi, W.G.; Li, H.; Liu, T.X.; Polle, A.; Peng, C.H.; Luo, Z.B. Exogenous abscisic acid alleviates zinc uptake and accumulation in *Populus* × *canescens* exposed to excess zinc. *Plant Cell Environ.* **2015**, *38*, 207–223. [CrossRef] [PubMed]

75. Valledor, L.; Castillejo, M.A.; Lenz, C.; Rodriguez, R.; Cañal, M.J.; Jorrín, J. Proteomic analysis of *Pinus radiata* needles: 2-DE map and protein identification by LC/MS/MS and substitution-tolerant database searching. *J. Proteome Res.* **2008**, *7*, 2616–2631. [CrossRef] [PubMed]

76. Lu, K.; Li, T.; He, J.; Chang, W.; Zhang, R.; Liu, M.; Yu, M.; Fan, Y.; Ma, J.; Sun, W.; et al. qPrimerDB: A thermodynamics-based gene-specific qPCR primer database for 147 organisms. *Nucleic Acids Res.* **2018**, *46*, D1229–D1236. [CrossRef]

77. Paolacci, A.R.; Tanzarella, O.A.; Porceddu, E.; Ciaffi, M. Identification and validation of reference genes for quantitative RT-PCR normalization in wheat. *BMC Mol. Biol.* **2009**, *10*, 11. [CrossRef]

78. Scholtz, J.J.; Visser, B. Reference gene selection for qPCR gene expression analysis of rust-infected wheat. *Physiol. Mol. Plant Pathol.* **2013**, *81*, 22–25. [CrossRef]

79. Livak, K.J.; Schmittgen, T.D. Analysis of relative gene expression data using real-time quantitative PCR and the $2^{-\Delta\Delta Ct}$ Method. *Methods* **2001**, *25*, 402–408. [CrossRef]

80. Valledor, L.; Jorrín, J. Back to the basics: Maximizing the information obtained by quantitative two dimensional gel electrophoresis analyses by an appropriate experimental design and statistical analyses. *J. Proteom.* **2011**, *74*, 1–18. [CrossRef]

81. Valledor, L.; Jorrín, J.V.; Rodriguez, J.L.; Lenz, C.; Meijon, M.; Rodriguez, R.; Canal, M.J. Combined proteomic and transcriptomic analysis identifies differentially expressed pathways associated to *Pinus radiata* needle maturation. *J. Proteome Res.* **2010**, *9*, 3954–3979. [CrossRef] [PubMed]

International Journal of
Molecular Sciences

Article

Proteomics Analysis of *E. angustifolia* Seedlings Inoculated with Arbuscular Mycorrhizal Fungi under Salt Stress

Tingting Jia [1,2], Jian Wang [1,2], Wei Chang [1,2], Xiaoxu Fan [1,2], Xin Sui [1,2] and Fuqiang Song [1,2,*]

[1] Engineering Research Center of Agricultural Microbiology Technology, Ministry of Education, Heilongjiang University, Harbin 150500, China; 18945081056@163.com (T.J.); 13796672755@163.com (J.W.); changwei77@126.com (W.C.); fan_xiao_xu@126.com (X.F.); xinsui_cool@126.com (X.S.)

[2] Heilongjiang Provincial Key Laboratory of Ecological Restoration and Resource Utilization for Cold Region, School of Life Sciences, Heilongjiang University, Harbin 150080, China

* Correspondence: 0431sfq@163.com

Received: 7 December 2018; Accepted: 1 February 2019; Published: 12 February 2019

Abstract: To reveal the mechanism of salinity stress alleviation by arbuscular mycorrhizal fungi (AMF), we investigated the growth parameter, soluble sugar, soluble protein, and protein abundance pattern of *E. angustifolia* seedlings that were cultured under salinity stress (300 mmol/L NaCl) and inoculated by *Rhizophagus irregularis* (RI). Furthermore, a label-free quantitative proteomics approach was used to reveal the stress-responsive proteins in the leaves of *E. angustifolia*. The result indicates that the abundance of 75 proteins in the leaves was significantly influenced when *E. angustifolia* was inoculated with AMF, which were mainly involved in the metabolism, signal transduction, and reactive oxygen species (ROS) scavenging. Furthermore, we identified chorismate mutase, elongation factor mitochondrial, peptidyl-prolyl cis-trans isomerase, calcium-dependent kinase, glutathione S-transferase, glutathione peroxidase, NADH dehydrogenase, alkaline neutral invertase, peroxidase, and other proteins closely related to the salt tolerance process. The proteomic results indicated that *E. angustifolia* seedlings inoculated with AMF increased the secondary metabolism level of phenylpropane metabolism, enhanced the signal transduction of Ca^{2+} and ROS scavenging ability, promoted the biosynthesis of protein, accelerated the protein folding, and inhibited the degradation of protein under salt stress. Moreover, AMF enhanced the synthesis of ATP and provided sufficient energy for plant cell activity. This study implied that symbiosis of halophytes and AMF has potential as an application for the improvement of saline-alkali soils.

Keywords: arbuscular mycorrhizal fungi; salt stress; *E. angustifolia*; proteomics

1. Introduction

Salt stress is one of the most important abiotic stresses and limiting factors for plant growth and agricultural production. It is a major abiotic stress in the world. Land salinization causes many ecological and environmental problems, such as soil erosion, land desertification, forest and grassland degradation, and biodiversity reduction [1]. At present, along with the increase in soil salinization and secondary salinization, it is estimated that 30% of the arable land in the world will disappear in the next 25 years, and 50% by the middle of the 21st century [2,3]. Hence, the question of how to treat saline alkali soil has attracted widespread attention around the world. In recent years, it was demonstrated that using biological means to treat soil salinization is highly efficient, and environmental and sustainable, thus providing a new breakthrough method for saline alkali land treatment.

E. angustifolia, a member of the family, Elaeagaceae, is a deciduous tree that is widespread in the vast desert and semidesert in the Northwest of China. A few varieties of the species, *E. angustifolia*, can survive in the Gobi, such as the desert and saline, and is called the "treasure tree" locally. It is important to further improve the salt tolerance of *E. angustifolia* using biotechnology under saline-alkali conditions. Arbuscular mycorrhizal fungi (AMF) exist widely in soil and form a mutualism system with most higher plants [4,5]. Plants are subjected to salt stress in the presence of a high salinity in the soil, which reduces the absorption and transport of water, inhibits the metabolic process, and affects nutrient absorption and the cell infiltration balance, resulting in the fragmentation of the horny layer of plants and leakage of the cell membrane. This leads to plant growth retardation. AMF can adapt to the saline soil habitat and survive in a heavy salt environment, indicating that AMF is adaptable to saline soil [6]. Previous studies have shown that symbiosis between AMF and plants under salt stress can promote plant growth and improve plant salt tolerance [7,8]. Therefore, the symbiosis of halophytes with AMF has great potential for the improvement of salt resistance and restoration of saline-alkali land, which has been a major research field globally.

Previous studies on the application of proteomics technology revealed the salt tolerance of plant leaves [9–15]. However, the response mechanism of mycorrhizal plants to salt stress needs to be further revealed. In this study, the symbiosis of AMF *Rhizophagus irregularis* (RI) and the salt-tolerant plant, *E. angustifolia*, was used as a breakthrough point. The stress-responsive proteomics in the leaves of *E. angustifolia* were detected under salt stress conditions. These results will provide more information for the understanding of the function of AMF in the improvement of plant salt tolerance.

2. Results

2.1. Growth of E. angustifolia under Salt Stress and Colonization of AMF in the Plant Roots

As shown in Figure 1A, both mycorrhizal and non-mycorrhizal seedlings grew well in the treatments lacking salt, but the mycorrhizal seedlings' leaves grew stronger than the non-mycorrhizal seedlings; some leaves of the mycorrhizal and non-mycorrhizal seedlings were yellow during salt stress, however, the number of withered leaves of the mycorrhizal plants was significantly less than that of the non-mycorrhizal plants.

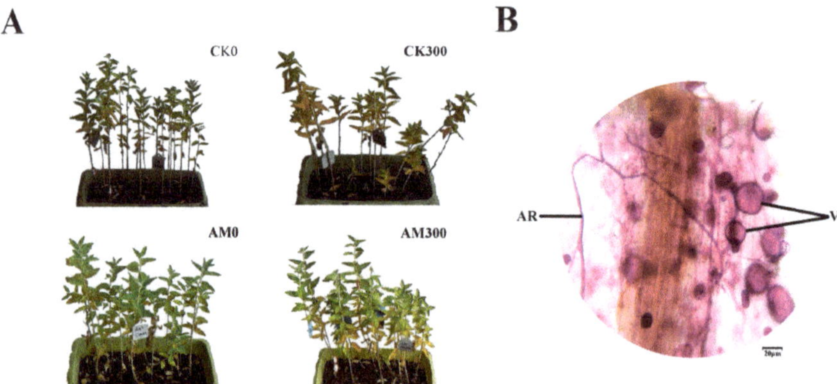

Figure 1. Growth of *E. angustifolia* inoculated AMF (arbuscular mycorrhizal fungi) under salt stress and a representative image of AMF colonization. Note: (**A**) represents the growth contrast in mycorrhizal and non-mycorrhizal *E. angustifolia* after salt stress. (**B**) represents a photomicrograph of the structural colonization of AMF in the root of *R. irregularis*. AM, mycorrhizal; CK, non-mycorrhizal; 0 mmol/L, without salt stress; 300 mmol/L, during salt stress; AR: Arbuscule; V: Vesicles. Scar bar: 20 µm.

The typical AMF morphological structure was detected in inoculated *E. angustifolia* roots, including vesicles and arbuscules (Figure 1B). The maximum AMF colonization percentage of the root reached more than 90% at approximately 100 at approximately 30 days after inoculation. The maximum

AM colonization percentage of the root reached more than 90% at approximately 30 days after salt stress. At the same time, no colonization was found in the non-inoculated seedlings. This result shows that *E. angustifolia* and *R. irregularis* established a vigorous symbiosis.

2.2. Effects of RI and CK on Height, Diameter, and Roots of E. angustifolia under Salt Stress

Salt stress decreased the height, diameter, length, and area, but mycorrhizal seedlings had a greater height, diameter, length, and area than non-mycorrhizal seedlings during salt stress (Table 1). During the 300 mmol/L NaCl treatment, the height, diameter, length, and area of the mycorrhizal seedlings increased by 9.1%, 20.8%, 17.4%, and 35.5%, respectively, compared with those of the non-mycorrhizal seedlings. AMF inoculation significantly enhanced the growth parameter of *E. angustifolia* seedlings in the presence of 300 mmol/L NaCl.

Table 1. Effects of RI and CK on the height, diameter, and roots of *E. angustifolia* under salt stress.

Level of Salinity/(mmol/L)	Different Treatment	Height/(cm)	Diameter/(mm)	Length/(cm)	Area/(cm^2)
0	CK	45.50 ± 0.24c	5.65 ± 0.17b	985.73 ± 27.80b	146.04 ± 5.98c
	RI	49.07 ± 0.54a	6.54 ± 0.20a	1256.7 ± 22.52a	213.07 ± 13.04a
	Significance	**	**	**	**
300	CK	39.57 ± 0.26f	3.99 ± 0.14e	763.64 ± 23.34e	93.68 ± 6.27e
	RI	43.17 ± 0.21de	4.82 ± 0.11d	896.56 ± 42.36bcd	126.96 ± 8.03cd
	Significance	**	**	*	**

RI, mycorrhizal; CK, non-mycorrhizal; 0 mmol/L, without salt stress; 300 mmol/L, during salt stress. Data are means ± SD (standard deviation) of six replicates. The same letter within each column shows no significant differences among treatments ($p < 0.05$). Levels of significance: * $p < 0.05$, ** $p < 0.01$.

2.3. Effects of RI and CK on the Soluble Sugar Content, Soluble Protein Content in the Leaves of E. angustifolia under Salt Stress

As shown in Figure 2A, salinity stress caused a significant decline in the leaf soluble protein content of mycorrhizal and non-mycorrhizal seedlings, while mycorrhizal seedlings had a higher leaf soluble protein content than non-mycorrhizal seedlings during salt stress treatments. As shown in Figure 2B, AMF inoculation significantly promoted the leaf soluble sugar content in the treatments lacking salt. The soluble sugar content in the leaves of mycorrhizal and non-mycorrhizal seedlings increased, but mycorrhizal seedlings had a higher leaf soluble sugar content than that of the non-mycorrhizal seedlings during salt stress.

2.4. Effect of RI on Protein Abundance under Salt Stress

In the CK, AM-NaCl, AM, and AM-NaCl groups, a total of 25,082 peptides and 4349 proteins were identified in the *E. angustifolia* seedlings. The number of proteins identified in the three replicates of each treatment group is shown in Figure 3. Quantifiable proteins were identified in at least two of the three replicates for further analysis. The significance of the differential proteins' abundances was filtered by the ratio > ±2 and p value < 0.05. The numbers of the differentially abundant proteins between treatments (NaCl vs CK, AM vs CK, AM + NaCl vs AM, and AM + NaCl vs NaCl) are shown in Table 2.

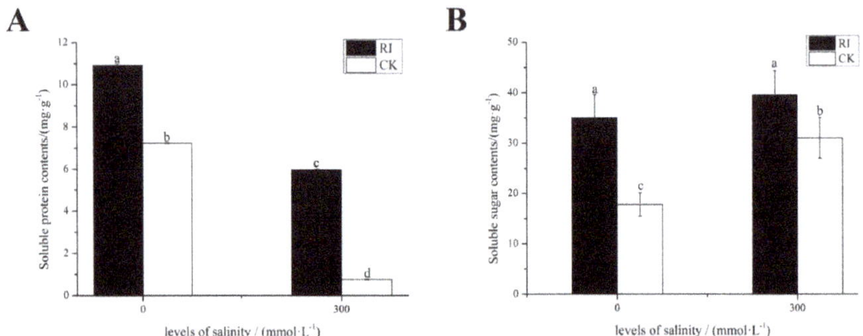

Figure 2. Effects of RI and CK on the soluble sugar content and soluble protein content in the leaves of *E. angustifolia* under salt stress. Note: (**A**) soluble protein, (**B**) soluble sugar. RI, mycorrhizal; CK, non-mycorrhizal; 0 mmol/L, without salt stress; 300 mmol/L, during salt stress. Columns represent the means for three replicates (n = 3). Error bars show the standard error. Columns with different letters indicate significant differences between the treatments at $p < 0.05$.

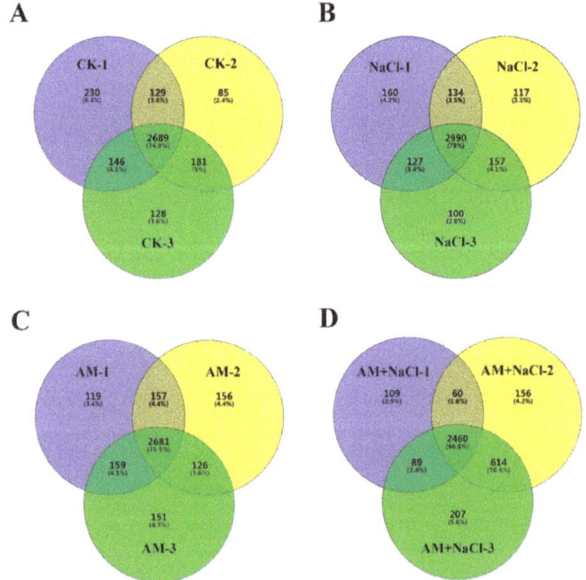

Figure 3. Statistics of the number of proteins identified in each treatment group. Note: (**A**) the number of proteins identified in the three replicates of CK group; (**B**) the number of proteins identified in the three replicates of NaCl group; (**C**) the number of proteins identified in the three replicates of AM group; (**D**) the number of proteins identified in the three replicates of AM + NaCl group.

Table 2. Differentially expressed proteins between treatments.

Treatments	Number of Differential Proteins
NaCl vs CK variation analysis	402 a + 335 b
AM vs CK variation analysis	35 a + 152 b
AM + NaCl vs AM variation analysis	166 a + 226 b
AM + NaCl vs NaCl variation analysis	62 a + 189 b

a: The number of proteins was the satisfied condition (ratio > ±2 and p value < 0.05); b: The number of proteins was only detected at CK or treatments (NaCl, AM, AM + NaCl).

2.5. Functional Classification of Proteins

2.5.1. Salt Tolerance-Related Proteins Induced by Symbiosis

As shown in Table 2, a total of 187 differentially expressed proteins were identified in the AM vs the CK group. The 186 proteins were compared with the AM + NaCl vs the AM group; a total of 112 were found in the AM + NaCl vs the AM group. Among the 112 proteins, four proteins were highly abundant, 25 were newly expressed proteins under salt stress, and 14 proteins were identified as symbiotic salt tolerance related proteins after referring to many academic documents, as shown in Table A1. These 14 proteins are beneficial to the maintenance of AMF-*E. angustifolia* symbiosis and improved the salt tolerance of the plant under salt stress.

2.5.2. Functional Classification of Salt Tolerance-Related Proteins Induced by Symbiosis

Blast2GO (Version 3.3.5) was used to annotate the biological functions of the targeted proteins. These proteins were divided into seven groups (Figure 4), including metabolism, signal transduction, redox, transport, cytoskeleton, protein synthesis, protein folding, and degradation (Table A1). Among them, the proportion of metabolic and protein folding related proteins were the largest, which was 22%. The second category was related to redox, transport, and cytoskeleton, which was 14%. The third category included signal transduction-related and protein synthesis proteins.

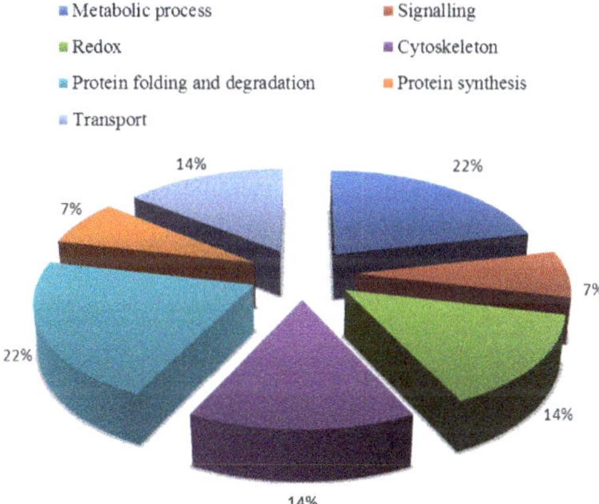

Figure 4. Biological functional classification of salt-tolerant proteins induced by symbiosis.

2.5.3. Salt-Induced Mycorrhizal Protein

As shown in Figure 5, after a comparison of differentially expressed proteins between NaCl vs CK and AM + NaCl vs AM groups by VENNY 2.1 (http://bioinfogp.cnb.csic.es/tools/venny/index.html), 121 out of 392 proteins in the AM + NaCl vs AM group were identical with those in the NaCl vs CK group. It is suggested that these 121 proteins are salt-tolerant related proteins of *E. angustifolia* under salt stress, and are not caused by mycorrhizal. However, 271 proteins not in the NaCl vs CK group were considered to be salt-induced mycorrhizal proteins. The mycorrhizal proteins produced by AMF-*E. angustifolia* symbiosis to adapt to salt stress under salt stimulation. Thus, the salt tolerance of plants can be improved.

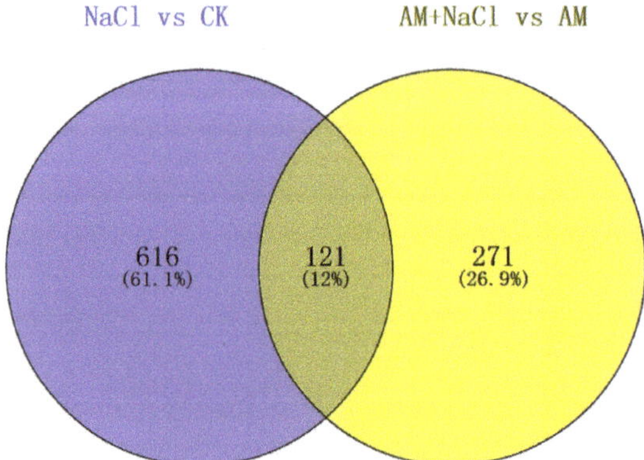

Figure 5. Venn diagram of the protein differential expression between the NaCl/CK group and AM + NaCl/AM group.

2.5.4. Functional Classification of Salt-Induced Mycorrhizal Protein

A total of 57 out of 271 proteins were previously reported to be key proteins for the salt response, as shown in Table A2. These were divided into 10 groups by Blast 2 GO analysis (Figure 6). In these functional groups, the first class (23%) are proteins related to metabolism. There are nine (accounting for 14%) different expressed proteins in the signal transduction pathway, which is the second class. Meanwhile, the other 10 functional groups are also involved in protein redox, protein synthesis, photosynthesis, energy, transport, the cytoskeleton, and stress response related proteins.

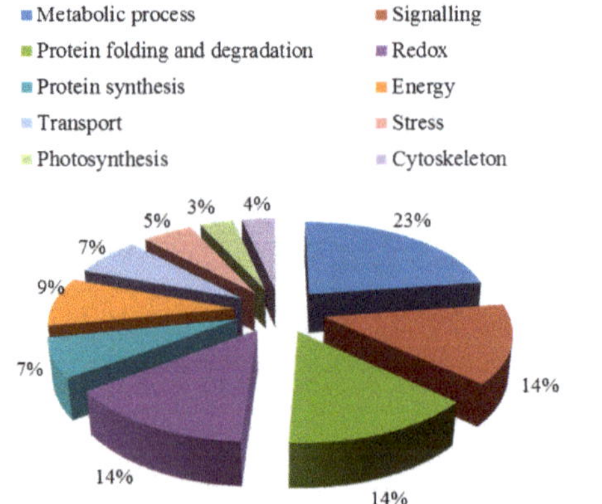

Figure 6. Biological functional classification of salt-induced mycorrhizal protein.

3. Discussion

AMF and salt affected the obvious physical appearance of *E. angustifolia* leaf, and both have a clear interaction. For this reason, we think that AMF has a great influence on salt habitats and that, moreover, salt stress is also a factor in this influence. On the one hand, compared with the AM vs CK group and the AM + NaCl vs AM group, the salt-tolerant proteins caused by the symbiosis could be identified; on the other hand, by comparing the differentially abundant proteins of the NaCl vs CK group and the AM + NaCl vs AM group, the protein for the symbiosis response caused by salt treatment could be identified. Therefore, these two ways were selected to discuss how symbiosis responds to salt stress at the protein level (Figure 7).

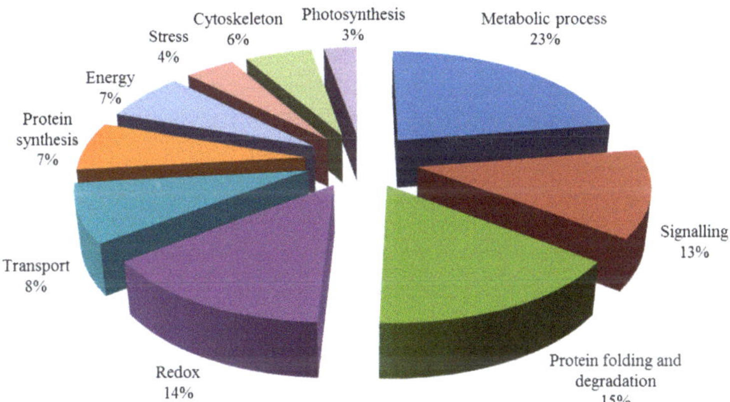

Figure 7. Biological functional classification of differential proteins.

3.1. Proteins Related to Metabolism

Metabolism, consisting of basic physiological processes, maintains a series of activities of a living organism. This research shows the most important factor in the leaves of the *E. angustifolia* after salt stress was the abundance of secondary metabolism-related proteins. In secondary metabolism, especially the metabolism of the protein, phenylpropane and flavonoids increased significantly under salt stress.

Studies have shown that secondary metabolites of plants change during the symbiosis of mycorrhizal fungi and plants [16]. These secondary metabolites play an important role in the symbiotic relationship between plants and mycorrhizal fungi [17]. For example, the content of lignin and soluble phenol in the tomato was increased by inoculation with Arbuscular mycorrhizal fungi. Flavonoids can promote spore germination and mycelium growth and increase the content of flavonoids after mycorrhizal formation [18,19]. In this study, there were two pathways involved in metabolism-related proteins, as shown in Figure 8.

1. In this study, we found a chorismate mutase (CM), which catalyzes the conversion of branched acid to prebenzoic acid. Prebenzoic acid can produce phenolic compounds through the phenylpropane metabolic pathway, including phenylalanine (Phe), tyrosine (Tyr), anthocyanin, and tannin [20,21]. Phenylpropane metabolism is indirectly generated by the shikimic acid pathway. This pathway might play an important role in plant stress defense. We found three proteins that relate to the phenylalanine metabolic pathways, including shikimate O-hydroxycinnamoyltransferase, cinnamyl alcohol dehydrogenase, and caffeoyl-CoA O-methyltransferase. Flavonoids are synthesized by the condensation of phenylpropane derivatives with malonate monoacyl coenzyme A. In addition, shikimic acid O-hydroxyacinnamate transferase and caffeoyl coenzyme A-O-methyltransferase were also involved.

2. In this study, we found that the phosphoribosyltransferase (APT) was up-regulated, which was the first key enzyme in the tryptophan production reaction of o-aminobenzoic acid. Phosphorylribosyltransferase activity of o-aminobenzoic acid was enhanced, which accelerated the synthesis of tryptophan in plants under salt stress. It is well known that tryptophan is a precursor of auxin (indole acetic acid) as well as protein synthesis in plants. Auxin response was also identified in the leaves of *E. angustifolia*. We deduced that these two pathways synthesize auxin to maintain the growth and metabolism of mycorrhizal plants under salt stress.

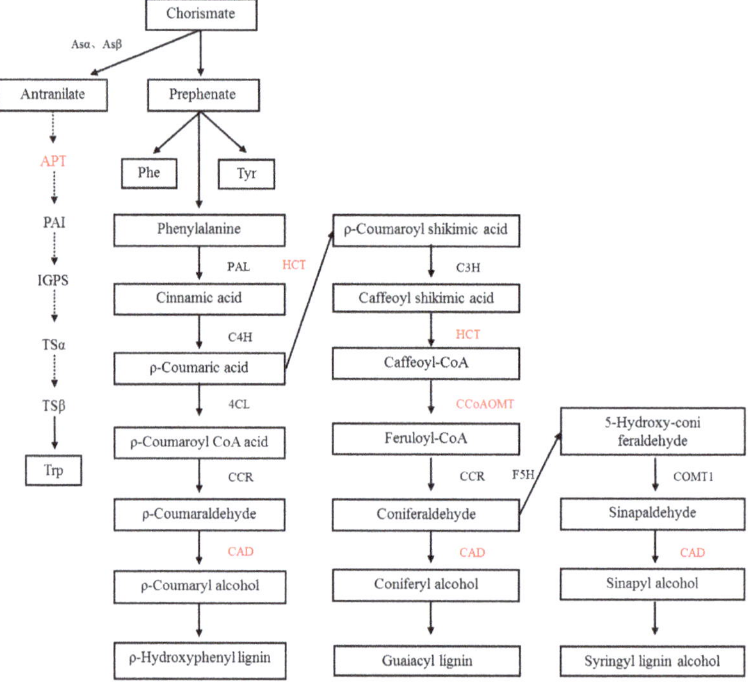

Figure 8. Secondary metabolism of the alleviation of salt stress in mycorrhizal *E. angustifolia* seedlings. Note: red color items represents proteins related to secondary metabolism in this study.CM: chorismate mutase; APT: anthranilate phosphoribosyltransferase; HCT: shikimate O-hydroxycinnamoyltransferase; CAD: cinnamyl alcohol dehydrogenase; CCoAOMT: caffeoyl-CoA O-methyltransferase.

Alkaline neutral invertase is involved in the decomposition of sucrose into glucose and fructose and plays an important role in plant growth and development. The study showed that NaCl and PEG (Polyethylene glycol) stress increased the differential expression of sugar cane *SoNIN1* in the root and leaf [22], and the alkaline neutral transformation enzyme was involved in the stress response. This study found that after inoculation of AMF to the *E. angustifolia* leaf in the treatment of salt stress, alkaline neutral invertase increased its expression, thereby improving the soluble sugar content of the *E. angustifolia* leaf, and providing more sugar for plant metabolism.

3.2. Protein Synthesis, Folding, and Degradation Related Proteins

Protein synthesis plays an important role in plant growth under abiotic stress. We found two proteins (ubiquitin-60S ribosomal L4 and 60S acidic ribosomal P1) were up-regulated in the mycorrhizal plant. Similar ribosomal proteins were also found in studies [23]. One study showed that mitochondrial elongation factors can extend peptide chains more [24], and this was also found to be up-regulated in this study. These proteins related to protein synthesis increase the tolerance of the mycorrhizal plant to salt stress by increasing the expression level.

Proteins can lose their biological functions due to denaturation under adverse conditions. Correct folding and degradation of proteins are key to the maintenance of normal cell functions. Molecular chaperones and folding enzymes play an important role in the maintenance of the natural conformation of proteins, which can help them fold correctly [23,25]. In this study, we found that four folding enzymes, peptide-based prolyl cis-trans isomerases, and four molecular chaperones were up-regulated under salt stress, including peptidyl-prolyl cis-trans isomerase FKBP12, FKBP-type peptidyl-prolyl cis-trans isomerase 5 isoform 1, peptidyl-prolyl cis-trans isomerase CYP18-1, peptidyl-prolyl cis-trans isomerase FKBP62, prefoldin subunit 1, prefoldin subunit 2, heat shock 70 kDa partial, and small heat shock protein 17.3 kDa. Therefore, the up-regulation of these four peptidyl prolyl cis-trans isomerases and four molecular chaperones completes the correct folding of proteins and helps the mycorrhizal *E. angustifolia* resist salt stress.

E3 ubiquitin ligase (UPL3) and ubiquitin-like 1-activating enzyme (E1 B) increased their expression under salt stress. Studies have shown that E3 ubiquitin protein ligase, ubiquitin activation enzyme E1, and ubiquitin ligase all catalyze ubiquitin to their target proteins. Moreover, the specificity of the ubiquitin pathway is controlled by E3, which is pertinent because it can provide the greatest response to environmental stress by regulating the transcription factor of the downstream stress response [26].

3.3. Signal Transduction-Related Proteins

Plants exposed to an adverse environment for a long time will produce a complex system of signal sensing and transduction. It is particularly important to understand how mycorrhizal plants perceive and transmit this stress signal, as well as method of improving the plant salt tolerance under salt stress.

In this study, a series of proteins related to signal transduction were screened out, including G protein, phospholipase C, plasma membrane Ca^{2+} transporter ATPase (PMCA), calcium-dependent protein kinase (CDPKs), calmodulin (CaM), and calcium binding (CML), and were up-regulated. They communicated with each other, thus completing the process of perceiving and transmitting stress signals.

G protein, also known as signal-converting protein or coupling protein, can specifically bind and recognize signals on the cell membrane, and produce intracellular signals with the medial membrane effector enzyme (phospholipase C), which plays a role in signal transduction. After transmembrane conversion, extracellular signals are further transmitted and expanded through Ca^{2+} (second messenger) signals, which eventually lead to a series of physiological and biochemical reactions in cells. Ca^{2+} participates in metabolic pathways that are mainly dependent on changes in the Ca^{2+} concentration [27], and this process is achieved by the various calcium transport systems distributed in the cell organelles and cell membranes [28]. The changes in the Ca^{2+} concentration in the various processes in this study depended on the plasma membrane Ca^{2+} transporter, ATPase (PMCA), which is the main Ca^{2+} transporter, transporting Ca^{2+} to the extracellular space at the cost of hydrolyzing an ATP molecule. Studies have shown that the calcium-dependent protein kinase (CDPKs), calmodulin, calmodulin, and the calcium phosphatase B protein, are involved in cell signal transduction and responses to specific stimuli [29,30]. In this study, the Ca^{2+} concentrations maintained a steady state in the cell wall, mitochondria, endoplasmic reticulum, and vacuole, while the concentration in the cytoplasm was low. After salt stimulation, the Ca^{2+} concentration in the cytoplasm increased significantly. On the one hand, Ca^{2+} directly binds to the calmodulin or the calmodulin, transfers the received signal to the protein kinase and stimulates its activity, or the activity of calcium depends on the protein kinases (CDPKs), which are directly stimulated by the Ca^{2+} signal, and directly participate in and cause subsequent physiological responses through these two processes. These two processes directly participate in and cause subsequent physiological responses.

3.4. ROS Scavenging-Related Proteins

Active oxygen species (ROS) are usually accumulated in plants under salt stress [31], which can be used as signal molecules to activate the plant stress defense system [32]. However, all ROS are very harmful to organisms at high concentrations, leading to cell membrane peroxidation, destruction of enzyme activity, and eventually leading to cell inactivation. Therefore, the removal of ROS can resist salt damage and improve the salt tolerance of plants. In this study, thioredoxin (TRX) and glutathione (GRX) were involved in ROS scavenging as redox enzymes. In these plants, based on the TRX redox system that runs in various kinds of organelles, including the cytoplasm, mitochondria, and chloroplasts [33–36], it was shown that TRX plays a key role in plant redox regulation. At present, many GRXs have been identified in different plants. For example, over-expression of the tomato

SLGRX1 gene can enhance the plant's resistance to oxidative stress, drought resistance, and salt pollution in *Arabidopsis*. In contrast, silencing this gene leads to an increased insensitivity to stress in the tomato [37]. Meanwhile, the gene silencing increases the membrane lipid peroxide level and the accumulation of ROS and suppresses the activity of antioxidant enzymes under high-temperature stress. This suggests that GRXs are involved in the regulation of the redox status and the response to high-temperature stress.

In our study, three glutathione s-transferase (GST) and two glutathione peroxidase (GPX) were found. Studies have reported that to prevent ROS damage, the amount of glutathione transferase (GST) in plants will increase, which can catalyze the removal of ROS in plants [38]. Glutathione peroxidase (GPX) is a sulfur-containing peroxidase, which can remove hydrogen peroxide, organic hydroperoxides, and lipid peroxides from the organism, and block further damage of ROS to the organism [39,40]. It has been shown that chloroplast glutathione transferase plays a very important role in the resistance of low concentrations in Stargrass seedlings [41]. When plants are exposed to high salt stress, the expression activity of GPX will be enhanced and the tolerance of plants to salt stress will be enhanced [42–44]. In addition, excessive expression of GST/GPX in transgenic tobacco under salt stress conditions for seed germination and seedling growth were improved more than for the control group, suggesting that GST/GPX increases the ROS removal in plants, protects plants from oxidative damage, and maintains the growth of plants [45,46]. These results suggest that the GST/GPX system is a key factor in the improvement of the salt tolerance of plants through its ROS scavenging ability under salt stress. Peroxidase (POD) is one of the key enzymes in plants under stress conditions in the enzymatic defense system. It cooperates with superoxide dismutase and catalase to remove excess ROS to improve plant resistance.

"Arbuscular Mycorrhizal Symbiosis Modulates Antioxidant Response and Ion Distribution in Salt-Stressed *Elaeagnus angustifolia* Seedlings"- Changes of SOD, CAT, POD, and APX activities in the leaves of *E. angustifolia* inoculated with AMF and those of non-inoculated *E. angustifolia* under 0 and 300 salt concentrations were analyzed [47].

As shown in Figure 9, mycorrhizal seedlings had a higher leaf SOD (superoxide dismutase), CAT (Catalase), POD (Peroxidase), and APX (ascorbate peroxidase) activity than that of the non-mycorrhizal seedlings during salt stress. The mycorrhization of the plants led to increased levels of leaf antioxidant defense systems during stress conditions. The results obtained at the protein level are consistent with the results of the apparent physiological indicators in this study. Therefore, we can conclude that TRX/GRX, GST/GPX, and POD are up-regulated in mycorrhizal plants under salt stress, which results in a significant increase of SOD, CAT, POD, and APX activities in plant leaves, thus improving the salt tolerance of mycorrhizal plants.

3.5. Energy-Related Protein

Mitochondria are the main site of oxidative phosphorylation and the synthesis of adenosine triphosphate (ATP) in cells, and they provide energy for cell activity. NADH dehydrogenase, cytochrome C oxidase, iron sulfur protein NADH dehydrogenase, and ATP synthase up-regulation were found in this study as complex compounds I, IV, and V participated in the mitochondrial electron transport. The mitochondrial electron transport chain is also called the respiratory chain, and the cells transfer electrons obtained from the oxidation of macromolecules via I, II, III, and IV complexes and the energy produced by the electron transfer maintains the proton gradient of the mitochondrial inner membrane, which is utilized by complex V (ATP synthase) to catalyze the formation of ATP. In this study, proteins involved in providing cellular energy were up-regulated to provide energy for mycorrhizal plants to resist salt stress.

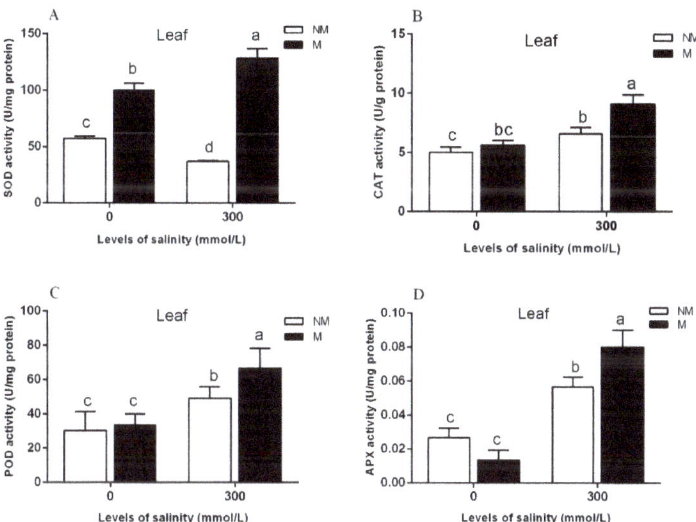

Figure 9. Effects of AMF inoculation on the superoxide dismutase (SOD) (**A**), catalase (CAT) (**B**), peroxidase (POD) (**C**), and ascorbate peroxidase (APX) (**D**) activities in the leaves during different salt conditions. M, mycorrhizal; NM, non-mycorrhizal; 0 mmol/L, without salt stress; 300 mmol/L, during salt stress. Columns represent the means for three plants ($n = 3$). Error bars show the standard error. Columns with different letters indicate significant differences between treatments at $p < 0.05$. Note: cited from "Arbuscular Mycorrhizal Symbiosis Modulates Antioxidant Response and Ion Distribution in Salt-Stressed *Elaeagnus angustifolia* Seedlings" [47].

3.6. Photosynthesis-Related Proteins

AMF and plant symbiosis can promote plant growth by increasing the photosynthesis of plants under salt stress, thereby increasing the ability of plants to resist the salt stress [48–53]. In this study, a photosystem II D1 precursor processing PSB27 was found to have upregulated expression. Photosystem II D1 protein is the core protein of photosystem II, which is synthesized in light and is injured by light or other adverse factors. Repairing D1 protein damage repairs PS II, maintaining its dynamic balance through continuous turnover [54,55]. In addition, the reactive center proteins, D1 and D2, are the binding sites of the auxiliary factors for all redox activities, which are necessary for PSII electron transport [54,56]. Thus, the photosystem II D1 protein plays an important role in the maintenance of the stability of the PSII reaction center. The previously mentioned thioredoxin (TRX), which is electronically reduced by the ferredoxin/thioredoxin system (Fd/TRX) in the chloroplasts of plants, is involved in the electron transport of plant photosynthesis. Guo [57] and other studies found that GRXs gene-silenced plants resulted in a significant decrease in ETR of photosystem II, and a significant increase in NPQ in varying degrees. Meanwhile, GRXs gene silencing results in a significant decrease in the maximum quantum efficiency (Fv/Fm) and the actual electronic yield (Φ PSII) under high temperature stress. This is consistent with the results of the chlorophyll fluorescence parameters in the early stage of this experiment [58], in which Fv/Fm, PSII, NPQ, and ETR in the leaves of *E. angustifolia* inoculated with AMF were higher than those of non-mycorrhizal plants under salt stress. In this study, TRX and GRX were up-regulated at the protein level, and Fv/Fm, PSII, NPQ, and ETR of mycorrhizal plants were significantly increased at the apparent level, thus alleviating the damage of salt stress to plants.

3.7. Network Interaction Predictions Based on Differential Expression

The protein–protein interaction information of the studied proteins was retrieved from the IntAct molecular interaction database (http://www.ebi.ac.uk/intact/) by their gene symbols or STRING software (http://string-db.org/). These interactions included the direct physical proteins and the indirect proteins correlated with indirect functions as shown in Figure 10. The results showed that ubiquitin-60S ribosomal was the most correlated protein, directly or indirectly, with connections to proteins, such as 60S acidic ribosomal, elongation factor mitochondrial, ubiquitin-activating enzyme E1, and E3 ubiquitin-ligase. Additionally, most of the target proteins were associated with ubiquitin-60S ribosomal and were in the protein synthesis, folding, and degradation pathway. Thus, this pathway might play an important role in stress tolerance. Ubiquitin-60S ribosomal interacts with various peroxidases, thereby participating in *E. angustifolia* protein synthesis, folding, and degradation biosynthetic processes, and modulates stress responses. These findings showed that the salt stress response is a multi-factor process involving many protein interactions.

The hypothesized mechanism of the improved salt tolerance of the mycorrhizal plant was revealed from the proteome.

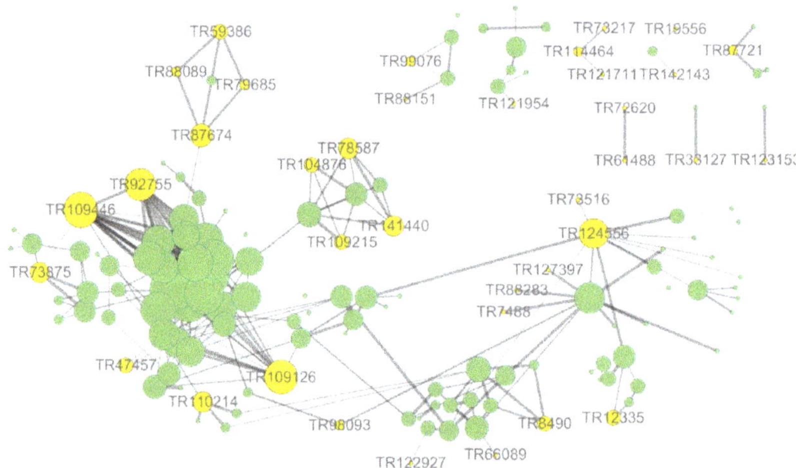

Figure 10. Protein—interaction network interactions for differentially expressed proteins. Note: shows the integrated network for all the differentially expressed proteins; yellow circle represents salt tolerance-related proteins induced by symbiosis and salt-induced mycorrhizal protein in this study; green circle represents other differentially expressed proteins in this study, respectively. The sizes represent the abundance of differentially expressed proteins.

In the current study, based on proteomic data analysis, it is suggested that AMF can improve the salt tolerance of *E. angustifolia* seedlings (Figure 11).

1. AMF accelerates the secondary metabolism of plants, mainly phenylpropanoid metabolism (shikimate O-hydroxycinnamoyltransferase, cinnamyl alcohol dehydrogenase, and caffeoyl-CoA O-methyltransferase), reducing salt damage to plants.

2. AMF enhances the signal transduction of the second messenger Ca^{2+} (G protein, phospholipase C, plasma membrane Ca^{2+} transporter ATPase (PMCA), calcium binding (CML), calcium-dependent kinases (CDPKs), and calmodulin (CaM), increasing the speed of sensing and transmitting of stress signals, allowing plants to follow up.

3. Among a variety of antioxidant pathways (TRX/GRX, GST/GPX, POD), AMF enhances the antioxidant capacity of plants by increasing ROS clearance.

4. AMF promotes protein biosynthesis, speeding up protein folding, and inhibiting protein degradation (ubiquitin-60S ribosomal, 60S acidic ribosomal, elongation factor mitochondrial, ubiquitin activating enzyme E1, and E3 ubiquitin- ligase).

5. In the chloroplast, AMF maintains the PSII reaction centre conformation stability and speeds up photosynthetic electron transport (TRX/GRX, photosystem II D1 precursor processing PSB27-chloroplastic); in mitochondria, AMF enhances the synthesis of ATP (NADH dehydrogenase, cytochrome C oxidase, iron sulfur protein NADH dehydrogenase, and ATP synthase), providing sufficient energy for cellular activities.

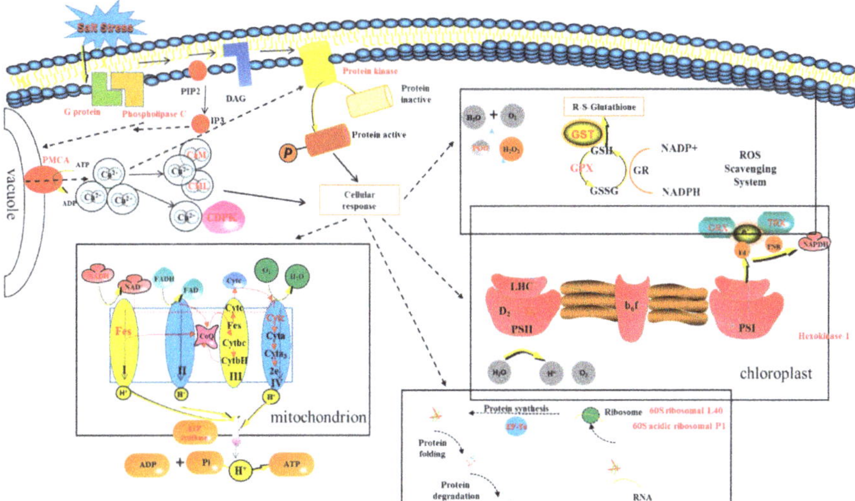

Figure 11. Mechanisms of mycorrhizal *E. angustifolia* seedlings to alleviate salt stress. Note: Protein folding and degradation: peptidyl-prolyl cis-trans isomerase FKBP12; FKBP-type peptidyl-prolyl cis-trans isomerase 5 isoform 1; peptidyl-prolyl cis-trans isomerase CYP18-1; peptidyl-prolyl cis-trans isomerase FKBP62; prefoldin subunit 1; prefoldin subunit 2; Heat shock 70 kDa partial; small heat shock protein 17.3 kDa; E3 ubiquitin ligase (UPL3); ubiquitin-like 1-activating enzyme (E1 B). PMCA: calcium-transporting ATPase; CaM: calmodulin; CML: calcium-binding CML20; CDPK: calcium-dependent kinase; POD: peroxidase; GST: glutathione S-transferase; GPX: glutathione peroxidase; GRX:glutaredoxin; TRX:thioredoxin; D1:photosystem II D1 precursor processing PSB27; EF-Tu: elongation factor mitochondrial; NADH: NADH dehydrogenase [ubiquinone] 1 beta subcomplex subunit 7; Fes: NADH dehydrogenase [ubiquinone] iron-sulfur mitochondrial; Cytc: cytochrome c oxidase subunit mitochondrial.

4. Materials and Methods

4.1. Experimental Materials and Salinity Treatments

E. angustifolia seeds were provided by Heilongjiang Jinxiudadi Biological Engineering Co. Ltd. (Haerbin, China). AM fungus *Rhizophagus irregularis* (RI) was propagated and preserved by Heilongjiang Provincial Key Laboratory of Ecological Restoration. The mycorrhizal inoculum containing approximately 25–30 AM propagules/g consisted of soil, spores, mycelia, and infected root fragments. The soil was collected from the Forest Botanical Garden of Heilongjiang Province, sieved (5 mm), and mixed with vermiculite (3:1, soil:vermiculite, *v/v*). The soil medium was pre-autoclaved at 121 °C for 2 h.

There were four different treatments as follows: *E. angustifolia* inoculated *Rhizophagus irregularis* without salt stress, *E. angustifolia* inoculated *Rhizophagus irregularis* under salt stress (300 mmol/L

NaCl), *E. angustifolia* non-inoculated *Rhizophagus irregularis* without salt stress, and *E. angustifolia* non-inoculated *Rhizophagus irregularis* under salt stress (300 mmol/L NaCl). Each treatment had six replicates. The inoculated dosage of mycorrhizal inoculum per pot was 10 g. The same amount of inactive mycorrhizal inoculum (121 °C, 2 h) was used in non-inoculated treatments. The 300 mmol/L NaCl was added into the pots after 4 months of being inoculated with mycorrhizal inoculum as described [47]. Seedlings continued to be cultivated for 30 days. The experiment was carried out under outdoor natural conditions.

Three seedlings were randomly selected from each pot and 6–7 round leaves were removed from each seedling. Leaves from two pots of the same treatment were combined as one sample. There were three biological repeats per treatment. The proteomic, AMF colonization, and growth parameter of the samples was detected.

4.2. Measurement of AMF Colonization and Growth Parameter

The AMF colonization rate of *E. angustifolia* was determined by the acid fuchsin staining method [59]. The height and diameter of *E. angustifolia* were measured by a vernier caliper. The root area and root length of *E. angustifolia* were measured by a root scanner.

4.3. Measurement of Soluble Sugar Content, Soluble Protein Content in the Leaves of E. angustifolia under Salt Stress

To assess the degree of stress, the contents of sugar and soluble protein were detected using the anthrone colorimetric method and coomassie brilliant blue G-250 method [60,61]. Samples were taken during the same growth period and from the same leaf positions.

4.4. Extraction and Quantification of Proteins

The leaves of all samples were frozen in liquid nitrogen and ground with a pestle and mortar. TCA/acetone (1:9) was added to the powder (1:5, *v/v*) and mixed by vortex. The mixture was placed at –20 °C for 4 h, and centrifuged at $6000\times g$ for 40 min at 4 °C. The supernatant was discarded. The pre-cooling acetone was added into the pellet and washed three times. The precipitation was air dried. SDT buffer (1:30, *v/v*) was added to 20–30 mg of powder, mixed, and boiled for 5 min. The supernatant was filtered with 0.22 μm filters. The filtrate was quantified with the BCA Protein Assay Kit (P0012, Beyotime, Shanghai, China).

4.5. FASP Digestion

Proteins of each sample (200 μg) were mixed with 200 μL UA buffer, and the mixture was filtered by a ultrafiltration centrifugal tube (10 kD) at $14,000\times g$ for 15 min, and the pellet was re-suspended and filtered. The 100 μL iodoacetamide (100 mM IAA in UA buffer) was added into each sample and incubated for 30 min in darkness. The filters were washed with 100 μL UA buffer three times and subsequently with 100 μL 25 mM NH_4HCO_3 buffer twice. The protein suspensions were digested with 4 μg trypsin in 40 μL 25 mM NH_4HCO_3 buffer overnight at 37 °C. The collection of the resulting peptides with a new collector and peptides were desalted on C18 Cartridges, freeze-dried, and reconstituted in 40 μL of 0.1% (*v/v*) formic acid. The peptide content was estimated by UV light spectral density at 280 nm.

4.6. LC-MS/MS Analysis

Proteins were separated by using an Easy nLC HPLC liquid phase system with an increasing flow rate. The chromatographic column was balanced by 95% A solution (0.1% formic acid aqueous solution). The sample was injected onto the C18 column (Thermo Scientific Acclaim PepMap100, 100 μm × 2 cm, nanoViper C18) by an automatic sample injector and separated by C18-a2 analytical column (Thermo scientific EASY column, 10 cm, ID75 μm, 3 μm, C18-A) at a flow rate of 300 mL/min. Solution B (0.1% formic acid acetonitrile aqueous solution) was then used for gradient elution.

The samples separated by chromatography were analyzed with a Q-Exactive mass spectrometer. The analysis time was 120 min, the detection method was the positive-ion mode, the parent-ion scanning range was 300–1800 m/z, the resolution of a first-order mass spectrometer was 70,000 at 200 m/z, the AGC target was 3e6, the first level maximum IT was 10 ms, the number of scan ranges was 1, and the dynamic exclusion was 40.0 s.

4.7. Database Search and Protein Quantification

The database used was P16440_Unigene.fasta.transdecoder_73797_20161212.fasta (Sequence 73797, self-building). Maxquant software 1.3.0.5 [62] was used to analyze the protein qualitatively and quantitatively in the original raw file. The maxquant software parameter table is shown in Table 3.

Table 3. Maxquant software parameter table.

Item	Value
Main search ppm	6
Missed cleavage	2
MS/MS tolerance ppm	20
De-Isotopic	True
Enzyme	Trypsin
Database	P16440_Unigene.fasta.transdecoder_73797_20161212.fasta
Fixed modification	Carbamidomethyl (C)
Variable modification	Oxidation(M), Acetyl (Protein N-term)
Decoy database pattern	reverse
LFQ	True
LFQ min. ratio count	1
Match between runs	2min
Peptide FDR	0.01
Protein FDR	0.01

For the proteins identified by mass spectrometry in the original data, differentially expressed proteins and differentially expressed proteins were screened by the screening criteria of Ratio > $+/-2$ and p value < 0.05. The quantified protein sequence information was extracted in batches from the UniProtKB database (version number: 201701).

4.8. Protein GO Functional Annotation and KEGG Pathway Annotation

Blast 2 GO was used to annotate the functions of the targeted proteins [63]. KASS software was used for pathway analysis. The target protein sequences were classified by KO compared to the KEGG GENES database, and then the pathway information involved in the target protein sequence was automatically acquired according to the KO classification.

4.9. Protein—Protein Interact Network (PPI)

The protein—protein interaction information of the studied proteins was retrieved from the IntAct molecular interaction database (http://www.ebi.ac.uk/intact/) by their gene symbols or STRING software (http://string-db.org/). The results were downloaded in the XGMML format and imported into Cytoscape5 software (http://www.cytoscape.org/, version 3.2.1) to visualize and further analyze the functional protein-protein interaction networks. Furthermore, the degree of each protein was calculated to evaluate the importance of the protein in the PPI network.

5. Conclusions

E. angustifolia seedlings' growth was significantly inhibited by salt stress, and growth was improved in mycorrhizal symbionts. Furthermore, mycorrhizal *E. angustifolia* had a higher leaf soluble sugar and soluble protein content than that of the non-mycorrhizal seedlings during salt stress. Additionally, it was found that AMF inoculated *E. angustifolia* seedlings increased secondary metabolism, enhanced Ca^{2+} signal transduction and ROS scavenging capacity, promoted protein biosynthesis, accelerated protein folding, and inhibited protein degradation compared with non-inoculated plants under salt stress. In addition, AMF maintained the conformation stability of the PS II reaction center, accelerated the photosynthetic electron transfer, enhanced ATP synthesis, and provided sufficient energy for cell activity. Overall, these findings show that AMF played an important role in easing salt stress in plants and contributed to saline alkali soil remediation.

Author Contributions: T.J. and J.W. contributed equally for this article. T.J. and J.W. did the experiment and analyzed data and wrote this paper. W.C. participated to analyze the data. X.F. and X.S. participated to do this experiment. F.S. designed this experiment. All authors read and approved the final manuscript.

Funding: This research was funded by the National Nature Science Foundation of China (No. 31570635), the Special Fund for Forest Scientific Research in the Public Welfare, (No. 201504409), Heilongjiang Natural Fund Team Project: Study on Microbial Control Technology of Agricultural Non-point Source Pollution in Cold Region, 100 Discipline Young Scholars Project of the Heilongjiang University.

Acknowledgments: We thank Shanghai Applied Protein Technology Co. Ltd. for the help with Label-free sequencing. This work was supported by Key Laboratory of Microbiology, College of Heilongjiang Province.

Conflicts of Interest: The authors declare no conflict of interest.

Appendix A

Table A1. The identification and quantitative analysis of the protein in leaves of *E. angustifolia* with AMF under salt stress (salt-tolerance-related proteins induced by symbiosis).

TR Number	Protein Name	Unique Peptides	Sequence Coverage [%]	Mol. Weight [kDa]	AM/CK	AM+NaCl/AM	Fold Change	t Test p Value
Metabolic process								
115007	Auxin response 4	2	5.7	52.521	under	over		
142773	purple acid phosphatase 2-like [64,65]	2	11.2	53.569	under	over		
71893	probable alkaline neutral invertase B [22]	2	4	65.249	under	over		
Protein folding and degradation								
101573	AAA-ATPase At5g17760-like [66]	2	8	58.706	under	over		
12335	Heat shock 70 kDa partial [67]	1	55.3	22.844	under	over		
143526	small heat shock protein 17.3 kDa	2	1.1	81.397	over	over	2.32	0.032
Protein synthesis								
109446	ubiquitin-60S ribosomal L40 [23]	1	41.4	14.643	under	over		
Redox								
127397	thioredoxin 1 [68,69]	1	10.5	20.973	under	over		
130845	glutaredoxin 3 [37]	2	24	13.046	under	over		
Cytoskeleton								
114464	actin-related 7 [70]	1	4.7	39.154	under	over		
Transport								
119356	transmembrane 147 [71]	2	10.5	25.299	under	over		
88151	vesicle transport v-SNARE 13 [72]	2	15.8	24.96	under	over		
Signalling								
136541	serine/threonine-protein kinase PRP4 [73,74]	1	5.1	28.709	under	over		

Table A2. The identification and quantitative analysis of the protein in leaves of *E. angustifolia* with AMF under salt stress (salt-induced mycorrhizal protein).

TR Number	Protein Name	Unique Peptides	Sequence Coverage [%]	Mol. Weight [kDa]	AM+NaCl/AM	Fold Change	*t* Test *p* Value
Metabolic process							
103310	beta-glucosidase 40 [75]	3	7.7	60.254	over		
142616	beta-glucosidase 42 [75]	7	18.8	57.009	over	2.16	0.037
105849	1,2-dihydroxy-3-keto-5-methylthiopentene dioxygenase [76]	6	32.2	23.919	over	2.42	0.014
121954	anthocyanidin reductase [77]	2	5.2	37.122	over		
123153	shikimate O-hydroxycinnamoyltransferase-like	2	5.5	47.978	over		
124556	acetyl-CoA carboxylase [78]	5	4.8	155.54	over		
47457	chorismate mutase chloroplastic-like [20,21]	1	8.7	14.428	over		
66089	probable cinnamyl alcohol dehydrogenase 1 [20,21]	9	31.5	38.574	over	2.22	0.006
99076	caffeoyl-CoA O-methyltransferase [20,21]	1	1.5	72.473	over		
73516	phosphoglucan phosphatase chloroplastic isoform X1	1	2	62.123	over		
98093	delta-1-pyrroline-5-carboxylate dehydrogenase mitochondrial-like [2]	9	35.5	37.633	over	2.07	0.001
99156	anthranilate phosphoribosyltransferase [20,21]	2	3.3	83.928	over	2.06	0.016
120341	UDP-glucose 4-epimerase [79]	15	49.7	43.428	over	2.68	0.007
Signalling							
87721	phospholipase C	6	14	59.562	over		
114971	calcium-dependent kinase 29 [30]	1	6	51.414	over		
142143	calcium-transporting ATPase plasma membrane-type [28]	1	11.4	15.308	over		
155255	glycine-rich 2-like [80]	2	32.8	11.363	over	2.17	0.018
19556	calcium-transporting ATPase plasma membrane-type [28]	1	8.8	18.439	over		
67342	probable calcium-binding CML20 [29]	3	15.8	22.997	over		
61926	calmodulin-7 [30]	7	46.7	21.009	over	2.7	0.046
24540	guanylate-binding family	1	7.8	13.089	over		
Protein folding and degradation							
110214	E3 ubiquitin- ligase UPL3 [26]	2	2.3	196.04	over		
122927	peptidyl-prolyl cis-trans isomerase FKBP12 [23,25]	2	21.4	12.076	over		
74495	FKBP-type peptidyl-prolyl cis-trans isomerase 5 isoform 1 [23,25]	1	8.9	33.514	over		
61488	prefoldin subunit 2 [23,25]	3	31.1	16.223	over		
72620	prefoldin subunit 1 [23,25]	3	26.4	14.922	over		
91034	peptidyl-prolyl cis-trans isomerase CYP18-1 [23,25]	1	8.8	17.474	over		
93048	peptidyl-prolyl cis-trans isomerase FKBP62-like [23,25]	1	20.8	13.805	over		
73875	ubiquitin-like 1-activating enzyme E1 B [26]	6	11.4	74.427	over	2.64	0.000
Protein synthesis							
92755	60S acidic ribosomal P1	2	36.9	14.968	over	2.73	0.013
33127	exportin-2 [81]	1	14.1	11.097	over		

Table A2. *Cont.*

TR Number	Protein Name	Unique Peptides	Sequence Coverage [%]	Mol. Weight [kDa]	AM+NaCl/AM	Fold Change	t Test p Value
106351	nuclear pore complex NUP98A [82,83]	1	3.8	50.09	over		
109126	elongation factor mitochondrial [24]	2	5.4	44.427	over		
Redox							
101431	monothiol glutaredoxin-mitochondrial	3	15.6	19.243	over	2.27	0.010
12964	peroxidase 16-like	1	7.6	13.952	over		
29124	probable linoleate 9S-lipoxygenase 5	2	19.3	17.097	over	3.06	0.036
59386	glutathione S-transferase L3-like [39,40]	11	42.6	29.735	over	2.65	0.003
79685	probable glutathione peroxidase 2 [45,46]	6	33.2	21.733	over	2.35	0.017
97566	2,4-D inducible glutathione S-transferase [39,40]	4	22.7	25.559	over		
87674	glutathione S-transferase T1 [39,40]	3	16.8	27.036	over		
88089	probable glutathione peroxidase 8 [45,46]	3	17	19.394	over	2.43	0.006
Energy							
109215	ATP synthase subunit mitochondrial	4	60.6	14.69	over	3.36	0.015
141440	NADH dehydrogenase [ubiquinone] 1 beta subcomplex subunit 7	5	40.6	15.264	over	2.59	0.010
28754	NADH dehydrogenase [ubiquinone] 1 beta subcomplex subunit 7	1	7.9	11.862	over		
78587	NADH dehydrogenase [ubiquinone] iron-sulfur mitochondrial	2	9.4	18.933	over	2.19	0.029
104876	cytochrome c oxidase subunit mitochondrial-like	2	30.3	11.185	over	2.27	0.050
Transport							
88283	aquaporin PIP2-1-like [84]	1	4.9	30.753	over		
7488	aquaporin PIP1-4 [84]	2	11.9	21.091	over		
87290	mechanosensitive ion channel 1 [85]	2	2.5	85.926	over		
88244	vacuolar sorting-associated 2 homolog 1 [86]	2	6.7	25.106	over		
Stress							
109624	GDSL esterase lipase 1 [87]	3	13.2	42.914	over		
93161	GDSL esterase lipase At3g27950-like [88]	2	8.9	30.265	over	2.07	0.029
58645	stress response NST1-like	1	2.8	35.36	over		
Photosynthesis							
71168	photosystem II D1 precursor processing PSB27-chloroplastic-like [54–56]	2	12.3	23.507	over		
8490	hexokinase-3 isoform X1 [89]	1	14.9	16.704	over	2.5	0.002
Cytoskeleton							
73217	profilin	2	13.1	19.413	over	2.45	0.039
121711	actin-depolymerizing factor 1 [90]	2	26.6	16.042	over		

References

1. Sun, Y.F.; Song, F.Q.; Chang, W.; Fan, X.X. The effects of the AMF on the growth and physiology of the *E. angustifolia* seedlings under the stress of saline. *Sci. Silvae Sin.* **2016**, *52*, 18–27.
2. Porcel, R.; Aroca, R.; Ruiz-Lozano, J.M. Salinity stress alleviation using arbuscular mycorrhizal fungi. *Agron. Sustain. Dev.* **2012**, *32*, 181–200. [CrossRef]
3. Kapoor, R.; Evelin, H.; Mathur, P.; Giri, B. *Plant Acclimation to Environmental Stress*; Springer: New York, NY, USA, 2012; pp. 359–401.
4. Li, X.L.; Feng, G. *Arbuscular Mycorrhizal Ecological Physiology*; Sino-Culture Press: Beijing, China, 2001.
5. Guo, X.Z.; Bi, G.C. *Mycorrhiza and Its Application Technology*; China Forestry Publishing House: Beijing, China, 1989.
6. Li, G.Z.; Chen, Z.C.; Li, X.C.; Sheng, J.D.; Huang, C.F.; Jin, J.X. Spatial distribution of arbuscular mycorrhizal fungi in the roots of *Phragmites australis* in Xinjiang. *Pr. Sci.* **2016**, *33*, 1267–1274.
7. Ruiz-Lozano, J.M.; Porcel, R.; Azcon, C.; Aroca, R. Regulationby arbuscular mycorrhizae of the integrated physiological responseto salinity in plants: New challenges in physiological andmolecular studies. *J. Exp. Bot.* **2012**, *5*, 1–12.
8. Kumar, A.; Sharma, S.; Mishra, S. Influence ofarbuscularmycorrhizal (AM) fungi and salinity on seedling growth, soluteaccumulation, and mycorrhizal dependency of *Jatrophacurcas* L. *J. Plant Growth Regul.* **2010**, *29*, 297–306. [CrossRef]
9. Li, M.N.; Zhang, K.; Long, G.; Sun, Y.; Kang, J.M.; Zhang, T.J.; Cao, S.H. iTRAQ-based comparative proteomic analysis reveals tissue-specific and novel early-stage molecular mechanisms of salt stress response in *Carex rigescens. Environ. Exp. Bot.* **2017**, *143*, 99–114. [CrossRef]
10. Ma, Q.L.; Kang, J.M.; Long, R.C.; Cui, Y.G.; Zhang, T.J.; Xiong, J.B.; Yang, Q.C.; Sun, Y. Proteomic analysis of salt and osmotic-drought stress in alfalfaseedlings. *J. Integr. Agric.* **2016**, *15*, 2266–2278. [CrossRef]
11. Yu, J.J.; Chen, S.X.; Dai, S.J. Physiological and proteomic analysis of salinity tolerance in *Puccinellia tenuiflora. J. Proteome Res.* **2011**, *10*, 3852–3870. [CrossRef] [PubMed]
12. Banaei-Asl, F.; Farajzadeh, D.; Bandehagh, A.; Komatsu, S. Comprehensive proteomic analysis of canola leaf inoculated with a plant growth-promoting bacterium, *Pseudomonas fluorescens*, under salt stress. *Biochim. Biophys. Acta* **2016**, *1864*, 1222–1236. [CrossRef] [PubMed]
13. Yin, Y.Q.; Yang, R.Q.; Han, Y.B.; Gu, Z.X. Comparative proteomic and physiological analysesreveal the protective effect of exogenous calcium on the germinating soybean response to salt stress. *J. Proteom.* **2015**, *113*, 110–126. [CrossRef] [PubMed]
14. Niu, X.P.; Xu, J.T.; Chen, T.; Tao, A.; Qi, J.M. Proteomic changes in kenaf (*Hibiscus cannabinus* L.) leaves under saltstress. *Ind. Crop. Prod.* **2016**, *91*, 255–263. [CrossRef]
15. Lv, D.W.; Zhu, G.R.; Zhu, D.; Bian, Y.W.; Liang, X.N.; Cheng, Z.W.; Deng, X.; Yan, Y.M. Proteomic and phosphoproteomic analysis reveals the response and defense mechanism in leaves of diploid wheat *T. monococcum* under salts tress and recovery. *J. Proteom.* **2016**, *143*, 93–105. [CrossRef] [PubMed]
16. Morandi, D. Occurrence of Phytoalexins and Phenolic com Pounds in endomyeorrhizalinteractions, and their Potential role in biological control. *Plant Soil* **1996**, *185*, 241–251. [CrossRef]
17. Akiyama, K.; Matsuoka, H.; Hayashi, H. Isolation and identification of a phosphate deficieny-induced C-glycosylflavonoid that stimulates arbuscular mycorrhiza formation in melon roots. *Mol. Plant-Microbe Interact.* **2002**, *15*, 334–340. [CrossRef] [PubMed]
18. Zhao, X.; Yan, X.F. Effects of arbuscular mycorrhizal fungi on plant secondary metabolism. *Chin. J. Plant Ecol.* **2006**, *30*, 514–521.
19. Larose, G.; Chnevert, R.; Moutoglis, P.; Gagne, S.; Piche, Y.; Vietheilig, H. Flavonoid Ievels in roots of Medicagosativa are modulated by the developmental stage of the symbiosis and the root colonizing arbuseular mycorrhizal fungus. *J. Plant Physiol.* **2002**, *159*, 1329–1339. [CrossRef]
20. Qian, X.; Yu, K.J.; Yuan, G.S.; Yang, H.Y.; Pan, Q.H.; Zhu, B.Q. Research progress in plant anthranilate synthase. *J. Trop. Biol.* **2015**, *6*, 504–511.
21. Pérez, E.; Rubio, M.B.; Cardoza, R.E.; Gutiérrez, S.; Bettiol, W.; Monte, E.; Hermosa, R. The importance of chorismate mutase in the biocontrol potential of *Trichoderma parareesei. Front. Microbiol.* **2015**, *6*, 1181. [CrossRef]

22. Niu, J.Q.; Wang, A.Q.; Huang, J.L.; Yang, L.T.; Li, Y.R. Cloning and expression analysis of sugarcane alkaline/neutral invertase gene SoNIN1. *Acta Agron. Sin.* **2014**, *40*, 253–263. [CrossRef]

23. Li, H.Y.; Yu, P.; Zhang, Y.X.; Wu, C.; Ma, C.Q.; Yu, B.; Zhu, N.; Koh, J.; Chen, S.X. Salt stress response of membrane proteome of sugarbeet monosomic addition line M14. *J. Proteom.* **2015**, *127*, 18–33. [CrossRef]

24. Suzuki, H.; Ueda, T.; Taguchi, H.; Takeuchi, N. Chaperoneproperties of mammalian mitochondrial translation elongationfactor Tu. *J. Biol. Chem.* **2007**, *282*, 4076–4084. [CrossRef] [PubMed]

25. Yu, J.J.; Chen, S.X.; Dai, S.J. Comparative proteomic analysis of *Puccinellia tenuiflora* leaves under Na_2CO_3 stress. *Int. J. Mol. Sci.* **2013**, *14*, 1740–1762. [CrossRef] [PubMed]

26. Lyzenga, W.J.; Stone, S.L. Abiotic stress tolerance mediated by protein ubiquitination. *J. Exp. Botary* **2012**, *63*, 599–616. [CrossRef] [PubMed]

27. Chen, S.S.; Lan, H.Y. Signal Transduction Pathways in Response to Salt Stress in Plants. *Plant Physiol. J.* **2011**, *47*, 119–128.

28. Kim, J.S.; Park, S.J.; Kwak, K.J.; Kim, Y.O.; Kim, J.K.; Song, J.; Jang, B.; Jung, S.H.; Kang, H. Cold shock domain proteins and glycine-rich RNA-binding proteins from *Arabidopsis thaliana* can promote the cold adaptation process in Escherichia coli. *Nucleic Acids Res.* **2007**, *35*, 505–516. [CrossRef]

29. Harmon, A.C. Calcium-regulated protein kinases of plants. *Gravit. Space Biol. Bull.* **2003**, *16*, 83–90. [PubMed]

30. Asano, T.; Hayashi, N.; Kikuchi, S.; Ohsugi, R. CDPK-mediated abiotic stress signaling. *Plant Signal. Behav.* **2012**, *7*, 817–821. [CrossRef] [PubMed]

31. Choudhury, S.; Panda, P.; Sahoo, L.; Panda, S.K. Reactive oxygen species signalingin plants under abiotic stress. *Plant Signal. Behav.* **2013**, *8*, e23681. [CrossRef]

32. Wang, L.X.; Pan, D.Z.; Li, J.; Tan, F.L.; Benning, S.H.; Liang, W.Y.; Chen, W. Proteomic analysis of changes in the *Kandelia candel* chloroplast proteins reveals pathways associated with salt tolerance. *Plant Sci.* **2015**, *231*, 159–172. [CrossRef] [PubMed]

33. Rouhier, N.; Gelhaye, E.; Jacquot, J.P. Redox control by dithiol–disulfide exchange in plants: II. The cytosolic and mitochondrial systems. *Ann. N. Y. Acad. Sci.* **2002**, *973*, 520–528. [CrossRef] [PubMed]

34. Scheibe, R.; Dietz, K.J. Reduction-oxidation network for flexible adjustment ofcellular metabolism in photoautotrophic cells. *Plant Cell Environ.* **2012**, *35*, 202–216. [CrossRef] [PubMed]

35. Jacquot, J.P.; Rouhier, N.; Gelhaye, E. Redox control by dithiol–disulfide exchange in plants: I. The chloroplastic systems. *Ann. N. Y. Acad. Sci.* **2002**, *973*, 508–519. [CrossRef] [PubMed]

36. Michelet, L.; Zaffagnini, M.; Morisse, S.; Sparla, F.; Pérez-Pérez, M.E.; Francia, F.; Danon, A.; Marchand, C.H.; Fermani, S.; Trost, P.; et al. Redox regulation of the Calvin–Benson cycle: Something old, something new. *Front. Plant Sci.* **2013**, *4*, 470. [CrossRef] [PubMed]

37. Guo, Y.; Huang, C.; Xie, Y.; Song, F.; Zhou, X. A tomato glutaredoxin gene SLGRX1 regulates plant responses to oxidative, drought and salt stresses. *Planta* **2010**, *232*, 1499–1509. [CrossRef] [PubMed]

38. Song, C.Q.; Miao, H.F.; Zhu, B. The role of plant glutathione-S-transferase in phytoremediation. *Anhui Agric. Bull.* **2010**, *16*, 1007–7731.

39. Mittler, R. Oxidative stress, antioxidants and stress tolerance. *Trends Plant Sci.* **2002**, *7*, 405–410. [CrossRef]

40. Margis, R.; Dunand, C.; Teixeira, F.K.; Pinheiro, M.M. Glutathione peroxidase family-an evolutionary overview. *FEBS J.* **2008**, *275*, 3959–3970. [CrossRef] [PubMed]

41. Sun, G.R.; Wang, J.B.; Cao, W.Z.; Du, K.; Zhang, B. GST activity and its related indexes in the Chloroplasts of *Puccinellia tenuiflora* seedlings under Na_2CO_3 stress. *Northwest Bot. Gaz.* **2005**, *25*, 2495–2501.

42. Avsian, K.O.; Gueta, D.Y.; Ben, H.G.; Eshdat, Y. Regulation of stress-induced phospholipid hydroperoxide glutathione peroxidase expression in citrus. *Planta* **1999**, *4*, 469–477. [CrossRef]

43. Beeor-Tzahar, T.; Ben-Hayyim, G.; Holland, D.; Faltun, Z.; Eshdat, Y. A stress-associated citrus protein is a distinct plant phospholipid hydroperoxide glutathione peroxidase. *FEBS Lett.* **1995**, *366*, 151–155. [CrossRef]

44. Kim, Y.J.; Jang, M.G.; Noh, H.Y.; Lee, H.J.; Sukweenadhi, J.; Kim, J.H.; Kim, S.Y.; Kwon, W.S.; Yang, D.C. Molecular characterization of two glutathione peroxidase genes of Panax ginseng and their expression analysis against environmental stresses. *Gene* **2014**, *1*, 33–41. [CrossRef] [PubMed]

45. Roxas, V.P.; Smith, R.K., Jr.; Allen, E.R.; Allen, R.D. Overexpression of glutathione S-transferase/glutathione peroxidase enhances growth of transgenic tobacco seedlings during stress. *Nat. Biotechnol.* **1997**, *15*, 988–991. [CrossRef] [PubMed]

46. Roxas, V.P.; Lodhi, S.A.; Garrett, D.K.; Mahan, J.R.; Allen, R.D. Stress Tolerance in Transgenic Tobacco Seedlings that Overexpress Glutathione S-Transferase/Glutathione Peroxidase. *Plant Cell Physiol.* **2000**, *41*, 1229–1234. [CrossRef] [PubMed]

47. Chang, W.; Sui, X.; Fan, X.X.; Jia, T.T. Arbuscular Mycorrhizal Symbiosis Modulates Antioxidant Response and Ion Distribution in Salt-Stressed *Elaeagnus angustifolia* Seedlings. *Front. Microbiol.* **2018**, *9*, 652. [CrossRef] [PubMed]

48. Talaat, N.B.; Shawky, B.T. Protective effects of arbuscular mycorrhizal fungi on wheat (*Triticum aestivum* L.) plants exposed to salinity. *Environ. Exp. Bot.* **2014**, *98*, 20–31. [CrossRef]

49. Yue, Y.N. *Effects of Arbuscular Mycorrhizal Fungi on Plant Salt Tolerance in Songnen Saline Grassland*; Northeast Forestry University: Haerbin, China, 2015.

50. Sheng, M.; Tang, M.; Chen, H.; Yang, B.W.; Zhang, F.F.; Huang, Y.H. Influence of arbuscular mycorrhizae on photosynthesis and water status of maize plants under salt stress. *Mycorrhiza* **2008**, *18*, 287–296. [CrossRef] [PubMed]

51. Lin, J.X.; Wang, Y.N.; Sun, S.N.; Mu, C.S.; Yan, X.F. Effects of arbuscular mycorrhizal fungi on the growth, photosynthesis and photosynthetic pigments of *Leymus chinensis* seedlings under salt-alkali stress and nitrogen deposition. *Sci. Total Environ.* **2017**, *576*, 234–241. [CrossRef]

52. Liu, H.G. *AMF to Improve the Mechanism of Salt Resistance of Lycium barbarum L.*; Northwest A & F University: Yangling, China, 2016.

53. Tang, J. *Effects of Mycorrhizal Fungi (AMF) on Salinity Tolerance in the Cultivation of Healthy Vegetables*; Sichuan Agricultural University: Yaan, China, 2016.

54. Järvi, S.; Suorsa, M.; Aro, E.M. Photosystem II repair in plant chloroplasts-Regulation, assisting proteins and shared components with photosystem II biogenesis. *Biochim. Biophys. Acta* **2015**, *1847*, 900–909. [CrossRef] [PubMed]

55. Greenberg, B.M.; Gaba, V.; Canaani, O.; Malkin, S.; Mattoo, A.K.; Edelman, M. Separate photosensitizers mediate degradation of the 32 kDa photosystem II reaction center protein in the visible and UV spectral regions. *Proc. Natl. Acad. Sci. USA* **1989**, *86*, 6617–6620. [CrossRef]

56. Zhu, S.Q.; Xia, S.L.; Chen, Q. Advances in the regulation mechanism of photosystem IID1 protein expression. *Guizhou Agric. Sci.* **2012**, *40*, 37–42.

57. Guo, X. Role of Glutathione GRX in Response to Temperature Stress in Tomato. Master's Thesis, Zhejiang University, Hangzhou, China, 2016.

58. Jia, T.T.; Chang, W.; Fan, X.X. Effects of Arbuscular mycorrhizal fungi on photosynthetic and chlorophyll fluorescence characteristics in Elaeagnus angustifolia seedlings under salt stress. *Acta Ecol. Sin.* **2018**, *38*, 1337–1347.

59. Phillips, J.M.; Hayman, D.S. Improved procedures for clearing roots and staining parasitic and vesicular-arbuscular mycorrhizal fungi for rapid assessment of infection. *Trans. Br. Mycol. Soc.* **1970**, *55*, 158–161. [CrossRef]

60. Wang, X.K.; Huang, S.L. *Plant Physiological and Biochemical Experimental Principles and Technology*, 3rd ed.; Higher Education Press: Beijing, China, 2015.

61. Zhang, Z.L.; Qu, W.J. *Plant Physiology Experimental Guide*, 3rd ed.; Higher Education Press: Beijing, China, 2003.

62. Wisniewski, J.R.; Zougman, A.; Nagaraj, N.; Mann, M. Universal sample preparation method for proteome analysis. *Nat. Methods* **2009**, *6*, 359–362. [CrossRef] [PubMed]

63. Cox, J.; Mann, M. MaxQuant enables high peptide identification rates, individualized p.p.b.-range mass accuracies and proteome-wide protein quantification. *Nat. Biotechnol.* **2008**, *6*, 1367–1372. [CrossRef] [PubMed]

64. Del Pozo, J.C.; Allona, I.; Rubio, V.; Leyva, A.; de la Peña, A.; Aragoncillo, C.; Paz-Ares, J. A type 5 acid phosphatase gene from *Arabidopsis thaliana* is induced by phosphate starvation and by some other types of phosphate mobilising/oxidativestress conditions. *Plant J.* **1999**, *19*, 579–589. [CrossRef] [PubMed]

65. Li, W.Y.; Shao, G.; Lam, H. Ectopic expression of GmPAP3 alleviates oxidative damage caused by salinity and osmotic stresses. *New Phytol.* **2008**, *178*, 80–91. [CrossRef] [PubMed]

66. Monroe, N.; Hill, C.P. Meiotic Clade AAA ATPases: Protein Polymer Disassembly Machines. *J. Mol. Biol.* **2016**, *428 Pt B*, 1897–1911. [CrossRef]

67. Ren, Y.; Pan, H.; Yang, Y.; Pan, B.; Bu, W. Molecular cloning, characterization and functional analysis of a heat shock protein 70 gene in *Cyclina sinensis*. *Fish Shellfish Immunol.* **2016**, *58*, 663–668. [CrossRef]

68. Wong, J.H.; Cai, N.; Balmer, Y.; Tanaka, C.K.; Vensel, W.H.; Hurkman, W.J.; Buchanan, B.B. Thioredoxin targets of developing wheat seeds ide ntified by complementary proteomic approaches. *Phytochemistry* **2004**, *65*, 1629–1640. [CrossRef]

69. Traverso, J.A.; Vignols, F.; Cazalis, R.; Pulido, A.; Sahrawy, M.; Cejudo, F.J.; Meyer, Y.; Chueca, A. PsTRXh1 and PsTRXh2 Are Both Pea h-Type Thioredoxins with Antagonistic Behavior in Redox Imbalances. *Plant Physiol.* **2007**, *143*, 300. [CrossRef]

70. Wang, J.; Zhang, S.B.; Ma, D.; Ma, G.Y. Advances in Plant Actin Research. *J. Anhui Agric. Sci.* **2007**, *35*, 2860–2863.

71. Song, J.H.; Zhang, L.X. Progress on the transmembrane protein in plant. *J. Biol.* **2009**, *26*, 62–64.

72. Weber, T.; Parlati, F.; McNew, J.A.; Johnston, R.J.; Westermann, B.; Söllner, T.H.; Rothman, J.E. SNAREpins are functionally resistant to dis-ruption by NSF and αSNAP. *J. Cell Biol.* **2000**, *149*, 1063–1072. [CrossRef]

73. Sun, X.L.; Yu, Q.Y.; Tang, L.L.; Ji, W.; Bai, X.; Cai, H.; Liu, X.F.; Ding, X.D.; Zhu, Y.M. GsSRK, a G-type lectin S-receptor-like serine/threonine protein kinase, is apositive regulator of plant tolerance to salt stress. *J. Plant Physiol.* **2013**, *170*, 505–515. [CrossRef] [PubMed]

74. Ge, R.C.; Chen, G.P.; Zhao, B.C.; Shen, Y.Z.; Huang, Z.J. Cloning and functional characterization of a wheat serine/threoninekinase gene (TaSTK) related to salt-resistance. *Plant Sci.* **2007**, *173*, 55–60. [CrossRef]

75. Lee, K.H.; Piao, H.L.; Kim, H.Y.; Choi, S.M.; Jiang, F.; Hartung, W.; Hwang, I.; Kwak, J.M.; Lee, I.J.; Hwang, I. Activation of glucosidase via stress-induced polymerization rapidly increases active pools of abscisic acid. *Cell* **2006**, *126*, 1109–1120. [CrossRef]

76. Ge, C.; Wan, D.; Wang, Z.; Ding, Y.; Wang, Y.; Shang, Q.; Ma, F.; Luo, S. A proteomic analysis of rice seedlings responding to 1,2,4-trichlorobenzene stress. *J. Environ. Sci.* **2008**, *20*, 309–319. [CrossRef]

77. Luo, P.; Shen, Y.X.; Jin, S.J. Overexpression of Rosa rugosa anthocyanidin reductase enhances tobacco tolerance to abiotic stress through increased ROS scavenging and modulation of ABA signaling. *Plant Sci.* **2016**, *245*, 35–39. [CrossRef]

78. Lee, J.Q.; Zheng, S.X.; Yu, Z.N. Acetyl-CoA carboxylase: Advances in research on key enzymes of fatty acid metabolism and gene cloning. *J. Appl. Environ. Biol.* **2011**, *17*, 753–758.

79. Zhang, Q.; Hrmova, M.; Shirley, N.J.; Lahnstein, J.; Fincher, G.B. Gene expression patterns and catalytic properties of UDP-D-glucose4-epimerases from barley *Hordeum vulgare* L. *Biochem. J.* **2006**, *394*, 115–124. [CrossRef]

80. Singh, S.; Virdi, A.S.; Jaswal, R. A temperature-responsive gene in sorghum encodes a glycine-rich protein that interacts with calmodulin. *Biochimie* **2017**, *137*, 115–123. [CrossRef]

81. Gupta, A.; Kailasam, S.; Bansall, M. Insights into the Structural Dynamics of Nucleocytoplasmic Transport of tRNA by Exportin-t. *Biophys. J.* **2016**, *110*, 1264–1279. [CrossRef]

82. Kohler, A.; Hurt, E. Exporting RNA from the nucleus to the cytoplasm. *Nat. Rev. Mol. Cell Biol.* **2007**, *8*, 761–773. [CrossRef]

83. Mattaj, I.W.; Englmeier, L. Nucleocytoplasmic transport: The soluble phase. *Annu. Rev. Biochem.* **1998**, *67*, 265–306. [CrossRef]

84. Aroca, R.; Porcel, R.; Ruiz-Lozano, J.M. How does arbuscular mycorrhizal symbiosis regulate root hydraulic properties and plasma membrane aquaporins in Phaseolus vulgaris under drought, cold or salinity stresses? *New Phytol.* **2007**, *73*, 808–816. [CrossRef]

85. Geelen, D.; Leyman, B.; Batoko, H.; Di Sansebastiano, G.P.; Moore, I.; Blatt, M.R. The abscisic acid-related SNARE homolog NtSyr1contributes to secretion and growth: Evidence from competition with its cytosolic domain. *Plant Cell* **2002**, *14*, 387–406. [CrossRef]

86. Iqbal, M.S.; Siddiqui, A.A.; Alam, A.; Goyal, M.; Banerjee, C.; Sarkar, S.; Mazumder, S.; De, R.; Nag, S.; Saha, S.J.; et al. Expression, purification and characterization of *Plasmodium falciparum* vacuolar protein sorting 29. *Protein Expr. Purif.* **2016**, *120*, 7–15. [CrossRef]

87. Hong, J.K.; Choi, H.W.; Hwang, I.S.; Kim, D.S.; Kim, N.H.; Choi, D.S.; Kim, Y.J.; Hwang, B.K. Function of a novel GDSL-type pepper lipase gene, CaGLIP1, in disease sus-ceptibility and abiotic stress tolerance. *Planta* **2008**, *227*, 539–558. [CrossRef]

88. Naranjo, M.; Forment, J.; Roldan, M.; Serrano, R.; Vicente, O. Overexpression of *Arabidopsis thaliana* LTL1, a salt-induced gene encod-ing a GDSL-motif lipase, increases salt tolerance in yeast and transgenic plants. *Plant Cell Environ.* **2006**, *29*, 1890–1900. [CrossRef]

89. Li, N.N.; Qian, W.J.; Wang, L.; Cao, H.L.; Hao, X.Y.; Yang, Y.J.; Wang, X.C. Isolation and expression features of hexose kinase genes under various abiotic stresses in the tea plant (*Camellia sinensis*). *J. Plant Physiol.* **2017**, *209*, 95–104. [CrossRef]

90. Van Gisbergen, P.A.; Bezanilla, M. Plant formins: Membrane anchors for actin polymerization. *Trends Cell Biol.* **2013**, *23*, 227–233. [CrossRef]

Article

Identification of Early Salinity Stress-Responsive Proteins in *Dunaliella salina* by isobaric tags for relative and absolute quantitation (iTRAQ)-Based Quantitative Proteomic Analysis

Yuan Wang [1,2,†], Yuting Cong [2,†], Yonghua Wang [3], Zihu Guo [4], Jinrong Yue [2], Zhenyu Xing [2], Xiangnan Gao [2] and Xiaojie Chai [2,*]

[1]   Key Laboratory of Hydrobiology in Liaoning Province's Universities, Dalian Ocean University, Dalian 116021, China; wangyuan@dlou.edu.cn

[2]   College of fisheries and life science, Dalian Ocean University, Dalian 116021, China; congyuting@dlou.edu.cn (Y.C.); yuejinrong0604@126.com (J.Y.); xxxzy0520@163.com (Z.X.); gao.xn@foxmail.com (X.G.)

[3]   Bioinformatics Center, College of Life Sciences, Northwest A&F University, Yangling 712100, China; yh_wang@nwsuaf.edu.cn

[4]   College of Life Sciences, Northwest University, Xi'an, Shaanxi 710069, China; guozihu2010@yahoo.com

*   Correspondence: cxj63@126.com

†   These authors contribute equally to the work.

Received: 22 November 2018; Accepted: 16 January 2019; Published: 30 January 2019

Abstract: Salt stress is one of the most serious abiotic factors that inhibit plant growth. *Dunaliella salina* has been recognized as a model organism for stress response research due to its high capacity to tolerate extreme salt stress. A proteomic approach based on isobaric tags for relative and absolute quantitation (iTRAQ) was used to analyze the proteome of *D. salina* during early response to salt stress and identify the differentially abundant proteins (DAPs). A total of 141 DAPs were identified in salt-treated samples, including 75 upregulated and 66 downregulated DAPs after 3 and 24 h of salt stress. DAPs were annotated and classified into gene ontology functional groups. The Kyoto Encyclopedia of Genes and Genomes pathway analysis linked DAPs to tricarboxylic acid cycle, photosynthesis and oxidative phosphorylation. Using search tool for the retrieval of interacting genes (STRING) software, regulatory protein–protein interaction (PPI) networks of the DAPs containing 33 and 52 nodes were built at each time point, which showed that photosynthesis and ATP synthesis were crucial for the modulation of early salinity-responsive pathways. The corresponding transcript levels of five DAPs were quantified by quantitative real-time polymerase chain reaction (qRT-PCR). These results presented an overview of the systematic molecular response to salt stress. This study revealed a complex regulatory mechanism of early salt tolerance in *D. salina* and potentially contributes to developing strategies to improve stress resilience.

Keywords: Salinity stress; *Dunaliella salina*; isobaric tags for relative and absolute quantitation; differentially abundant proteins; proteomics

---

1. Introduction

Salinity stress greatly affects plant growth and productivity, resulting in cellular energy depletion, redox imbalances, and oxidative damage [1,2]. Scientists have long sought to understand the mechanisms of salt tolerance in plants in order to improve the yield of economically important crops. The unicellular eukaryotic green alga *Dunaliella salina* has extreme salt tolerance, with a unique ability to adapt and grow in salt concentrations ranging from 0.05 to 5.5 M. This organism is an established model for studying plant adaptation to high salinity [3,4], and also has significant pharmaceutical and industrial value, mainly as food for marine aquacultures [5].

To cope with high salinity in its sessile existence, *D. Salina* has evolved a considerable degree of developmental plasticity, including adaptation via cascades of molecular networks [6], which regulate cellular homeostasis and promote survival [1,2,6]. For example, the mitogen-activated protein kinase (MAPK) pathway plays a key regulatory role in plant development as well as in numerous stress responses [2]. We previously cloned the MAPK kinase (MAPKK) cDNA from *D. salina*, and found that its expression was induced upon salt stress [7]. Although salt tolerance in *D. salina* has been studied extensively at the phenotypic, physiological, and genetic level, and many candidate genes associated with energy metabolism, signal transduction, transcription, protein biosynthesis and degradation have been identified [6], the underlying mechanisms remain unclear. Genes responsive to salt stress are regulated at the transcriptional, translational, and post-translational levels. Analysis of the proteome during stress response provides deeper insights into the molecular phenotype since unlike mRNA transcripts, the proteome reflects the actual response of the organism to environmental change [8–11]. It is necessary to identify the salt stress response proteins to further elucidate the salt tolerance mechanisms in *D. salina*.

Proteomics is a high-throughput approach to study the dynamic protein profile of a cell or organism, and therefore also the intricate molecular networks [2]. Plant cell responses to salinity depend on the tissue, and the severity and duration of the stress, which lead to various changes at the proteome level [12,13]. Recently, salt stress-induced changes have been reported in the protein profiles of rice roots and leaves [9], *Arabidopsis thaliana* roots [11], soybean leaves [13,14], hypocotyls and roots [14] and chloroplasts of diploid and tetraploid black locust [15]. Proteome analysis of *D. salina* subjected to salinity by 2-D gels [16,17] and blue native gels using plasma membrane [3] have shown changes in photosynthesis, protein and ATP biosynthesis, and signal transduction proteins in response to salt stress. Quantitative proteomics and phosphoproteomics have also been applied to explore palmella formation mechanism in *D. salina* under salt stress [4], and have identified several proteins and phosphoproteins as potential candidates for augmenting salt tolerance in this organism. However, previous studies focused on relatively late response to salinity, and did not clarify the early response to short-term salt stress in *D. salina*. The initial phases of stress response in organisms could reveal more profound differences at the proteomic level, as compared to later phases, which could help elucidate novel mechanisms of homeostasis between plant and environment [9,13,18,19]. Therefore, the main objectives of this study were to investigate the early salt stress-response proteome and identify the differentially abundant proteins under short-term salt stress in *D. salina*.

Isobaric tags for relative and absolute quantitation (iTRAQ) is currently one of the most robust techniques used for proteomic quantitation [13,20]. The iTRAQ approach has been used to compare salt tolerant and susceptible cultivars, and has helped identify many proteins that can potentially enhance salt resistance in plants [13,15,20,21]. In the present study, iTRAQ was used to assess proteomic changes and identify differentially abundant proteins (DAPs) in *D. salina* at the early stage of salt stress. NaCl was used in the culture medium to imitate environmental salt stress. We identified 65 and 102 proteins with significantly altered abundance after 3 and 24 h of salt stress respectively; these were then classified into gene ontology (GO) functional groups and pathways. This study advanced our understanding of early salt-responsive mechanisms in *D. salina*. Since *D. salina* is a unicellular organism, the findings reveal how cell can adapt to extreme salinity and provide potential molecular elements for enhancing salinity tolerance in crop plants.

2. Results

2.1. Identification of Dunaliella Salina *Differentially Abundant Proteins (DAPs) Using Isobaric Tags for Relative and Absolute Quantitation (iTRAQ)*

The optimal in-vitro salt stress stimulus with which to explore early responses to salt stress was determined to be 3-M NaCl [3,4] for 3 or 24 h. Proteomes of *D. salina* subjected to the above conditions were analyzed using iTRAQ and liquid chromatography-tandem mass spectrometry/ mass spectrometry (LC-MS/MS). A total of 23,461 spectra were generated from control and salt-treated *D. salina* and were analyzed using the Mascot search engine. 23,461 spectra matched known spectra of 2,283 unique peptides and 1140 proteins from control and salt-treated samples. Detailed annotation information including peptide sequences, accession numbers, matching criteria, unused scores, P-value and sequence coverage of total identified and differential protein species is provided in Supplementary Table S1. The overall changes in protein abundance in *D. salina* after 3 and 24 h salt stress are illustrated in Figure 1. Of the 1140 unique proteins of the control and treated *D. salina*, a total of 141 DAPs were observed in response to salt stress relative to control, with the threshold for upregulated expression >1.2 and downregulated expression <0.83 ($p < 0.05$) [22,23]. 75 of these DAPs were upregulated and 66 were downregulated in the cells after 3 and/or 24 h salt stress (Figure 1a, Supplementary Table S2). Details for each protein are also provided in Table S1. Of these 141 proteins in the control vs. salt treatment, there was an overlap in two categories, which means there were 26 proteins that changed in 3 and 24 h exposures (Figure 1a). Volcano plots show the overall changes to protein abundance in treated compared to control cultures at each time point, with significant difference ($p < 0.05$) in the expression of a few proteins (Figure 1b).

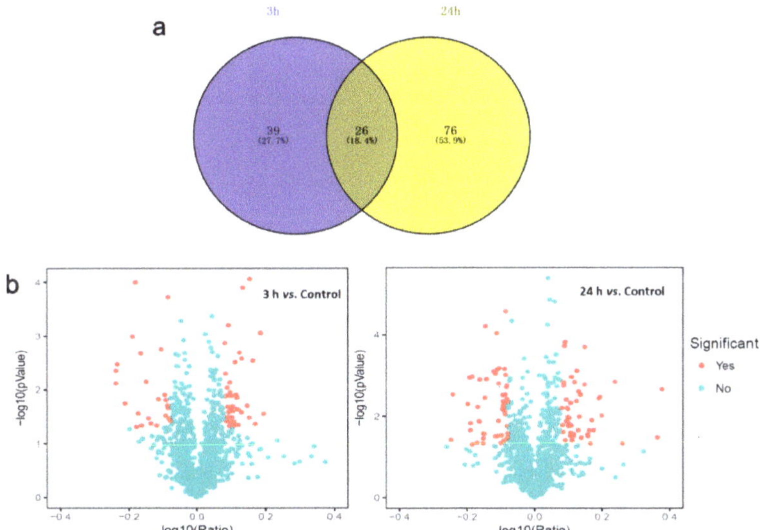

Figure 1. Differentially abundant proteins (DAPs) in response to salt stress in *Dunaliella salina*. (**a**) Venn diagram of 3 and 24 h specific DAPs with overlapping regions indicating the number of common proteins. (**b**) Volcano plots of the proteins quantified during iTRAQ analysis comparing control to 3 and 24 h salt treatments. Each point represents the difference in expression (fold-change) between the two groups plotted against the level of statistical significance. Proteins represented by a filled red square are those with expression that differs at a statistically significant level.

2.2. Gene Ontology (GO) Annotation of DAPs in Dunaliella Salina

To determine the putative functions of DAPs involved in early response to salt stress, they were subjected to GO enrichment analysis. The DAPs were categorized by three sets of ontologies: biological processes (BP), cellular components (CC), and molecular functions (MF) (Figure 2), and a GO term was considered significant at p-value < 0.05. As shown in Figure 2a, the significantly enriched GO terms in the BP category included regulation of protein-chromophore linkage, regulation of ATP synthesis coupled proton transport, regulation of ATP hydrolysis coupled proton transport, regulation of translation, regulation of photosynthesis, negative regulation of glycolytic process, negative regulation of nitrogen compound metabolic process, and negative regulation of cell redox homeostasis. In the CC category, integral component of membrane, photosystem I, photosystem II, chloroplast thylakoid membrane and proton-transporting ATP synthase complex were the most significantly enriched terms (Figure 2b). For MF, chlorophyll binding, structural constituent of ribosome, proton-transporting ATP synthase activity, and electron transporter-transferring electrons within the cyclic electron transport pathway of photosynthesis activity were the most significant terms. After 3 h of salt stress, six, seven and four proteins respectively enriched in chlorophyll binding, structural constituent of ribosome, and proton-transporting ATP synthase activity, respectively, were upregulated. After 3 and 24 h of stress, the downregulated DAPs were predominantly related to magnesium ion binding, and after 24 h salt stress the downregulated DAPs were also predominantly related to ribulose−bisphosphate carboxylase activity (Figure 2c).

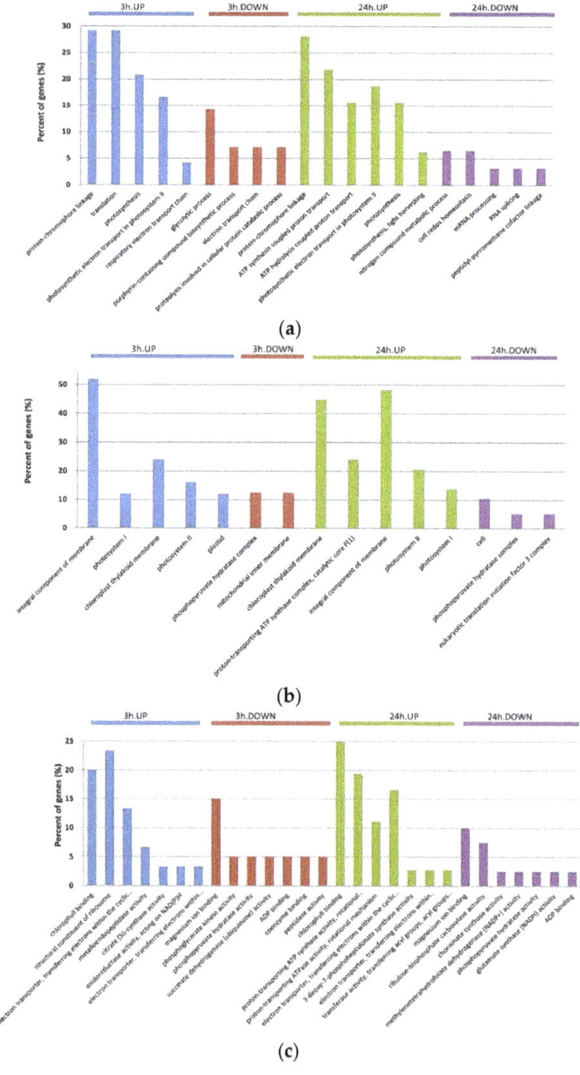

Figure 2. Gene ontology (GO) classification of the DAPs detected at each time point. GO terms in the biological process (**a**), cellular component (**b**), and molecular function (**c**) categories are presented.

2.3. Kyoto Encyclopedia of Genes and Genomes (KEGG) Analysis of DAPs in Dunaliella Salina

Kyoto Encyclopedia of Genes and Genomes (KEGG) enrichment analysis was used to determine any potential clustering of the DAPs in specific metabolic pathways. Using the KEGG database as a reference, 115 DAPs were annotated and classified into 25 different pathways (Figure 3 and Table S3). After 3 h salt stress, the main KEGG pathway classifications of the DAPs were photosynthesis and oxidative phosphorylation. Porphyrin and chlorophyll metabolism were among the pathways that were most significantly downregulated during the early response. DAPs contributing to photosynthesis, oxidative phosphorylation and metabolic pathways were also significantly enriched after 24 h salt stress, while the 'one carbon pool by folate' pathway was significantly downregulated after 24 h. All pathways are listed in Supplementary Table S3.

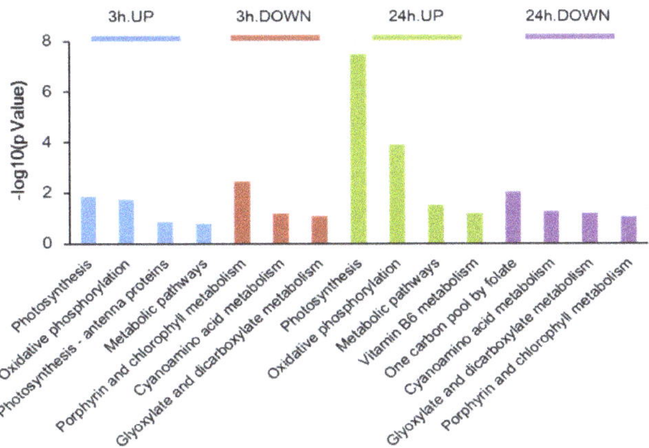

Figure 3. Kyoto Encyclopedia of Genes and Genomes (KEGG) pathway enrichment analysis of the DAPs in *Dunaliella salina* exposed to salt stress at 3 and 24 h. The *x*-axis shows representative enriched KEGG pathways, and the y-axis indicates corresponding p-values of enriched pathways ($-\log10$ *p*-value).

2.4. Search Tool for the Retrieval of Interacting Genes (STRING)-Based Protein-Protein Interaction (PPI) Analysis

To determine the regulatory mechanisms of the DAPs and their potential roles in salt stress, we built a regulatory network with the up- and downregulated proteins using STRING analysis; this revealed functional links among the DAPs that were significantly altered after salt stress (Figure 4). Among the 141 DAPs identified by iTRAQ, two regulatory PPI networks of the DAPs containing 33 and 52 nodes (3 and 24 h, respectively) were built. Considerable overlapping was seen among the major clusters, especially with the DAPs involved in photosynthesis, ATP synthesis and ribosome structure regulation pathways. Furthermore, 3 h salt stress induced several protein interactions, including FTSY-RPL13a-RPL18-EIF3A and chlL-chlN-rbcL-psaB-psaA-LHCB4-ATPvL1-atpI-cox1 (Figure 4a). After 24 h salt stress, ATPC, ATPvL1, atpA, atpB, atpE, psaA, psaB, psbB, psbC, HSP70B, rbl2 and EIF3A were the most important protein upregulation hubs, while FTSY, rbcL, HSP90A, LHCB5, EDP00988, SHMT3 and FTSH3 were the most important protein downregulation hubs in the constructed network (Figure 4b). These proteins were not separated and together they formed a related network in response to salt stress.

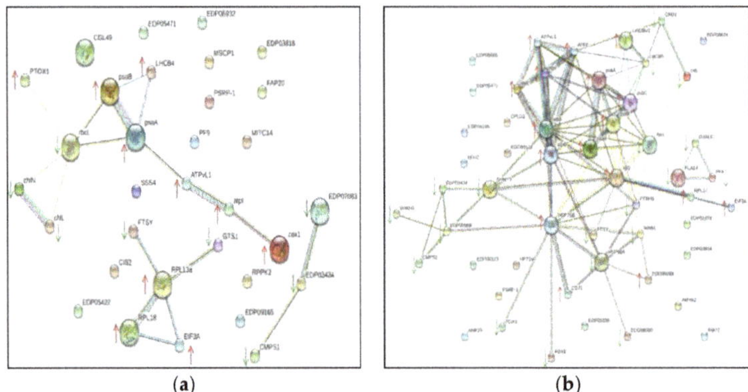

(a)   (b)

Figure 4. Search tool for the retrieval of interacting genes (STRING)-based protein–protein interaction (PPI) analysis of the DAPs in *Dunaliella salina* exposed to salt stress at each time point. (**a**) 3 h and (**b**) 24 h. The DAPs from *D. salina* were used for constructing PPI network using STRING software. The circles represent proteins while the straight lines represent the interactions between different proteins: gene fusion (red), neighborhood (green), co-occurrence across genomes (blue), protein homology (light green), co-mentioned in PubMed abstracts (yellow), experimentally determined interactions (purple), and interactions determined from curated databases (light blue). The small nodes represent proteins of unknown 3D structure, and large nodes are proteins of known or predicted 3D structure. Red arrows indicate the upregulated DAPs and green arrows indicate the downregulated DAPs.

2.5. Analysis of Transcripts Encoding Selected DAPs

To investigate whether the differences in protein abundances were reflected at the mRNA level and to validate the proteomic analysis, the quantitative real-time polymerase chain reaction (qRT-PCR) was used to verify the level of gene expression associated with DAPs between control and salt-treated groups. Four upregulated proteins and one downregulated protein in *D. salina* were selected for verification at the mRNA level. The fold changes of transcript abundances are provided in Figure 5. The transcript levels of four genes displayed the same trend with the abundance of the corresponding protein species, namely, *atpE*, *psaB*, *rps11*, and *rbcL*. In contrast, *psbB* showed the opposite trend with the abundance of the corresponding protein in *D. salina* under 3 or 24 h salt stress (Figure 5 and Supplementary Table S2). The discrepancy between the transcription level of the gene and the abundance of the corresponding protein probably resulted from various post-translational modifications [24] under salt stress.

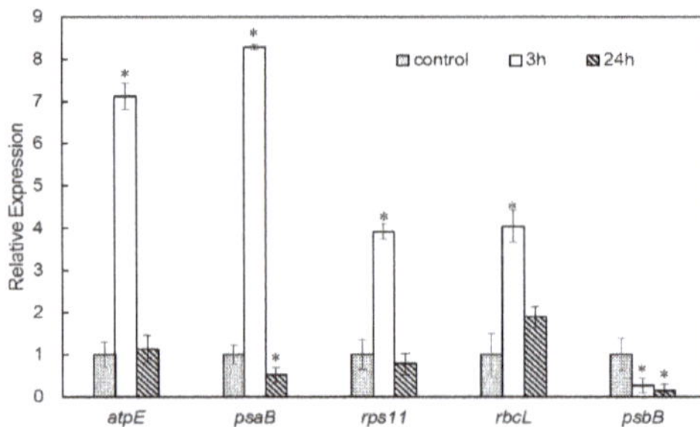

Figure 5. Analysis of transcript levels of the DAPs between salt stress and control conditions by quantitative real-time polymerase chain reaction (qRT-PCR). mRNA expression values were rescaled relative to the control. Statistical significant between experimental and control groups marked with asterisks: (*) p-value < 0.05.

3. Discussion

D. salina possesses an extraordinary ability to cope with salt stress, and is therefore a model organism for studying salt tolerance [3,4]. In recent years, *D. salina* proteome networks have been analyzed to reveal the molecular basis underlying this organism's tolerance to different stresses [4,16,25,26]. Previous publications on physiological and molecular changes under salt stress focused majorly on the long-term stress ($\geq$48 h) [3,16,17]. Notably, the response of plants to salinity stress can be determined by the rapid perception of stress shock that occurs within a few hours. The early stress response identified in the previous literature ranged from 1 to 24 h [9,13,18]. However, the adaptive mechanisms underlying *D. salina* in the early response to salt stress (<24 h) at the proteomic level were still unclear. The iTRAQ technology in combination with LC-MS/MS is an effective method for investigating altered protein profiles in plant cells during environmental stress [13,15,20,21]. In the present study, we used an iTRAQ comparative proteomic strategy to analyze the dynamic changes in the protein profiles of *D. salina* exposed to 3 M NaCl [3,4], which simulated environmental salt stress. The aim was to identify protein species to help elucidate the molecular mechanisms associated with early salinity response.

A comparison of the treated and control *D. salina* proteomes revealed 141 DAPs in response to salt stress, indicating a massive metabolic reprogramming. Of these DAPs, 75 were upregulated and 66 were downregulated after both 3 and 24 h of salt stress (Figure 1a). The abundance most DAPs we observed were less than twofold. A possible explanation is that proteins play major roles in most biological processes; as a consequence, protein expression levels are subject to diverse and complex control [27]. Functionally important proteins are subject to higher levels of constraint [28,29]. Functional annotations of these DAPs with GO enrichment analysis showed that most of the upregulated proteins were involved in respiratory metabolism, transport and photosynthesis, while the majority of the downregulated proteins accounted for glycolysis and nitrogen compound metabolic process, in terms of BP. This indicates that salt stress likely affects energy metabolism and ion transportation in *D. salina*, which is useful information for further research into the molecular mechanisms of salt tolerance. DAPs after 3 h of salt stress were predominantly binding proteins that were involved in cellular organization or biogenesis, while the response DAPs after 24 h of salt stress, also primarily binding proteins, were involved in aldehyde and organic acid metabolism. With regards to subcellular location, the significant protein species were mainly enriched in the chloroplast, photosystem, peroxisome, and ribosome (Figure 2). Therefore, our data indicate that *D. salina* is able to mount an early response (as early as 3 h) to salt stress owing to the active stimulation of the critical cell signaling pathways involved in classical stimuli response.

Proteins typically do not exert their functions independently, but rather coordinate with each other in biochemical and physiological processes. Pathway analysis can therefore help reveal cellular processes involved in early salt tolerance [15,20]. These DAPs were further investigated using KEGG database. The main responses to salt stress were seen in proteins that regulate metabolism and energy conversion, carbon fixation in photosynthetic organisms and transport (Figure 3). This is consistent with previous studies which showed that the adaptation of microalgae to stress conditions is accompanied by multiple changes in proteins involved in carbon and energy metabolism [2,9]. In this study, salt stress affected photosynthesis both in 3 and 24 h treated groups. Additionally, we observed an increase in the abundances of many proteins involved in mitochondrial oxidative phosphorylation, indicating an increased need for ATP and energy in response to salt stress.

Photosynthesis in chloroplasts is one of the primary processes that is affected by abiotic stress [30], and the rapid response of photosynthetic machinery and metabolism is pivotal for plants to cope with salt stress [15,30]. We observed that several crucial proteins related to photosynthesis and energy metabolism, such as atpl, ATPvL1, psaA, and psaB, were upregulated upon salt stress in *D. salina*. The ATP synthase alpha subunit (atpA) was upregulated after both 3 and 24 h of salt stress in *D. salina* (Figure 3 and Supplementary Table S2). The increased accumulation of atpA may enhance ATP synthesis to meet increasing energy demands for sustained salt resistance. Previous studies have also documented a regulated expression of multiple subunits of this protein complex under salinity [31,32]. PsaA and PsaB, which encode the large submits of the core complex in PSI that carries the cofactors of the electron transport chain [33], were also upregulated as part of the early response. In plants and green algae, photosystem II and photosystem I consist of a core complex and a light-harvesting complex (LHC) containing electron chain transport cofactors [34]. The LHC protein, together with chlorophyll, captures light energy and delivers it to the photosystems [30]. Under pressure conditions, light harvesting must be reduced to avoid the excessive excitation of and damage to photosystems [34]. We observed that LHCB4 was upregulated after 3 h, while LHCBM2 was upregulated and LHCB5 was downregulated after 24 h salt stress in *D. salina*. The response mechanisms of LHC proteins in early salt stress require further study.

Salt stress can increase the rate of protein unfolding; challenge cellular protein homeostasis for the available folding capacity becomes insufficient. Molecular chaperones or heat shock proteins are a large family of proteins that have the important function of helping other proteins fold and repair misfolding [35,36]. HSP70B and HSP90A play important roles in plant growth and responses to environmental stimuli [37]. In a previous proteomic study, post-translational modifications of HSP90A and HSP90C were speculated to be involved in salt stress responses [38]. In our study, HSP90A was downregulated while HSP70B was upregulated after 24 h salt stress, suggesting an involvement with increased degradation and reduced biosynthesis of proteins during salt stress. This is surprising given that the cooperation between HSP70 and HSP90 systems in chloroplasts has been suggested [39]. Thus, HSP90A may play different roles within cells exposed to salinity.

After 3 h of salt stress, the downregulated metabolic pathways in *D. salina* included not only those of primary but also secondary metabolism. The majority of DAPs related to central pathways, such as porphyrin and chlorophyll metabolism, and one carbon pool related to folate pathways, were downregulated after 24 h salt exposure (Figure 3). These findings indicate that salt stress fundamentally inhibited normal carbohydrate and energy metabolism in *D. salina* during the early stages of response. Both FTSY (a signal recognition particle docking protein) and rbcL (a Rubisco large subunit protein) were arrested in *D. salina* upon salt stress (Supplementary Table S2). This contradicts another study which showed upregulation of FTSY and rbcL and a strengthened glycolysis pathway, which could result in more energy for the generation of ATP and NADPH to resist salt stress [40]. The molecular response of organisms to salt stress may vary depending on species and stress levels.

The early response of *D.salina* to salt stress is a dynamic process. This organism can enhance the tolerance/resistance mechanism and establish cellular metabolic homeostasis under stress conditions. Based on functional and pathway analysis, we revealed the early response mechanisms to acclimatize to salinity in *D.salina*. Firstly, during initial exposure to salinity, changes in photosynthesis proteins may be related to the early response. Secondly, salt stress acclimatization is an energy consuming process. *D.salina* enhances the expression of oxidative phosphorylation-related proteins and the generation of production of ATP repairing of stress-induced damages. Thirdly, salt stress leads to cell instability leading to an increased risk of protein damage. Several heat shock proteins act as molecular chaperones to prevent denaturation and help denature proteins to restore their natural conformation. Furthermore, early stress responses are related to the activity of the protein synthesis system. *D.salina* enhances the processing and renewal of chloroplast and cytoplasmic proteins to cope with salt stress.

A regulatory PPI network was constructed for the DAPs using STRING (Figure 4), which showed considerable interactive networks among proteins involved in photosynthesis, ATP synthesis, and stress responsive signal transduction. Significant interaction was seen between several photosynthesis related proteins including photosystem components (psaA, psaB, psbB, psbC) and ATP synthase subunit proteins (atpA, atpB, atpE). Additionally, ATPC, ATPvL1, atpA, atpB, atpE, psaA, psaB, psbB, psbC, HSP70B, rbl2, and EIF3A, were the most important hubs orchestrating protein regulation in the constructed regulatory network. The downregulation of rbcL, HSP90A, and LHCB5 in the PPI network was consistent with previous findings [9,30,40]. We also observed that LHCB5 was downregulated in *D. salina* after salt treatment. Taken together, our findings indicate that salt stress affected multiple metabolic and physiological pathways in *D. salina*, predominantly photosynthesis, energy metabolism, carbon assimilation and metabolism and heat shock proteins. We also found variations in the DAPs in 3 h versus 24 h response to salt stress, indicating that the alga detected the extent of salt stress and alleviated it by modulating the expression of stress-responsive proteins.

In conclusion, we identified a number of novel proteins whose expression and abundance were significantly altered in the early response to salt stress. Multiple proteins mainly involved in photosynthesis, ATP synthesis, and oxidative phosphorylation, were putatively linked to early *D. salina* salt stress response. Furthermore, important metabolic pathways, including glycolysis, purine, and chlorophyll metabolism were compromised by salt treatment. PPI network analysis suggested that protein metabolism, energy supply, and photosynthesis work together to reconstitute cellular homeostasis under stress. ATPC, ATPvL1, atpA, atpB, atpE, psaA, psaB, psbB, psbC, HSP70B, rbl2, and EIF3A were the most important protein upregulation hubs. The identification of these stress-induced proteins can increase our knowledge of the molecular networks involved in plant salt tolerance, and help to mine more salt stress associated genes. This study provides a better understanding of the molecular mechanisms involved in stress response at the translational level.

4. Materials and Methods

4.1. Algae Culture

D. salina was obtained from the Hydrobiology Laboratory of the Dalian Ocean University (Dalian, China), and maintained at 50 mM photons/m^3 on alternate 12 h light–dark cycles. The algal cells were cultured in f/2 medium for several weeks in 1 M NaCl as previously described [3,4,41], with temperature and pH maintained at 25 ± 1 °C and 7.5 ± 0.2, respectively. Cells were seeded at a density of ~5 $\times$ 10^5 cells/mL, corresponding to the optical density at 630 nm (OD_{630}) of 0.06–0.08. When the cells reached the logarithmic growth phase (~2–3 $\times$ 10^6 cells/mL), they were transferred to fresh medium containing 3 M (salinity shock) NaCl [3,4]. Algal culture with 1 M NaCl addition (normal growth condition) was set as the control. Four replicates were made for the control and salt treatment groups. After 3 or 24 h of 3 M NaCl treatment, algal cells were harvested, and either used fresh or stored at −80 °C for later analyses.

4.2. Protein Extraction and Quantification

Proteins were extracted from 2 $\times$ 10^8 cells per sample using Plant Total Protein Lysis Buffer (7 M Urea, 1% CHAPS (3-[(3-cholamidopropyl) dimethylammonio]-1-propanesulfonate), 2 M Thiourea, 40 mM Tris-HCl pH8.5, 2 mM EDTA and 1 mM PMSF) and the cells were sonicated in the buffer for 60 s (0.2 s on, 2 s off, amplitude 25%). Homogenized samples were then incubated for 1 h at 25°C and the remaining debris was removed by centrifugation at 30,000$\times g$ at 4°C for 30 min. An aliquot of the supernatant was taken and the protein concentration was determined by Bio-Rad DC protein assay (Bio-Rad, Hercules, CA, USA) [42]. A total of 20 µg of protein per sample from cell lysate was subjected to sodium dodecyl sulfate-polyacrylamide gel electrophoresis (SDS-PAGE) to verify protein quality.

4.3. Protein Digestionand iTRAQ Labeling and Fractionation by Strong Cationic Exchange (SCX)

For each sample, a total of 200 µg of proteins were precipitated in 4 $\times$ volumes of cold acetone overnight at −20 °C. The protein pellets were dissolved in 1% SDS with 100 mM triethylammonium bicarbonate (pH 8.5) and sonicated in ice. Protein samples were reduced and digested with trypsin at 30:1 (*w/w*) for 16 h at 37 °C. Peptides were labeled with an iTRAQ Reagents 8-plex kit (AB Sciex Inc., Foster City, CA, USA) and incubated for 2 h at room temperature. The labeled peptide mixtures were then combined and dried by vacuum centrifugation. After labeling, the peptides were reconstituted in solvent A (25% acetonitrile, 25 mM NaH_2PO_4, pH 2.7) and then loaded into an Ultremex strong cationic exchange (SCX) column (Phenomenex, Torrence, CA, USA). The peptides were eluted using the SCX column to remove interfering substances such as excess iTRAQ reagents, organic solvents and SDS. The elution process was monitored by measuring the absorbance at 280 nm, and 12 fractions were collected [13].

4.4. Liquid Chromatography-Tandem Mass Spectrometry (LC-MS/MS) Analysis

Peptides fraction from each sample were analyzed using a nano-high-performance liquid chromatography (HPLC) system (Shimadzu LC-20AD, Kyoto, Japan) [25]. The 100-μL labeled peptides were resolved in solvent A containing 5% acetonitrile and 0.1% formic acid. Samples with individual volumes of 10 μL were loaded into a C18 trap column. Subsequently, solvent B (95% acetonitrile v/v, 0.1% formic acid) was used to separate the peptide with the following linear gradient conditions. Peptides were eluted with a flow rate of 0.6 mL/min. The elution peptide gradient was used from 5% solvent B to 35% solvent B for 35 min, then ramped up to 60% solvent B over five minutes, raised to 80% in two minutes and held for five minutes. The LC elute was then subjected to a Q Exactive MS (Thermo Fisher, NY, USA) coupled online to the HPLC. The applied electrospray voltage was 2.5 kV.

4.5. Analysis of Differentially Abundant Proteins

Protein identification and quantification was performed using the Mascot 2.3.02 search engine (Matrix Science, Boston, MA) against the UniProt database (http://www.uniprot.org). Viridiplantae (39,754 entries in UniProt) was chosen for taxonomic categorization. All DAPs were compared to the *D.salina* genome database (http://genomeportal.jgi.doe.gov) to further identify the annotated protein entries [43]. The protein mass were predicted using online software (http://www.expasy.ch/tools/) on the basis of the protein sequences. The peptide mass tolerance was set as ±20 ppm and the fragment mass tolerance was 0.1 Da. The results were filtered based on a false discovery rate (FDR) of no more than 0.01. To demonstrate the reproducibility of the replicates, protein abundances between various biological replicates were compared, and ratios for each protein comparison were normalized to 1. Only proteins with at least one unique peptide and unused value of >1.2 were considered for further analysis. When differences in protein expression between salt-treated and control groups were >1.2-fold or <0.83-fold [22,23], with a p-value < 0.05, the protein was considered to be differentially abundant. The mass spectrometry proteomics data have been deposited to the ProteomeXchange Consortium via the PRIDE [44] partner repository with the dataset identifier PXD010739.

4.6. Go, KEGG and STRING Enrichment Analyses

Functional analysis of DAPs was based on biological process, molecular function, and cellular components, using GO annotation and protein classification (http://www.geneontology.org) [45]. The DAPs were further assigned to the KEGG database (http://www.genome.jp/kegg) and the STRING database (http://www.string-db.org/). KEGG was used to predict the major metabolic and signal transduction pathways involved in the identified DAPs [46,47]. STRING was used for protein interaction analysis in order to identify protein interaction networks of *D.salina* under salinity stress conditions. The protein interaction (PPI) networks' responses to salinity stress were obtained [48].

4.7. RNA Extraction and qRT-PCR

Total RNA was extracted from *D. salina* using Trizol (Invitrogen, Carlsbad, CA, USA) according to the manufacturer's instructions. The RNA quality was analyzed with a NanoDrop 2000 spectrophotometer (Thermo, USA), after which cDNA was synthesized using the PrimeScript Reverse Transcriptase Kit (Takara, Japan), and cDNAs were amplified and detected using SYBR Green PCR Kit (Qiagen, Valencia, CA, USA). qRT-PCR was completed using the ABI 7500 Real-Time PCR system (Applied Biosystems, Foster City, CA, USA). The 18s-rRNA gene served as an endogenous control for normalization. The details of gene-specific qRT-PCR primers are listed in Supplementary Table S4. The primers were designed using the Primer-BLAST program (https://www.ncbi.nlm.nih.gov/tools/primer-blast) based on National Center for Biotechnology Information (NCBI) sequence data (http://www.ncbi.nlm.nih.gov/genbank/). The relative expression level was calculated as follows:

ratio = $2^{-\Delta\Delta Ct} = 2^{-(\Delta Ctt - \Delta Ctc)}$ (Ct: cycle threshold; Ctt: Ct of the target gene; Ctc: Ct of the control gene) [49].

Supplementary Materials: Supplementary materials can be found at http://www.mdpi.com/1422-0067/20/3/599/s1.

Author Contributions: Y.W., X.C. and Y.C. conceived and designed the experiments; Y.W., Z.G. and X.G. performed the experiments; J.Y., Z.X., Y.W. and Z.G. analyzed the data; Y.W. and X.C contributed reagents/materials/analysis tools; Y.W. and Y.C. wrote the paper.

Funding: This research was funded by the National Natural Science Fund of China, grant number 31472260.

Acknowledgments: We thank the PRIDE team for proteomics data processing and repository assistance.

Conflicts of Interest: The authors declare no conflicts of interest.

Abbreviations

iTRAQ	Isobaric Tags for Relative and Absolute Quantitation
DAPs	differentially abundant proteins
BP	biological process
MF	molecular function
CC	cellular components
GO	Gene Ontology
KEGG	The Kyoto Encyclopedia of Genes and Genomes
STRING	Search Tool for the Retrieval of Interacting Genes
PPI	predict protein interaction
LHC	light-harvesting complex
qRT-PCR	quantitative real-time polymerase chain reaction
SDS-PAGE	sodium dodecyl sulfate -polyacrylamide gel electrophoresis
SCX	strong cationic exchange
LC-MS/MS	liquid chromatography-tandem mass spectrometry

References

1. Zhang, H.; Han, B.; Wang, T.; Chen, S.; Li, H.; Zhang, Y.; Dai, S. Mechanisms of plant salt response: Insights from proteomics. *J. Proteome Res.* **2012**, *11*, 49–67. [CrossRef]
2. Parihar, P.; Singh, S.; Singh, R.; Singh, V.P.; Prasad, S.M. Effect of salinity stress on plants and its tolerance strategies: A review. *Environ. Sci. Pollut. Res. Int.* **2015**, *22*, 4056–4075. [CrossRef]
3. Katz, A.; Waridel, P.; Shevchenko, A.; Pick, U. Salt-induced changes in the plasma membrane proteome of the halotolerant alga *Dunaliella salina* as revealed by blue native gel electrophoresis and nano-LC-MS/MS analysis. *Mol. Cell. Proteom.* **2007**, *6*, 1459–1472. [CrossRef]
4. Wei, S.; Bian, Y.; Zhao, Q.; Chen, S.; Mao, J.; Song, C.; Cheng, K.; Xiao, Z.; Zhang, C.; Ma, W.; et al. Salinity-induced palmella formation mechanism in halotolerant algae *Dunaliella salina* revealed by quantitative proteomics and phosphoproteomics. *Front. Plant Sci.* **2017**, *8*, 810. [CrossRef]
5. Morowvat, M.H.; Ghasemi, Y. Culture medium optimization for enhanced β-carotene and biomass production by *Dunaliella salina* in mixotrophic culture. *Biocatal. Agric. Biotechnol.* **2016**, *7*, 217–223. [CrossRef]
6. Gong, W.-F.; Zhao, L.-N.; Hu, B.; Chen, X.-W.; Zhang, F.; Zhu, Z.-M.; Chen, D.-F. Identifying novel salt-tolerant genes from *Dunaliella salina* using a *Haematococcus pluvialis* expression system. *Plant Cell Tissue Organ* **2014**, *117*, 113–124. [CrossRef]
7. Chai, X.; Liu, Y.; Wang, Y.; Wu, T.; Liu, L. Prokaryotic Expression, Purification and Cloning of MAPKK Kinase Gene DsMAPKKK from *Dunaliella salina*. *Chin. Agric. Sci. Bull.* **2015**, *24*, 1085–1091.
8. Dani, V.; Simon, W.J.; Duranti, M.; Croy, R.R. Changes in the tobacco leaf apoplast proteome in response to salt stress. *Proteomics* **2005**, *5*, 737–745. [CrossRef]
9. Liu, C.W.; Chang, T.S.; Hsu, Y.K.; Wang, A.Z.; Yen, H.C.; Wu, Y.P.; Wang, C.S.; Lai, C.C. Comparative proteomic analysis of early salt stress responsive proteins in roots and leaves of rice. *Proteomics* **2014**, *14*, 1759–1775. [CrossRef]

10. Lv, D.W.; Subburaj, S.; Cao, M.; Yan, X.; Li, X.; Appels, R.; Sun, D.F.; Ma, W.; Yan, Y.M. Proteome and phosphoproteome characterization reveals new response and defense mechanisms of Brachypodium distachyon leaves under salt stress. *Mol. Cell. Proteom.* **2014**, *13*, 632–652. [CrossRef]

11. McLoughlin, F.; Arisz, S.A.; Dekker, H.L.; Kramer, G.; de Koster, C.G.; Haring, M.A.; Munnik, T.; Testerink, C. Identification of novel candidate phosphatidic acid-binding proteins involved in the salt-stress response of Arabidopsis thaliana roots. *Biochem. J.* **2013**, *450*, 573–581. [CrossRef]

12. Hossain, Z.; Khatoon, A.; Komatsu, S. Soybean proteomics for unraveling abiotic stress response mechanism. *J. Proteome Res.* **2013**, *12*, 4670–4684. [CrossRef]

13. Ji, W.; Cong, R.; Li, S.; Li, R.; Qin, Z.; Li, Y.; Zhou, X.; Chen, S.; Li, J. Comparative proteomic analysis of soybean leaves and roots by iTRAQ provides insights into response mechanisms to short-term salt stress. *Front Plant Sci.* **2016**, *7*, 573. [CrossRef]

14. Sobhanian, H.; Razavizadeh, R.; Nanjo, Y.; Ehsanpour, A.A.; Jazii, F.R.; Motamed, N.; Komatsu, S. Proteome analysis of soybean leaves, hypocotyls and roots under salt stress. *Proteome Sci.* **2010**, *8*, 19. [CrossRef]

15. Meng, F.; Luo, Q.; Wang, Q.; Zhang, X.; Qi, Z.; Xu, F.; Lei, X.; Cao, Y.; Chow, W.S.; Sun, G. Physiological and proteomic responses to salt stress in chloroplasts of diploid and tetraploid black locust (*Robinia pseudoacacia* L.). *Sci. Rep.* **2016**, *6*, 23098. [CrossRef]

16. Jia, Y.L.; Chen, H.; Zhang, C.; Gao, L.J.; Wang, X.C.; Qiu, L.L.; Wu, J.F. Proteomic analysis of halotolerant proteins under high and low salt stress in *Dunaliella salina* using two-dimensional differential in-gel electrophoresis. *Genet. Mol. Biol.* **2016**, *39*, 239–247. [CrossRef]

17. Liska, A.J.; Shevchenko, A.; Pick, U.; Katz, A. Enhanced photosynthesis and redox energy production contribute to salinity tolerance in Dunaliella as revealed by homology-based proteomics. *Plant Physiol.* **2004**, *136*, 2806–2817. [CrossRef]

18. Nam, M.H.; Huh, S.M.; Kim, K.M.; Park, W.J.; Seo, J.B.; Cho, K.; Kim, D.Y.; Kim, B.G.; Yoon, I.S. Comparative proteomic analysis of early salt stress-responsive proteins in roots of SnRK2 transgenic rice. *Proteome Sci.* **2012**, *10*, 25. [CrossRef]

19. Yin, X.; Komatsu, S. Comprehensive analysis of response and tolerant mechanisms in early-stage soybean at initial-flooding stress. *J. Proteom.* **2017**, *169*, 225–232. [CrossRef]

20. Chen, T.; Zhang, L.; Shang, H.; Liu, S.; Peng, J.; Gong, W.; Shi, Y.; Zhang, S.; Li, J.; Gong, J.; et al. iTRAQ-based quantitative proteomic analysis of cotton roots and Leaves reveals pathways associated with salt stress. *PLoS ONE* **2016**, *11*, e0148487. [CrossRef]

21. Lan, P.; Li, W.; Wen, T.N.; Shiau, J.Y.; Wu, Y.C.; Lin, W.; Schmidt, W. iTRAQ protein profile analysis of Arabidopsis roots reveals new aspects critical for iron homeostasis. *Plant Physiol.* **2011**, *155*, 821–834. [CrossRef]

22. Long, R.; Gao, Y.; Sun, H.; Zhang, T.; Li, X.; Li, M.; Sun, Y.; Kang, J.; Wang, Z.; Ding, W.; et al. Quantitative proteomic analysis using iTRAQ to identify salt-responsive proteins during the germination stage of two Medicago species. *Sci. Rep.* **2018**, *8*, 9553. [CrossRef]

23. Xia, F.; Yao, X.; Tang, W.; Xiao, C.; Yang, M.; Zhou, B. Isobaric tags for relative and absolute quantitation (iTRAQ)-based proteomic analysis of Hugan Qingzhi and its protective properties against free fatty acid-induced L02 hepatocyte Injury. *Front. Pharmacol.* **2017**, *8*, 99. [CrossRef]

24. Vedeler, A.; Hollas, H.; Grindheim, A.K.; Raddum, A.M. Multiple roles of annexin A2 in post-transcriptional regulation of gene expression. *Curr. Protein Pept. Sci.* **2012**, *13*, 401–412. [CrossRef]

25. Zhao, Y.; Hou, Y. Identification of NaHCO3 stress responsive proteins in *Dunaliella salina* HTBS using iTRAQ-based analysis. *J. Proteom. Bioinform.* **2016**, *9*, 137–143. [CrossRef]

26. Ge, Y.; Ning, Z.; Wang, Y.; Zheng, Y.; Zhang, C.; Figeys, D. Quantitative proteomic analysis of *Dunaliella salina* upon acute arsenate exposure. *Chemosphere* **2016**, *145*, 112–118. [CrossRef]

27. Laurent, J.M.; Vogel, C.; Kwon, T.; Craig, S.A.; Boutz, D.R.; Huse, H.K.; Nozue, K.; Walia, H.; Whiteley, M.; Ronald, P.C. Protein abundances are more conserved than mRNA abundances across diverse taxa. *Proteomics* **2010**, *10*, 4209–4212. [CrossRef]

28. Drummond, D.A.; Bloom, J.D.; Adami, C.; Wilke, C.O.; Arnold, F.H. Why highly expressed proteins evolve slowly. *Proc. Natl. Acad. Sci. USA* **2005**, *102*, 14338–14343. [CrossRef]

29. Vogel, C.; Marcotte, E.M. Insights into the regulation of protein abundance from proteomic and transcriptomic analyses. *Nat. Rev. Genet.* **2012**, *13*, 227. [CrossRef]

30. Zhao, P.; Cui, R.; Xu, P.; Wu, J.; Mao, J.L.; Chen, Y.; Zhou, C.Z.; Yu, L.H.; Xiang, C.B. ATHB17 enhances stress tolerance by coordinating photosynthesis associated nuclear gene and ATSIG5 expression in response to abiotic stress. *Sci. Rep.* **2017**, *7*, 45492. [CrossRef]

31. Asrar, H.; Hussain, T.; Hadi, S.; Gul, B.; Nielsen, B.; Khan, M. Salinity induced changes in light harvesting and carbon assimilating complexes of *Desmostachya bipinnata* (L.) Staph. *Environ. Exp. Bot.* **2017**, *135*, 86–95. [CrossRef]

32. Shu, S.; Yuan, Y.; Chen, J.; Sun, J.; Zhang, W.; Tang, Y.; Zhong, M.; Guo, S. The role of putrescine in the regulation of proteins and fatty acids of thylakoid membranes under salt stress. *Sci. Rep.* **2015**, *5*, 14390. [CrossRef]

33. Caffarri, S.; Tibiletti, T.; Jennings, R.; Santabarbara, S. A comparison between plant Photosystem I and Photosystem II architecture and functioning. *Curr. Protein Pept. Sci.* **2014**, *15*, 296–331. [CrossRef]

34. Natali, A.; Croce, R. Characterization of the major light-harvesting complexes (LHCBM) of the green alga *Chlamydomonas reinhardtii*. *PLoS ONE* **2015**, *10*, e0119211. [CrossRef]

35. Zhang, H.; Li, L.; Ye, T.; Chen, R.; Gao, X.; Xu, Z. Molecular characterization, expression pattern and function analysis of the OsHSP90 family in rice. *Biotechnol. Biotechnol. Equip.* **2016**, *30*, 669–676. [CrossRef]

36. Rutgers, M.; Muranaka, L.S.; Schulz-Raffelt, M.; Thoms, S.; Schurig, J.; Willmund, F.; Schroda, M. Not changes in membrane fluidity but proteotoxic stress triggers heat shock protein expression in *Chlamydomonas reinhardtii*. *Plant Cell Environ.* **2017**, *40*, 2987–3001. [CrossRef]

37. Tang, T.; Yu, A.; Li, P.; Yang, H.; Liu, G.; Liu, L. Sequence analysis of the Hsp70 family in moss and evaluation of their functions in abiotic stress responses. *Sci. Rep.* **2016**, *6*, 33650. [CrossRef]

38. Yokthongwattana, C.; Mahong, B.; Roytrakul, S.; Phaonaklop, N.; Narangajavana, J.; Yokthongwattana, K. Proteomic analysis of salinity-stressed Chlamydomonas reinhardtii revealed differential suppression and induction of a large number of important housekeeping proteins. *Planta* **2012**, *235*, 649–659. [CrossRef]

39. Hahn, A.; Bublak, D.; Schleiff, E.; Scharf, K.D. Crosstalk between Hsp90 and Hsp70 chaperones and heat stress transcription factors in tomato. *Plant Cell* **2011**, *23*, 741–755. [CrossRef]

40. Liu, A.; Xiao, Z.; Li, M.W.; Wong, F.L.; Yung, W.S.; Ku, Y.S.; Wang, Q.; Wang, X.; Xie, M.; Yim, A.K.; et al. Transcriptomic reprogramming in soybean seedlings under salt stress. *Plant Cell Environ.* **2018**, *12*, e0189159. [CrossRef]

41. Katz, A.; Avron, M. Determination of intracellular osmotic volume and sodium concentration in dunaliella. *Plant Physiol.* **1985**, *78*, 817–820. [CrossRef]

42. Bradford, M.M. A rapid and sensitive method for the quantitation of microgram quantities of protein utilizing the principle of protein-dye binding. *Anal. Biochem.* **1976**, *72*, 248–254. [CrossRef]

43. Polle, J.E.W.; Barry, K.; Cushman, J.; Schmutz, J.; Tran, D.; Hathwaik, L.T.; Yim, W.C.; Jenkins, J.; McKie-Krisberg, Z.; Prochnik, S.; et al. Draft Nuclear Genome Sequence of the Halophilic and Beta-Carotene-Accumulating Green Alga *Dunaliella salina* Strain CCAP19/18. *Genome Announc.* **2017**, *5*, e01105–e01117. [CrossRef]

44. Vizcaino, J.A.; Csordas, A.; del-Toro, N.; Dianes, J.A.; Griss, J.; Lavidas, I.; Mayer, G.; Perez-Riverol, Y.; Reisinger, F.; Ternent, T.; et al. 2016 update of the PRIDE database and its related tools. *Nucleic Acids Res.* **2016**, *44*, D447–D456. [CrossRef]

45. Conesa, A.; Gotz, S.; Garcia-Gomez, J.M.; Terol, J.; Talon, M.; Robles, M. Blast2GO: A universal tool for annotation, visualization and analysis in functional genomics research. *Bioinformatics* **2005**, *21*, 3674–3676. [CrossRef]

46. Kanehisa, M.; Goto, S. KEGG: Kyoto encyclopedia of genes and genomes. *Nucleic Acids Res.* **2000**, *28*, 27–30. [CrossRef]

47. Kanehisa, M.; Goto, S.; Sato, Y.; Kawashima, M.; Furumichi, M.; Tanabe, M. Data, information, knowledge and principle: Back to metabolism in KEGG. *Nucleic Acids Res.* **2014**, *42*, D199–D205. [CrossRef]

Int. J. Mol. Sci. **2019**, *20*, 599

48. Szklarczyk, D.; Franceschini, A.; Wyder, S.; Forslund, K.; Heller, D.; Huerta-Cepas, J.; Simonovic, M.; Roth, A.; Santos, A.; Tsafou, K.P.; et al. STRING v10: Protein-protein interaction networks, integrated over the tree of life. *Nucleic Acids Res.* **2015**, *43*, D447–D452. [CrossRef]
49. Livak, K.J.; Schmittgen, T.D. Analysis of relative gene expression data using real-time quantitative PCR and the 2(-Delta Delta C(T)) Method. *Methods* **2001**, *25*, 402–408. [CrossRef]

Article

Global Phosphoproteomic Analysis Reveals the Defense and Response Mechanisms of *Jatropha Curcas* Seedling under Chilling Stress

Hui Liu, Fen-Fen Wang, Xian-Jun Peng, Jian-Hui Huang and Shi-Hua Shen *

Key Laboratory of Plant Resources, Institute of Botany, Chinese Academy of Sciences, Beijing 100093, China; huiliu@ibcas.ac.cn (H.L.); wangfenfen@ibcas.ac.cn (F.-F.W.); pengxianjun@ibcas.ac.cn (X.-J.P.); jhhuang@ibcas.ac.cn (J.-H.H.)
* Correspondence: shshen@ibcas.ac.cn; Tel.: +86-10-6283-6545

Received: 24 November 2018; Accepted: 3 January 2019; Published: 8 January 2019

Abstract: As a promising energy plant for biodiesel, *Jatropha curcas* is a tropical and subtropical shrub and its growth is affected by one of major abiotic stress, chilling. Therefore, we adopt the phosphoproteomic analysis, physiological measurement and ultrastructure observation to illustrate the responsive mechanism of *J. curcas* seedling under chilling (4 °C) stress. After chilling for 6 h, 308 significantly changed phosphoproteins were detected. Prolonged the chilling treatment for 24 h, obvious physiological injury can be observed and a total of 332 phosphoproteins were examined to be significantly changed. After recovery (28 °C) for 24 h, 291 phosphoproteins were varied at the phosphorylation level. GO analysis showed that significantly changed phosphoproteins were mainly responsible for cellular protein modification process, transport, cellular component organization and signal transduction at the chilling and recovery periods. On the basis of protein-protein interaction network analysis, phosphorylation of several protein kinases, such as SnRK2, MEKK1, EDR1, CDPK, EIN2, EIN4, PI4K and 14-3-3 were possibly responsible for cross-talk between ABA, Ca^{2+}, ethylene and phosphoinositide mediated signaling pathways. We also highlighted the phosphorylation of HOS1, APX and PIP2 might be associated with response to chilling stress in *J. curcas* seedling. These results will be valuable for further study from the molecular breeding perspective.

Keywords: *Jatropha curcas*; phosphoproteomics; seedling; chilling stress; regulated mechanism

1. Introduction

As one of the most critical limiting factors, low temperature affects the plant growth and development broadly as well as yield, quality, postharvest life and geographic distribution [1]. Chilling tolerance is the ability of a plant to tolerate low temperature (0–15 °C) without injury or damage [2]. Although it is possible to enhance the physical and physiochemical tolerance according to cold acclimation [3], plant species origin from tropical and subtropical areas, such as *Oryza sativa*, *Zea mays* and *Lycopersicon esculentum*, are sensitive to chilling stress and easily damaged by chilling temperature [4]. The cellular structure and physiological characterization of tropical plants were both changed in response to chilling stress, especially the photosynthetic organelle-chloroplast [5–7]. Compared to PSI, PSII was likely more sensitive to chilling temperature under moderate light in tropical trees. In order to protect PSI at such stress situation, the PSII photoinhibition appeared and PSII reaction centers was closed [6]. Subsequently, the recovery of PSII from low temperature depended on its ability to maintain PsaA, Cyt b6/f and D1 protein at photoinhibitory conditions [8]. Low temperature stress induced the changes of a variety of protein kinases and transcription factors in plants [5,9,10]. It was reported that increases in the cytosolic transient calcium flux played a vital role in an early step of cold stress signaling [11–13]. ABA was responsible for the stomatal closure as well as low temperature stress responses [14] and SnRK2s were active protein kinases that participated in the regulation of ABA [15]. ABI5 and TRAB1 both were bZIP transcriptional factors and had been demonstrated to have the ability to mediated ABA signals [16–19]. HOS1 was another important negative regulator of cold stress signaling in plant cells [20], as well as acted as an E3 ligase to be required for the ubiquitination of ICE1 [21] and appeared to act upstream of CBF transcription.

As the major and reversible post-translational modification, protein phosphorylation was crucial for providing the basis for complex signaling networks and the regulation of diverse cellular functions in plants [22]. Phosphorylation mainly regulated kinases and phosphatases in a dynamic process, including signal transduction, homeostasis, protein degradation, metabolism and stress responses [23]. With the rapid development of proteomic technology, determination of phosphorylation sites mainly employed LC-MS/MS without gel-based [24]. Additionally, with the development of functional genomes and mass spectrum tools, it was practicable to identify and quantify the protein phosphorylation on a large scale [25]. Therefore, plant protein phosphorylation events played important roles in designing strategies to prevent crops from biotic and abiotic stresses, in particular, quantitative phosphoproteomics should be taken into consideration when illustrate the stress-induced related signal pathways [26]. It has been widely used for illustrating phosphorylation networks in *Arabidopsis thaliana*, *Brachypodium distachyon*, *Triticum aestivum*, *Broussonetia papyrifera* after hormonal and dehydration treatment, salt and chilling stress [27–30].

Jatropha curcas L., which originated from the tropics or subtropics, is a woody oil plant that belongs to the Euphorbiaceae family [31]. Due to the abundant oil content in its seeds (as high as 40%), *J. curcas* has been considered as the ideal material for biodiesel in the world [32]. Taking the advantage of the high-throughput technologies, a great body of information on *J. curcas* has been achieved with the application of genomic [33,34], transcriptomic [35,36] and proteomic [37–41] sequencing. The whole size of *J. curcas* was calculated as about 410 Mb and approximately 286 Mb consisting of 120,586 contigs and 29,831 singlets have been sequenced. These databases are valuable for further studies focusing on all aspects of molecular mechanisms. As an arisen economic woody species, *J. curcas* has been paid worldwide attentions. However, low temperature is still the major factor limited its distribution and affects the production of *J. curcas* seeds consequently. In consideration of the research and application, it is valuable to explore the metabolic network and regulated signal pathway of *J. curcas* under low temperature stress. Nevertheless, only few studies focused on molecular mechanism of chilling response in *J. curcas*. A previous study had employed transcriptome to analyze *J. curcas* under cold stress (12 °C), however, the result was sweeping without accurate mapping the regulated network basing on the experimental data, besides, the treatment temperature was also too moderate [42] to evaluate almost the limiting adverse effect under chilling stress. Another study only identified 8 photosynthesis related proteins significantly changed from *J. curcas* seedling under cold stress (4 °C) instead of high-throughput identification [43]. In this study, quantitative phosphoproteomics combining with traditional cellular observation, physiological measurement were employed to explore the chilling response and defense mechanism in *J. curcas* seedling at phosphorylation level. We aim to understand how sensitive is *J. curcas* to chilling stress and further unveil the specific phosphorylated proteins involved in potential pathways in *J. curcas* under chilling stress. The results will be benefit for providing clue to screen the chilling resistant *J. curcas* species.

2. Results

2.1. Physiological Changes of Leaves from J. curcas Seedling under Chilling Stress

The leaves of *J. curcas* seedling drooped and pseudostem tilted after 24 h of chilling treatment (4 °C) and mostly recovered after being returned to 28 °C for 24 h (Figure 1A). To evaluate the adverse effects of chilling stress quantitatively and determine the best time for sample collection after chilling stress for subsequent phosphoproteomic analysis, photosynthetic characteristics were measured. In response to the chilling treatment, the value of net photosynthetic rate (Pn) at each treatment stage (C0 h, C6 h,

C24 h and R24 h) displayed significant difference with each other. the Pn decreased to 65.9% and 4.5% of pre-treatment (C0 h) levels after chilling for 6 h (C6 h) and 24 h (C24 h), respectively but returned to 44.3% after recovery for 24 h (R24 h). Comparing to Pn, the intercellular CO_2 concentration (Ci) showed the opposite change tendency, the C6 h and C24 h of Ci increased to 160% and 408% respectively and almost returned to C0 h level after 24 h recovery. Conductance to H_2O (Cond) and transpiration rate (Trmmol) shared the similar change curve, only the R24 h of Cond and Trmmol showed the significant changes compared with C0 h (Figure 1B). These findings were slightly different with the studies for *O. sativa* and *Musa paradisiaca* under chilling stress (4 °C), the species also originated in the tropics [44,45].

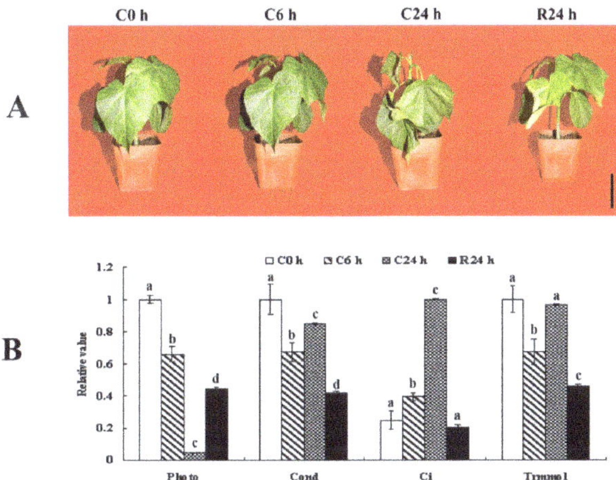

Figure 1. Morphological and physiological responses of *J. curcas* seedling to different treatment condition. Four-leaf stage seedling were treated at 4 °C for 0, 6 and 24 h (C0 h, C6 h and C24 h) and then allowed to recover for 24 h (R24 h) (**A**). The black bar equals 6 cm. The Pn, Cond, Ci and Trmmol were showed in (**B**). The values of relative % for each column are means ±S.D. of three biological replicates. The different lowercase letters labeled above columns indicate significant changes according to one-way ANOVA ($p < 0.05$).

2.2. Ultrastructure Change of J. curcas Seedling under Chilling Stress

To assess the damage of *J. curcas* seedling under chilling treatment, the fourth-leaves at each treatment (C0 h, C6 h, C24 h and R24 h) were prepared for ultrastructural observation. In the leaf cell from C0 h, the morphological structures were intact chloroplasts with numerous embedded starch granules, which mainly distributed closely to the cell membrane. Besides, intact vacuole and mitochorina could also be observed clearly (Figure 2, C0 h). As a universal symptom, the obvious manifestations of chilling stress were chloroplast swelling, a distortion of vacuole, a reduction in the size of starch granules (Figure 2, C6 h), prolonged chilling treatment leaded to grana unstacked and starch granules continued to diminish with time till disappeared completely (Figure 2, C24 h). However, the mitochorina from the *J. curcas* seedling showed relative stationary and was not as sensitive as chloroplast and vacuole in the whole chilling treatment process (Figure 2). After recovery for 24 h, the chloroplasts of *J. curcas* seedling returned to normal morphological status (Figure 2, R24 h).

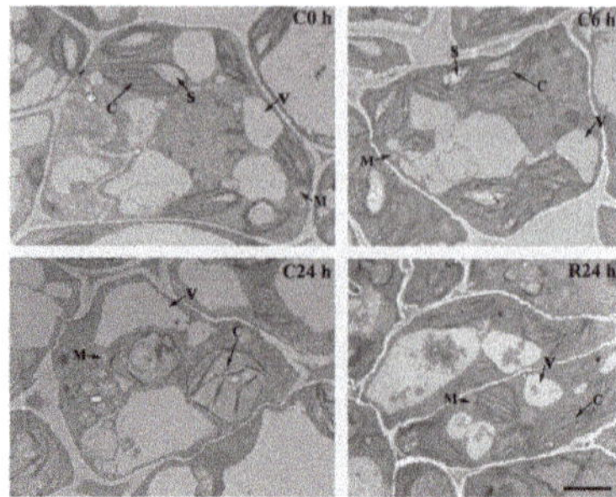

Figure 2. The ultrastructures of leaves at different treatment condition (C0 h, C6 h, C24 h, R24 h) from *J. curcas* seedling. The bar equals 2 μm. C, chloroplast; M, Mitochondria; S, starch granules; V, vacuole.

2.3. Phosphoprotein Identification and Phosphorylated Site Location

In total, 3101 phosphopeptides with 3101 phosphorylated sites corresponding to 1810 phosphoproteins were identified (Table S1) and the proportions of pS, pT and pY sites were calculated as 89.4%, 9.8% and 0.8%, respectively (Figure S1). The mass spectrometry proteomics data have been deposited in the ProteomeXchange Consortium (http://proteomecentral.proteomexchange.org) via the PRIDE partner repository [46] with the dataset identifier PXD011438. In order to evaluate analytical reproducibility, a range of quality control measures were taken for the three biological replicates of each condition before the comparison analysis of the phosphorylation levels between the chilling stress condition and the control. The result of Pearson correlation analysis showed that the four samples were individually clustered confidently with their replicates (Figure S2). Only the phosphopeptides identified from all biological repeats were used for further analysis.

2.4. Screening Phosphoproteins with Phosphorylation Level Significantly Changed

The intensity of each phosphopeptide was normalized as the Zhang et al. described [28]. According to the ANOVA analysis, 996 phosphorylated sites corresponding to 805 phosphoproteins were screened out (Table S2). Compared to untreated *J. curcas* seedling (C0 h), 308, 332 and 291 phosphorylation sites corresponding to 279, 313 and 270 phosphoproteins were differentially changed after chilling for 6, 24 h (C6 h, C24 h) and recovery for 24 h (R24 h), respectively (Tables S3–S5 and Figure 3A). There were 44, 76, 76 and 112 phosphorylation sites corresponding to induced, up regulated, down regulated and depressed regulation after 6 h chilling treatment, prolonged the chilling treatment time to 24 h, the number of phosphorylation site for each regulation has changed, which are 38, 94, 67 and 133, respectively (Figure 3A). To be mentioned, when recovery for 24 h, the number of induced phosphorylation sites increased obviously to 76 while the depressed ones decreased to 63 (Figure 3A). Further venny charts indicated that a number of differential phosphorylation sites were overlapped between C6 h and C24 h when compared to C0 h, however, only few differential phosphorylation sites were overlapped no matter the comparison between C6 h and R24 h or C24h and R24 h (Figure 3B).

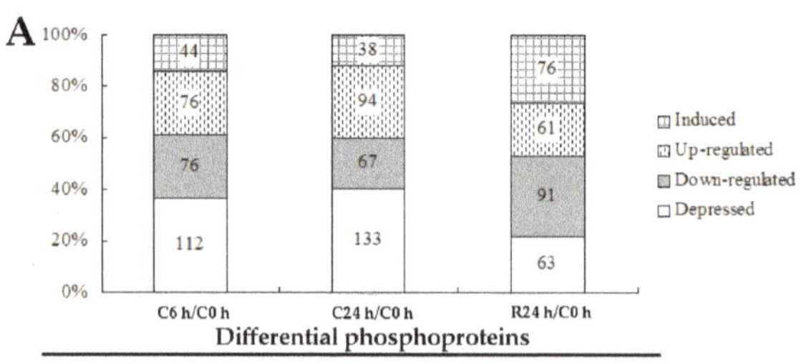

Figure 3. *Cont.*

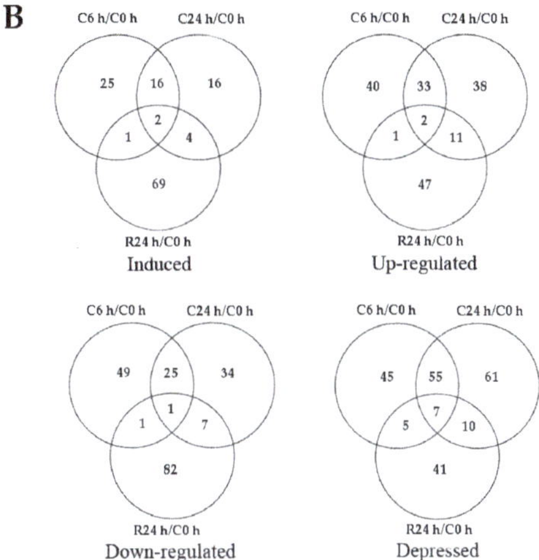

Figure 3. Statistical analysis of phosphoproteomics of *J. curcas* seedling under different treatment. The significantly changed phosphopeptides in each sample (C6 h, C24 h and R24 h) when compared to C0 h (**A**); Venn diagram of significantly changed phosphorylation sites and phosphoproteins distributed in each sample (C6 h, C24 h and R24 h) when compared to C0 h (**B**).

2.5. GO Annotation of Significant Changed Phosphoproteins

All the identified significant changed phosphoproteins (805) were used for GO annotation. The distribution pie charts for biological process, cellular component and molecular function are shown in Figure 4. In the whole treatment process, the cellular protein modification process, transport, cellular component organization and signal transduction were significantly overrepresented from the biological process perspective. Cytoskeleton and ribosome were significantly overrepresented from the cellular component perspective. Nucleotide binding, protein binding and kinase activity were significantly overrepresented from the molecular function perspective (Figure 4).

Figure 4. *Cont.*

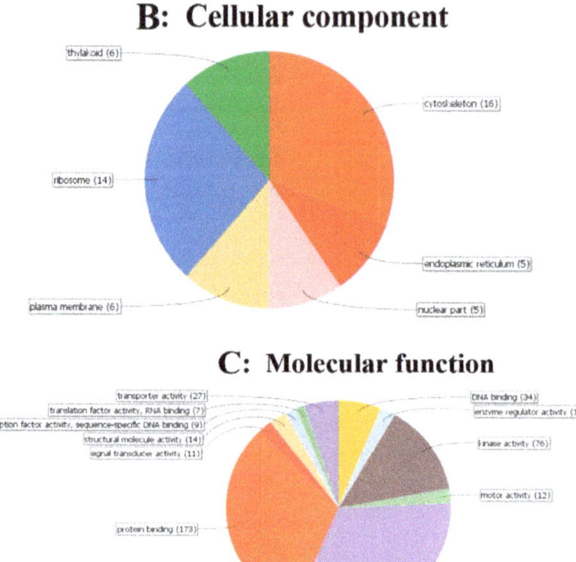

Figure 4. Functional classification of the significantly changed phosphoproteins from *J. curcas* seedling under different treatment by GO analysis. Over-represented GO terms were displayed graphically as pie charts for three GO vocabularies: (**A**) biological process; (**B**) cellular component; (**C**) molecular function. The number in brackets represents the phosphoprotein number within the group and the color of the pie chart represents the significance of enrichment.

2.6. Conservation Analysis of the Significant Changed Phosphoproteins

The sequences of the 805 significant changed phosphoproteins were used as queries to blast phosphoprotein databases that were constructed using data sets in the Plant Protein Phosphorylation DataBase (P3DB) [47] and PhoPhAt [48]. *O. sativa* and *A. thaliana* were compared against *J. curcas* to determine the degree of conservation of phosphorylation sites among different plant species. The thresholds were set as score ≥ 80, E-value $< 1 \times 10^{-10}$ and identity $\geq 30\%$. In all, 561 (69.7%) of the 805 phosphoproteins had phosphorylated orthologs in the two species, 163 (20.2%) had phosphorylated orthologs in only one species (Figure 5 and Table S6), 81 phosphoproteins had no phosphorylated orthologs in both of the two species (Table S6).

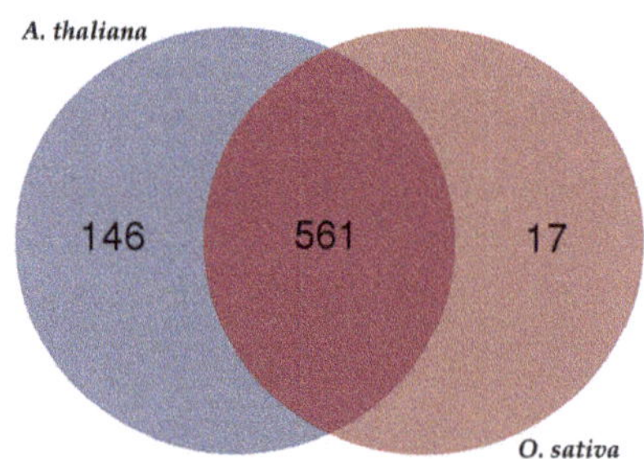

Figure 5. Conservation analysis of significantly changed phosphoproteins in *J. curcas* seedling under different treatment.

2.7. Analysis of Phosphorylation Motifs of Significant Changed Phosphopeptides

The kinase related phosphorylation motifs of the significantly changed phosphopeptides were identified by employing WebLogo and motif-X. Briefly, the significant changed phosphopeptides were centered at the phosphorylated amino acid residues of each experimental group (Table S7) and then were submitted for Weblogo analysis and phosphorylation motif extraction. Nine phosphorylation motifs were enriched from the four experimental groups (Figure 6). Those phosphorylation motifs were then searched in relevant databases to find the specific protein kinases. [sPxK] and [sPxR] were the CDK motifs [49], [sPxxxxR] motif resembles the sPxR motif which is recognized by CDK [50,51]. [sP] and [tP] motifs were the proline-directed motifs, which were potential substrates of MAPK [49]. [LxRxxs] was basic motif representative of CDPK substrate and the motif [Rxxs] was a potential substrate for CDPK-II, which was also the 14-3-3 binding motif [52–54]. The two motifs [Rxxs] and [Kxxs] could be assigned to motif [-(K/R)-x-x-(pS/pT)-] and were identified as a phosphorylation motif of the SnRK2 or CDPK in plants by previous study [55]. [sF] contained the phenylalanine residue and was the minimal MAPK target motif [30].

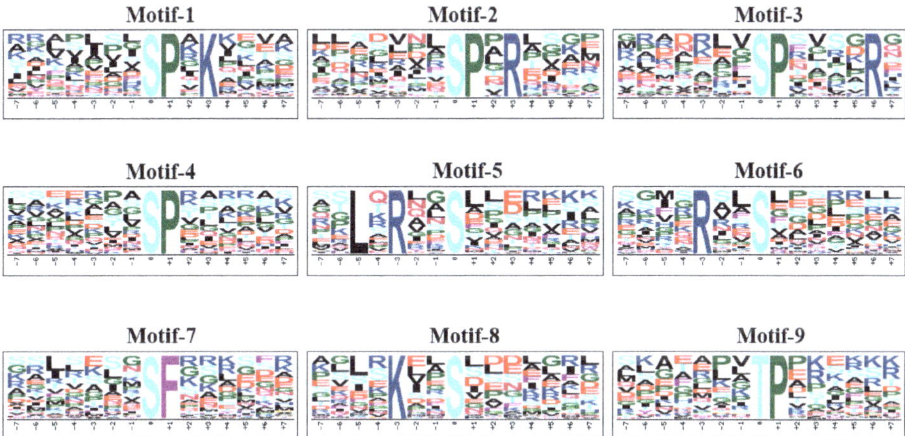

Figure 6. Phosphorylation motifs extracted from the phosphopeptides with the significantly changes by Motif-X from *J. curcas* seedling under different treatment.

2.8. Protein-Protein Interaction (PPI) Analysis of Significant Changed Phosphoproteins

The PPI network of the significant changed phosphoproteins identified in the current study were analyzed by STRING (http://string-db.org,version9.1). A total of 514 KOGs representing 610 phosphoproteins (Table S8) were used to construct the PPI network. In order to improve the reliability of the PPI analysis, the confidence score was set at the highest level ($\geq$0.900). Finally, a complex PPI network that contained 319 nodes and 1924 edges was displayed through Cytoscape (Figure S3). With the aim to further extract the key potential interacting proteins from the whole PPI network, 85 KOGs representing 183 significantly changed phosphoproteins, which related to signal transduction, posttranslational modification and intracellular trafficking, transport (Tables S9 and S10), were chosen and centered to construct the subnetwork (Figure 7). The result showed that KOG0583 (protein id: 897 and 2462), KOG0841 (protein id: 1724 and 2765) and KOG0070 (Protein id: 5) were the centered phosphoproteins in each functional group (Table S9 and Figure 7).

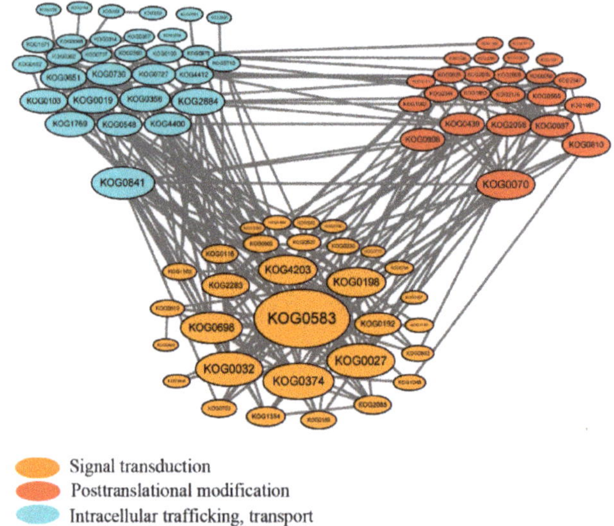

Figure 7. Protein-protein interaction (PPI) network of signal transduction, posttranslational modification and intracellular trafficking, transport related phosphoproteins by STRING. Nodes with orange, red and blue background color represent the KOGs of differential phosphoproteins related with signal transduction, posttranslational modification and intracellular trafficking, transport, respectively.

2.9. An Overview of Response and Defense Mechanisms of J. curcas Seedling under Chilling Stress

Based on the above results, 111 phosphoproteins (Table 1) with significant changes (with credible ANOVA analysis and fold change $\geq$ 2) were chosen to figure out a systematic chilling response and defense pathway in *J. curcas* seedling (Figure 8). The ion stress signal was transferred by chilling sensors on plasma membrane into cells and led to increase of Ca^{2+}. Subsequently, ABA, ethylene, MAPK, Phosphatidylinositol, CDPK and 14-3-3 signal pathways were activated by phosphorylation modification in *J. curcas* seedling under chilling stress. These signals were then transferred into nucleus and induced the expression changes of response and defense related genes. Under chilling stress, photosynthesis was depressed, which resulted in the proteins associated with PSI and PSII to be significantly changed at the phosphorylation level. Channels and transporters on the membrane which were associated with ion, auxin, H_2O_2 also regulated through phosphorylation or dephosphorylation. According to change at phosphorylation level, the misfolded proteins were possibly handled either by refolding or degradation. Therefore, an ubiquitination mediated degradation pathway centered on E3s was displayed in the schematic representation.

Table 1. 110 significantly changed phosphoproteins from *J. curcas* seedling in response to chilling.

ID	Proteins	Annotation	Modified Sequence	Change Pattern *
		Ions and Transport		
1	ABM54183.1	Aquaporin pip2-8	_ALGS(ph)FRS(ph)NPTN_	
2	ABM54183.1	Aquaporin pip2-8	_ALGSFRS(ph)NPTN_	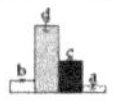
3151	KDP44363.1	Aquaporin PIP2-5	_VKDVAEQGS(ph)FSAK_	
555	KDP24122.1	Two pore calcium channel protein 1	_(ac)M(ox)EKPLLGETSSNAS(ph)FR_	
172	KDP21053.1	Sodium/calcium exchanger NCL2-like	_LVNDEGQVDVS(ph)CIKR_	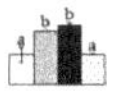
1832	KDP33816.1	K(+) efflux antiporter 4	_GDS(ph)FRADSAK_	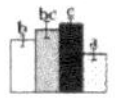
535	KDP23925.1	Auxin efflux carrier component 1c	_EDFS(ph)FGNR_	
536	KDP23925.1	Auxin efflux carrier component 1c	_LAVS(ph)PGKVEGQR_	
2342	KDP38049.1	Auxin efflux carrier component 3-like	_AIANAGDFPGEDFS(ph)FAGK_	
2343	KDP38049.1	Auxin efflux carrier component 3-like	_LGS(ph)SSTAELHPK_	
2352	KDP38049.1	Auxin efflux carrier component 3-like	_S(ph)LGPGS(ph)FSALTPR_	
1468	KDP31195.1	ATPase plasma membrane-type	_(ac)GDKS(ph)EVLEAVLK_	
2710	KDP40607.1	ATPase plasma membrane-type	_GHVES(ph)VIR_	
2711	KDP40607.1	ATPase plasma membrane-type	_S(ph)LQGLM(ox)AADLEFNGK_	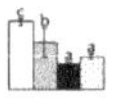
3274	KDP45466.1	Plasma membrane ATPase 1-like	_GS(ph)FNELNQMAEEAK_	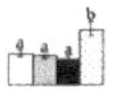
3755	KDP31195.1	ATPase plasma membrane-type	_T(ph)LHGLQPPDTK_	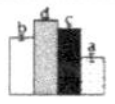
3794	KDP33445.1	Plasma membrane ATPase 4	_GLDIDTIQQHYT(ph)V_	
3836	KDP35383.1	Plasma membrane ATPase 4	_T(ph)LHGLQPPETASIFNEK_	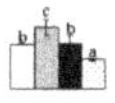
3941	KDP40607.1	ATPase plasma membrane-type	_NLDLNVIQGAHT(ph)V_	
		Phosphorylation Events Involved in Photoinhibition		
9	ACN72692.1	PsaA	_IIRS(ph)PEPEVK_	
508	KDP23672.1	Photosystem II reaction center W chloroplastic-like	_LATILPAAS(ph)FK_	
1690	KDP32802.1	Photosystem II 22 kDa chloroplastic	_FVDDPPTGIEGAVIPPGKS(ph)FR_	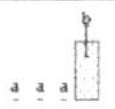

Table 1. *Cont.*

ID	Proteins	Annotation	Modified Sequence	Change Pattern *
3541	ACN72673.1	photosystem II protein D1	_(ac)T(ph)AILERR_	
3542	ACN72687.1	photosystem II protein D2	_(ac)T(ph)IALGKFTK_	
3544	ACN72707.1	photosystem II protein L	_(ac)T(ph)QSNPNEQNVELNR_	
3634	KDP24461.1	Chlorophyll a-b binding protein CP29.1 chloroplastic	_T(ph)ELADVK_	
4027	KDP44781.1	Chlorophyll a-b binding protein CP29.3 chloroplastic	_FGFPGFGT(ph)KK_	
3900	KDP38815.1	Ferredoxin-A-like	_LLT(ph)PEGEK_	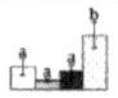
3091	KDP44002.1	Ru large subunit-binding protein subunit alpha	_NVVLDEFGS(ph)PK_	
98	KDP20412.1	Translocase of chloroplast chloroplastic-like	_LVNGS(ph)SEDIR_	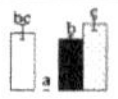
1008	KDP28139.1	Translocase of chloroplast chloroplastic	_M(ox)NEETEVLS(ph)GGNEK_	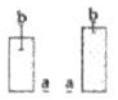
Chloroplast movement				
1231	KDP29354.1	Protein CHUP1, chloroplastic	_NAGETVAITS(ph)FGK_	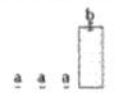
1233	KDP29354.1	Protein CHUP1, chloroplastic	_SFS(ph)GGSPR_	
2153	KDP36752.1	Plastid movement impaired1	_LTELDS(ph)IAQQIK_	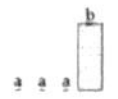
2154	KDP36752.1	Plastid movement impaired1	_MEDETES(ph)QRLDADEETVTR_	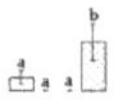
2419	KDP38475.1	J domain-containing protein required for chloroplast accumulation response 1	_VLS(ph)PGRPLPPR_	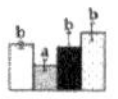
2248	KDP37156.1	Root phototropism protein 3	_S(ph)PNLGFEQPGGSISK_	
2767	KDP40950.1	Root phototropism protein 3	_LLEHFLVQEQTENSS(ph)PSR_	
3298	KDP45678.1	Root phototropism protein 2	_NQLQTDVS(ph)LIR_	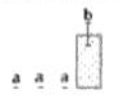
3118	KDP44148.1	Phototropin-1	_RNS(ph)ENVPSNR_	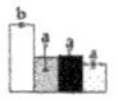
3490	KDP47050.1	Phototropin-2	_FAVDSTRTS(ph)EESEAGAFPR_	
ROS related				
15	ACV50426.1	Cytosolic ascorbate peroxidase-1	_HS(ph)AELAHAANTGLDIALR_	
16	ACV50426.1	Cytosolic ascorbate peroxidase-1	_LPGANEGS(ph)DHLR_	

Table 1. *Cont.*

ID	Proteins	Annotation	Modified Sequence	Change Pattern *
773	KDP26232.1	Pyrroline-5-carboxylate reductase	_DDVAS(ph)PGGTTIAGIHELEK_	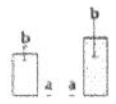
2948	KDP42714.1	Monodehydroascorbate reductase	_VIGAFLEGGS(ph)PDENQAIAK_	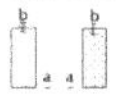
CDPK related signal				
133	KDP20751.1	Calcium-dependent protein kinase 21-like	_LGS(ph)KLSETEVK_	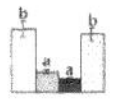
286	KDP21986.1	CDPK-related kinase 4-like	_WPLPPPS(ph)PAK_	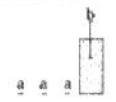
290	KDP22101.1	Calcium-dependent protein kinase 8-like	_ENPFFGNDYVVNNGS(ph)GR_	
580	KDP24440.1	Calcium-dependent protein kinase 13	_FNSLS(ph)MK_	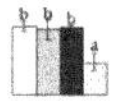
1253	KDP29499.1	Calcium-dependent protein kinase 1	_NS(ph)FSIGFR_	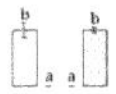
2491	KDP39132.1	Calcium-dependent protein kinase 33-like	_S(ph)PVQPTYQLPSQQPPIHVPR_	
3476	KDP47014.1	Calcium-dependent protein kinase 26	_LYQGINQPEEQSAASHS(ph)K_	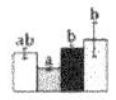
3477	KDP47014.1	Calcium-dependent protein kinase 26	_NS(ph)LNMSMR_	
2413	KDP38457.1	Calcium-binding protein CML41	_LITSS(ph)LPR_	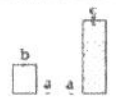
3546	KDP45998.1	Calmodulin-7	_M(ox)KDT(ph)DSEEELK_	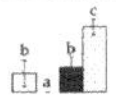
Ethylene and ABA related signal				
521	KDP23695.1	Protein EIN4	_VFSENGS(ph)EGKNDR_	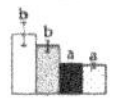
1204	KDP29130.1	Ethylene-insensitive protein 2	_M(ox)VPGSISSVTYDGPPS(ph)FR_	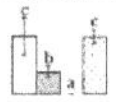
897	KDP27185.1	SNF1-related protein kinase catalytic subunit alpha kin10	_M(ox)HANEPTS(ph)PAVGHR_	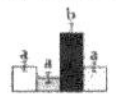
1412	KDP30923.1	SNF1-related protein kinase regulatory subunit gamma-1	_S(ph)PEANLGM(ox)K_	
2036	KDP35506.1	SNF1-related protein kinase regulatory subunit beta-2	_GEEEGLNS(ph)PSGSGGGGIDGGGGR_	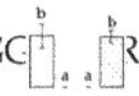
2462	KDP38831.1	Serine threonine-protein kinase SRK2a	_VKEAQAS(ph)GEFR_	
58	AIA57942.1	Abscisic acid-insensitive 5-like protein 2	_QAS(ph)LSLTSALSK_	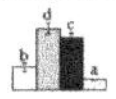
2400	KDP38330.1	bZIP transcription factor TRAB1	_QGS(ph)LTLPR_	
2703	KDP40550.1	E3 ubiquitin-protein ligase KEG	_VGFPGAS(ph)R_	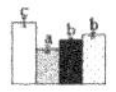
2644	KDP40139.1	Protein phosphatase 2c 70	_S(ph)GPEKDDLLESVPK_	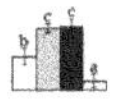

Table 1. *Cont.*

ID	Proteins	Annotation	Modified Sequence	Change Pattern *
3925	KDP40139.1	Protein phosphatase 2C 70	_VGQT(ph)LKR_	
MAPK related signal				
1678	KDP32742.1	Mitogen-activated protein kinase kinase kinase 1	_FHDM(ox)DS(ph)PR_	
2287	KDP37397.1	Protein-tyrosine-phosphatase MKP1	_S(ph)LDEWPK_	
2304	KDP37750.1	serine/threonine-protein kinase EDR1	_SIS(ph)MTPEIGDDIVR_	
2597	KDP39907.1	E3 ubiquitin-protein ligase HOS1	_IS(ph)PSSLADR_	
Phosphoinositide metabolism				
1558	KDP31989.1	Phosphatidylinositol-3,4,5-trisphosphate 3-phosphatase and dual-specificity protein phosphatase PTEN, putative	_LTS(ph)GFGLHFASAAPGPNESSK_	
1873	KDP33967.1	1-phosphatidylinositol-3-phosphate 5-kinase Fab1a	_S(ph)FGSGEYR_	
2004	KDP35146.1	Phosphoinositide phosphatase SAC9	_RAS(ph)FGGSVENDPCLHAR_	
2372	KDP38175.1	type IV inositol polyphosphate 5-phosphatase 9	_GPS(ph)LDLPR_	
2718	KDP40633.1	1-phosphatidylinositol-3-phosphate 5-kinase Fab1b	_TSSS(ph)FGSGEFR_	
2735	KDP40719.1	Phospholipid:diacylglycerol acyltransferase 1-like	_S(ph)REPTNSSTDLK_	
2911	KDP42390.1	Phosphoinositide phosphatase SAC1	_IGS(ph)GSDNLSSTMHR_	
3110	KDP44118.1	Phosphoinositide phospholipase C 2	_RLS(ph)LSEPQLEK_	
3180	KDP44771.1	Phosphoinositide phospholipase C 2-like	_GAS(ph)DEEAWGKEVSDLK_	
3364	KDP46231.1	phosphatidylinositol 4-kinase alpha 1	_LSGVGAAES(ph)K_	
3551	KDP24987.1	Diacylglycerol kinase 5-like	_KFGAAPT(ph)FR_	
3975	KDP41761.1	Phosphatidylinositol phosphatidylcholine transfer protein SFH8	_LT(ph)PVREESK_	
Protein misfolding and degradation				
126	KDP20698.1	Plant UBX domain-containing protein 4	_TLS(ph)DLNRR_	
239	KDP21542.1	NPL4-like protein 1	_TIAGPAIHPAGS(ph)FGR_	
1207	KDP29157.1	Cell division cycle protein 48 homolog	_DFS(ph)TAILER_	

Table 1. *Cont.*

ID	Proteins	Annotation	Modified Sequence	Change Pattern *
1583	KDP32077.1	Peptidyl-prolyl cis-trans isomerase cyp65	_S(ph)FTSTSFDPVTK_	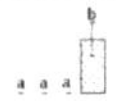
1216	KDP29225.1	E3 ubiquitin-protein ligase UPL7	_DLS(ph)LDFTVTEESFGKR_	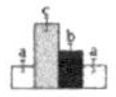
1276	KDP29647.1	E3 ubiquitin-protein ligase UPL3	_S(ph)SVNIGDAAR_	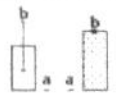
2820	KDP41504.1	E3 ubiquitin-protein ligase UPL2-like	_ANLGNVNAGS(ph)VHGK_	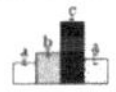
2907	KDP42227.1	E3 ubiquitin-protein ligase RING1	_NAGDRS(ph)PFNPVIVLR_	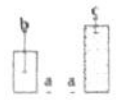
1279	KDP29680.1	Ubiquitin-conjugating enzyme E2 23	_LSPAAVSTS(ph)DSESAGELK_	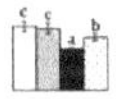
3548	KDP36146.1	ubiquitin-conjugating enzyme E2 variant 1D	_T(ph)LGSGGSSVVVPR_	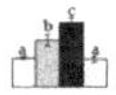
1272	KDP29632.1	26s proteasome non-atpase regulatory subunit 4 homolog	_VEEPSSTS(ph)QDATVVEK_	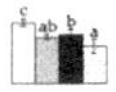
1742	KDP33301.1	26s protease regulatory subunit 6b homolog	_PVLEPLPS(ph)IPK_	
14-3-3 Related				
655	KDP25197.1	Serine threonine-protein kinase BLUS1	_ASANSLS(ph)APIK_	
657	KDP25197.1	Serine threonine-protein kinase BLUS1	_KLPS(ph)FSGPLM(ox)LPNR_	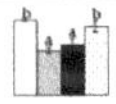
1724	KDP32938.1	14-3-3-like protein a	_MS(ph)PTETSR_	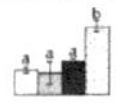
2765	KDP40925.1	14-3-3 protein 6	_(ac)AAGS(ph)PREDNVYMAK_	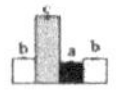
3118	KDP44148.1	Phototropin-1	_RNS(ph)ENVPSNR_	
HSPs				
1375	KDP30625.1	Chaperonin 60 subunit beta chloroplastic	_LAS(ph)KVDAIK_	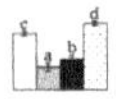
1658	KDP32613.1	Heat shock cognate protein 80	_EVSHEWS(ph)LVNK_	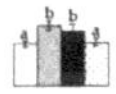
2898	KDP42168.1	kDa class I heat shock	_(ac)ALLPSFFGNS(ph)R_	
3205	KDP44981.1	Heat shock cognate 70 kDa protein 2-like	_RFS(ph)DASVQSDIK_	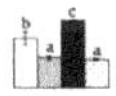
3206	KDP44981.1	Heat shock cognate 70 kDa protein 2-like	_FSDASVQS(ph)DIK_	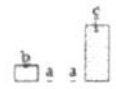
3984	KDP42448.1	Stromal 70 kDa heat shock-related chloroplastic	_LKT(ph)PVENSLR_	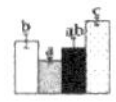
Others				
1743	KDP33317.1	Mitochondrial import receptor subunit TOM9-2	_TVS(ph)ESAVLNTAK_	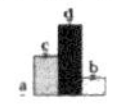

Table 1. *Cont.*

ID	Proteins	Annotation	Modified Sequence	Change Pattern *
1984	KDP34991.1	ATP synthase subunit mitochondrial	_GQNVLNTGS(ph)PITVPVGR_	
3239	KDP45196.1	Acyl-coenzyme a thioesterase mitochondrial	_(ac)MDFNSPS(ph)PR_	
3307	KDP45748.1	Malonyl-acyl carrier protein mitochondrial	_LEAALAATAIKS(ph)PR_	
385	KDP22798.1	Regulatory-associated protein of TOR 1	_PGEPTTSS(ph)PTTSLAGLAR_	
415	KDP23116.1	Autophagy-related protein 13	_GAPFTVNQPFGGS(ph)PPAYR_	

* The four columns from left to right were corresponding to C0 h, C6 h, C24 h and R24 h. The different lowercase letters labeled above columns indicate significant changes according to one-way ANOVA ($p < 0.05$).

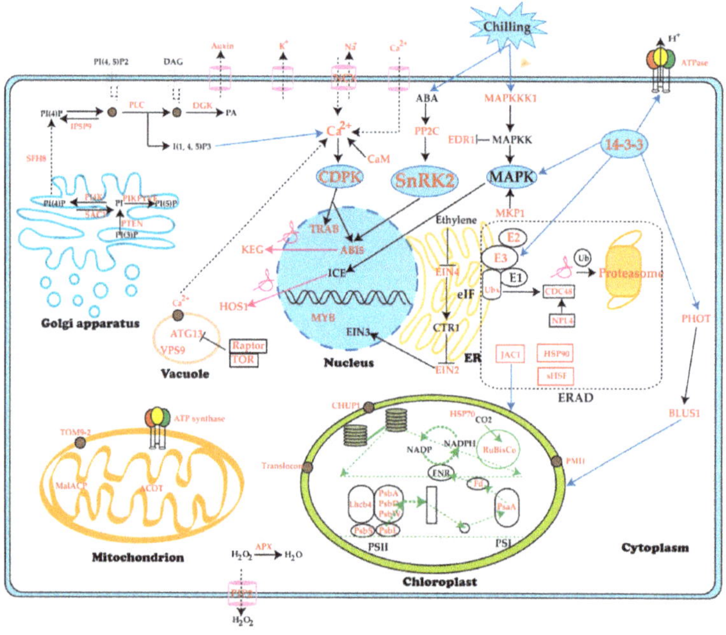

Figure 8. Schematic presentation of systematic chilling response and defense mechanisms in *J. curcas* seedling at phosphorylation level. The phosphorylation regulatory sites were shown in Table 1. Black solid lines with arrows represent catalytic reaction, while dotted lines with arrows represent material transport. Blue solid lines with arrows represent direct relationship, while magenta lines with arrows represent ubiquination. Black solid lines with stubs represent negative regulation. Green dotted lines with arrows in chloroplast represent transport between PSII and PSI. The proteins labeled with red colors represent significantly changed phosphoproteins among C0 h, C6 h, C24 h and R24 h. The area outlined with dotted square represents ER associated degradation (ERAD).

3. Discussion

We performed a comprehensive analysis of *J. curcas* seedling under chilling stress by employing the phosphoproteomic approach. Phosphorylation changes frequently occurred under chilling stress and the number of significantly changed phosphoproteins continually increased within the chilling duration (Figure 3 and Tables S2–S5). As a low temperature-sensitive species, *J. curcas* appeared to respond to and defend against chilling stress mainly through reversible protein phosphorylation, which was stimulated by Ca^{2+}, phytohormones and phosphoinositide. The whole regulated network illustrated in this study mainly included signal transduction, ion transport and protein ubiquitination. At the early chilling treatment period, protein phosphorylation might involve in the regulation of ion transport for homeostasis. Prolonging the time of chilling treatment, the physiological injury appeared, which resulted in the proteins related chloroplast movement were phosphorylated against chilling (Figure 2 and Table 1).

3.1. SnRKs Played a Central Role in the Chilling Responsive Signal Pathways

When plants were exposed to chilling stress, the osmotic stress mediated by ABA accompanied and the corresponding signal pathways were triggered [4]. According to the PPI analysis (Figure 7), KOG0583 represented SnRKs (Table S9, protein id: 897 and 2462) showed the central position, which involved in the ABA mediated signal pathway and seemed to act as a vital defense approach against chilling stress in *J. curcas* seedling. JcSRK2a (id 2462) were observed to be significantly up-regulated phosphorylated in the chilling treatment process and return to normal condition at the recovery process, which might indicate that the JcSRK2a phosphorylation was activated by osmotic stress caused by chilling treatment. Multiple sequence alignments between JcSRK2a and 10 subfamily of SnRK2 from *A. thaliana* showed the JcSRK2a shared the highest similarity (86.44%) with AtSnRK2.4 which could be activated by osmotic stress [56]. Furthermore, overexpression of *TaSnRK2.4* in *A. thaliana* enhanced the plant tolerance to drought, salinity and low temperature [57]. The JcSRK2a (KDP38831.1) shared 79% homology with TaSnRK2.4 (ACU65228.1) and the phosphorylation site Ser348 of JcSRK2a was relative conserved comparing to the site of TaSnRK2.4 (Figure S4). These results provided a possible clue for us to improve low temperature resistance of *J. curcas*. In other hand, as the member of the SnRK1 subfamily [58], KIN10 is one of central transcriptional integrators of stress and energy signaling [59,60]. We observed KIN10 (id 897) was also observed to be phosphorylated changed as the SnRK2a showed. In consideration of these results, we hypothesized the phosphorylation of SnRK2a and KIN10 mainly involved in defense against chilling stress and promoting catabolism respectively in *J. curcas* seedling.

ABI5 was highly inducible by either ABA treatment or stress [61]. In the presence of ABA, ABI5 accumulation was promoted by inducing KEG degradation, which might be accomplished by KEG self-phosphorylation [62]. In this study, the phosphorylation level of ABI5-2 (id 58) shared the similar change pattern as the SRK2a (id 2462) displayed (Figure S5), which might suggest the phosphorylation of ABI5-2 and SRK2a were both triggered by ABA. Additionally, ABI5 also underwent sumoylation at K391 residue to protect it from degradation [19], thereby possibly accelerating ABA-mediated growth inhibition in *J. curcas* seedling under chilling stress. Coincidently, as a negative regulator, JcKEG (id 2703) showed down regulation trend at phosphorylation level, which possibly indicated that ABA had induced KEG degradation. This assumption was supported by previous studies [18,62]. In response to ABA, TRAB1 phosphorylation at Ser 102 was essential and the TRAB1 phosphorylation level increased in a very short period and declined thereafter in *O. sativa* [16,63]. The JcTRAB1 (id 2400) phosphorylation level changed in a similar trend with OsTRAB1 (Table 1). Therefore, we compared the phosphorylation site of JcTRAB1 (KDP38330.1, id 2400) with OsTRAB1 (BAA83740.1). The result showed the phosphorylation site of JcTRAB1 was different with OsTRAB1, however, these two phosphorylation sites were very-close and both of them were in the conserved region II (Figure 9). Moreover, the phosphorylation motifs of JcTRAB1 and OsTRAB1 were [RxxS] and [LxRxxs] respectively, which were potential substrates of SnRK2 or CDPK in plants [55]. It was assumed the JcTRAB1 phosphorylation at Ser in this conserved region was possibly related to the ABA response.

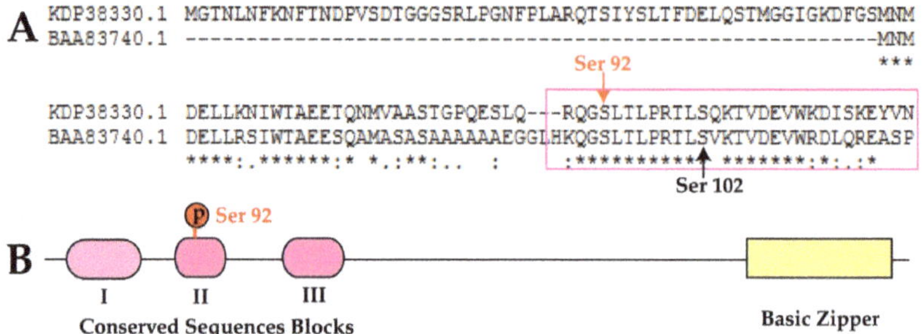

Figure 9. Sequence alignment and predicted structural composition of TRAB1 in *J. curcas*. (**A**) JcTRAB1 (KDP38330.1) sequence alignment with OsTRAB1 (BAA83740.1). The red line with arrow indicated the phosphorylation site of TRAB1 in *J. curcas* while black line with arrow indicated the phosphorylation site Ser 102 in *O. sativa*. The pink rectangle showed the conserved region II of TRAB1. (**B**) Predicted functional domain distribution of JcTRAB1.

The neighbors of SnRKs were chilling stress related protein kinases, which participated in multiple biological processes. Cytosolic Ca^{2+} transiently increased as the result from chilling stress [64]. CDPK played a key role in abiotic stress and mediated in Ca^{2+} signal transduction [65]. The phosphorylation of OsCDPK13 had been indicated as an important signal event in response to cold stress [66]. In this study, JcCDPK8 (id 290) was inducible phosphorylated at Ser 43 only under chilling stress (Table 1), which might indicate that JcCDPK8 functioned in the phosphorylation status in response to chilling stress. In ethylene mediated signal pathway, ethylene responses were negatively regulated by receptors family [67]. Two proteins EIN4 (id 521) and EIN2 (id 1204) were phosphorylated changed. EIN4 and EIN2 were responsible for the negative and positive regulations respectively in ethylene mediated signal [68]. Although it was unclear whether the protein expression level of EIN4 and EIN2 had changed, their changes at phosphorylation level might suggest they were likely related to the ethylene signaling. Additionally, phosphorylation changes were also implicated in MAPK mediated signaling. Phosphorylated proteins involved in MAPK signaling including EDR (id 2304), MKP1 (id 2287) and HOS1 (id 2597) were all significantly changed at phosphorylation level. MAPKs were implicated in response to low temperature condition [69–72], as well as interacted with MKP1 [73]. In this study, we did not detect the significantly change of MAPK at phosphorylation level. However, MKP1 (id 2287) showed an elevated phosphorylation level under chilling treatment (id 2287), suggesting a possible link with MAPK function in *J. curcas*. MAPK subsequently interacted with ICE1, which was degraded by an E3 ligase HOS1 [21]. The phosphorylation level of HOS1 (id 2597) was increasing under chilling stress, possibly resulting in ICE1 degradation. It might be a possible reason to explain why no significantly changed ICE1 at the phosphorylation level was detected in this study.

3.2. Phosphorylation Participated in the Regulation of Phosphoinositide Metabolism

As the primary lipid-derived signals, phosphoinositides were involved in various plant responses to surrounding environment [74]. A total of 13 participants in phosphoinositide metabolism were differentially phosphorylated in this study (Table 1). They participated in reactions leading to the generation of different phosphoinositide species. Among them, PLC2 was the decisive enzyme in phosphoinositide metabolism [75], it catalyzed the hydrolysis of PI(4,5)P2 into DAG and IP3, which subsequently acted as two important second messenger molecules and resulted in release of Ca^{2+} into the cytoplasm [76]. Recent studies showed the expressions of *JcPLC2* and *BpPLC2* were constitutive even under low temperature stress [27,77]. In *J. curcas* seedling, PLC2s (id 3110, 3180) were both differential phosphorylated at recovery stage instead of the chilling stage. Their phosphorylation sites were Ser312 and Ser278 respectively. The phosphorylation site Ser312 from PLC2 (KDP44118.1, id 3110) was located in the catalytic Y domain (Figure S6), which have been reported in AtPLC2

(NP_187464.1) [78,79]. It possibly suggested the regulation of PLC2 activity by phosphorylation in *J. curcas* seedling at the recovery stage. Besides, the present study showed that PI4K (id 3364) was significantly and continuously up-regulated at the phosphorylation level under chilling stress and returned to normal level at the recovery stage (Table 1). PI4K was responsible for the generation of PI4P from PI and it showed important role in trafficking from the trans-Golgi network to the prevacuolar compartment [76]. Meanwhile, the phosphorylation level of DAGK 5-like (id 3551) showed the completely opposite trend with PI4K (Figure S5). In *A. thaliana*, PI4K was indicatively activated during cold stress [80]. In other hand, PI4P could be dephosphorylated by SAC1-domain phosphoinositide phosphatases [81]. As the revisable phosphatase, SAC1 (id 2911) was down regulated at phosphorylation level (Table 1). By means of the PI4K, DAGK and SAC1 changes at phosphorylation level under chilling stress, we assumed that there was a potential straightforward metabolic pathway for generation of IP3 and the phosphorylation event was seemed to be decisive for phosphoinositide metabolism in *J. curcas* seedling under chilling stress.

3.3. Detoxification and Stress Defense

Chilling caused H_2O_2 accumulation and H_2O_2 could be diffused by aquaporins in phosphorylation status at the plasma membrane in the response to chilling stress in maize [82]. Aquaporins were proteinaceous pores that facilitated passive water transport through membranes of living cells [83]. Three aquaporins were examined as the significantly phosphorylated proteins in this study. The phosphorylation level of PIP2-8 (id 1 and id 2) showed the identical change trends (Figure S5), they both increased at the chilling treatment stage and return to normal level after recovery for 24 h. However, another one (PIP2-5, id 3151) maintained constant phosphorylation level even under chilling stress and increased at the recovery stage (Table 1). This might suggest phosphorylation was able to influence the activity of aquaporins and potentially related to H_2O_2 diffusion in *J. curcas* seedling under chilling stress. On the other hand, as the typical antioxidant enzyme, APX1 catalyzed the H_2O_2-dependent oxidation of ascorbate in plants [84]. APX1 (id 15) was significantly up-regulated phosphorylated in response to chilling, which possibly indicated the APX1 phosphorylated change was related to H_2O_2 detoxification in *J. curcas* seedling.

Serine threonine-protein kinase BLUS1 mediated a primary step for phototropin signaling in guard cells and it was directly phosphorylated by PHOT1. BLUS1 functioned as a PHOT substrate and primary regulator of stomatal control to enhance photosynthetic CO_2 assimilation under natural light conditions [85]. BLUS1 (id 655) showed down regulated trends at phosphorylation level, which was corresponding to the phosphorylated change pattern of PHOT1 (id 3118) (Figure S5). Meanwhile, CDPK 33 was also involved in the ABA-mediated regulation of stomatal closure and drought stress responses [86]. The phosphorylation level of JcCDPK33 (id 2491) increased greatly after chilling for 6 h, CDPK33 contained the motif [sP] and potentially combined with MAPK [49], as well as possibly interacted with 14-3-3 [87]. 14-3-3 protein 6 (id 2765) also contained [sP] motif, which was dramatically up-regulated (2.7-fold) at phosphorylation level after chilling for 6 h (Table 1). It was hypothesized that phosphorylation of PHOT, CDPK and 14-3-3 signaling related proteins was possibly involved in stomatal closure in *J. curcas* seedling under chilling stress.

Under chilling stress, many proteins were misfolded and these proteins could not be assembled and function as normal. There were two approaches to handle these misfolded proteins. One was refolding in the assistance of molecular chaperones. In this study, CPN60B2 (id 1375) and CHSP70 (id 3984) were both phosphorylated changed and might result in the activation of RuBisCo assembly [88–90]. Nevertheless, the phosphorylation level of sHSP (id 2898), Hsp90 (id 1658) and PPI cyp65 (id 1583) showed little changes under chilling treatment. These results possibly indicate the accumulation of misfolded proteins in ER [91]. Another was degradation of the ubiquitinated misfolded proteins by the 26S proteasome. 26S protease regulatory subunit 6b homolog (id 1742) showed obviously phosphorylated change after chilling for 24 h and remained the phosphorylation level after 24 h recovery (Table 1). We speculated that the misfolded proteins in *J. curcas* seedling were mainly handled through degradation by ubiquitin-proteasome system at continuous chilling (24 h) and recovery period.

3.4. Photoinhibition and Chloroplast Movement

In response to the chilling stress, changes could be observed at the physiological and cellular level in *J. curcas* seedling within 24 h chilling treatment (Figure 1; Figure 2), which was consistent with the previous description [5]. Chloroplast was the first and most severely affected organelle when plants were exposed to freezing stresses. The photosyntic rate of *J. curcas* seedling was dramatically decreased (Figure 1B) and the chloroplasts were swelling after 24 h chilling treatment (Figure 2), which indicated serious damage of the photosystem in *J. curcas* seedling. At chilling temperature, it was well known that photoinhibition of photosynthesis was enhanced [92] and the primary target of photoinhibition was PSII [93] and phosphorylation of PSII centers increased the stability of PSII complexes and concomitantly improved their survival under stress conditions [94]. In this study, as an important component of PSII, D1 (id 3541) was rapidly phosphorylated under chilling treatment for 6 h, which might result from an imbalance between energy supply and utilization in chloroplasts of chilling-sensitive *J. curcas* seedling. The rate of PSII repair was reduced by inhibiting D1 synthesis at the translation-elongation stage if the repair was not efficiently scavenged [95]. Coincidently, with the time of chilling treatment prolonged, the phosphorylation level of D1 dramatically decreased. It might suggest the phosphorylation of D1 was closely related to stability of PSII complexes. During recovery from chilling-induced photoinhibition in leaves of *J. curcas* seedling, we observed that the CP29 (id 3634) dephosphorylation matched very well with those of PSII recovery. These results might suggest that PSII reactivation from low temperature photoinhibition was closely related to CP29 dephosphorylation. Similar result had been illuminated at *O. sativa* during recovery from chilling induced photoinhibition [96].

Chloroplast movement was an efficient strategy to alleviate PSII damage and maintained maximal photosynthetic output [97]. In plants, temperature was an important factor in modulating chloroplast relocations and PHOTs were considered to mediate not only stomatal opening but also chloroplast movement [98]. The protein level of PHOT1 decreased while the PHOT2 slightly increased at low temperature in *A. thaliana* [99]. In this study, the phosphorylation level of PHOT1 (id 3118) was down-regulated while the PHOT2 (id 3490) was up-regulated after chilling treatment for 6 h and 24 h, which possibly indicated that phosphorylation related to the activity of PHOT1 and PHOT2. However, both of the PHOT1 and PHOT2 maintained the phosphorylation level previously even after the stimulus was removed (Table 1). Besides the PHOTs, the chloroplast movement was also associated with several proteins [100]. CHUP1 might function at the periphery of the chloroplast outer membrane and most likely represented an essential component for chloroplast movement. Similarly, the mutant *pmi1* exhibited severely attenuated chloroplast movements under low and high light intensities, indicating that PMI1 was necessary for both chloroplast accumulation and avoidance movements [101]. Nevertheless, we observed little change of CHUP1 (id 1231 and id 1233), PMI1 (id 2153 and id 2154) at the phosphorylation level after chilling treatment in *J. curcas* seedling, subsequently, all of them inducible regulated at the phosphorylation level in the recovery stage. To date, there were relatively few physiological and molecular evidences for explaining chloroplast movement through phosphorylation. Therefore, we speculated that the activities of CHUP1 and PMI1 might be regulated by other post-translational modification, such as glycosylation or sumoylation.

4. Materials and Methods

4.1. Plant Material and Chilling Treatment

The uniform mature *J. curcas* seeds with same provenance were collected from Xishuangbanna Tropical Botanical Garden, Chinese Academy of Sciences, Yunnan Province. The mature *J. curcas* seeds were sowed and grown in a mixed soil of peat and vermiculite (1:1) in a growth chamber under 28/25 °C (day/night) temperature regime, a photon flux density of 150 μmol m^{-2}·s^{-1} throughout a 14-h photoperiod and a relative humidity of 80%. Four-leaf stage seedlings were used in this experiment. Low temperature treatments were started at 11:00 AM on the first day by setting the temperature to 4 °C, which was reached about 50 min later. The temperature was returned to 28 °C after 24 h of chilling treatment. The fourth true leaves were detached from the top of *J. curcas* seedling at each time point (4 °C for 0, 6, 24 h and then returned to 28 °C for 24 h), which indicated as C0 h, C6 h, C24 h and R24 h respectively for each biological replicate. one part of the collected leaves were used for tissue preparation of transmission electron microscopy and other part of the collected leaves from the five plants were frozen in liquid nitrogen and stored at −80 °C until use. However, in consideration of the functional maturity, the first true leaves were chosen for measurements of photosynthesis, stomatal conductance and transpiration. The measurement was carried out by using a portable gas analysis system, LI-COR 6400 with a light-emitting diode light source (LICOR Inc., Lincoln, NE, USA). The measurement conditions were set as follows: block temperature 20 °C, photon flux density 1200 μmol m^{-2}·s^{-1}, humidity 80% and the CO_2 concentration was stabilized by 5 L surge flask.

4.2. Tissue Preparation for Transmission Electron Microscopy and Protein Preparation

The fourth true leaves from C0 h, C6 h, C24 h and R24 h were fixed in 2.5% glutaraldehyde in 100 mM phosphate buffer (pH 7.0) for 4 h at room temperature. Then the samples were handled as the previous operation [38]. Leaf total protein was extracted according to the method of Shen et al. [102] with minor modifications. Approximately 500 mg leaves of each sample was homogenized in 2 mL of the homogenization buffer containing 20 mM Tris-HCl (pH 7.5), 250 mM sucrose, 10 mM ethylene glycol tetra acetic acid, 1mM phenylmethylsulfonyl fluoride, 1% dithiothreitol (DTT) and 1% Triton X-100 on the ice. The homogenate was collected into an Eppendorf tube and centrifuged at 10,000× g for 10 min at 4 °C. The supernatant was transferred to a fresh tube and precipitated by adding 10% cold trichloroacetic acid on ice for over 30 min. The mixture was centrifuged at 15,000 g for 10 min at 4 °C and the supernatant was discarded. After washed three times with acetone containing 1% DTT, the pellet was collected by centrifugation, air-dried and then suspended in SDT buffer containing 4% SDS, 1 mM DTT, 100 mM Tris-HCl (pH 7.5), sonicated for 2 min (with cooling on ice) and boiled for 15 min. The protein mixtures were harvested via centrifugation at 15,000× g for 20 min at 4 °C to remove insoluble material. Protein concentrations of experimental samples were quantified according to Bradford method [103]. Albumin (A5503, Sigma, St. Louis, MO, USA) was used as a standard for protein quantification. Three independent biological replicates were performed independently for each experimental sample.

4.3. Phosphopeptide Enrichment

The same amount of extracted protein mixture in each sample was directly reduced with DTT, alkylated with iodoacetamide and subsequently digested with endoproteinase Lys-C and trypsin as previously described [104]. The enrichment procedure for the phosphopeptides was performed as followed. Tryptic peptides (5 mg) were dissolved in 400 μL of loading buffer containing 65% acetonitrile (ACN)/2% trifluoroaceticacid (TFA) saturated with glutamic acid and incubated with an appropriate amount (tryptic peptide: TiO$_2$ = 1:1, w/w) of TiO$_2$ beads (GL Sciences, Tokyo, Japan) for 40 min. The mixture was centrifuged at 10, 000× g for 3 min at 4 °C and the supernatant was discarded. After washing with 800 μL wash buffer (65% ACN/0.1% TFA) for 40 min, then centrifugation as above, then the phosphopeptides were eluted twice with 800 μL elution buffer (500 mM NH$_4$OH/60% ACN), centrifugation and the eluate was dried and reconstituted in 0.1% formic acid/H$_2$O for MS analysis.

4.4. LC-MS/MS Analysis

The enriched phosphopeptides were separated on a self-packed C18 reverse-phase column (70 μm inner diameter, 150-mm length, Column Technology, Fremont, CA, USA) that directly connected to the nanospray ion source to an Q Exactive mass spectrometer (Thermo Fisher Scientific, San Jose, CA, USA) running in the positive ion mode. The pump flow was split to achieve a flow rate of 1 μL/min for sample loading and 300 nL/min for MS analysis. The mobile phases consisted of 0.1% formic acid (A) and 0.1% formic acid and 90% ACN (B). A four-step linear gradient of 2% to 5% B in 5 min, 5% to 22% B in 55 min, 22% to 90% B in 5 min and keeping 90% B for 10 min. The spray voltage was set at 2.0 kV and the temperature of the heated capillary was 270 °C. For data acquisition, each MS scan was acquired at a resolution of 60,000 (at 400 m/z) with the lock mass option enabled. A lock mass function was used to obtain high mass accuracy. The 12 most intense precursor ions were selected for collision-induced fragmentation in the linear ion trap at normalized collision energy of 37%. The threshold for precursor ion selection was 500 and the mass window for precursor ion selection was 2.0 Da. The dynamic exclusion duration was 120 s, the repeat count was 1 and the repeat duration was 30 s. Three biological replicates were performed independently from sample collection to the phosphopeptide identification using LC-MS/MS.

4.5. Protein Identification

The raw files were processed with MaxQuant (version 1.3.0.5) and searched against the NCBI Jatropha protein database (update to 20180509, 41287 entries) concatenated with a decoy consisting of reversed sequences. The following parameters were used for database searches: trypsin/P was chosen as enzyme specificity, Carbamidomethyl (C) was selected as a fixed modification, Oxidation (M), Acetyl (Protein N-term) and Phospho (STY) were selected as variable modifications. Up to two missing cleavage points were allowed. The precursor ion mass tolerances were 7 ppm and the fragment ion mass tolerances for the MS/MS spectra were 20 ppm. The false discovery rate (FDR) was set to $\leq$0.01 for peptide and protein. The minimum peptide length was set to 6. The localization of the phosphorylation site was based on the PTM scores that assigned probabilities for each of the possible sites according to their site-determining ions. The software MaxQuant was used to calculate the PTM scores and PTM localization probabilities. Only when the phosphorylation site of localization probabilities ($p \geq 0.75$) and the PTM score ($\geq$5) was considered as the reliable. An FDR of 0.01 was used for phosphorylation site identification.

4.6. Label Free Quantification and Screening of Phosphopeptides with Significant Changes at the Phosphorylation Level

We integrated the ion intensities over its chromatographic elute profile and employed MaxQuant software to calculate the quantification of phosphopeptides on the basis of a label free approach. For each phosphopeptide, its intensity was normalized to the mean of the intensities of all phosphopeptides within each biological replicate. The screening terms of phosphopeptides with significant changed at phosphorylation level were listed as below: (1) phosphopeptides detected in all three biological replicates; (2) phosphopeptides with credible ANOVA analysis or Student's-test analysis (FDR<0.05); (3) phosphorylation localization probability $\geq$ 0.75 and (4) phosphorylation site score difference $\geq$ 5.

4.7. Bioinformatics

Protein function was annotated through Blast2GO software (http://www.blast2go.com/b2ghome), significantly enriched phosphorylation motifs were extracted from phosphopeptides with confidently identified phosphorylation residues using the Motif-X algorithm (http://motif-x.med.harvard.edu/). The phosphopeptides were centered at the phosphorylated amino acid residues and aligned and seven positions upstream and downstream of the phosphorylation sites were included. Because the upload restriction of Motif-X is 10 Mb, a FASTA format data set (nearly 10 Mb) containing the protein sequences from the *J. curcas* protein database was used as the background database to normalize the scores against the random distributions of amino acids. The occurrences threshold was set to 5% of the input data, set at a minimum of 20 peptides and the probability threshold was set to $p < 10^{-6}$. The phosphoproteins blasted by the National Center for Biotechnology Information (NCBI) were used to obtain the KOG numbers of those proteins by eggNOG (http://eggnogdb.embl.de/). A data set containing all the KOG numbers was then used for PPI by using the Search Tool for the Retrieval of Interaction Genes/Proteins (STRING) database (http://string-db.org/) and the PPI network was displayed by Cytoscape software (Version 3.6.3).

5. Conclusions

In summary, the woody oil plant species, *J. curcas* seedling was sensitive to chilling stress. It was studied by employing a comparative phosphoproteomic analysis, physiological measurement, ultrastructure observation under chilling stress and recovery. For the 805 significantly changed phosphopeptides, 9 phosphorylation motifs were extracted, which were mainly regulated by CDK, SnRK2, MAPK and CDPK. A complex PPI subnetwork were constructed and indicated the crucial roles of SnRK, 14-3-3 and ADP-ribosylation factor in the responsive network. Our results showed the phosphorylation was an essential event and responsible for chilling response and defense in *J. curcas* seedling. Consequently, Ca^{2+}, ABA, ethylene, phosphoinositide and 14-3-3 mediated signal pathways cross linked for chilling response. In response to chilling stress, the phosphorylation of SnRK2a, ABI5 and TRAB1 were possibly related to ABA mediated signal pathway. Additionally, the phosphorylation level of transport, photoinhibition and chloroplast movement related proteins were also significantly regulated. We also highlighted that the phosphorylation of JcHOS1, JcKEG, JcAPX, JcPIP2 and JcPI4K might activate the potential functions to defense against chilling stress. Finally, we depicted a schematic presentation including signal transduction, metabolism and ion transport on the basis of changes at phosphorylation level, which is valuable for us to understand the chilling response and defense network in *J. curcas* seedling.

Supplementary Materials: Supplementary materials can be found at http://www.mdpi.com/1422-0067/20/1/208/s1.

Author Contributions: H.L. and S.-H.S. conceived and designed the experiment; H.L. and F.-F.W. performed the experimental work; H.L. wrote the manuscript; S.-H.S., J.-H.H. and X.-J.P. jointly directed this work.

Funding: This research was supported by China Postdoctoral Science Foundation (2015M571154) and the National Natural Science Foundation of China (31870247).

Conflicts of Interest: The authors declare no conflict of interest.

Abbreviations

GO	Gene Ontology
PS	Photosystem
ABA	Abscisic acid
SnRK2	Sucrose non-fermenting 1-related protein kinase 2
ABI5	ABA Insensitive 5
HOS1	High expression of osmotically responsive gene 1
CBF	C-repeat binding factor
ICE1	Inducer of CBF expression 1
LC-MS	Liquid chromatograph-mass spectrometer
CDK	Cyclin-dependent kinase
MAPK	Mitogen-activated protein kinase
CDPK	Calmodulin-dependent protein kinase
KEG	RING-type E3 ligase KEEP ON GOING
KOG	Eukaryotic orthologous groups
ABI5	ABA-insensitive 5 protein
EIN	Ethylene insensitive protein
MKP	Mitogen-activated protein kinase phosphatase
EDR	Enhanced disease resistance
PLC	Phosphoinositide phospholipase C
PI(4,5)P2	Phosphatidylinositol 4,5-bisphosphate
DAG	Diacylglycerol
PI4K	phosphatidylinositol 4-kinase
DAGK	Diacylglycerol kinase 5
PI4P	Phosphatidylinositol 4-phosphate
IP3	Inositol 1,4,5-trisphosphate
APX	Cytosolic ascorbate peroxidase
PHOT	Phototropin
CPN60B2	Chaperonin 60 subunit beta 2, chloroplastic
CHSP70	Stromal 70 kDa heat shock-related chloroplastic
Hsp90	Heat shock cognate protein 80
sHSP	17.3 kDa class I heat shock protein
PPI cyp65	Peptidyl-prolyl cis-trans isomerase cyp65
ER	Endoplasmic Reticulum
CHUP1	Chloroplast unusual positioning1
PMI	Plastid movement impaired

References

1. Thakur, P.; Kumar, S.; Malik, J.A.; Berger, J.D.; Nayyar, H. Cold stress effects on reproductive development in grain crops: An overview. *Environ. Exp. Bot.* **2010**, *67*, 429–443. [CrossRef]
2. Levitt, J. *Chilling, Freezing and High Temperature Stresses. Responses of Plants to Environmental Stress*, 2nd ed.; Academic Press: London, UK; New York, NY, USA, 1980; Volume 1, p. 497. ISBN 0-12-445501-8.
3. Dong, C.H.; Hu, X.; Tang, W.; Zheng, X.; Kim, Y.S.; Lee, B.H.; Zhu, J.K. A putative Arabidopsis nucleoporin, AtNUP160, is critical for RNA export and required for plant tolerance to cold stress. *Mol. Cell. Biol.* **2006**, *26*, 9533–9543. [CrossRef] [PubMed]
4. Zhu, J.; Dong, C.H.; Zhu, J.K. Interplay between cold-responsive gene regulation, metabolism and RNA processing during plant cold acclimation. *Curr. Opin. Plant. Biol.* **2007**, *10*, 290–295. [CrossRef] [PubMed]
5. Thomashow, M.F. Plant Cold Acclimation: Freezing tolerance genes and regulatory mechanisms. *Annu. Rev. Plant Physiol. Plant Mol. Biol.* **1999**, *50*, 571–599. [CrossRef] [PubMed]
6. Huang, W.; Zhang, S.B.; Cao, K.F. The different effects of chilling stress under moderate light intensity on photosystem II compared with photosystem I and subsequent recovery in tropical tree species. *Photosynth. Res.* **2010**, *103*, 175–182. [CrossRef]

7. Kratsch, H.A.; Wise, R.R. The ultrastructure of chilling stress. *Plant Cell Environ.* **2000**, *23*, 337–350. [CrossRef]
8. Bascunan-Godoy, L.; Sanhueza, C.; Cuba, M.; Zuniga, G.E.; Corcuera, L.J.; Bravo, L.A. Cold-acclimation limits low temperature induced photoinhibition by promoting a higher photochemical quantum yield and a more effective PSII restoration in darkness in the Antarctic rather than the Andean ecotype of Colobanthus quitensis Kunt Bartl (Cariophyllaceae). *BMC Plant Biol.* **2012**, *12*, 114.
9. Shinozaki, K.; Yamaguchi-Shinozaki, K. Molecular responses to drought and cold stress. *Curr. Opin. Biotechnol.* **1996**, *7*, 161–167. [CrossRef]
10. Ishitani, M.; Xiong, L.; Stevenson, B.; Zhu, J.K. Genetic analysis of osmotic and cold stress signal transduction in Arabidopsis: Interactions and convergence of abscisic acid-dependent and abscisic acid-independent pathways. *Plant Cell* **1997**, *9*, 1935–1949. [CrossRef]
11. Knight, M.R.; Campbell, A.K.; Smith, S.M.; Trewavas, A.J. Transgenic plant aequorin reports the effects of touch and cold-shock and elicitors on cytoplasmic calcium. *Nature* **1991**, *352*, 524–526. [CrossRef]
12. Knight, H.; Trewavas, A.J.; Knight, M.R. Cold calcium signaling in Arabidopsis involves two cellular pools and a change in calcium signature after acclimation. *Plant Cell* **1996**, *8*, 489–503. [CrossRef] [PubMed]
13. Monroy, A.F.; Dhindsa, R.S. Low-temperature signal transduction: Induction of cold acclimation-specific genes of alfalfa by calcium at 25 °C. *Plant Cell* **1995**, *7*, 321–331. [PubMed]
14. Yamaguchi-Shinozaki, K.; Shinozaki, K. Transcriptional regulatory networks in cellular responses and tolerance to dehydration and cold stresses. *Annu. Rev. Plant Biol.* **2006**, *57*, 781–803. [CrossRef] [PubMed]
15. Kobayashi, Y.; Yamamoto, S.; Minami, H.; Kagaya, Y.; Hattori, T. Differential activation of the rice sucrose nonfermenting1-related protein kinase2 family by hyperosmotic stress and abscisic acid. *Plant Cell* **2004**, *16*, 1163–1177. [CrossRef] [PubMed]
16. Kagaya, Y.; Hobo, T.; Murata, M.; Ban, A.; Hattori, T. Abscisic acid-induced transcription is mediated by phosphorylation of an abscisic acid response element binding factor, TRAB1. *Plant Cell* **2002**, *14*, 3177–3189. [CrossRef] [PubMed]
17. Hobo, T.; Kowyama, Y.; Hattori, T. A bZIP factor, TRAB1, interacts with VP1 and mediates abscisic acid-induced transcription. *Proc. Natl. Acad. Sci. USA* **1999**, *96*, 15348–15353. [CrossRef] [PubMed]
18. Finkelstein, R.R.; Lynch, T.J. The Arabidopsis abscisic acid response gene ABI5 encodes a basic leucine zipper transcription factor. *Plant Cell* **2000**, *12*, 599–609. [CrossRef]
19. Miura, K.; Lee, J.; Jin, J.B.; Yoo, C.Y.; Miura, T.; Hasegawa, P.M. Sumoylation of ABI5 by the Arabidopsis SUMO E3 ligase SIZ1 negatively regulates abscisic acid signaling. *Proc. Natl. Acad. Sci. USA* **2009**, *106*, 5418–5423. [CrossRef]
20. Ishitani, M.; Xiong, L.; Lee, H.; Stevenson, B.; Zhu, J.K. HOS1, a genetic locus involved in cold-responsive gene expression in arabidopsis. *Plant Cell* **1998**, *10*, 1151–1161. [CrossRef]
21. Dong, C.H.; Agarwal, M.; Zhang, Y.; Xie, Q.; Zhu, J.K. The negative regulator of plant cold responses, HOS1, is a RING E3 ligase that mediates the ubiquitination and degradation of ICE1. *Proc. Natl. Acad. Sci. USA* **2006**, *103*, 8281–8286. [CrossRef]
22. Kline-Jonakin, K.G.; Barrett-Wilt, G.A.; Sussman, M.R. Quantitative plant phosphoproteomics. *Curr. Opin. Plant Biol.* **2011**, *14*, 507–511. [CrossRef] [PubMed]
23. Tichy, A.; Salovska, B.; Rehulka, P.; Klimentova, J.; Vavrova, J.; Stulik, J.; Hernychova, L. Phosphoproteomics: Searching for a needle in a haystack. *J. Proteomics* **2011**, *74*, 2786–2797. [CrossRef] [PubMed]
24. Kersten, B.; Agrawal, G.K.; Iwahashi, H.; Rakwal, R. Plant phosphoproteomics: A long road ahead. *Proteomics* **2006**, *6*, 5517–5528. [CrossRef] [PubMed]
25. Kersten, B.; Agrawal, G.K.; Durek, P.; Neigenfind, J.; Schulze, W.; Walther, D.; Rakwal, R. Plant phosphoproteomics: An update. *Proteomics* **2009**, *9*, 964–988. [CrossRef]
26. Rampitsch, C.; Bykova, N.V. The beginnings of crop phosphoproteomics: Exploring early warning systems of stress. *Front Plant Sci.* **2012**, *3*, 144. [CrossRef] [PubMed]
27. Pi, Z.; Zhao, M.L.; Peng, X.J.; Shen, S.H. Phosphoproteomic analysis of paper mulberry reveals phosphorylation functions in chilling tolerance. *J. Proteome Res.* **2017**, *16*, 1944–1961. [CrossRef] [PubMed]
28. Zhang, M.; Ma, C.Y.; Lv, D.W.; Zhen, S.M.; Li, X.H.; Yan, Y.M. Comparative Phosphoproteome Analysis of the Developing Grains in Bread Wheat (*Triticum aestivum* L.) under Well-Watered and Water-Deficit Conditions. *J. Proteome Res.* **2014**, *13*, 4281–4297. [CrossRef]

29. Lv, D.W.; Subburaj, S.; Cao, M.; Yan, X.; Li, X.H.; Appels, R.; Sun, D.F.; Ma, W.J.; Yan, Y.M. Proteome and phosphoproteome characterization reveals new response and defense mechanisms of *Brachypodium distachyon* leaves under salt stress. *Mol. Cell. Proteom.* **2014**, *13*, 632–652. [CrossRef] [PubMed]

30. Umezawa, T.; Sugiyama, N.; Takahashi, F.; Anderson, J.C.; Ishihama, Y.; Peck, S.C.; Shinozaki, K. Genetics and phosphoproteomics reveal a protein phosphorylation network in the abscisic acid signaling pathway in *Arabidopsis thaliana*. *Sci. Signal.* **2013**, *6*. [CrossRef] [PubMed]

31. Heller, J. *Physic Nut. Jatropha curcas L. Promoting the Conservation and Use of Underutilized and Neglected Crops*; IPGRI: Zschortau, Germany, 1996; pp. 13–42.

32. Fairless, D. Biofuel: The little shrub that could—Maybe. *Nature* **2007**, *449*, 652–655. [CrossRef] [PubMed]

33. Sato, S.; Hirakawa, H.; Isobe, S.; Fukai, E.; Watanabe, A.; Kato, M.; Kawashima, K.; Minami, C.; Muraki, A.; Nakazaki, N.; et al. Sequence analysis of the genome of an oil-bearing tree, *Jatropha curcas* L. *DNA Res.* **2010**, *18*, 65–76. [CrossRef] [PubMed]

34. Carvalho, C.R.; Clarindo, W.R.; Praça, M.M.; Araújo, F.S.; Carels, N. Genome size, base composition and karyotype of *Jatropha curcas* L., an important biofuel plant. *Plant Sci.* **2008**, *174*, 613–617. [CrossRef]

35. Jiang, H.; Wu, P.; Zhang, S.; Song, C.; Chen, Y.; Li, M.; Jia, Y.; Fang, X.; Chen, F.; Wu, G. Global analysis of gene expression profiles in developing physic nut (*Jatropha curcas* L.) seeds. *PLoS ONE* **2012**, *7*, e36522. [CrossRef] [PubMed]

36. Zhang, L.; Zhang, C.; Wu, P.; Chen, Y.; Li, M.; Jiang, H.; Wu, G. Global analysis of gene expression profiles in physic nut (*Jatropha curcas* L.) seedlings exposed to salt stress. *PLoS ONE* **2014**, *9*, e97878. [CrossRef]

37. Liu, H.; Liu, Y.J.; Yang, M.F.; Shen, S.H. A comparative analysis of embryo and endosperm proteome from seeds of *Jatropha curcas*. *J. Integr. Plant Biol.* **2009**, *51*, 850–857. [CrossRef] [PubMed]

38. Liu, H.; Wang, C.; Chen, F.; Shen, S. Proteomic analysis of oil bodies in mature *Jatropha curcas* seeds with different lipid content. *J. Proteom.* **2015**, *113*, 403–414. [CrossRef] [PubMed]

39. Liu, H.; Wang, C.; Komatsu, S.; He, M.; Liu, G.; Shen, S. Proteomic analysis of the seed development in *Jatropha curcas*: From carbon flux to the lipid accumulation. *J. Proteom.* **2013**, *91*, 23–40. [CrossRef] [PubMed]

40. Liu, H.; Yang, Z.; Yang, M.; Shen, S. The differential proteome of endosperm and embryo from mature seed of *Jatropha curcas*. *Plant Sci.* **2011**, *181*, 660–666. [CrossRef] [PubMed]

41. Yang, M.F.; Liu, Y.J.; Liu, Y.; Chen, H.; Chen, F.; Shen, S.H. Proteomic analysis of oil mobilization in seed germination and postgermination development of *Jatropha curcas*. *J. Proteome Res.* **2009**, *8*, 1441–1451. [CrossRef] [PubMed]

42. Wang, H.; Zou, Z.; Wang, S.; Gong, M. Global analysis of transcriptome responses and gene expression profiles to cold stress of *Jatropha curcas* L. *PLoS ONE* **2013**, *8*, e82817. [CrossRef] [PubMed]

43. Liang, Y.; Chen, H.; Tang, M.J.; Yang, P.F.; Shen, S.H. Responses of *Jatropha curcas* seedlings to cold stress: Photosynthesis-related proteins and chlorophyll fluorescence characteristics. *Physiol. Plant* **2007**, *131*, 508–517. [CrossRef]

44. Yan, S.P.; Zhang, Q.Y.; Tang, Z.C.; Su, W.A.; Sun, W.N. Comparative proteomic analysis provides new insights into chilling stress responses in rice. *Mol. Cell. Proteom.* **2006**, *5*, 484–496. [CrossRef] [PubMed]

45. Yang, Q.S.; Wu, J.H.; Li, C.Y.; Wei, Y.R.; Sheng, O.; Hu, C.H.; Kuang, R.B.; Huang, Y.H.; Peng, X.X.; McCardle, J.A.; et al. Quantitative proteomic analysis reveals that antioxidation mechanisms contribute to cold tolerance in plantain (*Musa paradisiaca* L.; ABB Group) seedlings. *Mol. Cell. Proteom.* **2012**, *11*, 1853–1869. [CrossRef]

46. Vizcaino, J.A.; Cote, R.G.; Csordas, A.; Dianes, J.A.; Fabregat, A.; Foster, J.M.; Griss, J.; Alpi, E.; Birim, M.; Contell, J.; et al. The Proteomics Identifications (PRIDE) database and associated tools: Status in 2013. *Nucleic Acids Res.* **2013**, *41*, D1063–D1069. [CrossRef] [PubMed]

47. Yao, Q.; Bollinger, C.; Gao, J.; Xu, D.; Thelen, J.J. P^{3}DB: An integrated database for plant protein phosphorylation. *Front Plant Sci.* **2012**, *3*, 206. [CrossRef] [PubMed]

48. Heazlewood, J.L.; Durek, P.; Hummel, J.; Selbig, J.; Weckwerth, W.; Walther, D.; Schulze, W.X. PhosPhAt: A database of phosphorylation sites in *Arabidopsis thaliana* and a plant-specific phosphorylation site predictor. *Nucleic Acids Res.* **2008**, *36*, 1015–1021. [CrossRef]

49. Schwartz, D.; Gygi, S.P. An iterative statistical approach to the identification of protein phosphorylation motifs from large-scale data sets. *Nat. Biotechnol.* **2005**, *23*, 1391–1398. [CrossRef]

50. Amanchy, R.; Periaswamy, B.; Mathivanan, S.; Reddy, R.; Tattikota, S.G.; Pandey, A. A curated compendium of phosphorylation motifs. *Nat. Biotechnol.* **2007**, *25*, 285–286. [CrossRef] [PubMed]

51. Prasad, T.S.K.; Goel, R.; Kandasamy, K.; Keerthikumar, S.; Kumar, S.; Mathivanan, S.; Telikicherla, D.; Raju, R.; Shafreen, B.; Venugopal, A.; et al. Human protein reference database-2009 update. *Nucleic Acids Res.* **2009**, *37*, D767–D772. [CrossRef]

52. Villen, J.; Beausoleil, S.A.; Gerber, S.A.; Gygi, S.P. Large-scale phosphorylation analysis of mouse liver. *Proc. Natl. Acad. Sci. USA* **2007**, *104*, 1488–1493. [CrossRef]

53. Ku, N.O.; Liao, J.; Omary, M.B. Phosphorylation of human keratin 18 serine 33 regulates binding to 14-3-3 proteins. *EMBO J.* **1998**, *17*, 1892–1906. [CrossRef] [PubMed]

54. Zhang, S.H.; Kobayashi, R.; Graves, P.R.; PiwnicaWorms, H.; Tonks, N.K. Serine phosphorylation-dependent association of the band 4.1-related protein-tyrosine phosphatase PTPH1 with 14-3-3 beta protein. *J. Biol. Chem.* **1997**, *272*, 27281–27287. [CrossRef]

55. Vlad, F.; Turk, B.E.; Peynot, P.; Leung, J.; Merlot, S. A versatile strategy to define the phosphorylation preferences of plant protein kinases and screen for putative substrates. *Plant J.* **2008**, *55*, 104–117. [CrossRef] [PubMed]

56. Boudsocq, M.; Barbier-Brygoo, H.; Lauriere, C. Identification of nine sucrose nonfermenting 1-related protein kinases 2 activated by hyperosmotic and saline stresses in *Arabidopsis thaliana*. *J. Biol. Chem.* **2004**, *279*, 41758–41766. [CrossRef]

57. Mao, X.; Zhang, H.; Tian, S.; Chang, X.; Jing, R. TaSnRK2.4, an SNF1-type serine/threonine protein kinase of wheat (*Triticum aestivum* L.), confers enhanced multistress tolerance in Arabidopsis. *J. Exp. Bot.* **2010**, *61*, 683–696. [CrossRef] [PubMed]

58. Hrabak, E.M. The Arabidopsis CDPK-SnRK superfamily of protein kinases. *Plant Physiol.* **2003**, *132*, 666–680. [CrossRef] [PubMed]

59. Baena-Gonzalez, E.; Rolland, F.; Thevelein, J.M.; Sheen, J. A central integrator of transcription networks in plant stress and energy signalling. *Nature* **2007**, *448*, 938–942. [CrossRef] [PubMed]

60. Hardie, D.G.; Carling, D.; Carlson, M. The AMP-activated/SNF1 protein kinase subfamily: Metabolic sensors of the eukaryotic cell? *Annu. Rev. Biochem.* **1998**, *67*, 821–855. [CrossRef]

61. Qin, F.; Shinozaki, K.; Yamaguchi-Shinozaki, K. Achievements and challenges in understanding plant abiotic stress responses and tolerance. *Plant Cell Physiol.* **2011**, *52*, 1569–1582. [CrossRef]

62. Liu, H.; Stone, S.L. Abscisic acid increases Arabidopsis ABI5 transcription factor levels by promoting KEG E3 ligase self-ubiquitination and proteasomal degradation. *Plant Cell* **2010**, *22*, 2630–2641. [CrossRef]

63. Kobayashi, Y.; Murata, M.; Minami, H.; Yamamoto, S.; Kagaya, Y.; Hobo, T.; Yamamoto, A.; Hattori, T. Abscisic acid-activated SnRK2 protein kinases function in the gene-regulation pathway of ABA signal transduction by phosphorylating ABA response element-binding factors. *Plant J.* **2005**, *44*, 939–949. [CrossRef] [PubMed]

64. Sanders, D.; Brownlee, C.; Harper, J.F. Communicating with calcium. *Plant Cell* **1999**, *11*, 691–706. [CrossRef]

65. Li, W.G.; Komatsu, S. Cold stress induced calcium-dependent protein kinase(s) in rice (*Oryza sativa* L.) seedling stem tissues. *Theor. Appl. Genet.* **2000**, *101*, 355–363. [CrossRef]

66. Khan, M.; Takasaki, H.; Komatsu, S. Comprehensive phosphoproteome analysis in rice and identification of phosphoproteins responsive to different hormones/stresses. *J. Proteome Res.* **2005**, *4*, 1592–1599. [CrossRef]

67. Hua, J.; Meyerowitz, E.M. Ethylene responses are negatively regulated by a receptor gene family in *Arabidopsis thaliana*. *Cell* **1998**, *94*, 261–271. [CrossRef]

68. Zheng, Y.; Zhu, Z. Relaying the ethylene signal: New roles for EIN2. *Trends Plant Sci.* **2016**, *21*, 2–4. [CrossRef] [PubMed]

69. Solanke, A.U.; Sharma, A.K. Signal transduction during cold stress in plants. *Physiol. Mol. Biol. Plants* **2008**, *14*, 69–79. [CrossRef] [PubMed]

70. Yadav, S.K. Cold stress tolerance mechanisms in plants. A review. *Agron. Sustain. Dev.* **2010**, *30*, 515–527. [CrossRef]

71. Chinnusamy, V.; Zhu, J.K.; Sunkar, R. Gene regulation during cold stress acclimation in plants. *Methods Mol. Biol.* **2010**, *639*, 39–55.

72. Jonak, C.; Kiegerl, S.; Ligterink, W.; Barker, P.J.; Huskisson, N.S.; Hirt, H. Stress signaling in plants: A mitogen-activated protein kinase pathway is activated by cold and drought. *Proc. Natl. Acad. Sci. USA* **1996**, *93*, 11274–11279. [CrossRef]

73. Keyse, S.M. Protein phosphatases and the regulation of mitogen-activated protein kinase signalling. *Curr. Opin. Cell Biol.* **2000**, *12*, 186–192. [CrossRef]

74. Boss, W.F.; Im, Y.J. Phosphoinositide signaling. *Annu. Rev. Plant Biol.* **2012**, *63*, 409–429. [CrossRef] [PubMed]

75. Kanehara, K.; Yu, C.Y.; Cho, Y.; Cheong, W.F.; Torta, F.; Shui, G.; Wenk, M.R.; Nakamura, Y. Arabidopsis AtPLC2 is a primary phosphoinositide-specific Phospholipase C in phosphoinositide metabolism and the endoplasmic reticulum stress response. *PLoS Genet.* **2015**, *11*, e1005511. [CrossRef] [PubMed]

76. Kim, D.H.; Eu, Y.J.; Yoo, C.M.; Kim, Y.W.; Pih, K.T.; Jin, J.B.; Kim, S.J.; Stenmark, H.; Hwang, I. Trafficking of phosphatidylinositol 3-phosphate from the trans-Golgi network to the lumen of the central vacuole in plant cells. *Plant Cell* **2001**, *13*, 287–301. [CrossRef]

77. Hirayama, T.; Mitsukawa, N.; Shibata, D.; Shinozaki, K. *AtPLC2*, a gene encoding phosphoinositide-specific phospholipase C, is constitutively expressed in vegetative and floral tissues in *Arabidopsis thaliana*. *Plant Mol. Biol.* **1997**, *34*, 175–180. [CrossRef] [PubMed]

78. Chen, Y.; Hoehenwarter, W.; Weckwerth, W. Comparative analysis of phytohormone-responsive phosphoproteins in Arabidopsis thaliana using TiO_2-phosphopeptide enrichment and mass accuracy precursor alignment. *Plant J.* **2010**, *63*, 1–17. [CrossRef]

79. Singh, A.; Bhatnagar, N.; Pandey, A.; Pandey, G.K. Plant phospholipase C family: Regulation and functional role in lipid signaling. *Cell Calcium* **2015**, *58*, 139–146. [CrossRef]

80. Delage, E.; Ruelland, E.; Guillas, I.; Zachowski, A.; Puyaubert, J. Arabidopsis type-III phosphatidylinositol 4-kinases beta1 and beta2 are upstream of the phospholipase C pathway triggered by cold exposure. *Plant Cell Physiol.* **2012**, *53*, 565–576. [CrossRef]

81. Delage, E.; Puyaubert, J.; Zachowski, A.; Ruelland, E. Signal transduction pathways involving phosphatidylinositol 4-phosphate and phosphatidylinositol 4,5-bisphosphate: Convergences and divergences among eukaryotic kingdoms. *Prog. Lipid Res.* **2013**, *52*, 1–14. [CrossRef]

82. Aroca, R.; Amodeo, G.; Fernandez-Illescas, S.; Herman, E.M.; Chaumont, F.; Chrispeels, M.J. The role of aquaporins and membrane damage in chilling and hydrogen peroxide induced changes in the hydraulic conductance of maize roots. *Plant Physiol.* **2005**, *137*, 341–353. [CrossRef]

83. Chrispeels, M.J.; Agre, P. Aquaporins: Water channel proteins of plant and animal cells. *Trends Biochem. Sci.* **1994**, *19*, 421–425. [CrossRef]

84. Raven, E.L. Understanding functional diversity and substrate specificity in haem peroxidases: What can we learn from ascorbate peroxidase? *Nat. Prod. Rep.* **2003**, *20*, 367–381. [CrossRef] [PubMed]

85. Takemiya, A.; Sugiyama, N.; Fujimoto, H.; Tsutsumi, T.; Yamauchi, S.; Hiyama, A.; Tada, Y.; Christie, J.M.; Shimazaki, K.I. Phosphorylation of BLUS1 kinase by phototropins is a primary step in stomatal opening. *Nat. Commun.* **2013**, *4*, 2094. [CrossRef] [PubMed]

86. Li, C.L.; Wang, M.; Wu, X.M.; Chen, D.H.; Lv, H.J.; Shen, J.L.; Qiao, Z.; Zhang, W. THI1, a thiamine thiazole synthase, interacts with Ca^{2+}-dependent protein kinase CPK33 and modulates the S-type anion channels and stomatal closure in Arabidopsis. *Plant Physiol.* **2016**, *170*, 1090–1104. [CrossRef] [PubMed]

87. Roberts, M.R.; Salinas, J.; Collinge, D.B. 14-3-3 proteins and the response to abiotic and biotic stress. *Plant Mol. Biol.* **2002**, *50*, 1031–1039. [CrossRef]

88. Viitanen, P.V.; Schmidt, M.; Buchner, J.; Suzuki, T.; Vierling, E.; Dickson, R.; Lorimer, G.H.; Gatenby, A.; Soll, J. Functional characterization of the higher plant chloroplast chaperonins. *J. Biol. Chem.* **1995**, *270*, 18158–18164. [CrossRef]

89. Gething, M.J.; Sambrook, J. Protein folding in the cell. *Nature* **1992**, *355*, 33–45. [CrossRef]

90. Salvucci, M.E. Association of RuBisCo activase with chaperonin-60beta: A possible mechanism for protecting photosynthesis during heat stress. *J. Exp. Bot.* **2008**, *59*, 1923–1933. [CrossRef]

91. Deng, Y.; Srivastava, R.; Howell, S.H. Endoplasmic reticulum (ER) stress response and its physiological roles in plants. *Int. J. Mol. Sci.* **2013**, *14*, 8188–8212. [CrossRef]

92. Sonoike, K. Various aspects of inhibition of photosynthesis under light/chilling stress: "Photoinhibition at chilling temperatures" versus "chilling damage in the light". *J. Plant Res.* **1998**, *111*, 121–129. [CrossRef]

93. Powles, S.B. Photoinhibition of photosynthesis induced by visible light. *Annu. Rev. Plant Physiol.* **1984**, *35*, 15–44. [CrossRef]

94. Salonen, M.; Aro, E.M.; Rintamaki, E. Reversible phosphorylation and turnover of the D1 protein under various redox states of Photosystem II induced by low temperature photoinhibition. *Photosynth. Res.* **1998**, *58*, 143–151. [CrossRef]

95. Nishiyama, Y.; Allakhverdiev, S.I.; Yamamoto, H.; Hayashi, H.; Murata, N. Singlet oxygen inhibits the repair of photosystem II by suppressing the translation elongation of the D1 protein in synechocystis sp. PCC 6803. *Biochemistry* **2004**, *43*, 11321–11330. [CrossRef] [PubMed]

96. Hong, J.H.; Chang, C.X.; Moon, B.Y.; Lee, C.H. Recovery from low-temperature photoinhibition is related to dephosphorylation of phosphorylated CP29 rather than zeaxanthin epoxidation in rice leaves. *J. Plant Biol.* **2003**, *46*, 122–129.

97. Davis, P.A.; Hangarter, R.P. Chloroplast movement provides photoprotection to plants by redistributing PSII damage within leaves. *Photosynth. Res.* **2012**, *112*, 153–161. [CrossRef] [PubMed]

98. Sakai, T.; Kagawa, T.; Kasahara, M.; Swartz, T.E.; Christie, J.M.; Briggs, W.R.; Wada, M.; Okada, K. Arabidopsis nph1 and npl1: Blue light receptors that mediate both phototropism and chloroplast relocation. *Proc. Natl. Acad. Sci. USA* **2001**, *98*, 6969–6974. [CrossRef] [PubMed]

99. Labuz, J.; Hermanowicz, P.; Gabrys, H. The impact of temperature on blue light induced chloroplast movements in *Arabidopsis thaliana*. *Plant Sci.* **2015**, *239*, 238–249. [CrossRef] [PubMed]

100. Banas, A.K.; Aggarwal, C.; Labuz, J.; Sztatelman, O.; Gabrys, H. Blue light signalling in chloroplast movements. *J. Exp. Bot.* **2012**, *63*, 1559–1574. [CrossRef]

101. Christie, J.M. Phototropin blue-light receptors. *Annu. Rev. Plant Biol.* **2007**, *58*, 21–45. [CrossRef]

102. Shen, S.; Jing, Y.; Kuang, T. Proteomics approach to identify wound-response related proteins from rice leaf sheath. *Proteomics* **2003**, *3*, 527–535. [CrossRef]

103. Ramagli, L.S. Quantifying protein in 2-D PAGE solubilization buffers. *Methods Mol. Biol.* **1999**, *112*, 99–103. [PubMed]

104. Olsen, J.V.; Blagoev, B.; Gnad, F.; Macek, B.; Kumar, C.; Mortensen, P.; Mann, M. Global, in vivo and site-specific phosphorylation dynamics in signaling networks. *Cell* **2006**, *127*, 635–648. [CrossRef] [PubMed]

 International Journal of
Molecular Sciences

Article

Phosphoproteomic Analysis of Two Contrasting Maize Inbred Lines Provides Insights into the Mechanism of Salt-Stress Tolerance

Xiaoyun Zhao [†], Xue Bai [†], Caifu Jiang and Zhen Li *

State Key Laboratory of Plant Physiology and Biochemistry, College of Biological Sciences,
China Agricultural University, Beijing 100193, China; xiaoyunzhao@cau.edu.cn (X.Z.);
xuebai0827@163.com (X.B.); cfjiang@cau.edu.cn (C.J.)
* Correspondence: lizhenchem@cau.edu.cn; Tel.: 86-10-62731128
† These authors contributed equally to this work.

Received: 25 February 2019; Accepted: 9 April 2019; Published: 16 April 2019

Abstract: Salinity is a major abiotic stress that limits maize yield and quality throughout the world. We investigated phosphoproteomics differences between a salt-tolerant inbred line (Zheng58) and a salt-sensitive inbred line (Chang7-2) in response to short-term salt stress using label-free quantitation. A total of 9448 unique phosphorylation sites from 4116 phosphoproteins in roots and shoots of Zheng58 and Chang7-2 were identified. A total of 209 and 243 differentially regulated phosphoproteins (DRPPs) in response to NaCl treatment were detected in roots and shoots, respectively. Functional analysis of these DRPPs showed that they were involved in carbon metabolism, glutathione metabolism, transport, and signal transduction. Among these phosphoproteins, the expression of 6-phosphogluconate dehydrogenase 2, pyruvate dehydrogenase, phosphoenolpyruvate carboxykinase, glutamate decarboxylase, glutamate synthase, L-gulonolactone oxidase-like, potassium channel AKT1, high-affinity potassium transporter, sodium/hydrogen exchanger, and calcium/proton exchanger CAX1-like protein were significantly regulated in roots, while phosphoenolpyruvate carboxylase 1, phosphoenolpyruvate carboxykinase, sodium/hydrogen exchanger, plasma membrane intrinsic protein 2, glutathione transferases, and abscisic acid-insensitive 5-like protein were significantly regulated in shoots. Zheng58 may activate carbon metabolism, glutathione and ascorbic acid metabolism, potassium and sodium transportation, and the accumulation of glutamate to enhance its salt tolerance. Our results help to elucidate the mechanisms of salt response in maize seedlings. They also provide a basis for further study of the mechanism underlying salt response and tolerance in maize and other crops.

Keywords: maize; phosphoproteomics; salt tolerance; label-free quantification; root and shoot

1. Introduction

Salt stress is a major factor that limits plant growth and development throughout the world. A thorough understanding of salt tolerance and response mechanisms is essential for the planting and breeding of crops [1,2]. Increased salt concentrations in plant roots and shoots cause ion toxicity, hyperosmotic stress, and oxidative damage, impair metabolic processes, and decrease photosynthetic efficiency in crops [2,3]. One mechanism that alleviates salt stress involves removal of sodium (Na$^+$) from the cytoplasm by transporting Na$^+$ into the vacuole or out of the cell [4]. This transportation is executed by Na/H exchangers (antiporters) and induced by salt stress [5–7]. Salt stress leads to the production of reactive oxygen species (ROS) in mitochondria, chloroplasts, and peroxisomes in plants, which causes oxidative damage to proteins, DNA, and lipids [8]. Elimination of excessive ROS via the glutathione–ascorbate cycle and maintaining tolerable salt levels inside the plant cells through exportation or compartmentalization are generally accepted as two major strategies used by plants to survive salinity stress [9]. In addition, it is reported that plants can synthesize and accumulate small molecules, such as proline, betaine, soluble sugars, and amino acids, to protect themselves from salt stress [10–12].

Mass spectrometry (MS)-based proteomics have played an indispensable role in large-scale characterization of complex protein mixtures. It can identify thousands of proteins and their modifications in a single experiment [13]. It has been widely used in proteomic and phosphoproteomic analysis on Arabidopsis [14–17], maize [18–21], rice [22–24], soybean [25–27], and sorghum [28,29]. A comparative phosphoproteomic and proteomic analysis on roots of soybean seedlings identified 2692 phosphoproteins and 5509 phosphorylation sites. Eighty-nine differentially abundant proteins were discovered, and a novel salt tolerance pathway involving chalcone metabolism, which was mediated by the phosphorylation of MYB transcription factors, was proposed [30]. A proteomics-based approach has also been used to identify salt-responsive proteins in leaves of salt-resistant Chinese wheat cultivar. Fifty-two differentially abundant proteins were identified, including H^+-ATPases, glutathione S-transferase, ferritin, and triosephosphateisomerase, which were found to be upregulated under salt stress [31]. A proteomic comparison between maize seedling roots from two different inbred lines under NaCl treatment for 2 days was reported [32]. Twenty-eight differentially abundant proteins were identified, such as 14-3-3 proteins, plasma membrane intrinsic proteins (PIP1 and PIP2), ribosomal protein S8, and 60S ribosomal protein L3-1. These proteins were mainly involved in signal processing, water conservation, protein synthesis, and abiotic stress-tolerance.

Protein phosphorylation plays critical roles in various biological processes, including cell growth, proliferation, metabolism, signal transduction, and apoptosis [33,34]. Many stress-related signaling pathways are involved in protein phosphorylation [35–38]. It is estimated that more than 30% of the proteins are phosphorylated at any given time in a living cell [39]. The SOS (Salt Overly Sensitive) pathway is a well-studied salt response signaling pathway in plants [38,40–44]. Salt stress elicits a calcium signal in the cytoplasm, and the calcium-binding protein encoded by SOS3 senses and binds to these Ca^{2+} to activate SOS2, a serine/threonine protein kinase. The SOS3–SOS2 complex activates SOS1 (plasma membrane Na^+/H^+ antiporter) or other transporters. The SOS signaling pathway is conserved in rice and maize [45,46]. Moreover, several protein-kinase- and protein-phosphatase-related signal pathways are also involved in salt response [47]. Therefore, identification and quantification of phosphorylated peptides under salt stress are highly desirable in understanding the salt resistance/tolerance mechanism in plants.

Maize (*Zea mays* L.) is a glycophyte plant that is hypersensitive to salt stress [48,49]. Salt stress not only results in weak seeding, a short radicle, and a low survival rate [50], but also impairs the photosynthesis system and leads to decreased stomatal conductance and an increased intercellular carbon dioxide concentration [2]. Therefore, characterizing the salt-tolerance mechanism is essential for increasing corn yield. Different maize inbred lines exhibit different salt tolerance characteristics, and comparing the proteomics and phosphoproteomics changes between salt-tolerant and salt-sensitive varieties under saline conditions is an effective means to explore the mechanisms responsible for the diversity in tolerance to salt stress. At present, studies on the phosphorylation of maize under salt stress are still very limited [51,52]. Most studies limit themselves to a few specific proteins, and most of these proteomic and phosphoproteomics studies focus on long-term saline conditions (e.g., over 48 h of NaCl treatment) [53,54]. Under salt stress, receptors located on the cell membrane quickly sense changes in NaCl content in the external environment, and second messengers, such as Ca^{2+}, inositol phosphate, ROS, and phytohormones, are rapidly produced in the cytoplasm to transduce and amplify the salt stress signals, which, in turn, induce immediate phosphorylation of protein kinases and their downstream substrates [55,56]. Therefore, in-depth analysis of phosphoproteomics between salt-sensitive and salt-tolerant maize inbred lines under short-term salt treatment (0.5 to 2 h after salt treatment) is critical to obtain a comprehensive understanding of the salt response and salt-tolerance mechanism in maize.

In this study, an MS-based label-free quantitative phosphoproteomics analysis was employed to identify differentially regulated phosphoproteins in two maize cultivars: the salt-sensitive Chang7-2 and the salt-tolerant Zheng58. Here, 209 and 243 phosphoproteins were found to be significantly regulated in roots and shoots of the two maize lines after 100 mM NaCl treatment for 0.5 h and 2 h. Physiological parameters, such as hydrogen peroxide content, proline content, and metal ion contents, were examined in roots and shoots (Figure S1). The results of this study help to elucidate the salt response and regulation mechanism in maize and provide a theoretical basis for the cultivation of salt-tolerant varieties.

2. Results

2.1. Physiological Assays of Chang7-2 and Zheng58 in Response to Salt Stress

2.1.1. Proline and H_2O_2 Contents under Salt Treatment

To investigate the difference in salt tolerance between Chang7-2 and Zheng58, we measured the hydrogen peroxide (H_2O_2) content, the proline content, and the metal ion contents (Na^+, K^+, and Mg^{2+} etc.) in the roots and shoots of the two cultivars. It can be seen that the proline content in roots increased significantly in both inbred lines after salt treatment for 0.5 h and the extent of increase was greater in Zheng58 (Figure 1a). The proline content dropped to the original level 2 h after treatment. The proline content in shoots showed the opposite trend, that is, the proline content in shoots decreased after 0.5 h of salt treatment and recovered after 2 h of treatment (Figure 1b). These results suggest that, under short-term salt stress, plants preferentially transport existing proline from shoots to roots instead of de novo synthesis [57,58]. Salt stress usually induces excessive amounts of H_2O_2 in plants. We found that the H_2O_2 content in roots of Zheng58 increased significantly after exposure to salt for 2 h but was unaffected in Chang7-2 (Figure 1c). The H_2O_2 content in shoots decreased significantly in both Chang7-2 and Zheng58 after exposure to salt for 0.5 h, and almost stayed the same at 2 h (Figure 1d). Thus, salt treatment may induce rapid accumulation of H_2O_2 in salt-tolerant plants to stimulate the ROS scavenging system.

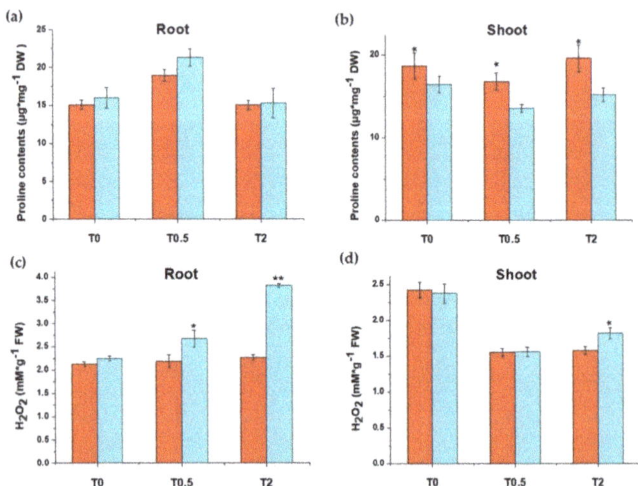

Figure 1. Physiological analysis of Chang7-2 and Zheng58 seedlings under saline conditions (100 mM NaCl treatment for 0.5 h and 2 h). Proline content of root (**a**) and shoot (**b**); Hydrogen peroxide content of root (**c**) and shoot (**d**). * indicates significant difference at 0.05; ** indicates a difference at the 0.01 level determined by Student's *t*-test. Red, Chang7-2; Light green, Zheng58.

2.1.2. Metal Ions Response to Salt Stress in Chang7-2 and Zheng58

The fluctuation of metal elements in Chang7-2 and Zheng58 under salt treatment was measured by ICP-OES. K^+ contents were significantly higher in shoots than in roots, while Mn^{2+} and Zn^{2+} contents were lower than in roots (Table S1). Principal component analysis (PCA) of all eight metal elements (K^+, Na^+, Ca^{2+}, Mg^{2+}, Mn^{2+}, Fe^{2+}, Zn^{2+}, and Cu^{2+}) in roots showed that there were distinct differences in ion contents between the two inbred lines, even before salt treatment. For Zheng58, the treatment groups at 0.5 h and 2 h can be clearly separated from the control group, but the difference between the two salt treatment groups was not significant. For Chang7-2, the control group and the treatment groups could not be completely separated on the PCA score plot (Figure 2a), suggesting that salt stress influences the ion content of Zheng58 to a higher extent than that of Chang7-2. K^+, Ca^{2+}, and Fe^{2+} had the highest loading in PC1, while Mg^{2+} and Mn^{2+} contributed most to PC2 (Figure S2a). Similar to the results from roots, salt treatment induced significant changes on metal ion contents in shoots of Zheng58 but not in Chang7-2 (Figure 2b). K^+, Na^+, and Ca^{2+} were the key elements responsible for such differences (Figure S2b).

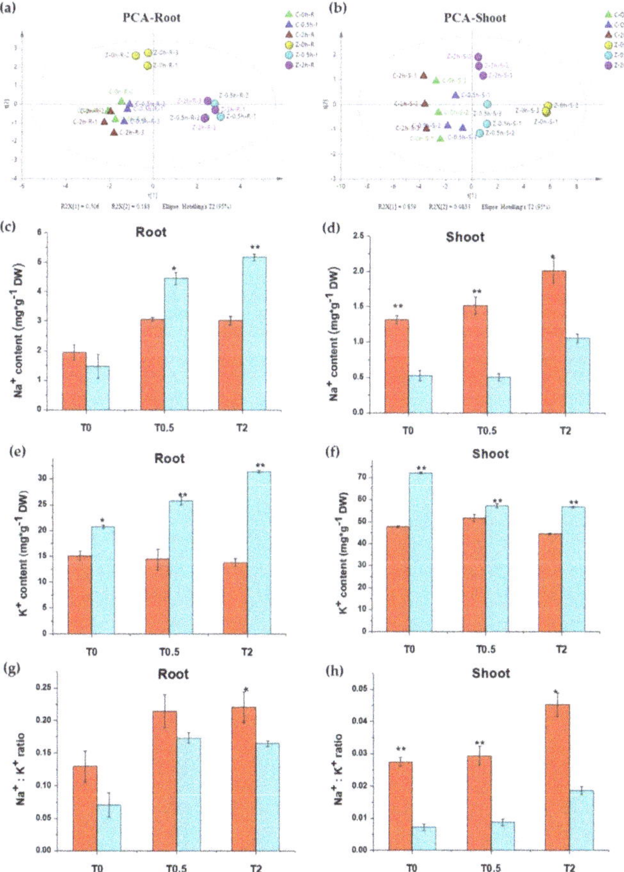

Figure 2. Metal ion contents in Chang7-2 and Zheng58 after salt treatment for 0.5 h and 2 h. Principal Component Analysis (PCA) score plot for root (**a**) and shoot (**b**); Na^+ content (**c,d**); K^+ content (**e,f**); and Na^+/K^+ ratio (**g,h**) for root and shoot. * indicates significant difference at 0.05; ** indicates a difference at the 0.01 level determined by Student's *t*-test. Red, Chang7-2; Light green, Zheng58.

Ionomic data showed that the K^+ and Na^+ content in both roots and shoots changed significantly after salt treatment. One interesting finding was that Na^+ contents in the roots of Zheng58 were much higher than those of Chang7-2 after salt treatment, while in shoots, it was Chang7-2 that had the higher Na^+ content (Figure 2c,d). Such a finding indicates that, under salt stress, Zheng58 greatly reduced Na^+ transportation from roots to shoots, which is an effective strategy for salt-tolerant plants to resist salt stress [59,60]. Also, the K^+ content increased significantly in the roots of Zheng58 but was unaffected in Chang7-2 under salt treatment (Figure 2e), and the Na^+/K^+ ratio was much lower in Zheng58 than Chang7-2 in both roots and shoots (Figure 2g,h). Taken together, when faced with salt stress, Zheng58 maintained a lower Na^+ concentration and Na^+/K^+ ratio in shoots to remain viable [60].

2.2. Overview of Phosphoproteins Identified in Maize Seedlings

Label-free quantitation of phosphoproteins after salt treatment was realized with Progenesis QI for Proteomics and protein identification was realized by the Mascot search engine. In roots and shoots, 36,635 and 40,119 peptide spectrum matches (PSMs) were identified separately at a 1% false discovery rate (FDR). In roots, 9772 unique phosphopeptides assigning to 3084 phosphoproteins were quantified and 11,062 unique phosphopeptides mapping to 2986 phosphoproteins were quantified in shoots. Altogether, 9448 unique phosphorylation sites from 4116 phosphoproteins were detected, and more than 7000 phosphorylation sites can be detected from each sample (Table S2). We detected 2211 and 2454 proteins in roots and shoots, respectively, with at least two unique phosphopeptides, accounting for more than one half of the identified phosphoproteins (Figure 3a). Most of the identified peptides contained one phosphorylation site, and 4225 and 4030 peptides in roots and shoots, respectively, were identified with multiple phosphorylation sites (Figure 3c,d).

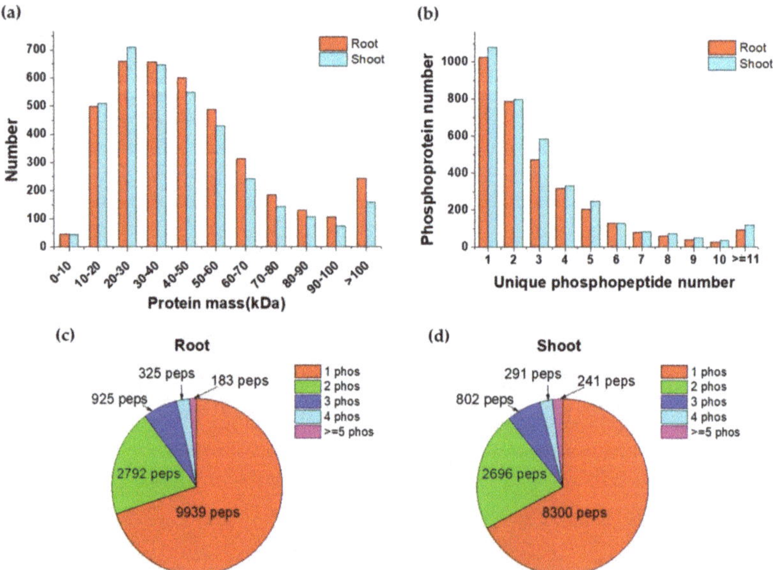

Figure 3. Overview of phosphoprotein identification. Distribution of protein mass (**a**) and numbers of unique peptides (**b**) of identified proteins. The phosphorylation status of the identified phosphopeptides in roots (**c**) and shoots (**d**).

2.3. Differentially Regulated Phosphoproteins in Chang7-2 and Zheng58 in Response to Salt Treatment

Differentially regulated phosphoproteins were defined as phosphoproteins with more than 2-fold of the changes in intensity after 0.5 h or 2 h of salt treatment when compared with their respective control group ($p < 0.05$). For the roots, 129 phosphoproteins (99 increased and 30 decreased) from Chang7-2 and 131 phosphoproteins (42 increased and 89 decreased) from Zheng58 met the above criteria (Figure 4a and Figure S3a,b). Fifty-one phosphoproteins were identified in both lines, and 78 and 80 phosphoproteins were unique to Chang7-2 and Zheng58, respectively. (Figure 4a). For the shoots, 134 phosphoproteins (74 upregulated and 60 downregulated) from Chang7-2 and 205 phosphoproteins (136 upregulated and 69 downregulated) from Zheng58 were significantly changed after salt treatment (Figure 4b and Figure S3c,d). Among them, 96 phosphoproteins were shared between Chang7-2 and Zheng58 shoots, and 38 and 109 phosphoproteins were unique to Chang7-2 and Zheng58, respectively. (Figure 4b).

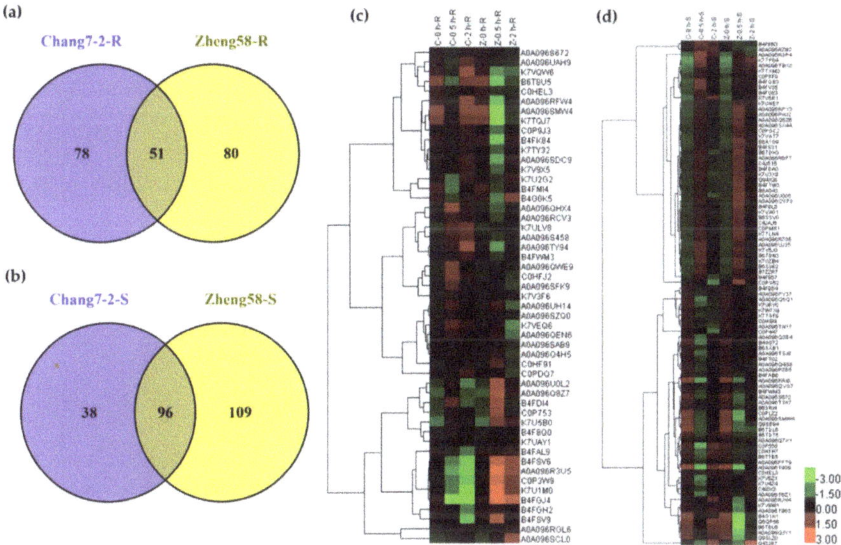

Figure 4. Differently regulated phosphoproteins (DRPPs) in salt-treated Chang7-2 and Zheng58 in roots (**a**) and shoots (**b**); clustering analysis of DRPPs detected in both Chang7-2 and Zheng58 seedlings ((**c**), roots and (**d**), shoots).

Of the 51 differentially regulated phosphoproteins (DRPPs) detected in the roots of both Chang7-2 and Zheng58, hierarchical clustering analysis showed that they exhibited different abundance patterns in different inbred lines (Figure 4c), indicating different response mechanisms of these two maize lines under salt stress. Among the 51 DRPPs, 29 DRPPs were upregulated in Chang7-2 but downregulated in Zheng58. These phosphoproteins included kinases, transcription factors, and ion-transport-related phosphoproteins. Five DRPPs were decreased in Chang7-2 but increased in Zheng58, which consisted of the 6-phosphogluconate dehydrogenase family protein, ribosomal protein, pyruvate dehydrogenase complex E1 alpha subunit, phosphoenolpyruvate carboxykinase, and putative L-gulonolactone oxidase. The remaining 17 DRPPs showed the same trend in the two lines. These 96 DRPPs in shoots shared between the two lines are clustered in Figure 4d. The trends of phosphorylation of these proteins in Chang 7-2 and Zheng 58 were also diverse.

2.4. Clustering Analysis of DRPPs Unique to Chang7-2 or Zheng58 in Maize Roots

Clustering analysis of DRPPs unique to Chang7-2 or Zheng58 were conducted using the Mfuzz algorithm embedded in the Wukong data analysis platform [61]. DRPPs from Chang7-2 can be grouped into six categories (Figure 5A). The phosphorylation level of proteins in cluster 1 was significantly upregulated after 2 h of salt treatment. There were 23 phosphoproteins in this cluster. They were involved in metal ion transport, phosphatidylinositol phosphorylation, signal transduction, transcription, regulation of lipid kinase activity, and the cellulose biosynthetic process. Similar to cluster 1, DRPPs in cluster 2 were upregulated in response to salt stress, but the extent of the increase was greater than that in cluster 1. Proteins in this cluster included: chloride channel, putative casein kinase protein, and proteasomal ubiquitin receptor ADRM1-like. DRPPs that functioned in the cell surface receptor signaling pathway and intracellular protein transport were only significantly upregulated after 2 h treatment, and were assigned to cluster 3. Cluster 4 contained proteins that were downregulated after exposure to salt stress. DRPPs in cluster 5 and 6 were induced after 0.5 h of treatment, and recovered after 2 h. Proteins that played roles in protein transport, response to oxidative stress, and retrograde-vesicle-mediated transport were assigned to cluster 5, and proteins involved in signal transduction, the lipid metabolic process, and defense response to fungus were assigned to cluster 6 (Figure 5a).

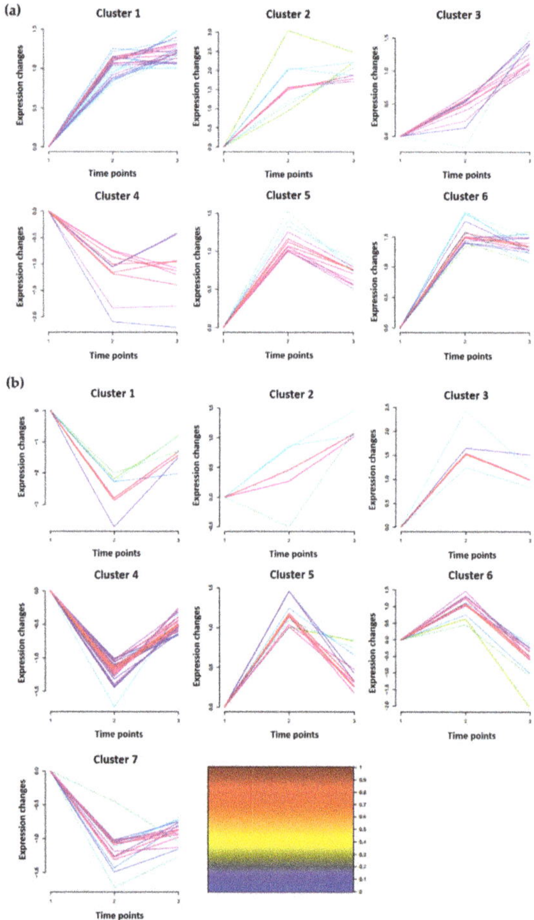

Figure 5. MFUZZ clustering analysis of DRPPs unique to Chang7-2 (**a**) and Zheng58 (**b**).

Eighty DRPPs that were unique to Zheng58 roots after salt stress were clustered into seven categories (Figure 5b). DRPPs in cluster 1, 4, and 7 were all significantly downregulated at 0.5 h after salt treatment and recovered at 2 h. Proteins in cluster 1 decreased to a higher extent than those in clusters 4 and 7, and seven proteins, including plasma membrane ATPase (K7TX67) and UBP1-associated protein (C0P558), were in this cluster. Twenty-three DRPPs were in cluster 4. They were involved in such biological processes as cell redox homeostasis, transport, pentose-phosphate shunt, microtubule cytoskeleton organization, the glutamate metabolic process, rRNA processing, the auxin-activated signaling pathway, and regulation of transcription. The DRPPs in cluster 7 were proteins related to transcription initiation, nucleotide binding, and calcium ion binding. The DRPPs in clusters 3 and 5 were both upregulated at 0.5 h and then declined at 2 h. Serine/threonine-protein kinase SRK2A, mitogen-activated protein kinase, and ribosomal proteins were included in cluster 3. Potassium channel 5, glutamate synthase, uridine-cytidine kinase C-like, and several uncharacterized proteins were included in cluster 5. Twelve DPRRs that increased in abundance at 0.5 h and declined at 2 h were grouped into cluster 6, including sodium/hydrogen exchanger 2-like (A0A096Q7K1), high-affinity potassium transporter (W5U5W2), molybdate transporter 2-like (A0A096TR23), and protein NRT1/PTR family 8.3-like (A0A096TGV7).

2.5. Gene Ontology Analysis of Salt-Responsive DRPPs in the Two Inbred Lines

Gene ontology (GO) enrichment of salt-responsive DRPPs in Chang7-2 and Zheng58 roots were performed using the Wukong platform [61]. GO functional analysis revealed that most of the differentially phosphorylated proteins under salt stress in the two lines were involved in the same biological processes or were endowed with the same molecular functions, although some unique GO terms were also enriched for each individual line. As shown in Figure 6, 14 GO terms and 17 GO terms were enriched in Chang7-2 and Zheng58, respectively, in the biological process category. One GO term, carbohydrate biosynthetic process (GO: 0016051), was unique to Chang7-2. Four GO terms, including monovalent inorganic cation homeostasis (GO: 0055067), pyrimidine-containing compound biosynthetic process (GO: 0072528), potassium ion homeostasis (GO: 0055075), and regulation of intracellular transport (GO: 0032386), were unique in Zheng58. Among these, potassium ion homeostasis was upregulated and regulation of intracellular transport was downregulated in Zheng58. Moreover, two GO terms, aldonic acid metabolic process and D-gluconate metabolic process, were upregulated in Zheng58 but downregulated in Chang7-2. For the molecular function category, nine GO terms and 12 GO terms were enriched in Chang7-2 and Zheng58, respectively. Three GO terms were unique to Chang7-2, namely, organic cyclic compound binding (GO: 0097159), nucleic acid binding (GO: 0003676), and sequence-specific DNA binding (GO: 0043565). Six GO terms, including sodium ion transmembrane transporter activity (GO:0015081), protein homodimerization activity (GO: 0042803), nucleoside-triphosphatase regulator activity (GO: 0060589), enzyme activator activity (GO: 0008047), uridine kinase activity (GO: 0004849), and nucleoside kinase activity (GO: 0019206), were specifically enriched in Zheng58. Among these, sodium ion transmembrane transporter activity, uridine kinase activity, and nucleoside kinase activity were upregulated in Zheng58, and nucleoside-triphosphatase regulator activity and enzyme activator activity were downregulated in Zheng58 (Figure 6).

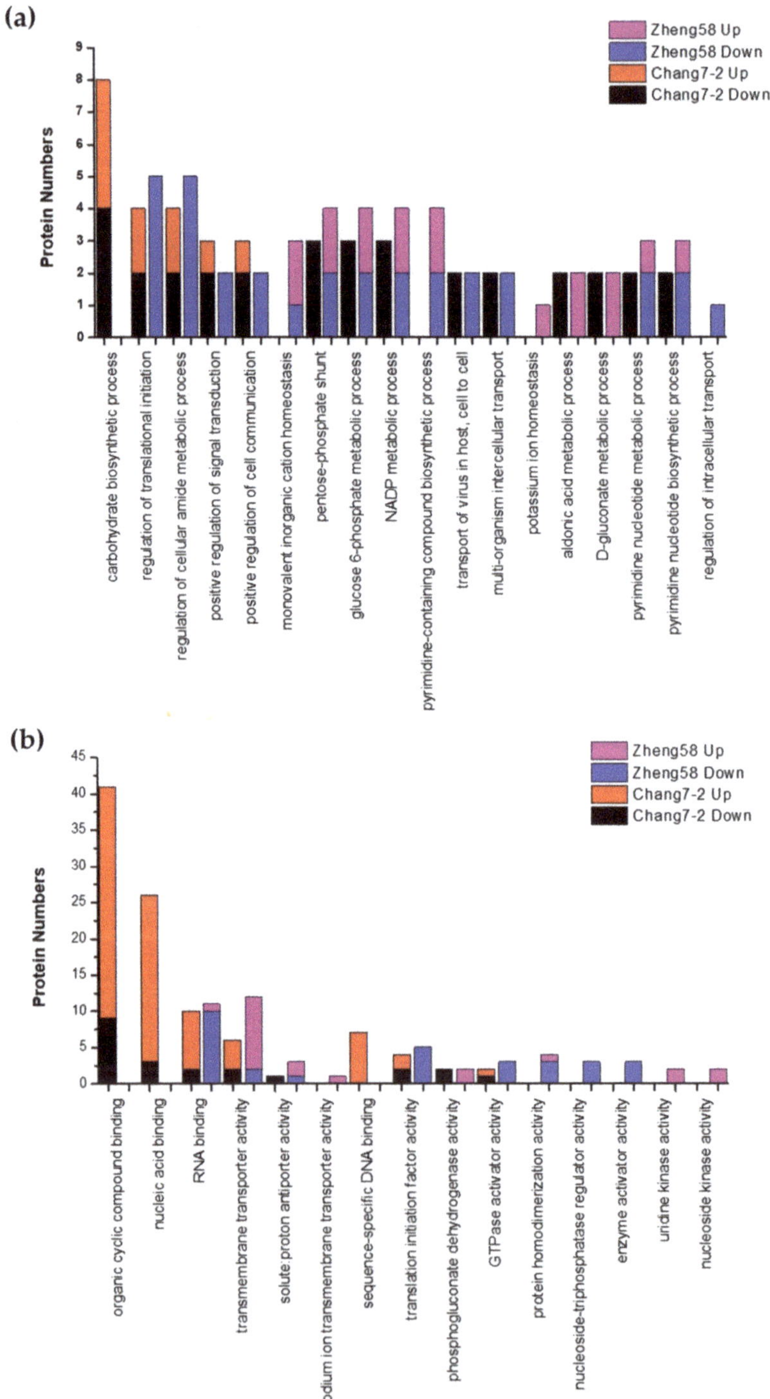

Figure 6. Gene Ontology (GO) analysis of DRPPs in roots of Chang7-2 and Zheng58. GO terms in biological processes (**a**) and molecular functions (**b**) are shown.

2.6. Protein–Protein Interaction Analysis

To investigate the transduction of salt stress signals in maize root cells, DRPPs identified in Chang7-2 and Zheng58 roots were further analyzed using the String 10.5 database (https://string-db.org/) [62]. Protein–protein interactions with confidence scores higher than 0.7 were shown. Five groups of interactions were significantly enriched in Zheng58 (Figure 7). Proteins including sodium/hydrogen exchanger (A0A096Q7K1, Zm.82157), sugar transport protein (C0P753, ERD6), glutamate synthase (A0A096RAH5, Zm.24266), and one uncharacterized protein (transport activity, B4FC34, GRMZM2G082184) constituted the first group. These proteins are related to sodium transport, amino acid transport/metabolism, and sugar transport. The second group was comprised of fructose-bisphosphate aldolase (B4FAL9, ALD1), pyruvate dehydrogenase E1 component subunit alpha (B4FGJ4, Zm.95858), and phosphoenolpyruvate carboxylase (C0P3W9, GRMZM2G001696). Proteins in this group are involved in carbon metabolism. In the third interaction group, uridine kinase (K7UTD0, Zm.84726) interacted with uridine-cytidine kinase C-like (A0A096SG98, Zm.2141) and shaggy-related protein kinase alpha-like (A0A096S672, Zm.155457). Proteins in this group are related to pyrimidine metabolism. The fourth group contained proteins involved in protein translation, including ribosomal protein S10 (J7LC26, Zm.16861), 40S ribosomal protein S27 (Q9ZQX9, GRMZM2G066222), 40S ribosomal protein S10-1 (B4FW06, GRMZM2G446960), and 60S ribosomal protein L24 (A0A096R3U5, GRMZM2G142640). Finally, calcium-binding protein (C0P445, IDP554), which is involved in plant–pathogen interaction, interacted with myosin1 (A0A096PFT4, Zm.442). They are connected with host defenses against pathogens. Additionally, five protein interaction groups were predicted for Chang7-2 roots as shown in Figure S4.

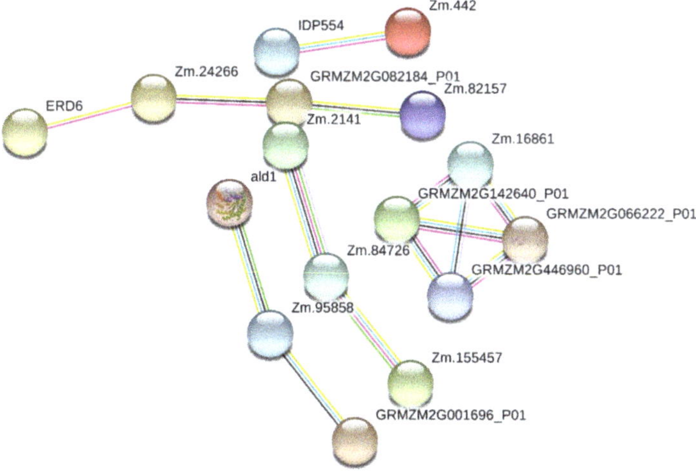

Figure 7. The protein interaction network of DRPPs in Zheng58 roots.

3. Discussion

Saline soil is a major threat to maize cultivation and yield enhancement. Therefore, characterizing the mechanism of salt tolerance in different maize varieties is critical for maize breeding and production. In this study, we compared the phosphoproteomics of two maize inbred lines with distinct salt tolerances in response to short-term salt treatment to explore the salt tolerance mechanism of maize. Most of the previous studies focus on proteomics and phosphoproteomics profiling under long-term salt treatment, i.e., proteomics studies were performed several days after salt treatment [54]. It is well-known that salt stress induces many processes related to protein phosphorylation in plants, which occurs in the time scale of seconds to minutes [38,47].

Therefore, to complement the salt-tolerant mechanism in maize, we investigated the short-term phosphoproteomics response of maize under salt stress conditions. We compared the phosphorylation levels of two contrasting inbred lines, one salt-tolerant (Zheng58) and one salt-sensitive (Chang7-2), in response to short-term (0.5 and 2 h) salt stress. Physiological changes in line with the phosphorylation events were also assessed to explore the salt tolerance mechanism in maize. Roots and shoots of seedlings from the two inbred lines were collected and assayed separately. A summary of the differences in physiology and phosphoproteomics in response to salinity stress in roots and shoots of Chang7-2 and Zheng58 is listed in Figure 8.

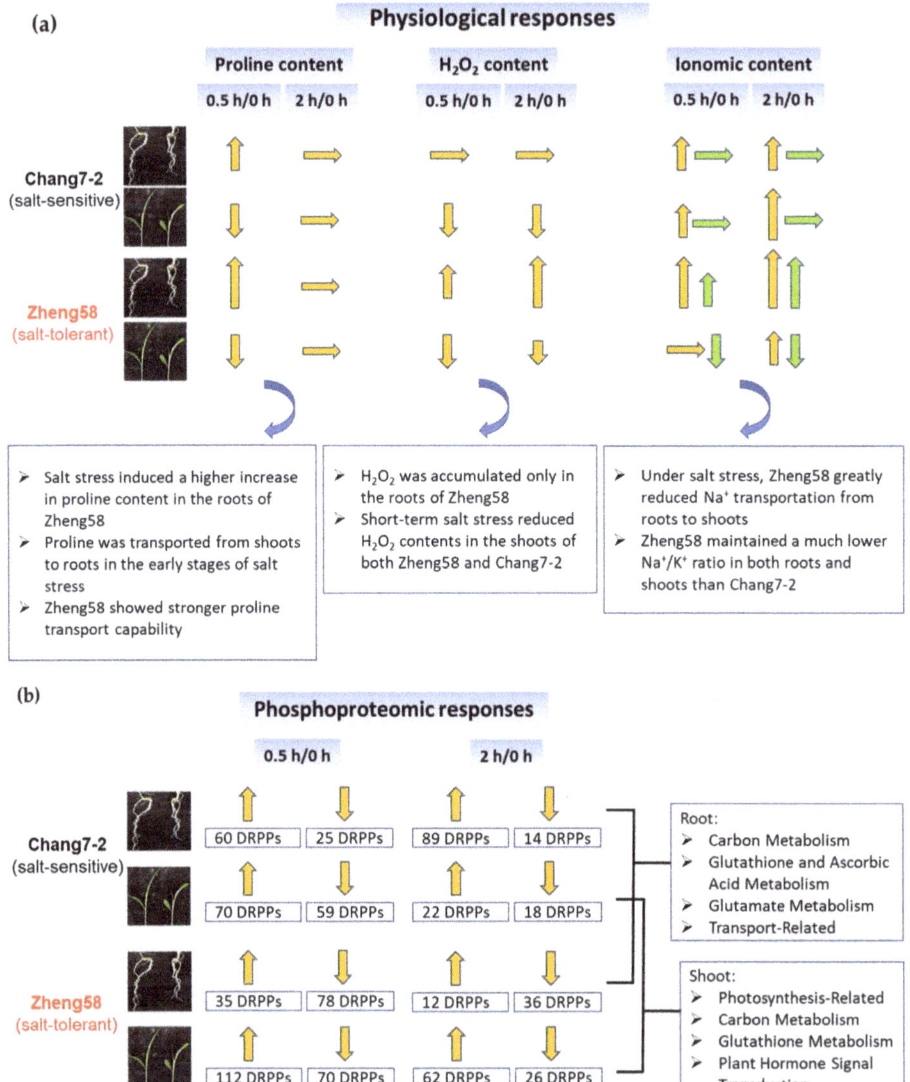

Figure 8. A summary of differences in physiology (**a**) and phosphoproteomics (**b**) in response to salinity stress in roots and shoots of Chang7-2 and Zheng58. The up arrows indicate upregulation and the down arrows indicate downregulation; The lengths of the arrows indicate the degree of changes; In the ionomic content section, the yellow arrows represent Na^+ and the green arrows indicate K^+.

3.1. Physiological Difference between Zheng58 and Chang7-2 under Salt Stress

The physiological responses of Chang7-2 and Zheng58 to short-term salt treatment were evaluated to investigate their distinct salt endurance characteristics. In the roots of the two inbred lines, the proline content increased at 0.5 h of salt treatment and returned to the original levels at 2 h; the opposite trends were observed in the shoots (Figure 1a,b). Salt stress induced a larger increase in proline content in Zheng58, suggesting that Zheng58 mobilized the defense mechanism to cope with salt stress more quickly. The proline content of both inbred lines decreased to normal levels after 2 h of treatment, indicating that other defense strategies were initiated by the plants to adapt to the osmotic stress caused by salt stress after a certain period of salt treatment. In shoots, the proline content decreased at 0.5 h of salt treatment, and slightly rebounded at 2 h, which was the exact opposite of the change in proline content in roots. Thus, we proposed that the proline accumulated in roots was mainly transported from the shoots to initiate a quick response in the early stages of salt stress. Zheng58 showed a stronger proline transport capability; supporting it was its salt-tolerant nature.

The H_2O_2 contents in shoots were higher than those in roots for both maize lines under control conditions. Short-term salt stress induced H_2O_2 accumulation only in the roots of Zheng58. In the meanwhile, decreased H_2O_2 contents were observed in the shoots of both Zheng58 and Chang7-2 after salt treatment (Figure 1c,d). These results indicate that the roots of the salt-tolerant plant Zheng58 accumulated more H_2O_2 under short-term salt stress, and it had a higher tolerance to elevated H_2O_2. The H_2O_2 content decreased in shoots in both inbred lines after salt treatment, indicating immediate removal of H_2O_2 in the shoots.

Metal ions assay of the two inbred lines revealed that the metal ions contents in the two inbred lines were statistically different even before salt treatment, as indicated by principal component analysis (Figure 2a,b). The metal ions contents in the two lines showed different responses to salt stress. The metal ions contents in Zheng58 changed significantly after salt treatment, while the metal ions contents in Chang7-2 did not show notable changes under saline conditions. Zheng58 greatly reduced the transportation of Na^+ from roots to shoots and maintained a much lower Na^+/K^+ ratio than Chang7-2 under salt stress (Figure 2g,h), demonstrating that Zheng58 had a better salt tolerance capability than Chang7-2.

3.2. Functional Analysis of DRPPs in Roots of Zheng58 and Chang7-2 under Salt Stress

3.2.1. Carbon Metabolism

6-phosphogluconate dehydrogenase 2 (B4FSV6) catalyzes the dehydrogenation of 6-phosphate-gluconate to form 5-phosphate-ribulose; it is the key enzyme in the pentose phosphate pathway [63]. This reaction is accompanied by the production of NADPH, which is a reducing agent for various synthetic reactions in living systems. Moreover, NADPH is the cofactor required for oxidized glutathione to be converted to reduced glutathione. Glutathione is vital for redox homeostasis and is one of the main antioxidants involved in eliminating reactive oxygen species (ROS) [64]. The phosphorylation of 6-phosphogluconate dehydrogenase 2 was downregulated in Chang7-2 (0.37-fold and 0.31-fold at 0.5 and 2 h treatment, respectively, Table 1) but upregulated in Zheng58 (2.05-fold at 2 h treatment), suggesting that Zheng58 can produce more NADPH under salt stress. It has been reported that 6-phosphogluconate dehydrogenase was upregulated in salt-treated Jing724, a salt-tolerant maize line, but not in the salt-sensitive maize line D9H, consistent with our results [54].

Table 1. Details of differentially regulated phosphoproteins related to salt tolerance in Zheng58 roots.

Accession	Annotation	p-Value	C-0.5h fc [1]	C-2h fc	Z-0.5h fc [2]	Z-2h fc
Carbon Metabolism						
B4FSV6	6-phosphogluconate dehydrogenase 2	0.0095	0.37	0.31	1.88	2.05
B4FGJ4	Pyruvate dehydrogenase	1.1×10^{-6}	0.08	0.06	3.60	2.02
C0P3W9	phosphoenolpyruvate carboxykinase	0.0124	0.38	0.38	2.16	1.58
Glutathione and Ascorbic Acid Metabolism						
B4FK84	Glutathione S-transferase 3	0.0427	0.47	1.59	0.04	0.53
K7U1M0	L-gulonolactone oxidase-like	0.0054	0.32	0.19	2.57	1.37
Glutamate Metabolism						
B4F972	Glutamate decarboxylase	0.0203	-	-	0.30	0.80
A0A096RAH5	glutamate synthase	0.0002	1.01	0.79	2.00	1.53
Transport-Related Proteins						
K7V3Z4	Potassium channel AKT1	0.0092	1.28	0.99	2.06	1.39
W5U5W2	high-affinity potassium transporter	0.0095	1.07	0.95	2.49	0.78
A0A096Q7K1	Sodium/hydrogen exchanger	4.29×10^{-6}	1.55	1.64	2.13	0.71
B4F910	Calcium/proton exchanger CAX1-like protein	1.01×10^{-5}	1.17	0.99	1.16	0.35

[1] C-0.5h fc indicates a fold change at 0.5 h of salt treatment compared with the control group in Chang7-2. [2] Z-0.5h fc indicates a fold change at 0.5 h of salt treatment compared with the control group in Zheng58.

B4FGJ4 (pyruvate dehydrogenase), B4FAL9 (aldolase 1), and C0P3W9 (phosphoenolpyruvate carboxykinase) are all involved in carbon metabolism. Pyruvate dehydrogenase catalyzes the dehydrogenation and decarboxylation of pyruvate to form acetyl-CoA, which is the core reaction of the Tricarboxylic acid (TCA) cycle [65]. Phosphoenolpyruvate carboxykinase functions as the catalytic enzyme in converting oxaloacetate to phosphoenolpyruvate, an intermediate product in glycolysis [65]. The increased phosphorylation of these two proteins (a 2.16 to 3.60-fold increase) in response to salt stress in Zheng58 helped the plant resist salt stress by producing more energy.

3.2.2. Glutathione and Ascorbic Acid Metabolism

Glutathione S-transferase 3 (B4FK84) is the key enzyme that catalyzes the binding reaction between glutathione and various electrophilic compounds [66]. The phosphorylation level of this protein decreased at 0.5 h of salt treatment and then returned afterwards at 2 h in Chang7-2, but remained at a lower level in Zheng58 (Table 1). The lower phosphorylation level of glutathione S-transferase 3 was beneficial to the accumulation of glutathione, which is an important antioxidant to scavenge ROS and protect sulfhydryl groups in many proteins and enzymes [64]. Thus, the reduced phosphorylation of glutathione S-transferase 3 in Zheng58 suggests that the plant may maintain a higher level of glutathione in the case of salt stress, which may account for the salt-tolerant trait of the plant.

L-gulonolactone oxidase-like (K7U1M0) was downregulated in Chang7-2 (0.32-fold) but upregulated in Zheng58 (2.57-fold) after 0.5 h of salt treatment. L-gulonolactone oxidase is the key enzyme in the synthesis of ascorbic acid [67]. Ascorbic acid has a strong reducing capability; it is, like glutathione, an important antioxidant. We observed increased H_2O_2 content in response to salt treatment in Zheng58 roots (Figure 1c), which required more antioxidant to remove H_2O_2, consistent with the phosphoproteomics results. The abundance of ascorbate peroxidase was reported to be upregulated in a salt-tolerant maize line [54], suggesting that enhanced synthesis of glutathione and ascorbic acid for scavenging ROS is a common strategy for salt-tolerant plants.

3.2.3. Glutamate Metabolism

Glutamate decarboxylase (B4F972) promotes the decarboxylation of glutamate to produce γ-aminobutyric acid (GABA) and CO_2 [68]. Salt treatment (0.5 h) induced a significant decrease in phosphorylation of the protein in Zheng58, though it slightly recovered at 2 h. The decomposition of glutamate was inhibited when the phosphorylation of glutamate decarboxylase was decreased, leading to the accumulation of glutamate. Glutamate is the major substrate for the synthesis of proline under stress conditions [69]. Increased proline content is an adaptive response of plants to salt stress [12,70,71]. In addition, glutamate synthase (A0A096RAH5), which is involved in glutamate synthesis, was upregulated in Zheng58 but unchanged in Chang7-2. These results indicate that Zheng58 could maintain a higher level of glutamate under salt stress conditions, which can be used for the synthesis of proline, and this may contribute to its salt tolerance trait. A similar result was found in the comparative proteomics study between other maize lines [54].

3.2.4. Transport-Related Proteins

Under salt stress, sodium and potassium ions are selectively absorbed by plants; potassium instead of sodium is the preferred metal ion for absorption [5]. A previous study reported that salt-tolerant plants have higher concentrations of potassium in the extended tissues [72]. Potassium channel AKT1 (K7V3Z4) enables the transmembrane transfer of potassium ions by a voltage-gated channel and plays vital roles in K^+ uptake [73]. The phosphorylation of potassium channel AKT1 was upregulated in Zheng58 (2.06-fold at 0.5 h treatment), which enhanced its ability to transport potassium, resulting in its salt tolerance. In Chang7-2, the phosphorylation of this protein remained unchanged (a 1.28-fold change at 0.5 h treatment). Also, the phosphorylation of high-affinity potassium transporter (W5U5W2) was increased in Zheng58 under salt stress, hinting that it is closely related to the salt tolerance of Zheng58.

Sodium/hydrogen exchanger (A0A096Q7K1) has sodium/proton antiporter activity and pumps sodium ion from the cytoplasm to the vacuole, which not only reduces the damage that Na^+ exerts on enzymes and the membrane system, but also alleviates osmotic stress imposed by an elevated Na^+ concentration [7]. Maintaining a lower Na^+/K^+ ratio in the cytoplasm by pumping Na^+ into the vacuole is an important mechanism for plants to cope with salt stress. Two V-type proton ATPases were found to be upregulated after salt treatment in sugar beet to enhance the Na^+ compartmentalization into vacuoles and reduce Na^+ accumulation in the cytosol [74]. The phosphorylation level of the sodium/hydrogen exchanger remained unchanged in Chang7-2 but was significantly increased in Zheng58 at 0.5 h treatment (Table 1), evidencing that Zheng58 was more salt-tolerant.

Calcium is a signaling molecule ubiquitously existing in plants and is considered to play important roles in response to salt stress. Salt stress causes an increase in intracellular calcium concentration, and this signal is sensed and passed down by related proteins and eventually results in activation of target kinase and transcription factors and the expression of downstream regulating genes to initiate the salt-resistance mechanism [75–77]. Calcium/proton exchanger CAX1-like protein (B4F910) translocates Ca^{2+} and other metal ions into vacuoles using the proton gradient formed by H^+-ATPase and H^+-pyrophosphatase [78]. Phosphorylation of this protein was decreased at 2 h of salt treatment in Zheng58 but remained unchanged in Chang7-2 (Table 1), so more Ca^{2+} can be accumulated in the cytoplasm of Zheng58 under saline conditions, which facilitated the quick initiation of signaling transduction to resist salt stress.

3.3. Functional Analysis of DRPPs in Shoots of Zheng58 and Chang7-2 under Salt Stress

3.3.1. Photosynthesis-Related Proteins

Many photosynthesis-associated proteins have been reported to be accumulated in *Sorghum bicolor* in response to salt stress [28]. In our study, photosystem I reaction center subunit IV A (B6TH55) and Photosystem II reaction center protein H (P24993) were both upregulated in phosphorylation in Zheng58 (2.13-fold and 2.44-fold, respectively) but remained unchanged in Chang7-2 (Table 2). This indicated that photosynthesis was enhanced in Zheng58 under short-term salt stress, suggesting that salt stress could stimulate photosynthesis in salt-tolerant plants to supply more energy for the plants to resist salt stress.

Table 2. Details of differentially regulated phosphoproteins related to salt tolerance in Zheng58 shoots.

Accession	Annotation	p-Value	C-0.5h fc	C-2h fc	Z-0.5h fc	Z-2h fc
		Photosynthesis-Related Proteins				
B6TH55	Photosystem I reaction center subunit IV A	2.45×10^{-8}	1.56	1.21	2.13	1.60
P24993	Photosystem II reaction center protein H	8.67×10^{-8}	1.32	1.16	0.53	2.44
		Carbon Metabolism				
Q43267	phosphoenolpyruvate carboxylase 1 (PEP1)	8.42×10^{-9}	0.51	4.72	0.39	6.46
Q9SLZ0	Phosphoenolpyruvate carboxykinase	1.65×10^{-11}	0.26	0.90	0.08	0.27
A0A096T909	UDP-glucose 6-dehydrogenase	8.58×10^{-12}	0.05	0.77	0.04	0.26
		Glutamate Metabolism				
B4FTF8	glutathione S-transferase	1.12×10^{-7}	1.18	1.40	2.05	1.74
B6T7H0	glutathione S-transferase 6	9.47×10^{-7}	1.10	0.75	2.07	1.25
		Plant Hormone Signal Transduction				
B4F831	Abscisic acid-insensitive 5-like protein	9.29×10^{-9}	2.05	1.52	2.36	2.50
B6TNQ7	Ninja-family protein 6	6.63×10^{-9}	1.93	1.88	2.20	2.53

3.3.2. Proteins Involved in Carbon Metabolism

Carbon metabolism is greatly affected in plants under saline conditions [28,74]. Phosphoenolpyruvate carboxylase 1 (PEP1) (Q43267) catalyzes the reaction between phosphoenolpyruvate and carbon dioxide to form oxaloacetate, which is involved in the TCA cycle to provide energy for plants [79]. The phosphorylation of PEP1 was decreased in both inbred lines at 0.5 h of salt treatment and then significantly increased at 2 h; the increase was greater in Zheng58 than in Chang7-2 (6.46-fold versus 4.72-fold); thus, in Zheng58, carbon metabolism was enhanced to produce more energy for salt stress relief. Phosphoenolpyruvate carboxykinase (Q9SLZ0) functions oppositely to PEP1: it converts oxaloacetate to phosphoenolpyruvate and carbon dioxide [80]. The phosphorylation of phosphoenolpyruvate carboxykinase was significantly downregulated in Zheng58 shoots after salt treatment (0.08-fold and 0.27-fold at 0.5 and 2 h treatment, respectively), leading to the accumulation of oxaloacetate that eventually got into the TCA cycle. UDP-glucose 6-dehydrogenase (A0A096T909) is responsible for turning UDP-glucose into UDP-glucuronic acid. UDP-glucuronic acid is one of the most important compounds for polysaccharide synthesis [81]. Studies have shown that polysaccharide acts as an osmotic regulator to protect plants from stress caused by high salt conditions [82,83]. The phosphorylation of UDP-glucose 6-dehydrogenase was downregulated after 0.5 h of salt stress and slightly recovered after 2 h in Chang7-2; it maintained significant downregulation in Zheng58 within the two hours of salt treatment (Table 2). These results suggest that short-term salt treatment did not induce but rather reduced polysaccharide synthesis. Thus, we proposed that, in the early stage of salt stress, plants produced more energy through the TCA cycle and reduced energy consumption in the synthesis of polysaccharide, thus providing more energy for salt relief processes, such as ion transport and synthesis of substances for cell growth.

3.3.3. Glutathione Metabolism

Two glutathione transferases, glutathione S-transferase (B4FTF8) and glutathione S-transferase 6 (B6T7H0), were identified as DRPPs in shoots. These two glutathione transferases were both increased in phosphorylation in Zheng58 at 0.5 h of salt treatment and slightly decreased at 2 h, while they both remain unchanged in Chang7-2 (Table 2). The abundance of glutathione S-transferase was also upregulated under salt stress in wheat [31]. Glutathione transferases catalyze the binding reaction of glutathione and various electrophilic compounds [66]. We suspected that the phosphorylation of glutathione transferases helped to eliminate some toxic substances that were produced in the shoot under salt stress.

3.3.4. Plant Hormone Signal Transduction

Abscisic acid-insensitive 5-like protein (B4F831) is a bZIP transcription factor involved in the ABA signaling pathway [84]. ABA is a stress-signaling molecule that plays pivotal roles in stressful environments, such as drought, cold, and salt [85–87]. It has been documented that BnaABF2, a bZIP transcription factor from rapeseed, enhanced the salt tolerance in transgenic Arabidopsis [88]. The phosphorylation of B4F831 was upregulated initially and slightly decreased afterwards in Chang7-2; however, in Zheng58, it showed sustained significant upregulation. The large increase of B4F831 in Zheng58 may contribute to its salt tolerance. The increased phosphorylation of this protein enhanced the downstream expression of stress-related genes. In addition, an RNA-binding protein involved in signal transduction, Ninja-family protein 6 (B6TNQ7), was identified as a DRPP and was also upregulated in Zheng58 but showed no significant change in Chang7-2 after NaCl treatment.

4. Materials and Methods

4.1. Plant Materials and NaCl Treatment

Maize recombinant inbred lines (RILs) Zheng58 and Chang7-2 were kindly provided by Prof. Yan He from the China Agricultural University. Seeds were germinated in wet sand at room temperature (about 22–25 °C) with 40–60% humidity. Seven-day-old seedlings were uprooted from sand, and their roots were rinsed with distilled water and transferred to pots filled with Hoagland nutrient solution (Macronutrients: K_2SO_4, $Ca(NO_3)_2 \cdot 4H_2O$, KH_2PO_4, $MgSO_4 \cdot 7H_2O$, Fe-EDTA, KCl; micronutrients: H_3BO_3, $MnSO_4 \cdot H_2O$, $ZnSO_4 \cdot 7H_2O$, $CuSO4 \cdot 5H_2O$, $(NH_4)_6Mo_7O_{24} \cdot 4H_2O$). The nutrient solution was replaced every other day. Seedlings at the trifoliate stage were treated with 100 mM NaCl [54–56] for 0 h, 0.5 h, and 2 h, and the roots and shoots were collected afterwards. Parts of the plant were used to measure physiological parameters immediately after treatment, and the remaining samples were frozen in liquid nitrogen and stored at −80 °C for further analysis.

4.2. Physiological Parameter Measurements

Roots and shoots were dried at 80 °C to a constant weight for 24 h. The dry weights of the samples were measured. The dried roots and shoots were incinerated in a muffle furnace at 300 °C for 3 h and 575 °C for 6 h. The ashes were dissolved in 10 cm^3 5% nitric acid and diluted with 5% nitric acid accordingly. The concentrations of macroelements (Na, K, Ca, and Mg) and microelements (Cu, Fe, Mn, and Zn) in the digested solution were determined using a 4100-MP ICP-OES (Agilent, Santa Clara, CA, USA) [89]. Three biological replicates were tested.

Proline contents in fresh plant tissues were measured using the AccQ-Tag derivatization kit (Waters, Milford, MA, USA) with LC-MS, and three biological replicates were examined. In brief, 20 mg lyophilized plant tissue powder was extracted with 1 cm^3 water under sonication. The extract was derivatized with the Waters AccQ-Tag amino acid derivatization kit and analyzed with LC-MS [90].

H_2O_2 contents were measured using the Amplex Red Hydrogen Peroxide/Peroxidase Assay Kit (Thermo Scientific, Waltham, MA, USA) according to manufacturer's instructions. Each sample was subjected to three biological replicates and two technical replicates.

4.3. Protein Extraction and Digestion

Protein extraction was adopted from a published method with a slight modification [91]. Two biological replicates and two technical replicates were measured. Two grams of frozen plant samples were ground to a fine powder with a mortar in liquid nitrogen. Extraction buffer (6 cm^3) (500 mM Tris–HCl, 500 mM EDTA, 700 mM sucrose, 100 mM KCl, pH 8.0, 1% protease inhibitor cocktail, 1% phospho-STOP) was added, and the powder was ground for 10 min. Then, 6 cm^3 Tris saturated phenol was added, and the powder was ground for another 10 min. The phenol layer was collected after centrifugation and the protein was precipitated with 0.1 M ammonia acetate in methanol. The protein precipitate was washed with cold acetone and dried under a vented hood.

The protein was dissolved in 7 M urea and 2 M thiourea. The concentration of the protein solution was measured by the Bradford assay. The protein was digested with trypsin using a modified FASP method [92]. In short, the protein solution was loaded onto an ultrafiltration device (30 kDa, 500 mm^3,

Sartorius, Gottingen, Germany), washed with 50 mM NH_4HCO_3, reduced with 200 mM DTT at 56 °C, and alkylated with 200 mM iodoacetamide in the dark. The protein was digested with trypsin at 37 °C overnight (enzyme: protein ratio = 1:50). Parts of the digested peptides were diluted with 0.1% FA for nano LC-MS analysis. The remaining peptides were lyophilized for phosphopeptide enrichment.

4.4. Phosphopeptide Enrichment

Phosphopeptides were enriched using TiO_2-tips (Thermo Scientific, Waltham, MA, USA) according to the manufacturer's protocol. The phosphopeptides were reconstituted in 0.1% FA for nano LC-MS analysis.

4.5. Mass Spectrometry Analysis

Nano LC-MS analysis was performed on a Waters nano-Acquity nano HPLC (Waters, Milford, MA, USA) coupled with a Thermo Q-Exactive high resolution mass spectrometer (Thermo Scientific, Waltham, MA, USA). A peptide solution (7 mm^3) was loaded onto a trap column (AcclaimPepMap100, 75 μm × 2 mm, Thermo Scientific, Waltham, MA, USA). The analytical column was a 100 μm I.D. fused silica capillary filled with 20 cm of C_{18} stationary phase (Aqua C_{18}, 3 μm, 125 Å, Phenomenex, Los Angeles, CA, USA). A 125 min gradient was used to elute the peptides. Mobile phase A was 0.1% FA in water, B was 0.1% FA in acetonitrile, and the flow rate was 400 nL/min. The nano ESI spray voltage was 2.0 kV and a full mass scan in the range of m/z 300–1800 was obtained with a resolution of 70,000 at m/z 200. The 10 most intensive peptide signals from the full scan were selected for an MS/MS scan with a resolution of 17,500 at m/z 200. The dynamic exclusion time was 20 s.

4.6. Protein Identification and Quantification

Raw data were imported into Progenesis QI for Proteomics (build 2.0, Nonlinear Dynamics, Newcastle, UK) for peak alignment and peak picking. The obtained MGF files were submitted to the Mascot search engine (version 2.5.0, Matrix Science, London, UK) and searched against the Uniprot maize database (58,418 sequences, 2015-12, https://www.uniprot.org/). Trypsin was selected as the specific enzyme with a maximum of two missed cleavages. Fixed modification was carbamidomethyl (C). Oxidation (M) and phosphorylations (S, T, and Y) were set as variable modifications. The MS mass tolerance was 10 ppm and the MS/MS mass tolerance was 0.02 Da. A decoy database was used, and peptides were filtered at 1% FDR. Peptides with a mascot score higher than 29 were selected for further analysis. The mass spectrometric data were deposited at Integrated Proteome Resources with the project No. IPX0001523000 (https://www.iprox.org).

4.7. Bioinformatics Analysis

The GO enrichment analysis and MFUZZ analysis were conducted with the help of the Wukong platform (http://www.omicsolution.org/wu-kong-beta-linux/main/). Protein–protein interactions were analyzed by STRING 10.5 (http://string-db.org/) [62] using the DRPPs as input.

4.8. Statistical Analysis

Hydrogen peroxide and proline contents were measured in six and three biological replicates, respectively, and metal ion contents were determined in three biological replicates. Statistical analysis was performed using Student's *t*-test in Excel. Significant differences are indicated by an asterisk. One asterisk indicates a significant difference at the 0.05 level and two asterisks indicate a difference at the 0.01 level. Four biological replicates were performed in the phosphoproteomic experiment, and ANOVA was used to calculate the *p*-values. Proteins were considered to have significant changes in abundance when a *p*-value of <0.05 was reached, with a fold change of >2. Hierarchical cluster analysis of DRPPs was performed using cluster 3.0 (https://cluster.updatestar.com/en/edit) and Treeview software (http://jtreeview.sourceforge.net/). PCA analysis was conducted using SIMCA 13.0 software (Umetrics, Umeå, Sweden).

5. Conclusions

In conclusion, we explored the mechanism of salt tolerance in maize by comparing differentially regulated phosphoproteins in the maize inbred lines Chang7-2 and Zheng58 under short-term salt stress. Zheng58 had higher Na^+ content in roots, lower Na^+ content in shoots, and a lower Na^+/K^+ ratio after salt treatment, which were crucial for maintaining salt tolerance. The H_2O_2 content in the roots of Zheng58 was significantly increased after salt stress, and the accumulation of H_2O_2 acted as a stimulator to enhance Zheng 58's salt tolerance. Proline, an osmotic regulating compound, was increased in both inbred lines after 0.5 h of salt treatment, and the increase was greater in Zheng58 than in Chang7-2. The content of proline in shoots changed to the opposite direction. These results indicate that, under short-term salt stress, proline that is already in the shoots is recruited to the roots to cope with salt stress. Moreover, Zheng58 was able to transport proline more efficiently from shoots to roots to cope with salt stress. Several phosphoproteins related to salt tolerance were found to be significantly regulated in Zheng58. In roots, 6-phosphogluconate dehydrogenase 2, pyruvate dehydrogenase, and phosphoenolpyruvate carboxykinase provided NADPH and energy for the plants to resist salt stress. Glutathione S-transferase 3 and L-gulonolactone oxidase-like were related to glutathione and ascorbic acid metabolism, which offered antioxidants to remove ROS, such as H_2O_2. The phosphorylation changes in glutamate decarboxylase and glutamate synthase in Zheng58 under NaCl treatment were favorable for the accumulation of glutamate, which acts as a major substrate of proline synthesis under stress conditions. Potassium channel AKT1, High-affinity potassium transporter, and sodium/hydrogen exchanger were found to be related to potassium and sodium transportation. The upregulation of these proteins in Zheng58 under salt stress resulted in a lower Na^+ content in shoots and a lower Na^+/K^+ ratio in both roots and shoots in Zheng58. Calcium/proton exchanger CAX1-like protein was found to be related to calcium ions transportation. The downregulation of this protein in Zheng58 after salt treatment resulted in accumulation of Ca^{2+} in the cytoplasm to initiate the salt signal transduction pathway. In shoots, photosystem I reaction center subunit IV A and photosystem II reaction center protein H, which are involved in photosynthesis, supplied more energy to Zheng58. PEP1 and phosphoenolpyruvate carboxykinase, which are involved in carbon metabolism, promote the degradation of carbohydrates to produce more energy. The downregulation of UDP-glucose 6-dehydrogenase was helpful to save more energy for salt relief processes. Two glutathione transferases that are involved in eliminating some toxic substances produced during salt stress were found to be upregulated in Zheng58. Abscisic acid-insensitive 5-like protein, which is involved in the ABA signaling pathway, enhanced the expression of stress-related genes in Zheng58 under salt stress.

Supplementary Materials: Supplementary Materials can be found at http://www.mdpi.com/1422-0067/20/8/1886/s1.

Author Contributions: Conceptualization, Z.L. and C.-F.J.; methodology, X.-Y.Z.; formal analysis, Z.L.; X.-Y.Z., and X.B.; investigation, X.B.; resources, C.-F.J.; data curation, Z.L. and X.-Y.Z.; writing—original draft preparation, X.-Y.Z.; writing—review and editing, Z.L.; supervision, Z.L.; project administration, Z.L.; funding acquisition, Z.L. All authors read and approved the final manuscript.

Funding: This research was funded by The State Key Laboratory of Plant Physiology and Biochemistry at China Agricultural University.

Acknowledgments: We are very thankful to Yan He (Professor, College of Agriculture and Biotechnology, China Agricultural University) for providing the two maize inbred lines, and we would like to thank Yiting Shi (Associate Professor, State Key Laboratory of Plant Physiology and Biochemistry, College of Biological Sciences, China Agricultural University) for reading the manuscript and providing thoughtful insights.

Conflicts of Interest: The authors declare no conflict of interest.

Abbreviations

bZIP	Basic region-leucine zipper
DRPPs	Differential regulated phosphoproteins
FDR	false discovery rate
GABA	γ-aminobutyric acid
GO	Gene ontology
ICP-OES	Inductively Coupled Plasma Optical Emission Spectrometry
PCA	Principal component analysis
PEP1	Phosphoenolpyruvate carboxylase 1
PIP	plasma membrane intrinsic proteins
PSMs	Peptide spectrum matches
RILs	Recombinant inbred lines
ROS	Reactive oxygen species
SOS	Salt Overly Sensitive
STRING	Search Tool for the Retrieval of Interacting Genes
TCA	Tricarboxylic acid

References

1. Tester, M.; Davenport, R. Na$^+$ tolerance and Na$^+$ transport in higher plants. *Ann. Bot.* **2003**, *91*, 503–527. [CrossRef] [PubMed]
2. Munns, R.; Tester, M. Mechanisms of salinity tolerance. *Annu. Rev. Plant Biol.* **2008**, *59*, 651–681. [CrossRef] [PubMed]
3. Zhang, L.C.; Zhao, G.Y.; Jia, J.Z.; Liu, X.; Kong, X.Y. Molecular characterization of 60 isolated wheat MYB genes and analysis of their expression during abiotic stress. *J. Exp. Bot.* **2012**, *63*, 203–214. [CrossRef] [PubMed]
4. Zhu, J.K. Regulation of ion homeostasis under salt stress. *Curr. Opin. Plant Biol.* **2003**, *6*, 441–445. [CrossRef]
5. Blumwald, E.; Aharon, G.S.; Apse, M.P. Sodium transport in plant cells. *Biochim. Biophys. Acta* **2000**, *1465*, 140–151. [CrossRef]
6. Blumwald, E. Sodium transport and salt tolerance in plants. *Curr. Opin. Cell Biol.* **2000**, *12*, 431–434. [CrossRef]
7. Hasegawa, P.M.; Bressan, R.A.; Zhu, J.K.; Bohnert, H.J. Plant cellular and molecular responses to high salinity. *Annu. Rev. Plant Physiol. Plant Mol. Biol.* **2000**, *51*, 463–499. [CrossRef]
8. Apel, K.; Hirt, H. Reactive oxygen species: Metabolism, oxidative stress, and signal transduction. *Annu Rev Plant Biol* **2004**, *55*, 373–399. [CrossRef]
9. Galvan-Ampudia, C.S.; Testerink, C. Salt stress signals shape the plant root. *Curr. Opin. Plant Biol.* **2011**, *14*, 296–302. [CrossRef]
10. Kant, S.; Kant, P.; Raveh, E.; Barak, S. Evidence that differential gene expression between the halophyte, *Thellungiella halophila*, and *Arabidopsis thaliana* is responsible for higher levels of the compatible osmolyte proline and tight control of Na$^+$ uptake in *T. halophila*. *Plant Cell Environ.* **2010**, *29*, 1220–1234. [CrossRef]
11. Storey, R.; Jones, R.G.W. Quaternary ammonium compounds in plants in relation to salt resistance. *Phytochemistry* **1977**, *16*, 447–453. [CrossRef]
12. Jouve, L.; Hoffmann, L.; Hausman, J.F. Polyamine, carbohydrate, and proline content changes during salt stress exposure of Aspen (*Populus tremula* L.): Involvement of oxidation and ssmoregulation metabolism. *Plant Biol* **2010**, *6*, 74–80.
13. Erickson, B.K.; Jedrychowski, M.P.; McAlister, G.C.; Everley, R.A.; Kunz, R.; Gygi, S.P. Evaluating multiplexed quantitative phosphopeptide analysis on a hybrid quadrupole mass filter/linear ion trap/orbitrap mass spectrometer. *Anal. Chem.* **2015**, *87*, 1241–1249. [CrossRef] [PubMed]
14. Umezawa, T.; Sugiyama, N.; Takahashi, F.; Anderson, J.C.; Ishihama, Y.; Peck, S.C.; Shinozaki, K. Genetics and phosphoproteomics reveal a protein phosphorylation network in the abscisic acid signaling pathway in *Arabidopsis thaliana*. *Sci. Signal* **2013**, *6*, rs8. [CrossRef]
15. Kline, K.G.; Barrett-Wilt, G.A.; Sussman, M.R. In planta changes in protein phosphorylation induced by the plant hormone abscisic acid. *Proc. Natl. Acad. Sci. USA* **2010**, *107*, 15986–15991. [CrossRef]

16. Minkoff, B.B.; Stecker, K.E.; Sussman, M.R. Rapid phosphoproteomic effects of abscisic acid (ABA) on wild-type and ABA receptor-deficient *A. thaliana* mutants. *Mol. Cell Proteomics* **2015**, *14*, 1169–1182. [CrossRef]

17. Wang, X.; Bian, Y.Y.; Cheng, K.; Gu, L.F.; Ye, M.L.; Zou, H.F.; Sun, S.S.M.; He, J.X. A large-scale protein phosphorylation analysis reveals novel phosphorylation motifs and phosphoregulatory networks in *Arabidopsis*. *J. Proteomics* **2013**, *78*, 486–498. [CrossRef] [PubMed]

18. Wei, S.S.; Wang, X.Y.; Jiang, D.; Dong, S.T. Physiological and proteome studies of maize (*Zea mays* L.) in response to leaf removal under high plant density. *BMC Plant Biol.* **2018**, *18*, 378. [CrossRef]

19. Chao, Q.; Gao, Z.F.; Wang, Y.F.; Li, Z.; Huang, X.H.; Wang, Y.C.; Mei, Y.C.; Zhao, B.G.; Li, L.; Jiang, Y.B. The proteome and phosphoproteome of maize pollen uncovers fertility candidate proteins. *Plant Mol. Biol.* **2016**, *91*, 287–304. [CrossRef] [PubMed]

20. Zhao, Y.L.; Wang, Y.K.; Yang, H.; Wang, W.; Wu, J.Y.; Hu, X.L. Quantitative proteomic analyses identify ABA-related proteins and signal pathways in maize leaves under drought conditions. *Front. Plant Sci.* **2016**, *7*, 1827. [CrossRef]

21. Wang, X.Y.; Shan, X.H.; Wu, Y.; Su, S.Z.; Li, S.P.; Liu, H.K.; Han, J.Y.; Xue, C.M.; Yuan, Y.P. iTRAQ-based quantitative proteomic analysis reveals new metabolic pathways responding to chilling stress in maize seedlings. *J. Proteomics* **2016**, *146*, 14–24. [CrossRef]

22. Hou, Y.X.; Qiu, J.H.; Wang, Y.F.; Li, Z.Y.; Zhao, J.; Tong, X.H.; Lin, H.Y.; Zhang, J. A quantitative proteomic analysis of brassinosteroid-induced protein phosphorylation in rice (*Oryza sativa* L.). *Front. Plant Sci.* **2017**, *8*, 514. [CrossRef]

23. Qiu, J.H.; Hou, Y.X.; Wang, Y.F.; Li, Z.Y.; Zhao, J.; Tong, X.H.; Lin, H.Y.; Wei, X.J.; Ao, H.J.; Zhang, J. A comprehensive proteomic survey of ABA-induced protein phosphorylation in rice (*Oryza sativa* L.). *Int. J. Mol. Sci.* **2017**, *18*, 60. [CrossRef]

24. Zhong, M.; Li, S.F.; Huang, F.L.; Qiu, J.H.; Zhang, J.; Sheng, Z.H.; Tang, S.Q.; Wei, X.J.; Hu, P.S. The phosphoproteomic response of rice seedlings to cadmium stress. *Int. J. Mol. Sci.* **2017**, *18*, 2055. [CrossRef]

25. Kamal, A.H.M.; Rashid, H.; Sakata, K.; Komatsu, S. Gel-free quantitative proteomic approach to identify cotyledon proteins in soybean under flooding stress. *J. Proteomics* **2015**, *112*, 1–13. [CrossRef]

26. Oh, M.W.; Nanjo, Y.; Komatsu, S. Identification of nuclear proteins in soybean under flooding stress using proteomic technique. *Protein Pept. Lett.* **2014**, *21*, 458–467. [CrossRef]

27. Yin, Y.Q.; Qi, F.; Gao, L.; Rao, S.Q.; Yang, Z.Q.; Fang, W.M. iTRAQ-based quantitative proteomic analysis of dark-germinated soybeans in response to salt stress. *RSC Adv.* **2018**, *8*, 17905–17913. [CrossRef]

28. Ngara, R.; Ndimba, R.; Borch-Jensen, J.; Jensen, O.N.; Ndimba, B. Identification and profiling of salinity stress-responsive proteins in Sorghum bicolor seedlings. *J. Proteomics* **2012**, *75*, 4139–4150. [CrossRef]

29. Roy, S.K.; Cho, S.W.; Kwon, S.J.; Kamal, A.H.M.; Lee, D.G.; Sarker, K.; Lee, M.S.; Xin, Z.G.; Woo, S.H. Proteome characterization of copper stress responses in the roots of sorghum. *Biometals* **2017**, *30*, 765–785. [CrossRef]

30. Pi, E.; Qu, L.Q.; Hu, J.W.; Huang, Y.Y.; Qiu, L.J.; Lu, H.F.; Jiang, B.; Liu, C.; Peng, T.T.; Zhao, Y.; et al. Mechanisms of soybean roots' tolerances to salinity revealed by proteomic and phosphoproteomic comparisons between two cultivars. *Mol. Cell Proteomics* **2016**, *15*, 266–288. [CrossRef]

31. Gao, L.; Yan, X.; Li, X.; Guo, G.; Hu, Y.; Ma, W.; Yan, Y. Proteome analysis of wheat leaf under salt stress by two-dimensional difference gel electrophoresis (2D-DIGE). *Phytochemistry* **2011**, *72*, 1180–1191. [CrossRef] [PubMed]

32. Cui, D.Z.; Wu, D.D.; Liu, J.; Li, D.T.; Xu, C.Y.; Li, S.; Li, P.; Zhang, H.; Liu, X.; Jiang, C.; et al. Proteomic analysis of seedling roots of two maize inbred lines that differ significantly in the salt stress response. *PLoS ONE* **2015**, *10*, e0116697. [CrossRef] [PubMed]

33. Macek, B.; Mann, M.; Olsen, J.V. Global and site-specific quantitative phosphoproteomics: Principles and applications. *Annu. Rev. Pharmacool. Toxicol.* **2009**, *49*, 199–221. [CrossRef]

34. Morandell, S.; Stasyk, T.; Grosstessner-Hain, K.; Roitinger, E.; Mechtler, K.; Bonn, G.K.; Huber, L.A. Phosphoproteomics strategies for the functional analysis of signal transduction. *Proteomics* **2006**, *6*, 4047–4056. [CrossRef]

35. Fujii, H.; Chinnusamy, V.; Rodrigues, A.; Rubio, S.; Antoni, R.; Park, S.Y.; Cutler, S.R.; Sheen, J.; Rodriguez, P.L.; Zhu, J.K. In vitro reconstitution of an abscisic acid signalling pathway. *Nature* **2009**, *462*, 660–664. [CrossRef]

36. Fujii, H.; Zhu, J.K. Osmotic stress signaling via protein kinases. *Cell. Mol. Life Sci.* **2012**, *69*, 3165–3173. [CrossRef] [PubMed]

37. Shi, Y.T.; Ding, Y.L.; Yang, S.H. Molecular regulation of CBF signaling in cold acclimation. *Trends Plant Sci.* **2018**, *23*, 623–637. [CrossRef] [PubMed]

38. Zhu, J.K. Salt and drought stress signal transduction in plants. *Annu. Rev. Plant Biol.* **2002**, *53*, 247–273. [CrossRef] [PubMed]

39. Zolnierowicz, S. Type 2A protein phosphatase, the complex regulator of numerous signaling pathways. *Biochem. Pharmacol.* **2000**, *60*, 1225–1235. [CrossRef]

40. Liu, J.P.; Ishitani, M.; Halfter, U.; Kim, C.S.; Zhu, J.K. The *Arabidopsis thaliana* SOS2 gene encodes a protein kinase that is required for salt tolerance. *Proc. Natl. Acad. Sci. USA* **2000**, *97*, 3730–3734. [CrossRef]

41. Halfter, U.; Ishitani, M.; Zhu, J.K. The Arabidopsis SOS2 protein kinase physically interacts with and is activated by the calcium-binding protein SOS3. *Proc. Natl. Acad. Sci. USA* **2000**, *97*, 3735–3740. [CrossRef]

42. Liu, J.P.; Zhu, J.K. A calcium sensor homolog required for plant salt tolerance. *Science* **1998**, *280*, 1943–1945. [CrossRef]

43. Ishitani, M.; Liu, J.P.; Halfter, U.; Kim, C.S.; Shi, W.M.; Zhu, J.K. SOS3 function in plant salt tolerance requires N-myristoylation and calcium binding. *Plant Cell* **2000**, *12*, 1667–1677. [CrossRef]

44. Knight, H.; Trewavas, A.J.; Knight, M.R. Calcium signalling in *Arabidopsis thaliana* responding to drought and salinity. *Plant J.* **2010**, *12*, 1067–1078. [CrossRef]

45. MartiNez-Atienza, J.; Jiang, X.Y.; Garciadeblas, B.; Mendoza, I.; Zhu, J.K.; Pardo, J.M.; Quintero, F.J. Conservation of the salt overly sensitive pathway in rice. *Plant Physiol.* **2007**, *143*, 1001–1012. [CrossRef]

46. Wang, M.Y.; Gu, D.; Liu, T.S.; Wang, Z.Q.; Guo, X.Y.; Hou, W.; Bai, Y.F.; Chen, X.P.; Wang, G.Y. Overexpression of a putative maize calcineurin B-like protein in Arabidopsis confers salt tolerance. *Plant Mol. Biol.* **2007**, *65*, 733–746. [CrossRef]

47. Xu, Q.; Fu, H.H.; Gupta, R.; Luan, S. Molecular characterization of a tyrosine-specific protein phosphatase encoded by a stress-responsive gene in Arabidopsis. *Plant Cell* **1998**, *10*, 849–857. [CrossRef]

48. Geilfus, C.M.; Zoerb, C.; Mühling, K.H. Salt stress differentially affects growth-mediating β-expansins in resistant and sensitive maize (*Zea mays* L.). *Plant Physiol. Biochem.* **2010**, *48*, 993–998. [CrossRef]

49. Muhammad, F.; Mubshar, H.; Abdul, W.; Kadambot, H.M.S. Salt stress in maize: Effects, resistance mechanisms, and management. A review. *Agron. Sustain. Dev.* **2015**, *35*, 461–481.

50. Katerji, N.; van Hoorn, J.W.; Hamdy, A.; Mastrorilli, M. Salinity effect on crop development and yield, analysis of salt tolerance according to several classification methods. *Agric. Water Manag.* **2003**, *62*, 37–66. [CrossRef]

51. Nakagami, H.; Sugiyama, N.; Mochida, K.; Daudi, A.; Yoshida, Y.; Toyoda, T.; Tomita, M.; Ishihama, Y.; Shirasu, K. Large-scale comparative phosphoproteomics identifies conserved phosphorylation sites in plants. *Plant Physiol.* **2010**, *153*, 1161–1174. [CrossRef]

52. Bi, Y.D.; Wang, H.X.; Lu, T.C.; Li, X.H.; Shen, Z.O.; Chen, Y.B.; Wang, B.C. Large-scale analysis of phosphorylated proteins in maize leaf. *Planta* **2011**, *233*, 383–392. [CrossRef] [PubMed]

53. Soares, A.L.C.; Geilfus, C.M.; Carpentier, S.C. Genotype-specific growth and proteomic responses of maize toward salt stress. *Front. Plant Sci.* **2018**, *9*, 661. [CrossRef]

54. Luo, M.J.; Zhao, Y.X.; Wang, Y.D.; Shi, Z.; Zhang, P.P.; Zhang, Y.X.; Song, W.; Zhao, J.R. Comparative proteomics of contrasting maize genotypes provides insights into salt-stress tolerance mechanisms. *J. Proteome Res.* **2018**, *17*, 141–153. [CrossRef]

55. Zhu, J.K. Abiotic stress signaling and responses in plants. *Cell* **2016**, *167*, 313–324. [CrossRef]

56. Xiong, L.M.; Schumaker, K.S.; Zhu, J.K. Cell signaling during cold, drought, and salt stress. *Plant Cell* **2002**, *14* (Suppl 1), S165–S183. [CrossRef] [PubMed]

57. Girousse, C.; Bournoville, R.; Bonnemain, J.L. Water deficit-induced changes in concentrations in proline and some other amino acids in the Phloem Sap of Alfalfa. *Plant Physiol.* **1996**, *111*, 109–113. [CrossRef] [PubMed]

58. Ueda, A.; Shi, W.M.; Sanmiya, K.; Shono, M.; Takabe, T. Functional analysis of salt-inducible proline transporter of Barley roots. *Plant Cell Physiol.* **2001**, *42*, 1282–1289. [CrossRef]

59. Matsushita, N.; Matoh, T. Function of the shoot base of salt-tolerant reed (*Phragmites communis* Trinius) plants for Na$^+$ exclusion from the shoots. *Soil Sci. Plant Nutr.* **1992**, *38*, 565–571. [CrossRef]

60. Zhang, M.; Cao, Y.B.; Wang, Z.P.; Wang, Z.Q.; Shi, J.P.; Liang, X.Y.; Song, W.B.; Chen, Q.J.; Lai, J.S.; Jiang, C.F. A retrotransposon in an HKT1 family sodium transporter causes variation of leaf Na$^+$ exclusion and salt tolerance in maize. *New Phytol.* **2017**, *217*, 1161–1176. [CrossRef] [PubMed]

61. Wang, S.S.; Chen, X.L.; Du, D.; Zheng, W.; Hu, L.Q.; Yang, H.; Cheng, J.Q.; Gong, M. MetaboGroupS: A group entropy-based web platform for evaluating normalization methods in blood metabolomics data from maintenance hemodialysis patients. *Anal. Chem.* **2018**, *90*, 11124–11130. [CrossRef]

62. Szklarczyk, D.; Franceschini, A.; Wyder, S.; Forslund, K.; Heller, D.; Huerta-Cepas, J.; Simonovic, M.; Roth, A.; Santos, A.; Tsafou, K.P.; et al. STRING v10: Protein-protein interaction networks, integrated over the tree of life. *Nucleic Acids Res.* **2015**, *43*, D447–D452. [CrossRef] [PubMed]

63. Schnarrenberger, C.; Oeser, A.; Tolbert, N.E. Two isoenzymes each of glucose-6-phosphate dehydrogenase and 6-phosphogluconate dehydrogenase in spinach leaves. *Arch. Biochem. Biophys.* **1973**, *154*, 438–448. [CrossRef]

64. Diaz, V.P.; Wolff, T.; Markovic, J.; Pallardó, F.V.; Foyer, C.H. A nuclear glutathione cycle within the cell cycle. *Biochem. J.* **2010**, *431*, 169–178.

65. Lea, P.J.; Leegood, R.C. Plant biochemistry and molecular biology. *Plant Mol. Biol.* **1993**, *19*, 169–191.

66. Edwards, R.; Dixon, D.P.; Walbot, V. Plant glutathione S -transferases: Enzymes with multiple functions in sickness and in health. *Trends Plant Sci.* **2000**, *5*, 193–198. [CrossRef]

67. Lara, H.; Vogeli, P.; Stoll, P.; Kramer, S.S.; Stranzinger, G.; Neuenschwander, S. Intragenic deletion in the gene encoding L-gulonolactone oxidase causes vitamin C deficiency in pigs. *Mamm. Genome* **2004**, *15*, 323–333.

68. Hyun, T.K.; Eom, S.H.; Jeun, Y.C.; Han, S.H.; Kim, J.S. Identification of glutamate decarboxylases as a γ-aminobutyric acid (GABA) biosynthetic enzyme in soybean. *Ind. Crops Prod.* **2013**, *49*, 864–870. [CrossRef]

69. Hare, P.D.; Cress, W.A.; Staden, J.V. Proline synthesis and degradation: A model system for elucidating stress-related signal transduction. *J. Exp. Bot.* **1999**, *50*, 413–434. [CrossRef]

70. Demiral, T.; Türkan, I. Exogenous glycinebetaine affects growth and proline accumulation and retards senescence in two rice cultivars under NaCl stress. *Environ. Exp. Bot.* **2006**, *56*, 72–79. [CrossRef]

71. Hmida-Sayari, A.; Gargouri-Bouzid, R.; Bidani, A.; Jaoua, L.; Savouré, A.; Jaoua, S. Overexpression of Δ1-pyrroline-5-carboxylate synthetase increases proline production and confers salt tolerance in transgenic potato plants. *Plant Sci.* **2005**, *169*, 746–752. [CrossRef]

72. Khatun, S.; Flowers, T.J. Effects of salinity on seed set in rice. *Plant Cell Environ.* **2010**, *18*, 61–67. [CrossRef]

73. Li, J.; Long, Y.; Qi, G.N.; Li, J.; Xu, Z.J.; Wu, W.H.; Wang, Y. The Os-AKT1 channel is critical for K^+ uptake in rice roots and is modulated by the rice CBL1-CIPK23 complex. *Plant Cell* **2014**, *26*, 3387–3402. [CrossRef]

74. Wu, G.Q.; Wang, J.L.; Feng, R.J.; Li, S.J.; Wang, C.M. iTRAQ-based comparative proteomic analysis provides insights into molecular mechanisms of salt tolerance in Sugar Beet (*Beta vulgaris* L.). *Int. J. Mol. Sci.* **2018**, *19*, 3866. [CrossRef] [PubMed]

75. Mahajan, S.; Pandey, G.K.; Tuteja, N. Calcium- and salt-stress signaling in plants: Shedding light on SOS pathway. *Arch. Biochem. Biophys.* **2008**, *471*, 146–158. [CrossRef] [PubMed]

76. Sánchez-Barrena, M.J.; Martínez-Ripoll, M.; Zhu, J.K.; Albert, A. The structure of the *Arabidopsis thaliana* SOS3: Molecular mechanism of sensing calcium for salt stress response. *J. Mol. Biol.* **2005**, *345*, 1253–1264. [CrossRef] [PubMed]

77. Bressan, R.A.; Hasegawa, P.M.; Pardo, J.M. Plants use calcium to resolve salt stress. *Trends Plant Sci.* **1998**, *3*, 411–412. [CrossRef]

78. Shigaki, T.; Rees, I.; Nakhleh, L.; Hirschi, K.D. Identification of three distinct phylogenetic groups of CAX cation/proton antiporters. *J. Mol. Evol.* **2006**, *63*, 815–825. [CrossRef]

79. Melzer, E.; O'Leary, M.H. Anapleurotic CO_2 fixation by phosphoenolpyruvate carboxylase in C3 plants. *Plant Physiol.* **1987**, *84*, 58–60. [CrossRef] [PubMed]

80. Leegood, R.C.; Ap, R.T. Phosphoenolpyruvate carboxykinase and gluconeogenesis in cotyledons of Cucurbita pepo. *Biochim. Biophys. Acta* **1978**, *524*, 207–218. [CrossRef]

81. Roman, E. Studies on the Role of UDP-Glucose Dehydrogenase in Polysaccharide Biosynthesis. Ph.D. Thesis, Acta Universitatis Upsaliensis, Uppsala, Sweden, 2004.

82. Arora, M.; Kaushik, A.; Rani, N.; Kaushik, C.P. Effect of cyanobacterial exopolysaccharides on salt stress alleviation and seed germination. *J. Environ. Biol.* **2010**, *31*, 701–704.

83. Ghanem, M.E.; Han, R.M.; Classen, B.; Quetin-Leclerq, J.L.; Mahy, G.; Ruan, C.J.; Qin, P.; Pérez-Alfocea, F.; Lutts, S. Mucilage and polysaccharides in the halophyte plant species *Kosteletzkya virginica*: Localization and composition in relation to salt stress. *J. Plant Physiol.* **2010**, *167*, 382–392. [CrossRef]

84. Brocard, I.M.; Lynch, T.J.; Finkelstein, R.R. Regulation and role of the Arabidopsis abscisic acid-insensitive 5 gene in abscisic acid, sugar, and stress response. *Plant Physiol.* **2002**, *129*, 1533–1543. [CrossRef]

85. Setter, T.L.; Yan, J.B.; Warburton, M.; Ribaut, J.M.; Xu, Y.B.; Sawkins, M.; Buckler, E.S.; Zhang, Z.W.; Gore, M.A. Genetic association mapping identifies single nucleotide polymorphisms in genes that affect abscisic acid levels in maize floral tissues during drought. *J. Exp. Bot.* **2011**, *62*, 701–716. [CrossRef]

86. Singh, K.; Singla-Pareek, S.L.; Pareek, A. Dissecting out the crosstalk between salinity and hormones in roots of Arabidopsis. *OMICS* **2011**, *15*, 913–924. [CrossRef]

87. Knight, H.; Zarka, D.G.; Okamoto, H.; Thomashow, M.F.; Knight, M.R. Abscisic acid induces CBF gene transcription and subsequent induction of cold-regulated genes via the CRT promoter element. *Plant Physiol.* **2004**, *135*, 1710–1717. [CrossRef]

88. Zhao, B.Y.; Hu, Y.F.; Li, J.J.; Yao, X.; Liu, K.D. BnaABF2, a bZIP transcription factor from rapeseed (*Brassica napus* L.), enhances drought and salt tolerance in transgenic Arabidopsis. *Bot. Stud.* **2016**, *57*, 12. [CrossRef]

89. Xu, J.; Li, H.D.; Chen, L.Q.; Wang, Y.; Liu, L.L.; He, L.; Wu, W.H. A protein kinase, interacting with two calcineurin B-like proteins, regulates K^+ transporter AKT1 in Arabidopsis. *Cell* **2006**, *125*, 1347–1360. [CrossRef]

90. Yin, B.J.; Li, T.T.; Zhang, S.R.; Li, Z.; He, P.L. Sensitive analysis of 33 free amino acids in serum, milk, and muscle by ultra-high performance liquid chromatography-quadrupole-orbitrap high resolution mass spectrometry. *Food Anal. Methods* **2016**, *9*, 2814–2823. [CrossRef]

91. Carpentier, S.C.; Witters, E.; Laukens, K.; Deckers, P.; Swennen, R.; Panis, B. Preparation of protein extracts from recalcitrant plant tissues: An evaluation of different methods for two-dimensional gel electrophoresis analysis. *Proteomics* **2005**, *5*, 2497–2507. [CrossRef]

92. Wisniewski, J.R.; Zougman, A.; Nagaraj, N.; Mann, M. Universal sample preparation method for proteome analysis. *Nat. Methods* **2009**, *6*, 359–362. [CrossRef]

International Journal of
Molecular Sciences

Article

The Proteomic Analysis of Maize Endosperm Protein Enriched by Phos-tag^tm Reveals the Phosphorylation of Brittle-2 Subunit of ADP-Glc Pyrophosphorylase in Starch Biosynthesis Process

Guowu Yu [1,†], Yanan Lv [1,†], Leiyang Shen [1], Yongbin Wang [1], Yun Qing [1], Nan Wu [1], Yangping Li [1], Huanhuan Huang [1], Na Zhang [2], Yinghong Liu [3], Yufeng Hu [1], Hanmei Liu [4], Junjie Zhang [4] and Yubi Huang [1,3,*]

[1] College of Agronomy, Sichuan Agricultural University, Huimin Road 211#, Wenjiang District, Chengdu 611130, Sichuan, China; 2002ygw@163.com (G.Y.); lvyn@shhrp.com (Y.L.); sly1832633910@163.com (L.S.); wwyybb007@163.com (Y.W.); 18328652715@163.com (Y.Q.); 15982862780@163.com (N.W.); yangpingli103@gmail.com (Y.L.); hh820423@163.com (H.H.); 13958@sicau.edu.cn (Y.H.)
[2] College of Science, Sichuan Agricultural University, Huimin Road 211#, Wenjiang District, Chengdu 611130, Sichuan, China; 03118663@163.com
[3] Maize Research Institute of Sichuan Agricultural University, Huimin Road 211#, Wenjiang District, Chengdu 611130, Sichuan, China; sclydx@163.com
[4] College of Life Science, Sichuan Agricultural University, Xingkang Road 46#, Ya'an 625014, Sichuan, China; hanmeil@163.com (H.L.); junjiezh@163.com (J.Z.)
[*] Correspondence: 10024@sicau.edu.cn; Tel.: +86-028-8629-0868
[†] These authors contributed equally to this work.

Received: 22 January 2019; Accepted: 18 February 2019; Published: 24 February 2019

Abstract: AGPase catalyzes a key rate-limiting step that converts ATP and Glc-1-p into ADP-glucose and diphosphate in maize starch biosynthesis. Previous studies suggest that AGPase is modulated by redox, thermal and allosteric regulation. However, the phosphorylation of AGPase is unclear in the kernel starch biosynthesis process. Phos-tag^{TM} technology is a novel method using phos-tag^{TM} agarose beads for separation, purification, and detection of phosphorylated proteins. Here we identified phos-tag^{TM} agarose binding proteins from maize endosperm. Results showed a total of 1733 proteins identified from 10,678 distinct peptides. Interestingly, a total of 21 unique peptides for AGPase sub-unit Brittle-2 (Bt2) were identified. Bt2 was demonstrated by immunoblot when enriched maize endosperm protein with phos-tag^{TM} agarose was in different pollination stages. In contrast, Bt2 would lose binding to phos-tag^{TM} when samples were treated with alkaline phosphatase (ALP). Furthermore, Bt2 could be detected by Pro-Q diamond staining specifically for phosphorylated protein. We further identified the phosphorylation sites of Bt2 at Ser^{10}, Thr^{451}, and Thr^{462} by iTRAQ. In addition, dephosphorylation of Bt2 decreased the activity of AGPase in the native gel assay through ALP treatment. Taking together, these results strongly suggest that the phosphorylation of AGPase may be a new model to regulate AGPase activity in the starch biosynthesis process.

Keywords: maize; AGPase; phosphorylation; brittle-2; phos-tag^{TM}

---

1. Introduction

Maize is one of the three crops with the highest production in the world [1]. Maize starch, accounting for about 70% of maize kernels, is not only widely used as main energy source to support the majority of the world's population, but is also used as a feedstock for the production of industrial material and bioethanol [2]. Kernel starch is a crucial factor in determining the yield and quality of maize. It is a main goal of breeders to increase starch content and improve the quality of starch in maize.

Maize is generally an excellent source of starch which is synthesized by four enzymes: ADP-Glc pyrophosphorylase (AGPase; EC 2.7. 7.27), starch synthase (SS; EC 2.4.1.21), starch branch enzyme (SBE; EC 2.4.1.18) and starch debranch enzyme (SDBE; EC 3.2.1.68) [3]. AGPase catalyzes the synthesis of ADP-glucose and diphosphate from ATP and Glc-1-p [4], and the product ADP-glucose serves as the activated glucosyl donor in α-1, 4-glucan synthesis [5,6]. This reaction is a key rate-limiting step in maize endosperm starch synthesis. AGPase widely exists in plant leaves and cereal endosperm in a heterotetrameric $\alpha2\beta2$ format and is composed of two identical large subunits (Sh2) and two identical small subunits (Bt2) in maize endosperm [7,8].

Previous studies suggest that AGPase activity is mainly modulated by allosteric regulation, thermal inactivation and redox regulation [9–12]. Some small molecule effectors are thought to regulate the activity of AGPase in an allosteric regulation fashion, such as 3-phosphoglycerate (3-PGA) [13], inorganic phosphate (Pi) [14], and sugars [15]. 3-phosphoglycerate (3-PGA) and inorganic phosphate (Pi) are demonstrated as activator and inhibitor, respectively, in most cereals [16–18]. 3-PGA activates AGPase by increasing the affinity for substrate G-1-P, but Pi inhibits the activity of AGPase by reversing the effect of 3-PGA [19]. The heat lability of AGPase is another regulation mechanism [9]. AGPase from maize endosperm will lose 95% of activity when it is heated at 57 °C for 5 min [7]. The heat lability of AGPase leads to grain loss during hot weather. It is an aim of breeders to generate new maize varieties with enhanced heat stability in terms of AGPase for increasing yields [20–22]. In addition, AGPase is also subject to redox-dependent regulation in potato tubers [23], potato leaves, arabidopsis, and peas [24].

Phosphorylation of proteins is involved in various cellular processes and plays a central role in signal transduction [25]. For a protein molecule, phosphorylation can alter enzyme activity, protein conformation and protein–protein interaction, and modulate localization of the protein and its stability [26]. Phosphorylation of proteins can change a protein's complex formation [27]. In wheat endosperm, phosphorylation of SBEI, SBEIIb and SP can form a protein complex to enhance the synthesis of starch while dephosphorylation of phosphoproteins complex can disturb protein complex formation and reduce the activity of SBEIIa and SBEIIb [28,29]. In the process of starch synthesis, Liu reports that the phosphorylation of SBEI and SBEIIb is involved in heterogenic complexes of proteins [30]. The phosphorylation of SSI and GBSS is detected by phos-tag^{TM} technology [31]. Maize endosperm SBEIIb is phosphorylated on multiple sites [32]. In maize seed development, recent progress in large-scale maize phosphoproteomics has shown that phosphorylation of starch biosynthetic enzymes might play an important role in starch synthesis [33]. However, with AGPase, as a key enzyme in the starch synthetic pathway, its phosphorylation is unclear in the maize kernel starch synthesis process.

Here, we analyzed proteomic data from 08-641 maize endosperm protein enriched by phos-tag^{TM} agarose, which is specific for phosphorylated protein at maize 20 days endosperm lysate after pollination. The proteomic results showed that a total of 1733 proteins were identified from 10,678 distinct peptides with at least one unique peptide. The majority of binding proteins were attributed to the metabolic process and cellular process proteins. A total of 21 unique peptides for Bt2 were identified by mass spectrometry. Bt2 was further demonstrated at 15, 20, and 27 days endosperm by immunoblot. In contrast, Bt2 lost binding when incubated with alkaline phosphatase, which can remove phosphate from protein. Furthermore, the phosphorylation of AGPase subunit Bt2 protein band was further demonstrated by Pro-Q diamond dye staining technology. The phosphorylation sites at Bt2-Ser^{10}, Bt2-Thr^{451}, and Bt2-Thr^{462} were identified by iTRAQ. In addition, dephosphorylating of AGPase subunit Bt2 decreased its activity in the native gel assay. Taking together, these results suggested that the phosphorylation of AGPase may be a new model for regulating AGPase activity in the starch biosynthesis process.

2. Results

2.1. Proteomic Analysis of Maize Endosperm Protein Enriched by Phos-tag^{TM} Agarose

To analyze the phosphorylation of protein in maize endosperm at the starch accumulation process, maize endosperm samples were collected at 20 days after pollination from the 08-641 maize inbred line, a high starch content material widely used in southwest China. Maize endosperm protein was extracted from the 08-641 inbred line and then enriched with phos-tag^{TM} agarose that was specific for phosphorylated protein. Protein from phos-tag^{TM} agarose enrichment was performed SDS-PAGE electrophoresis and stained with commassie brilliant blue. The gel was cut for in-gel tryptic digestion for mass spectrometry. The results showed that a total of 1733 proteins were identified from 10,678 peptides with at least one unique peptide (Figure 1A and Table S1). GO term analysis results showed that these proteins identified were annotated into biological process, cellular component and molecular function (Table S2). Of these proteins, 46.49% and 41.69% were annotated into catalytic activity and binding activity, respectively, in the molecular function (Figure 1B). A total of 35.21% and 33.53% of the proteins were cell processes and metabolic processes, respectively, in the biological process (Figure 1C). We also found that 24.64% and 23.87% of the proteins were cell and cell part, respectively (Figure 1D). These results potentially suggested that the phosphorylation of proteins was involved in various biological processes and the molecular function in maize endosperm during the starch synthesis process. Interestingly, the representative peptides of AGPase small subunit Bt2 were identified by mass spectrometry in proteomic data (Figure 1E). The data showed that a total of 21 unique peptides for Bt2 were identified. The cover percentage of peptides for Bt2 was 53.89% (Table S3). Therefore, we speculated that Bt2 might be phosphorylated by kinase; thus, it could be enriched with phos-tag^{TM} agarose.

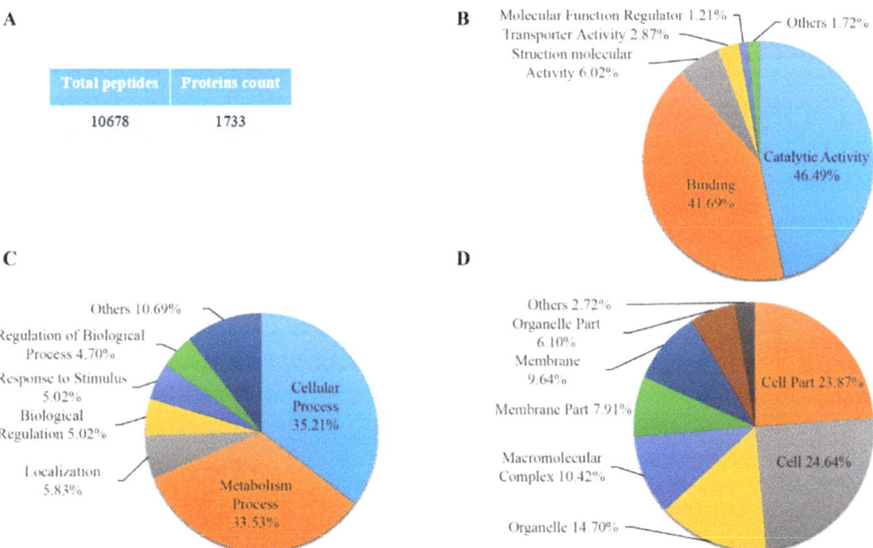

Figure 1. *Cont.*

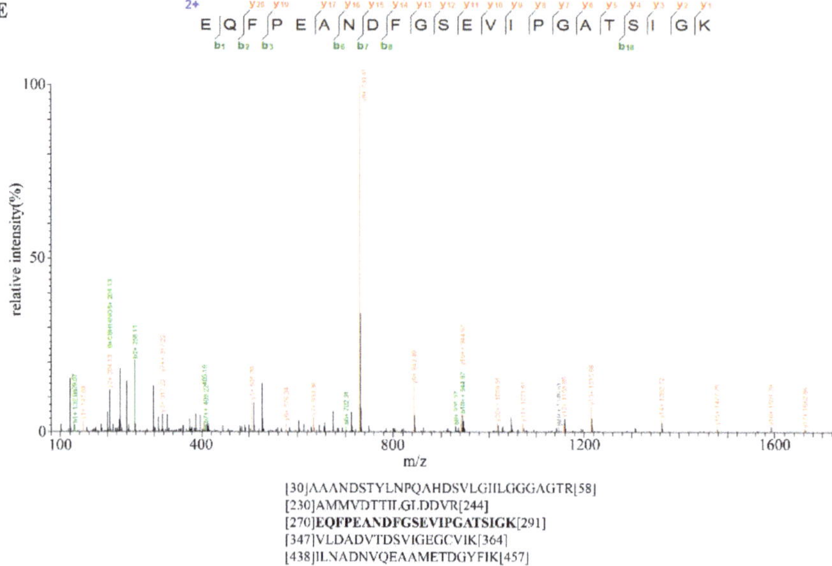

Figure 1. The proteomic analysis of maize endosperm proteins enriched by phos-tag™ agarose. (**A**) Summary of peptides and proteins identified. (**B**) Gene ontology analysis of molecular function level. (**C**) Gene ontology analysis of biological process level. (**D**) Gene ontology analysis of cellular component level. Percentages shown indicated GO term protein accounted for total protein number. (**E**) The representative mass spectrometry diagram and peptides of Bt2 protein. The mass spectrometry diagram shows the bold type of peptide. The numbers show the position of amino acid in Bt2 protein.

2.2. Antibody Preparation and Evaluation of AGPase Small Subunit Bt2

In order to further investigate the phosphorylation of AGPase subunit Bt2, we prepared the antibody Bt2. We purified GST-Bt2 proteins as the antigen for immunization of New Zealand rabbits to prepare the antibody. Through molecule cloning, we successfully constructed pGEX-6-t-GST-Bt2 vectors. The fusion protein could be cut by TEV enzyme between the GST and target proteins (Figure 2A). After induction of protein expression in BL21 bacterial cells with 0.2 mM IPTG, we purified GST-Bt2 proteins with GST beads (Figure 2B). According to the antibody preparation procedure, antiserum was harvested and purified for assessing to detect target protein. In order to evaluate the specificity of antibodies of Bt2, we cut GST-Bt2 proteins with TEV enzyme at different time intervals from 0h to 6 h. Immunoblot results showed Bt2 antibody could recognize Bt2 and GST (Figure 2C). These results showed that Bt2 antiserum could specifically recognize Bt2 proteins in vitro.

2.3. Co-Immunoprecipitation of Stromal Protein and Expression Analysis of Bt2 in Maize Endosperm

Specific Bt2 antibody was prepared; however, whether it could recognize Bt2 in maize endosperm in vivo was unclear. In order to further demonstrate the specificity of antibody of Bt2 in vivo, first, we performed immunoblot to detect Bt2 protein from 20 days maize endosperm after pollination with Bt2 antibody (Figure 3B). However, we could not detect Bt2 protein using pre-antiserum (Figure 3A). These results showed that Bt2 antibody was prepared successfully. Second, we used 27 days endosperm after pollination to lysate for immunoprecipitation with Bt2 antibody. The product of immunoprecipitation with Bt2 antibody was identified by immunoblot. In addition, Bt2 protein was immunoprecipited with Bt2 antibody and run the gel and digested with trypsin and analyzed by mass spectrometry. The result showed that Bt2 antibody could immunoprecipite Bt2 protein very specifically and that IgG control could not (Figure 3C). mass spectrometry results showed the band is Bt2 (Figure 3E). Third, we performed the expression analysis of Bt2 proteins in endosperm from 5 days to 40 days after pollination. According to previous reports, the expression of Bt2 is increasing in endosperm from 0 to 20 days after pollination and is decreasing from 20 to 40 days after pollination at mRNA level [34]. We further demonstrated the result by immunoblot using Bt2 antibodies. The result showed that Bt2 protein levels were increasing in endosperm from 0 to 20 days after pollination and then decreased from 25 to 40 days after pollination (Figure 3D). These results showed that Bt2 antibodies could be used for immunoblot, IP assay in vivo and in vitro.

2.4. Enrichment of Bt2 Phosphorylation Protein by Phos-tag^{TM}

In order to demonstrate the AGPase subunits phosphorylation, we extracted protein from the different development process of endosperm cells in the maize for experiments. After enriching all phosphoproteins from the lysate using phos-tag^{TM} technology and removing nonspecific binding protein by washing, we used the specific Bt2 antibodies to detect Bt2 proteins by immunoblot in phos-tag^{TM} binding protein. Although phos-tag^{TM} enrichment step is hardly get 100% phosphoproteins, we could clearly detect bands of Bt2 protein. However, Bt2 bands would disappear when we added alkaline phosphatase to treat the samples from 15, 20, and 27 days 08-641 maize endosperm after pollination (Figure 4). The result showed that Bt2 might be phosphorylated by kinases in the starch biosythesis process.

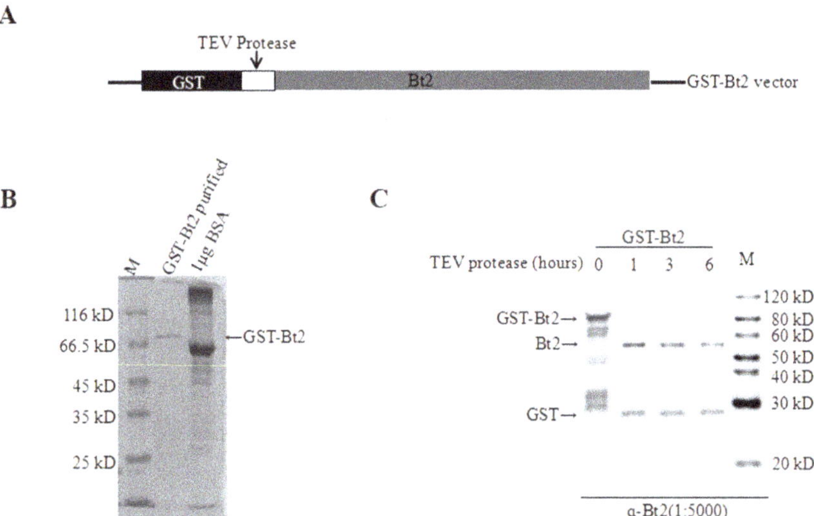

Figure 2. Evaluation of Brittle-2 (Bt2) antibodies using antigen protein purified from GST in vitro. (**A**) Structural diagram of GST-Bt2 vectors. The white region showed the TEV protease site between GST tag and protein interested. (**B**) GST-Bt2 protein purified and BSA standard were detected by coomassie brilliant blue G-250 staining method. The protein marker ladder is indicated in the left column. (**C**) GST-Bt2 digested with TEV protease for 1–6 h by immunoblot. 0.5 μg protein was loaded in each well. Ratio of antibody dilution was 1:5000.

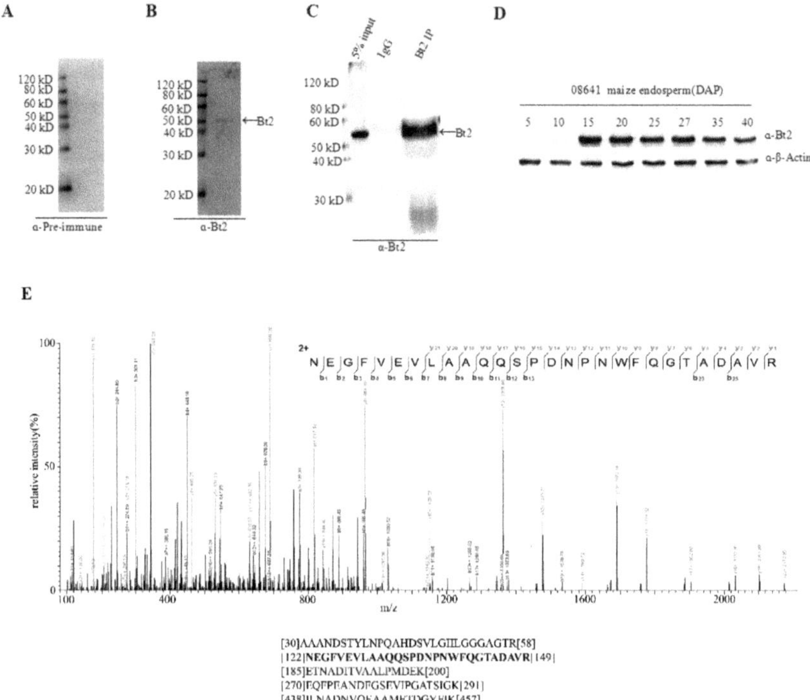

Figure 3. Immunoprecipitation of stromal proteins and expression analysis of Bt2 in maize endosperm. (**A**) Pre-immune serum was evaluated as control with 20 DAP maize endosperm. Ratio of antibodies dilution was 1:5000. (**B**) Bt2 antibody was evaluated by immunoblot with 20 DAP maize endosperm. Ratio of antibodies dilution was 1:5000. Arrow shows Bt2 protein. (**C**) Bt2 antibody was used as immunoprecipitation to detect maize endosperm Bt2 protein by immunoblot. Arrow shows Bt2 protein. (**D**) Expression analysis of Bt2 protein in days indicated after pollination in maize 08-641 endosperm. Ratio of antibodies dilution was 1:5000 for Bt2 and 1:1000 for actin. (**E**) Identification of maize endosperm Bt2 peptides after immunoprecipitation using Bt2 antibody. The protein was precipitated, run the gel, digested with trypsin and analyzed by mass spectrometry. The mass spectrometry diagram showed the bold type of peptide. The numbers show the location of peptides in Bt2 protein.

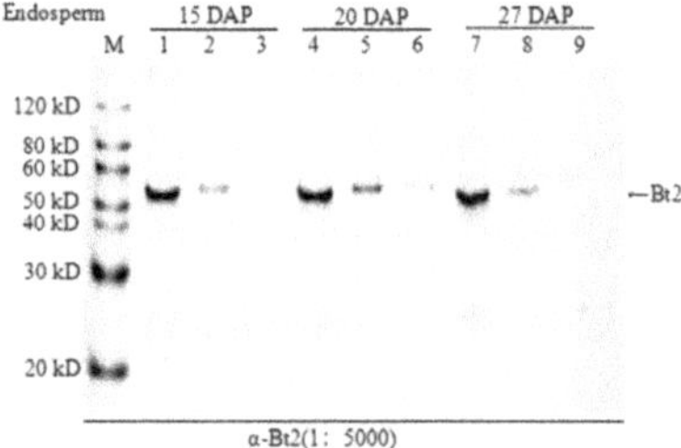

Figure 4. Enrichment of phosphorylated AGPase Bt2. 1, 4, and 7 were 20 μg total protein extracted from maize endosperm 15, 20, and 27 days after pollination, respectively; 2, 5, and 8 were phosphorylated protein enriched from 200 μg total protein by phos-tag^TM agarose from maize endosperm 15, 20, and 27 days after pollination, respectively; 3, 6, and 9 were protein from maize endosperm 15, 20, and 27 days after pollination 200 μg total protein treated with alkaline phosphatase and then enriched by phos-tag^TM agarose, respectively

2.5. Detection of Bt2 Phosphorylation Protein by Diamond Q Staining Technology

Using commercially available phosphorylated proteins as control proteins, phosphorylation-specific staining of maize endosperm proteins was performed to further detect the phosphorylation of Bt2 by Pro-Q diamond dye technology, which is specific for phosphylated proteins [35,36]. Immunoblot was performed to demonstrate Bt2 protein by specific antibody. We harvested maize 15, 20, and 27 days endosperm after pollination and divided it into three groups for assay. The first group was used for enriching phosphoproteins by phos-tag^{TM} technology, the second group was used as control for immunoblot and the third group was used for immunoprecipitation assay using Bt2 antibody. Commercially available phosphorylated proteins were used as a control marker. The results showed that Bt2 was obviously phosphorylated in the maize endosperm, although it appears much loss of phospho-Bt2 through enrichement and transfer-menbrane steps (Figure 5A). Using the specific Bt2 antibody made before, we did a immunoblot test to further demonstrate that the stain band was Bt2 (Figure 5B). These results obviously showed Bt2 was phosphorylated in maize kernel starch biosynthesis.

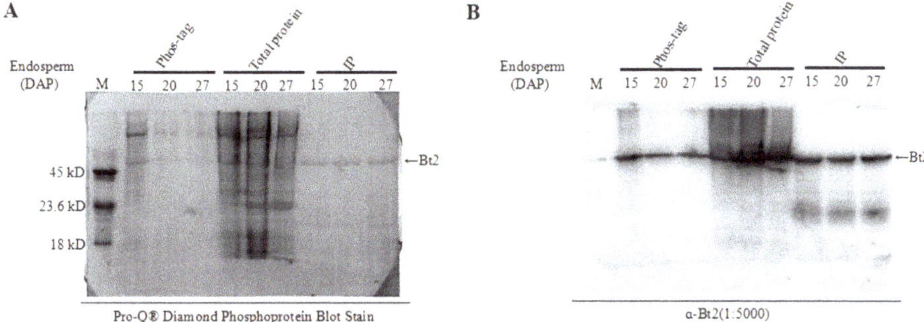

Figure 5. Detection of Bt2 phosphorylation proteins by Diamond Q staining technology. (**A**) Phosphoprotein Pro-Q staining of maize endosperm protein in indicated days after pollination; phos-tag^{TM} group 200 µg total protein was enriched by phos-tag^{TM} agarose with maize endosperm protein; 20 µg total protein was the control, and the IP group 200 µg total protein was immunoprecipitated with the Bt2 antibody; M was a protein marker phosphorylated from Fisher Scientific Company. (**B**) The stained Bt2 protein band was demonstrated by immunoblot. Arrow shows Bt2 band.

2.6. Identification of Bt2 Phosphorylation Sites by iTRAQ

The previous results showed that Bt2 was phosphorylated during maize kernel starch biosynthesis. However, the phosphorylation site of Bt2 protein was still unknown. In order to identify the specific phosphorylation sites of Bt2, we collected 08-641 maize 15 DAP endosperm and a transgenic maize to perform iTRAQ assay. The results showed that Bt2-Ser^{10} and Bt2-Thr^{451} and Bt-Thr^{462} were phosphorylated (Figure 6A,B and Table S4). Bioinformatics analysis showed that Bt2-Thr^{451} and Bt2-Thr^{462} were conserved in selected plants (Figure 6C). Instead, Bt2-Ser^{10} was specific for maize, potato, and tomato in 10 selected plants (Figure 6D). These results show that three sites were phosphorylated for Bt2 protein in maize endosperm and that phosphorylation regulation of Bt2 may be complicated in the starch synthesis process.

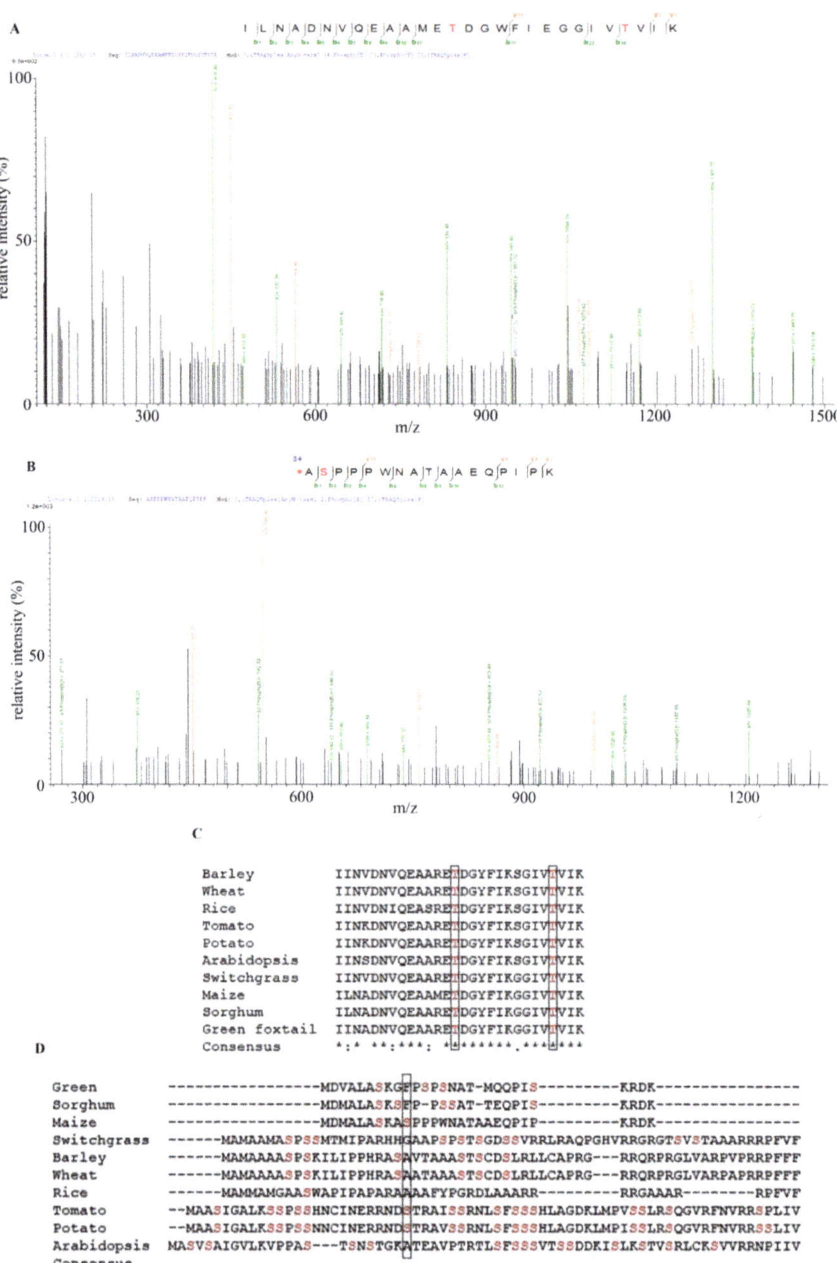

Figure 6. Identification of Bt2 phosphorylation sites by iTRAQ^{TM}. (**A,B**) Mass spectrometry diagram of Bt2 phosphopeptides. Red letters S and T show the phosphorylated serine and threonine. (**C,D**) Multiple alignment analysis of phosphorylated sites of maize Bt2 at Ser[10], Thr[451], and Thr[462] in indicated species by ClustalX 2.1. software. The frame shows the phosphorylated sites identified in maize Bt2 position and putative position in other species. Asterisk shows the consensus of the protein sequence.

2.7. Enzyme Characteristics of AGPase Phosphorylation and a Potential Regulatory Model

Previous results suggested that Bt2 could be phosphorylated in maize endosperm during the starch accumulation process. Next, we wanted to know whether the phosphorylation of Bt2 might affect activity of AGPase. We detected phosphorylated AGPase format and dephosphorylated AGPase format by adding alkaline phosphatase to remove the phosphate group. Native gel assay was performed to examine activity of AGPase according to previous reports [34,37]. The active band could be observed in native gel when soaked in the solution, which adds substrate glucose-1-phosphate and ATP in the no-alkaline phosphatase column; however, there are no bands in the alkaline phosphatase or no ATP column (Figure 7A). The results revealed that dephosphorylating of AGPase small subunit Bt2 may abolished or declined the active bands that AGPase catalyzed to form. Therefore, the phosphorylation of Bt2 might be important for AGPase activity. Based on these results, we proposed a potential regulation mode in which phosphorylation of AGPase catalyzed by kinases would enhance its activity. However, dephosphorylation of AGPase catalyzed by alkaline phosphatase would inhibit its activity (Figure 7B). In all, phosphorylation of AGPase subunits may be a new model to regulate the activity of AGPase in the maize kernel starch synthesis process.

A   **B**

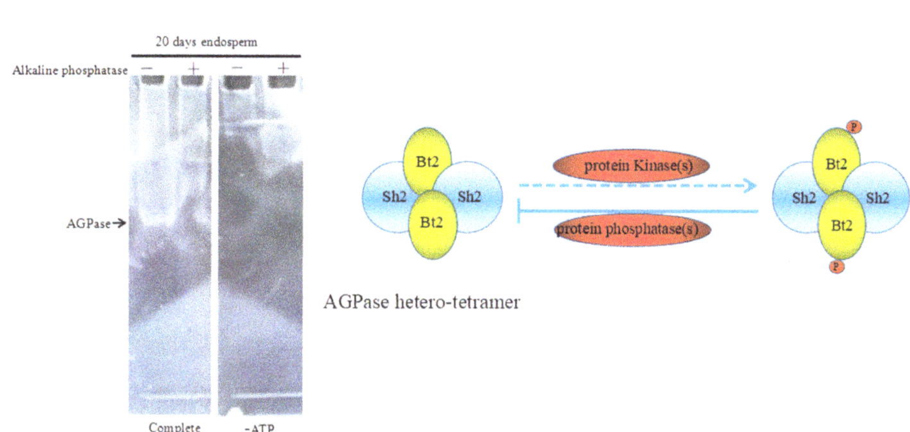

Figure 7. Enzyme characteristics of AGPase dephosphorylation and a potential regulatory model. (**A**) Native gel was performed with 40 µg total maize endosperm protein and protein treated with alkaline phosphatase at 20 days after pollination. Gel was incubated in reaction buffer including ATP and Glc-1-P(Complete) in the presence of divolent cations or in the same buffer lacking ATP (-ATP). Arrow show that the white band is native AGPase band. (**B**) Putative potential regulatory model of phosphorylation of AGPase. Sh2: large subunits.

3. Discussion

AGPase is a key enzyme for catalyzing ATP and Glc-1-p to form ADP-glucose and diphosphate. As the precursor of starch, ADP-glucose is used as a material for incorporating glucosyl units into starch [5]. Genes encoding AGPase activity are mutated in maize, barley, and rice, which leads to a reduction in total endosperm starch in the range of 20–70% of normal. The activity of AGPase is one of determinants for the content of starch in maize kernel. Therefore, it is very important for the regulation of activity of AGPase. To date, there are three main ways reported to regulate the activity of AGPase: small effector molecules for allosteric regulation, thermal regulation and redox regulation [9–12,38]. In this study, we used phos-tag^{TM} technology, Pro-Q diamond staining and iTRAQ^{TM} to investigate phosphorylation of AGPase. The results show that AGPase subunits Bt2-Ser^{10}, Bt2-Thr^{451}, and Bt2-Thr^{462} were phosphorylated in the kernel starch accumulation process.

In addition, based on the current data, it seems important for the activity of AGPase regulated by phosphorylation. Removal of phosphate groups from AGPase subunits could reduce its activity. Therefore, phosphorylation of AGPase may be an important means of regulating its activity.

The phosphorylation of AGPase leads to binding phos-tag^{TM} agarose. Phos-tag^{TM} agarose is a dinuclear metal complex and acts as a selective phosphate-binding tag molecule specifically for phosphorylated proteins [39–41]. According to our current data, phos-tag^{TM} agarose could enrich phosphorylated AGPase, including Bt2. Using SDS to unfold and denature protein, we could still detect Bt2 by phos-tag^{TM} agarose (data not shown). Thus, the result is reliable for the binding of phos-tag^{TM} agarose by the phosphorylation of AGPase. Dephosphorylation of AGPase leads to losing the binding between AGPase subunits and phos-tag^{TM} agarose. Alkaline phosphatase is widely used as a dephosphorylation agent to dephosphorylate various proteins [42,43]. In our experimental system, we used alkaline phosphatase to remove the phosphate group from AGPase, and the binding of phosphorylated AGPase subunit Bt2 with phos-tag^{TM} agarose significantly decreased or disappeared. In this study, we also used a commercially available phosphorylated protein marker as a positive control and Pro-Q diamond staining technology, which is specific for detecting the phosphorylated Bt2 protein in denatured condition by boiling the protein sample. We demonstrated the Bt2 protein by immunoblot using specific Bt2 antibody made in the current experiment. Also, we detected the phosphorylated Bt2 protein. We analyzed AGPase subunit Bt2 peptides mass spectrometry data from 15 DAP maize endosperm by iTRAQ^{TM}. Bt2-Ser^{10}, Bt2-Thr^{451}, and Bt2-Thr^{462} were phosphorylated from mass spectrometry. Therefore, it could be concluded that the phosphorylation of Bt2 actually occurred in vivo from this indirect and direct evidence.

The phosphorylation of AGPase may be complicated and exist in different regulation mechanisms in maize and other species. Walley reported maize AGPase subunit Sh2-Ser^{95} and Bt2-Ser^{10} and Bt2-Ser^{104} were phosphorylated in vivo from proteomic data [33]. Nakagami reported *At*APS1-Thr^{231} and *At*APS1-Thr^{236} were phosphorylated by large-scale comparative phosphoproteomics research in the plant *Arabidopsis thaliana* [44]. Our current data show that Bt2-Ser^{10}, Bt2-Thr^{451}, and Bt2-Thr^{462} were phosphorylated in maize endosperm. The phopsphorylation of Bt2-Ser^{10} may regulate the activity of AGPase because this site is in N-term of Bt2 [45]. Bt2-Thr^{451} and Bt2-Thr^{462} may affect the structure of AGPase because these sites are in C-term of Bt2 [46,47]. In addition, Bt2-Ser^{10} is a common phosphorylated site for B73 and our inbred maize line. This result suggests that there is common regulatory mechanism in different inbred maize lines. Instead, Bt2-Ser^{104} was phosphorylated in B73, Bt2-Thr^{451} and Bt2-Thr^{462} were phosphorylated in our inbred line. This result also suggests that there are specific regulatory mechanisms in different inbred maize lines. In addition, in different plants, Bt2-Thr^{451} and Bt2-Thr^{462} were very conservative; however, Bt2-Ser^{10} was comparatively specific for maize, potato, and tomato. Thus, it is possible that there are common and specific regulatory mechanisms in different species. Phosphorylation regulation of Bt2 could be complicated in starch synthesis.

The phosphorylation of Bt2 might change the activity and stability of AGPase. In general, phosphorylation of a protein will change the activity or function of enzyme, localization, and binding specificity of target proteins [27]. In order to answer the question of whether it activates or inactivates AGPase after Bt2 phosphorylation, we tried to detect activity of AGPase phosphorylation through native gel assay. Our current results clearly show that the native band disappeared or was non-detectable when alkaline phosphatase was present in the protein sample. We speculate that the result of dephosphorylation of AGPase might inhibit its activity because alkaline phosphatase removing phosphate group from AGPase will lead to a free phosphate group. As previous reported, the free phosphate will inhibit the activity of AGPase [11]. In addition, phosphorylation of SBEI, SBEIIb, and SP is important for stability and activity of the protein complex formed with SBI, SBEI, and SBEIIb. Dephosphorylation of the SBEI-SBEIIb-SP complex will disturb the protein complex and decrease its activity [28,29]. As a heterotetrameric AGPase, which is composed of two identical large Sh2 subunits and two identical small Bt2 subunits, it is potentially possible that like the

SBEI-SBEIIb-SP complex, the phosphorylation of AGPase would increase the enzyme stability and activity, the dephosphorylation of AGPase subunits would cause a reduction or loss of its activity and stability.

4. Materials and Methods

4.1. Plant Materials

Seeds for 08-641 inbred maize line were provided by the maize research institute of Sichuan Agricultural University and grown at the school farm in the summer of 2013–2017. Developing kernels from self-pollinated ears were collected 10 DAP, 15 DAP, 20 DAP, 27 DAP, 30 DAP, 35 DAP, and 40 DAP and were quickly frozen in liquid nitrogen and stored at $-80\,^{\circ}$C until use. For phos-tag^{TM} enrichment assay, three independent biology repeated maize endosperm samples collected at the same time were mixed as a pool for proteomic analysis.

4.2. GST-Gene Fusion System Protein Expression and Purification

GST-gene fusion system protein expression vector pGEX-6t-1-Bt2 were constructed by adding Bt2 genes into the pGEX-6t-1 vector. The cloning primers of Bt2 were as follows: Bt2 Forward: 5′-CG ggatccATGGACTGGCTTTGGCGTCTA-3′, Reverse: 5′-CAGctcgagTCATATAACTGTTCCACTAG GGAG-3′. The lowercase letters indicate the introduced base to create an BamHI and XhoI, respectively. The protein expression and purification of Bt2 were performed according to the GST gene fusion system handbook from GE Healthcare (Piscataway, NY, USA).

4.3. Rabbit Breeding, Anti-serum Preparation, and Antibody Purification

New Zealand white rabbits were provided by Da Shuo experimental animal company. Rabbits were maintained in the Animal Core Facility following procedures approved by the Animal Care and Use Committee of Sichuan Agricultural University (no 20160320, Chengdu, China). After one week of acclimation, rabbits were injected with an antigen mixed with adjuvant every two weeks and their venous blood harvested after three injections. Polyclonal rabbit antisera targeted to maize Bt2 were raised against the antigen, in which GST-Bt2 were purified by GST-gene fusion system protein expression assay. Antiserum containing the polyclonal maize antibody was applied to the column with 1 cm^3 50% protein A and 50% protein G and washed with 10 cm^3 ice-cold TBS (50 mM Tris-HCl, pH 7.4, 150 mM NaCl, and 0.05% sodium azide). The antiserum was then thawed in ice water and clarified by centrifugation at $15,000\times g$ for 5 min at 4 $^{\circ}$C. 3.6 cm^3 of the clarified antiserum was added to the column and the column was washed with 36 cm^3 TBS buffer. 2.5 cm^3 elution buffer with pH 2.7 and pH 1.9 (100 mM glycine pH 2.7 and 100 mM glycine pH 1.9, respectively) was gently added to the column. Roughly 0.4 cm^3 fractions were collected in the tubes above and neutralized with NB buffer (1 M Tris-HCl, pH 8.0; 1.5 M NaCl; 1 mM EDTA; 0.5% sodium azide) to adjust the pH to approximately 7.4. Pure antibodies were used in immunoblot and immunoprecipitation. Pre-immune sera for each of the antibodies used above were employed as negative controls, and showed no cross-reaction with proteins from maize endosperm lysates and co-immunoprecipitation experiments.

4.4. Plant Protein Extraction and Protein Determination

Maize kernels were harvested and quickly frozen in liquid nitrogen and stored at $-80\,^{\circ}$C until use. Maize endosperm was dissected using pre-chilled tweezers and a mortar on ice, and total endosperm proteins were extracted from endosperm tissues flash frozen in liquid nitrogen. Two grams of endosperm tissue were pulverized using a mortar and pestle under liquid nitrogen. 6 cm^3 of native protein extraction buffer (100 mM Tris-HCl, pH 7, 10 mM MgCl$_2$, 100 mM KCl, 15% glycerol and DDT, 40 mM β-mercaptoethanol, 1 mM PMSF and phosphatase inhibitor cocktail (Sigma) added freshly) were added followed by further shaking in a vortex. The homogenates were centrifuged ($16,000\times g$, 10 min, 4 $^{\circ}$C) and supernatants stored at $-80\,^{\circ}$C until use. Protein concentration was determined using the Bio-Rad protein assay (Bio-Rad, Hercules, CA, USA) according to the manufacturer's instructions and with BSA as a standard.

4.5. SDS-PAGE and Immunoblotting

Zn^{2+}-phos-tag^{TM} agarose was used to enrich phosphoproteins according to the manufacturer's instructions (Wako Pure Chemical Industries Ltd., Hiroshima, Japan). Briefly, the total maize endosperm cell lysate sample containing 200 µg protein was used for enriching phosphoproteins with 200 mm^3 Zn^{2+}-phos-tag^{TM} agarose. For dephosphorylation assay, we treated cell lysate by adding alkaline phosphatase. The binding assay was performed for 4 h at 4 °C, and then washed 3 times with washing buffer (0.1 M Tris-CH_3COOH, 1.0 M CH_3COONa, pH 7.5). Elution buffer (0.1 M Tris-CH_3COOH, 1.0 M NaCl, 10 mM NaH_2PO_4-NaOH, pH 7.5) was used for the elution of phosphoproteins. Zn^{2+}-phos-tag^{TM} agarose binding proteins were separated by electrophoresis. In addition, 20 µg maize endosperm cell lysates were subjected to SDS-PAGE as a control. The SDS polyacrylamide gels consisted of an 8% or 10% acrylamide separation gel and a 4% stacking gel. The resolved proteins were electrophoretically transferred to nitrocellulose membranes (GE Healthcare Life Sciences, Cat: 10600003). The membranes were incubated with anti-Bt2 antibody diluted 1:5000 and anti-actin antibody diluted 1:1000 for 2 h at room temperature. The membranes were then incubated with horseradish peroxidase conjugated with anti-rabbit IgG (Kangwei Company) diluted 1:5000 for 30 min, and the immunoreactive bands were detected using the chemiluminescent substrate, Lumi-Light immunoblotting Substrate (Thermo Scientific, Cat: PI208186 and PJ209602, Rockford, IL, USA). All immunoblot assays were independently performed at least three times.

4.6. Mass Spectrometry and Data Processing

The binding protein was collected from phos-tag^{TM} agarose incubated with endosperm protein. Similarly, the production of IP was collected. Protein was performed SDS-PAGE electrophoresis and stained commassie brilliant blue. The gel was cut for in-gel tryptic digestion, gel pieces were destained in 400 mm^3 50 mM NH_4HCO_3 in 50% acetonitrile (v/v) until clear. Gel pieces were dehydrated with 100 mm^3 of 100% acetonitrile for 5 min, the liquid was removed and the gel pieces were rehydrated in 10 mM dithiothreitol and incubated at 56 °C for 60 min. Gel pieces were again dehydrated in 100% acetonitrile, liquid was removed and gel pieces were rehydrated with 55 mM iodoacetamide. Samples were incubated at room temperature in the dark for 45 min. Gel pieces were washed with 50 mM NH_4HCO_3 and dehydrated with 100% acetonitrile. Gel pieces were rehydrated with 10 ng/mm^3 trypsin resuspended in 50 mM NH_4HCO_3 on ice for 1 h. Excess liquid was removed and gel pieces were digested with trypsin at 37 °C overnight. Peptides were extracted with 50% acetonitrile and 5% formic acid, followed by 100% acetonitrile. Peptides were dried and re-suspended in 0.1% formic acid.

The tryptic peptides were dissolved in 0.1% formic acid (solvent A), directly loaded onto a homemade reversed-phase analytical column (15 cm in length, 75 µm i.d.). The gradient was comprised of an increase from 4% to 50% solvent B (0.1% formic acid in 98% acetonitrile) in 50 min, 50% to 100% in 4 min, then holding at 100% for the last 6 min, all at a constant flow rate of 400 nl/min on an EASY-nLC 1000 UPLC system. The peptides were subjected to NSI source followed by tandem mass spectrometry (MS/MS) in Q Exactive^{TM} Plus (Thermo) coupled online to the UPLC. The electrospray voltage applied was 2.0 kV. The m/z scan range was 350 to 1800 for full scan, and intact peptides were detected in the Orbitrap at a resolution of 70,000. Peptides were then selected for MS/MS using NCE setting as 28 and the fragments were detected in the Orbitrap at a resolution of 17,500. This was a data-dependent procedure that alternated between one MS scan followed by 20 MS/MS scans with 15.0 s dynamic exclusion. Automatic gain control (AGC) was set at 5E4. The resulting MS/MS data were processed using a Proteome Discoverer 1.3.6 Tandem mass spectra were searched using MASCOT (Matrix Science, London, UK; version 2.2) and against an uniprot Zea mays database (https://www.uniprot.org; 132356 sequences, download on 1 August 2018). Trypsin/P (or other enzymes if any) was specified as the cleavage enzyme, allowing up to 2 missing cleavages. Mass error was set to 10 ppm for precursor ions and 0.02 Da for fragment ions. Peptide confidence was set at high, and the peptide ion score was set >20. The GO term protein function classification in levels 1 and 2 were analyzed by http://www.ebi.ac.uk/GOA/.

4.7. Immunoprecipitation (IP) and Co-immunoprecipitation (Co-IP)

Co-immunoprecipitation experiments were conducted using the methods described by Fushan Liu with some modifications [30]. Purified Bt2 antibodies (each approximately 10 μg) were individually used for the immunoprecipitation and co-immunoprecipitation experiments with 27 days endosperm after pollination cell lysates (1 cm^3, 0.5 mg/cm^3 proteins). The mixture of antibody and cell lysate was incubated at 4 °C on a rotator for 4 h and precipitation of the antibody performed by adding 20 mm^3 of Protein A/G-Sepharose (Biorad) made up as a 50% (*w/v*) slurry with phosphate buffered saline (PBS, 137 mM NaCl, 10 mM Na$_2$HPO$_4$, 2.7 mM KCl, 1.8 mM KH$_2$PO$_4$, pH 7.4) at 4 °C for 4 h. The Protein A-Sepharose/antibody/protein complex was centrifuged at 2000× *g* for 5 min at 4 °C in a refrigerated microfuge and the supernatant was discarded. The pellet was washed five times (1.0 cm^3 each) with PBS, followed by washing five times with a buffer containing 10 mM HEPES-NaOH, pH 7.5, and 150 mM NaCl. Washed pellets were boiled in SDS loading buffer and separated by SDS-PAGE, followed by immunoblot analysis.

4.8. Pro-Q Diamond Phosphoprotein Staining

For the control group, 20 μg total protein were extracted from 08-641 maize endosperm 15 days, 20 days and 27 days after pollination for running gel. For the phos-tag^{TM} group, a total of 200 μg maize endosperm cell lysate protein was enriched with 20 mm^3 slurry made up as Zn^{2+}-phos-tag^{TM} agarose and a 50% (*v/v*) suspension buffer containing 20 mM Tris-CH$_3$COOH pH 7.4, 20% (*v/v*) 3-propanol. For the immunoprecipitation (IP) group, 200 μg protein from endosperm was incubated with the control IgG and Bt2, respectively. Then precipitation of the antibody was performed by adding 20 mm^3 of Protein A/G-Sepharose (Biorad) made up as a 50% (*w/v*) slurry with phosphate buffered saline (PBS, 137 mM NaCl, 10 mM Na$_2$HPO$_4$, 2.7 mM KCl, 1.8 mM KH$_2$PO$_4$, pH 7.4) at 4 °C on a rotator for 4 h. The Protein A-Sepharose/antibody/protein complex was centrifuged at 2000× *g* for 5 min at 4 °C in a refrigerated microfuge, and the supernatant was discarded. The pellet was washed five times (1.0 cm^3 each) with PBS buffer. Washed pellets were boiled in SDS loading buffer and separated by SDS-PAGE. For all protein samples, transferring protein to a polyvinylidene difluoride (PVDF) membrane (GE Healthcare Life Sciences, Cat: 10600023) was used for Pro-Q diamond phosphoprotein dye staining according to manufacturer's instruction from Invitrogen (Cat: MP33300). A commercially available phosphorylated protein marker was used as a positive control and was purchased from Thermo Fisher Scientific (Cat: MP33350). In order to exclude the possibility that the signal of Pro-Q diamond phosphoprotein dye interfered with the signal from immunoblotting, after destaining of the PVDF membrane with Pro-Q diamond phosphoprotein dye, the PVDF membrane with the protein sample was used for Bt2 immunoblotting.

4.9. iTRAQ^{TM} Labeling and Mass Spectrometry Analysis

Phosphorylated protein was identified by the iTRAQ^{TM} method described by and Ma [48]. Briefly, three biological repeats maize 15 DAP kernels and three dependent transgene samples protein were extracted and digested by sequencing-grade trypsin. Peptides were collected and lyophilized. The samples were reconstituted in TEAB buffer and three independent samples labelled by iTRAQ reagents (8 PLEX multiplex kit, AB Sciex, Cat: 4381663). Three independent WT samples were respectively labeled with iTRAQ tag 115,116,117. Three independent transgenic samples were respectively labeled with iTRAQ tag 118,119,121. TiO$_2$ was used to enrich the phosphopeptides for mass spectrometry analysis. All analyses were performed by a Triple TOF 5600 Mass Spectrometer (SCIEX, USA) equipped with a Nanospray III source (SCIEX, Framingham, MA, USA). Samples were loaded by a capillary C18 trap column (3 cm × 100 μm) and then separated by a C18 column (15 cm

$\times$ 75 μm) on an Eksigent nanoLC-1D plus system (SCIEX, USA). The flow rate was 300 nL/min and the linear gradient was 90 min (from 5–85% B over 67 min; mobile phase A = 2%ACN/0.1%FA and B = 95%ACN/0.1%FA). Data were acquired with a 2.4 kV ion spray voltage, 35 psi curtain gas, 5 psi nebulizer gas, and an interface heater temperature of 150 °C. The MS scanned between 400 and 1500 with an accumulation time of 250 ms. For IDA, 30 MS/MS spectra (80 ms each, mass range 100–1500) were acquired with MS peaks above intensity 260 and a charge state of between 2 and 5. A rolling collision energy voltage was used for CID fragmentation for MS/MS spectra acquisitions. Mass was dynamically excluded for 22 s.

4.10. Zymograms of Native PAGE

AGPase zymograms modified method was used for in-gel assay of AGPase activity as described by Huang et al. [34]. Five endosperm samples were ground in liquid nitrogen using a pre-chilled mortar and pestle, and an extraction buffer (100 mM Tris-HCl, pH 7, 40 mM β-mercaptoethanol added freshly, 10 mM MgCl$_2$, 100 mM KCl, and 15% glycerol) was added in 1.5 cm^3 EP tubes. The homogenate was centrifuged at 16,000$\times$ *g* for 30 min at 4°C, and the supernatant was stored at -80 °C or used immediately. Equal amounts of protein (40 μg) were loaded onto a 7.5% (*w/v*) native acrylamide gel and electrophoresed in Laemmli buffer lacking SDS at 4 °C at 90 V for 2 to 4 h. The gel was then incubated overnight at 37 °C in 100 mM Tris-HCl, pH 8, 5 mM β-mercaptoethanol, 5 mM CaCl$_2$, 10 mM MgCl$_2$, 5 mM Glc-1-P, 5 mM ATP, and 10 mM 3-PGA. Gels were photographed on a dark background to visualize white precipitate bands. Control assays omitted ATP.

5. Conclusions

We provide the evidence that AGPase subunit Bt2 was phosphorylated in the maize endosperm during starch synthesis process. Proteomic and mass spectrometry analysis showed the peptides of AGPase subunit Bt2 were identified from product of phos-tag^{TM} agarose binding maize endosperm proteins. Phosphorylation of AGPase subunits could bind phos-tag^{TM} agarose. However, dephosphorylation of AGPase subunits would lose the binding treating with ALP. We further demonstrated the result by immunoblot through specific Bt2 antibodies. Pro-Q diamond staining further demonstrated the phosphorylation of Bt2. The specific phosphorylation sites of Bt2 at Ser^{10}, Thr^{451}, and Thr^{462} were identified by iTRAQ^{TM}. Protein sequence multiple alignment analysis showed Bt2-Thr^{451} and Bt2-Thr^{462} were very conserved sites in different species. Bt2-Ser^{10} is specific for maize, potato and tomato. In native gel assay, removing phosphate group from Bt2 with alkaline phosphatase, the activity of AGPase was abolished. In all, our data potentially suggest that phosphorylation of Bt2 may be a new model to regulate activity of AGPase.

Supplementary Materials: Supplementary materials can be found at http://www.mdpi.com/1422-0067/20/4/986/s1.

Author Contributions: G.Y., Y.H. (Yubi Huang) designed the experiments. G.Y., Y.L. (Yanan Lv), L.S., Y.Q., N.W., Y.L. (Yangping Li), H.H., N.Z. performed the experiments; Y.W., Y.L. (Yinghong Liu), Y.H. (Yufeng Hu), H.L., J.Z. analyzed the data. G.Y. and Y.H. (Yubi Huang) wrote the paper. All authors have given approval of the final version of the manuscript.

Funding: This work was supported by the Natural Science Foundation of China (No: 31501322), Postdoctoral Special Foundation of Sichuan Province (No: 03130104), Overseas Scholar Science and Technology Activities Project Merit Funding (No: 00124300) and the National Key Basic Research Program of China (No: 2014CB138200).

Acknowledgments: We thank Zhengli Chen and Li Tang for animals technology support.

Conflicts of Interest: The authors declare no conflict of interest.

Abbreviations

ALP	Alkaline phosphatase
Co-IP	Co-immunoprecipitation
DAP	Days After Pollination
GBSS	Granule bound starch synthase
IP	immunoprecipitation
SBEI	Starch Branch Enzyme I
SBEIIa	Starch Branch Enzyme IIa
SP	Starch Phosphorylase
SSI	Starch synthase I
SSII	Starch synthase II

References

1. Chen, J.; Yi, Q.; Cao, Y.; Wei, B.; Zheng, L.; Xiao, Q.; Xie, Y.; Gu, Y.; Li, Y.; Huang, H.; et al. ZmbZIP91 regulates expression of starch synthesis-related genes by binding to ACTCAT elements in their promoters. *J. Exp. Bot.* **2016**, *67*, 1327–1338. [CrossRef] [PubMed]
2. Hannah, L.C.; James, M. The complexities of starch biosynthesis in cereal endosperms. *Curr. Opin. Biotechnol.* **2008**, *19*, 160–165. [CrossRef] [PubMed]
3. Smith, A.M.; Denyer, K.; Martin, C. The Synthesis of the Starch Granule. *Annu. Rev. Plant Physiol. Plant Mol. Biol.* **1997**, *48*, 67–87. [CrossRef] [PubMed]
4. Georgelis, N.; Braun, E.L.; Shaw, J.R.; Hannah, L.C. The two AGPase subunits evolve at different rates in angiosperms, yet they are equally sensitive to activity-altering amino acid changes when expressed in bacteria. *Plant Cell* **2007**, *19*, 1458–1472. [CrossRef] [PubMed]
5. Zeeman, S.C.; Kossmann, J.; Smith, A.M. Starch: Its metabolism, evolution, and biotechnological modification in plants. *Annu. Rev. Plant Biol.* **2010**, *61*, 209–234. [CrossRef] [PubMed]
6. Tiessen, A.; Nerlich, A.; Faix, B.; Hummer, C.; Fox, S.; Trafford, K.; Weber, H.; Weschke, W.; Geigenberger, P. Subcellular analysis of starch metabolism in developing barley seeds using a non-aqueous fractionation method. *J. Exp. Bot.* **2012**, *63*, 2071–2087. [CrossRef] [PubMed]
7. Hannah, L.C.; Tuschall, D.M.; Mans, R.J. Multiple forms of maize endosperm adp-glucose pyrophosphorylase and their control by shrunken-2 and brittle-2. *Genetics* **1980**, *95*, 961–970. [PubMed]
8. Hannah, L.C.; Shaw, J.R.; Giroux, M.J.; Reyss, A.; Prioul, J.L.; Bae, J.M.; Lee, J.Y. Maize genes encoding the small subunit of ADP-glucose pyrophosphorylase. *Plant Physiol.* **2001**, *127*, 173–183. [CrossRef] [PubMed]
9. Greene, T.W.; Hannah, L.C. Enhanced stability of maize endosperm ADP-glucose pyrophosphorylase is gained through mutants that alter subunit interactions. *Proc. Natl. Acad. Sci. USA* **1998**, *95*, 13342–13347. [CrossRef] [PubMed]
10. Linebarger, C.R.; Boehlein, S.K.; Sewell, A.K.; Shaw, J.; Hannah, L.C. Heat stability of maize endosperm ADP-glucose pyrophosphorylase is enhanced by insertion of a cysteine in the N terminus of the small subunit. *Plant Physiol.* **2005**, *139*, 1625–1634. [CrossRef] [PubMed]
11. Boehlein, S.K.; Shaw, J.R.; Stewart, J.D.; Hannah, L.C. Studies of the kinetic mechanism of maize endosperm ADP-glucose pyrophosphorylase uncovered complex regulatory properties. *Plant Physiol.* **2010**, *152*, 1056–1064. [CrossRef] [PubMed]
12. Boehlein, S.K.; Shaw, J.R.; Georgelis, N.; Hannah, L.C. Enhanced heat stability and kinetic parameters of maize endosperm ADPglucose pyrophosphorylase by alteration of phylogenetically identified amino acids. *Arch. Biochem. Biophys.* **2014**, *543*, 1–9. [CrossRef] [PubMed]
13. Burger, B.T.; Cross, J.M.; Shaw, J.R.; Caren, J.R.; Greene, T.W.; Okita, T.W.; Hannah, L.C. Relative turnover numbers of maize endosperm and potato tuber ADP-glucose pyrophosphorylases in the absence and presence of 3-phosphoglyceric acid. *Planta* **2003**, *217*, 449–456. [CrossRef] [PubMed]
14. Cross, J.M.; Clancy, M.; Shaw, J.R.; Boehlein, S.K.; Greene, T.W.; Schmidt, R.R.; Okita, T.W.; Hannah, L.C. A polymorphic motif in the small subunit of ADP-glucose pyrophosphorylase modulates interactions between the small and large subunits. *Plant J.* **2005**, *41*, 501–511. [CrossRef] [PubMed]

15. Boehlein, S.K.; Shaw, J.R.; Stewart, J.D.; Hannah, L.C. Heat stability and allosteric properties of the maize endosperm ADP-glucose pyrophosphorylase are intimately intertwined. *Plant Physiol.* **2008**, *146*, 289–299. [CrossRef] [PubMed]

16. Sowokinos, J.R. Pyrophosphorylases in Solanum tuberosum: II. Catalytic properties and regulation of ADP-glucose and UDP-glucose pyrophosphorylase activities in potatoes. *Plant Physiol.* **1981**, *68*, 924–929. [CrossRef] [PubMed]

17. Cross, J.M.; Clancy, M.; Shaw, J.R.; Greene, T.W.; Schmidt, R.R.; Okita, T.W.; Hannah, L.C. Both subunits of ADP-glucose pyrophosphorylase are regulatory. *Plant Physiol.* **2004**, *135*, 137–144. [CrossRef] [PubMed]

18. Sowokinos, J.R.; Preiss, J. Pyrophosphorylases in Solanum tuberosum: III. Purification, Physical, and Catalytic Properties of Adpglucose Pyrophosphorylase in Potatoes. *Plant Physiol.* **1982**, *69*, 1459–1466. [CrossRef] [PubMed]

19. Boehlein, S.K.; Shaw, J.R.; McCarty, D.R.; Hwang, S.K.; Stewart, J.D.; Hannah, L.C. The potato tuber, maize endosperm and a chimeric maize-potato ADP-glucose pyrophosphorylase exhibit fundamental differences in Pi inhibition. *Arch. Biochem. Biophys.* **2013**, *537*, 210–216. [CrossRef] [PubMed]

20. Hannah, L.C.; Futch, B.; Bing, J.; Shaw, J.R.; Boehlein, S.; Stewart, J.D.; Beiriger, R.; Georgelis, N.; Greene, T. A shrunken-2 transgene increases maize yield by acting in maternal tissues to increase the frequency of seed development. *Plant Cell* **2012**, *24*, 2352–2363. [CrossRef] [PubMed]

21. Li, N.; Zhang, S.; Zhao, Y.; Li, B.; Zhang, J. Over-expression of AGPase genes enhances seed weight and starch content in transgenic maize. *Planta* **2011**, *233*, 241–250. [CrossRef] [PubMed]

22. Sweetlove, L.J.; Burrell, M.M.; ap Rees, T. Starch metabolism in tubers of transgenic potato (Solanum tuberosum) with increased ADPglucose pyrophosphorylase. *Biochem. J.* **1996**, *320*, 493–498. [CrossRef] [PubMed]

23. Tiessen, A.; Hendriks, J.H.; Stitt, M.; Branscheid, A.; Gibon, Y.; Farre, E.M.; Geigenberger, P. Starch synthesis in potato tubers is regulated by post-translational redox modification of ADP-glucose pyrophosphorylase: A novel regulatory mechanism linking starch synthesis to the sucrose supply. *Plant Cell* **2002**, *14*, 2191–2213. [CrossRef] [PubMed]

24. Hendriks, J.H.; Kolbe, A.; Gibon, Y.; Stitt, M.; Geigenberger, P. ADP-glucose pyrophosphorylase is activated by posttranslational redox-modification in response to light and to sugars in leaves of Arabidopsis and other plant species. *Plant Physiol.* **2003**, *133*, 838–849. [CrossRef] [PubMed]

25. Wu, X.; Gong, F.; Cao, D.; Hu, X.; Wang, W. Advances in crop proteomics: PTMs of proteins under abiotic stress. *Proteomics* **2016**, *16*, 847–865. [CrossRef] [PubMed]

26. Burnell, J.N.; Hatch, M.D. Activation and inactivation of an enzyme catalyzed by a single, bifunctional protein: A new example and why. *Arch. Biochem. Biophys.* **1986**, *245*, 297–304. [CrossRef]

27. Pesaresi, P.; Pribil, M.; Wunder, T.; Leister, D. Dynamics of reversible protein phosphorylation in thylakoids of flowering plants: The roles of STN7, STN8 and TAP38. *Biochim. Biophys. Acta* **2011**, *1807*, 887–896. [CrossRef] [PubMed]

28. Tetlow, I.J.; Wait, R.; Lu, Z.; Akkasaeng, R.; Bowsher, C.G.; Esposito, S.; Kosar-Hashemi, B.; Morell, M.K.; Emes, M.J. Protein phosphorylation in amyloplasts regulates starch branching enzyme activity and protein-protein interactions. *Plant Cell* **2004**, *16*, 694–708. [CrossRef] [PubMed]

29. Tetlow, I.J.; Morell, M.K.; Emes, M.J. Recent developments in understanding the regulation of starch metabolism in higher plants. *J. Exp. Bot.* **2004**, *55*, 2131–2145. [CrossRef] [PubMed]

30. Liu, F.; Makhmoudova, A.; Lee, E.A.; Wait, R.; Emes, M.J.; Tetlow, I.J. The amylose extender mutant of maize conditions novel protein-protein interactions between starch biosynthetic enzymes in amyloplasts. *J. Exp. Bot.* **2009**, *60*, 4423–4440. [CrossRef] [PubMed]

31. Liu, F.; Ahmed, Z.; Lee, E.A.; Donner, E.; Liu, Q.; Ahmed, R.; Morell, M.K.; Emes, M.J.; Tetlow, I.J. Allelic variants of the amylose extender mutation of maize demonstrate phenotypic variation in starch structure resulting from modified protein-protein interactions. *J. Exp. Bot.* **2012**, *63*, 1167–1183. [CrossRef] [PubMed]

32. Makhmoudova, A.; Williams, D.; Brewer, D.; Massey, S.; Patterson, J.; Silva, A.; Vassall, K.A.; Liu, F.; Subedi, S.; Harauz, G.; et al. Identification of multiple phosphorylation sites on maize endosperm starch branching enzyme IIb, a key enzyme in amylopectin biosynthesis. *J. Biol. Chem.* **2014**, *289*, 9233–9246. [CrossRef] [PubMed]

33. Walley, J.W.; Shen, Z.; Sartor, R.; Wu, K.J.; Osborn, J.; Smith, L.G.; Briggs, S.P. Reconstruction of protein networks from an atlas of maize seed proteotypes. *Proc. Natl. Acad. Sci. USA* **2013**, *110*, E4808–E4817. [CrossRef] [PubMed]

34. Huang, B.; Hennen-Bierwagen, T.A.; Myers, A.M. Functions of multiple genes encoding ADP-glucose pyrophosphorylase subunits in maize endosperm, embryo, and leaf. *Plant Physiol.* **2014**, *164*, 596–611. [CrossRef] [PubMed]

35. Goodman, T.; Schulenberg, B.; Steinberg, T.H.; Patton, W.F. Detection of phosphoproteins on electroblot membranes using a small-molecule organic fluorophore. *Electrophoresis* **2004**, *25*, 2533–2538. [CrossRef] [PubMed]

36. Schulenberg, B.; Goodman, T.N.; Aggeler, R.; Capaldi, R.A.; Patton, W.F. Characterization of dynamic and steady-state protein phosphorylation using a fluorescent phosphoprotein gel stain and mass spectrometry. *Electrophoresis* **2004**, *25*, 2526–2532. [CrossRef] [PubMed]

37. Dinges, J.R.; Colleoni, C.; Myers, A.M.; James, M.G. Molecular structure of three mutations at the maize sugary1 locus and their allele-specific phenotypic effects. *Plant Physiol.* **2001**, *125*, 1406–1418. [CrossRef] [PubMed]

38. Wakuta, S.; Shibata, Y.; Yoshizaki, Y.; Saburi, W.; Hamada, S.; Ito, H.; Hwang, S.K.; Okita, T.W.; Matsui, H. Modulation of allosteric regulation by E38K and G101N mutations in the potato tuber ADP-glucose pyrophosphorylase. *Biosci. Biotechnol. Biochem.* **2013**, *77*, 1854–1859. [CrossRef] [PubMed]

39. Kinoshita, E.; Kinoshita-Kikuta, E.; Kubota, Y.; Takekawa, M.; Koike, T. A Phos-tag SDS-PAGE method that effectively uses phosphoproteomic data for profiling the phosphorylation dynamics of MEK1. *Proteomics* **2016**, *16*, 1825–1836. [CrossRef] [PubMed]

40. Kinoshita, E.; Kinoshita-Kikuta, E.; Koike, T. Phosphate-affinity polyacrylamide gel electrophoresis for SNP genotyping. *Meth. Mol. Biol.* **2009**, *578*, 183–192.

41. Kinoshita-Kikuta, E.; Kinoshita, E.; Yamada, A.; Endo, M.; Koike, T. Enrichment of phosphorylated proteins from cell lysate using a novel phosphate-affinity chromatography at physiological pH. *Proteomics* **2006**, *6*, 5088–5095. [CrossRef] [PubMed]

42. Pesaresi, P.; Hertle, A.; Pribil, M.; Schneider, A.; Kleine, T.; Leister, D. Optimizing photosynthesis under fluctuating light: The role of the Arabidopsis STN7 kinase. *Plant Signal. Behav.* **2010**, *5*, 21–25. [CrossRef] [PubMed]

43. Pribil, M.; Pesaresi, P.; Hertle, A.; Barbato, R.; Leister, D. Role of plastid protein phosphatase TAP38 in LHCII dephosphorylation and thylakoid electron flow. *PLoS Biol.* **2010**, *8*, e1000288. [CrossRef] [PubMed]

44. Nakagami, H.; Sugiyama, N.; Mochida, K.; Daudi, A.; Yoshida, Y.; Toyoda, T.; Tomita, M.; Ishihama, Y.; Shirasu, K. Large-scale comparative phosphoproteomics identifies conserved phosphorylation sites in plants. *Plant Physiol.* **2010**, *153*, 1161–1174. [CrossRef] [PubMed]

45. Batra, R.; Saripalli, G.; Mohan, A.; Gupta, S.; Gill, K.S.; Varadwaj, P.K.; Balyan, H.S.; Gupta, P.K. Comparative Analysis of AGPase Genes and Encoded Proteins in Eight Monocots and Three Dicots with Emphasis on Wheat. *Front. Plant Sci.* **2017**, *8*, 19. [CrossRef] [PubMed]

46. Jin, X.; Ballicora, M.A.; Preiss, J.; Geiger, J.H. Crystal structure of potato tuber ADP-glucose pyrophosphorylase. *EMBO J.* **2005**, *24*, 694–704. [CrossRef] [PubMed]

47. Comino, N.; Cifuente, J.O.; Marina, A.; Orrantia, A.; Eguskiza, A.; Guerin, M.E. Mechanistic insights into the allosteric regulation of bacterial ADP-glucose pyrophosphorylases. *J. Biol. Chem.* **2017**, *292*, 6255–6268. [CrossRef] [PubMed]

48. Ma, Q.; Wu, M.; Pei, W.; Li, H.; Li, X.; Zhang, J.; Yu, J.; Yu, S. Quantitative phosphoproteomic profiling of fiber differentiation and initiation in a fiberless mutant of cotton. *BMC Genom.* **2014**, *15*, 466. [CrossRef] [PubMed]

International Journal of
Molecular Sciences

Article

Dynamic TMT-Based Quantitative Proteomics Analysis of Critical Initiation Process of Totipotency during Cotton Somatic Embryogenesis Transdifferentiation

Haixia Guo [†], Huihui Guo [†], Li Zhang, Yijie Fan, Yupeng Fan, Zhengmin Tang and Fanchang Zeng *

State Key Laboratory of Crop Biology, College of Agronomy, Shandong Agricultural University, Tai'an 271018, China; diya_haixiaguo@163.com (H.G.); hhguo@sdau.edu.cn (H.G.); 15610418001@163.com (L.Z.); yjfan@sdau.edu.cn (Y.F.); fanyupeng@chnu.edu.cn (Y.F.); 17861500710@163.com (Z.T.)
* Correspondence: fczeng@sdau.edu.cn; Tel.: +86-538-824-1828
† These authors contributed equally to this work.

Received: 21 February 2019; Accepted: 2 April 2019; Published: 4 April 2019

Abstract: The somatic embryogenesis (SE) process of plants, as one of the typical responses to abiotic stresses with hormone, occurs through the dynamic expression of different proteins that constitute a complex regulatory network in biological activities and promotes plant totipotency. Plant SE includes two critical stages: primary embryogenic calli redifferentiation and somatic embryos development initiation, which leads to totipotency. The isobaric labels tandem mass tags (TMT) large-scale and quantitative proteomics technique was used to identify the dynamic protein expression changes in nonembryogenic calli (NEC), primary embryogenic calli (PEC) and globular embryos (GEs) of cotton. A total of 9369 proteins (6730 quantified) were identified; 805, 295 and 1242 differentially accumulated proteins (DAPs) were identified in PEC versus NEC, GEs versus PEC and GEs versus NEC, respectively. Eight hundred and five differentially abundant proteins were identified, 309 of which were upregulated and 496 down regulated in PEC compared with NEC. Of the 295 DAPs identified between GEs and PEC, 174 and 121 proteins were up- and down regulated, respectively. Of 1242 differentially abundant proteins, 584 and 658 proteins were up- and down regulated, respectively, in GEs versus NEC. We have also complemented the authenticity and accuracy of the proteomic analysis. Systematic analysis indicated that peroxidase, photosynthesis, environment stresses response processes, nitrogen metabolism, phytohormone response/signal transduction, transcription/posttranscription and modification were involved in somatic embryogenesis. The results generated in this study demonstrate a proteomic molecular basis and provide a valuable foundation for further investigation of the roles of DAPs in the process of SE transdifferentiation during cotton totipotency.

Keywords: cotton; somatic embryogenesis; transdifferentiation; quantitative proteomics; regulation and metabolism; molecular basis; concerted network

---

1. Introduction

Somatic embryogenesis (SE) is a notable illustration of cell totipotency as one of the typical responses to abiotic stresses with hormone, which processes the developmental reprogramming of somatic cells toward the embryogenesis pathway. Cotton (*Gossypium hirsutum* L.), as the foremost natural fiber source [1] and one of the most important economic crops worldwide, has a global socioeconomic impact worth approximately $56 billion [2]. However, plant regeneration in SE is still a limiting method for transgenic development in cotton [1,3]. Somatic embryogenesis represents a unique phenomenon in the plant kingdom [4]. This developmental pathway is one of the most striking examples of plant cell developmental plasticity [5,6]. It includes a series of characteristic events, including somatic dedifferentiation, cell division activation, metabolism alterations and gene expression pattern reprogramming [4]. During SE, the development of somatic cells is reprogrammed to the embryogenic pathway, and SE forms the basis of cellular totipotency in higher plants [7]. Each transformed cell has the potential to produce a plant from the callus [8]. Somatic embryogenesis and subsequent plant regeneration have been reported in most major crop varieties [9]. Soybeans and cotton have proven to be the most difficult to regenerate [10].

In cotton, only a few percent of somatic embryos are able to mature and regenerate into plantlets. Most embryos develop abnormally, redifferentiate in to calli, or become necrotic and die [11]. Sakhanokho and Rajasekaran [12] obtained a variety of factors affecting cotton during in vitro regeneration, including plant growth regulators, explants, compositions of media and environmental conditions. The transition of somatic cells into embryogenic cells is the most intriguing and the least understood part of somatic embryogenesis [13–15]. Now, it is generally accepted that stress and hormones play a crucial role in collectively inducing cell dedifferentiation and initiation of the embryogenic program in plants with responsive genotypes [16–18].

Since the first observations of somatic embryo formation in suspension cultures of carrot cells by Stewards [7] and Reinert [19], the potential for SE has been demonstrated to be characteristic of extensive tissue culture systems from both dicotyledonous and monocotyledonous plants [20,21]. Considerable efforts have been expended in identifying the various factors that control SE [22,23]. An important gene that marks embryonic cells is the transcription factor gene *WUSCHEL* (*WUS*) [24]. Using a genetic gain-of-function screening approach, Zuo et al. [25] found that overexpression of *WUS* in roots, leaf petioles, stems, or leaves of *Arabidopsis* can induce the formation of somatic embryos. These results indicate that *WUS* participates in the promotion and/or maintenance of totipotent embryogenic stem cells. However, the *wus* mutants are still able to produce somatic embryos, suggesting that multiple alternative pathways can lead to the expression of totipotent potential. *WUS* is the only transcription factor that has been found to be involved in regulating meristematic stem cells (pluripotent) and embryogenic stem cells (totipotent) [15]. In addition, when auxin biosynthesis rates were manipulated in *Arabidopsis* embryos, polar auxin transport activity apparently buffered the normal distribution of auxin, suggesting a compensatory mechanism for buffering auxin gradients in the embryo, with *PIN1* and *PIN4* being the most important genes [26]. The results by Su [27] suggested that the establishment of auxin gradients and the polar distribution of *PIN1* are critical for the regulation of *WUS* expression during somatic embryogenesis. *ERF* plays an important role in hormone signal transduction and interconnecting different hormone pathways [28]. Inhibition of gibberellin (GA) biosynthesis increases the fraction of *lec1-tnp* seedlings displaying the mutant phenotype, suggesting that reduced GA levels enhance maturation processes induced by *LEAFY COTYLEDON 1* (*LEC1*) [29]. The plant hormone abscisic acid (ABA) regulates many important plant developmental processes and induces epigenetic reprogramming against tolerance to different stresses, including drought, salinity, low temperature and some pathogens [30,31]. ABA serves as a critical chemical messenger for stress responses. The roles of several genes in somatic embryogenesis masses (SEM) have been well-characterized, including *Arabinogalactan protein 1* (*AGP1*) [32], *Glutathione-S-transferase* (*GST*) [33], *SOMATIC EMBRYO RELATED FACTOR1* (*MtSERF1*) [34], *BABY BOOM* (*BBM*) [35], *Agamous-like 15* (*AGL15*) [36,37] and *SOMATIC EMBRYOGENESIS RECEPTOR-LIKE KINASE* (*SERK*) from *Daucus carota*, which was the first identified marker gene with a crucial role in SEM [38].

At present, a great number of SE-related genes and transcription factors have been identified at the transcription level. For example, Zeng [39], with the suppression subtractive hybridization (SSH) technique, identified 671 cDNAs in the initial period of SE in cotton. Nonetheless, reports on the identification of cotton SE at high-throughput proteins levels are still insufficient, especially during the initial stage of SE transdifferentiation. Proteomics is a powerful approach aimed at systematic studies of protein structure, function, interaction, and dynamics [4]. To further investigate the molecular regulatory mechanisms of somatic embryogenesis, the protein dynamics of NEC, PEC and GE were identified by TMT quantitative proteomics techniques. The highly sensitive proteomic platform based on the isobaric labels tandem mass tags was recently developed as one of the most robust proteomics techniques [40,41]. Through identification and annotation of DAPs, we uncovered the key genes/proteins and pathways involved in cotton SE transdifferentiation. The results generated in this study provide a valuable foundation for further investigation of the roles of DAPs in cotton SE.

2. Results

2.1. Somatic Embryogenesis in Cotton

PEC were formed from NEC after approximately 3 months of culture. The initial development period of GEs from embryogenic callus was the most restrictive step during cotton SEM (Figure 1). To identify proteins related to SE and morphogenesis in cotton, we sampled the critical representative periods of NEC, PEC and GEs for protein preparation and TMT-based quantitative proteomics analyses.

(a)	(b)	(c)

Figure 1. Samples used for proteomic assays: (**a**) Nonembryogenic calli; (**b**) Primary embryogenic calli; (**c**) Globular embryos. Bar (**a**,**b**) = 2.5 mm; bar (**c**) = 0.5 mm.

2.2. TMT-Based Quantitative Proteomic Basis Data Analysis and Overall Protein Identification

TMT-based quantitative proteomics was conducted to assess protein changes among NEC, PEC and GEs in cotton. Pair wise Pearson's correlation coefficients displayed sufficient reproducibility of this experiment (Figure 2a). After quality validation, a total of 360,720 (74,579 matched) spectra were obtained. Of these spectra, 45062 identified peptides (27,673 unique peptides) and 9369 identified proteins (6730 quantified proteins) were detected (Table 1), and the average peptides mass error was <10 ppm, indicating a high mass accuracy of the MS data (Figure 2b). The lengths of most identified peptides were 8 to 20 amino acid residues (Figure 2c), suggesting that our sampling met the required standard. The detail information of identified proteins, including protein accession, protein description, gene name, peptide number, matching scores, carried charges and delta mass, is shown in Supplementary Table S1.

To further understand their functions, all identified proteins were annotated according to different categories, including subcellular localizations, Gene Ontology (GO) terms, Kyoto Encyclopedia of Genes and Genomes (KEGG) pathways, predicted functional domains and other data. The detailed information of all identified proteins is listed in Supplementary Table S1.

Pearson's correlation of quantitation									
	NEC1	**NEC2**	**NEC3**	**PEC1**	**PEC2**	**PEC3**	**GE1**	**GE2**	**GE3**
NEC1	1	0.94	0.94	-0.37	-0.31	-0.41	-0.77	-0.76	-0.75
NEC2	0.94	1	0.94	-0.37	-0.31	-0.42	-0.77	-0.76	-0.75
NEC3	0.94	0.94	1	-0.36	-0.3	-0.4	-0.78	-0.77	-0.76
PEC1	-0.37	-0.37	-0.36	1	0.85	0.88	0.13	0.13	0.11
PEC2	-0.31	-0.31	-0.3	0.85	1	0.86	0.05	0.05	0.03
PEC3	-0.41	-0.42	-0.4	0.88	0.86	1	0.18	0.18	0.15
GE1	-0.77	-0.77	-0.78	0.13	0.05	0.18	1	0.93	0.92
GE2	-0.76	-0.76	-0.77	0.13	0.05	0.18	0.93	1	0.92
GE3	-0.75	-0.75	-0.76	0.11	0.03	0.15	0.92	0.92	1

(**a**)

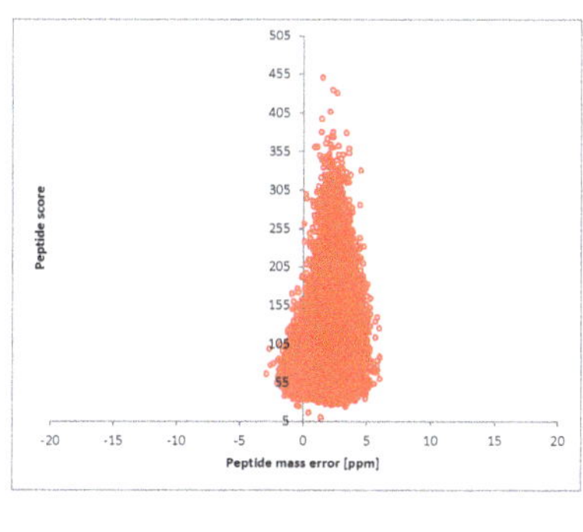

(**b**)

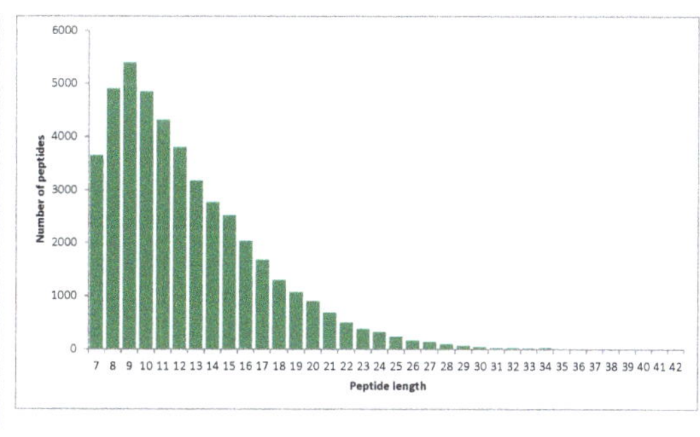

(**c**)

Figure 2. Experimental strategy for quantitative proteome analysis and quality control validation of MS data: (**a**) Mass delta of all identified peptides; (**b**) Average peptide mass error; (**c**) Length distribution of all identified peptides. NEC: Nonembryogenic calli; PEC: Primary embryogenic calli; GE: Globular embryos. Each staged sample was prepared for three biological replicates.

Table 1. MS/MS spectrum database search analysis summary.

Total Spectrum	Matched Spectrum	Peptides	Unique Peptides	Identified Proteins	Quantifiable Proteins
360,720	74,579 (20.7%)	45,062	27,673	9369	6730

2.3. Enrichment of the Chloroplast Subcellular Location and GO Functional Classification of All Identified Proteins

To characterize the subcellular locations and functions of the identified differential proteins among NEC, PEC and GEs in cotton, subcellular locations and GO functional classification were performed (Figure 3; Supplementary Table S1). Subcellular distribution predictions (Figure 3a) showed that the identified proteins were distributed predominantly in chloroplast (31.10%), cytoplasm (25.75%) and nucleus (23.07%) during the transformation periods of somatic embryogenesis. Significantly, the highest proportion of differential proteins was enriched in chloroplasts, highlighting that this organelle plays an important role in cotton SE.

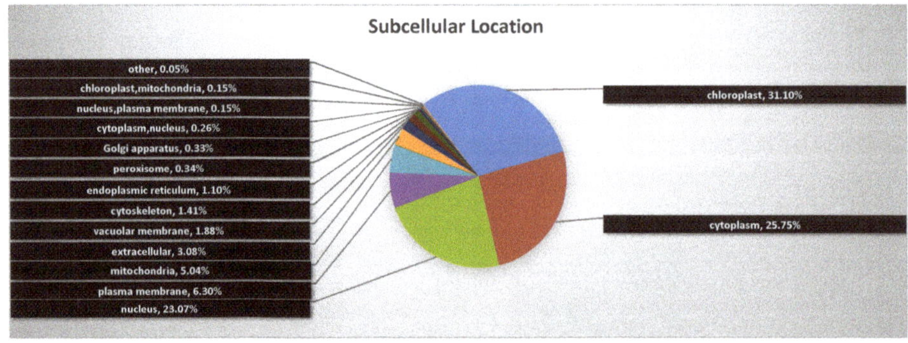

(**a**)

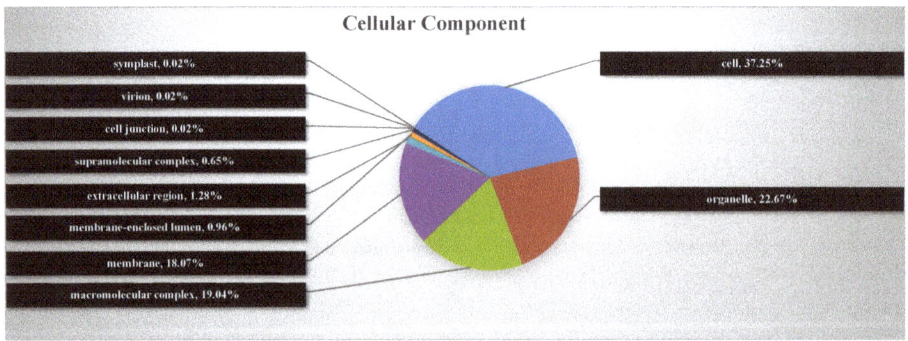

(**b**)

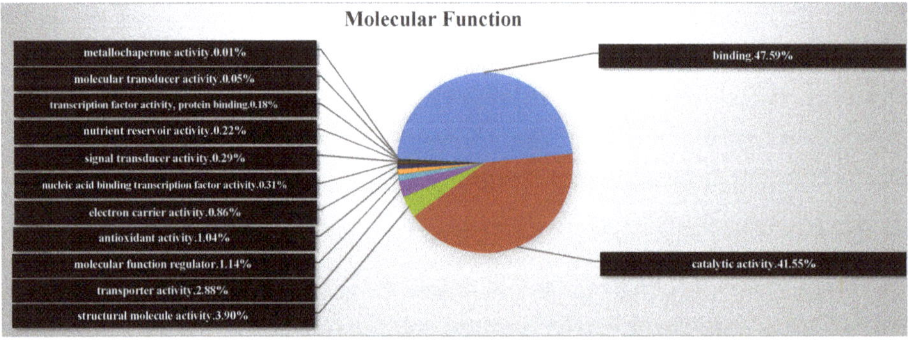

(**c**)

Figure 3. *Cont.*

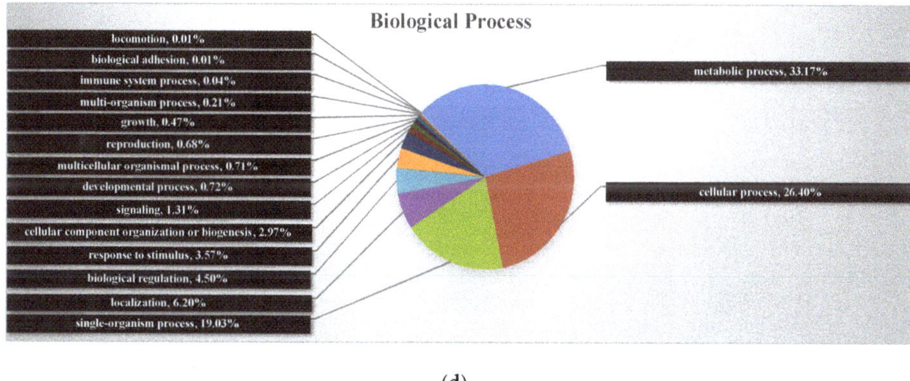

(d)

Figure 3. Subcellular functional annotation and GO functional classification of identified proteins. (a) Subcellular locations of identified proteins; (b) GO annotation in terms of cellular component; (c) GO annotation in terms of molecular function; (d) GO annotation in terms of biological process. GO: Gene Ontology.

The results of cellular component analysis further revealed that 22.67% of the identified proteins were catalogued in organelles, 19.04% with macromolecular complexes and 18.07% with the membrane (Figure 3b). Regarding molecular function, the largest two GO categories, binding and catalytic activity, accounted for 47.59 and 41.55% of the identified proteins, respectively (Figure 3c). At the biological process level, proteins involved in the metabolic process, cellular process and single-organism process accounted for 33.17, 26.40 and 19.03% of identified proteins, respectively (Figure 3d). These results demonstrated that the identified proteins are found in multiple cellular components, have diversified molecular functions, and are involved in a variety of biological processes.

2.4. Identification of Differentially Abundant Proteins

Differentially abundant proteins were defined as those with a $\geq$2-fold or $\leq$0.5-fold change in relative abundance ($p < 0.05$) between PEC and NEC, GEs and PEC, and GEs and NEC. In total, 805, 295 and 1242 DAPs were identified in comparing PEC versus NEC, GEs versus PEC and GEs versus NEC, respectively. In PEC compared with NEC, 805 proteins differentially accumulated were identified, 309 of which were up regulated and 496 of which were down-regulated. Of the 295 DAPs identified between GEs and PEC, 174 and 121 proteins were up- and down regulated in GEs, respectively. Of 1242 proteins differentially accumulated in GEs compared to NEC, 584 and 658 proteins were up- and down regulated, respectively (Figure 4a; Supplementary Table S2).

To identify the commonly and specifically changed proteins between PEC and NEC, GEs and PEC or between GEs and NEC, a Venn diagram was generated (Figure 4b). It clearly showed that 122 and 29 proteins were specifically expressed in PEC and GE processes, respectively, and 85 common proteins (25 and 60) were involved in both PEC and GE.

To investigate the overall dynamics of proteome changes in SEM, we performed eight types of protein expression pattern analyses for the DAPs identified in NEC, PEC and GEs (Figure 4c). These analyses suggested protein expression patterns including down- to up regulation, up- to down regulation, down- to down regulation, up- to up regulation, down- to constant-regulation, up- to constant-regulation, constant- to down regulation and constant- to up regulation.

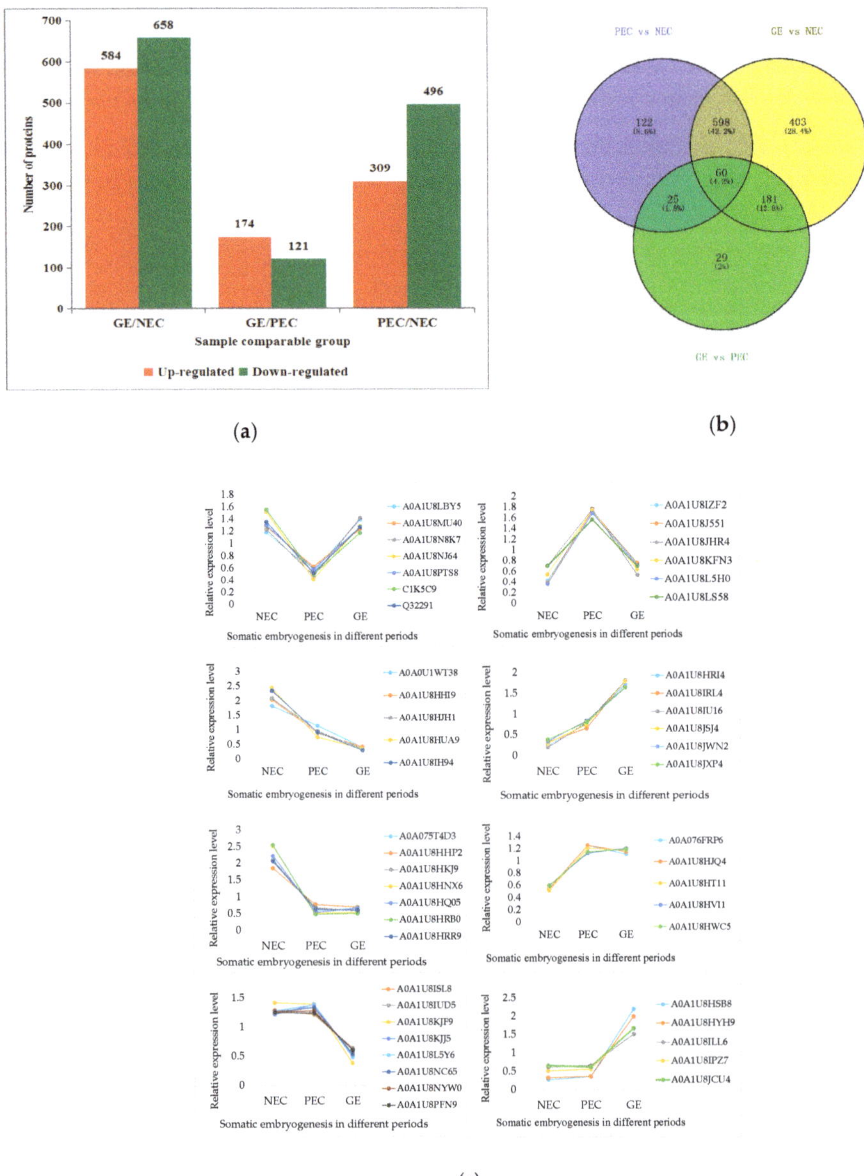

(a)

(b)

(c)

Figure 4. Distribution of differentially accumulated proteins (DAPs): (**a**) Number of up(red)- and down(green)-regulated DAPs in GE vs. NEC, GE vs. PEC and PEC vs. NEC; (**b**) Venn diagram to show the distribution of DAPs between PEC vs. NEC (blue circle), GE vs. PEC (yellow circle) and GE vs. NEC (green circle). (**c**) Expression patterns of DAPs. NEC: Nonembryogenic calli; PEC: Primary embryogenic calli; GE: Globular embryos.

2.5. Enrichment Analysis of DAPs in GO, KEGG and Protein Domain

In total, 1418 proteins (nonrepetitive DAPs) were differentially accumulated and significantly regulated by the NEC, PEC and GEs under the given culture conditions (Supplementary Table S2/Total DAPs). The biological functions of the DAPs could also be identified by their GO terms, KEGG pathways and protein domain enrichment, as summarized in Figures 5–7.

2.5.1. Enrichment Cluster Analysis of DAPs between the Groups in GO Terms

In the different GO functional classifications, we carried out comparative cluster analysis between the sample groups, indicating the change of the co-expression trends of different proteins between the groups.

The Enzyme Metabolism Activity of Molecular Function Category in Cotton SE

For up-regulated DAPs, 'protein dimerization activity' and 'protein heterodimerization activity' showed a certain degree of enrichment in PEC versus NEC, GE versus PEC and GE versus NEC, especially in GE versus NEC. 'DNA helicase activity' and 'helicase activity' were functional categories which lower in GE versus NEC compared to PEC versus NEC. The up regulation of 'peroxiredoxin activity' and 'nutrient reservoir activity' in GE versus PEC was greater than in GE versus NEC, indicating that the enzyme and nutritional protein activities were higher in GE (Figure 5a). Most of the differential proteins are clearly clustered among the down regulated proteins. Four DAPs of peptidase-related protein activity, proteins from 'hydrolase activity' to 'hydrolase activity, hydrolyzing O-glycosyl compounds' and other enzyme activity were significantly down regulated in PEC versus NEC, but the enrichment of these DAPs was not significant in GE versus PEC; there were various degree of enrichment in GE versus NEC. Additionally, 'glutamate dehydrogenase (NAD+) activity', two proteins of 'oxidoreductase activity' and 'phosphoenolpyruvate carboxykinase (ATP) activity' accumulated to a certain extent in GE versus PEC (Figure 5a).

The results above indicated that enzyme metabolism activity affected the SE of cotton, with dynamic features in NEC, PEC, and GE.

The Photosynthesis-Related Proteins of the Cellular Component Category in Cotton SE

In the cellular component category of PEC versus NEC and GE versus PEC, a large number of DAPs were clustered in photosynthesis-related cellular components and proteins from 'plastid thylakoid' to 'photosystem I', indicating a significant decrease of photosynthesis in PEC versus NEC. Furthermore, in GE versus PEC, the corresponding photosynthetic cell components showed slightly up regulated enrichment. However, the photosynthetic effect of GE was not higher than NEC; this result showed that the photosynthesis-related DAPs of the cell component classification in GE versus NEC were concentrated in the down regulated expression region (Figure 5b). Additionally, the DAPs' of expression pattern from 'photosynthetic membrane' to 'photosystem II', photosynthesis-related proteins' was similar to the above results in that there was significant down regulation in PEC versus NEC (Figure 5b).

The analysis above showed that photosynthesis is a critical process involved in SE of cotton, which is consistent with our subcellular localization results.

The Regulation, Response and Metabolism-related Proteins of the Biological Process Category in Cotton SE

Proteins related to 'lipid transport', 'reproductive system development', 'DNA metabolic process' and 'regulation' were up regulated with different degrees of enrichment in PEC versus NEC and GE versus NEC. In addition, we also found that from 'amide biosynthetic process' to 'cellular protein metabolic process' proteins in GE versus NEC were uniformly enriched in up regulation (Figure 5c). Furthermore, 'monosaccharide metabolic process', 'hexose metabolic process', glycometabolism related proteins and the DAPs from 'aminoglycan catabolic process' to 'cell wall macromolecule metabolic process' were down regulated to different degrees in the three sample groups. Other proteins involved in response regulation were also enriched in the down regulated region in PEC versus NEC and GE versus NEC (Figure 5c).

Interestingly and consistently, the photosynthesis-related proteins in 'photosynthesis, light harvesting' were down regulated in PEC versus NEC and up regulated in GE versus PEC (Figure 5c). The above results demonstrated that SE of cotton might frequently involve proteins associated with environmental stress response, biological regulation, central metabolic processes, and photosynthetic metabolism.

(a)

Figure 5. *Cont.*

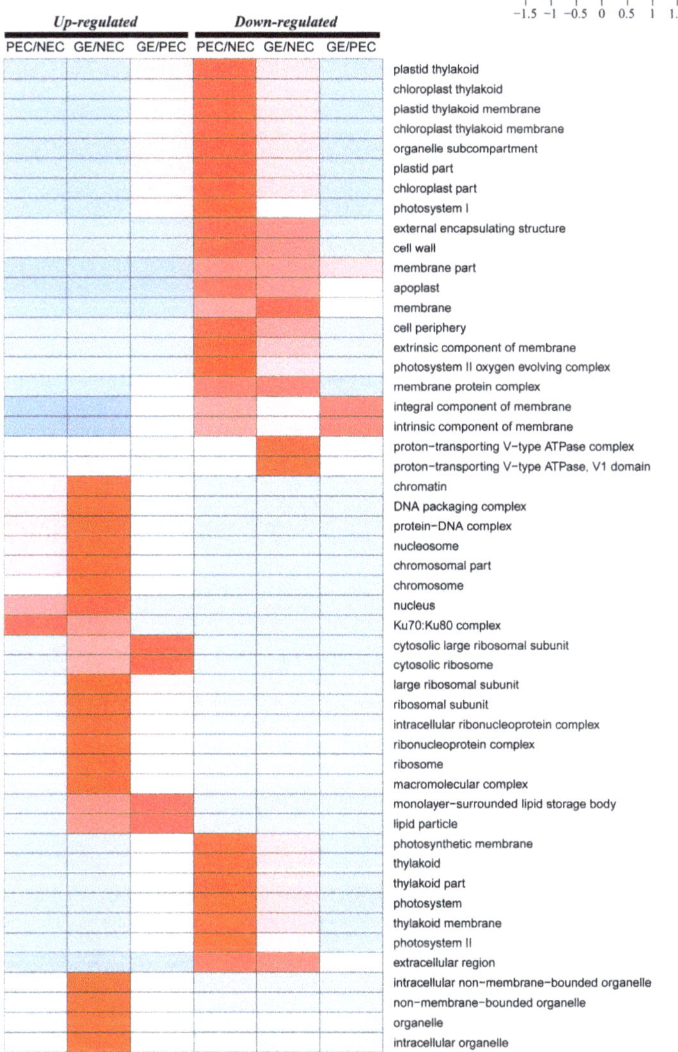

(**b**)

Figure 5. *Cont.*

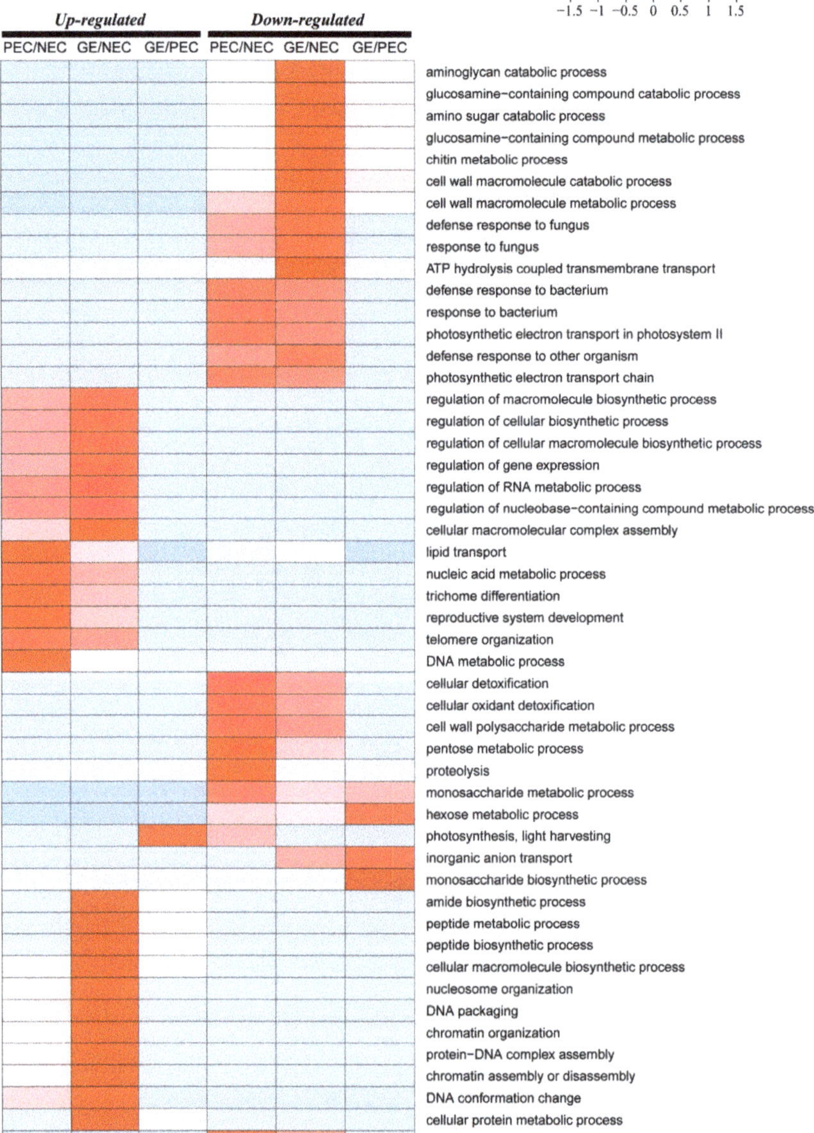

(c)

Figure 5. GO functional cluster of differentially accumulated proteins (DAPs) in PEC vs. NEC, GE vs. PEC and GE vs. NEC: (**a**) GO functional cluster of DAPs in the molecular function; (**b**) GO cluster of DAPs in the cellular component; (**c**) GO functional cluster of DAPs in the biological process. NEC: Nonembryogenic calli; PEC: Primary embryogenic calli; GE: Globular embryos; GO: Gene Ontology.

2.5.2. Enrichment Analysis in KEGG of the DAPs Involved in Phenylpropanoid Biosynthesis, Nitrogen Metabolism, Photosynthesis and Other Related Biological Processes

Enrichment Analysis in KEGG Clusters of Related Biological Processes among Groups

To further understand the function of SE-related proteins, we analyzed the differences and dynamic changes among groups of rich clustering classes in KEGG pathways. Cluster analysis of up regulated expression pathways showed that protein from 'ribosome biogenesis in eukaryotes' and 'DNA replication' were slightly enriched to varying degrees in PEC versus NEC and GE versus NEC (Figure 6a). In the down regulated enrichment region, multiple types of proteins were enriched in different levels in different sample groups, of which 'phenylpropanoid biosynthesis' and 'nitrogen metabolism' were enriched significantly in PEC versus NEC and GE versus PEC, respectively (Figure 6a). The categories 'photosynthesis—antenna proteins', 'glycolysis/gluconeogenesis' and 'carbon fixation in photosynthetic organisms' were down regulated in PEC versus NEC and up regulated in GE versus PEC (Figure 6a).

Phenylpropanoid biosynthesis and nitrogen metabolism were significantly enriched and photosynthesis was re-enriched. The study suggesting the above biological processes possible involvement in cotton SE transformation.

KEGG Pathway Enrichment Analysis of Related Biological Processes within the Sample Groups

In PEC versus NEC, KEGG pathway enrichment analysis demonstrated that the 'phenylpropanoid biosynthesis', 'photosynthesis', 'glutathione metabolism' and 'glycolysis/gluconeogenesis' were the most significantly affected pathways. A great deal of peroxidase proteins of the phenylpropanoid biosynthesis pathway were down regulated in PEC and NEC. Furthermore, a variety of proteins related to 'photosynthesis' and a slight number of DAPs of the 'glycolysis/gluconeogenesis' pathway were also down-regulated in PEC and NEC. Additionally, the 'glutathione metabolism' pathway was down regulated in PEC and NEC (Figure 6b).

KEGG analysis of GE versus PEC indicated that the enriched pathways of the DAPs were most remarkably associated with 'glutathione metabolism' and 'nitrogen metabolism' pathways, which were both down regulated. Once again, 'photosynthesis-antenna proteins' and 'photosynthesis' were significantly enriched up regulated pathways (Figure 6c).

In addition, DAPs of GE versus NEC were classified into various KEGG pathways, of which 6 metabolic pathways were significantly enriched. Interestingly, the largest number of DAPs was once more enriched in 'phenylpropanoid biosynthesis' and 'photosynthesis' were down-regulated. Moreover, 'ribosome' was up regulated pathways that were also enriched (Figure 6d).

The results of the KEGG pathway analysis further indicated that the three metabolic pathways, including phenylpropanoid biosynthesis, nitrogen metabolism and photosynthesis, are essential for the process of cotton SE transformation.

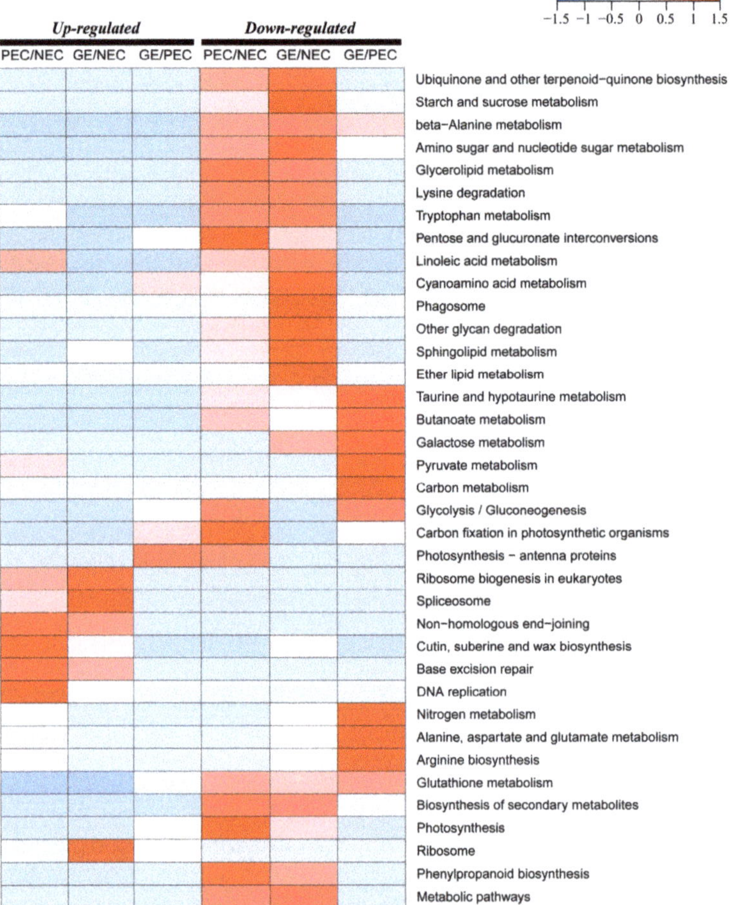

(**a**)

Figure 6. *Cont.*

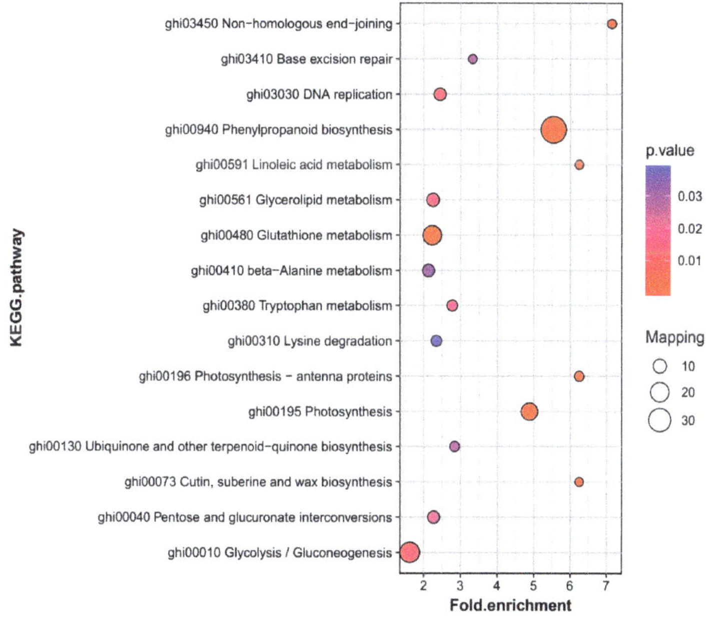

(**b**)

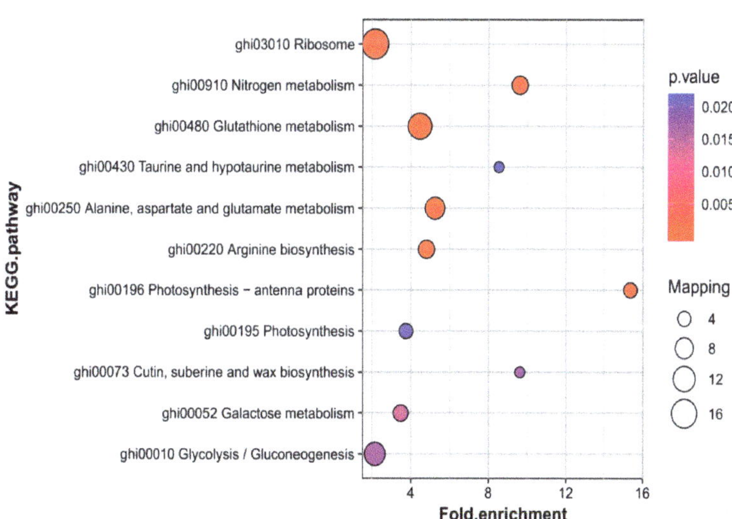

(**c**)

Figure 6. *Cont.*

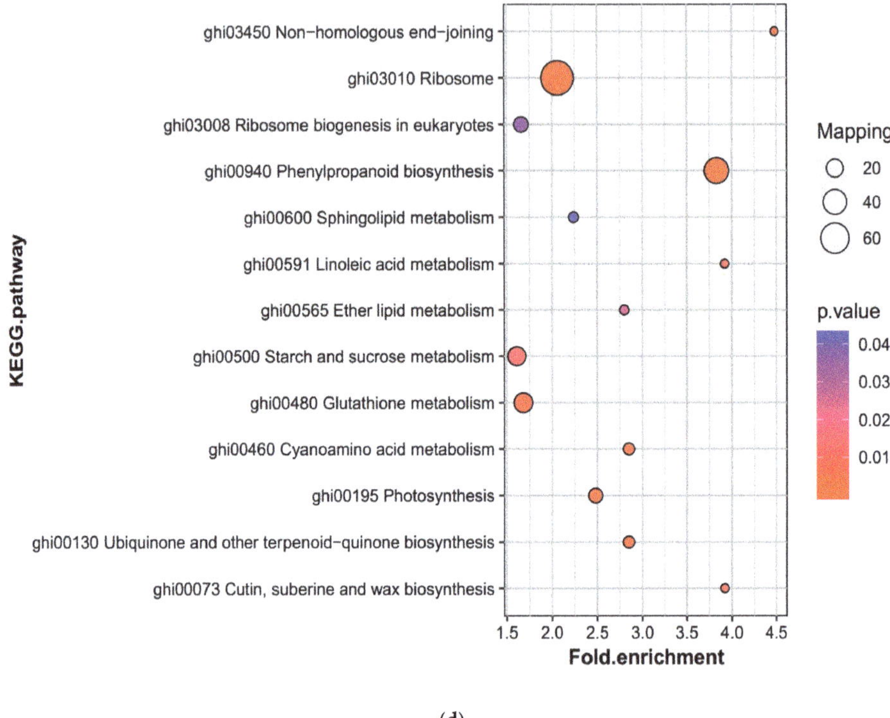

(**d**)

Figure 6. KEGG cluster and pathway enrichment analysis of DAPs: (**a**) KEGG clusters in PEC vs. NEC, GE vs. PEC and GE vs. NEC; (**b**) Pathway enrichment in PEC vs. NEC; (**c**) Pathway enrichment in GE vs. PEC; (**d**) Pathway enrichment in GE vs. NEC. The pathway enrichment statistical analysis was performed by Fisher's exact test. The X-axis is folded enrichment; the y-axis is enrichment pathway. The mapping is the protein number. NEC: Nonembryogenic calli; PEC: Primary embryogenic calli; GE: Globular embryos; KEGG: Kyoto encyclopedia of genes and genomes

2.5.3. Enrichment Cluster Analysis of Differential Proteins Functional Domain

Domain enrichment analysis of up regulated proteins revealed that 'histone'-related domain, 'ribosomal'-related domain, 'seed maturation protein', 'translation protein SH3−like domain', 'RmlC−like jelly roll fold', 'Cupin 1' and 'RmlC−like cupin domain' were enriched in the three sample groups with different dynamic expression patterns. The degree of enrichment of 'rmlC-like jelly roll fold' and 'cupin 1', seed storage protein-related domain, was extremely high in GE versus PEC (Figure 7). This result indicates that the development of GE requires storage proteins to provide nutrients for the regeneration of somatic embryos. For the down regulated expression region, domains including the 'aspartic peptidase domain' and the 'START−like domain' were abundant in three or two samples groups, including glutathione S−transferase domain. Furthermore, domains related to 'haem peroxidase, plant/fungal/bacterial', 'secretory peroxidase', 'aquaporin−like' and 'glycoside hydrolase' were equally enriched in PEC versus NEC and GE versus NEC (Figure 7).

The cluster analysis of dynamic enrichment changes through different functional domains showed that the 'rmlC−like cupin domain', seed maturation protein, glutathione S−transferase and peroxidase-related domain were driving diverse tasks in different development processes of cotton SE.

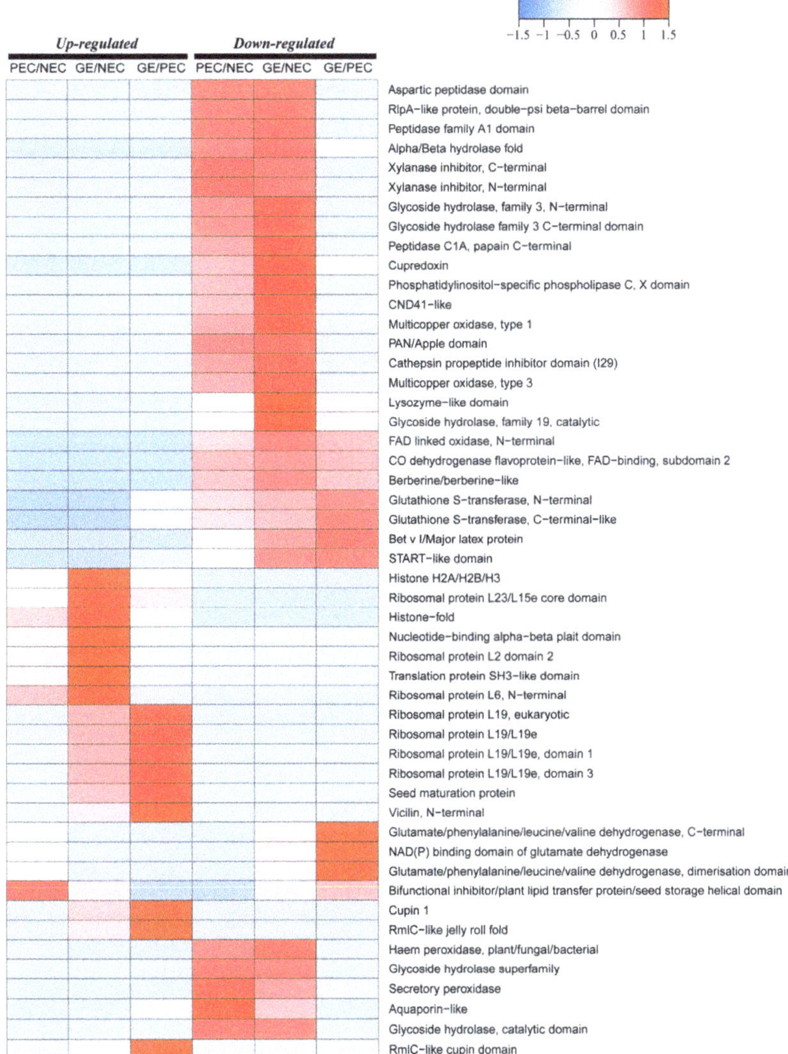

Figure 7. Protein domain enrichment analysis of the differentially accumulated proteins (DAPs) in PEC vs. NEC, GE vs. PEC and GE vs. NEC. NEC: Nonembryogenic calli; PEC: Primary embryogenic calli; GE: Globular embryos.

2.6. Enrichment Analysis of the Major Biological Process between Different Comparison Groups

Above all, the significantly enriched GO terms of the biological process in different comparison groups were investigated. In PEC and NEC, the top 5 GO terms were peroxidase-related, further demonstrated the above results that significant enrichment of the phenylpropanoid biosynthesis pathway was observed during the initiation process of SE. In addition, 'photosynthesis' term showed significant changes in the PEC differentiation, promoting cotton SE. In GE and PEC, environmental response and photosynthesis related proteins were presented in the top 12 GO terms, indicating that the abundant abiotic stress and photosynthesis responsive proteins might regulate the maturity and development of globular embryos. In GE and NEC, the top 8 GO terms were peroxidase, photosynthesis and environmental response related proteins, being part of an important biological process in the process of cotton somatic embryo transformation. What's more, glutamate dehydrogenase (GDH) is significantly enriched in the molecular function classification of PEC and GE. GDH is one of the main enzymes of nitrogen metabolism and participates in important biological processes of plant SE. These important biological processes throughout the development process suggested that complex regulatory networks are involved in the cotton SE process (Figure 8).

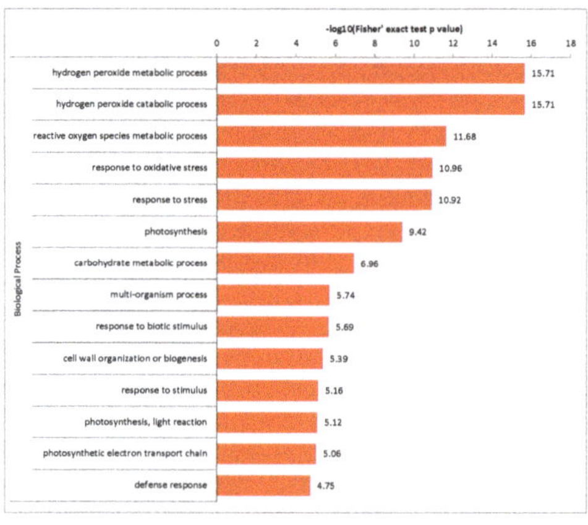

(**a**)

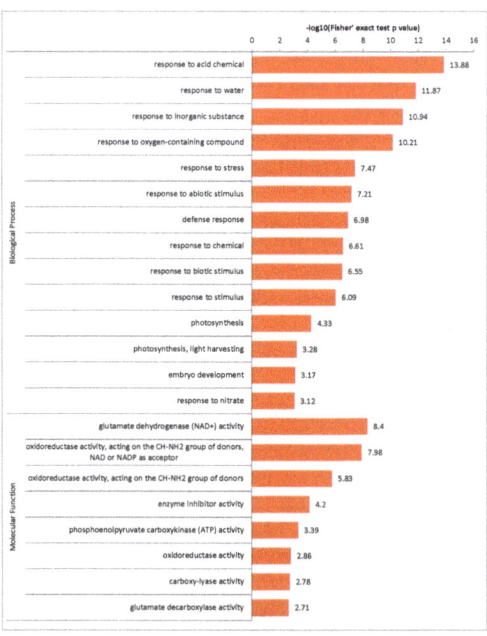

(**b**)

Figure 8. *Cont.*

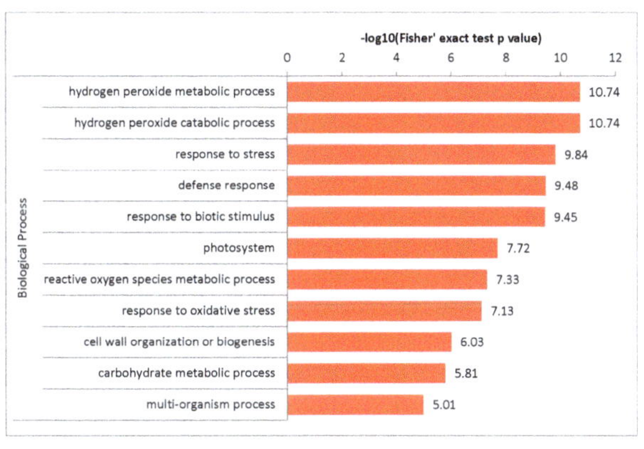

(**c**)

Figure 8. GO terms of differentially accumulated proteins (DAPs) in the biological process and the molecular function: (**a**) The biological process of DAPs in PEC vs. NEC; (**b**) The biological process and the molecular function of DAPs in GE vs. PEC; (**c**) The biological process of DAPs in GE vs. NEC. NEC: Nonembryogenic calli; PEC: Primary embryogenic calli; GE: Globular embryos; GO: Gene Ontology.

2.7. Several Major DAPs are Associated with SE Regulation and Modification

SE of cotton is regulated by many factors. The GO and KEGG enrichment analysis showed that peroxidase, photosynthesis-related proteins, stress-responsive proteins, amino acid metabolism-related proteins and other energy metabolism enriched proteins all play an important role in the process of cotton SE. Furthermore, we comprehensively explored and analyzed differentially abundant proteins in cotton SE involving hormone signal response/signaling transduction, transcription/posttranscription and modification regulation (Table 2).

Table 2. Significantly representative SE regulatory DAPs in PEC vs. NEC, GE vs. PEC and GE vs. NEC.

Gene ID	Gene Name	Protein ID	Protein Description	Pathway Annotation	PEC/NEC Ratio	GE/PEC Ratio	GE/NEC Ratio
LOC107907377	PIN2	A0A120KAE0	Auxin efflux carrier component	Auxin signal	2.36	—	—
LOC107909506	GH3.17	A0A1U8JQJ4	Indole-3-acetic acid-amido synthetase GH3.17-like isoform X2	Auxin signal	2.14	—	—
LOC107948437	ETR1	A0A1U8NHA4	ethylene receptor-like isoform X1	Ethylene signal	2.02	—	3.705
LOC107938108	GASL1	M1GN42	GA-stimulated transcript-like protein 1	GA signal	0.25	0.27	0.068
LOC107955576	GASL4	M1GMV2	GA-stimulated transcript-like protein 4	GA signal	3.00		—
LOC107950128	PYR1	A0A1U8NR07	Abscisic acid receptor PYR1-like	ABA signal	—	2.31	2.361
LOC107893363	At5g01020	A0A1U8I6G4	serine/threonine-protein kinase At5g01020-like	Signal transduction	0.50	—	0.482
LOC107897915	—	A0A1U8IRF6	A-kinase anchor protein 12-like isoform X2	Signal transduction	0.44	—	0.459
LOC107909143	—	A0A1U8JP78	leucine-rich repeat receptor-like protein kinase PXC2	Signal transduction	0.45	—	0.36
LOC107945188	BAM3	A0A1U8N9I0	leucine-rich repeat receptor-like Ser/Thr -protein kinase BAM3	Signal transduction	2.03	—	—

Table 2. *Cont.*

Gene ID	Gene Name	Protein ID	Protein Description	Pathway Annotation	PEC/NEC Ratio	GE/PEC Ratio	GE/NEC Ratio
LOC107935259	PCKA	A0A1U8M980	phosphoenolpyruvate carboxykinase [ATP]-like	Signal transduction	2.08	0.38	—
LOC107943515	At1g56140	A0A1U8N4H5	probable LRR receptor-like serine/threonine-protein kinase At1g56140	Signal transduction	0.29	—	0.215
LOC107931208	TPK1	A0A1U8LYM7	thiamine pyrophosphokinase 1-like isoform X1	Signal transduction	0.46	—	0.469
LOC107905700	PFK	A0A1U8JGW8	ATP-dependent 6-phosphofructokinase	Signal transduction	—	2.55	2.504
LOC107943957	PV42A	A0A1U8N623	SNF1-related protein kinase regulatory subunit gamma-like PV42a	Signal transduction	—	2.83	2.757
LOC107937641	CPK11	A0A1U8MGW7	calcium-dependent protein kinase 11-like	Signal transduction	2.25	—	3.84
LOC107930954	CML27	A0A1U8LUL1	probable calcium-binding protein CML27	Signal transduction	4.16	—	2.88
LOC107916423	RHN1	A0A1U8KFK5	ras-related protein RHN1-like	Signal transduction	—	0.47	0.333
LOC107889787	—	A0A1U8HV05	Embryonic protein DC-8-like	Somatic embryogenesis related proteins	—	3.58	4.522
LOC107937048	Lea2A-A	Q03791	Embryogenesis abundant protein	Somatic embryogenesis related proteins	—	4.42	4.746
LOC107941722	WOX9	A0A1U8MVD7	WUSCHEL-related homeobox 9-like	Transcription factor	2.58	—	—
LOC107905698	NFYB6	A0A1U8JC47	Nuclear transcription factor Y subunit B-6	Transcription factor	3.22	2.07	6.661
	bHLH4	W5XUY9	BHLH4 transcription factor	Transcription factor	2.16	—	—
LOC107920272	NFYB9	A0A1U8KSD1	nuclear transcription factor Y subunit B-9-like	Transcription factor	4.07	—	2.509
LOC107931333		A0A1U8LVZ2	transcription factor HBP-1b (C38)-like	Transcription factor	2.40	—	—
LOC107924015	PHL1	A0A1U8L8P3	Protein PHR1-LIKE 1-like	Transcription factor	0.15	—	0.166
LOC107891610	At1g07170	A0A1U8I119	PHD finger-like domain-containing protein 5B	Zinc finger	3.49	—	4.21
LOC107909066	NERD	A0A1U8JUI6	zinc finger CCCH domain-containing protein 19-like isoform X2	Zinc finger	2.03	—	—
LOC107927097	TAF15B	A0A1U8LG36	transcription TFIID subunit 15b-like	Zinc finger	0.44	—	—
LOC107962890	ZHD5	A0A1U8PUW9	zinc-finger homeodomain protein 5-like	Zinc finger	—	2.93	5.554
LOC107890886	AGO1	A0A1U8HY77	protein argonaute 1-like isoform X2	Posttranscriptional regulation	8.44	—	6.734
LOC107906203	AGO4	A0A1U8JEA7	protein argonaute 4-like	Posttranscriptional regulation	2.38	—	2.426
LOC107962954	HEN1	A0A1U8PXD5	small RNA 2′-O-methyltransferase-like isoform X4	Posttranscriptional regulation	2.06	—	—
LOC107891032	IDM1	A0A1U8HYR9	increased DNA methylation 1-like isoform X4	Modification-related protein	2.62	0.49	—
LOC107906306	MMT1	A0A1U8JEK1	Methionine S-methyltransferase	Modification-related protein	0.46	—	0.458
LOC107948568	SUVH4	A0A1U8NHS0	histone-lysine N-methyltransferase, H3 lysine-9 specific SUVH4-like isoform X2	Modification-related protein	0.28	—	0.196
LOC107943854	CCOAOMT	A0A1U8N5R4	caffeoyl-CoA O-methyltransferase -like	Modification-related protein	0.35	—	0.363
LOC107953938	EMB1691	A0A1U8P3T9	methyltransferase-like protein 1	Modification-related protein	2.33	—	2.774

Table 2. *Cont.*

Gene ID	Gene Name	Protein ID	Protein Description	Pathway Annotation	PEC/NEC Ratio	GE/PEC Ratio	GE/NEC Ratio
LOC107916882	IAMT1	A0A1U8KGS5	indole-3-acetate O-methyltransferase 1	Modification-related protein	2.32	—	—
LOC107958653	—	A0A1U8PI31	chromatin modification-related protein MEAF6-like isoform X3	Modification-related protein	—	3.48	5.641
LOC107926365	—	A0A1U8LDK7	RNA cytidine acetyltransferase	Modification-related protein	2.17	—	2.333
LOC107960303	—	A0A1U8PLN7	Acetyltransferase component of pyruvate dehydrogenase complex	Modification-related protein	0.48	—	0.404
LOC107947121	UBR7	A0A1U8NDR8	putative E3 ubiquitin-protein ligase UBR7	Modification-related protein	2.90	—	2.679
LOC107959749	RUB2	A0A1U8PJK7	ubiquitin-NEDD8-like protein RUB2	Modification-related protein	0.26	—	0.286
LOC107938100	—	A0A1U8MII2	Phosphotransferase	Modification-related protein	0.42	—	0.476
LOC107898863	CYP86B1	A0A1U8IP72	Cytochrome P450 86B1-like	Fatty acid	4.51	—	2.388
LOC107922796	CYP86A8	A0A1U8L513	Cytochrome P450 86A8-like	Fatty acid	2.02	—	—
LOC107915850	PIP2-5	A0A1U8KIL6	probable aquaporin PIP2-5	Aquaporins	—	2.58	—
LOC107898442	TIP3-2	A0A1U8IU16	probable aquaporin TIP3-2	Aquaporins	—	2.29	9.086
LOC107963873	GhPIP2;10	D8FSK4	Aquaporin PIP210	Aquaporins	0.14	—	0.102
—	GhTIP1;4	D8FSK6	Aquaporin TIP14	Aquaporins	0.20	—	0.195
LOC107934987 LOC107944588	PIP1;4	G8XV51	PIP protein	Aquaporins	0.22	—	0.109

DAPs: differentially accumulated proteins; NEC: Nonembryogenic calli; PEC: Primary embryogenic calli; GE: Globular embryos; GO: Gene Ontology.

2.8. Comparative and Complementary Proteome of the Candidate DAPs

To complement the changes in abundance at the transcriptional level and confirm the authenticity and accuracy of the proteomic analysis, we analyzed ten candidate DAPs in NEC and PEC. Eight out of ten genes under this analysis showed positive correlation between the expression levels of protein and mRNA, indicating that most proteins were regulated directly at the transcription level. For the other two DAPs, negative correlation between their expression levels of protein and mRNA was observed, suggesting that their protein levels might be depended not only on the transcript level but also on the post-translational level (Figure 9).

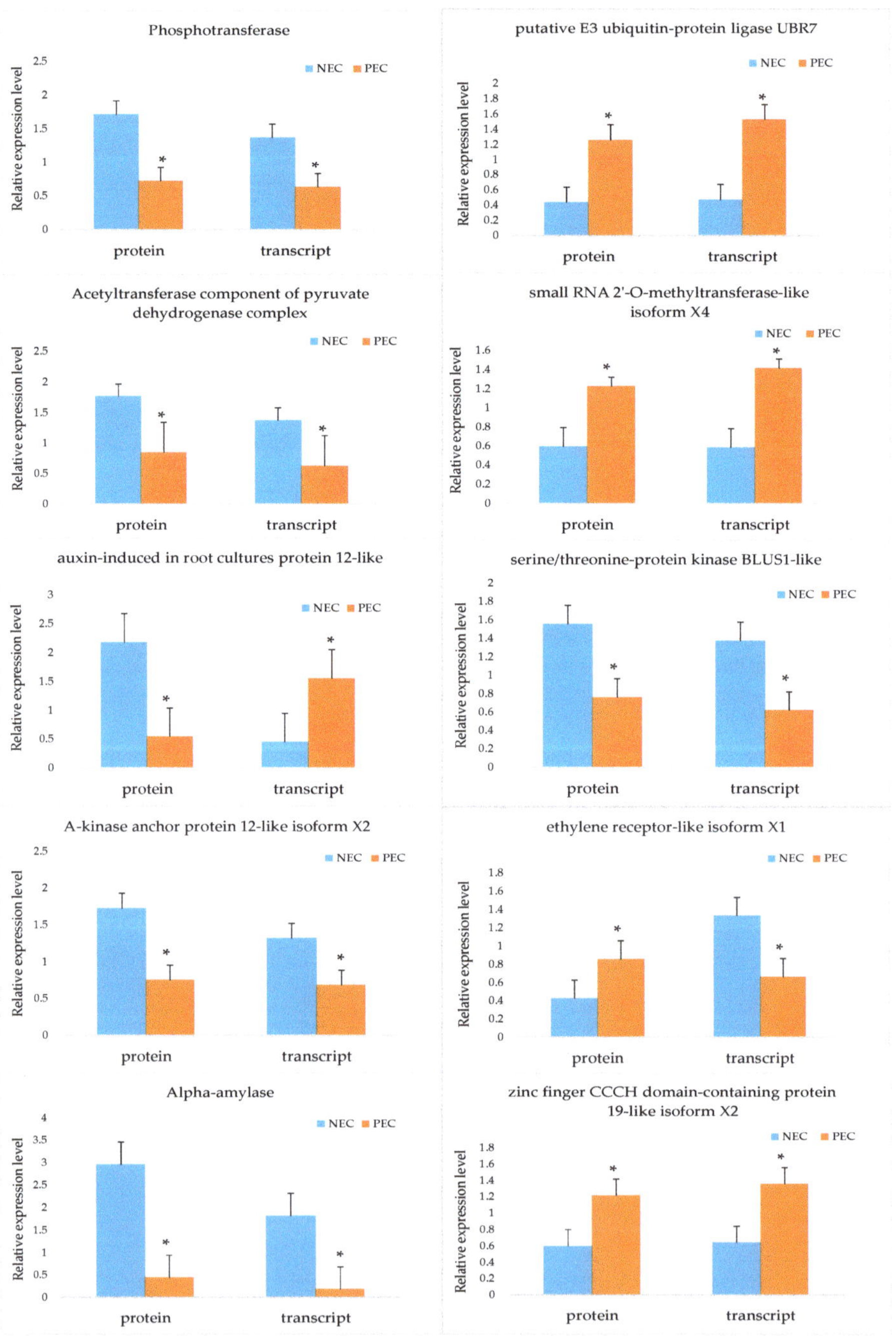

Figure 9. Comparative and complementary proteome of the candidate DAPs in stage of NEC and PEC. Significant differences in expression level were indicated by "*". DAPs: differentially accumulated proteins; NEC: Nonembryogenic calli; PEC: Primary embryogenic calli; GE: Globular embryos.

3. Discussion

Proteomics analyses have long been recognized as a useful tool to dissect the molecular mechanisms of SE. The effectiveness of this technique is strongly dependent on the applied technique of the proteomics analysis system and the experimental system. Proteomics analyses for the somatic embryogenesis in *Pinus nigra Arn.* [42] and *Phoenix dactylifera* L. [43] have been previously performed. In this study, we performed quantitative proteomics analysis using the advanced EASY-nLC 1000 UPLC system based on the high-throughput TMT-labeling quantitative detection technique, and we consequently identified 9369 proteins for our samples. Thus, our new study significantly improved the resolution. TMT is advantageous and offers greater sensitivity for the analysis of cotton SE proteomic dynamics than previous methods. Furthermore, most previous studies have focused on the molecular mechanisms of regulation in the late stage of somatic embryogenesis, and little is known about protein regulation and metabolism in the early stage of embryogenesis. In the isobaric tags for relative and absolute quantitation (iTRAQ) proteomics analysis of Ge [11], 6318 proteins were identified in somatic spherical embryo and cotyledon embryo. Subsequently, Zhu [44] identified 5892 proteins related to SE through iTRAQ technology. In our study, a total of 9369 proteins were identified in stages of NEC, PEC and somatic embryos' initial development period of GEs by TMT-labeling quantitative detection technique. Through identification and annotation of DAPs, we uncovered the key genes/proteins and pathways involved in the critical initial stage of cotton SE. The results generated in this study provide a valuable foundation for further investigation of the roles of DAPs during the expression of totipotency in cotton SE.

3.1. DAPs Enriched in Crucial Biological Processes Associated with Cotton SE

3.1.1. Peroxidase Proteins Involved in Phenylpropanoid Biosynthesis Affect SE

Recently, a proteomic analysis of the somatic embryogenesis induction stage of *Medicago truncatula* revealed that peroxidase accumulates by day 5 after the induction of somatic embryogenesis and increases four-fold by day 14 [45]. Peroxidase is known to take part in diverse plant processes, such as auxin metabolism, cell wall elongation and stiffening [46]. However, in the present study, our data showed that peroxidase is involved in 'phenylpropanoid biosynthesis' pathways (Figure 6a,b; Figure 7), and the expression pattern was different from that of Almeida [45]. We presume that peroxidase might participate in different metabolic pathways with different expression patterns to regulate somatic embryogenesis in different species. Peroxidase is a phenol oxidase and is highly representative in date palm, and polyphenoloxidase are involved in oxidative browning in date palm [47,48]. Abohatem [49] demonstrates that the low rate of successive transfer culture (every 15 or 20 days) reduced the increase in phenolic contents and peroxidase activities in plant tissue leading to an enhancement of tissues/cells browning and then to a decrease in embryonic cell proliferation. Fresh culture medium every 7 days can significantly reduce the oxidative browning of tissue/cells, which is related to the reduction of phenolic compounds and peroxidase activity, thus increasing the proliferation of embryonic protocells. Based on our proteomic profile results, down regulation of many peroxidase DAPs in PEC versus NEC indicates that peroxidase protein activity is weak during SE, which prevents callus browning and promotes embryogenic differentiation.

3.1.2. Photosynthesis in Cotton SE

The photosynthetic potential of cotyledon embryos has been reported in previous studies. For example, in *Coffea × arabusta* cotyledons photosynthetic capacity and germinated embryos, the cotyledon embryo stage is the earliest photoautotrophic stage to ensure plant development [50]. Rival [51] reported that the maximum photochemical activity of photosystem II is extremely low in proliferating embryos of oil palm and strongly increases at later developmental stages. In this study, the result is that the expression pattern of DAPs in the 'photosynthetic' pathway is down- to up regulation. According to our results, 'Photosystem I P700 chlorophyll an apoprotein', 'oxygen-evolving enhancer protein 2', 'ATP synthase gamma chain' and 'Chlorophyll a–b binding protein' of the 'Photosynthesis' pathway were down regulated in PEC in comparison to GEs (Figure 5b,c; Figure 6a–c). The results indicate that photosynthesis organs and photosynthetic capacity gradually developed from the GE stage for future autotrophy in cotton.

3.1.3. Response to Environment Stresses during SE of Cotton

Stresses are the factors that have been increasingly recognized as having important role in the induction of SE [52]. Embryogenic competence of in vitro cultured somatic cells can be stimulated by various factors, such as phytohormone [53,54], dehydration [55,56], explant wounding [57], heavy metal ions [17,56], high osmotic pressure [58,59], etc. Our data showed that a series of stress responsive biological processes were significantly enriched in PEC and GE, including response to acid chemical, water, inorganic substance, oxygen-containing compound, chemical, abiotic stimulus and biotic stimulus, indicating that callus was protected from the external environmental stress through complex regulatory networks to ensure the embryos development of cotton during the PEC-to-embryos transition of cotton (Figure 5c; Figure 8b). Previous studies have shown that decrease of water availability stimulated a shift from proliferation of cells and early embryos to the production of cotyledonary embryos in the developmental program of the culture [60]. In our study, the water content gradually decreased with the callus culture time increased prompted the tissue cells to respond to dry stress and initiate defense mechanisms to ensure embryo development. In the carrot somatic embryogenesis, 2,4-dichlorophenoxyacetic acid (2,4-D) functions as a stress chemical as well as an auxin [61]. During Arabidopsis somatic embryos induction, cells in the shoot apical meristem (SAM) of wild-type seedlings acquired pluripotency or embryogenic potential under initial stress, and then these cells form somatic embryos on 2,4-D treatment [17]. In our culture system, 2,4-D is also used to induce SE, but stress treatment is necessary before exposure to 2,4-D. From the datas above, we can conclude that stress responses are the indispensable biological processes in plant embryos induction.

3.1.4. Effects of Nitrogen Metabolism Related to SE

Glutamate dehydrogenase (GDH) is one of the enzymes directly related to nitrogen metabolism. Induction of various GDH isoenzymes may suggest their varied anabolic and catabolic functions [62]. GDH, which is directly involved in the oxidation of amino acids, protect tissues from the toxicity of ammonium [63]. Ganced [64,65] strongly suggested that sugar could control the expression of the GDH gene through catabolic inhibition, which has been described in bacteria and yeast. In this study, GDH in nitrogen metabolism was significantly enriched in GE versus PEC and showed a down regulated trend (Figure 6a,c). We speculate that GDH may be more important during the cotton PEC redifferentiation period.

3.2. Other DAPs of Regulatory Factors Associated with Cotton SE

3.2.1. Phytohormone Response Related Proteins

Hormones are the most likely candidates in the regulation of developmental switches [16]. Auxin is the main growth regulator in plants, which is involved in the regulation of cell division and differentiation, as well as the existence of other growth regulators such as abscisic acid [66], ethylene [67], gibberellin [68].

In cotton primary embryogenic calli, the DAPs of 'Auxin efflux carrier component' (AP2) and 'indole-3-acetic acid-amido synthetase GH3. 17-like' (GH3) were up regulated (Table 2). Auxin-related proteins are essential for initiating dedifferentiation and cell division in already differentiated cells before they can express embryogenic competence [44]. The *PIN* gene is believed to be the coding element that regulates the auxin efflux mechanism of the polar auxin transport, which is concluded by the polarity localization of the PIN membrane protein and auxin absorption experiment [69]. Blilou [70] pointed to polar auxin transport as a major factor in organ formation. The *GH3* gene is one of several sequences screened by differential hybridization of auxin-induced cDNA sequences extracted from auxin-treated soybean tissues [71]. Expression of the *GH3* gene has been shown to be rapidly and specifically induced by the application of auxin [72,73]. These conclusions also explained the up regulation of AP2 and GH3 in PEC due to the trend of auxin polar transport organ formation and the key regulation of induced somatic embryo formation.

Gallie [74] identified two ethylene receptor gene families in maize. In developing embryos, the expression levels of members of the two ethylene receptor families were significantly increased, which indicated that embryonic development was involved in ethylene synthesis. In this study, DAPs of 'ethylene receptor-like isoform X1' (Table 2) were involved in the ethylene signaling pathway and were up regulated in PEC compared to NEC, suggesting that ethylene receptor may positively regulate SE initiation in cotton.

In the present study, the differentially accumulated 'GA-stimulated transcript-like protein 1' (GASL1) was down regulated in PEC and GE (Table 2). Ge [11] demonstrated that, at 10 days after GA treatment, 95% of the embryos showed an aberrant structure, large size, and light-green color. Therefore, we presume that GA negatively affects somatic embryo production and growth via regulation of the GA signaling pathway.

Treatment with ABA improves the efficiency of somatic embryo maturation of *Panax ginseng* [75] and promotes sugar cane embryo growth [76]. Somatic embryos treated with ABA generate the highest yield of plantlets in *Picea abies* [77]. In our results, the 'abscisic acid receptor PYR1-like' DAP, which is involved in the ABA signal pathway, was up regulated in GE compared to PEC (Table 2); this suggests that the 'abscisic acid receptor PYR1-like protein' promotes embryo maturation in GE.

3.2.2. Signal Transduction Related Proteins

In this study, we also found that a large number of important signal regulators, including kinases, calcium signals and GTP-binding related proteins (Table 2), were significantly differential expressed in cotton SE. Ca^{2+} has an important role in the establishment of cellular polarity during embryogenesis in plants [78]. It as a secondary messenger may trigger various signal transduction pathways [79] in plants SE. Calmodulin or Ca^{2+}-dependent protein kinase may be involved in the regulation of Ca^{2+} levels in the proembryogenic cell mass (PEM) of sandalwood to promote embryo development [80]. In our TMT profile, the calcium-related proteins 'calcium-dependent protein kinase 11-like' and 'probable calcium-binding protein CML27' were simultaneously up regulated in PEC versus NEC. We presume that calcium-related proteins involved in signal transduction and regulation of calcium balance, thereby establishing polarity to promote cotton SE.

Phosphoenolpyruvate carboxylase (PEPC), a glycolytic protein and CO_2-fixing enzyme [81], participates in TCA cycle under non-photosynthesis conditions [82] and signal transduction of plant embryo development [83]. Therefore, this metabolic pathway maintains the carbon residue pool necessary for oil and storage protein biosynthesis that occurred the later in embryonic development [84]. In our study, the protein 'phosphoenolpyruvate carboxykinase [ATP]-like' was up regulated in PEC versus NEC and down regulated in GE versus PEC (Table 2). Thus, we presume that PEPC may not only participate in TCA cycle and fixation of CO_2 but also in signal transduction in PEC, providing energy for cotton SE transformation process.

As a result, calcium-related proteins, PEPC and other signal transduction proteins may be involved in a variety of biological processes to promote the transformation process of cotton SE.

3.2.3. SE Associated Proteins of Aquaporins and Fatty Acid Metabolism

In our TMT profile, a large number of aquaporins, including PIP and TIPs, were found to be significantly involved in PEC and GE initiation during SE. Aquaporin was involved in water transport by osmosis to prevent dryness and abortion during embryonic development, which was of great significance in the development of *Picea asperata* somatic embryos [85]. Here, we found that the aquaporins (PIP210, TIP14 and PIP type) were significantly down regulated during PEC; the aquaporins (PIP2-5, TIP3-2) were significantly up regulated during GE (Table 2). Therefore, we suspect that these proteins are sensitive to water content and light induction during cotton SE initiation process of transdifferentiation. Similarly, previous studies have shown that aquaporin, under stress conditions, forms a 'tunnel' to regulate the water transport in the cell membrane [86], and can be induced by light [87]. In addition to reducing the activation energy of water transport, aquaporin also enhanced the permeability of the plasma membrane [88].

Many fatty acid biosynthesis- and metabolism-related proteins, such as 'cytochrome P450 86B1-like' and 'cytochrome P450 86A8-like', were differentially accumulated. In the fatty acid biosynthesis pathway, they were up regulated in PEC (Table 2). A previous study reported that fatty acids, which affect cell function and growth patterns, appear to be a part of the TDZ action pattern and may play an important role in inducing regeneration [89]. These results implied that aquaporins and perturbations of fatty acid metabolism contribute to the initiation of SE in cotton.

3.2.4. Regulation of SE-related Proteins, Transcription, Posttranscription and Modification

In this study, two types of SE-related proteins and multiple types of transcription factors, zinc finger domains, microRNAs and modification-related proteins were identified (Table 2). Late embryogenesis abundant protein (LEA) was first found in cotton (*Gossypium hirsutum*) seeds and accumulated in the late stages of embryogenesis, which played a crucial role in cell dehydration tolerance [90]. In current study, two types of SE-related proteins, 'embryonic protein DC-8-like' and 'embryogenesis abundant protein', were up regulated in GE. It indicating that they may positively regulate somatic embryo maturation and involve in cell dehydration tolerance in cotton.

WUSCHEL (*WUS*) is a vital transcription factor for labeling embryonic cells [24]. The capability of promoting the vegetative to embryonic transition by *WUS*, and eventually forming somatic embryos, suggesting that the homeodomain protein also plays a critical role during embryogenesis, in addition to its function in meristem development [91]. In this result, '*WUSCHEL*-related homeobox 9-like' was up regulated in PEC, indicating that the *WUSCHEL*-related protein is essential for the initiation of somatic embryogenesis. In addition to *WUS*, we also found other transcription factors, zinc finger domains related to SE (Table 2).

Members of the Argonaute protein family are key players in the small RNA-directed gene silencing pathway. Various types of small RNA and Argonaute proteins played important roles in embryonic development, cell differentiation and transposon silencing in all higher eukaryotes [92]. In our data, 'protein argonaute 1-like isoform X2', 'protein argonaute 4-like' and 'small RNA 2'-O-methyltransferase-like isoform X4' were significantly up regulated during PEC versus NEC (Table 2), suggesting that RNA-mediated post-transcriptional regulation played an important role in the process of cotton SE transformation as reported in other plants [93–95].

In current study, we identified several types of modification-related DAPs, including DNA methylation, chromatin modification, acetylation, ubiquitination and phosphorylation. In the process of carrot SE, the removal of auxin led to the loss of DNA methylation, so that the embryo continued to develop [96]. Six types of methylation-related proteins were identified, which dynamically regulated to cotton SE by different expression patterns. Efficient modification of chromatin structure was crucial in the epigenetic regulation of genes [97]. In our data, 'chromatin modification-related protein MEAF6-like isoform X3' promoted maturation and development of cotton GEs by epigenetic regulation. In addition, during the PEC period, we also identified acetylation, ubiquitination and phosphorylation-related proteins with different expression patterns (Table 2), indicating that they played diverse and indispensable functions in regulating cotton SE.

The different expression patterns of SE-related proteins, multitudinal transcription, posttranscription and modification showed that they played a pivotal role in the process of cotton SE [98–103]. The results of our high-through put proteomics assay, the large scale of proteins associated with SE, and their complex expression patterns suggest that SE is a concerted process involving multiple molecular pathways controlled by a complicated gene regulatory network.

4. Materials and Methods

4.1. Plant Materials and Culture Conditions

Seeds of cotton cultivar Yuzao 1 (Institute of Cotton Research of CAAS), uncovering the coats, were imbibed in 0.1% (w/v) $HgCl_2$ for 10 min, then rinsed four times by sterile distilled water and germinated on Murashige and Skoog (MS) medium containing 3% (w/v) sucrose and 0.3% (w/v) Phytegel. Hypocotyl explants (0.5–1.0 cm) taken from 5–7 d old seedlings and non-embryogenic callus initiation of Yuzao 1 was done, as described by Wu [104], in MS medium plus B5 vitamins medium (MSB) containing 0.46 μmol L–1 kinetin and 0.45 μmol L–1 2,4-D. Calli were subcultured in MSB medium without any hormone to induce embryogenic calli (EC), as described by Zeng [39]. Samples were collected from the following three stages of SEM (1) NEC from explants cultured in MSB, induced (2) PEC and (3) GEs. The collected samples were immediately frozen in liquid nitrogen, and stored at $-80\,^{\circ}C$ before protein extraction. Each staged sample was prepared for three biological replicates.

4.2. Protein Extraction and Identification

4.2.1. Protein Samples Preparation

The three biological replicates of sample were first ground by liquid nitrogen, then the powder was transferred to 5 cm^3 centrifuge tube and sonicated three times on ice using a high intensity ultrasonic processor (Scientz, Ningbo, China) in 4-fold volume phenol extraction buffer (including 10 mM dithiothreitol, 1% Protease Inhibitor Cocktail and 2 mM EDTA). The equal amount of trisaturated phenol (pH 8.0) was added; then, the mixture was further vortexed for 5 min. After centrifugation (4 °C, 10 min, 5000× g), the upper stage of phenol was transferred to a new centrifuge tube. Proteins were precipitated by adding at five volume of 0.1 M ammonium sulfate-saturated methanol to precipitate overnight. After centrifugation at 4 °C for 10 min, the supernatant was discarded. The remaining precipitate was washed with ice-cold methanol once, followed by ice-cold acetone for three times. The protein was redissolved in 8 M urea and the protein concentration was determined with BCA kit (Beyotime Biotechnology, Shanghai, China) according to the manufacturer's instructions.

4.2.2. Trypsin Digestion

Protein solution with dithiothreitol makes its final concentration of 5 mM, 56 °C for 30 min. After that, acetamide was added to make the final concentration 11 mM and incubated in the dark at room temperature for 15 min. Finally, the sample urea concentration was diluted to less than 2 M. With 1:50 (w/w) quality ratio (trypsin: protein) to join the pancreatic enzyme, 37 °C enzyme solution for the night. Then the trypsin (Promega, Madison, WI, USA) was added at a mass ratio of 1:100 (trypsin: protein), and the enzymatic hydrolysis continued for 4 h.

4.2.3. TMT Labeling

After trypsin digestion, the peptide was desalted by Strata X C18 SPE column (Phenomenex, Torrance, CA, USA) and vacuum-dried. The peptide was reconstituted in 0.5 M TEAB (Sigma, St. Louis, MI, USA) and processed according to the manufacturer's protocol for TMT kit (Thermo Fisher Scientific, Waltham, MA, USA). Briefly, one unit of TMT reagent was thawed and reconstituted in acetonitrile. The peptide mixtures were then incubated for 2 h at room temperature and pooled, desalted and dried by vacuum centrifugation.

4.2.4. HPLC Fractionation and LC-MS/MS Analysis

The tryptic peptides were fractionated into fractions by high pH reverse-phase HPLC using Agilent 300Extend C18 column (5 μm particles, 4.6 mm ID, 250 mm length, Agilent, Santa Clara, USA). Briefly, peptides were first separated with a gradient of 8% to 32% acetonitrile (pH 9.0) over 60 min into 60 fractions. Then, the peptides were combined into 9 fractions and dried by vacuum centrifuging.

The tryptic peptides were dissolved in 0.1% formic acid (solvent A), directly loaded onto a home-made reversed-phase analytical column (15-cm length, 75 μm i.d.). The gradient was comprised of an increase from 8% to 23% solvent B (0.1% formic acid in 90% acetonitrile) over 38 min, 23% to 35% in 14 min and climbing to 80% in 4 min then holding at 80% for the last 4 min, all at a constant flow rate of 450 nL/min on an EASY-nLC 1000 UPLC system (Thermo Fisher Scientific, Waltham, MA, USA).

The peptides were subjected to NSI source followed by tandem mass spectrometry (MS/MS) in Q Exactive™ HF-X (Thermo Fisher Scientific, Waltham, MA, USA) coupled online to the UPLC. The electrospray voltage applied was 2.0 kV. The m/z scan range was 350–1600 *m/z* for a full scan, and intact peptides were detected in the Orbitrap (Thermo Fisher Scientific, Waltham, MA, USA) at a resolution of 60,000. Peptides were then selected for MS/MS using NCE setting as 28 and the fragments were detected in the Orbitrap at a resolution of 17,500. A data-dependent procedure that alternated between one MS scan followed by 20 MS/MS scans with 30 s dynamic exclusion. Automatic gain control (AGC) was set at 1E5. The fixed first mass was set as 100 *m/z*. The HPLC Fractionation and LC-MS/MS Analysis in our research is supported by Jingjie PTM BioLabs (Hangzhou, China).

4.2.5. Database Search

The resulting MS/MS data were processed using Maxquant search engine (v.1.5.2.8). Tandem mass spectra were searched against the UniProt 14.1 (2009)—*Gossypium hirsutum* database (78,387 sequences) concatenated with reverse decoy database. Trypsin/P was specified as cleavage enzyme allowing up to 2 missing cleavages. The first search was 20 ppm, the main search was 5 ppm, and the fragment ion mass tolerance was 0.02 Da.

4.3. Bioinformatics

4.3.1. Annotation Methods

GO Annotation

Gene Ontology (GO) annotation proteome was derived from the UniProt-GOA database (http://www.ebi.ac.uk/GOA/). Firstly, converting identified protein ID to UniProt ID and then mapping to GO IDs by protein ID. If some identified proteins were not annotated by UniProt-GOA database, the InterProScan soft would be used to annotated protein's GO functional based on protein sequence alignment method. Then proteins were annotated according to the biological process, cellular component and molecular function of the three categories of protein Gene Ontology annotation.

Domain Annotation

Identified proteins domain functional description were annotated by InterProScan (a sequence analysis application) based on protein sequence alignment method, and the InterPro domain database was used. InterPro (http://www.ebi.ac.uk/interpro/) is a database that integrates diverse information about protein families, domains and functional sites, and makes it freely available to the public via Web-based interfaces and services. At the heart of the database are diagnostic models, known as signatures, from which protein sequences can be searched for potential functions. InterPro has applications in large-scale genome-wide and metagenomic analyses, and the characterization of individual protein sequences.

KEGG Pathway Annotation

KEGG connects known information on molecular interaction networks, such as pathways and complexes (the "Pathway" database), information about genes and proteins generated by genome projects (including the gene database) and information about biochemical compounds and reactions (including compound and reaction databases). These databases are different networks, known as the "protein network", respectively, and the "chemical universe", respectively. Efforts are being made to increase KEGG knowledge, including information on orthographic clustering in the KEGG orthographic database. KEGG Pathways mainly including: Metabolism, Genetic Information Processing, Environmental Information Processing, Cellular Processes, Rat Diseases, Drug Development. Kyoto Encyclopedia of Genes and Genomes (KEGG) database (http://www.genome.jp/kegg/) was used to annotate protein pathway. Firstly, using KEGG online service tools KAAS (http://www.genome.jp/kaas-bin/kaas_main) to annotate protein's KEGG database description. Then mapping the annotation result on the KEGG pathway database using KEGG online service tools KEGG mapper (http://www.kegg.jp/kegg/mapper.html).

Subcellular Localization Prediction

Eukaryotic cells are elaborately subdivided into membrane-bound chambers with unique functions. Some major constituents of eukaryotic cells are extracellular space, cytoplasm, nucleus, mitochondria, Golgi apparatus, endoplasmic reticulum (ER), peroxisome, vacuoles, cytoskeleton, nucleoplasm, nucleolus, nuclear matrix, and ribosomes. There, we used wolfpsort (http://www.genscript.com/psort/wolf_psort.html) a subcellular localization predication soft to predict subcellular localization. Wolfpsort is an updated version of PSORT/PSORT II for the prediction of eukaryotic sequences.

4.3.2. Functional Enrichment

Enrichment of Gene Ontology Analysis

Through GO annotation, proteins are divided into biological process, cellular compartment, and molecular function. For each class, we used a double-tailed Fisher's precision test to detect the enrichment of differentially abundant proteins relative to all identified proteins. GO with a revised p value of <0.05 is considered significant.

Enrichment of Pathway Analysis

KEGG database identified enrichment pathways by double-tailed Fisher's precision test to detect the enrichment of differentially abundant proteins against all identified proteins. The pathway with a corrected p-value < 0.05 was considered significant. According to the KEGG website, these paths are classified into hierarchical categories.

Enrichment of Protein Domain Analysis

For each category protein, InterPro (a resource that provides a functional analysis of protein sequences by classifying them into families and predicting the presence of domains and important sites) database was researched and a two-tailed Fisher's exact test was employed to test the enrichment of the differentially abundant proteins against all identified proteins. Protein domains with a p-value < 0.05 were considered significant.

Enrichment-Based Clustering

For further hierarchical clustering based on different protein functional classification (such as GO, Domain, Pathway, Complex). We first collated all the categories obtained after enrichment along with their p values, and then filtered for those categories which were at least enriched in one of the clusters with p value <0.05. This filtered P value matrix was transformed by the function x = $-\log10$ (p value). Finally, these x values were z-transformed for each functional category. These z scores were then clustered by one-way hierarchical clustering (Euclidean distance, average linkage clustering) in Genesis. Cluster membership was visualized by a heat map using the "heatmap.2" function from the "gplots" R-package.

Supplementary Materials: Supplementary materials can be found at http://www.mdpi.com/1422-0067/20/7/1691/s1.

Author Contributions: Data curation, H.G. (Haixia Guo), H.G. (Huihui Guo), L.Z., Y.F. (Yijie Fan), Y.F. (Yupeng Fan) and Z.T.; Formal analysis, H.G. (Haixia Guo), H.G. (Huihui Guo), Z.T. and F.Z.; Funding acquisition, F.Z.; Investigation, H.G. (Haixia Guo), H.G. (Huihui Guo), L.Z., Y.F. (Yijie Fan) and Y.F. (Yupeng Fan); Methodology, Z.T.; Project administration, F.Z.; Resources, H.G. (Haixia Guo), H.G. (Huihui Guo), L.Z., Y.F. (Yijie Fan) and Y.F. (Yupeng Fan); Supervision, F.Z.; Writing—original draft, H.G. (Haixia Guo), H.G. (Huihui Guo) and L.Z.; Writing—review & editing, H.G. (Haixia Guo) and F.Z.

Funding: This work was supported by National Key Research and Development Program (2018YFD0100303; 2016YFD0100306), Taishan Scholar Talent Project from PRC (TSQN20161018), the National Natural Science Foundation of China (31401428) and Fok Ying-Tong Foundation (151024).

Acknowledgments: We are grateful for the technique support provided by Jingjie Biotechnology Co., Ltd., Hangzhou, China.

Conflicts of Interest: The authors declare no conflict of interest.

Abbreviations

ABA	Abscisic acid
DAPs	Differentially accumulated proteins
EC	Embryogenic calli
GA	Gibberellin
GEs	Globular embryos
GO	Gene Ontology
iTRAQ	Isobaric tags for relative and absolute quantitation
KEGG	Kyoto encyclopedia of genes and genomes
MS	Murashige and Skoog
MSB	Vitamins medium
NEC	Nonembryogenic calli
PEC	Primary embryogenic calli
PEM	Proembryogenic cell mass
SE	Somatic embryogenesis
SEM	Somatic embryogenesis masses
SSH	Suppression subtractive hybridization
TMT	The isobaric labels tandem mass tags

References

1. Wilkins, T.A.; Rajasekaran, K.; Anderson, D.M. Cotton biotechnology. *Crit. Rev. Plant Sci.* **2000**, *19*, 511–550. [CrossRef]

2. Brookes, G.; Barfoot, P. *GM Crops: Global Socio-Economic and Environmental Impacts 1996–2010*; PG Economics, Ltd.: Dorchester, UK, 2012; Available online: http://www.pgeconomics.co.uk/page/33/global-impact-2012 (accessed on 13 January 2013).

3. Zhao, X.Y.; Su, Y.H.; Cheng, Z.J.; Zang, X.S. Cell fate switch during in vitro plant organogenesis. *J. Integr. Plant Biol.* **2008**, *50*, 816–824. [CrossRef] [PubMed]

4. Mujib, A. *Somatic Embryogenesis in Ornamentals and Its Applications*; Springer: New Delhi, India, 2016; pp. 67–86.

5. Fehér, A. The initiation phase of somatic embryogenesis: What we know and what we don't. *Acta Biol. Szeged.* **2008**, *52*, 53–56.

6. Fehér, A. Somatic embryogenesis—stress-induced remodeling of plant cell fate. *BBA-Gene Regul. Mech.* **2015**, *1849*, 385–402.

7. Stewards, F.C.; Mapbs, M.O.; Mears, K. Growth and organized development of cultured cells I. Organization in cultures grown from freely suspended cells. *Am. J. Bot.* **1958**, *45*, 705–708. [CrossRef]

8. Rao, A.Q.; Hussain, S.S.; Shahzad, M.S.; Bokhari, S.Y.; Raza, M.H.; Rakha, A.; Majeed, A.; Shahid, A.A.; Saleem, Z.; Husnain, T.; Riazuddin, S. Somatic embryogenesis in wild relatives of cotton (*Gossypium* Spp.). *J. Zhejiang Univ. Sci. B* **2006**, *7*, 291–298. [CrossRef] [PubMed]

9. Evans, D.A.; Sharp, W.R.; Flick, C.E. *Growth and Behavior of Cell Cultures: Embryogenesis and Organogenesis. Plant Tissue Cultures*; Trevor, A.T., Ed.; Academic Press: New York, NY, USA, 1981; pp. 45–113.

10. Scowcroft, W.R. *Genetic Variability in Tissue Culture: Impact on Germplasm Conservation and Utilization*; IBPGR: Roma, Italy, 1984; p. 41. Available online: http://www.sidalc.net/cgi-bin/wxis.exe/?IsisScript=bac.xis&method=post&formato=2&cantidad=1&expresion=mfn=048729 (accessed on 3 April 2019).

11. Ge, X.; Zhang, C.; Wang, Q.; Yang, Z.; Wang, Y.; Zhang, X.; Wu, Z.; Hou, Y.; Wu, J.; Li, F. iTRAQ protein profile differential analysis between somatic globular and cotyledonary embryos reveals stress, hormone, and respiration involved in increasing plantlet regeneration of *Gossypium hirsutum* L. *J. Proteome Res.* **2014**, *14*, 268–278. [CrossRef] [PubMed]

12. Sakhanokho, H.F.; Rajasekaran, K. Cotton regeneration in vitro. In *Fiber Plants*; Ramawat, K.G., Ed.; Springer: Cham, Switzerland, 2016; pp. 87–110.

13. Fehér, A. Why somatic plant cells start to form embryos? In *Somatic Embryogenesis*; Mujib, A., Ed.; Springer: Berlin/Heidelberg, Germany, 2005; pp. 85–101.

14. Karami, O.; Aghavaisi, B.; Pour, A.M. Molecular aspects of somatic-to-embryogenic transition in plants. *J. Chem. Biol.* **2009**, *2*, 177–190. [CrossRef] [PubMed]

15. Elhiti, M.; Stasolla, C.; Wang, A. Molecular regulation of plant somatic embryogenesis. *In Vitro Cell. Dev. Biol. Plant* **2013**, *49*, 631–642. [CrossRef]

16. Fehér, A.; Pasternak, T.P.; Dudits, D. Transition of somatic plant cells to an embryogenic state. *Plant Cell Tissue Org.* **2003**, *74*, 201–228. [CrossRef]

17. Ikeda-Iwai, M.; Umehara, M.; Satoh, S.; Kamada, H. Stress-induced somatic embryogenesis in vegetative tissues of *Arabidopsis thaliana*. *Plant J.* **2003**, *34*, 107–114. [CrossRef] [PubMed]

18. Rose, R.J.; Nolan, K.E. Invited review: Genetic regulation of somatic embryogenesis with particular reference to *Arabidopsis thaliana* and *Medicago truncatula*. *In Vitro Cell. Dev. Biol. Plant* **2006**, *42*, 473–481. [CrossRef]

19. Reinert, J. Über die Kontrolle der Morphogenese und die Induktion von Adventivembryonen an Gewebekulturen aus Karotten. *Planta* **1959**, *53*, 318–333. [CrossRef]

20. Quiroz-Figueroa, F.R.; Rojas-Herrera, R.; Galaz-Avalos, R.M.; Loyola-Vargas, V.M. Embryo production through somatic embryogenesis can be used to study cell differentiation in plants. *Plant Cell Tissue Org.* **2006**, *86*, 285. [CrossRef]

21. Mathieu, M.; Lelu-Walter, M.A.; Blervacq, A.S.; David, H.; Hawkins, S.; Neutelings, G. Germin-like genes are expressed during somatic embryogenesis and early development of conifers. *Plant Mol. Biol.* **2006**, *61*, 615–627. [CrossRef] [PubMed]

22. Thorpe, T.A. In Vitro Embryogenesis in Plants. Springer Science & Business Media: Dordrecht, The Netherlands, 1995; pp. 267–308.

23. Lakshmanan, P.; Taji, A. Somatic embryogenesis in leguminous plants. *Plant Biol.* **2000**, *2*, 136–148. [CrossRef]

24. Elhiti, M.; Tahir, M.; Gulden, R.H.; Khamiss, K.; Stasolla, C. Modulation of embryo-forming capacity in culture through the expression of *Brassica* genes involved in the regulation of the shoot apical meristem. *J. Exp. Bot.* **2010**, *61*, 4069–4085. [CrossRef] [PubMed]

25. Zuo, J.; Niu, Q.W.; Frugis, G.; Chua, N. The *WUSCHEL* gene promotes vegetative-to-embryonic transition in *Arabidopsis*. *Plant J.* **2002**, *30*, 349–359. [CrossRef]

26. Weijers, D.; Sauer, M.; Meurette, O.; Friml, J.; Ljung, K.; Sandberg, G.; Hooykaas, P.; Offringa, R. Maintenance of embryonic auxin distribution for apical–basal patterning by *PIN-FORMED*–dependent auxin transport in *Arabidopsis*. *Plant Cell* **2005**, *17*, 2517–2526. [CrossRef]

27. Su, Y.H.; Zhao, X.Y.; Liu, Y.B.; Zhang, C.L.; O'Neill, S.D.; Zhang, X.S. Auxin-induced *WUS* expression is essential for embryonic stem cell renewal during somatic embryogenesis in *Arabidopsis*. *Plant J.* **2009**, *59*, 448–460. [CrossRef]

28. Vogler, H.; Kuhlemeier, C. Simple hormones but complex signalling. *Curr. Opin. Plant Biol.* **2003**, *6*, 51–56. [CrossRef]

29. Braybrook, S.A.; Harada, J.J. *LECs* go crazy in embryo development. *Trends Plant Sci.* **2008**, *13*, 624–630. [CrossRef] [PubMed]

30. Giraudat, J.; Parcy, F.; Bertauche, N.; Gosti, F.; Leung, J.; Morris, P.C.; Bouvier-Durand, M.; Vartanian, N. Current advances in abscisic acid action and signalling. In *Signals and Signal Transduction Pathways in Plants*; Palme, K., Ed.; Springer: Dordrecht, The Netherlands, 1994; pp. 321–341.

31. Ton, J.; Mauch-Mani, B. β-amino-butyric acid-induced resistance against necrotrophic pathogens is based on ABA-dependent priming for callose. *Plant J.* **2004**, *38*, 119–130. [CrossRef] [PubMed]

32. Letarte, J.; Simion, E.; Miner, M.; Kasha, K.J. Arabinogalactans and arabinogalactan-proteins induce embryogenesis in wheat (*Triticum aestivum* L.) microspore culture. *Plant Cell* **2006**, *24*, 691. [CrossRef] [PubMed]

33. Singla, B.; Tyagi, A.K.; Khurana, J.P.; Khurana, P. Analysis of expression profile of selected genes expressed during auxin-induced somatic embryogenesis in leaf base system of wheat (*Triticum aestivum*) and their possible interactions. *Plant Mol. Biol.* **2007**, *65*, 677–692. [CrossRef] [PubMed]

34. Mantiri, F.R.; Kurdyukov, S.; Lohar, D.P.; Sharopova, N.; Saeed, N.A.; Wang, X.D.; Vandenbosch, K.A.; Rose, R.J. The transcription factor *MtSERF1* of the *ERF* subfamily identified by transcriptional profiling is required for somatic embryogenesis induced by auxin plus cytokinin in *Medicago truncatula*. *Plant Physiol.* **2008**, *146*, 1622–1636. [CrossRef] [PubMed]

35. Kulinska-Lukaszek, K.; Tobojka, M.; Adamiok, A.; Kurczynska, E.U. Expression of the *BBM* gene during somatic embryogenesis of *Arabidopsis thaliana*. *Biol. Plant.* **2012**, *56*, 389–394. [CrossRef]

36. Yang, Z.; Li, C.; Wang, Y.; Zhang, C.; Wu, Z.; Zhang, X.; Liu, C.; Li, F. *GhAGL15*s, preferentially expressed during somatic embryogenesis, promote embryogenic callus formation in cotton (*Gossypium hirsutum* L.). *Mol. Genet. Genom.* **2014**, *289*, 873–883. [CrossRef]

37. Zheng, Q.; Zheng, Y.; Perry, S.E. *AGAMOUS-Like15* Promotes Somatic Embryogenesis in *Arabidopsis thaliana* and *Glycine max* in Part by Control of Ethylene Biosynthesis and Response. *Plant Physiol.* **2013**, *161*, 2113–2127. [CrossRef]

38. Schmidt, E.D.; Guzzo, F.; Toonen, M.A.; de Vries, S.C. A leucine-rich repeat containing receptor-like kinase marks somatic plant cells competent to form embryos. *Development* **1997**, *124*, 2049–2062.

39. Zeng, F.; Zhang, X.; Zhu, L.; Tu, L.; Guo, X.; Nie, Y. Isolation and characterization of genes associated to cotton somatic embryogenesis by suppression subtractive hybridization and macroarray. *Plant Mol. Biol.* **2006**, *60*, 167–183. [CrossRef] [PubMed]

40. Thompson, A.; Schäfer, J.; Kuhn, K.; Kienle, S.; Schwarz, J.; Schmidt, G.; Neumann, T.; Johnstone, R.; Mohammed, A.K.; Hamon, C. Tandem mass tags: A novel quantification strategy for comparative analysis of complex protein mixtures by MS/MS. *Anal. Chem.* **2003**, *75*, 1895–1904. [CrossRef] [PubMed]

41. Pagel, O.; Loroch, S.; Sickmann, A.; Zahedi, R.P. Current strategies and findings in clinically relevant post-translational modification-specific proteomics. *Expert Rev. Proteom.* **2015**, *12*, 235–253. [CrossRef] [PubMed]

42. Klubicová, K.; Uváčková, L.; Danchenko, M.; Nemecek, P.; Skultéty, L.; Salaj, J.; Salaj, T. Insights into the early stage of *Pinus nigra Arn.* somatic embryogenesis using discovery proteomics. *J. Proteom.* **2017**, *169*, 99–111. [CrossRef] [PubMed]

43. Sghaier-Hammami, B.; Drira, N.; Jorrín-Novo, J.V. Comparative 2-DE proteomic analysis of date palm (*Phoenix dactylifera* L.) somatic and zygotic embryos. *J. Proteom.* **2009**, *73*, 161–177. [CrossRef] [PubMed]

44. Zhu, H.G.; Cheng, W.H.; Tian, W.G.; Li, Y.J.; Liu, F.; Xue, F.; Zhu, Q.H.; Sun, Y.Q.; Sun, J. iTRAQ-based comparative proteomic analysis provides insights into somatic embryogenesis in *Gossypium hirsutum* L. *Plant Mol. Biol.* **2018**, *96*, 89–102. [CrossRef] [PubMed]

45. Almeida, A.M.; Parreira, J.R.; Santos, R.; Duque, A.S.; Francisco, R.; Tomé, D.F.; Ricardo, C.P.; Coelho, A.V.; Fevereiro, P. A proteomics study of the induction of somatic embryogenesis in *Medicago truncatula* using 2DE and MALDI-TOF/TOF. *Physiol. Plant.* **2012**, *146*, 236–249. [CrossRef]

46. Passardi, F.; Theiler, G.; Zamocky, M.; Cosioa, C.; Rouhier, N.; Teixera, F.; Margis-Pinheiroe, M.; Ioannidis, V.; Penel, C.; Falquet, L.; Dunand, C. PeroxiBase: The peroxidase database. *Phytochemistry* **2007**, *68*, 1605–1611. [CrossRef]

47. Baaziz, M.; Saaidi, M. Preliminary identification of date palm cultivars by esterase isoenzymes and peroxidase activities. *Can. J. Bot.* **1988**, *66*, 89–93. [CrossRef]

48. Baaziz, M. The activity and preliminary characterization of peroxidases in leaves of cultivars of date palm, *Phoenix dactylifera* L. *New Phytol.* **1989**, *111*, 403–411. [CrossRef]

49. Abohatem, M.; Zouine, J.; El Hadrami, I. Low concentrations of BAP and high rate of subcultures improve the establishment and multiplication of somatic embryos in date palm suspension cultures by limiting oxidative browning associated with high levels of total phenols and peroxidase activities. *Sci. Hortic. Amst.* **2011**, *130*, 344–348. [CrossRef]

50. Afreen, F.; Zobayed, S.M.A.; Kozai, T. Photoautotrophic culture of *Coffea arabusta* somatic embryos: Photosynthetic ability and growth of different stage embryos. *Ann. Bot.* **2002**, *90*, 11–19. [CrossRef] [PubMed]

51. Rival, A.; Beulé, T.; Lavergne, D.; Nato, A.; Havaux, M.; Puard, M. Development of photosynthetic characteristics in oil palm during in vitro micropropagation. *J. Plant Physiol.* **1997**, *150*, 520–527. [CrossRef]

52. Karami, O.; Saidi, A. The molecular basis for stress-induced acquisition of somatic embryogenesis. *Mol. Biol. Rep.* **2010**, *37*, 2493–2507. [CrossRef] [PubMed]

53. Cui, K.; Xing, G.; Zhou, G.; Liu, X.; Wang, Y. The induced and regulatory effects of plant hormones in somatic embryogenesis. *Hereditas* **2000**, *22*, 349–354.

54. Jiménez, V.M. Regulation of in vitro somatic embryogenesis with emphasis on to the role of endogenous hormones. *Rev. Bras. Fisiol. Veg.* **2001**, *13*, 196–223. [CrossRef]

55. Kumria, R.; Sunnichan, V.G.; Das, D.K.; Gupta, S.K.; Reddy, V.S.; Bhatnagar, R.K.; Leelavathi, S. High-frequency somatic embryo production and maturation into normal plants in cotton (*Gossypium hirsutum*) through metabolic stress. *Plant Cell Rep.* **2003**, *21*, 635–639. [PubMed]

56. Patnaik, D.; Mahalakshmi, A.; Khurana, P. Effect of water stress and heavy metals on induction of somatic embryogenesis in wheat leaf base cultures. *Indian J. Exp. Biol.* **2005**, *43*, 740–745. [PubMed]

57. Cheong, Y.H.; Chang, H.S.; Gupta, R.; Wang, X.; Zhu, T.; Luan, S. Transcriptional profiling reveals novel interactions between wounding, pathogen, abiotic stress, and hormonal responses in *Arabidopsis*. *Plant Physiol.* **2002**, *129*, 661–677. [CrossRef]

58. Karami, O.; Deljou, A.; Esna-Ashari, M.; Ostad-Ahmadi, P. Effect of sucrose concentrations on somatic embryogenesis in carnation (*Dianthus caryophyllus* L.). *Sci. Hortic.* **2006**, *110*, 340–344. [CrossRef]

59. Karami, O.; Deljou, A.; Kordestani, G.K. Secondary somatic embryogenesis of carnation (*Dianthus caryophyllus* L.). *Plant Cell* **2008**, *92*, 273–280. [CrossRef]

60. Klimaszewska, K.; Cyr, D.R.; Sutton, B.C.S. Influence of gelling agents on culture medium gel strength, water availability, tissue water potential, and maturation response in embryogenic cultures of *Pinus strobus* L. *In Vitro Cell Dev. Biol. Plant* **2000**, *36*, 279–286. [CrossRef]

61. Kiyosue, T. Somatic embryogenesis in higher plants. *J. Plant Res.* **1993**, *3*, 75–82.

62. Lehmann, T.; Polcyn, W.; Ratajczak, L. Glutamate dehydrogenase isoenzymes in yellow lupine root nodules. III. Affinity for ammonia. *Acta Physiol. Plant* **1990**, *12*, 259–263.

63. Yamaya, T.; Oaks, A. Synthesis of glutamate by mitochondria–an anaplerotic function for glutamate dehydrogenase. *Physiol. Plant.* **1987**, *70*, 749–756. [CrossRef]

64. Gancedo, J.M. Yeast carbon catabolite repression. *Microbiol. Mol. Biol. R.* **1998**, *62*, 334–361.

65. Saier, M.H. Protein phosphorylation and allosteric control of inducer exclusion and catabolite repression by the bacterial phosphoenolpyruvate: Sugar phosphotransferase system. *Microbiol. Mol. Biol. Rev.* **1989**, *53*, 109–120.

66. Nishiwaki, M.; Fujino, K.; Koda, Y.; Masuda, K.; Kikuta, Y. Somatic embryogenesis induced by the simple application of abscisic acid to carrot (*Daucus carota* L.) seedlings in culture. *Planta* **2000**, *211*, 756–759. [CrossRef]

67. Jin, L.G.; Li, H.; Liu, J.Y. Molecular characterization of three ethylene responsive element binding factor genes from cotton. *J. Integr. Plant Biol.* **2010**, *52*, 485–495. [CrossRef]

68. Ho, S.L.; Huang, L.F.; Lu, C.A.; He, S.L.; Wang, C.C.; Yu, S.P.; Chen, J.; Yu, S.M. Sugar starvation-and GA-inducible calcium-dependent protein kinase 1 feedback regulates GA biosynthesis and activates a 14-3-3 protein to confer drought tolerance in rice seedlings. *Plant Mol. Biol.* **2013**, *81*, 347–361. [CrossRef]

69. Friml, J.; Benková, E.; Blilou, I.; Wisniewska, J.; Hamann, T.; Ljung, K.; Woody, S.; Sandberg, G.; Scheres, B.; Jürgens, G.; Palme, K. *AtPIN4* mediates sink-driven auxin gradients and root patterning in *Arabidopsis*. *Cell* **2002**, *108*, 661–673. [CrossRef]

70. Blilou, I.; Xu, J.; Wildwater, M.; Willemsen, V.; Paponov, I.; Friml, J.; Heidstra, R.; Aida, M.; Palme, K.; Scheres, B. The *PIN* auxin efflux facilitator network controls growth and patterning in *Arabidopsis* roots. *Nature* **2005**, *433*, 39. [CrossRef] [PubMed]

71. Hagen, G.; Kleinschmidt, A.; Guilfoyle, T. Auxin-regulated gene expression in intact soybean hypocotyl and excised hypocotyl sections. *Planta* **1984**, *162*, 147–153. [CrossRef] [PubMed]

72. Hagen, G.; Guilfoyle, T.J. Rapid induction of selective transcription by auxins. *Plant Mol. Biol.* **1985**, *5*, 1197–1203. [CrossRef]

73. Yang, X.Y.; Zhang, X.L. Regulation of somatic embryogenesis in higher plants. *Crit. Rev. Plant Sci.* **2010**, *29*, 36–57. [CrossRef]

74. Gallie, D.R.; Young, T.E. The ethylene biosynthetic and perception machinery is differentially expressed during endosperm and embryo development in maize. *Mol. Genet. Genom.* **2004**, *271*, 267–281. [CrossRef] [PubMed]

75. Langhansova, L.; Konradova, H.; Vaněk, T. Polyethylene glycol and abscisic acid improve maturation and regeneration of *Panax ginseng* somatic embryos. *Plant Cell Rep.* **2004**, *22*, 725–730. [CrossRef]

76. Nieves, N.; Martinez, M.E.; Castillo, R.; Blanco, M.A.; González-Olmedo, J.L. Effect of abscisic acid and jasmonic acid on partial desiccation of encapsulated somatic embryos of sugarcane. *Plant Cell Tissue Org.* **2001**, *65*, 15–21. [CrossRef]

77. Von, A.S.; Hakman, I. Regulation of somatic embryo development in *Picea abies* by abscisic acid (ABA). *J. Plant Physiol.* **1988**, *132*, 164–169.

78. Jürgens, G. Apical–basal pattern formation in *Arabidopsis* embryogenesis. *EMBO J.* **2001**, *20*, 3609–3616. [CrossRef]

79. Zhu, J.K. Abiotic stress signaling and responses in plants. *Cell* **2016**, *167*, 313–324. [CrossRef] [PubMed]

80. Anil, V.S.; Rao, K.S. Calcium-mediated signaling during sandalwood somatic embryogenesis. Role for exogenous calcium as second messenger. *Plant Physiol.* **2000**, *123*, 1301–1312. [CrossRef] [PubMed]

81. Noah, A.M.; Niemenak, N.; Sunderhaus, S.; Haase, C.; Omokolo, D.N.; Winkelmann, T.; Braun, H.P. Comparative proteomic analysis of early somatic and zygotic embryogenesis in *Theobroma cacao* L. *J. Proteom.* **2013**, *78*, 123–133. [CrossRef] [PubMed]

82. O'Leary, B.; Park, J.; Plaxton, W.C. The remarkable diversity of plant PEPC (phosphoenolpyruvate carboxylase): Recent insights into the physiological functions and post-translational controls of non-photosynthetic PEPCs. *Biochem. J.* **2011**, *436*, 15–34. [CrossRef]

83. Aivalakis, G.; Dimou, M.; Flemetakis, E.; Plati, F.; Katinakis, P.; Drossopoulos, J.B. Immunolocalization of carbonic anhydrase and phosphoenolpyruvate carboxylase in developing seeds of *Medicago sativa*. *Plant Physiol. Biochem.* **2004**, *42*, 181–186. [CrossRef] [PubMed]

84. Rolletschek, H.; Borisjuk, L.; Radchuk, R.; Miranda, M.; Heim, U.; Wobus, U.; Weber, H. Seed-specific expression of a bacterial phosphoenolpyruvate carboxylase in Vicia narbonensis increases protein content and improves carbon economy. *Plant Biotechnol. J.* **2004**, *2*, 211–219. [CrossRef]

85. Jing, D.; Zhang, J.; Xia, Y.; Kong, L.; OuYang, F.; Zhang, S.; Zhang, H.; Wang, J. Proteomic analysis of stress-related proteins and metabolic pathways in *Picea asperata* somatic embryos during partial desiccation. *Plant Biotechnol. J.* **2017**, *15*, 27–38. [CrossRef] [PubMed]

86. Boursiac, Y.; Chen, S.; Luu, D.T.; Sorieul, M.; van den Dries, N.; Maurel, C. Early effects of salinity on water transport in *Arabidopsis* roots. Molecular and cellular features of aquaporin expression. *Plant Physiol.* **2005**, *139*, 790–805. [CrossRef] [PubMed]

87. Loqué, D.; Ludewig, U.; Yuan, L.; von Wirén, N. Tonoplast intrinsic proteins AtTIP2;1 and AtTIP2;3 facilitate NH3 transport into the vacuole. *Plant Physiol.* **2005**, *137*, 671–680. [CrossRef] [PubMed]

88. Leitão, L.; Prista, C.; Moura, T.F.; Loureiro-Dias, M.C.; Soveral, G. Grapevine aquaporins: Gating of a tonoplast intrinsic protein (TIP2;1) by cytosolic pH. *PLoS ONE* **2012**, *7*, e33219. [CrossRef] [PubMed]

89. Murch, S.J.; Saxena, P.K. Modulation of mineral and free fatty acid profiles during thidiazuron mediated somatic embryogenesis in peanuts (*Arachis hypogeae* L.). *J. Plant Physiol.* **1997**, *151*, 358–361. [CrossRef]

90. Hundertmark, M.; Hincha, D.K. LEA (late embryogenesis abundant) proteins and their encoding genes in *Arabidopsis thaliana*. *BMC Genom.* **2008**, *9*, 118. [CrossRef] [PubMed]

91. Palovaara, J.; Hakman, I. Conifer *WOX*-related homeodomain transcription factors, developmental consideration and expression dynamic of *WOX2* during *Picea abies* somatic embryogenesis. *Plant Mol. Biol.* **2008**, *66*, 533–549. [CrossRef] [PubMed]

92. Höck, J.; Meister, G. The Argonaute protein family. *Genome Biol.* **2008**, *9*, 210. [CrossRef] [PubMed]

93. Szyrajew, K.; Bielewicz, D.; Dolata, J.; Wójcik, A.M.; Nowak, K.; Szczygieł-Sommer, A.; Szweykowska-Kulinska, Z.; Jarmolowski, A.; Gaj, M.D. MicroRNAs are intensively regulated during induction of somatic embryogenesis in *Arabidopsis*. *Front. Plant Sci.* **2017**, *8*, 18. [CrossRef]

94. Takahata, K. Isolation of carrot Argonaute1 from subtractive somatic embryogenesis cDNA library. *Biosci. Biotechnol. Biochem.* **2008**, *72*, 900–904. [CrossRef] [PubMed]

95. Tahir, M.; Law, D.A.; Stasolla, C. Molecular characterization of *PgAGO*, a novel conifer gene of the ARGONAUTE family expressed in apical cells and required for somatic embryo development in spruce. *Tree Physiol.* **2006**, *26*, 1257–1270. [CrossRef] [PubMed]

96. Munksgaard, D.; Mattsson, O.; Okkels, F.T. Somatic embryo development in carrot is associated with an increase in levels of S-adenosylmethionine, S-adenosylhomocysteine and DNA methylation. *Physiol. Plant.* **1995**, *93*, 5–10. [CrossRef]

97. Jarillo, J.A.; Piñeiro, M.; Cubas, P.; Martínez-Zapater, J.M. Chromatin remodeling in plant development. *Int. J. Dev. Biol.* **2009**, *53*, 1581–1596. [CrossRef] [PubMed]

98. Cheng, W.H.; Zhu, H.G.; Tian, W.G.; Zhu, S.H.; Xiong, X.P.; Sun, Y.Q.; Zhu, Q.H.; Sun, J. De novo transcriptome analysis reveals insights into dynamic homeostasis regulation of somatic embryogenesis in upland cotton (*G. hirsutum* L.). *Plant Mol. Biol.* **2016**, *92*, 279–292. [CrossRef]

99. Cao, A.; Zheng, Y.; Yu, Y.; Wang, X.; Shao, D.; Sun, J.; Cui, B. Comparative Transcriptome Analysis of SE initial dedifferentiation in cotton of different SE capability. *Sci. Rep.* **2017**, *7*, 8583. [CrossRef] [PubMed]

100. Xiao, Y.; Chen, Y.; Ding, Y.; Wu, J.; Wang, P.; Yu, Y.; Wei, X.; Wang, Y.; Zhang, C.; Li, F.; Ge, X. Effects of *GhWUS* from upland cotton (*Gossypium hirsutum* L.) on somatic embryogenesis and shoot regeneration. *Plant Sci.* **2018**, *270*, 157–165. [CrossRef] [PubMed]

101. Li, J.; Wang, M.; Li, Y.; Zhang, Q.; Lindsey, K.; Daniell, H.; Jin, S.; Zhang, X. Multi-omics analyses reveal epigenomics basis for cotton somatic embryogenesis through successive regeneration acclimation process. *Plant Biotechnol. J.* **2019**, *17*, 435–450. [CrossRef] [PubMed]

102. da Cunha Soares, T.; da Silva, C.R.C.; Chagas Carvalho, J.M.F.; Cavalcanti, J.J.V.; de Lima, L.M.; de Albuquerque Melo Filho, P.; Severino, L.S.; Dos Santos, R.C. Validating a probe from *GhSERK1* gene for selection of cotton genotypes with somatic embryogenic capacity. *J. Biotechnol.* **2018**, *270*, 44–50. [CrossRef] [PubMed]

103. Xu, J.; Yang, X.; Li, B.; Chen, L.; Min, L.; Zhang, X. *GhL1L1* affects cell fate specification by regulating *GhPIN1*-mediated auxin distribution. *Plant Biotechnol. J.* **2019**, *17*, 63–74. [CrossRef] [PubMed]

104. Wu, J.; Zhang, X.; Nie, Y.; Jin, S.X.; Liang, S.G. Factors affecting somatic embryogenesis and plant regeneration from a range of recalcitrant genotypes of Chinese cottons (*Gossypium hirsutum* L.). *In Vitro Cell. Dev. Biol. Plant* **2004**, *40*, 371–375. [CrossRef]

International Journal of
Molecular Sciences

Review

The Latest Studies on Lotus (*Nelumbo nucifera*)-an Emerging Horticultural Model Plant

Zhongyuan Lin [1,2,3], Cheng Zhang [1], Dingding Cao [2,3], Rebecca Njeri Damaris [1] and Pingfang Yang [1,*]

[1] State Key Laboratory of Biocatalysis and Enzyme Engineering, School of Life Sciences, Hubei University, Wuhan 430062, China
[2] Key Laboratory of Plant Germplasm Enhancement and Specialty Agriculture, Wuhan Botanical Garden, Chinese Academy of Sciences, Wuhan 430074, China
[3] University of Chinese Academy of Sciences, Beijing 100039, China
* Correspondence: yangpf@hubu.edu.cn; Tel.: +86-27-88661237; Fax: +86-27-87700860

Received: 16 June 2019; Accepted: 20 July 2019; Published: 27 July 2019

Abstract: Lotus (*Nelumbo nucifera*) is a perennial aquatic basal eudicot belonging to a small family *Nelumbonaceace*, which contains only one genus with two species. It is an important horticultural plant, with its uses ranging from ornamental, nutritional to medicinal values, and has been widely used, especially in Southeast Asia. Recently, the lotus obtained a lot of attention from the scientific community. An increasing number of research papers focusing on it have been published, which have shed light on the mysteries of this species. Here, we comprehensively reviewed the latest advancement of studies on the lotus, including phylogeny, genomics and the molecular mechanisms underlying its unique properties, its economic important traits, and so on. Meanwhile, current limitations in the research of the lotus were addressed, and the potential prospective were proposed as well. We believe that the lotus will be an important model plant in horticulture with the generation of germplasm suitable for laboratory operation and the establishment of a regeneration and transformation system.

Keywords: *Nelumbo nucifera*; phylogeny; genomics; molecular mechanisms; model plant

---

1. Introduction

Lotus is a perennial aquatic plant. It belongs to the small family of *Nelumbonaceae*, comprising of only one genus Nelumbo with two species: *Nelumbo nucifera* Gaertn. and *Nelumbo lutea* Pear., which are popularly named as Asian lotus and American lotus, respectively [1]. Generally, lotus refers to Asian lotus and mainly distributes in Asia and the north of Oceania, while the American lotus primarily occurs in the eastern and southern parts of North America, as well as the north of South America [1–4] (Figure 1). Being separated by the Pacific Ocean, these two species differ in their external morphologies, such as petal color and shape, leaf shape and plant size [5] (Figure 1). In spite of this, both of them have the same chromosome number (2n = 16), and show a similar life style, with about five months of life span for each generation. Crossing between these two species could generate an F1 population, which is totally infertile. Although there are only two species of lotus in taxonomy, very abundant germplasms exist all over the world, which display variable genetic backgrounds and phenotypes, especially in Asia. In addition, the lotus is a basal eudicot, which makes it a very important species in plant phylogenetic and evolution studies.

Asian lotus is also named as sacred lotus because of its significance in the religions of Buddhism and Hinduism [5]. It is a very good symbol in Chinese traditional culture. All of these make sacred lotus a very popular ornamental plant. In addition, it is also a popular vegetable and traditional medicinal plant with great economic value in South-East Asia. China is regarded as one of the major centers in lotus cultivation and breeding, with over several thousands of years of cultivation history [1].

As the result of the long period of breeding, domestication and cultivation, large amounts of lotus cultivars have been obtained, showing variable morphology and other traits. The cultivated lotus is generally divided into three categories, namely, rhizome, seed and flower lotus, according to their usage in reality. The lotus rhizome and seed could not only be consumed as vegetables, but are also used for lotus propagation, whereas, the flower lotus is mainly applied in ornamentation and environmental improvement. Based on the climatic regions they are accustomed to, sacred lotus could also be classified into two ecotypes, which are temperate lotus and tropical lotus. The temperate lotus has an enlarged rhizome occurring after flowering and its leaves wither. In contrast, the tropical lotus has a whip-like rhizome with a longer green period and flowering time [1,2].

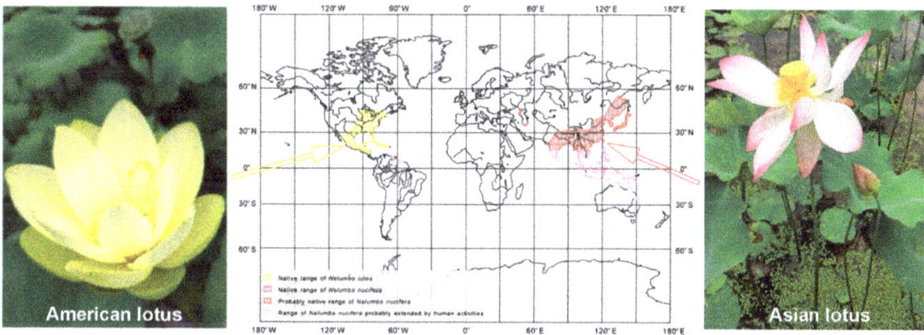

Figure 1. Overview of the lotus species and their global distribution. The left and right panels show the flowers of American and Asian lotus, respectively. The yellow and red shadow areas in the world map of the middle panel show the distributions of American and Asian lotus, respectively. (Figure revised from Li, Y. et al. [4]).

Because of its importance in horticulture, medicinal usage and plant phylogeny, the sacred lotus has gained increasing interests from the scientific community. It will undoubtedly enhance the breeding and application of lotus to obtain enough fundamental knowledge about this plant. Recently, the genome of two sacred lotus germplasms were sequenced and released [6,7], which facilitates further study on this species. Up-to-date, there are nearly 1000 research publications focusing on different aspects of the lotus, half of which were published in the last decade. In this review, we summarized the latest advancement of studies on the sacred lotus in order to provide a comprehensive insight into the basic biology and economic usage of this important plant, which might also contribute to future studies on lotus breeding and germplasm enhancement.

2. Phylogeny and Genomic Studies

Taxonomically, lotus belongs to the genus of *Nelumbo*, which is the only existing genus of the Nelumbonaceae family. Cretaceous fossils have been assigned to Nelumbonaceae. Analysis on these fossils indicate that the family of Nelumbonaceae might have more than 100 Ma years of history, and showed considerable morphological stasis. [8]. Determination of lotus classification in taxonomy took a long time. Because of its superficial similarities in the flowers and vegetative body with the waterlily, Nelumbo used to be regarded as one genus of the Nymphaeaceae family in the old classification system. In the Cronquist system, the Nelumbonaceae family was recognized, but still placed in the order of Nymphaeales [9]. In both the Dahlgren system [10] and the Thorne system [11], the Nelumbonaceae family was placed in its own order, Nelumbonales. Takhtajan [12] removed Nelumbonaceae from Nymphaeales, but placed them alone in the subclass of the Nelumbonidae. With the increasing accumulation of evidence at the molecular level, The Angiosperm Phylogeny Group (APG) has placed it into the basal eudicot order of Proteales, which is outside of the core eudicots (http://www.mobot.org/MOBOT/research/APweb/, last accessed date: 23 June, 2019) [13].

Except for Nelumbonaceae, Proteales contains three other families, including Platanaceae, Proteaceae, and Sabiaceae, of which the former two are the closest relatives of the lotus, and are mainly shrubs and woody trees [14], indicating the possibility of the lotus being a land plant adapted to aquatic environments. Interestingly, the family of Nelumbonaceae is still classified within the order of Nymphaeales on the USDA webpage (https://plants.sc.egov.usda.gov/core/profile?symbol=NENU2, last accessed date: 22 July, 2019). Furthermore, studies also showed that the gene expression patterns in the floral organs of *Nymphaea* and *Nelumbo* are remarkably similar to each other [15]. It would be interesting to understand the evolutionary convergences between Nymphaeales and the lotus.

From the genetic point of view, both species of lotus are diploid with the number of chromosomes 2n = 16. The predicted size of the lotus genome is 929 Mb, which is based on the flow cytometry analysis [16]. In 2013, the draft genomes of two lotus wild germplasms 'China Antique' and 'Chinese Tai-zi' were sequenced, assembled and released [6,7], which made lotus into a model angiosperm along with the other 22 species (http://www.mobot.org/MOBOT/research/APweb/trees/modeltreemap.html, last accessed date: 23 June, 2019). The assembled genome size of 'China Antique' is 804 Mb and the sequencing data shows that this genome contains 26,685 protein-coding genes [6]. Recently, a more comprehensive transcriptomic analysis increased the number of protein-coding genes to 32,121 in 'China Antique' [17]. The assembled genome of 'Chinese Tai-zi' is 792 Mb with 36,385 protein-coding genes [7,18]. Based on their data, it seems that the lotus genome contains a high content of repeat sequences, with transposable elements (TEs) accounting for about 50% of the genome sequence. The availability of these two genomes will facilitate further studies on the different biological features of lotus, including agronomic and horticultural traits, and might contribute to the knowledge of flowering plant evolution. Wang et al. [19] combined the lotus genome and transcriptome data of 'China Antique' and constructed the public accessible lotus genome database (http://lotus-db.wbgcas.cn, last accessed date: 20 March, 2015), which makes further molecular and genetic studies on this species more convenient among the scientific community. Additionally, the lotus chloroplast and mitochondrion genome were also sequenced, which have been applied in optimizing the genetic maps and analyzing the evolution of the lotus [20,21]. Because of the availability of abundant genome information, phylogenetic and evolution analysis of lotus at the molecular level was also conducted, which showed the functional divergence of miRNAs in temperate and tropical lotus [22]. Based on the study, 57 pre-eudicot miRNA families from different evolutionary stages were predicted. Combining the miRNA data and the lotus genome information, it was revealed that the loss of miRNA families in descendent plants is associated with that in duplicated genomes [22]. However, because of the high percentage of repetitive sequences (>47%), the assembly of the lotus genome, especially for 'China Antique', is still far behind completion, although a study has been conducted aiming at anchoring the megascaffolds into eight chromosomes [23]. The nine anchored megascaffolds, which have a combined size of 543.4 Mb, just account for 67.6% of the lotus genome. The advent of a third generation sequencing technique has been successfully applied in many other species, which will be able to improve the assembly of this lotus genome in the near future.

3. Unique Properties of Lotus

Biologically, lotus has not only the common aquatic plant features, but also certain unique features that distinguish it from other plant species. These features include seed longevity, leaf ultrahydrophobicity and floral thermoregulation. Understanding of the mechanisms that lead to the formation of these unique properties is important, for not only the basic plant biology, but also their great usage potential through bionics.

Lotus fruit is famous for its longevity. It was reported that lotus fruits buried underground over 1300 years in the Northeast of China could still be germinated [24]. Understanding the underlying mechanism of lotus seed longevity may contribute to enhancing seed storage in agriculture, and even in the healthcare of human beings.

Previous studies have shown that the first factor contributing to this feature might be the chemical compositions of lotus fruit wall, which contains high contents of polysaccharides (galactose, mannose) and tannins [25]. These compounds might help to prevent any negative effects from the environment. Recently, another study showed that the polyphenols content in lotus seed epicarp increased along with the ripening, and showed strong anti-oxidation activity [26], which might also be helpful. Besides of the physical factors, several thermo-proteins, which showed high stability under high temperature, were also indicated to be helpful. These proteins include CuZn-SOD, 1-CysPRX, dehydrin, Cpn20, Cpn60, HSP80, EF-1α, Enolase1, vicilin, Met-Synthase and PIMT [27]. The functions of some genes involved in seed thermos-tolerance and germination vigor, including *NnANN1* and *NnPER1* (*Peroxiredoxin 1*), were verified in transgenic *Arabidopsis* [28,29]. To achieve this, the lotus genome contains multi-copies for most of the antioxidative genes [6,7]. Recent study showed that small RNA might also be involved in the regulation of lotus seed longevity [30]. How these different factors cooperatively function to contribute to the lotus seed longevity is still elusive, but worthy of studying. More importantly, it is very interesting to know if these factors also work in other systems.

Lotus leaves exhibit ultra-hydrophobicity, which is also known as the "lotus effect" [31]. This characteristic of ultra-hydrophobicity could ensure that the leaf upper epidermis is not covered by water, thus maintaining the normal function of its stomata [32]. Because of this, ultrahydrophobicity is believed to be an advantage in the evolution of the lotus. Studies have shown that it is achieved by a special dense layer of waxy papillae on the lotus leaf surface [33,34]. Further studies showed that the easily rolling water droplets could help to remove the dirt particles adhering on the leaf surface and result in a self-cleaning phenomenon, which is heavily dependent upon the contact angle [35]. Two wax biosynthesis-related genes (*NnCER2* and *NnCER2-LIKE*) were cloned from the lotus, and transformed in *Arabidopsis*, which resulted in an alteration of the cuticle wax structure in inflorescence stems, and proved their function in the biosynthesis of the extra-long fatty acids [36]. More studies on the lotus leaf chemical compositions and structure might be very helpful in producing materials with super-hydrophobicity and self-cleaning features.

In addition, floral organ thermogenesis is another unique feature of the lotus, which independently occurs at receptacle, stamen and petal, respectively [37]. This property has been proven to be the results of a cyanide-resistant alternative oxidase pathway conducted in the floral organs [38–40], which initiated extensive studies on alternative oxidases (AOXs) and plant uncoupling mitochondrial proteins (PUMPs) [41]. This feature of thermogenesis seems to be ecologically important for the sexual reproduction of the lotus through attracting insect pollinators [42]. Studies have shown that the generated heat could either provide a warm environment to the thermo-sensitive pollinators or help to release the volatile compounds to attract the flying insects, mainly beetles [37,43,44]. Generation of heat only occurs before anthesis, which ends with pollination and a fertilized ovary. After anthesis, there is no need to attract the pollinators any more, and the main function of the floral organs, especially the receptacle, transits into photosynthesis [45,46]. It will be very important to explore the mechanism that controls this kind of metabolism transition.

4. Genetic and Molecular Studies on the Horticultural Traits of the Lotus

As mentioned above, a lotus is also a popular ornamental, vegetable and medicinal plant, with great potential of utilization in reality, based on which, three types of lotus, named as flower, seed, and rhizome lotus, were defined. Each type of lotus shows notable abundant variable phenotypes (Figure 2), which provide suitable germplasm for its breeding and further study on different traits. Recently, a number of studies have been conducted focusing on the genetic and molecular mechanisms underlying the formation of different traits of lotus flower, seed and rhizome. These traits could largely determine the economic value of the lotus, hence becoming the main factors selected in its breeding. Several genetic maps have been constructed through crossing between different germplasms with contrasting phenotypes in some of the economic traits, based on which a number of molecular markers associated with the target traits were developed, including ISSR, AFLP, SSR, RAPD, and SRAP [47–50].

Figure 2. The diversified phenotypes of the Asian lotus germplasm. (**A**) Flower lotus germplasm showing different flower color and shape. (**B**) Seed lotus germplasm showing different size and shape of seed and seedpod. (**C**) Rhizome lotus germplasm showing different branching, elongation and expansion of the rhizome.

Meanwhile, whole genome re-sequencing on the natural germplasm also identified abundant SNPs and Indels [51–53]. Together, these data will undoubtedly facilitate the lotus breeding.

4.1. The Flower of Lotus

Lotus flower is among the top ten traditional famous flowers in China, and was chosen as the national flower in India and Vietnam. It is widely cultivated for its aesthetic value, which is largely attributed to its gorgeous color and its diversified form and shape (Figure 2). For ornamental plants, flower color and shape are the major two factors that determine their ornamental value. The lotus petals show three major colors; white, red and yellow, with the former two existing only in Asian lotus and the later one only in American lotus. Through breeding and artificial selection, many cultivars with mixed colors have been obtained on the purpose of increasing its ornamental value (Figure 2). A large-scale analysis on the pigment composition of different germplasm has shown that the yellow and red color is mainly determined by the contents of carotenoids and anthocyanins, respectively [54]. Genome-wide analysis of the *MYB* gene family indicated that there is a similar anthocyanin biosynthesis regulatory system in lotus and *Arabidopsis* [55], based on which an overexpression of *NnMYB5* in *Arabidopsis* resulted in the accumulation of anthocyanin in immature seeds and flower stalks [56]. In spite of this similarity, a comparative proteomics study between white and red cultivars showed that the expression of the *ANS* gene might be the major reason for the absence of anthocyanin biosynthesis in the white flower lotus [57]. Further analysis found that different levels of methylation occur in the promoter regions of *ANS* gene between the two cultivars, which indicates the epigenetic regulation on expression of this gene. However, the gene that lead to the different methylation level on the promoter of *ANS* gene between the red and white lotus cultivars is still unknown. In addition, there are cultivars showing genetic constant spotted color (Figure 3), which is still not understood. It will be very important not only to the breeding of flower lotus, but also to enriching our knowledge on the coloration of plant flowers to explore the mechanism underlying the regulation of spotted color in lotus.

Figure 3. Lotus cultivar with spotted color flower.

In addition to color, flower shape is also important for the economic value of ornamental plants. Based on different purposes of breeding, lotus cultivars with diversified flower shapes were obtained, including few-petalled, semidouble-petalled, double-petalled, duplicate-petalled and all-double-petalled cultivars [2]. For the semidouble-petalled, double-petalled shapes, they are usually the resultants of stamen petaloid. Comparative transcriptomic studies among petal, stamen petaloid and stamen through RNA-seq were conducted, which identified several candidate genes involved in stamen petaloid, especially some MADS-box genes [58]. Their study revealed 11 MADS-box genes and one *APETALA2* (*AP2*) gene being involved in the stamen petaloid phenomenon. Among them, *AGL15*, *AGL80* and *AGAMOUS* genes are positively related to, and *AGL6* is negatively related to the stamen petaloid [58]. Meanwhile, a genome-wide DNA methylation analysis was also conducted among these three tissues, which indicates the potential involvement of epigenetic regulation on the stamen petaloid [59]. However, this study did not detect any obvious changes of the methylation on the MADS-box genes [59]. There also exist pistil petaloid cultivars (Figure 2), in which the stamen petaloid also occurs. How these are coordinately regulated is still unknown in the lotus. Furthermore, it is well known that lotus bloom in the summer days, which brings some challenges for its wide utilization in ornamentation. It will be very important to make it bloom either earlier or later for ornamental purposes. Hence, unveiling the mechanism controlling the time of flowering is also important. A transcriptomic analysis has been conducted aiming at exploring the candidate genes that control the time of flowering, which indicate the existence of a complicated regulatory network [60]. Their data indicate that the differential regulation of some photoperiod related genes, such as *COP1*, *CCA1*, *LHY*, *CO-LIKE*, and *FT*, the vernalization gene *VIN3* and the gibberellic acid-related gene *GAI*, might be involved in the regulation of early flowering in lotus. Specifically, several isoforms of the *FT* gene were found to be differentially expressed [60].

4.2. Rhizome and Seeds

As mentioned above, lotus is not only an ornamental plant, but also a vegetable because of its edible rhizome and seeds. Lotus has a morphologically modified subterraneous stem. Especially for the temperate ecotype, its subterraneous stem is enlarged in autumn, which is known as rhizome (Figure 2). The rhizome contains abundant starch, proteins and vitamins, making it a popular edible vegetable. Enlargement of lotus rhizome could largely determine its economic value. In addition, the enlarged rhizome could also help the lotus to survive from winter during its bud dormancy, and provide substrates and energy for its asexual propagation. This phenomenon is very similar with the tuberization of the potato, which has been proven to be regulated through a very intricate genetic network. Being a significant feature distinguishing between the temperate and tropical lotus, it may also facilitate in understanding the evolution and domestication of the lotus [2,61]. It seems that rhizome enlargement is tightly related to the flowering in a lotus. Usually, the enlargement occurs after flowering. For the purposes of increasing its yield in agricultural production, genetic and transcriptomic studies focusing on the enlargement of this rhizome have been conducted.

Gene expressions during the rhizome development were analyzed through RNA-Seq, which identified the specific candidate genes for rhizome enlargement [62]. The results also indicated the role of SNPs and alternative splicing (AS) events in Asian lotus rhizome development [61,63]. Similar with the yield traits in many crops, the enlargement of the lotus rhizome is a quantitative trait. Developing a suitable genetic population and constructing high density genetic map will be very helpful to elucidate the mechanism underlying rhizome development and enlargement.

Besides its longevity, lotus seed is also edible either fresh or dry matured, with an additional medicinal versatility resulting from compounds like alkaloids, flavonoids and certain micronutrients [3,5]. Both the size and number of the seeds per seedpod vary among different lotus cultivars (Figure 2). It is very important to increase its nutrition as well as its yield in lotus seed production. To achieve this, comparative proteomics and metabolomics studies were conducted on lotus seeds during its development, which not only deepen the understanding on the development of lotus seed, but also determine candidate genes crucial for lotus seed size [62]. In addition, comparative transcriptomic analysis was also conducted between two lotus germplasms with contrasting phenotypes in both seed size and seed number per seedpod [64]. Similar to rhizome, the yield of seed is also a quantitative trait, which requires more study at the genetic aspect. Meanwhile, because of its medicinal usage, it is necessary to conduct a comprehensive analysis on it metabolites during seed development.

4.3. Secondary Metabolites and Medicinal Usage of Lotus

Lotus is a traditional herb, of which nearly each tissue has a medicinal usage [65–67]. It has been used as a traditional Chinese medicine for over a thousand years. This might ascribe to its abundant content of secondary metabolites, including flavonoids, phenolic acids and alkaloids [65–67]. Systematic studies were conducted in optimizing the method to extract these metabolites from different tissues of the lotus [26,54,68–77]. Meanwhile, distributions of different secondary metabolites in different tissues of lotus were profiled [26,54,68–77]. Furthermore, assessment of the lotus germplasm with different origins was also performed by these established methods [65–67], which helped in screening of the germplasm with a high content of specific secondary metabolites. These candidate germplasms might be used for either the breeding or for further study on the biosynthesis of different metabolites in the lotus. In addition, the potential medicinal usage of different lotus secondary metabolites was also assessed [65–67]. However, the exact compounds that function in each medicinal usage are still unknown, which seems to be the general challenge for most traditional Chinese medicine.

Specifically, the leaf of a lotus is a very important traditional Chinese herbal medicine, which has been widely used in controlling the blood lipids and treating hyperlipidemia [78]. In the last decade, it is becoming more and more popular as weight-losing tea in China to reduce the level of lipids in the human body [79]. Studies have shown that alkaloids are the major bioactive compounds in lotus leaves, with nuciferine and N-nornuciferine being the major two [80–82]. To evaluate the biosynthesis pathway of alkaloids and its regulation in lotus leaf, several transcriptomic studies were performed [83,84], which revealed that a benzylisoquinoline alkaloids (BIA) biosynthetic pathway and its transcriptional regulation differ in high BIAs lotus compared with low BIAs lotus [84]. Several genes encoding the enzymes involved in the BIA biosynthetic pathway were proposed based on sequence similarity analysis [85]. Further functional analysis of these genes will be necessary to obtain comprehensive knowledge on the biosynthesis of these bioactive compounds.

4.4. Studies on the Establishment of Lotus Regeneration and Transformation System

To be a model horticultural plant, it might be necessary to establish a transformation system, which will facilitate the studies on the functions of different genes in the lotus. A study was conducted to induce the formation of a callus from different explants of the lotus, in which somatic embryo cultivated in suitable medium containing a combination of different growth regulators was proposed [86]. To obtain more in-depth understanding, a proteomic analysis was conducted to identify the key proteins that might be critical for the induction of callus from developing cotyledon [87].

Directly inducing the formation of a shoot from the bud has also been successfully performed [88]. Based on this system, various studies have been conducted to transform the lotus. It seems that the induced shoot from the embryo apical bud could be successfully transformed through a particle bombardment device with a pCAMBIA2301 vector [89]. This method not only succeeded with the GUS reporter gene, but also with the anti-sense of two anthocyanin biosynthesis genes *dihydroflavonol 4-reductase* (anti-*DFR*) and *Chalcone synthase* (anti-*CHS*) [89,90]. Except for the group from Thailand, there are still no other studies conducted successfully on the transformation of the lotus, although a lot of researchers are working on this. It seems there are still challenges on the reproducibility and the efficiency of the transformation, as well as the selection of a suitable cultivar.

5. Conclusions and Perspectives

Because of its significance in the ordinary life of the population in South and East Asia, as well as in horticultural and medicinal usage, lotus is attracting more and more attention from the scientific community. A large number of studies have been conducted on nearly all aspects of this plant, including phylogeny and evolution, genomics, genetics and breeding and medicinal usage. With the release of its genome information, -omics and molecular genetics studies, focusing on the economic traits of this plant have stepped into the center, which undoubtedly will contribute a lot to the lotus breeding. Unfortunately, there are still some limitations that constrain the studies, especially the molecular biology study, on this species. The first one might be the assembly and annotation of its genome, which still needs to be improved further. Secondly, there is no universally recognized lotus cultivar or germplasm that is commonly used for the basic biology studies in the scientific community. Among all the germplasm, the sequenced one 'China Antique' might be an ideal candidate because of its genetic homozygosity. Thirdly, the low efficiency of the regeneration and transformation system seriously prevents the molecular genetic studies on the lotus, which is a prerequisite for gene function study. The fourth, but not the last, is the indeterminate growth and long life span (~5 months per generation) of the lotus plant, which limits the cultivation of lotus in small space. Through artificial selection, a number of cultivars with small plant architecture and short life span (~3 months' generation time) were obtained in lotus, which are very popular in the ornamental market, and named as 'Wan Lian' (bowl lotus). To cross these bowl lotus with 'China Antique', and then subject to backcrossing breeding, it might be possible to obtain germplasm with both small plant size and the 'China Antique' genetic background. This type of germplasm might be suitable for cultivation in the lab, and hence for further studies at molecular level. In conclusion, lotus could be regarded as an emerging model of horticultural plants, and be capable for the utilization in studying many aspects of unique features in plants.

Author Contributions: Original draft preparation, writing, review and art work, Z.L. and C.Z.; review and editing, D.C. and R.N.D.; editing, review, conceptualization, supervision, P.Y.

Funding: This work was supported by distinguished talents project to Pingfang Yang from Hubei University.

Acknowledgments: We thank to all the colleagues who have been involved in the studies on lotus. It is their great achievements that have provided abundant data for this review. Due to space limitation, we could not include and cite all the available literatures on lotus, and we apologize for this. We are grateful to Chen Jingxing for kindly providing image in graphic abstract.

Conflicts of Interest: The authors declare no conflict of interest.

Abbreviations

AFLP	Amplified Fragment Length Polymorphism
AGL	AGAMOUS-like
ANN	Annexin
ANS	anthocyanin synthase
AOX	alternative oxidases
APG	The Angiosperm Phylogeny Group

AS	alternative splicing
BIA	benzylisoquinoline alkaloid
CCA	CIRCADIAN CLOCK ASSOCIATED
CER	ECERIFERUM
CHS	Chalcone synthase
CO-LIKE	CONSTANS-like
COP	CONSTITUTIVELY PHOTOMORPHOGENIC
Cpn	Chaperonin
DFR	dihydroflavonol 4-reductase
EF	elongation factor
HSP	Heat shock protein
FT	FLOWERING LOCUS T
GAI	gibberellic acid insensitive
ISSR	inter-simple sequence repeat
LHY	LATE ELONGATED HYPOCOTYL
MADS-box	MINICHROMOSOME MAINTENANCE 1 (MCM1), AGAMOUS (AG), DEFICIENS (DEF), and SERUM RESPONSE FACTOR (SRF) domain
PIMT	Protein L-isoaspartyl methyltransferase
PRX	Peroxiredoxin
PUMP	plant uncoupling mitochondrial protein
RAPD	Random Amplified Polymorphic DNA
SNP	single nucleotide polymorphism
SOD	superoxide demutase
SRAP	Sequence—related amplified polymorphism
SSR	Simple Sequence Repeats
TE	transposable element
VIN3	vernalization

References

1. Wang, Q.; Zhang, X. *Colored Illustration of Lotus Cultivars in China*; China Forestry Publishing House: Beijing, China, 2005.
2. Zhang, X.; Chen, L.; Wang, Q. *New Lotus Flower Cultivars in China*; China Forestry Publishing House: Beijing, China, 2011.
3. Zhang, Y.; Lu, X.; Zeng, S.; Huang, X.; Guo, Z.; Zheng, Y.; Tian, Y.; Zheng, B. Nutritional composition, physiological functions and processing of lotus (*Nelumbo nucifera* Gaertn.) seeds: A review. *Phytochem. Rev.* **2015**, *14*, 321–334. [CrossRef]
4. Li, Y.; Smith, T.; Svetlana, P.; Yang, J.; Jin, J.-H.; Li, C.-S. Paleobiogeography of the lotus plant (Nelumbonaceae: Nelumbo) and its bearing on the paleoclimatic changes. *Palaeogeogr. Palaeoclimatol. Palaeoecol.* **2014**, *399*, 284–293. [CrossRef]
5. Shen-Miller, J. Sacred lotus, the long-living fruits of China Antique. *Seed Sci. Res.* **2002**, *12*, 131–143. [CrossRef]
6. Ming, R.; Vanburen, R.; Liu, Y.; Yang, M.; Han, Y.; Li, L.T.; Zhang, Q.; Kim, M.J.; Schatz, M.C.; Campbell, M. Genome of the long-living sacred lotus (*Nelumbo nucifera* Gaertn.). *Genome Biol.* **2013**, *14*, R41. [CrossRef] [PubMed]
7. Wang, Y.; Fan, G.; Liu, Y.; Sun, F.; Shi, C.; Liu, X.; Peng, J.; Chen, W.; Huang, X.; Cheng, S.; et al. The sacred lotus genome provides insights into the evolution of flowering plants. *Plant J.* **2013**, *76*, 557–567. [CrossRef] [PubMed]
8. Gandolfo, M.A.; Cuneo, R.N. Fossil Nelumbonaceae from the La Colonia Formation (Campanian-Maastrichtian, Upper Cretaceous), Chubut, Patagonia, Argentina. *Rev. Palaeobot. Palynol.* **2005**, *133*, 169–178. [CrossRef]
9. Cronquist, A. *An Integrated System of Classification of Flowering Plants*; Columbia University Press: New York, NY, USA, 1981.
10. Dahlgren, G. An updated angiosperm classification. *Bot. J. Linn. Soc.* **1989**, *100*, 197–203. [CrossRef]

11. Thorne, R.F. An Updated Phylogenetic Classification of the Flowering Plants. *Aliso* **1992**, *13*, 365–389. [CrossRef]

12. Takhtajan, A. *Diversity and Classification of Flowering Plants*; Columbia University Press: New York, NY, USA, 1997.

13. Stevens, P.F. (2001 onwards). *Angiosperm Phylogeny Website*, Version 14; July 2017. Available online: http://www.mobot.org/MOBOT/research/APweb/ (accessed on 23 June 2019).

14. Byng, J.; Chase, M.; Christenhusz, M.; Fay, M.; Judd, W.; Mabberley, D.; Sennikov, A.; Soltis, D.; Soltis, P.; Stevens, P. An update of the Angiosperm Phylogeny Group classification for the orders and families of flowering plants: APG IV. *Bot. J. Linn. Soc.* **2016**, *181*, 1–20.

15. Yoo, M.J.; Soltis, P.S.; Soltis, D.E. Expression of floral MADS-box genes in two divergent water lilies: Nymphaeales and Nelumbo. *Int. J. Plant Sci.* **2010**, *171*, 121–146. [CrossRef]

16. Diao, Y.; Chen, L.; Yang, G.; Zhou, M.; Song, Y. Nuclear DNA C-values in 12 species in Nymphaeales. *Caryologia* **2006**, *59*, 25–30. [CrossRef]

17. Zhang, Y.; Nyong, A.T.; Shi, T.; Yang, P. The complexity of alternative splicing and landscape of tissue-specific expression in lotus (Nelumbo nucifera) unveiled by Illumina- and single-molecule real-time-based RNA-sequencing. *DNA Res.* **2019**. [CrossRef] [PubMed]

18. Gui, S.; Peng, J.; Wang, X.; Wu, Z.; Cao, R.; Salse, J.; Zhang, H.; Zhu, Z.; Xia, Q.; Quan, Z.; et al. Improving *Nelumbo nucifera* genome assemblies using high-resolution genetic maps and BioNano genome mapping reveals ancient chromosome rearrangements. *Plant J.* **2018**, *94*, 721–734. [CrossRef] [PubMed]

19. Wang, K.; Deng, J.; Damaris, R.N.; Yang, M.; Xu, L.; Yang, P. LOTUS-DB: An integrative and interactive database for *Nelumbo nucifera* study. *Database* **2015**, *2015*, bav023. [CrossRef] [PubMed]

20. Gui, S.; Wu, Z.; Zhang, H.; Zheng, Y.; Zhu, Z.; Liang, D.; Ding, Y. The mitochondrial genome map of *Nelumbo nucifera* reveals ancient evolutionary features. *Sci. Rep.* **2016**, *6*, 30158. [CrossRef]

21. Wu, Z.; Gui, S.; Quan, Z.; Pan, L.; Wang, S.; Ke, W.; Liang, D.; Ding, Y. A precise chloroplast genome of *Nelumbo nucifera* (Nelumbonaceae) evaluated with Sanger, Illumina MiSeq, and PacBio RS II sequencing platforms: Insight into the plastid evolution of basal eudicots. *BMC Plant Biol.* **2014**, *14*, 289. [CrossRef]

22. Shi, T.; Wang, K.; Yang, P. The evolution of plant microRNAs: Insights from a basal eudicot sacred lotus. *Plant J.* **2017**, *89*, 442–457. [CrossRef] [PubMed]

23. Meng, Z.; Hu, X.; Zhang, Z.; Li, Z.; Lin, Q.; Yang, M.; Yang, P.; Ming, R.; Yu, Q.; Wang, K. Chromosome Nomenclature and Cytological Characterization of Sacred Lotus. *Cytogenet. Genome Res.* **2017**, *153*, 223–231. [CrossRef]

24. Shen-Miller, J.; Mudgett, M.B.; Schopf, J.W.; Clarke, S.; Berger, R. Exceptional seed longevity and robust growth: Ancient Sacred Lotus from China. *Am. J. Bot.* **1995**, *82*, 1367–1380. [CrossRef]

25. Van Bergen, P.F.; Hatcher, P.G.; Boon, J.J.; Collinson, M.E.; de Leeuw, J.W. Macromolecular composition of the propagule wall of *Nelumbo nucifera*. *Phytochemistry* **1997**, *45*, 601–610. [CrossRef]

26. Liu, Y.; Ma, S.S.; Ibrahim, S.A.; Li, E.H.; Yang, H.; Huang, W. Identification and antioxidant properties of polyphenols in lotus seed epicarp at different ripening stages. *Food Chem.* **2015**, *185*, 159–164. [CrossRef] [PubMed]

27. Shen-Miller, J.; Lindner, P.; Xie, Y.; Villa, S.; Wooding, K.; Clarke, S.G.; Loo, R.R.; Loo, J.A. Thermal-stable proteins of fruit of long-living Sacred Lotus *Nelumbo nucifera* Gaertn var. China Antique. *Trop. Plant Biol.* **2013**, *6*, 69–84. [CrossRef] [PubMed]

28. Chu, P.; Chen, H.; Zhou, Y.; Li, Y.; Ding, Y.; Jiang, L.; Tsang, E.W.; Wu, K.; Huang, S. Proteomic and functional analyses of *Nelumbo nucifera* annexins involved in seed thermotolerance and germination vigor. *Planta* **2012**, *235*, 1271–1288. [CrossRef] [PubMed]

29. Chen, H.H.; Chu, P.; Zhou, Y.L.; Ding, Y.; Li, Y.; Liu, J.; Jiang, L.W.; Huang, S.Z. Ectopic expression of NnPER1, a Nelumbo nucifera 1-cysteine peroxiredoxin antioxidant, enhances seed longevity and stress tolerance in Arabidopsis. *Plant J.* **2016**, *88*, 608–619. [CrossRef] [PubMed]

30. Hu, J.; Jin, J.; Qian, Q.; Huang, K.; Ding, Y. Small RNA and degradome profiling reveals miRNA regulation in the seed germination of ancient eudicot Nelumbo nucifera. *BMC Genom.* **2016**, *17*, 684. [CrossRef]

31. Darmanin, T.; Guittard, F. Superhydrophobic and superoleophobic properties in nature. *Mater. Today* **2015**, *18*, 273–285. [CrossRef]

32. Ensikat, H.J.; Ditsche-Kuru, P.; Neinhuis, C.; Barthlott, W. Superhydrophobicity in perfection: The outstanding properties of the lotus leaf. *Beilstein J. Nanotechnol.* **2011**, *2*, 152–161. [CrossRef]

33. Zhang, Y.; Wu, H.; Yu, X.; Chen, F.; Wu, J. Microscopic Observations of the Lotus Leaf for Explaining the Outstanding Mechanical Properties. *J. Bionic Eng.* **2012**, *9*, 84–90. [CrossRef]

34. Marmur, A. The Lotus effect: Superhydrophobicity and metastability. *Langmuir* **2004**, *20*, 3517–3519. [CrossRef]

35. Bhushan, B.; Jung, Y.C.; Koch, K. Self-Cleaning Efficiency of Artificial Superhydrophobic Surfaces. *Langmuir* **2009**, *25*, 3240–3248. [CrossRef]

36. Yang, X.; Wang, Z.; Feng, T.; Li, J.; Huang, L.; Yang, B.; Zhao, H.; Jenks, M.A.; Yang, P.; Lü, S. Evolutionarily conserved function of the sacred lotus (*Nelumbo nucifera* Gaertn.) CER2-LIKE family in very-long-chain fatty acid elongation. *Planta* **2018**, *248*, 715–727. [CrossRef] [PubMed]

37. Li, J.K.; Huang, S.Q. Flower thermoregulation facilitates fertilization in Asian sacred lotus. *Ann. Bot.* **2009**, *103*, 1159–1163. [CrossRef] [PubMed]

38. Watling, J.; Robinson, S.; Seymour, R. Contribution of the Alternative Pathway to Respiration during Thermogenesis in Flowers of the Sacred Lotus. *Plant Physiol.* **2006**, *140*, 1367–1373. [CrossRef] [PubMed]

39. Seymour, R.; Schultze-Motel, P.; Lamprecht, I. Heat production by sacred lotus flowers depends on ambient temperature, not light cycle. *J. Exp. Bot.* **1998**, *49*, 1213–1217. [CrossRef]

40. Grant, N.; Miller, R.; Watling, J.; Robinson, S. Distribution of thermogenic activity in floral tissues of Nelumbo nucifera. *Funct. Plant Biol.* **2010**, *37*, 1085–1095. [CrossRef]

41. Zhu, Y.; Lu, J.; Wang, J.; Chen, F.; Leng, F.; Li, H. Regulation of thermogenesis in plants: The interaction of alternative oxidase and plant uncoupling mitochondrial protein. *J. Integr. Plant Biol.* **2011**, *53*, 7–13. [CrossRef] [PubMed]

42. Wang, R.; Zhang, Z. Floral thermogenesis: An adaptive strategy of pollination biology in Magnoliaceae. *Commun. Integr. Biol.* **2015**, *8*, e992746. [CrossRef]

43. Wagner, A.M.; Krab, K.; Wagner, M.J.; Moore, A.L. Regulation of thermogenesis in flowering Araceae: The role of the alternative oxidase. *Biochim. Biophys. Acta* **2008**, *1777*, 993–1000. [CrossRef]

44. Dieringer, G.; Leticia Cabrera, R.; Mottaleb, M. Ecological relationship between floral thermogenesis and pollination in Nelumbo lutea (Nelumbonaceae). *Am. J. Bot.* **2014**, *101*, 357–364. [CrossRef]

45. Miller, R.E.; Watling, J.R.; Robinson, S.A. Functional transition in the floral receptacle of the sacred lotus (*Nelumbo nucifera*): From thermogenesis to photosynthesis. *Funct. Plant Biol.* **2009**, *36*, 471–480. [CrossRef]

46. Grant, N.M.; Miller, R.E.; Watling, J.R.; Robinson, S.A. Synchronicity of thermogenic activity, alternative pathway respiratory flux, AOX protein content, and carbohydrates in receptacle tissues of sacred lotus during floral development. *J. Exp. Bot.* **2008**, *59*, 705–714. [CrossRef] [PubMed]

47. Li, Z.; Liu, X.Q.; Gituru, R.W.; Juntawong, N.; Zhou, M.Q.; Chen, L.Q. Genetic diversity and classification of Nelumbo germplasm of different origins by RAPD and ISSR analysis. *Sci. Hortic.* **2010**, *125*, 724–732. [CrossRef]

48. Hu, J.; Pan, L.; Liu, H.; Wang, S.; Wu, Z.; Ke, W.; Ding, Y. Comparative analysis of genetic diversity in sacred lotus (*Nelumbo nucifera* Gaertn.) using AFLP and SSR markers. *Mol. Biol. Rep.* **2012**, *39*, 3637–3647. [CrossRef] [PubMed]

49. Yang, M.; Han, Y.; VanBuren, R.; Ming, R.; Xu, L.; Han, Y.; Liu, Y. Genetic linkage maps for Asian and American lotus constructed using novel SSR markers derived from the genome of sequenced cultivar. *BMC Genom.* **2012**, *13*, 653. [CrossRef] [PubMed]

50. Yang, M.; Han, Y.N.; Xu, L.M.; Zhao, J.R.; Liu, Y.L. Comparative analysis of genetic diversity of lotus (Nelumbo) using SSR and SRAP markers. *Sci. Hortic.* **2012**, *142*, 185–195. [CrossRef]

51. Huang, L.; Yang, M.; Li, L.; Li, H.; Yang, D.; Shi, T.; Yang, P. Whole genome re-sequencing reveals evolutionary patterns of sacred lotus (*Nelumbo nucifera*). *J. Integr. Plant Biol.* **2018**, *60*, 2–15. [CrossRef] [PubMed]

52. Zhao, M.; Yang, J.-X.; Mao, T.-Y.; Zhu, H.-H.; Xiang, L.; Zhang, J.; Chen, L.-Q. Detection of Highly Differentiated Genomic Regions Between Lotus (*Nelumbo nucifera* Gaertn.) with Contrasting Plant Architecture and Their Functional Relevance to Plant Architecture. *Front. Plant Sci.* **2018**, *9*, 1219. [CrossRef] [PubMed]

53. Hu, J.; Gui, S.; Zhu, Z.; Wang, X.; Ke, W.; Ding, Y. Genome-Wide Identification of SSR and SNP Markers Based on Whole-Genome Re-Sequencing of a Thailand Wild Sacred Lotus (*Nelumbo nucifera*). *PLoS ONE* **2015**, *10*, e0143765. [CrossRef]

54. Deng, J.; Chen, S.; Yin, X.J.; Wang, K.; Liu, Y.L.; Li, S.H.; Yang, P.F. Systematic qualitative and quantitative assessment of anthocyanins, flavones and flavonols in the petals of 108 lotus (*Nelumbo nucifera*) cultivars. *Food Chem.* **2013**, *139*, 307–312. [CrossRef]

55. Deng, J.; Li, M.; Huang, L.; Yang, M.; Yang, P. Genome-wide analysis of the R2R3 MYB subfamily genes in lotus (*Nelumbo nucifera*). *Plant Mol. Biol. Report.* **2016**, *34*, 1016–1026. [CrossRef]

56. Sun, S.-S.; Gugger, P.F.; Wang, Q.-F.; Chen, J.-M. Identification of a R2R3-MYB gene regulating anthocyanin biosynthesis and relationships between its variation and flower color difference in lotus (*Nelumbo* Adans.). *PeerJ* **2016**, *4*, e2369. [CrossRef] [PubMed]

57. Deng, J.; Fu, Z.; Chen, S.; Damaris, R.N.; Wang, K.; Li, T.; Yang, P. Proteomic and Epigenetic Analyses of Lotus (*Nelumbo nucifera*) Petals Between Red and White cultivars. *Plant Cell Physiol.* **2015**, *56*, 1546. [CrossRef] [PubMed]

58. Lin, Z.; Damaris, R.N.; Shi, T.; Li, J.; Yang, P. Transcriptomic analysis identifies the key genes involved in stamen petaloid in lotus (*Nelumbo nucifera*). *BMC Genom.* **2018**, *19*, 554. [CrossRef] [PubMed]

59. Lin, Z.; Liu, M.; Damaris, R.N.; Nyong'a, T.M.; Cao, D.; Ou, K.; Yang, P. Genome-Wide DNA Methylation Profiling in the Lotus (Nelumbo nucifera) Flower Showing its Contribution to the Stamen Petaloid. *Plants* **2019**, *8*, 135. [CrossRef] [PubMed]

60. Yang, M.; Zhu, L.; Xu, L.; Pan, C.; Liu, Y. Comparative transcriptomic analysis of the regulation of flowering in temperate and tropical lotus (*Nelumbo nucifera*) by RNA-Seq. *Ann. Appl. Biol.* **2014**, *165*, 73–95. [CrossRef]

61. Yang, M.; Zhu, L.; Pan, C.; Xu, L.; Liu, Y.; Ke, W.; Yang, P. Transcriptomic Analysis of the Regulation of Rhizome Formation in Temperate and Tropical Lotus (*Nelumbo nucifera*). *Sci. Rep.* **2015**, *5*, 13059. [CrossRef]

62. Kim, M.-J.; Nelson, W.; Soderlund, C.A.; Gang, D.R. Next-Generation Sequencing-Based Transcriptional Profiling of Sacred Lotus 'China Antique'. *Trop. Plant Biol.* **2013**, *6*, 161–179. [CrossRef]

63. Yang, M.; Xu, L.; Liu, Y.; Yang, P. RNA-Seq Uncovers SNPs and Alternative Splicing Events in Asian Lotus (*Nelumbo nucifera*). *PLoS ONE* **2015**, *10*, e0125702. [CrossRef]

64. Li, J.; Shi, T.; Huang, L.; He, D.; Nyong'A, T.M.; Yang, P. Systematic transcriptomic analysis provides insights into lotus (Nelumbo nucifera) seed development. *Plant Growth Regul.* **2018**, *86*, 339–350. [CrossRef]

65. Chen, S.; Fang, L.; Xi, H.; Guan, L.; Fang, J.; Liu, Y.; Wu, B.; Li, S. Simultaneous qualitative assessment and quantitative analysis of flavonoids in various tissues of lotus (*Nelumbo nucifera*) using high performance liquid chromatography coupled with triple quad mass spectrometry. *Anal. Chim. Acta* **2012**, *724*, 127–135. [CrossRef]

66. Limwachiranon, J.; Huang, H.; Shi, Z.; Li, L.; Luo, Z. Lotus Flavonoids and Phenolic Acids: Health Promotion and Safe Consumption Dosages. *Compr. Rev. Food Sci. Food Saf.* **2018**, *17*, 458–471. [CrossRef]

67. Sharma, B.R.; Gautam, L.N.; Adhikari, D.; Karki, R. A Comprehensive Review on Chemical Profiling of *Nelumbo Nucifera*: Potential for Drug Development. *Phytother. Res.* **2017**, *31*, 3–26. [CrossRef] [PubMed]

68. Chen, S.; Xiang, Y.; Deng, J.; Liu, Y.L.; Li, S.H. Simultaneous Analysis of Anthocyanin and Non-Anthocyanin Flavonoid in Various Tissues of Different Lotus (Nelumbo) Cultivars by HPLC-DAD-ESI-MSn. *PLoS ONE* **2013**, *8*, e62291. [CrossRef] [PubMed]

69. Chen, S.; Zheng, Y.; Fang, J.B.; Liu, Y.L.; Li, S.H. Flavonoids in lotus (Nelumbo) leaves evaluated by HPLC-MSn at the germplasm level. *Food Res. Int.* **2013**, *54*, 796–803. [CrossRef]

70. Li, S.S.; Wu, J.; Chen, L.G.; Du, H.; Xu, Y.J.; Wang, L.J.; Zhang, H.J.; Zheng, X.C.; Wang, L.S. Biogenesis of C-Glycosyl Flavones and Profiling of Flavonoid Glycosides in Lotus (*Nelumbo nucifera*). *PLoS ONE* **2014**, *9*, e108860. [CrossRef] [PubMed]

71. Zhao, X.; Shen, J.; Chang, K.J.; Kim, S.H. Comparative Analysis of Antioxidant Activity and Functional Components of the Ethanol Extract of Lotus (*Nelumbo nucifera*) from Various Growing Regions. *J. Agric. Food Chem.* **2014**, *62*, 6227–6235. [CrossRef] [PubMed]

72. Zhu, M.Z.; Wu, W.; Jiao, L.L.; Yang, P.F.; Guo, M.Q. Analysis of Flavonoids in Lotus (*Nelumbo nucifera*) Leaves and Their Antioxidant Activity Using Macroporous Resin Chromatography Coupled with LC-MS/MS and Antioxidant Biochemical Assays. *Molecules* **2015**, *20*, 10553–10565. [CrossRef]

73. Feng, C.Y.; Li, S.S.; Yin, D.D.; Zhang, H.J.; Tian, D.K.; Wu, Q.; Wang, L.J.; Su, S.; Wang, L.S. Rapid determination of flavonoids in plumules of sacred lotus cultivars and assessment of their antioxidant activities. *Ind. Crop. Prod.* **2016**, *87*, 96–104. [CrossRef]

74. Guo, Y.J.; Chen, X.; Qi, J.; Yu, B.Y. Simultaneous qualitative and quantitative analysis of flavonoids and alkaloids from the leaves of *Nelumbo nucifera* Gaertn. using high-performance liquid chromatography with quadrupole time-of-flight mass spectrometry. *J. Sep. Sci.* **2016**, *39*, 2499–2507. [CrossRef]

75. Zhu, M.Z.; Liu, T.; Zhang, C.Y.; Guo, M.Q. Flavonoids of Lotus (*Nelumbo nucifera*) Seed Embryos and Their Antioxidant Potential. *J. Food Sci.* **2017**, *82*, 1834–1841. [CrossRef]

76. Tian, W.Y.; Zhi, H.; Yang, C.; Wang, L.K.; Long, J.T.; Xiao, L.M.; Liang, J.Z.; Huang, Y.; Zheng, X.; Zhao, S.Q.; et al. Chemical composition of alkaloids of Plumula nelumbinis and their antioxidant activity from different habitats in China. *Ind. Crop. Prod.* **2018**, *125*, 537–548. [CrossRef]
77. Zheng, J.X.; Tian, W.Y.; Yang, C.; Shi, W.P.; Cao, P.H.; Long, J.T.; Xiao, L.M.; Wu, Y.; Liang, J.Z.; Li, X.B.; et al. Identification of flavonoids in Plumula nelumbinis and evaluation of their antioxidant properties from different habitats. *Ind. Crop. Prod.* **2019**, *127*, 36–45. [CrossRef]
78. Commission, C.P. *Pharmacopoeia of the People's Republic of China*; China Medical Science and Technology Press: Beijing, China, 2015.
79. Huang, B.; Ban, X.; He, J.; Tong, J.; Tian, J.; Wang, Y. Hepatoprotective and antioxidant activity of ethanolic extracts of edible lotus (Nelumbo nucifera Gaertn.) leaves. *Food Chem.* **2010**, *120*, 873–878. [CrossRef]
80. Paudel, K.R.; Panth, N. Phytochemical Profile and Biological Activity of *Nelumbo nucifera*. *Evid. -Based Complement. Altern. Med.* **2015**, *2015*, 789124. [CrossRef] [PubMed]
81. Ye, L.H.; He, X.X.; Kong, L.T.; Liao, Y.H.; Pan, R.L.; Xiao, B.X.; Liu, X.M.; Chang, Q. Identification and characterization of potent CYP2D6 inhibitors in lotus leaves. *J. Ethnopharmacol.* **2014**, *153*, 190–196. [CrossRef] [PubMed]
82. Ye, L.-H.; He, X.-X.; You, C.; Tao, X.; Wang, L.-S.; Zhang, M.-D.; Zhou, Y.-F.; Chang, Q. Pharmacokinetics of Nuciferine and N-Nornuciferine, Two Major Alkaloids From *Nelumbo nucifera* Leaves, in Rat Plasma and the Brain. *Front. Pharmacol.* **2018**, *9*, 902. [CrossRef]
83. Yang, M.; Zhu, L.; Li, L.; Li, J.; Xu, L.; Feng, J.; Liu, Y. Digital Gene Expression Analysis Provides Insight into the Transcript Profile of the Genes Involved in Aporphine Alkaloid Biosynthesis in Lotus (*Nelumbo nucifera*). *Front. Plant Sci.* **2017**, *8*, 80. [CrossRef] [PubMed]
84. Deng, X.; Zhao, L.; Fang, T.; Xiong, Y.; Ogutu, C.; Yang, D.; Vimolmangkang, S.; Liu, Y.; Han, Y. Investigation of benzylisoquinoline alkaloid biosynthetic pathway and its transcriptional regulation in lotus. *Hortic. Res.* **2018**, *5*, 29. [CrossRef]
85. Menendez-Perdomo, I.M.; Facchini, P.J. Benzylisoquinoline Alkaloids Biosynthesis in Sacred Lotus. *Molecules* **2018**, *23*, 2899. [CrossRef]
86. Arunyanart, S.; Chaitrayagun, M. Induction of somatic embryogenesis in lotus (Nelumbo nucifera Geartn.). *Sci. Hortic.* **2005**, *105*, 411–420. [CrossRef]
87. Liu, Y.; Chaturvedi, P.; Fu, J.; Cai, Q.; Weckwerth, W.; Yang, P. Induction and quantitative proteomic analysis of cell dedifferentiation during callus formation of lotus (*Nelumbo nucifera* Gaertn. spp. *baijianlian*). *J. Proteom.* **2016**, *131*, 61–70. [CrossRef]
88. Shou, S.Y.; Miao, L.X.; Zai, W.S.; Huang, X.Z.; Guo, D.P. Factors influencing shoot multiplication of lotus (*Nelumbo nucifera*). *Biol. Plant.* **2008**, *52*, 529–532. [CrossRef]
89. Buathong, R.; Saetiew, K.; Phansiri, S.; Parinthawong, N.; Arunyanart, S. Tissue culture and transformation of the antisense DFR gene into lotus (*Nelumbo nucifera* Gaertn.) through particle bombardment. *Sci. Hortic.* **2013**, *161*, 216–222. [CrossRef]
90. Kanjana, S.; Prissadang, A.; Nanglak, P.; Sumay, A. Transformation of Antisense Chalcone Synthase (CHS) Gene into Lotus (*Nelumbo Nucifera* Gaertn.) by Particle Bombardment. *Open Biotechnol. J.* **2017**, *11*, 1–8. [CrossRef]

Article

Physiological Analysis and Proteome Quantification of Alligator Weed Stems in Response to Potassium Deficiency Stress

Li-Qin Li, Cheng-Cheng Lyu, Jia-Hao Li, Zhu Tong, Yi-Fei Lu, Xi-Yao Wang, Su Ni, Shi-Min Yang, Fu-Chun Zeng and Li-Ming Lu *

College of Agronomy, Sichuan Agriculture University, Chengdu 611130, China; liliqin88@163.com (L.-Q.L.); chengchengLyu@163.com (C.-C.L.); 18889589812@139.com (J.-H.L.); t173131021222@163.com (Z.T.); sarklu@126.com (Y.-F.L.); wxyrtl@163.com (X.-Y.W.); ns13@163.com (S.N.); yangshimin1@163.com (S.-M.Y.); zengfuchun78@163.com (F.-C.Z.)
* Correspondence: luliming@sicau.edu.cn

Received: 14 November 2018; Accepted: 27 December 2018; Published: 8 January 2019

Abstract: The macronutrient potassium is essential to plant growth, development and stress response. Alligator weed (*Alternanthera philoxeroides*) has a high tolerance to potassium deficiency (LK) stress. The stem is the primary organ responsible for transporting molecules from the underground root system to the aboveground parts of the plant. However, proteomic changes in response to LK stress are largely unknown in alligator weed stems. In this study, we investigated the physiological and proteomic changes in alligator weed stems under LK stress. First, the chlorophyll and soluble protein content and SOD and POD activity were significantly altered after 15 days of LK treatment. The quantitative proteomic analysis suggested that a total of 296 proteins were differentially abundant proteins (DAPs). The functional annotation analysis revealed that LK stress elicited complex proteomic alterations that were involved in oxidative phosphorylation, plant-pathogen interactions, glycolysis/gluconeogenesis, sugar metabolism, and transport in stems. The subcellular locations analysis suggested 104 proteins showed chloroplastic localization, 81 proteins showed cytoplasmic localization and 40 showed nuclear localization. The protein–protein interaction analysis revealed that 56 proteins were involved in the interaction network, including 9 proteins involved in the ribosome network and 9 in the oxidative phosphorylation network. Additionally, the expressed changes of 5 DAPs were similar between the proteomic quantification analysis and the PRM-MS analysis, and the expression levels of eight genes that encode DAPs were further verified using an RT-qPCR analysis. These results provide valuable information on the adaptive mechanisms in alligator weed stems under LK stress and facilitate the development of efficient strategies for genetically engineering potassium-tolerant crops.

Keywords: *Alternanthera philoxeroides*; proteomic; stem; potassium; stress

---

1. Introduction

Potassium (K^+) is the most important and abundant nutrient ion in living plant cells, and it plays crucial roles in many physiological and biochemical processes, such as photosynthesis, protein synthesis, enzyme activation, osmotic regulation, ion homeostasis, and stomata movement [1]. Under optimum growth conditions, the cytoplasmic K^+ concentration in plant cells is ~100 mM, but concentrations are very low in soils near roots and change continuously, varying from 0.1 to 1.0 mM [2]. Thus, most plants will face low K^+ stress during growth. Most plants can resist low K^+ stress, mainly because they have established a strategy to acclimate to this stress. Most plants have both high-affinity and low-affinity K^+ transport systems to sense the K^+ concentrations in the soil, and the high-affinity system functions when the K$^+$ concentrations are below 0.2 mM. The low-affinity system operates at K$^+$ concentrations above 0.5 mM; thus, these two transport systems are vital to help plants survive under LK stress [3,4].

Alligator weed (*Alternanthera philoxeroides*) is a dicotyledonous perennial herb that grows worldwide due to its high adaptability to harsh environments. Gao et al. reported that this adaptability is predominantly attributed to genome-wide DNA methylation and epigenetic regulation [5]. This plant was found to have a great ability to accumulate potassium, and three K$^+$ transporter (ApKUP1, ApKUP2 and ApKUP3) expression levels were up-regulated by low potassium, abscisic acid (ABA) and polyethylene glycol (PEG) treatments [6]. Comprehensive transcriptome analysis results suggest that 121 transcription factors, 108 kinases, 136 transporters and 178 genes were changed in response to LK stress in alligator weed roots [7].

Proteomics is considered one of the most robust methods; it presents an effective approach to pursue a systems-based perspective of how proteins change and thus how organisms adapt to various abiotic environments [8]. In recent years, gel-free quantitative proteomic methods (e.g., TMT and iTRAQ) have been used as precise and sensitive multiplexed peptide/protein quantification techniques with mass spectrometry [9]. Zeng et al. reported that 129 proteins were related to low K tolerance and were mainly involved in defense, transcription, signal transduction, energy, and protein synthesis in barley [10]. In cotton seedlings xylem sap, 258 peptides were qualitatively identified, of which, 90.31% were secreted proteins. LK significantly decreased the expression of most environmental-stress-related proteins [11]. The proteomics results suggest that most of the differentially abundant proteins were related to jasmonic acid (JA) synthesis, suggesting the importance of JA in the potassium deficiency response in wheat and rice [12]. Physiological and quantitative proteomic analyses reveal the proteins responsible for energy metabolism, signal sensing and transduction, and protein degradation that played crucial roles in alligator weed roots under LK stress [13]. Root-to-shoot translocation and shoot homeostasis of potassium determine the nutrient balance, growth, and stress tolerance of vascular plants. However, the proteomic profile of alligator weed stems under low potassium stress has not yet been studied. In this study, we used a quantitative proteomic analysis method to investigate the protein expression changes and metabolic process networks in alligator weed stems after LK stress. These results will provide critical insights into the potassium tolerance mechanisms of alligator weed.

2. Results

2.1. Effect of LK Stress on the Physiology of Alligator Weed Stems

The exposure of alligator weed seedlings to LK stress resulted in various physiological changes in the stem after 15 days of treatment. To determine the physiological responses of the stems after treatment, the chlorophyll and soluble protein content and SOD and POD activity were measured. First, the total chlorophyll content increased by 20% after treatment (Figure 1A), the soluble protein content showed greatly increased (Figure 1B), superoxide dismutase (SOD) and peroxidase (POD) were vitally protective enzymes in plant cells during stress. In the present study, the activity of SOD and POD were both significantly decreased after 15 days of treatment (Figure 1C,D).

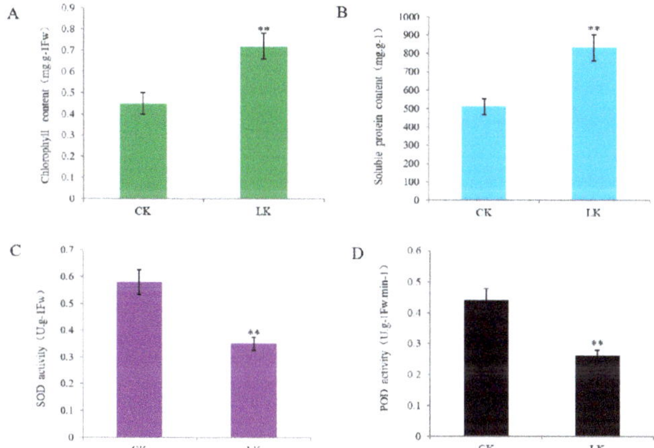

Figure 1. Stem physiology performance were listed. Note (**A**) measurements the total chlorophyll content; (**B**) measurements of soluble protein content; (**C**) measurements of superoxide dismutase (SOD) activity; (**D**) measurements of peroxidase (POD) activity. ** mean that data was significantly different among samples in the same treatment ($p \leq 0.05$) using the SPSS statistical software. Bars represent the mean $\pm$ SE (n = 5).

2.2. Protein Responses to LK Stress Revealed by the Proteomic Analysis

We employed TMT and LC-MS/MS to characterize the proteomic profiles of the stems. The proteome data showed that LK dramatically changed the protein abundance in the stem. The six samples that were analyzed included 3 replicates each of CK and LK stems. A change over 1.2-fold or a cut-off of less than 0.83-fold were considered statistically significant. A total of 296 proteins were altered in expression were regarded as DAPs. Among them, 152 proteins were up-regulated and 144 proteins were down-regulated; their information is shown (Table S1). Among them, 27 DAPs related to transport process were listed (Table 1).

Table 1. 27 differentially abundant proteins (DAPs) related to transport process were listed.

Protein Accession Number	Protein Annotation	Protein Score	Percentage of Protein Sequence Coverage %	Peptides Count	Number of Unique Peptides	Fold Change	*p*-Value
Gene.10143	ABC transporter F family member 5	52.24	8.1	5	5	1.34	0.0425
Gene.15326	ABC transporter B family member 19	91.58	10	11	8	1.27	0.00104
Gene.11683	ABC transporter F family member 4	45.75	6	3	3	1.25	0.0222
Gene.22137	ABC transporter B family member 1	72.34	9.9	11	3	1.23	0.00932
Gene.37843	ABC transporter C family member 10	7.88	3.8	1	1	1.21	0.0478
Gene.39686	ABC transporter D family member 2	30.70	3.1	2	2	1.21	0.0289
Gene.14866	ABC transporter B family member 2	107.13	11.8	12	10	1.21	0.0108
Gene.22512	Protein sieve element occlusion	30.17	5.9	4	4	1.29	0.00538
Gene.1036	Protein sieve element occlusion	19.51	26.9	3	3	1.29	0.00456
Gene.16899	Protein sieve element occlusion	323.31	35.3	31	29	1.27	0.000702
Gene.29767	Patellin-3	286.67	34.9	24	10	1.39	0.0226
Gene.42322	Patellin-3	25.04	25.7	4	4	1.34	0.0464
Gene.25481	Patellin-4	7.54	2.1	1	1	1.29	0.0105
Gene.42892	Patellin-5	12.17	8.9	3	2	1.85	0.0142
Gene.22040	Sugar transporter ERD6-like 6	46.34	3.8	1	1	1.47	0.000758
Gene.19069	Chloride channel protein CLC-b	13.79	2.4	2	2	1.46	0.042
Gene.44796	Two pore calcium channel protein 1	80.37	7.5	5	5	1.39	0.0196
Gene.41564	Vesicle-associated membrane protein	32.77	41.5	4	2	1.26	0.0172
Gene.23148	Transmembrane 9 superfamily member	14.87	8.8	6	2	1.21	0.0113
Gene.36797	Kinesin-like protein KIN-UA	9.03	1.8	1	1	1.21	0.0246
Gene.1641	Vacuolar membrane proton pump	25.00	12.9	9	1	0.81	0.0441
Gene.51508	Chloroplastic lipocalin	41.09	17.7	5	5	0.78	0.784
Gene.38083	Stem-specific protein TSJT1	48.56	18.5	4	4	0.78	0.0021
Gene.14846	Protein NRT1/PTR FAMILY 8.3	34.84	6.5	4	2	0.74	0.0466
Gene.1247	Syntaxin-61	6.42	4.5	1	1	0.78	0.0021
Gene.1250	Syntaxin-61	6.32	3.7	1	1	0.67	0.0248
Gene.11391	V-type proton ATPase subunit G	155.18	62.7	8	8	0.7	0.0448

2.3. GO and KEGG Analysis of DAPs

All DAPs were annotated and classified according to biological process (BP), molecular function (MF), and cellular component (CC) according to the GO database. The primary categories in BP were metabolic processes, cellular processes, and single-organism processes; the prominent MF categories were catalytic activity, binding, and transporter activity; and the most abundant categories in CC were cell, membrane, and macromolecular complex (Figure 2). Next, the biological metabolic pathways related to 119 DAPs were investigated using KEGG analysis. The results suggested that most represented DAPs were associated with carbohydrate metabolism (20.1%), energy metabolism (15.1%) and amino acid metabolism (15.1%), while the fourth and fifth category was lipid metabolism and transport; every group included 12 to 9 DAPs (Figure 3).

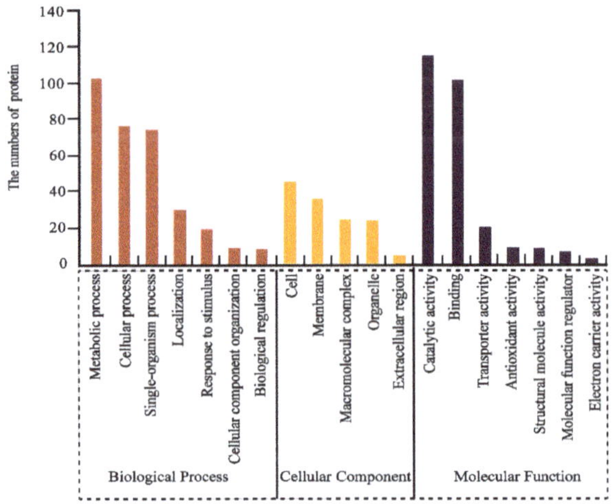

Figure 2. Functional classification of differentially abundant proteins.

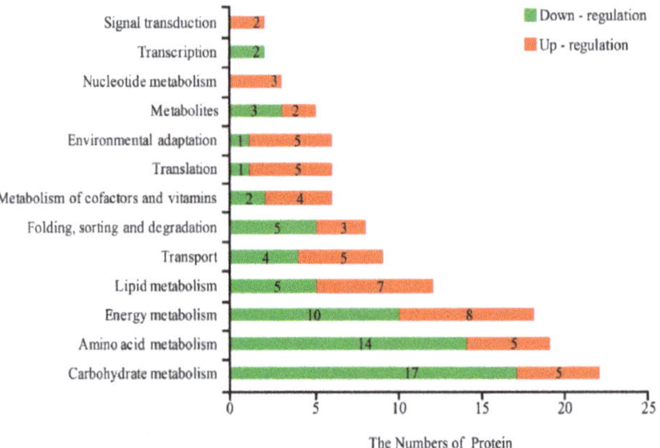

Figure 3. Differentially abundant proteins by KEGG analysis. Note number mean protein number.

2.4. Subcellular Location and Domain Analysis of DAPs

The subcellular locations of 296 identified DAPs identified were predicted by Target P1.1 software. The result suggested that 104 proteins showed chloroplastic localization, of which 53 were increased, 81 proteins showed cytoplasmic localization, of which 38 proteins were increased, and 40 showed nuclear localization, of which 25 were increased. These were the top three groups. Localization in the peroxisome, golgi apparatus and cytoskeleton were the lowest, with protein numbers of 2, 1, and 1, respectively (Figure 4). With the protein domain analysis, the top three protein groups were Bet v I/Major latex proteins, ABC transporter-like proteins and START-like domain proteins, and they all mapped 7 proteins. Among them, 7 ABC transporter-like proteins were all up-regulated, and 6 manganese/iron superoxide dismutases (3 N-terminal, 3 C-terminal) were all down-regulated (Figure 5).

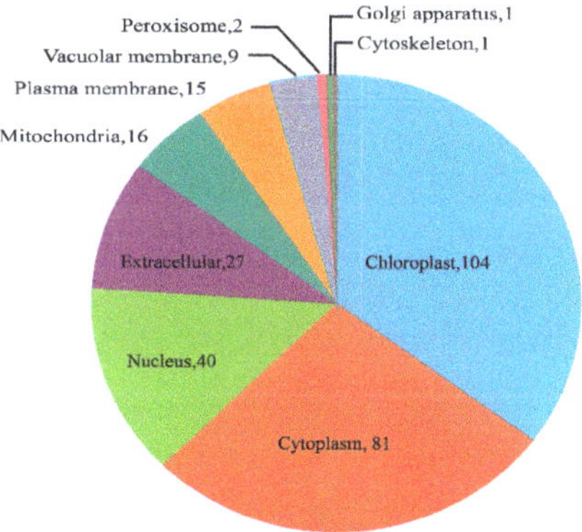

Figure 4. Subcellular locations analysis of differentially abundant proteins. Note number mean protein number in this group.

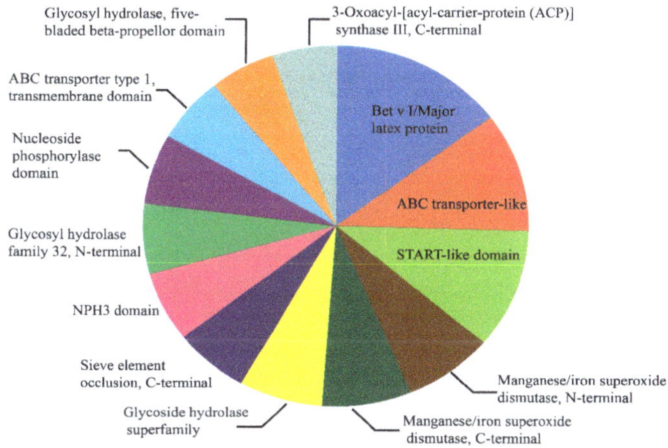

Figure 5. Protein domain analysis of differentially abundant proteins.

2.5. Interaction Network Analysis of the DAPs

A total of 56 related protein interaction networks were completed, among which 28 were up-regulated and 28 were down-regulated (Figure 6). Nine interaction proteins belonged to the ribosome network. These proteins included 7 up-regulated proteins, such as the 30S ribosomal protein S8 (RPS8), 60S ribosomal protein L3 (EMB2207), 30S ribosomal protein S4 (RPS4), 60S ribosomal protein L5 (RPL5A), chaperonin 60 (CPN60A), heat shock protein 90 (CR88) and heat shock 70 kDa protein 15 (Hsp70-15), and two down-regulated proteins, such as heat shock cognate 70 (Hsp70) and H/ACA ribonucleo protein complex subunit 2 (RP). Nine belonged to the oxidative phosphorylation network, including seven down-regulated proteins, such as Protein SGT1 (SGT1B), chaperonin (CRT1b), cytochrome b-c1 complex subunit 7-2 (CYDB), ATP synthase subunit O (ATP5), cytochrome b-c1 complex subunit Rieske-2 (UQCRFS), two ATP synthase subunit delta (ATPQ, ATPase), and 2 up-regulated proteins, such as endoplasmin (SHD) and heat shock protein 83 (HSP90.1). Other information about proteins is shown in Table S2.

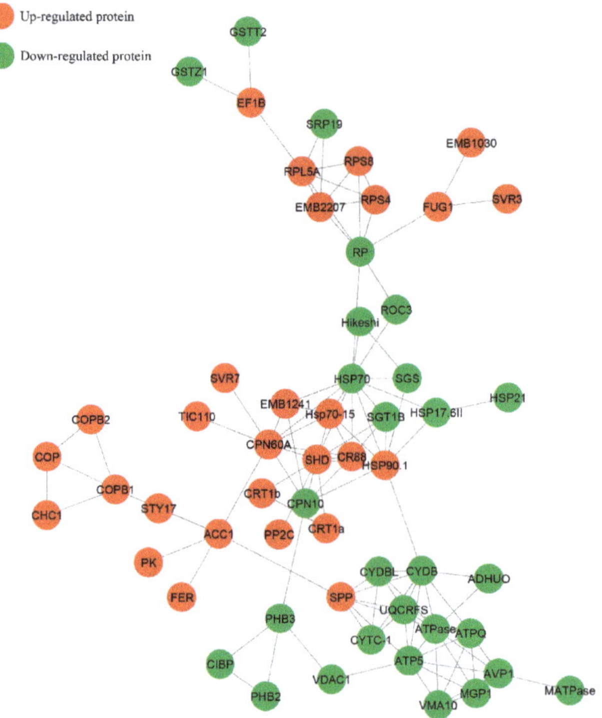

Figure 6. Functional correlation network of differentially abundance protein. Note: RPS8, 30S ribosomal protein S8; EMB2207, 60S ribosomal protein L3; ATP5, ATP synthase subunit O; ATPQ, ATP synthase subunit delta; RPS4, 30S ribosomal protein S4; COPB1, Coatomer subunit beta-1; COPB2, Coatomer subunit beta′-2; COP, Coatomer subunit gamma-2; PHB3, Prohibitin-3; PHB2, Prohibitin-2; RPL5A, 60S ribosomal protein L5; ATPase, ATP synthase subunit delta; CPN60A, Chaperonin 60 subunit alpha 1; CPN10, 10 kDa chaperonin; HSP90.1, Heat shock protein 83; SGT1B, Protein SGT1; SHD, Endoplasmin; CYDB, Cytochrome b-c1 complex subunit 7-2; UQCRFS, Cytochrome b-c1 complex subunit Rieske-2; CYTC-1, Cytochrome c; CR88, Heat shock protein 90-5; HSP70, Heat shock cognate 70 kDa protein; CYDBL, Cytochrome c oxidase subunit 6a; MGP1, Probable ATP synthase 24 kDa subunit; SPP, Stromal processing peptidase; Hsp70-15, Heat shock 70 kDa protein 15; RP, Aribonucleo protein complex subunit 2; CRT1b, Calreticulin; EMB1241, Uncharacterized protein; CRT1a, Calreticulin; ACC1, Acetyl-CoA carboxylase 1; ROC3, Peptidyl-prolyl cis-trans isomerase; SRP19, Signal recognition particle 19 kDa protein; FER, Receptor-like protein kinase; AVP1, Pyrophosphate-energized vacuolar membrane proton pump; PK, Probable receptor-like protein kinase; STY17, Serine/threonine-protein kinase STY17; CIBP, Uncharacterized protein; VMA10, V-type proton ATPase subunit G; CHC1, Clathrin heavy chain 1; SVR3, Translation factor GUF1; FUG1, Translation initiation factor IF-2; GSTT2, Glutathione S-transferase T2; EF1B, Elongation factor 1-delta; ADHUO, NADH dehydrogenase; HSP17.6II, 17.9 kDa class II heat shock protein; Hikeshi, Uncharacterized protein; TIC110, Protein TIC110; VDAC1, Mitochondrial outer membrane protein porin of 36 kDa; SVR7, Pentatricopeptide repeat-containing protein; GSTZ1, Glutathione S-transferase; HSP21, Small heat shock protein; EMB1030, Alanine—tRNA ligase; SGS, Uncharacterized protein; MATPase, mitochondrial ATP synthase; PP2C, Protein phosphatase 2C.

2.6. PRM-MS Quantification of DAPs

The expression levels of 5 DAPs were chosen for quantification by PRM-MS analysis to verify the proteomic results (Table 2). As this assay requires the signature peptide of the target protein to be unique, we only selected proteins with a unique signature peptide sequence for the PRM analysis. In the stem, five DAPs were sieve element occlusion, patellin 3, ATP synthase, NAD(P)H dehydrogenase, and glycine-rich RNA-binding protein 5. In general, the fold changes for these detected proteins were in agreement with the findings of proteomic analysis. Our PRM assay illustrated that the proteomic results were credible for further analysis.

Table 2. Confirmation of DAPs s in proteomic analysis using PRM analysis.

Description	Change in TMT	*p*-Value in TMT	Change in RPM	*p*-Value in RPM
Sieve element occlusion	1.27	0.000702	1.28	0.005
Patellin-3	1.39	0.0226	1.49	0.03
ATP synthase	0.83	0.00086	0.74	0.0015
NAD(P)H dehydrogenase	0.81	0.00398	0.88	0.01
Glycine-rich RNA-binding protein 5	0.83	0.0457	0.67	0.002

2.7. Complementation of the Proteomic Results via qRT-PCR

A total of eight proteins were randomly selected to complement the accuracy of the proteomics data using quantitative real-time PCR (qRT-PCR). There were 7 gene expression patterns that showed the same tendencies as those for protein expression, including receptor-like protein kinase, ubiquitin-conjugating enzyme E2, sugar transporter ERD6, U-box domain-containing protein 44, LRR receptor-like serine/threonine-protein kinase, serine/threonine-protein kinase STY17, and glycine-rich RNA-binding protein 5. Only one, ABC transporter B family member 19, showed an opposite expression pattern (Figure 7). The primer sequences for eight genes were listed (Table S3).

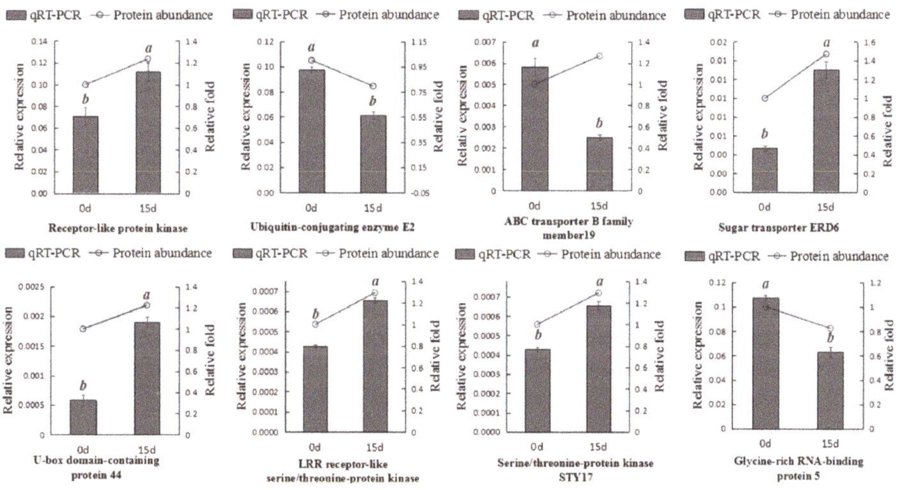

Figure 7. Complementation of proteomic results by qRT-PCR.

3. Discussion

3.1. LK Affected DAPs Involved in Transport Physiological Process

Membrane proteins fulfill critical functions in the transport of ions and organic molecules, which are an essential part of cellular stress responses. The ABC superfamily proteins mediate the transport across biological membranes not only as ATP-dependent pumps but also as ion channels and channel regulators [14]. Previous reports have suggested that ABC transporters in soybean and tomato are involved in salt stress responses and auxin transport [15,16], Seven ABC transporter proteins increased in expression in our study, suggesting that a high expression of ABC transporters in alligator weed stems are required for potassium transport regulation and LK responses [17,18]. However, further research is required to clarify the details of their function.

Patellin (PATL1) is a membrane trafficking-related protein. In *Arabidopsis*, PATL1 negatively modulates PM Na^+/H^+ antiport activity and modulates cellular redox homeostasis during salt stress [19]; in our study, four PATL1s, whose expressions were enhanced, were also identified, showing that they possibly had a vital function in modulating cellular redox homeostasis in alligator weed stems facing LK stress. The sieve element occlusion (SEO) protein is localized to phloem filaments and is required for phloem filament formation and to seal the phloem in wounded tobacco [20,21]; a high SEO expression would protect the stem from damage to maintain the optimum growth of alligator weed. Ca^{2+} wave propagation is channelled through two pore calcium channel protein 1 (TPC1), and the Ca^{2+} wave/TPC1 system likely elicits a systemic molecular response to plant stress tolerance [22]. Moreover, annexin functions as a Ca^{2+}-permeable channel in the plasma membrane to mediate the radical-activated plasma membrane Ca^{2+}- and K^+-permeable conductance in root cells [23]. In our study, the TPC1 and annexin proteins were both up-regulated, suggesting that fluxes in the transport of ions from the roots to leaves or post-phloem sugar transport to the root tip help plants rapidly manage stress [24].

One sugar transporter from *Dianthus spiculifolius* affects sugar metabolism and confers osmotic and oxidative stress tolerance in *Arabidopsis* [25]. The overexpression of the MdSUT2.2 gene (sucrose transporter) increased salt tolerance in transgenic apple, and further research suggest that MdSUT2.2 can be phosphorylated by MdCIPK13 and MdCIPK22 to enhance its stability and transport activity [26,27]. In our study, sugar transporter ERD6-like 6 (ERD6) increased in abundance, which may be a vital factor to help alligator weeds improve LK tolerance. Future research is needed to identify the interactions among CIPK proteins. Potassium and nitrogen are essential macronutrients and have a positive impact on crop yield. Previous studies have indicated that the absorption and translocation of K^+ and NO_3^- are correlated with each other in plants. A lack of NPF7.3/NRT1.5 resulted in K deficiency in shoots under low NO_3^- conditions by affecting xylem loading and root-to-shoot K^+ translocation through SKOR channel [28]. Further research suggest that NRT1.5 functions as a proton-coupled H^+/K^+ antiporter, plays a crucial role in K^+ translocation from the root to shoot and is also involved in the coordination of K^+/NO_3^- distribution in plants [29]. Thus, the down-regulation of NRT1/PTR FAMILY 8.3 in our study would decrease nitrate and potassium transport in the root-to-shoot process [30]. These findings provide a basis for the relationship between potassium and nitrogen nutrition in plants.

Syntaxin is a member of the SNARE (soluble *N*-ethylmaleimide-sensitive fusion protein attachment protein receptor) family. Arabidopsis R-SNARE VAMP721 interacts with the inward-rectifying K^+ channels KAT1 and KC1 and then suppresses the activities of two K^+ channels [31]. In our study, the decreased abundance of syntaxin-61 could improve KAT1 and KC1 activity to absorb and utilize more K^+. Recent reports have suggested that the overexpression of two SNARE proteins resulted in high tolerance to drought and salt stress in two kinds of plants [24,32]. Therefore, the SNARE proteins in different plants have multiple functions in regard to abiotic stress.

3.2. LK Affected DAPs Related to Carbohydrate and Energy Metabolism

In plants, carbohydrate and energy metabolism, which not only meet the energy demand but also afford many essential cofactors and substrates for other metabolisms and many transport processes, are dependent on the proton motive force that is achieved largely through the H^+ gradient across membranes afforded by the vacuolar H^+-ATPase (V-ATPase) [33]. In this study, one V-ATPase was decreased, but its response was different from the root results under LK conditions [13]. In plants, CYTc is also a multi-functional signaling molecule that influences the balance between life and death or triggers programmed cell death [34]. According to the present data, reduced cytochrome c oxidase activity may have reduces cell death in alligator weeds as it does in *Arabidopsis* [35].

Sucrose synthase (Sus) is a key enzyme in sucrose metabolism. One sucrose synthase was observed to be up-regulated, and the same results were found in alligator weed and Arabidopsis under K^+ deficient conditions [13,36]; a high expression of Sus may play a role in regulating energy metabolism in response to nutrition changes. Uridine-diphospho-(UDP)-glucose 4-epimerase (OsUGE-1) and nitrate reductase (NADH) increased in our study, and recent research has shown that overexpression OsUGEO lines maintain proportionally more galactose than glucose under low N conditions [37]. Nitrate reductase is also necessary under low nitrate stress [38], so we hypothesized that a high abundance of the two proteins could improve K tolerance by increasing N utilization in alligator weed shoots.

Pyruvate kinase (PK) is a glycolysis enzyme that catalyses the conversion of phosphoenolpyruvate (PEP) to pyruvate by transferring a phosphate from PEP to ADP; it has an absolute requirement for K^+, and a previous study showed that pyruvate kinase has protein kinase activity and plays a role in promoting tumor cell proliferation [39]. Two PKs identified in the present study were up-regulated, possibly having similar functions in plants to promote stem cell proliferation to improve lodging resistance in alligator weeds. Most represented DAPs were associated with carbohydrate and energy metabolism (35.2%) by KEGG analysis (Figure 3), this result was similar to *Arabidopsis* proteomic data [40], meanwhile, nine interaction proteins belonged to the oxidative phosphorylation network (Figure 6), these results supported the change of carbohydrate and energy metabolism were an adjustment mechanism of alligator weed to reduce LK damage.

3.3. LK Affected DAPs Related to Photosynthesis

Photosynthesis serves as the major energy source of plants and is directly affected by potassium deficiency. Magnesium chelatase is the first enzyme in the chlorophyll biosynthesis pathway and consists of 3 subunits that include ChlI, ChlD, and ChlH in plants. It is worth mentioning that ChlD and ChlH are related to abscisic acid (ABA) stress in *Arabidopsis*, and over-expression lines of the CHLD gene show more ABA sensitivity than do wild type [41]. ChlH, as an ABA receptor, can be phosphorylated by SnRK2.6 protein kinase [42]. Therefore, the accumulation of these two proteins in our study indicated that ABA pathway-related genes might also play positive roles in enabling plants to adapt LK stress.

Fructose-bisphosphate aldolase is involved in the calvin cycle. Cai et al. reported that the levels of superoxide anions and hydrogen peroxide were increased under low temperature and low-light intensity growth conditions in RNAi tomato seedlings [43]. Because H_2O_2 is a vital signal molecule in sensing LK stress [44], a low expression of this enzyme was hypothesized to increase the H_2O_2 content and contribute to alligator weed survival in stress.

Three other photosynthesis-related proteins, including one carbonic anhydrase (CA) and two oxygen-evolving enhancer protein 3 (OEE3), were down-regulated. CA plays a crucial role in the CO_2-concentrating mechanism (CCM), and recent research has shown that OEE is also an excellent antioxidant [45]. Additionally, *hcar* mutants (7-hydroxymethyl chlorophyll a reductase) showed an accelerated cell death phenotype due to excessive accumulation of singlet oxygen in rice and Arabidopsis, but HCAR-overexpressing plants were more tolerant to reactive oxygen species than were the *hcar* mutants [46]. HCAR and ribulose bisphosphate carboxylase were decreased in our study; therefore, it may be assumed that the down-regulation of photosynthesis-related proteins are associated with the LK stress response in the stems. The subcellular locations analysis revealed.

104 proteins were chloroplastic localization (Figure 4), the possible reason was that more proteins synthesized by the leaves were transported to the stems, or the stems cell synthesized more proteins for photosynthesis under LK stress for survival in alligator weed.

3.4. LK Affected DAPs Related to Common Stress Responses

LK stress may disturb cellular redox homeostasis and promote the production of reactive oxygen species (ROS); ROS can be scavenged by plant antioxidant defense systems consisting of a series of enzymes, such as superoxide dismutase (SOD), peroxidases (POD), glutathione-s-transferase (GST) and glutathione peroxidase (GPX). The expression of these enzymes were found to be changed in our proteomic data. Under LK conditions, these proteins are involved in detoxifying ROS to maintain the homeostasis of these molecules in the cytosol. Pattanayak et al. reported that the overexpression of protochlorophyllide reductase (PORCx) regulates oxidative stress in *Arabidopsis* and that overexpression reduced the generation of $^1O_2$ to reduce plasma membrane damage [47]. One PORCx was increased in our study. Thus, it may be assumed that this protein was helpful in facing LK stress responses. In summary, the expression changes of these proteins implied that the antioxidative defense system was provoked in the stems to protect alligator weed. Major latex protein (MLP)-like protein, as a positive regulator of downstream signaling, mainly responds to defense or abiotic stimuli [48]. In our study, two MLP-like proteins were remarkably increased, one was decreased. Previous research has suggested that the over-expression of MLP protein enhances the salt and drought tolerance of Arabidopsis [49]; however, the specific biological function of MLP related to LK was unknown, so it may be a novel LK-stress-responsive protein in alligator weed plants. In the current study, three heat shock proteins were found to be significantly up-regulated, while three were down-regulated. The expression differences imply that the gene family members probably have diverse functions to cope with various stresses [50].

Meanwhile, one chaperonin 60 showed enhanced expression, and the Oscpn60α1 mutant had a pale-green phenotype at the seedling stage. Further analysis indicated that the level of the rubisco large subunit (rbcL) was severely reduced in the mutant, so OsCpn60α1 is required for the folding of rbcL [51]. Therefore, a high expression of chaperonin 60 may maintain the correct protein folding under LK stress in the stem. Universal stress-related proteins are also redox-dependent chaperones, and a high expression of these proteins enhances plant tolerance to heat shock and oxidative stress in *Arabidopsis* [52]. Three stress-related proteins were found to be significantly down-regulated. In addition, other stress proteins, such as remorins, phenylalanine ammonialyase (PAL) and cellulose synthase, were all up-regulated. Taken together, these results suggest that the stress response proteins all play significant roles in improving the ability of alligator weed to cope with abiotic stress.

3.5. Comparative Analysis of Low-K^+ Responses between Alligator Weed and Arabidopsis

Phospholipids play crucial roles in regulating development and signal transduction in higher plants, protein phosphatase 2C (PP2C) physically interacts with CBL-CIPK and can activate the AKT1 channel in *Arabidopsis* [53]. AtPP2A plays vital roles for root auxin transport, gravity response, and lateral root development [54]. In our study, two protein phosphatase 2C and one phospholipase A were found to be up-regulated, this is consistent with the proteomic results in *Arabidopsis* seedlings after LK treatment [40]. These results indicated that high expression PP2C and PP2A maybe induce root growth and absorb more K^+ during low potassium conditions, Therefore, it is clear that phospholipids signal regulation pathway is conservative in plant. Meanwhile, the calcium signaling pathway is vital in phosphorylation signal transduction responses under stress conditions. Calreticulin (CRT) is a multifunctional protein that participates in plant growth and development as an important molecular partner on the endoplasmic network. CRT mutant exhibit more sensitivity to water stress in *Arabidopsis* [55]. In Arabidopsis CRT1 and CRT2 are critical components in the accumulation of VESICLE-ASSOCIATED MEMBRANE PROTEIN 721 (VAMP721) and VAMP722 during ER stress responses [56]. Arabidopsis VAMP721 could assemble with SYP121 to drive membrane fusion and bind to the KAT1 K^+ channel to control channel-gating [57]. In our data, two calreticulin (CRT) increased abundance but this protein was not reported in proteomic analysis under LK stress before, so this result raised CRT maybe maintain intracellular calcium homeostasis or interacts with VAMP721 to improve K^+ channels activity to increase LK tolerance in alligator weed.

The receptor-like protein kinases (RLKs) are involved in plant growth and development, hormonal responses, cellular differentiation, stress responses and pathogen recognition [58]. The heterologous overexpression of one serine/threonine-protein kinase (STK) from *Bambusa balcooa* in transgenic tobacco induce higher cellulose synthesis and high cellulose deposition in the xylem fibres, increase a greater number of xylary fibres and enhance mechanical strength in transgenic tobacco [59]. In our study, the expression of one STK was enhanced in the stem. At the same time, cellulose synthase was also increased, so we hypothesize that high STK expression can maintain the stem mechanical strength under LK stress by improving cellulose synthase activity. Previous research suggested RLKs can recognized CLE (CLAVATA3/Endosperm surrounding region-related) to form CLE-RLK module in order to convey extracellular and intracellular signaling in long distance [60], in our study, three RLKs increased abundance, maybe these protein kinase can interaction with CLE peptides to re-establish new optimal growth state under LK stress, this result could partly explain the high LK tolerance of alligator weeds because of good root to shoot signaling.

Some protein involved in protein synthesis was founded to be increased abundance in the present data, including two 30S ribosomal proteins, two 60S ribosomal proteins, three translation initiation factors and one translation factor. The overexpression of elongation factors in plants allow them to be more tolerant to high temperature stress [61], so it has been noticed that the accumulation of elongation factors could play a role in splicing, polyadenylation, mRNA stabilization and localization, and translation [62]. The expression of four RNA-binding protein were significantly increased. Nine interaction proteins belonged to the ribosome network after analysis (Figure 6), these results showed that protein synthesis and post-transcriptional regulation processes were improved in stems after LK treatment for 15 days compare to *Arabidopsis*. RBPs that consist of both an RNA-recognition motif (RRM) and a glycine-rich region termed GRPs [63]. Kim et al. reported GRPs are functionally conserved during the cold adaptation process in rice and *Arabidopsis* as RNA chaperones [64]. Glycine-rich RNA-binding protein (GRP) expression has been shown to be regulated by different abiotic stresses, including low temperature, water deficit and high salinity stress, and over-expressing ZjGRP from *Zoysia japonica* weaken salinity tolerance in *Arabidopsis*, affecting ion transportation, osmosis, and antioxidation [65]. The expression of GRP5 was reduced, and the result was confirmed by qRT-PCR and PRM analysis. Therefore, it can be hypothesized that the low expression of this protein may play a positive role in improving LK tolerance in alligator weeds. It was different to *Arabidopsis*.

Ubiquitination is a post-translational modification of proteins process consisting ubiquitin activating enzyme (E1), ubiquitin-conjugating enzyme (E2), and ubiquitin ligase enzyme (E3), which regulates a wide range of biological processes, including adaptation to drought, salinity, cold, and nutrient deprivation [66], which the overexpression of the soybean ubiquitin-conjugating enzyme gene GmUBC2 in *Arabidopsis* enhance its drought and salt tolerance by modulating abiotic stress-responsive gene expression [67]. It was worth noting that three E2 were decreased, which was similar to the proteomic results in *Arabidopsis* under LK stress for 7 days [40], and the role of E2s may be similar to those of AtUBC32, AtUBC33, and AtUBC34, because the suppression of three E2 expression level result in increased abscisic acid-mediated stomata closure and tolerance to drought stress [68]. Therefore, the three E2s in the stems play negative roles under LK stress, and this result can also provide clues for subsequent in-depth research.

In conclusion, much new information about the protein profile in alligator weed stems was obtained with respect to LK stress (Table S4), the physiological analysis and expression of these proteins were a clear response to potassium deficiency in stem (Figure 8). These proteins worked together to establish a new steady-state balance of metabolic processes to enable alligator weed to grow normally after enduring LK stress for 15 days.

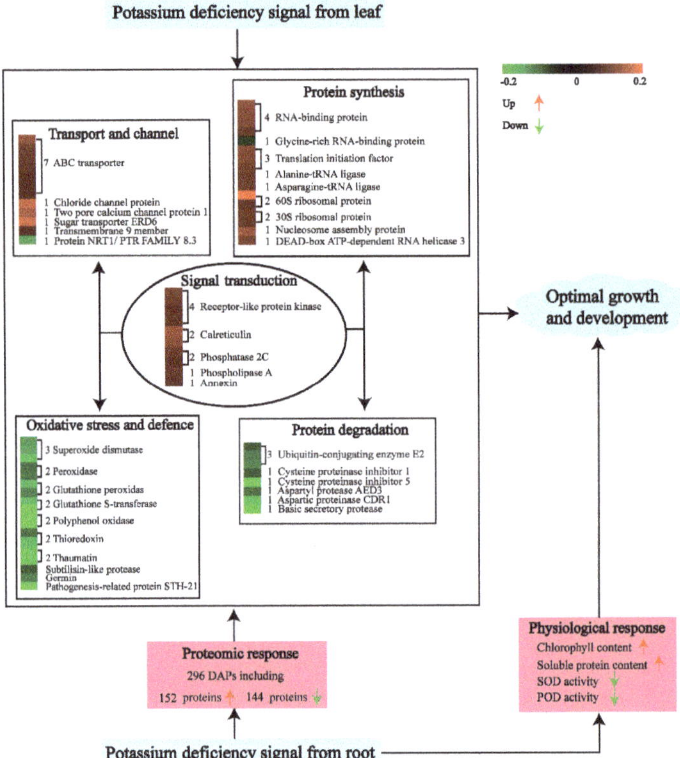

Figure 8. Model showed the physiological and proteomic responses of alligator weed stem under potassium deficiency (LK) stress.

4. Materials and Methods

4.1. Physiological Experiments of Alternanthera philoxeroides Stem

Naturally grown *Alternanthera philoxeroides* shoots were collected from a test field of Sichuan Agricultural University (Chengdu, China) and cultured hydroponically in a growth chamber for 10 day to induce root growth. The greenhouse was maintained under a 16 h/8 h day/night light cycle and a 28 °C/25 °C day/night temperature cycle.

The nutrient solution was refreshed every 2 day. The nutrient solution was prepared as described before [7]. The alligator weed plants had grown strong roots after 10 day of culture. Half of the plants were then transferred to a low potassium nutrient solution (lacking K_2SO_4) as the LK sample. The other half continued to grow in the solution with optimum K concentration as the CK sample. Then, LK and CK alligator weed stem samples were collected after 15 day of LK treatment for the physiological and molecular measurements. The chlorophyll and soluble protein content and SOD and POD activity were measured. The total chlorophyll and soluble protein content and SOD and POD activity were measured as described previously [15,50].

4.2. Protein Extraction

Six stem samples were ground in liquid nitrogen to cell powder and then transferred to a 5-mL centrifuge tube. Then, four volumes of lysis buffer (8 M urea, 1% Triton-100, 10 mM dithiothreitol, and 1% protease inhibitor cocktail) were added to the cell powder, followed by sonication three times on ice using a high intensity ultrasonic processor. The remaining debris was removed by centrifugation at 20,000× g at 4 °C for 10 min, and finally, the protein was precipitated with cold 20% TCA for 2 h at −20 °C. After centrifugation at 12,000× g at 4 °C for 10 min, the supernatant was discarded, the remaining precipitate was washed with cold acetone three times, the protein was redissolved in 8 M urea and the protein concentration was determined with a BCA kit according to the manufacturer's instructions.

4.3. Trypsin Digestion and TMT Labelling

For digestion, the protein solution was reduced with 5 mM dithiothreitol for 30 min at 56 °C and alkylated with 11 mM iodoacetamide for 15 min at room temperature in the dark. The protein sample was then diluted by adding 100 mM TEAB to obtain a urea concentration less than 2 M. Finally, trypsin was added at a 1:50 trypsin-to-protein mass ratio for the first digestion overnight and a 1:100 trypsin-to-protein mass ratio for a second 4 h digestion. After trypsin digestion, the peptides were desalted using a Strata X C18 SPE column (Phenomenex) and vacuum dried. The peptide were reconstituted in 0.5 M TEAB and processed according to the manufacturer's protocol for the TMT kit. Briefly, one unit of TMT reagent was thawed and reconstituted in acetonitrile. The peptide mixtures were then incubated for 2 h at room temperature, pooled, desalted and dried by vacuum centrifugation.

4.4. HPLC Fractionation

The tryptic peptides were fractionated by high pH reverse-phase HPLC using an Agilent 300Extend C18 column (5 μm particles, 4.6 mm ID, and 250 mm length). Briefly, the peptides were first separated with a gradient of 8% to 32% acetonitrile (pH 9.0) over 60 min into 60 fractions. Then, the peptides were combined into 18 fractions and dried by vacuum centrifugation.

4.5. LC-MS/MS Analysis

The tryptic peptides were dissolved in 0.1% formic acid (solvent A) and directly loaded onto a homemade reversed-phase analytical column. The gradient was comprised of an increase from 7% to 25% solvent B (0.1% formic acid in 98% acetonitrile) over 38 min and 25% to 40% over 14 min and then rose to to 80% for 4 min, followed by holding at 80% for the last 4 min, all at a constant flow rate of 700 nL/min on an EASY-nLC 1000 UPLC system. The peptides were subjected to an NSI source followed by tandem mass spectrometry (MS/MS) by using a Q Exactive^{TM} Plus (Thermo, Waltham, MA, USA) coupled online to the UPLC. The electrospray voltage applied was 2.0 kV. The m/z scan range was 350 to 1000 for the full scan, and the intact peptides were detected in the Orbitrap at a resolution of 70,000. The peptides were then selected for MS/MS using the NCE setting at 27, and the fragments were detected in the Orbitrap at a resolution of 17,500. A data-independent procedure that alternated between one MS scan followed by 20 MS/MS scans was performed. The automatic gain control (AGC) was set at 3E6 for the full MS and 1E5 for MS/MS. The maximum IT was set at 20 s for the full MS and auto for MS/MS. The isolation window for MS/MS was set at 2.0 m/z.

4.6. Database Search

The resulting MS/MS data were processed using the Maxquant search engine (v.1.5.2.8). The tandem mass spectra were searched against alligator weed transcription data. Trypsin/P was specified as a cleavage enzyme allowing up to two missing cleavages. The mass tolerance for the precursor ions was set as 20 ppm in the first search and 5 ppm in the main search, and the mass tolerance for the fragmented ions was set as 0.02 Da. Carbamidomethyl on Cys was specified as the fixed modification, and oxidation on Met was specified as the variable modifications. FDR was adjusted to <1%, and the minimum score for the peptides was set to >40.

4.7. DAPs Functional Analysis

The DAPs were assigned to the NCBI non-redundant (Nr) protein database using the Blast 2GO program to obtain their functional annotation. The Gene Ontology (GO) annotation proteome was derived from the UniProt-GOA database (available online: http://www.ebi.ac.uk/GOA/). The Kyoto Encyclopedia of Genes and Genomes (KEGG) database was used to annotate the protein metabolic pathways.

GO terms or KEGG pathways with p-values < 0.05 were regarded as significantly enriched. The DAPs subcellular location prediction was conducted followed by TargetP1.1 (available online: http://www.cbs.dtu.dk/services/TargetP/). Finally, the protein–protein interactions were analyzed for the identified proteins using the STRING v10 database (available online: http://string-db.org) to determine their functions and pathways.

4.8. Parallel Reaction Monitoring PRM-MS Analysis

The protein expression change obtained using the proteomic analysis were confirmed by a PRM-MS analysis carried out at Jingjie PTM-Biolab Co., Ltd. (Hang Zhou, China). The proteins (60 μg) from the stem sample were prepared, reduced, alkylated, and digested with trypsin following the protocol for the TMT analysis. The obtained peptide mixtures were introduced into the mass spectrometer via a C18 trap column (0.10×20 mm; 3 μm) and then via a C18 column (0.15×120 mm; 1.9 μm). The raw data obtained were then analyzed using Proteome Discoverer 1.4 (Thermo Fisher Scientific). The FDR was set to 0.01 for the proteins and peptides. Skyline 2.6 software (Download from the MacCoss Lab at the University of Washington) was used for the quantitative data processing and proteomic analysis.

4.9. Quantitative Reverse Transcription PCR (qRT-PCR) Analysis

For the qRT-PCR analysis, total RNA was extracted from the stem after 15 days of LK treatment using TRIzol reagent (Invitrogen, Carlsbad, CA, USA). The isolated total RNA was used to generate cDNA with a reverse transcriptase kit (Thermo, Tokyo, Japan). The relative quantification of the candidate genes by qRT-PCR was carried out using a 7500 Real Time PCR System machine (Bio-Ras, Life Technologies, Carlsbad, CA, USA) following the manufacturer's protocols. The formula $2^{-\Delta\Delta Ct}$ was used to calculate the relative gene expression levels. Actin2/8 expression was used as the internal control, and three replicates were conducted. All data are shown as the mean $\pm$ SD ($n = 3$).

4.10. Statistical Analysis

For all generated data, at least three biological replicates were performed for the chlorophyll and soluble protein content measurements and SOD and POD activity measurements. The data were subjected to unpaired student's t-tests at levels of $p \leq 0.01$ and $p \leq 0.05$. The data are shown as the mean $\pm$ SE ($n = 3$). Excel and the SPSS 14.0 statistical software package were used for the statistical analyses of the data. The statistical results are reported as the mean $\pm$ SD.

Supplementary Materials: The following are available online at http://www.mdpi.com/1422-0067/20/1/221/s1.

Author Contributions: L.-Q.L., X.-Y.W. and L.-M.L. designed the experiments. C.-C.L., J.-H.L., Z.T., Y.-F.L. performed the experiments, S.N., S.-M.Y., F.-C.Z. analyzed the data. L.-Q.L. and L.-M.L. wrote the paper. All authors read and approved the final manuscript.

Acknowledgments: This research was financially supported by the open funds of the State Key Laboratory of Plant Physiology and Biochemistry (SKLPPBKF1801). We thank to Manager Yuanhong Deng of Jingjie PTM-Biolab (Hang Zhou 310018, China) Co., Ltd. for giving the suggestions on differentially abundant protein analysis.

Conflicts of Interest: The authors declare that the research was conducted in the absence of any commercial or financial relationships that could be construed as a potential conflict of interest.

References

1. Kanai, S.; Ohkura, K.; Adu-Gyamfi, J.J.; Mohapatra, P.K.; Nguyen, N.T.; Saneoka, H.; Fujita, K. Depression of sink activity precedes the inhibition of biomass production in tomato plants subjected to potassium deficiency stress. *J. Exp. Bot.* **2007**, *58*, 2917–2928. [CrossRef] [PubMed]
2. Maathuis, F.J. Physiological functions of mineral macronutrients. *Curr. Opin. Plant Biol.* **2009**, *12*, 250–258. [CrossRef] [PubMed]
3. Wang, Y.; Wu, W.H. Potassium transport and signaling in higher plants. *Annu. Rev. Plant Biol.* **2013**, *64*, 451–476. [CrossRef] [PubMed]
4. Chérei, I.; Lefoulon, C.; Boeglin, M.; Sentenac, H. Molecular mechanisms involved in plant adaptation to low K$^+$ availability. *J. Exp. Bot.* **2014**, *65*, 833–848. [CrossRef] [PubMed]
5. Gao, L.X.; Geng, Y.P.; Li, B.; Chen, J.K.; Yang, J. Genome-wide DNA methylation alterations of *Alternanthera philoxeroidesin* natural and manipulated habitats: Implications for epigenetic regulation of rapid responses to environmental fluctuation and phenotypic variation. *Plant Cell Environ.* **2010**, *33*, 1820–1827. [CrossRef] [PubMed]
6. Song, Z.Z.; Su, Y.H. Distinctive potassium-accumulation capability of alligator weed (*Alternanthera philoxeroides*) links to high-affinity potassium transport facilitated by K$^+$-uptake systems. *Weed Sci.* **2013**, *6*, 77–84. [CrossRef]
7. Li, L.Q.; Xu, L.; Wang, X.Y.; Pan, G.; Lu, L.M. De novo characterization of the alligator weed (*Alternanthera philoxeroides*) transcriptome illuminates gene expression under potassium deprivation. *J. Genet.* **2015**, *94*, 95–104. [CrossRef]
8. Tomanek, L. Proteomics to study adaptations in marine organisms to environmental stress. *J. Proteom.* **2014**, *105*, 92–106. [CrossRef]
9. Savitski, M.M.; Mathieson, T.; Zinn, N.; Sweetman, G.; Doce, C.; Becher, I.; Pachl, F.; Kuster, B.; Bantscheff, M. Measuring and managing ratio compression for accurate iTRAQ/TMT quantification. *J. Proteome Res.* **2013**, *12*, 3586–3598. [CrossRef]
10. Zeng, J.B.; He, X.Y.; Quan, X.Y.; Cai, S.G.; Han, Y.; Nadira, U.A.; Zhang, G.P. Identification of the proteins associated with low potassium tolerance in cultivated and Tibetan wild barley. *J. Proteom.* **2015**, *126*, 1–11. [CrossRef]
11. Zhang, Z.; Chao, M.; Wang, S.; Bu, J.; Tang, J.; Li, F. Proteome quantification of cotton xylem sap suggests the mechanisms of potassium-deficiency-induced changes in plant resistance to environmental stresses. *Sci. Rep.* **2016**, *6*, 21060–21075. [CrossRef]
12. Li, G.Z.; Wu, Y.F.; Liu, G.Y.; Xiao, X.H.; Wang, P.F.; Gao, T. A Large-scale proteomics combined with transgenic experiments demonstrates an important role of jasmonic acid in potassium deficiency response in wheat and rice. *Mol. Cell. Proteom.* **2017**, *16*, 1889–1905. [CrossRef]
13. Li, L.Q.; Liu, L.; Zhuo, W.; Chen, Q.; Hu, S.; Peng, S.; Wang, X.Y.; Lu, Y.F.; Lu, L.M. Physiological and quantitative proteomic analyses unraveling potassium deficiency stress response in alligator weed (*Alternanthera philoxeroides* L.) root. *Plant Mol. Biol.* **2018**, *97*, 265–278. [CrossRef] [PubMed]
14. Theodoulou, F.L. Plant ABC transporters. *Biochim. Biophys. Acta* **2000**, *1465*, 79–103. [CrossRef]
15. Ji, W.; Cong, R.; Li, S.; Li, R.; Qin, Z.W.; Li, Y.J.; Zhou, X.L.; Chen, S.; Li, J. Comparative proteomic analysis of soybean leaves and roots by iTRAQ provides insights into response mechanisms to short-term salt stress. *Front. Plant Sci.* **2016**, *7*, 573. [CrossRef] [PubMed]
16. Ofori, P.A.; Geisler, M.; di Donato, M.; Pengchao, H.; Otagaki, S.; Matsumoto, S.; Shiratake, K. Tomato ATP-binding cassette transporter SlABCB4 is involved in auxin transport in the developing fruit. *Plants* **2018**, *7*, 65. [CrossRef] [PubMed]
17. Lin, D.X.; Tang, H.; Wang, E.T.; Chen, W.X. An ABC transporter is required for alkaline stress and potassium transport regulation in *Sinorhizobium meliloti*. *FEMS Microbiol.* **2009**, *293*, 35–41. [CrossRef]
18. Wang, C.; Chen, H.F.; Hao, Q.N.; Sha, A.H.; Shan, Z.H.; Chen, L.M.; Zhou, R.; Zhi, H.J.; Zhou, X.A. Transcript profile of the response of two soybean genotypes to potassium deficiency. *PLoS ONE* **2012**, *7*, e39856. [CrossRef]
19. Zhou, H.P.; Wang, C.W.; Tan, T.H.; Cai, J.Q.; He, J.X.; Lin, H.H. Patellin1 negatively modulates salt tolerance by regulating PM Na$^+$/H$^+$ antiport activity and cellular redox homeostasis in Arabidopsis. *Plant Cell Physiol.* **2018**, *59*, 1630–1642. [CrossRef]

20. Zhang, B.; Tolstikov, V.; Turnbull, C.; Hicks, L.M.; Fiehn, O. Divergent metabolism and proteome suggest functional independence of dual phloem transport systems in cucurbits. *Proc. Natl. Acad. Sci. USA* **2010**, *107*, 13532–13537. [CrossRef]

21. Ernst, A.M.; Jekat, S.B.; Zielonka, S.; Müller, B.; Neumann, U.; Rüping, B.; Twyman, R.M.; Krzyzanek, V.; Prüfer, D.; Noll, G.A. Sieve element occlusion (SEO) genes encode structural phloem proteins involved in wound sealing of the phloem. *Proc. Natl. Acad. Sci. USA* **2012**, *109*, 11084–11085. [CrossRef] [PubMed]

22. Choi, W.G.; Masatsugu Toyota, M.; Kim, S.H.; Hilleary, R.; Gilroy, S. Salt stress-induced Ca^{2+} waves are associated with rapid, long-distance root-to-shoot signaling in plants. *Proc. Natl. Acad. Sci. USA* **2014**, *111*, 6497–6502. [CrossRef] [PubMed]

23. Laohavisit, A.; Shang, Z.L.; Rubio, L.; Cuin, T.A.; Véry, A.A.; Wang, A.H.; Mortimer, J.C.; Macpherson, N.; Coxon, K.M.; Battey, N.H.; et al. Arabidopsis annexin1 mediates the radical-activated plasma membrane Ca^{2+}- and K^+-permeable conductance in root cells. *Plant Cell* **2012**, *24*, 1522–1533. [CrossRef]

24. Wang, P.; Sun, Y.; Pei, Y.K.; Li, X.C.; Zhang, X.Y.; Li, F.G.; Hou, Y.X. GhSNAP33, a t-SNARE protein from *Gossypium hirsutum*, mediates resistance to *Verticillium dahlia* infection and tolerance to drought stress. *Front. Plant Sci.* **2018**, *9*, 896. [CrossRef] [PubMed]

25. Zhou, A.M.; Ma, H.P.; Feng, S.; Gong, S.F.; Wang, J.G. A novel sugar transporter from *Dianthus spiculifolius*, DsSWEET12, affects sugar metabolism and confers osmotic and oxidative stress tolerance in *Arabidopsis*. *Int. J. Mol. Sci.* **2018**, *19*, 497. [CrossRef] [PubMed]

26. Ma, Q.J.; Sun, M.H.; Kang, H.; Lu, J.; You, C.X.; Hao, Y.J. A CIPK protein kinase targets sucrose transporter MdSUT2.2 at Ser 254 for phosphorylation to enhance salt tolerance. *Plant Cell Environ.* **2018**. [CrossRef] [PubMed]

27. Ma, Q.J.; Sun, M.H.; Lu, J.; Kang, H.; You, C.X.; Hao, Y.J. An apple sucrose transporter MdSUT2.2 is a phosphorylation target for protein kinase MdCIPK22 in response to drought. *Plant Biotechnol. J.* **2018**. [CrossRef] [PubMed]

28. Drechsler, N.; Zheng, Y.; Bohner, A.; Nobmann, B.; Wirén, N.; Kunze, R.; Rausch, C. Nitrate-dependent control of shoot K homeostasis by the nitrate transporter1/peptide transporter family member NPF7.3/NRT1.5 and the stelar K^+ outward rectifier SKOR in Arabidopsis. *Plant Physiol.* **2015**, *169*, 2832–2847.

29. Li, H.; Yu, M.; Du, X.Q.; Wang, Z.F.; Wu, W.H.; Quintero, F.J.; Jin, X.H.; Li, H.D.; Wang, Y. NRT1.5/NPF7.3 functions as a proton-coupled H^+/K^+ antiporter for K+ loading into the xylem in *Arabidopsis*. *Plant Cell* **2017**, *29*, 2016–2026. [CrossRef]

30. Lin, S.H.; Kuo, H.F.; Canivenc, G.; Lin, C.S.; Lepetit, M.; Hsu, P.K.; Tillard, P.; Lin, H.L.; Wang, Y.Y.; Tsai, C.B.; et al. Mutation of the Arabidopsis NRT1.5 nitrate transporter causes defective root-to-shoot nitrate transport. *Plant Cell* **2008**, *20*, 2514–2528. [CrossRef]

31. Zhang, B.; Karnik, R.; Wang, Y.; Wallmeroth, N.; Blatt, M.R.; Grefen, C. The *Arabidopsis* R-SNARE VAMP721 interacts with KAT1 and KC1 K^+ channels to moderate K^+ current at the plasma membrane. *Plant Cell* **2015**, *27*, 1697–1717. [CrossRef] [PubMed]

32. Singh, D.; Yadav, N.S.; Tiwari, V.; Agarwal, P.K.; Jha, B. A SNARE-like superfamily protein SbSLSP from the *Halophyte Salicornia brachiata* confers salt and drought tolerance by maintaining membrane stability, K^+/Na^+ ratio, and antioxidant machinery. *Front. Plant Sci.* **2016**, *7*, 737. [CrossRef] [PubMed]

33. Pittman, J.K. Multiple transport pathways for mediating intracellular pH homeostasis: The contribution of H^+/ion exchangers. *Front. Plant Sci.* **2012**, *3*, 11. [CrossRef] [PubMed]

34. Welchen, E.; Daniel, H.G. Cytochrome c, a hub linking energy, redox, stress and signaling pathways in mitochondria and other cell compartments. *Physiol. Plant.* **2016**, *157*, 310–321. [CrossRef]

35. Chivasa, S.; Tome, D.F.; Hamilton, J.M.; Slabas, A.R. Proteomic analysis of extracellular ATP-regulated proteins identifies ATP synthase beta-subunit as a novel plant cell death regulator. *Mol. Cell. Proteom.* **2011**, *10*, 1074–1087. [CrossRef] [PubMed]

36. Armengaud, P.; Sulpice, R.; Miller, A.J.; Stitt, M.; Amtmann, A.; Gibon, Y. Multilevel analysis of primary metabolism provides new insights into the role of potassium nutrition for glycolysis and nitrogen assimilation in *Arabidopsis* roots. *Plant Physiol.* **2009**, *150*, 772–785. [CrossRef] [PubMed]

37. Guevara, D.R.; El-Kereamy, A.; Yaish, M.W.; Mei-Bi, Y.; Rothstein, S.J. Functional characterization of the rice UDP-glucose 4-epimerase 1, OsUGE1: A potential role in cell wall carbohydrate partitioning during limiting nitrogen conditions. *PLoS ONE* **2014**, *9*, e96158. [CrossRef] [PubMed]

38. Reyes, T.H.; Scartazza, A.; Pompeiano, A.; Ciurli, A.; Lu, Y.; Guglielminetti, L.; Yamaguchi, J. Nitrate reductase modulation in response to changes in C/N balance and nitrogen source in *Arabidopsis*. *Plant Cell Physiol.* **2018**, *59*, 1248–1254. [CrossRef] [PubMed]

39. Gao, X.; Wang, H.; Yang, J.J.; Liu, X.; Liu, Z.R. Pyruvate kinase M2 regulates gene transcription by acting as a protein kinase. *Mol. Cell* **2012**, *45*, 598–609. [CrossRef] [PubMed]

40. Kang, J.G.; Pyo, Y.J.; Cho, J.W.; Cho, M.H. Comparative proteome analysis of differentially expressed proteins induced by K$^+$ deficiency in *Arabidopsis thaliana*. *Proteomics* **2004**, *4*, 3549–3559. [CrossRef]

41. Du, S.Y.; Zhang, X.F.; Lu, Z.; Xin, Q.; Wu, Z.; Jiang, T.; Lu, Y.; Wang, X.F.; Zhang, D.P. Roles of the different components of magnesium chelatase in abscisic acid signal transduction. *Plant Mol. Biol.* **2012**, *80*, 519–537. [CrossRef] [PubMed]

42. Liang, S.; Lu, K.; Wu, Z.; Jiang, S.C.; Yu, Y.T.; Bi, C.; Xin, Q.; Wang, X.F.; Zhang, D.P. A link between magnesium-chelatase H subunit and sucrose nonfermenting 1(SNF1)-related protein kinase SnRK2.6/OST1 in Arabidopsis guard cell signaling in response to abscisic acid. *J. Exp. Bot.* **2015**, *66*, 6355–6369. [CrossRef] [PubMed]

43. Cai, B.B.; Li, Q.; Liu, F.J.; Bi, H.G.; Ai, X.Z. Decreasing fructose-1,6-bisphosphate aldolase activity reduces plant growth and tolerance to chilling stress in tomato seedlings. *Physiol. Plant.* **2018**, *163*, 247–258. [CrossRef] [PubMed]

44. Kim, M.J.; Ciani, S.; Schachtman, D.P. A peroxidase contributes to ROS production during Arabidopsis root response to potassium deficiency. *Mol. Plant* **2010**, *3*, 420–427. [CrossRef] [PubMed]

45. Kim, E.Y.; Choi, Y.H.; Lee, J.I.; Kim, I.H.; Nam, T.J. Antioxidant activity of oxygen evolving enhancer protein 1 purified from *Capsosiphon fulvescens*. *J. Food Sci.* **2015**, *80*, H1412–H1417. [CrossRef] [PubMed]

46. Piao, W.L.; Han, S.H.; Sakuraba, Y.; Paek, N.C. Rice 7-hydroxymethyl chlorophyll a reductase is involved in the promotion of chlorophyll degradation and modulates cell death signaling. *Mol. Cells* **2017**, *40*, 773–786.

47. Pattanayak, G.K.; Tripathy, B.C. Overexpression of protochlorophyllide oxidoreductase C regulates oxidative stress in Arabidopsis. *PLoS ONE* **2011**, *6*, e26532. [CrossRef]

48. Chen, J.Y.; Dai, X.F. Cloning and characterization of the *Gossypium hirsutum* major latex protein gene and functional analysis in *Arabidopsis thaliana*. *Planta* **2010**, *231*, 861–873. [CrossRef]

49. Wang, Y.P.; Yang, L.; Chen, X.; Ye, T.T.; Zhong, B.; Liu, R.J.; Wu, Y.; Chan, Z.L. Major latex protein-like protein 43 (MLP43) functions as a positive regulator during abscisic acid responses and confers drought tolerance in *Arabidopsis thaliana*. *J. Exp. Bot.* **2016**, *67*, 421–434. [CrossRef]

50. Zhang, C.M.; Shi, S.L. Physiological and proteomic responses of contrasting alfalfa (*Medicago sativa* L.) varieties to PEG-induced osmotic stress. *Front. Plant Sci.* **2018**, *9*, 242. [CrossRef]

51. Kim, S.R.; Yang, J.; An, G. OsCpn60α1, encoding the plastid chaperonin 60α subunit, is essential for folding of rbcL. *Mol. Cells* **2013**, *35*, 402–409. [CrossRef] [PubMed]

52. Jung, Y.J.; Melencion, S.M.; Lee, E.S.; Park, J.H.; Alinapon, C.V.; Oh, H.T.; Yun, D.J.; Chi, Y.H.; Lee, S.Y. Universal stress protein exhibits a redox-dependent chaperone function in *Arabidopsis* and enhances plant tolerance to heat shock and oxidative Stress. *Front. Plant Sci.* **2015**, *6*, 1141. [CrossRef] [PubMed]

53. Lan, W.Z.; Lee, S.C.; Che, Y.F.; Jiang, Y.Q.; Luan, S. Mechanistic analysis of AKT1 regulation by the CBL–CIPK–PP2CA interactions. *Mol. Plant* **2011**, *4*, 527–536. [CrossRef] [PubMed]

54. Rashotte, A.M.; DeLong, A.; Muday, G.K. Genetic and chemical reductions in protein phosphatase activity alter auxin transport, gravity response, and lateral root growth. *Plant Cell* **2001**, *13*, 1683–1697. [CrossRef]

55. Kim, J.H.; Nguyen, N.H.; Nguyen, N.T.; Hong, S.W.; Lee, H. Loss of all three calreticulins, CRT1, CRT2 and CRT3, causes enhanced sensitivity to water stress in *Arabidopsis*. *Plant Cell Rep.* **2013**, *32*, 1843–1853. [CrossRef] [PubMed]

56. Kim, S.; Choi, Y.J.; Kwon, C.; Yun, H.S. Endoplasmic reticulum stress-induced accumulation of VAMP721/722 requires CALRETICULIN 1 and CALRETICULIN 2 in *Arabidopsis*. *J. Integr. Plant Biol.* **2018**. [CrossRef] [PubMed]

57. Zhang, B.; Karnik, R.; Waghmare, S.; Donald, N.; Blatt, M.R. VAMP721 conformations unmask an extended motif for K+ channel binding and gating control. *Plant Physiol.* **2017**, *173*, 536–551. [CrossRef] [PubMed]

58. Morillo, S.A.; Tax, F.E. Functional analysis of receptor-like kinases in monocots and dicots. *Curr. Opin. Plant Biol.* **2006**, *9*, 460–469. [CrossRef] [PubMed]

59. Ghosh, J.S.; Chaudhuri, S.; Dey, N.; Pal, A. Functional characterization of a serine-threonine protein kinase from *Bambusa balcooa* that implicates in cellulose overproduction and superior quality fiber formation. *BMC Plant Biol.* **2013**, *13*, 128–145. [CrossRef] [PubMed]

60. Wang, G.; Zhang, G.; Wu, M. CLE peptide signaling and crosstalk with phytohormones and environmental stimuli. *Front. Plant Sci.* **2016**, *6*, 1211. [CrossRef] [PubMed]

61. Bita, C.; Gerats, T. Plant tolerance to high temperature in a changing environment: Scientific fundamentals and production of heat stress-tolerant crops. *Front Plant Sci.* **2013**, *4*, 273. [CrossRef] [PubMed]

62. Ciuzan, O.; Hancock, J.; Pamfil, D.; Wilson, I.; Ladomery, M. The evolutionarily conserved multifunctional glycine-rich RNA binding proteins play key roles in development and stress adaptation. *Physiol. Plant.* **2015**, *153*, 1–11. [CrossRef] [PubMed]

63. Lorkovic, Z.J. Role of plant RNA-binding proteins in development, stress response and genome organization. *Trends Plant Sci.* **2009**, *14*, 229–236. [CrossRef] [PubMed]

64. Kim, J.Y.; Kim, W.Y.; Kwak, K.J.; Oh, S.H.; Han, Y.S.; Kang, H. Glycine-rich RNA-binding proteins are functionally conserved in *Arabidopsis thaliana* and *Oryza sativa* during cold adaptation process. *J. Exp. Bot.* **2010**, *61*, 2317–2325. [CrossRef] [PubMed]

65. Teng, K.; Tan, P.H.; Xiao, G.Z.; Han, L.B.; Chang, Z.H.; Chao, Y.H. Heterologous expression of a novel *Zoysia japonica* salt induced glycine-rich RNA-binding protein gene, ZjGRP, caused salt sensitivity in Arabidopsis. *Plant Cell Rep.* **2017**, *36*, 179–191. [CrossRef] [PubMed]

66. Stone, S.L. The role of ubiquitin and the 26S proteasome in plant abiotic stress signaling. *Front. Plant Sci.* **2014**, *5*, 1–10. [CrossRef] [PubMed]

67. Zhou, G.A.; Chang, R.Z.; Qiu, L.J. Overexpression of soybean ubiquitin-conjugating enzyme gene GmUBC2 confers enhanced drought and salt tolerance through modulating abiotic stress-responsive gene expression in Arabidopsis. *Plant Mol. Biol.* **2010**, *72*, 357–367. [CrossRef] [PubMed]

68. Ahn, M.Y.; Oh, T.R.; Seo, D.H.; Kim, J.H.; Cho, N.H.; Kim, W.T. *Arabidopsis* group XIV ubiquitin-conjugating enzymes AtUBC32, AtUBC33, and AtUBC34 play negative roles in drought stress response. *J. Plant Physiol.* **2018**, *230*, 73–79. [CrossRef]

International Journal of
Molecular Sciences

Article

Metabolome and Transcriptome Association Analysis Reveals Dynamic Regulation of Purine Metabolism and Flavonoid Synthesis in Transdifferentiation during Somatic Embryogenesis in Cotton

Huihui Guo [1,2,†], Haixia Guo [2,†], Li Zhang [2], Zhengmin Tang [2], Xiaoman Yu [2], Jianfei Wu [2] and Fanchang Zeng [2,*]

[1] State Key Laboratory of Crop Biology, College of Horticulture Science and Engineering, Shandong Agricultural University, Tai'an 271018, China; hhguo@sdau.edu.cn
[2] State Key Laboratory of Crop Biology, College of Agronomy, Shandong Agricultural University, Tai'an 271018, China; diya_haixiaguo@163.com (H.G.); 15610418001@163.com (L.Z.); 17861500710@163.com (Z.T.); 15890636826@163.com (X.Y.); jfwu@sdau.edu.cn (J.W.)
* Correspondence: fczeng@sdau.edu.cn; Tel.: +86-538-824-1828
† These authors contributed equally to this paper.

Received: 13 March 2019; Accepted: 24 April 2019; Published: 26 April 2019

Abstract: Plant regeneration via somatic embryogenesis (SE) is a key step during genetic engineering. In the current study, integrated widely targeted metabolomics and RNA sequencing were performed to investigate the dynamic metabolic and transcriptional profiling of cotton SE. Our data revealed that a total of 581 metabolites were present in nonembryogenic staged calli (NEC), primary embryogenic calli (PEC), and initiation staged globular embryos (GE). Of the differentially accumulated metabolites (DAMs), nucleotides, and lipids were specifically accumulated during embryogenic differentiation, whereas flavones and hydroxycinnamoyl derivatives were accumulated during somatic embryo development. Additionally, metabolites related to purine metabolism were significantly enriched in PEC vs. NEC, whereas in GE vs. PEC, DAMs were remarkably associated with flavonoid biosynthesis. An association analysis of the metabolome and transcriptome data indicated that purine metabolism and flavonoid biosynthesis were co-mapped based on the Kyoto encyclopedia of genes and genomes (KEGG) database. Moreover, purine metabolism-related genes associated with signal recognition, transcription, stress, and lipid binding were significantly upregulated. Moreover, several classic somatic embryogenesis (SE) genes were highly correlated with their corresponding metabolites that were involved in purine metabolism and flavonoid biosynthesis. The current study identified a series of potential metabolites and corresponding genes responsible for SE transdifferentiation, which provides a valuable foundation for a deeper understanding of the regulatory mechanisms underlying cell totipotency at the molecular and biochemical levels.

Keywords: cotton; somatic embryogenesis; transdifferentiation; widely targeted metabolomics; purine metabolism; flavonoid biosynthesis; molecular and biochemical basis; transcript-metabolite network

---

1. Introduction

Somatic embryogenesis (SE) plays a crucial role in the genetic transformation of plants and in their in vitro rapid propagation. SE refers to the process of transformation from the somatic to the embryogenic state and is a unique phenomenon similar to what occurs during the development of zygotic embryos. The SE process can be artificially controlled in vitro conditions. It is also a classic example of cell totipotency. The underlying mechanisms of SE are significant for revealing important scientific theoretical problems that involve cell development, differentiation, and morphogenesis [1–4].

Nic-Can et al. [5] reported that SE is a complex process of molecular regulation in which somatic cells acquire totipotency to transform into embryonic cells. How does a somatic cell become a whole plant? This question has been reported to be one of the 25 questions most important within the scientific community [6]. However, the regulatory networks that are involved in the transition from the nonembryogenic staged callus (NEC) to the somatic embryo during SE remain poorly understood [7]. In addition to environmental factors, external stimuli and hormones, many specifically expressed genes also participate in the SE process and play a decisive role [8–11]. Therefore, it is particularly important to identify and isolate key genes that regulate SE. Candidate genes that may regulate SE were identified using bioinformatics tools [12]. The results showed that *auxin response factor* (*ARF*) [11,13,14], *leafy cotyledon* (*LEC*) [10,15], *Wuschel* (*WUS*) [16–18], *somatic embryogenesis receptor kinase* (*SERK*) [8], *shoot meristemless* (*STM*), and *baby boom* (*BBM*) [10,19,20] were involved in SE. In addition, the salicylic acid (SA) and jasmonic acid (JA) signaling pathways were also predicted to regulate SE [12,21–27].

During the process of cotton SE, Hu et al. [28] found that interference of the high-mobility group box 3 (*GhHmgb3*) gene enhanced the proliferation and differentiation of embryogenic callus. Poon et al. [29] found that embryogenic callus could secrete a specific AGP protein (GhPLA1), which was extracted and added to the medium to enhance SE. Min et al. [7] showed that the process of SE was negatively regulated by a casein kinase gene (*GhCKI*) via a complex regulatory network. Several differentially expressed small RNAs and their target genes were identified via small RNA sequencing and degradation sequencing, which indicated that the SE process in cotton was regulated by complex gene expression networks [30]. Genes related to stress, such as *SERK1, abscisic aldehyde synthesis enzyme 2* (*ABA2*), *abscisic acid insensitive 3* (*ABI3*), *jasmonate ZIM-domain 1* (*JAZ1*), *late embriogenesis abundant protein 1* (*LEA1*), and transcription factors (NACs, WRKYs, MYBs, ERFs, Zinc finger family proteins) were also involved in callus induction [31–37]. Transcriptome sequencing was carried out during callus dedifferentiation and redifferentiation. Auxin and stress response molecules were found to be involved in SE in cotton [38]. Xu et al. [39] also found that plant growth regulators played an important regulatory role in SE.

In recent years, metabolomics, which involves the collection of all low molecular weight metabolites that are present in a certain organism, cell, or tissue during a specific physiological period, has attracted a great deal of attention. It mainly involves the examination of endogenous small molecules with molecular weights less than 1000 Da. Plants can produce 200,000 to 1 million kinds of metabolites, which can be divided into primary metabolites and secondary metabolites [40].

Metabolites are the ultimate result of gene transcription and protein expression in organisms that are under the influence of internal and external factors. They form a bridge between genes and phenotypes and can directly reflect the physiological phenomena within plants. At the same time, metabolites can regulate gene transcription and protein expression. As the final products of cellular regulatory processes, metabolites represent the building blocks of macromolecules and are also an essential component of cellular energy pathways [41]. Metabolomic technologies enable the examination and identification of endogenous biochemical reaction products and thereby reveal information about the metabolic pathways and processes occurring within a living cell [42].

Metabolomics data can provide a wealth of information about the biochemical status of tissues, and the interpretation of such data offers an effective approach that can be used for the functional characterization of genes [43–45]. Transcriptomic and proteomic studies are only able to predict changes at the gene expression and protein levels, respectively, while metabolomics studies investigate the changes in functioning exhibited by these genes or proteins [46]. Compared with the genome, transcriptome, or proteome, the metabolome more accurately reflects the phenotype of the organism, and the minor changes in the genome and proteome can be reflected and amplified by the metabolome [47]. The applicability of metabolomics to the SE process has been demonstrated in many plant genera [48–52].

The development of 'omics' technology has allowed for comprehensive analysis of the SE process at the transcript, protein, and metabolite levels [50,52]. The combination of modern omics technologies such as transcriptomics and metabolomics provides a great opportunity to acquire a deeper understanding of the mechanisms of cell totipotency at the molecular and biochemical levels [52]. To date, there have been few studies that have used transcriptomics and metabolomics techniques to identify potential key factors involved in the SE process [48,49,51,52]. Additionally, the relationship between callus tissues and the media used to culture them is not well understood at the biochemical level in terms of nutrient uptake by callus from the medium or release of chemicals into the medium [52]. Integrated metabolomic and transcriptomic network analyses can elucidate the functioning of a series of secondary metabolites along with changes in content, as well as the corresponding differentially expressed genes, which can broaden the global view of SE regulation [52].

In this study, we comparatively investigated the dynamic metabolomic and transcriptomic profiles of two transdifferentiation processes, embryogenic differentiation and somatic embryo development, during the SE process in cotton. By interactively comparing metabolomic and transcriptomic data, we were able to identify potential metabolites and the corresponding differentially expressed genes at the molecular and biochemical levels, as well as draw a comprehensive picture of the events underpinning SE development to provide a valuable foundation for uncovering the regulatory mechanisms underlying plant cell totipotency.

2. Results

2.1. Transdifferentiation Staged Cultures Derived from Cotton Somatic Embryogenesis

We established an efficient cotton SE system and obtained cultures from different developmental stages, including nonembryogenic staged calli (NEC), primary embryogenic calli (PEC), and initiation staged embryos with globular-like enriched (GE) based on our previous approach as published recently [53] (Figure 1). These representative staged samples would be highly enriched and collected to establish the metabolic and transcriptional profiles.

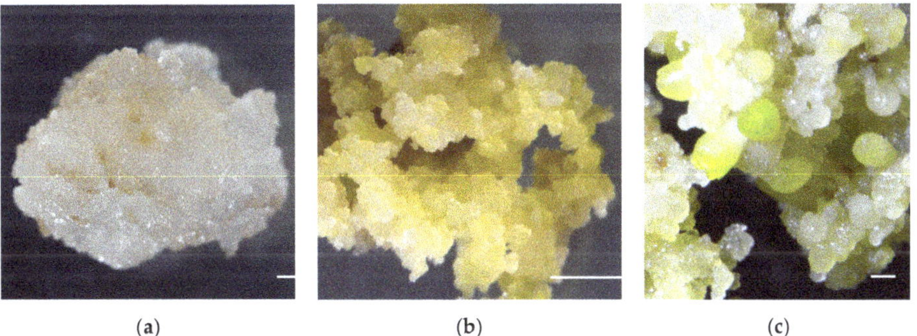

(a)   (b)   (c)

Figure 1. Cultures derived from different developmental stages of cotton somatic embryogenesis (SE) for the purposes of the metabolome and transcriptome assays. (**a**) Nonembryogenic staged calli (NEC); (**b**) primary embryogenic calli (PEC); (**c**) initiation staged embryos with globular-like enriched (GE). Bars = 1 mm.

2.2. UPLC-MS/MS-Based Quantitative Metabolomic Analysis and Overall Metabolite Identification

Ultra-performance liquid chromatography (UPLC) and tandem mass spectrometry (MS/MS) were conducted to assess the dynamic metabolite changes in NEC, PEC, and GE in cotton. During the process of instrumental analysis, one quality control (QC) sample was injected after every ten samples to monitor the repeatability of the analysis process. The repeatability of metabolite extraction and detection can be judged using overlapping analysis of the total ion current (TIC) in the different QC samples (Figure 2). The results showed the overlap in the TIC curves during metabolite detection. The retention times and peak intensities were consistent, which indicated that the signal was stable when an identical sample was detected at a different time. The instrumental stability provided an important guarantee of the repeatability and reliability of our metabonomic data.

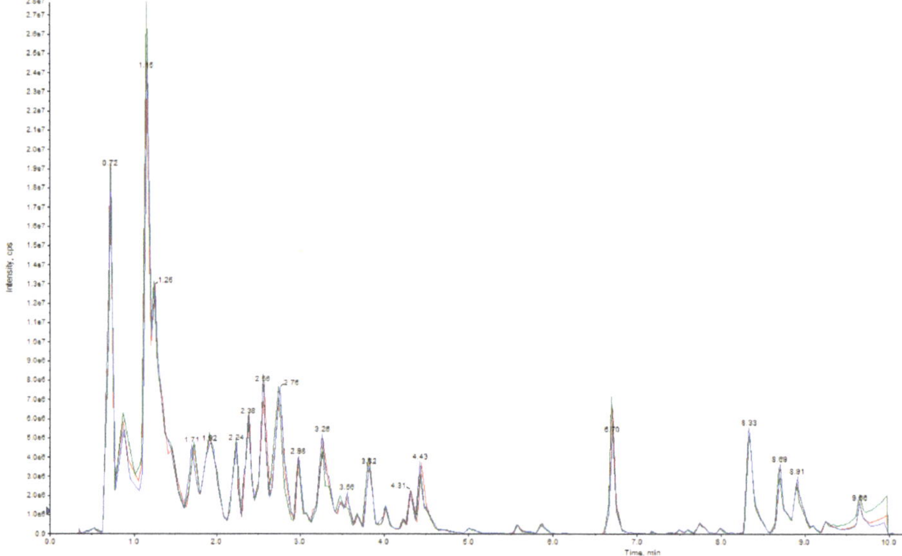

Figure 2. Overlapping analysis of the total ion current (TIC) in different quality control (QC) samples. The abscissa represents the retention time (min) of metabolite detection, and the ordinate represents the intensity of the ion current (cps: count per second).

The determination of Pearson's correlation coefficient reflected the repeatability among the intragroup samples (Figure 3a). Principal component analysis (PCA) was used to determine the rate of contribution of the first two primary components (67.49%). The three period materials were obviously separated, and each formed a cluster (Figure 3b). These results suggested that there was sufficient reproducibility of the materials, which made them suitable for the following qualitative and quantitative analyses.

After quality validation, a total of 581 metabolites with known structures were identified in NEC, PEC and GE, each of which was analyzed using three biological replicates (Table S1). Of the 581 metabolites, amino acids (15%), flavones (15%), organic acids (12%), lipids (11%), and nucleotides (10%) accounted for a large proportion (Figure 3c). Detailed information about the identified metabolites, including the compounds, classes, molecular weights (Da), ionization models, Kyoto Encyclopedia of Genes and Genomes (KEGG) pathways, and quantities for each of the three periods is shown in Table S1.

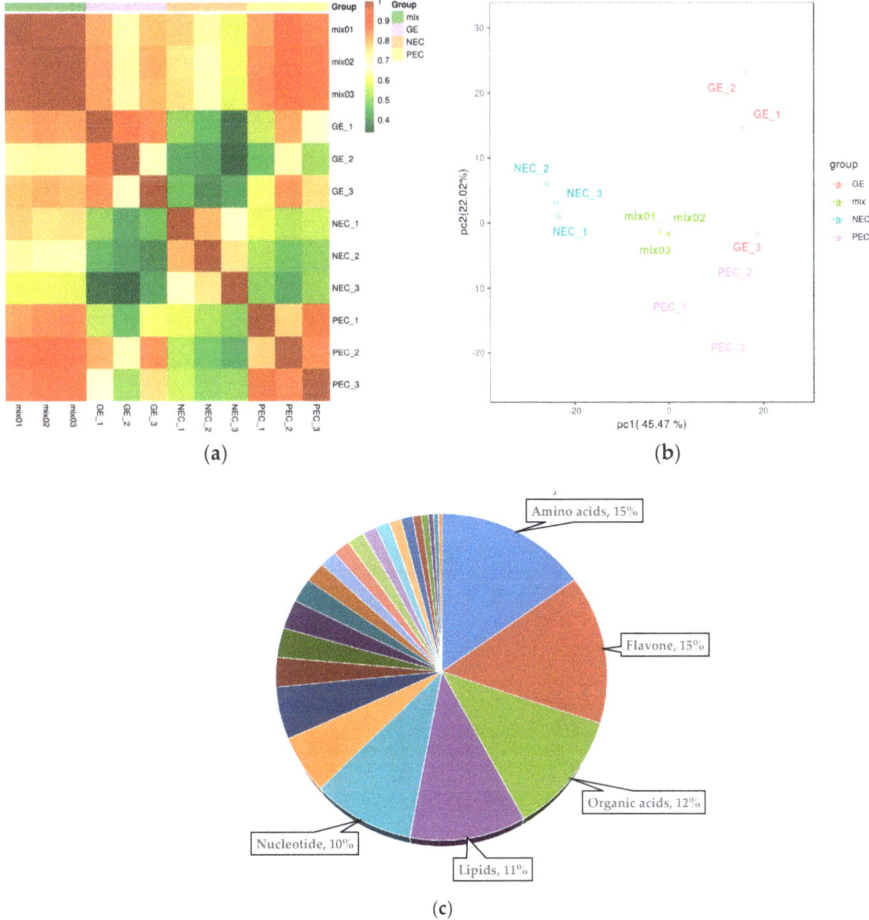

Figure 3. Overall qualitative and quantitative analysis of the metabolomics data. (**a**) Pearson's correlation coefficients among the three samples (NEC, PEC, and GE) and quality control samples (mix); (**b**) PCA analysis of the three samples (NEC, blue; PEC, purple; GE, red) and quality control samples (mix, green); the x-axis represents the first principal component and the y-axis represents the second principal component. (**c**) Component analysis of the identified metabolites. The top five metabolites are shown beside the graph. NEC, nonembryogenic staged calli; PEC, primary embryogenic calli; GE, initiation staged embryos with globular-like enriched.

2.3. Identification of Differentially Accumulated Metabolites (DAMs)

Differentially accumulated metabolites (DAMs) were defined as those exhibiting a fold change ≥2 or a fold change ≤0.5 and a variable importance in project (VIP) ≥1 between PEC versus NEC, GE versus PEC, or GE versus NEC ($p < 0.05$). In total, 156, 139, and 159 DAMs were identified, respectively (Table 1 and Table S2). For PEC vs. NEC, 124 of 156 (79.5%) were upregulated and 32 of 156 (20.5%) were downregulated. Of the 139 DAMs identified between GE and PEC, 86 (61.9%) and 53 (38.1%) metabolites were upregulated and downregulated, respectively. Of the 159 metabolites differentially accumulated in GE compared to NEC, 128 (80.5%) and 31 (19.5%) metabolites were upregulated and downregulated, respectively. Volcano plots were generated to represent the significant differences between PEC vs. NEC, GE vs. PEC, and GE vs NEC (Figure 4a–c). A hierarchical cluster analysis was also performed to assess the DAM accumulation patterns (Figure 4d–f).

Of these DAMs, amino acids (17%), organic acids (13%), flavones (12%), nucleotides (12%), and hydroxycinnamoyl derivatives (6%) accounted for a large proportion in PEC compared to NEC (Figure 4g). Flavones (23%), amino acids (12%), hydroxycinnamoyl derivatives (9%), organic acids (9%), and other compounds (9%) were shown to be differentially accumulated in GE and PEC (Figure 4h). Of the differentially accumulated metabolites in GE vs. NEC, flavones (16%), organic acids (15%), amino acids (14%), nucleotides (11%), and other metabolites (8%) had the highest representation (Figure 4i). Differential metabolites commonly and specifically accumulated in PEC vs. NEC, GE vs. PEC, and GE vs. NEC were also investigated (Figure 5).

Table 1. Summary of differentially accumulated metabolites (DAMs) among NEC, PEC, and GE.

Group Name	Number of Differential Metabolites	Number Upregulated	Number Downregulated
PEC vs. NEC	156	124	32
GE vs. PEC	139	86	53
GE vs. NEC	159	128	31

NEC, nonembryogenic staged calli; PEC, primary embryogenic calli; GE, initiation staged embryos with globular-like enriched.

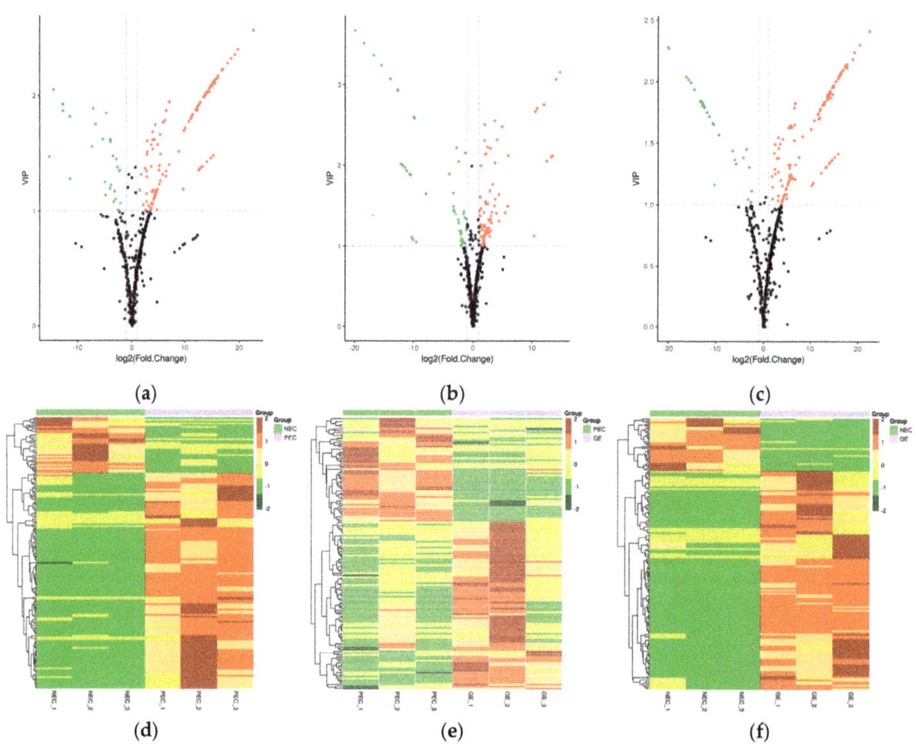

Figure 4. *Cont.*

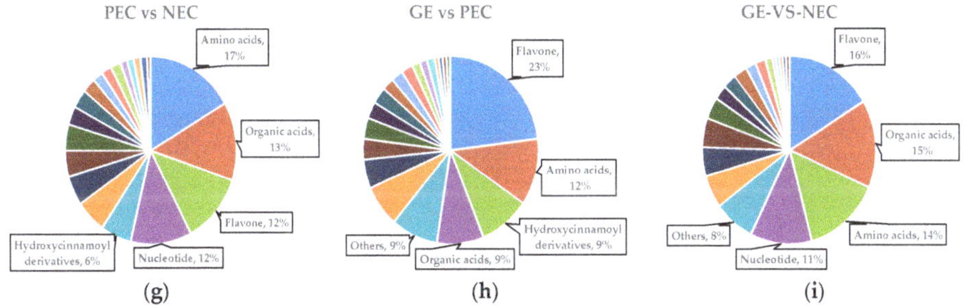

Figure 4. Differentially accumulated metabolites (DAMs) among NEC, PEC, and GE. (**a**) Volcano plot showing the differential metabolites in PEC vs. NEC; (**b**) volcano plot showing the differential metabolites in GE vs. PEC; (**c**) volcano plot showing the differential metabolites in GE vs. NEC. The red spots represent upregulated DAMs, the green dots represent downregulated DAMs, and the black dots represent nondifferentially accumulated metabolites; (**d**) Heat map representing the hierarchical cluster analysis in PEC vs. NEC; (**e**) heat map representing the hierarchical cluster analysis in GE vs. PEC; (**f**) heat map representing the hierarchical cluster analysis in GE vs. NEC. Red indicates high abundance, whereas the relatively low-abundance metabolites are shown in green; (**g**) component analysis of DAMs in PEC vs. NEC; (**h**) component analysis of DAMs in GE vs. PEC; (**i**) component analysis of DAMs in GE vs. NEC. The top five DAMs are noted beside each graph. NEC, nonembryogenic staged calli; PEC, primary embryogenic calli; GE, initiation staged embryos with globular-like enriched.

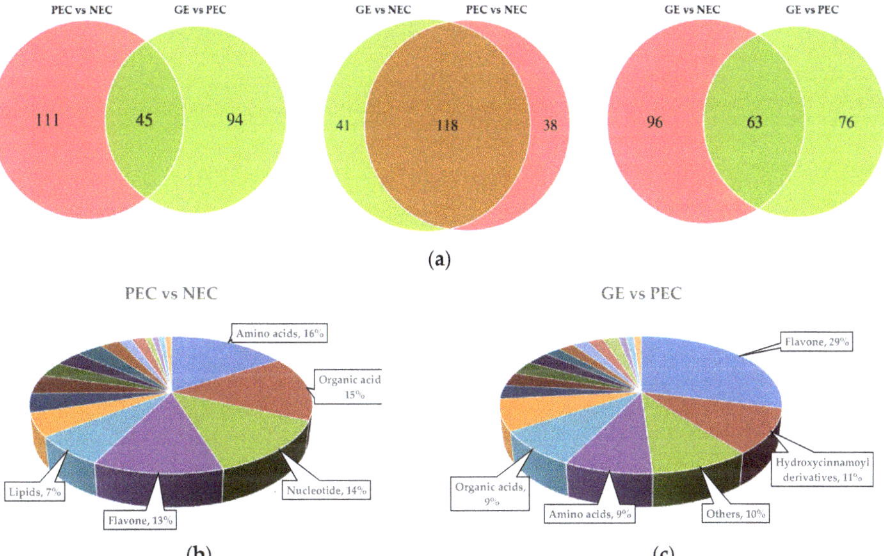

Figure 5. Differential metabolites commonly and specifically accumulated in PEC vs. NEC, GE vs. PEC, and GE vs. NEC. (**a**) Venn diagrams showing the distribution of DAMs in PEC vs. NEC, GE vs. PEC, and GE vs. NEC; (**b**) component analysis of DAMs accumulated specifically in PEC vs. NEC; (**c**) component analysis of DAMs accumulated specifically in GE vs. PEC. The top five specific DAMs are noted beside each graph. DAMs, differentially accumulated metabolites; NEC, nonembryogenic staged calli; PEC, primary embryogenic calli; GE, initiation staged embryos with globular-like enriched.

2.4. Functional Annotation and KEGG Enrichment Analysis of DAMs

Differentially accumulated metabolites in PEC vs. NEC, GE vs. PEC, and GE vs. NEC were functionally annotated using the Kyoto Encyclopedia of Genes and Genomes (KEGG) database, which was summarized in Table S4. To further investigate the functioning of SE-related metabolites, we analyzed the differences and dynamic changes of biological processes among the three groups. For PEC vs. NEC (embryogenic differentiation), the KEGG enrichment analysis showed that the terms 'purine metabolism', 'cGMP-PKG signaling pathway', and 'olfactory transduction' were significantly enriched (Figure 6a). However, the terms 'phenylpropanoid biosynthesis', 'flavone and flavonol biosynthesis', 'flavonoid biosynthesis', and 'biosynthesis of phenylpropanoids' were enriched in GE vs. PEC (somatic embryo development) (Figure 6b). Meanwhile, for GE vs. NEC, the DAMs were most strongly associated with the terms 'purine metabolism', 'microbial metabolism in diverse environments', 'biosynthesis of plant secondary metabolites', and 'cGMP-PKG signaling pathway' (Figure 6c).

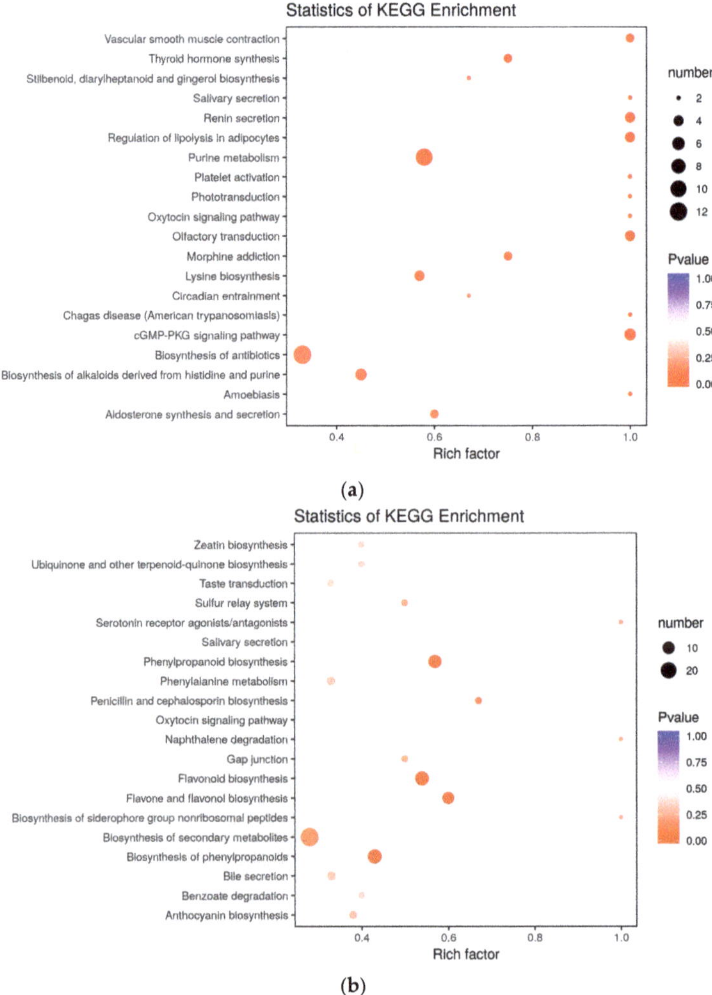

Figure 6. *Cont.*

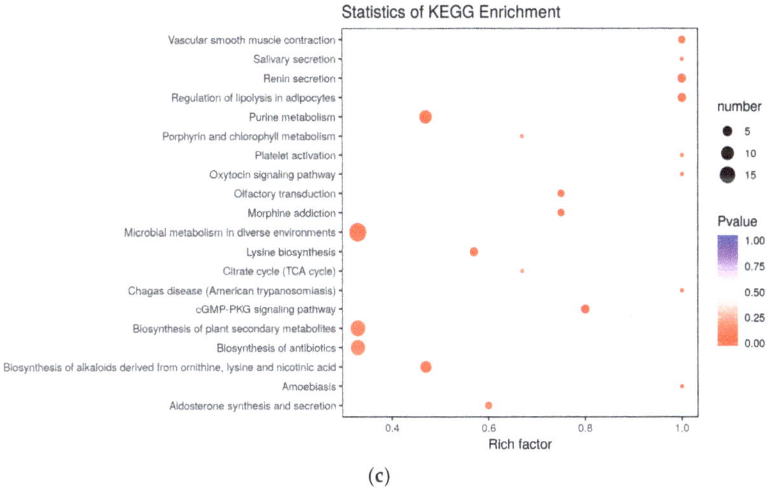

(**c**)

Figure 6. KEGG pathway enrichment analysis of DAMs. (**a**) Pathway enrichment in PEC vs. NEC; (**b**) pathway enrichment in GE vs. PEC; (**c**) pathway enrichment in GE vs. NEC. The x-axis represents the enrichment factor, while the y-axis represents the enrichment pathway. The dot sizes represent the number of differentially enriched metabolites. The statistical analysis of the pathway enrichment was performed using Fisher's exact test. DAMs, differentially accumulated metabolites; NEC, nonembryogenic staged calli; PEC, primary embryogenic calli; GE, initiation staged embryos with globular-like enriched.

2.5. Association Analysis of the Metabolome and Transcriptome Data Sets

To provide an overview of the system-wide changes that occur during SE in cotton, transcriptomic profiling based on RNA-seq was performed. Significant differentially expressed genes (DEGs) were detected among NEC, PEC, and GE (unpublished data). By interactively comparing the metabolomic and transcriptomic data, we were able to identify potential metabolites and the corresponding differentially expressed genes at the molecular and biochemical levels.

2.5.1. KEGG Enrichment Analysis of DAMs and DEGs During Cotton SE

To investigate the relationships between genes and metabolites involved in the two transdifferentiation processes during SE in cotton, the DAMs and DEGs in the two compared groups (PEC vs. NEC and GE vs. PEC) were mapped according to the KEGG database. There were 96 and 55 co-mapped pathways in PEC vs. NEC and GE vs. PEC, respectively (Table S5).

Interestingly, of these co-mapped pathways, 'purine metabolism' (ko00230) was significantly enriched in PEC vs. NEC. Meanwhile, 'phenylpropanoid biosynthesis' (ko00940), 'flavonoid biosynthesis' (ko00941), and 'flavone and flavonol biosynthesis' (ko00944) showed significant accumulation in GE vs. PEC. DAMs involved in purine metabolism during embryogenic differentiation and in flavonoid biosynthesis during somatic embryo development are listed in Tables 2 and 3, respectively.

Table 2. DAMs involved in purine metabolism during embryogenic differentiation (PEC vs. NEC).

Index	Compound	Class	Fold Change	Log$_2$ (FC)	Type
pmb0981	Adenosine 5′-monophosphate	Nucleotide and its derivates	3.92×10^4	15.3	up
pmb2684	Cyclic AMP	Nucleotide and its derivates	4.79×10^4	15.5	up
pmb2948	Adenosine 3′-monophosphate	Nucleotide and its derivates	2.54×10^3	11.3	up
pmb4344	Guanosine 5′-monophosphate	Nucleotide and its derivates	2.19×10^4	14.4	up
pmc0066	2′-Deoxyinosine-5′-monophosphate	Nucleotide and its derivates	1.29×10^4	13.7	up
pmd0023	Adenosine	Nucleotide and its derivates	1.36×10^2	7.08	up
pme1175	Guanosine	Nucleotide and its derivates	93.1	6.54	up
pme1181	Deoxyguanosine	Nucleotide and its derivates	17.1	4.10	up
pme1296	Xanthosine	Nucleotide and its derivates	1.37×10^2	7.10	up
pme3835	Guanosine 3′,5′-cyclic monophosphate	Nucleotide and its derivates	1.64×10^4	14.0	up
pme3960	Deoxyadenosine	Nucleotide and its derivates	7.32×10^4	16.2	up

DAMs, differentially accumulated metabolites; NEC, nonembryogenic staged calli; PEC, primary embryogenic calli.

Table 3. DAMs involved in flavonoid biosynthesis during somatic embryo development (GE vs. PEC).

Index	Compound	Class	Fold Change	Log$_2$ (FC)	Type
pme1580	Eriodictyol	Flavanone	4.17	2.06	up
pme2319	Hesperetin	Flavanone	2.43	1.28	up
pmb0605	Apigenin 7-O-glucoside (Cosmosiin)	Flavone	0.145	−2.79	down
pme0088	Luteolin	Flavone	3.56	1.83	up
pme2459	Luteolin 7-O-glucoside (Cynaroside)	Flavone	3.30	1.72	up
pme3300	Tricetin	Flavone	2.60	1.38	up
pme0199	Quercetin	Flavonol	56.9	5.83	up
pme0200	Kaempferol	Flavonol	4.09×10^3	12.0	up
pme0202	Quercetin 3-O-rutinoside (Rutin)	Flavonol	15.9	3.99	up
pme1478	Myricetin	Flavonol	44.2	5.46	up
pme1521	Dihydroquercetin (Taxifolin)	Flavonol	4.60	2.20	up
pme2898	Dihydromyricetin	Flavonol	2.92×10^{-6}	−18.4	down
pme3212	Quercetin 3-O-glucoside (Isotrifoliin)	Flavonol	6.63	2.73	up
pme3268	Kaempferol 3-O-galactoside (Trifolin)	Flavonol	6.26	2.65	up

DAMs, differentially accumulated metabolites; PEC, primary embryogenic calli; GE, initiation staged embryos with globular-like enriched.

2.5.2. Correlation Analysis of DAMs and DEGs during SE in Cotton

Log$_2$ conversion data for the differentially accumulated metabolites and differentially expressed genes were selected that had a Pearson's correlation coefficient (PCC) > 0.8. To obtain a systematic view of the variations in metabolites and their corresponding genes with PCC > 0.8, nine quadrant diagrams were generated for the embryogenic differentiation and somatic embryo development processes (Figure 7).

The black dotted lines divide each graph into 9 quadrants. Upregulated metabolites and downregulated genes are displayed in quadrant 1, upregulated metabolites and unchanged genes are displayed in quadrant 2, upregulated metabolites and upregulated genes are displayed in quadrant 3, unchanged metabolites and downregulated genes are displayed in quadrant 4, unchanged metabolites and unchanged genes are displayed in quadrant 5, unchanged metabolites and upregulated genes are displayed in quadrant 6, downregulated metabolites and downregulated genes are displayed in quadrant 7, downregulated metabolites and unchanged genes are displayed in quadrant 8, and downregulated metabolites and upregulated genes are displayed in quadrant 9. Moreover, the DAMs and DEGs shown in quadrant 3 and quadrant 7 are positively correlated and have similar consistent patterns, while the DAMs and DEGs shown in quadrant 1 and quadrant 9 are negatively correlated and have opposite patterns.

Notably, in quadrant 3, several representative factors that are related to receptor-like protein kinases, signal recognition, transcription, stress regulation, lipid binding, hormone responses, and histone modification were significantly activated during embryogenic differentiation process (Table 4).

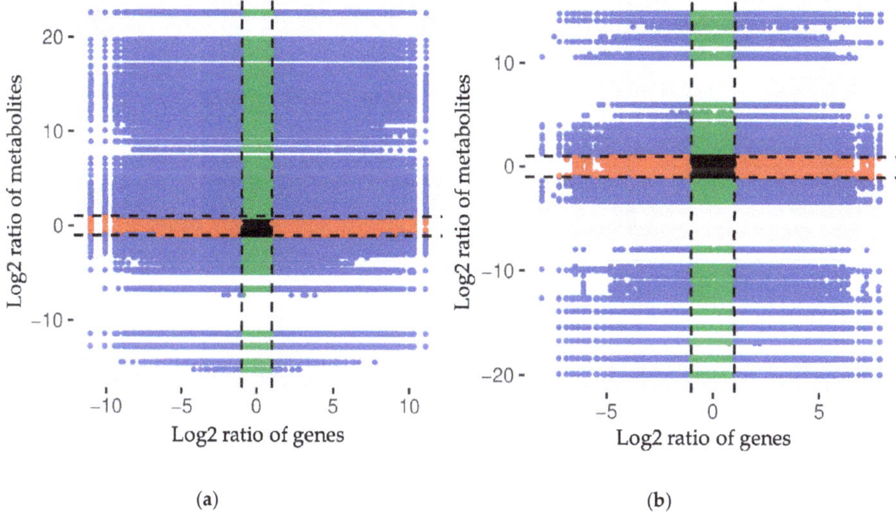

(a)   (b)

Figure 7. Quadrant diagrams representing association of metabolomic and transcriptomic variation during SE in cotton. (**a**) Overview of metabolomic and transcriptomic variation in PEC vs. NEC; (**b**) overview of metabolomic and transcriptomic variation in GE vs. PEC. The x-axis represents the fold changes in the transcriptome data; the y-axis represents the fold changes in the metabolome data. The black dotted lines represent the differential thresholds. Outside of the threshold lines, there were significant differences in the genes/metabolites, and within the threshold lines are shown the unchanged genes/metabolites. Each point represents a gene/metabolite. Black dots represent the unchanged genes/metabolites; green dots represent differentially accumulated metabolites with unchanged genes; red dots represent differentially expressed genes with unchanged metabolites; blue dots represent both differentially expressed genes and differentially accumulated metabolites. NEC, nonembryogenic staged calli; PEC, primary embryogenic calli; GE, initiation staged embryos with globular-like enriched.

Table 4. Significant representative differentially expressed genes (DEGs) involved in purine metabolism during embryogenic differentiation (PEC vs. NEC).

Gene ID	Gene Name	Description	Log$_2$ (FC)	Meta ID	Log$_2$ (FC)	PCC
Gh_D05G1280	*PERK13*	Proline-rich receptor-like protein kinase	3.61	pme1175	6.54	1
				pme1296	7.10	1
Gh_D10G1867	*IRK*	LRR receptor-like protein kinase	3.52	pmb0981	15.26	1
				pmb2948	11.31	1
				pmc0066	13.65	1
Gh_A03G1831	*CAO*	Signal recognition protein	1.99	pme1175	6.54	1
				pme1296	7.10	1
Gh_A07G2239	*ABI3*	B3 transcription factor	5.70	pmb0981	15.26	1
				pmb2948	11.31	1
				pmc0066	13.65	1
Gh_A08G2488	*WRKY2*	WRKY transcription factor 2	1.77	pmb0981	15.26	1
				pmb2948	11.31	1
				pmc0066	13.65	1
Gh_A11G2618	*NFYC2*	Nuclear transcription factor	3.07	pmb0981	15.26	1
				pmb2948	11.31	1
				pmc0066	13.65	1
Gh_D04G0086	*NFYA3*	Nuclear transcription factor	2.29	pmb0981	15.26	1
				pmb2948	11.31	1
				pmc0066	13.65	1
Gh_A06G1368	*SERP2*	Stress-associated protein	4.36	pme1175	6.54	1
			4.36	pme1296	7.10	1
Gh_D04G1347	*ARR4*	Two-component response regulator	2.55	pme1175	6.54	1
				pme1296	7.10	1

Table 4. *Cont.*

Gene ID	Gene Name	Description	Log$_2$ (FC)	Meta ID	Log$_2$ (FC)	PCC
Gh_D12G2180	*AIR1*	Putative lipid-binding protein	3.55	pme1175	6.54	1
			3.55	pme1296	7.10	1
Gh_D12G1336	*HMGB7*	High mobility group B protein	3.75	pmb0981	15.26	1
			3.75	pmb2948	11.31	1
			3.75	pmc0066	13.65	1
Gh_D11G0232	*HMGB13*	High mobility group B protein	5.55	pme1175	6.54	1
			5.55	pme1296	7.10	1
Gh_A06G0239	*IAA9*	Auxin-responsive protein	3.87	pmb2948	11.31	1
Gh_D09G0145	*GASA4*	Gibberellin-regulated protein	6.53	pmb0981	15.26	1
				pmb2948	11.31	1
				pmc0066	13.65	1
Gh_A10G0472	*B34*	Histone H3.2	3.80	pmb0981	15.26	1
			3.80	pmb2948	11.31	1
Gh_A13G1866	*B34*	Histone H3.2	3.46	pmb2948	11.31	1
Gh_D08G1979	*HIS2A*	Histone H2AX	4.15	pmb0981	15.26	1
Gh_D08G0034	*HIS2A*	Histone H2A	3.15	pmb2948	11.31	1

NEC, nonembryogenic staged calli; PEC, primary embryogenic calli.

2.5.3. Transcript-Metabolite Correlation Network Representing DAMs and DEGs during SE in Cotton

To model the synthetic and regulatory characteristics of DAMs and DEGs, subnetworks were constructed to determine transcript-metabolite correlations. Only the correlation pairs with a correlation coefficient > 0.8 were included in the analysis.

In PEC vs. NEC, the differential metabolites involved in purine metabolism are listed in Table 2. Several classic key genes that have been shown to be specifically activated during embryogenic differentiation were included in the correlation test, including *BBM* [10,19,20], *SERK* [8], *LEC* [10,15], *ARF* [11,13,14], *AGL15* [54], *AGP* [29], *GLP* [55], *AGO1* [56], *LTP*, and *AMY* [57]. The visualized network revealed that a total of 22 nodes were connected by 81 edges. Meanwhile, 53 pairs showed a positive correlation, and 28 pairs were negatively correlated (Figure 8a). Detailed information about the gene-metabolite pairs that are involved in purine metabolism during embryogenic differentiation is listed in Table 5.

The differential metabolites involved in flavonoid biosynthesis in GE vs. PEC are listed in Table 3. Genes that control somatic embryo development, including *wuschel* (*WUS*) [16–18], *clavata1* (*CLV1*) [16–18], *cup-shaped cotyledon 2* (*CUC2*) [58], *short-root* (*SHR*), and *scarecrow* (*SCW*) [59], were subjected to correlation tests. A transcript-metabolite correlation network was built that consisted of 12 nodes and 13 edges. Four pairs showed a positive correlation, and nine pairs were negatively correlated (Figure 8b). Detailed information about the gene-metabolite pairs involved in flavonoid biosynthesis during somatic embryo development is listed in Table 6.

These results indicated that several classic SE-related genes were highly correlated with their corresponding metabolites involved in purine metabolism and flavonoid biosynthesis, which reconfirmed the specific importance of purine metabolism during embryogenic differentiation, as well as the importance of flavonoid biosynthesis during somatic embryo development. The transcriptome data validated the authenticity and accuracy of the metabolic analysis.

Table 5. Classic gene-metabolite pairs involved in purine metabolism during embryogenic differentiation (PEC vs. NEC).

Gene ID	Gene Name	Description	Log$_2$ (FC)	Meta ID	Log$_2$ (FC)	PCC
				pmc0066	13.65	−1
				pmd0023	7.08	−1
				pmb2948	11.31	−1
				pmb0981	15.26	−1
Gh_A08G2008	*BBM*	Baby boom	6.54	pme3960	16.20	−1
				pme1181	4.10	−1
				pme1175	6.54	−1
				pme3835	14.00	−1
				pme1296	7.10	−1

Table 5. *Cont.*

Gene ID	Gene Name	Description	Log$_2$ (FC)	Meta ID	Log$_2$ (FC)	PCC
Gh_D06G1184	SERK1	Somatic embryogenesis receptor kinase	1.25	pmc0066	13.65	1
				pmd0023	7.08	1
				pmb2948	11.31	1
				pmb0981	15.26	1
				pmb2684	15.50	1
				pme3960	16.20	1
				pme1181	4.10	1
				pme1175	6.54	1
				pme3835	14.00	1
				pme1296	7.10	1
Gh_D13G1387	LEC1	Leafy cotyledon	6.44	pmc0066	13.65	1
				pmd0023	7.08	1
				pmb2948	11.31	1
				pmb0981	15.26	1
				pme3960	16.20	1
				pme3835	14.00	1
Gh_D07G0476	ARF2	Auxin response factor 2	1.89	pmc0066	13.65	1
				pmd0023	7.08	1
				pmb2948	11.31	1
				pmb0981	15.26	1
				pme3960	16.20	1
				pme1181	4.10	1
				pme3835	14.00	1
Gh_A12G0910	AGL15	Agamous-like 15	6.47	pmc0066	13.65	−1
				pmd0023	7.08	−1
				pmb2948	11.31	−1
				pmb0981	15.26	−1
				pme3960	16.20	−1
				pme1181	4.10	−1
				pme1175	6.54	−1
				pme3835	14.00	−1
				pme1296	7.10	−1
Gh_D07G0323	AGP1	Arabinogalactan protein1	5.93	pmc0066	13.65	1
				pmd0023	7.08	1
				pmb2948	11.31	1
				pmb0981	15.26	1
				pmb2684	15.50	1
				pme3960	16.20	1
				pme1181	4.10	1
				pme3835	14.00	1
Gh_A12G1076	GLP2	Germin-like protein 2	−1.13	pmb2684	15.50	−1
				pme1181	4.10	−1
Gh_A09G1557	AGO1	Argonaute 1	1.19	pmc0066	13.65	1
				pmd0023	7.08	1
				pmb2948	11.31	1
				pmb0981	15.26	1
				pmb2684	15.50	1
				pme3960	16.20	1
				pme1181	4.10	1
				pme3835	14.00	1
Gh_D05G1937	AMY1	Alpha-amylase 1	−4.34	pmc0066	13.65	−1
				pmd0023	7.08	−1
				pmb2948	11.31	−1
				pmb0981	15.26	−1
				pmb2684	15.50	−1
				pme3960	16.20	−1
				pme1181	4.10	−1
				pme3835	14.00	−1

Table 5. *Cont.*

Gene ID	Gene Name	Description	Log$_2$ (FC)	Meta ID	Log$_2$ (FC)	PCC
Gh_A12G0504	*LTP2*	Lipid transfer protein 2	3.58	pmc0066	13.65	1
				pmd0023	7.08	1
				pmb2948	11.31	1
				pmb0981	15.26	1
				pmb2684	15.50	1
				pme3960	16.20	1
				pme1181	4.10	1
				pme3835	14.00	1
				pmb4344	14.40	1
Gh_D12G0517	*LTP2-like*	Lipid transfer protein 2-like	3.03	pmc0066	13.65	1
				pmb0981	15.26	1
				pme3960	16.20	1
				pme1181	4.10	1
				pmb4344	14.40	1

NEC, nonembryogenic staged calli; PEC, primary embryogenic calli.

Table 6. Classic gene-metabolite pairs involved in flavonoid biosynthesis during somatic embryo development (GE vs. PEC).

Gene ID	Gene Name	Description	Log$_2$ (FC)	Meta ID	Log$_2$ (FC)	PCC
Gh_A10G0884	*WUS*	Wusche 1	2.12	pme3268	2.65	−1
				pme2898	−18.40	1
Gh_A02G0853	*CLV1*	Clavata 1	3.00	pme0200	12.00	−1
				pme3268	2.65	−1
				pme2459	1.72	−1
Gh_D01G0448	*CUC2*	Cup-shaped cotyledon 2	2.77	pme0202	3.99	−1
				pme2459	1.72	−1
				pme3212	2.73	−1
Gh_D02G0017	*SCW*	Scarecrow	2.55	pme2898	−18.40	−1
				pme0200	12.00	1
				pme3268	2.65	1
Gh_A12G0710	*SHR*	Short-root	1.90	pme2459	1.72	1
				pmb0605	−2.79	−1

PEC, primary embryogenic calli; GE, initiation staged embryos with globular-like enriched.

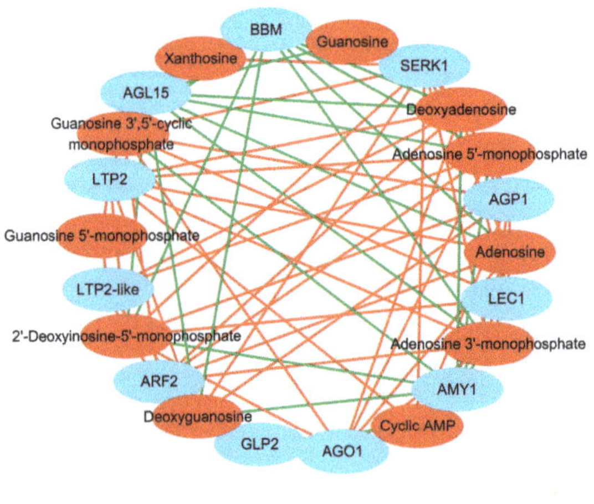

(**a**)

Figure 8. *Cont.*

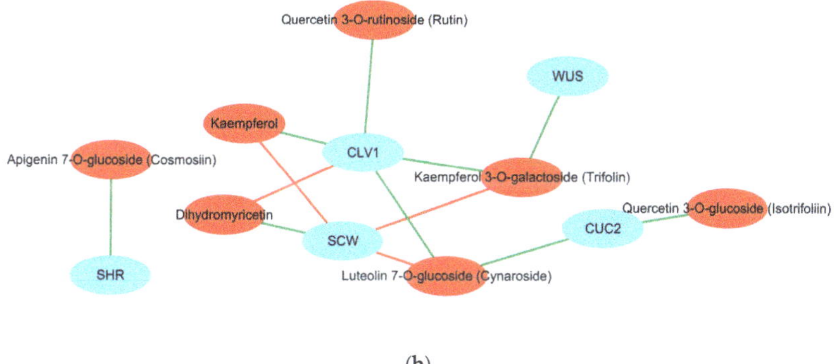

(**b**)

Figure 8. Transcript-metabolite correlation network representing DAMs and DEGs involved in PEC vs. NEC and GE vs. PEC. (**a**) Pearson correlation network representing factors associated with purine metabolism in PEC vs. NEC; (**b**) Pearson correlation network representing factors associated with flavonoid biosynthesis in GE vs. PEC. The gene-metabolite pairs were connected within the network by edges. Blue nodes represent genes and red nodes represent metabolites. Red edges represent positive correlations, and green edges represent negative correlations. DAMs, differentially accumulated metabolites; DEGs, differentially expressed genes; NEC, nonembryogenic staged calli; PEC, primary embryogenic calli; GE, initiation staged embryos with globular-like enriched.

3. Discussion

Metabolomics is an emerging omics technology that, like genomics and proteomics, can be used to qualify and quantify all metabolites of small molecular weight within the cells of an organism. Plant metabolomics has been widely applied to the investigation of patterns of metabolite accumulation and their underlying genetic basis via the identification of genes involved in metabolism, which is currently a topic of interest in modern plant biology. As the final products of genome expression, metabolites directly define the biochemical characteristics of a cell or tissue, which allows metabolomic data to be used to explain the biochemical mechanisms underlying SE. Meanwhile, integrated transcriptomic and metabolomic analysis allows for the more precise representation of gene-to-metabolite networks and thus will be an effective method that can be used to decipher the mechanisms involved in SE regulation in cotton. In the current study, we combined transcriptome and metabolome analyses to generate dynamic profiles of two transdifferentiation processes, embryogenic differentiation, and somatic embryo development, during SE in cotton, with the aim to provide a better understanding of the processes of SE transdifferentiation that resulted in cell totipotency at the molecular and biochemical levels.

3.1. Accumulated DAMs Specifically Involved in Two SE Transdifferentiation Processes in Cotton

In this study, a hierarchical cluster analysis (HCA) was performed to assess the patterns of accumulation of the metabolites among the different samples. Results showed that PEC and GE can be clustered together while NEC forms a separate cluster, which suggests that there is significant differential accumulation of metabolites between the two transdifferentiation processes, embryogenic differentiation (PEC-NEC) and somatic embryo developmental initiation (GE-PEC).

To investigate the differential metabolites specifically accumulated in (PEC-NEC) vs. (GE-PEC) vs. (GE-NEC), a Venn diagram was generated (Figure 5a; Table S3). The results showed that 26 metabolites were accumulated in common among (PEC-NEC) vs. (GE-PEC) vs. (GE-NEC). Meanwhile, 45 DAMs were accumulated in both (PEC-NEC) vs. (GE-PEC), 118 DAMs in both (PEC-NEC) vs. (GE-NEC), and 63 DAMs in both (GE-PEC) vs. (GE-NEC).

We examined the metabolites that were specifically accumulated between the two transdifferentiation processes, embryogenic differentiation (PEC-NEC), and somatic embryo development (GE-PEC). The results showed that there were 111 DAMs that were specifically accumulated in PEC vs. NEC, of which amino acids, organic acids, nucleotides, flavones, and lipids were the most represented (Figure 5b). Meanwhile, 94 DAMs were accumulated specifically in GE vs. PEC, of which flavones hydroxycinnamoyl derivatives, other metabolites, amino acids and organic acids accounted for a large proportion (Figure 5c). Specifically, nucleotides and lipids were more greatly accumulated during embryogenic differentiation, whereas greater amounts of flavones and hydroxycinnamoyl derivatives were accumulated during the somatic embryo development process. These data indicated that nucleotides and lipids may play important and special roles during embryogenic differentiation, whereas flavonoids are more important during the embryo maturation process.

3.2. Enrichment of Purine Metabolism in Embryogenic Differentiation

Purine metabolism refers to the metabolic pathways that synthesize and break down the purines that are present in many organisms. In our study, to investigate the functioning of SE-related metabolites, we analyzed the differences and dynamic metabolite changes among the three groups, PEC vs. NEC, GE vs. PEC, and GE vs. NEC. KEGG enrichment analysis showed that the 'purine metabolism' pathway was significantly enriched during embryogenic differentiation (Figure 6b). All of the differentially accumulated metabolites related to purine metabolism were upregulated. Meanwhile, enrichment analysis of DAMs and DEGs showed that 'purine metabolism' was co-mapped based on results from the KEGG database (Table S5).

In our study, several major DEGs of regulatory factors were identified and significantly associated with purine metabolism in embryogenic differentiation, including receptor-like protein kinases, signal recognition, transcription, stress regulation, lipid binding, hormone responses, and histone modification were significantly upregulated during the embryogenic differentiation process (Table 4).

The role of purine metabolism in the SE process has been demonstrated in many plants. A comparative omic analysis of tree fern *Cyathea delgadii* explants that were undergoing direct SE was performed. The results revealed that the differentially regulated proteins adenine phosphoribosyl transferase 3 and adenosine kinase 2 were assigned to the purine metabolism category and were associated with direct SE in *C. delgadii* [60]. To understand the molecular mechanisms that regulate early SE in *Eleutherococcus senticosus* Maxim, a high-throughput RNA-seq technology was used to investigate its transcriptome. The initiation of SE affected gene expression in many KEGG pathways but predominantly affected expression in metabolic pathways and those related to the biosynthesis of secondary metabolites. Other unigenes were classified into the purine metabolism pathway [61]. Similar results were obtained in cotton (*Gossypium hirsutum* L.). RNA-seq was performed to analyze the genes expressed during SE and their expression dynamics using RNAs isolated from nonembryogenic callus (NEC), embryogenic callus (EC), and somatic embryos (SEs). The differentially expressed genes in NEC, EC, and SEs were identified, annotated, and classified. Partial DEGs were identified that were related to purine metabolism [62]. A possible critical role for purines during embryogenesis in geranium hypocotyl tissues (*Pelargonium* x *hortorum*) has also been reported [63]. The above results revealed the significant role of purine metabolism in EC proliferation.

3.3. Enrichment of Flavonoid Biosynthesis in Somatic Embryo Developmental Initiation

Flavonoids are synthesized by the phenylpropanoid metabolic pathway in which the amino acid phenylalanine is used to produce 4-coumaroyl-CoA. The metabolic pathway continues through a series of enzymatic modifications to yield flavanones, dihydroflavonols, and then anthocyanins. Along this pathway, many products can be formed, including flavonols, flavan-3-ols, proanthocyanidins (tannins) and a host of other various polyphenolics. In the current study, 139 differentially accumulated metabolites were identified in GE vs. PEC, and flavone was largely detected (Figure 4h). 94 DAMs accumulated specifically in somatic embryo development, of which flavone accounted for a large proportion (Figure 5c). KEGG enrichment analysis of DAMs suggested that 'phenylpropanoid biosynthesis', 'flavone and flavonol biosynthesis', 'flavonoid biosynthesis', and 'biosynthesis of phenylpropanoids' were enriched in GE vs. PEC (Figure 6b). Meanwhile, in co-mapped pathways between genes and metabolites, 'phenylpropanoid biosynthesis' (ko00940), 'flavonoid biosynthesis' (ko00941), and 'flavone and flavonol biosynthesis' (ko00944) showed significant accumulation in GE vs. PEC (Table S5).

The significant role of flavonoid biosynthesis in the somatic embryo development process has been investigated. In citrus cell cultures, there was no detectable accumulation of flavonoid in the undifferentiated calli, but flavonoid accumulated after the morphological changes to embryos. Two chalcone synthase (*CHS*) genes differentially expressed during citrus SE and *CHS* gene may regulate the accumulation of flavonoid [64].

With the goal to better understand SE development and to improve the efficiency of SE conversion in *Theobroma cacao* L., gene expression differences between zygotic and somatic embryos were examined using a whole genome microarray. Expression levels of genes involved in fatty acid metabolism and flavonoid biosynthesis were differentially expressed in the two types of embryos. The relatively higher expression of flavonoid related genes during SE suggested that the developing tissues may be experiencing high levels of stress during SE maturation caused by the in vitro environment [65].

Moreover, Wang et al. reported that overexpression of *GhSPL10*, a target of *GhmiR157a*, activated the flavonoid biosynthesis pathway, and promoted initial cellular dedifferentiation and callus proliferation [66]. The synthesis of flavonoids like anthocyanin in several plant tissues has been associated with increased phenylalanine ammonia lyase (PAL) activity. The increased PAL level was detected in samples collected from different growth phases during SE in *Silybum marianum*. As intermediary products of phenylpropanoid metabolism, flavonoids may stimulate differentiation and create a situation that is more favorable for embryogenesis [67]. These conclusions demonstrated that flavonoids biosynthesis might be frequently associated with somatic embryo development during cotton SE transdifferentiation.

4. Materials and Methods

4.1. Plant Materials and Culture Conditions

Upland cotton (*Gossypium hirsutum* cv. YZ-1) seeds were sterilized in 0.1% $HgCl_2$ (*w/v*) for 8 min and were then rinsed 3–4 times with distilled water. The seeds were then germinated in Murashige and Skoog (MS) medium supplemented with 3% (*w/v*) sucrose and 0.25% (*w/v*) phytagel. Hypocotyl explants (0.5–1.0 cm) from 7-day-old seedlings were cultured in MS plus B_5 vitamin (MSB) medium containing 0.45 μmol·L^{-1} 2,4-dichlorophenoxyacetic acid (2,4-D) and 0.46 μmol·L^{-1} kinetin (KT). NEC were maintained in MSB medium for 6 weeks at 28 °C in a 16/8 h light/dark photoperiod and were then subcultured in fresh MSB medium without hormones. Following an additional 3–4 weeks of growth, the somatic-to-embryogenic transition progressed to the point where it induced the development of PEC and GE. Based on our previous approach as published recently [53], these representative staged samples of NEC, PEC, and GE could be highly enriched and collected, frozen immediately in liquid nitrogen, and stored at −80 °C for subsequent metabolic and transcriptomic profiling. The sample from each stage was prepared using three biological replicates.

4.2. Sample Preparation and Extraction for Widely Targeted Metabolic Profiling

Chemical extraction was carried out on nine samples (three biological replicates for each of three developmental stages) for the purposes of metabolic analyses. Each freeze-dried sample was crushed using a mixer mill (MM 400, Retsch, Haan, Germany) with a zirconia bead for 1.5 min at 30 Hz. One hundred milligrams of powder was weighed and extracted overnight at 4 °C in 1.0 mL 70% aqueous methanol. Following centrifugation at 10000 g for 10 min, the extracts were absorbed using a CNWBOND Carbon-GCB SPE Cartridge (250 mg, 3 mL; ANPEL, Shanghai, China, www.anpel.com.cn/cnw) and filtered with a SCAA-104 filter (0.22 μm pore size; ANPEL, Shanghai, China, http://www.anpel.com.cn/) prior to LC-MS analysis.

4.3. HPLC Conditions

The sample extracts were analyzed using an LC-ESI-MS/MS system (HPLC: Shim-pack UFLC SHIMADZU CBM30A system, www.shimadzu.com.cn/; MS: Applied Biosystems 6500 Q TRAP, www.appliedbiosystems.com.cn/). The analytical conditions were as follows: HPLC column, Waters ACQUITY UPLC HSS T3 C18 (1.8 μm, 2.1 × 100 mm); solvent system, water (0.04% acetic acid) and acetonitrile (0.04% acetic acid); gradient program, 95:5 *v/v* at 0 min, 5:95 *v/v* at 11.0 min, 5:95 *v/v* at 12.0 min, 95:5 *v/v* at 12.1 min, 95:5 *v/v* at 15.0 min; flow rate, 0.40 mL/min; temperature, 40 °C; injection volume, 2 μL. The effluent was alternatively connected to an ESI-triple quadrupole-linear ion trap (Q TRAP)-MS.

4.4. ESI-Q TRAP-MS/MS

LIT and triple quadrupole (QQQ) scans were acquired using a triple quadrupole-linear ion trap mass spectrometer (Q TRAP) API 6500 Q TRAP LC/MS/MS System equipped with an ESI Turbo ion-spray interface, operated in positive ion mode and controlled using Analyst 1.6 software (AB Sciex). The ESI source operation parameters were as follows: ion source, turbo spray; source temperature, 500 °C; ion spray voltage (IS), 5500 V; ion source gas I (GSI), gas II (GSII), and curtain gas (CUR), 55, 60, and 25.0 psi, respectively; the collision gas (CAD) was set to high. The instrument tuning and mass calibration were performed using 10 and 100 μmol/L polypropylene glycol solutions in QQQ and LIT modes, respectively. QQQ scans were acquired during MRM experiments in which the collision gas (nitrogen) was set to 5 psi. DP and CE for individual MRM transitions were conducted after further DP and CE optimization. A specific set of MRM transitions were monitored during each period based on the metabolites that eluted during this period [68,69].

4.5. Widely Targeted Metabolic Profiling

The samples were analyzed using a metabolomic platform that combined ultra-performance liquid chromatography (UPLC) and tandem mass spectrometry (MS/MS). Metabolite identification was performed using the MWDB metware database (Metware Biotechnology Co., Ltd. Wuhan, China) and other public databases according to standard metabolic operating procedures.

4.6. Statistical Analysis

The metabolite abundances were quantified using the peak areas. The data obtaining from the metabolite profiling were normalized for the principal component analysis (PCA) and partial least squares-discriminant analysis (PLS-DA) [70,71]. Metabolites with significant differences in content were defined as having a variable importance in project (VIP) ≥ 1 and a fold change ≥ 2 or ≤ 0.5. Fisher's exact test was applied to identify the significant KEGG pathways with a false discovery rate (FDR) < 0.05 [72]. Gene-metabolite pairs with a Pearson's correlation coefficient > 0.8 were used to construct the transcript-metabolite network.

5. Conclusions

Somatic embryogenesis is the developmental reprogramming of somatic cells toward the embryogenesis pathway. In this study, the dynamic metabolomic and transcriptomic profiling of cotton SE transdifferentiation processes, embryogenic differentiation, and somatic embryo development, were comparatively investigated.

During embryogenic differentiation (PEC vs. NEC), nucleotides and lipids were specifically accumulated. And the metabolome wide DAMs significantly enriched in purine metabolism (Table 2).

In addition, purine metabolism-related genes associated with signal recognition, transcription, stress, and lipid binding were remarkably activated (Table 4). Moreover, several classic SE genes that specifically activated during embryogenic differentiation, including *BBM, SERK1, LEC1, ARF2, AGL15, AGP1, GLP2, AGO1, LTP2,* and *AMY1,* were highly correlated with the corresponding metabolites that were involved in purine metabolism (Table 5).

During somatic embryo development (GE vs. PEC), flavones and hydroxycinnamoyl derivatives were largely accumulated. DAMs were most significantly associated with flavonoid biosynthesis (Table 3). Classic SE genes that control somatic embryo development, including *WUS, CLV1, CUC2, SHR,* and *SCW,* were highly correlated with the corresponding metabolites that were involved in flavonoid biosynthesis (Table 6).

By interactively comparing metabolomic and transcriptomic data, we identified a series of potential metabolites and the corresponding differentially expressed genes candidate for a relevant role in SE transdifferentiation. The findings in our work provide new insights into the underlying molecular and biochemical basis associated with embryogenic competence acquisition underpinning cotton SE development.

Supplementary Materials: Supplementary materials can be found at http://www.mdpi.com/1422-0067/20/9/2070/s1.

Author Contributions: Data curation, H.G. (Huihui Guo), H.G. (Haixia Guo), L.Z., Z.T., X.Y., J.W. and F.Z.; Formal analysis, H.G. (Huihui Guo), Z.T. and F.Z.; Funding acquisition, F.Z.; Investigation, H.G. (Huihui Guo), H.G. (Haixia Guo), L.Z. and X.Y.; Methodology, H.G. (Huihui Guo), H.G. (Haixia Guo), L.Z., Z.T. and J.W.; Project administration, F.Z.; Resources, H.G. (Haixia Guo), L.Z. and X.Y.; Software, J.W.; Supervision, F.Z.; Validation, H.G. (Huihui Guo), H.G. (Haixia Guo) and Z.T.; Writing—original draft, H.G. (Huihui Guo) and H.G. (Haixia Guo); Writing—review & editing, H.G. (Huihui Guo), X.Y. and F.Z.

Funding: This work was supported by the National Natural Science Foundation of China (31401428), Fok Ying-Tong Foundation (151024), Taishan Scholar Talent Project from PRC (TSQN20161018).

Conflicts of Interest: The authors declare no conflict of interest.

Abbreviations

ABA2	Abscisic aldehyde synthesis enzyme 2
ABI3	Abscisic acid insensitive 3
AGL15	Agamous-like 15
AGO1	Argonaute 1
AGP	Arabinogalactan protein
AMY	Alpha-amylase
ARF	Auxin response factor
BBM	Baby boom
CLV1	Clavata 1
CUC2	Cup-shaped cotyledon 2
DAMs	Differentially accumulated metabolites
DEGs	Differentially expressed genes
GE	Initiation staged embryos with globular-like enriched
GLP	Germin-like protein
HCA	Hierarchical cluster analysis
JAZ1	Jasmonate ZIM-domain 1
KEGG	Kyoto encyclopedia of genes and genomes
LEA1	Late embryogenesis abundant protein 1
LEC	Leafy cotyledon
LTP	Lipid transfer protein
MS/MS	Tandem mass spectrometry
NEC	Nonembryogenic staged calli
PCA	Principal component analysis
PCC	Pearson correlation coefficient
PEC	Primary embryogenic calli

QC	Quality control
SCW	Scarecrow
SE	Somatic embryogenesis
SERK	Somatic embryogenesis receptor kinase
SHR	Short-root
STM	Shoot meristemless
TIC	Total ions current
UPLC	Ultra-performance liquid chromatography
VIP	Variable importance in project
WUS	Wuschel

References

1. Zeng, F.; Zhang, X.; Cheng, L.; Hu, L.; Zhu, L.; Cao, J.; Guo, X. A draft gene regulatory network for cellular totipotency reprogramming during plant somatic embryogenesis. *Genomics* **2007**, *90*, 620–628. [CrossRef]
2. Loyolavargas, V.; Delapeña, C.; Galazávalos, R.; Quirozfigueroa, F. Plant tissue culture. In *Molecular Biomethods Handbook*, 2nd ed.; Walker, J., Rapley, R., Eds.; Humana Press: New York, NY, USA, 2008; pp. 875–904.
3. Ikeuchi, M.; Sugimoto, K.; Iwase, A. Plant callus: Mechanisms of induction and repression. *Plant Cell* **2013**, *25*, 3159–3173. [CrossRef] [PubMed]
4. Sakhanokho, H.; Rajasekaran, K. Cotton regeneration in vitro. In *Fiber Plants*; Ramawat, K.G., Ahuja, M.R., Eds.; Springer: Berlin, Germany, 2016; pp. 87–110.
5. Nic-Can, G.; De la Peña, C. Epigenetic advances on somatic embryogenesis of agronomical and important crops. In *Epigenetics in Plants of Agronomic Importance: Fundamentals and Applications*, 6th ed.; Alvarez-Venegas, R., Ed.; Springer International Publishing: Berlin, Germany, 2014; pp. 91–109.
6. Vogel, G. How does a single somatic cell become a whole plant? *Science* **2005**, *309*, 86. [CrossRef] [PubMed]
7. Min, L.; Hu, Q.; Li, Y.; Xu, J.; Ma, Y.; Zhu, L.; Yang, X.; Zhang, X. Leafy cotyledon1-casein kinase I-TCP15-phytochrome interacting factor 4 network regulates somatic embryogenesis by regulating auxin homeostasis. *Plant Physiol.* **2015**, *169*, 2805–2821. [PubMed]
8. Hu, H.; Xiong, L.; Yang, Y. Rice *SERK1* gene positively regulates somatic embryogenesis of cultured cell and host defense response against fungal infection. *Planta* **2005**, *222*, 107–117. [CrossRef] [PubMed]
9. Ikeuchi, M.; Iwase, A.; Rymen, B.; Harashima, H.; Shibata, M.; Ohnuma, M.; Breuer, C.; Morao, A.; Lucas, M.; Veylder, L.; et al. PRC2 represses dedifferentiation of mature somatic cells in *Arabidopsis*. *Nat. Plants* **2015**, *1*, 15089–15095. [CrossRef] [PubMed]
10. Rupps, A.; Raschke, J.; Rümmler, M.; Linke, B.; Zoglauer, K. Identification of putative homologs of *Larix decidua* to baby boom (*BBM*), leafy cotyledon1 (*LEC1*), wuschel-related homeobox2 (*WOX2*) and somatic embryogenesis receptor-like kinase (*SERK*) during somatic embryogenesis. *Planta* **2016**, *243*, 473–488. [CrossRef]
11. Zhai, L.; Xu, L.; Wang, Y.; Zhu, X.; Feng, H.; Li, C.; Luo, X.; Everlyne, M.; Liu, L. Transcriptional identification and characterization of differentially expressed genes associated with embryogenesis in radish (*Raphanus sativus* L.). *Sci. Rep.* **2016**, *6*, 21652–21664. [CrossRef]
12. Kumar, V.; Van Staden, J. New insights into plant somatic embryogenesis: An epigenetic view. *Acta Physiol. Plant* **2017**, *39*, 194–211. [CrossRef]
13. Lin, Y.; Lai, Z.; Tian, Q.; Lin, L.; Lai, R.; Yang, M.; Zhang, D.; Chen, Y.; Zhang, Z. Endogenous target mimics down-regulate miR160 mediation of *ARF10*, *-16*, and *-17* cleavage during somatic embryogenesis in *Dimocarpus longan* Lour. *Front. Plant Sci.* **2015**, *6*, 956–971. [CrossRef]
14. Su, Y.; Liu, Y.; Zhou, C.; Li, X.; Zhang, X. The microRNA167 controls somatic embryogenesis in Arabidopsis through regulating its target genes *ARF6* and *ARF8*. *Plant Cell Tiss. Org. Cult.* **2016**, *124*, 405–417. [CrossRef]
15. Harada, J. Role of *Arabidopsis LEAFY COTYLEDON* genes in seed development. *J. Plant Physiol.* **2001**, *158*, 405–409. [CrossRef]
16. Su, Y.H.; Zhang, X.S. Auxin gradients trigger de novo formation of stem cells during somatic embryogenesis. *Plant Signal. Behav.* **2009**, *4*, 574–576. [CrossRef]

17. Chen, S.K.; Kurdyukov, S.; Kereszt, A.; Wang, X.D.; Gresshoff, P.M.; Rose, R.J. The association of homeobox gene expression with stem cell formation and morphogenesis in cultured *Medicago truncatula*. *Planta* **2009**, *230*, 827–840. [CrossRef]

18. Elhiti, M.; Tahir, M.; Gulden, R.H.; Khamiss, K.; Stasolla, C. Modulation of embryo-forming capacity in culture through the expression of *Brassica* genes involved in the regulation of the shoot apical meristem. *J. Exp. Bot.* **2010**, *61*, 4069–4085. [CrossRef]

19. Kulinska-Lukaszek, K.; Tobojka, M.; Adamiok, A.; Kurczynska, E. Expression of the *BBM* gene during somatic embryogenesis of *Arabidopsis thaliana*. *Biol. Plant.* **2012**, *56*, 389–394. [CrossRef]

20. Florez, S.; Erwin, R.; Maximova, S.; Guiltinan, M.; Curtis, W. Enhanced somatic embryogenesis in *Theobroma cacao* using the homologous *BABY BOOM* transcription factor. *BMC Plant Biol.* **2015**, *15*, 121–132. [CrossRef]

21. Ahmadi, B.; Shariatpanahi, M.E.; Teixeira da Silva, J.A. Efficient induction of microspore embryogenesis using abscisic acid, jasmonic acid and salicylic acid in *Brassica napus* L. *Plant Cell Tiss. Org. Cult.* **2014**, *116*, 343–351. [CrossRef]

22. Pérez-Jiménez, M.; Cantero-Navarro, E.; Acosta, M.; Cos-Terrer, J. Relationships between endogenous hormonal content and direct somatic embryogenesis in *Prunus persica* L. Batsch cotyledons. *Plant Growth Regul.* **2013**, *71*, 219–224. [CrossRef]

23. Pérez-Jiménez, M.; Cantero-Navarro, E.; Pérez-Alfocea, F.; Le-Disquet, I.; Guivarc'hc, A.; Cos-Terrera, J. Relationship between endogenous hormonal content and somatic organogenesis in callus of peach (*Prunus persica* L. Batsch) cultivars and *Prunus persica* × *Prunus dulcis* rootstocks. *J. Plant Physiol.* **2014**, *171*, 619–624. [CrossRef]

24. Nieves, N.; Martínez, M.E.; Castillo, R.; Blanco, M.A.; González-Olmedo, J.L. Effect of abscissic acid and jasmonic acid on partial desiccation of encapsulated somatic embryos of sugarcane. *Plant Cell Tiss. Org. Cult.* **2001**, *65*, 15–21. [CrossRef]

25. Mira, M.M.; Wally, O.S.D.; Elhiti, M.; El-Shanshory, A.; Reddy, D.S.; Hill, R.D.; Stasolla, C. Jasmonic acid is a downstream component in the modulation of somatic embryogenesis by *Arabidopsis* Class 2 phytoglobin. *J. Exp. Bot.* **2016**, *67*, 2231–2246. [CrossRef] [PubMed]

26. Ma, J.; He, Y.H.; Hu, Z.Y.; Kanakala, S.; Xu, W.T.; Xia, J.X.; Guo, C.H.; Lin, S.Q.; Chen, C.J.; Wu, C.H.; et al. Histological analysis of somatic embryogenesis in pineapple: *AcSERK1* and its expression validation under stress conditions. *J. Plant Biochem. Biot.* **2016**, *25*, 49–55. [CrossRef]

27. Ruduś, I.; Weiler, E.W.; Kępczyńska, E. Do stress-related phytohormones, abscisic acid and jasmonic acid play a role in the regulation of *Medicago sativa* L. somatic embryogenesis? *Plant Growth Regul.* **2009**, *59*, 159–169. [CrossRef]

28. Hu, L.; Yang, X.; Yuan, D.; Zeng, F.; Zhang, X. *GhHmgB3* deficiency deregulates proliferation and differentiation of cells during somatic embryogenesis in cotton. *Plant Biotechnol. J.* **2011**, *9*, 1038–1048. [CrossRef]

29. Poon, S.; Heath, R.; Clarke, A. A chimeric arabinogalactan protein promotes somatic embryogenesis in cotton cell culture. *Plant Physiol.* **2012**, *160*, 684–695. [CrossRef]

30. Yang, X.; Wang, L.; Yuan, D.; Lindsey, K.; Zhang, X. Small RNA and degradome sequencing reveal complex miRNA regulation during cotton somatic embryogenesis. *J. Exp. Bot.* **2013**, *64*, 1521–1536. [CrossRef]

31. Jin, F.; Hu, L.; Yuan, D.; Xu, J.; Gao, W.; He, L.; Yang, X.; Zhang, X. Comparative transcriptome analysis between somatic embryos (SEs) and zygotic embryos in cotton: Evidence for stress response functions in SE development. *Plant Biotechnol. J.* **2014**, *12*, 161–173. [CrossRef]

32. Pandey, D.K.; Chaudhary, B. Oxidative stress responsive *SERK1* gene directs the progression of somatic embryogenesis in cotton (*Gossypium hirsutum* L. cv. Coker 310). *Am. J. Plant Sci.* **2014**, *5*, 80–102. [CrossRef]

33. Karami, O.; Saidi, A. The molecular basis for stress-induced acquisition of somatic embryogenesis. *Mol. Biol. Rep.* **2010**, *37*, 2493–2507. [CrossRef]

34. Fehér, A. Somatic embryogenesis-Stress-induced remodeling of plant cell fate. *BBA-Gene Regul. Mech.* **2015**, *1849*, 385–402. [CrossRef]

35. Ma, J.; He, Y.; Hu, Z.; Xu, W.T.; Xia, J.X.; Guo, C.H.; Lin, S.Q.; Cao, L.; Chen, C.J.; Wu, C.H.; et al. Characterization and expression analysis of *AcSERK2*, a somatic embryogenesis and stress resistance related gene in pineapple. *Gene* **2012**, *500*, 115–123. [CrossRef]

36. Cabrera-Ponce, J.L.; López, L.; León-Ramírez, C.G.; Jofre-Garfias, A.E.; Verver-y-Vargas, A. Stress induced acquisition of somatic embryogenesis in common bean *Phaseolus vulgaris* L. *Protoplasma* **2014**, *252*, 559–570. [CrossRef]

37. Salo, H.M.; Sarjala, T.; Jokela, A.; Häggman, H.; Vuosku, J. Moderate stress responses and specific changes in polyamine metabolism characterize Scots pine somatic embryogenesis. *Tree Physiol.* **2016**, *36*, 392–402. [CrossRef]

38. Yang, X.; Zhang, X.; Yuan, D.; Jin, F.; Zhang, Y.; Xu, J. Transcript profiling reveals complex auxin signalling pathway and transcription regulation involved in dedifferentiation and redifferentiation during somatic embryogenesis in cotton. *BMC Plant Biol.* **2012**, *12*, 110–128. [CrossRef]

39. Xu, Z.; Zhang, C.; Zhang, X.; Liu, C.; Wu, Z.; Yang, Z.; Zhou, K.; Yang, X.; Li, F. Transcriptome profiling reveals auxin and cytokinin regulating somatic embryogenesis in different sister lines of cotton cultivar CCRI24. *J. Integr. Plant Biol.* **2013**, *55*, 631–642. [CrossRef]

40. Dixon, R.A.; Strack, D. Phytochemistry meets genome analysis, and beyond. *Phytochemistry* **2003**, *62*, 815–816. [CrossRef]

41. Fiehn, O. Metabolomics-the link between genotypes and phenotypes. *Plant Mol. Biol.* **2002**, *48*, 155–171. [CrossRef]

42. Panopoulos, A.D.; Yanes, O.; Ruiz, S.; Kida, Y.S.; Diep, D.; Tautenhahn, R.; Herrerías, A.; Batchelder, E.M.; Plongthongkum, N.; Lutz, M.; et al. The metabolome of induced pluripotent stem cells reveals metabolic changes occurring in somatic cell reprogramming. *Cell Res.* **2012**, *22*, 168–177. [CrossRef]

43. Fiehn, O.; Kopka, J.; Dörmann, P.; Altmann, T.; Trethewey, R.N.; Willmitzer, L. Metabolite profiling for plant functional genomics. *Nat. Biotechnol.* **2000**, *18*, 1157–1161. [CrossRef]

44. Sumner, L.W.; Mendes, P.; Dixon, R.A. Plant metabolomics: Large-scale phytochemistry in the functional genomics era. *Phytochemistry* **2003**, *62*, 817–836. [CrossRef]

45. Weckwerth, W. Metabolomics in systems biology. *Annu. Rev. Plant Biol.* **2003**, *54*, 669–689. [CrossRef]

46. Liu, Q.; Wang, X.; Tzin, V.; Romeis, J.; Peng, Y.; Li, Y. Combined transcriptome and metabolome analyses to understand the dynamic responses of rice plants to attack by the rice stem borer *Chilo suppressalis* (Lepidoptera: Crambidae). *BMC Plant Biol.* **2016**, *16*, 259–275. [CrossRef]

47. Taylor, J.; King, R.D.; Altmann, T.; Fiehn, O. Application of metabolomics to plant genotype discrimination using statistics and machine learning. *Bioinformatics* **2002**, *18*, 241–248. [CrossRef]

48. Robinson, A.R.; Dauwe, R.; Ukrainetz, N.K.; Cullis, I.F.; White, R.; Mansfield, S.D. Predicting the regenerative capacity of conifer somatic embryogenic cultures by metabolomics. *Plant Biotechnol. J.* **2009**, *7*, 952–963. [CrossRef]

49. Dowlatabadi, R.; Weljie, A.M.; Thorpe, T.A.; Yeung, E.C.; Vogel, H.J. Metabolic footprinting study of white spruce somatic embryogenesis using NMR spectroscopy. *Plant Physiol. Biochem.* **2009**, *47*, 343–350. [CrossRef]

50. Palama, T.L.; Menard, P.; Fock, I.; Choi, Y.H.; Bourdon, E.; Govinden-Soulange, J.; Bahut, M.; Payet, B.; Verpoorte, R.; Kodja, H. Shoot differentiation from protocorm callus cultures of *Vanilla planifolia* (Orchidaceae): Proteomic and metabolic responses at early stage. *BMC Plant Biol.* **2010**, *10*, 82–99. [CrossRef]

51. Businge, E.; Brackmann, K.; Moritz, T.; Egertsdotter, U. Metabolite profiling reveals clear metabolic changes during somatic embryo development of Norway spruce (*Picea abies*). *Tree Physiol.* **2012**, *32*, 232–244. [CrossRef]

52. Mahmud, I.; Thapaliya, M.; Boroujerdi, A.; Chowdhury, K. NMR-based metabolomics study of the biochemical relationship between sugarcane callus tissues and their respective nutrient culture media. *Anal. Bioanal. Chem.* **2014**, *406*, 5997–6005. [CrossRef]

53. Guo, H.; Wu, J.; Chen, C.; Wang, H.; Zhao, Y.; Zhang, C.; Jia, Y.; Liu, F.; Ning, T.; Chu, Z.; et al. Identification and characterization of cell cultures with various embryogenic/regenerative potential in cotton based on morphological, cytochemical, and cytogenetical assessment. *J. Integr. Agr.* **2019**, *18*, 1–8. [CrossRef]

54. Zheng, Q.; Zheng, Y.; Perry, S. *AGAMOUS-Like15* promotes somatic embryogenesis in *Arabidopsis* and Soybean in part by the control of ethylene biosynthesis and response. *Plant Physiol.* **2013**, *161*, 2113–2127. [CrossRef]

55. Mathieu, M.; Lelu-Walter, M.A.; Blervacq, A.S.; David, H.; Hawkins, S.; Neutelings, G. Germin-like genes are expressed during somatic embryogenesis and early development of conifers. *Plant Mol. Biol.* **2006**, *61*, 615–627. [CrossRef]

56. Schlögl, P.; Santos, A.; Vieira, L.; Floh, E.; Guerra, M. Gene expression during early somatic embryogenesis in Brazilian pine (*Araucaria angustifolia* (Bert) O. Ktze). *Plant Cell Tiss. Org. Cult.* **2012**, *108*, 173–180. [CrossRef]

57. Guo, H.; Guo, H.; Zhang, L.; Fan, Y.; Fan, Y.; Zeng, F. *SELTP* assembled battery drives totipotency of somatic plant cell. *Plant Biotechnol. J.* **2019**. [CrossRef]

58. Aida, M.; Ishida, T.; Tasaka, M. Shoot apical meristem and cotyledon formation during *Arabidopsis* embryogenesis: Interaction among the cup-shaped cotyledon and shoot meristemless genes. *Development* **1999**, *126*, 1563–1570.

59. Di Laurenzio, L.; Wysocka-Diller, J.; Malamy, J.E.; Pysh, L.; Helariutta, Y.; Freshour, G.; Hahn, M.G.; Feldmann, K.A.; Benfey, F.N. The scarecrow gene regulates an asymmetric cell division that is essential for generating the radial organization of the *Arabidopsis* root. *Cell* **1996**, *8*, 423–433. [CrossRef]

60. Domżalska, L.; Kędracka-Krok, S.; Jankowska, U.; Grzyb, M.; Sobczak, M.; Rybczyński, J.J.; Mikuła, A. Proteomic analysis of stipe explants reveals differentially expressed proteins involved in early direct somatic embryogenesis of the tree fern *Cyathea delgadii* Sternb. *Plant Sci.* **2017**, *258*, 61–76. [CrossRef]

61. Tao, L.; Zhao, Y.; Wu, Y.; Wang, Q.; Yuan, H.; Zhao, L.; Guo, W.; You, X. Transcriptome profiling and digital gene expression by deep sequencing in early somatic embryogenesis of endangered medicinal *Eleutherococcus senticosus* Maxim. *Gene* **2016**, *578*, 17–24. [CrossRef]

62. Cheng, W.H.; Zhu, H.G.; Tian, W.G.; Zhu, S.H.; Xiong, X.P.; Sun, Y.Q.; Zhu, Q.H.; Sun, J. De novo transcriptome analysis reveals insights into dynamic homeostasis regulation of somatic embryogenesis in upland cotton (*G. hirsutum* L.). *Plant Mol. Biol.* **2016**, *92*, 279–292. [CrossRef]

63. Hutchinson, M.J.; Saxena, P.K. Role of purine metabolism in thidiazuron-induced somatic embryogenesis of geranium (*Pelargonium× hortorum*) hypocotyl cultures. *Physiol. Plant.* **1996**, *98*, 517–522. [CrossRef]

64. Moriguchi, T.; Kita, M.; Tomono, Y.; EndoInagaki, T.; Omura, M. One type of chalcone synthase gene expressed during embryogenesis regulates the flavonoid accumulation in citrus cell cultures. *Plant Cell Physiol.* **1999**, *40*, 651–655. [CrossRef]

65. Maximova, S.N.; Florez, S.; Shen, X.; Niemenak, N.; Zhang, Y.; Curtis, W.; Guiltinan, M.J. Genome-wide analysis reveals divergent patterns of gene expression during zygotic and somatic embryo maturation of *Theobroma cacao* L., the chocolate tree. *BMC Plant Biol.* **2014**, *14*, 185–201. [CrossRef] [PubMed]

66. Wang, L.; Liu, N.; Wang, T.; Li, J.; Wen, T.; Yang, X.; Lindsey, K.; Zhang, X. The *GhmiR157a-GhSPL10* regulatory module controls initial cellular dedifferentiation and callus proliferation in cotton by modulating ethylene-mediated flavonoid biosynthesis. *J. Exp. Bot.* **2017**, *69*, 1081–1093. [CrossRef] [PubMed]

67. Khan, M.A.; Abbasi, B.H.; Ali, H.; Ali, M. Temporal variations in metabolite profiles at different growth phases during somatic embryogenesis of *Silybum marianum* L. *Plant Cell Tiss. Org. Cult.* **2015**, *120*, 127–139. [CrossRef]

68. Chen, W.; Gong, L.; Guo, Z.; Wang, W.; Zhang, H.; Liu, X.; Yu, S.; Xiong, L.; Luo, J. A novel integrated method for large-scale detection, identification, and quantification of widely targeted metabolites: Application in the study of rice metabolomics. *Mol. Plant.* **2013**, *6*, 1769–1780. [CrossRef] [PubMed]

69. Fraga, C.G.; Clowers, B.H.; Moore, R.J.; Zink, E.M. Signature-discovery approach for sample matching of a nerve-agent precursor using liquid chromatography-mass spectrometry, XCMS, and chemometrics. *Anal. Chem.* **2010**, *82*, 4165–4173. [CrossRef] [PubMed]

70. Chen, Y.; Zhang, R.; Song, Y.; He, J.; Sun, J.; Bai, J.; An, Z.; Dong, L.; Zhan, Q.; Abliz, Z. RRLC-MS/MS-based metabonomics combined with in-depth analysis of metabolic correlation network: Finding potential biomarkers for breast cancer. *Analyst* **2009**, *134*, 2003–2011. [CrossRef] [PubMed]

71. Thévenot, E.A.; Roux, A.; Xu, Y.; Ezan, E.; Junot, C. Analysis of the human adult urinary metabolome variations with age, body mass index, and gender by implementing a comprehensive workflow for univariate and OPLS statistical analyses. *J. Proteome Res.* **2015**, *14*, 3322–3335. [CrossRef]

72. Kanehisa, M.; Goto, S. KEGG: Kyoto encyclopedia of genes and genomes. *Nucleic Acids Res.* **2000**, *28*, 27–30. [CrossRef] [PubMed]

International Journal of
Molecular Sciences

Article

Transcriptomic Analysis of Leaf Sheath Maturation in Maize

Lei Dong [1,†], Lei Qin [2,†], Xiuru Dai [2], Zehong Ding [3], Ran Bi [4], Peng Liu [4], Yanhui Chen [1], Thomas P. Brutnell [1], Xianglan Wang [2,*] and Pinghua Li [2,*]

[1] College of Agronomy, Synergetic Innovation Centre of Henan Grain Crops and National Key Laboratory of Wheat and Maize Crop Science, Henan Agricultural University, Zhengzhou 450002, China; dongleixyz@126.com (L.D.); chenyanhui@henau.edu.cn (Y.C.); tom@viridisgenomics.com (T.P.B.)

[2] State Key Laboratory of Crop Biology, College of Agronomic Sciences, Shandong Agricultural University, Tai'an 271018, China; qinlei2012shandong@163.com (L.Q.); daixiuru1993@163.com (X.D.)

[3] The Institute of Tropical Bioscience and Biotechnology (ITBB), Chinese Academy of Tropical Agricultural Sciences (CATAS), Haikou 571101, China; dingzehong@itbb.org.cn

[4] Department of Statistics, Iowa State University, Ames, IA 50011, USA; biran@iastate.edu (R.B.); pliu@iastate.edu (P.L.)

* Correspondence: wangxianglan@sdau.edu.cn (X.W.); pinghuali@sdau.edu.cn (P.L.)

† These authors contributed equally to this work.

Received: 2 April 2019; Accepted: 17 May 2019; Published: 19 May 2019

Abstract: The morphological development of the leaf greatly influences plant architecture and crop yields. The maize leaf is composed of a leaf blade, ligule and sheath. Although extensive transcriptional profiling of the tissues along the longitudinal axis of the developing maize leaf blade has been conducted, little is known about the transcriptional dynamics in sheath tissues, which play important roles in supporting the leaf blade. Using a comprehensive transcriptome dataset, we demonstrated that the leaf sheath transcriptome dynamically changes during maturation, with the construction of basic cellular structures at the earliest stages of sheath maturation with a transition to cell wall biosynthesis and modifications. The transcriptome again changes with photosynthesis and lignin biosynthesis at the last stage of sheath tissue maturation. The different tissues of the maize leaf are highly specialized in their biological functions and we identified 15 genes expressed at significantly higher levels in the leaf sheath compared with their expression in the leaf blade, including the *BOP2* homologs *GRMZM2G026556* and *GRMZM2G022606*, *DOGT1* (*GRMZM2G403740*) and transcription factors from the B3 domain, C2H2 zinc finger and homeobox gene families, implicating these genes in sheath maturation and organ specialization.

Keywords: leaf sheath; maturation; transcriptional dynamics; transcriptome

---

1. Introduction

Maize (*Zea mays* L.) is one of the most important grain crops globally. It is not only used as a grain food and feed but is also preserved as corn silage and as a bioenergy source. Regardless of its final use, the accumulation of maize biomass is important. The leaf is the primary source of photosynthate; thus, its morphological development greatly influences plant biomass and yield. A typical mature maize leaf is made up of three parts: The leaf blade, the ligular region and the leaf sheath. The blade is the distal part of the maize leaf and the major organ for photosynthesis, the ligular region is a wedge-shaped structure connecting the leaf blade and leaf sheath that acts as a hinge to project the leaf blade away from the stem, and the sheath wraps around the stem and provides strength for the growth and development of the leaf blade. The shape and development of these three regions control the architecture of the maize leaf, which is important for photosynthesis and ultimately maize yields [1–5].

The development of grass leaves proceeds basipetally, with the distal cells differentiating and maturing first while the basal cells divide and expand. Several recent studies have portrayed a very dynamic biochemical differentiation process along the maize leaf blade. Here, basic cellular functions—such as DNA synthesis and cell wall synthesis—become enriched in the basal region of the leaf sink, which transitions to secondary cell wall biosynthesis and the establishment of the photosynthetic machinery in the source-sink transition zone and the dominant photosynthetic reactions in the distal part of the leaf [6,7]. Also, several mutants have been identified that influence the morphogenesis of the leaf blade in maize, e.g., *ns* (narrow sheath) and *rld1* (rolled leaf 1) [8,9], and produce abnormal leaf blade development in maize. Several genes that control ligule development in maize have been identified. These include *LIGULELESS1* (*LG1*), which encodes a Squamosa binding protein (SBP) family transcription factor; *LIGULELESS2* (*LG2*), encoding a basic leucine zipper (bZIP) family transcription factor; and *LIGULELESS NARROW* (*LGN*), a putative serine-threonine kinase that controls early ligule formation in maize leaves [10–14]. Compared to studies on the leaf blade and ligule tissues, very few studies have focused on the leaf sheath, and our understanding of sheath development is limited.

The leaf sheath wraps around the stem in maize and is believed to provide strength for the growth and development of the leaf blade. Hatfield and colleagues compared changes in the cell wall component in maize leaf blades [15], midribs and sheaths from nodes 9 to 14 and concluded that sheath and midrib tissues always accumulate more neutral sugars, lignin and total phenolics than blade tissues. These metabolite measurements may explain the higher mechanical strength of leaf sheaths; however, information at the molecular level is still limited.

Most mutations that affect the blade-sheath boundary appear to extend the sheath tissue into the blade, but the functions of genes that regulate the development of the sheath tissues are poorly understood, and almost no genes have been identified that distinguish the leaf sheath from the blade and ligule tissues. To investigate the transcriptional dynamics associated with the maturation of maize sheath tissues, we conducted RNA sequencing (RNA-seq) profiling experiments to capture the progressive stages of sheath maturation. These data reveal a dynamic transcriptional process during leaf sheath maturation. In particular, we have defined transcriptional regulators that may specify unique activities in sheath tissue relative to those in the leaf blade. This dataset serves as a foundation for future studies of maize sheath development and maturation.

2. Results

2.1. Defining the Leaf Sheath Transcriptome

In the mature maize leaf, the proximal sheath and distal blade tissues are separated by the leaf ligule (Figure 1A,D). Relative to the leaf blade, epidermal cells in the sheath tissue are smaller and contain fewer cell files, as determined by counting the number of cell files between the same veins across the sheath and blade (Figure 1A,B). To investigate developmental and maturation changes in the sheath tissue, we planted maize seeds every day for 13 days and then harvested the leaf blades and sheaths from the third leaf on the same day, when seedlings ranged in age from 9 to 13 days after planting in order to minimize the environmental variation at the time of harvest. Two millimeter-wide sections of sheath and blade tissues just below or above the ligule were harvested from the third leaf at 10 (stage 1, S1), 11 (stage 2, S2), 12 (stage 3, S3), and 13 (stage 4, S4) days after planting. The sheath tissue was poorly defined on the third leaf at 9 days after planting (stage 0), so we could not harvest tissue from this stage (Figure 1C,D). Although the specification of sheath cell fate is complete in the leaf primordium [16], the sheath tissues we harvested from stage 1 to 4 were still undergoing differentiation and maturation (Figure 1C,D). In total, 221.9 million high-quality reads were obtained from the RNA-seq analysis, of which 164.0 million (73.9%) were uniquely mapped to the maize reference genome. We detected 24,704, 25,225, 25,224, and 24,498 expressed genes from stages 1 to 4, respectively, and 23,557 genes were expressed in all four stages during sheath maturation.

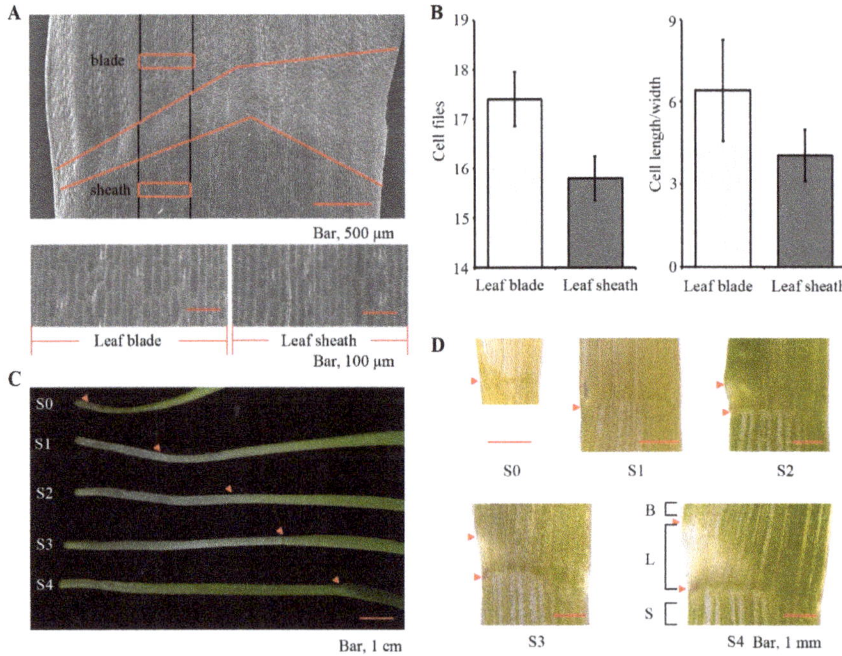

Figure 1. Observation of the leaf sheath and blade tissues. (**A**) Scanning electron micrograph of a mature maize leaf. B73 seedlings were grown for 13 days (stage 4). The red line indicates the ligular region, the tissues above the line are the blade and the tissues below the line are sheath tissues. The black line indicates the veins across the sheath, ligule and blade tissues of maize. (**B**) Cell file numbers and the ratio between the cell length and width were calculated in the leaf blade and sheath tissue between two lateral veins. (**C**) The blade and sheath tissues above or below the ligule of the third leaf from seedlings 10 (stage 1), 11 (stage 2), 12 (stage 3) and 13 (stage 4) days after planting were harvested at the same time. The arrowheads point to the leaf ligule. (**D**) Microscopic view of the blade (B), ligule (L) and sheath tissues (S) from stage 0 to stage 4 plants. The arrowheads point to the leaf ligule.

2.2. Maturation Gradient in the Leaf Sheath of Maize

We identified 7918 differentially expressed (DE) genes in the leaf sheath at four maturation stages with a false discovery rate (FDR) controlled at 0.001 (Supplemental Table S1), which accounted for approximately 31% of the sheath transcriptome. We then assigned genes to functional categories and grouped the genes by developmental dynamics using the k-means clustering algorithm. We identified nine clusters (K1–K9; Figure 2; Supplemental Figure S1 and Supplemental Table S1) from the leaf sheath and eight main clusters (K1–K8) that accounted for approximately 99% of the DE genes in four stages of development and maturation. As shown in Figure 2A,B, genes encoding proteins involved in cellular organization (e.g., actin and tubulin), vesicular transport, (e.g., syntaxin and snare), and DNA synthesis/chromatin structure were greatly enriched in clusters K1 and K2 and represented genes required for the installation of the basic cellular infrastructure. These genes are expressed at their highest levels in the early stage of sheath maturation (stage 1). Differences also existed between the K1 and K2 clusters. Namely, genes encoding fatty acid synthesis and elongation-related products, such as 3-ketoacyl-CoA synthase 4, 3-ketoacyl-CoA synthase 6, 3-ketoacyl-CoA synthase 9, 3-ketoacyl-CoA synthase 11 and 3-ketoacyl-CoA synthase 20 were differentially expressed [17]. The expression of genes related to respiration (including glycolysis, the tricarboxylic acid cycle and mitochondrial electron transport); signaling genes, especially G-proteins and LRR receptor kinases; and some transcription factors were greatly enriched in cluster K1 and slightly enriched in cluster K2, and genes involved in protein synthesis and targeting were only enriched in K2. The genes in clusters K3 and K4 were expressed at their highest levels in stage 2 and mainly participate in cell wall biosynthesis/modification and secondary metabolism. The genes that were highly expressed in stage 3 were included in clusters K5–K7 and are mainly involved in pathways related to photosynthesis (K6 and K7), hormone metabolism (K6 and K7), and redox and transcription regulation (K6 and K7), indicating that the sheath tissues at stage 3 begin building photosynthetic machinery. The genes in the K8 cluster were enriched in the functional categories of photosynthesis, sulfate assimilation, and secondary metabolism, e.g., lignin biosynthesis to strengthen the sheath, as well as cell organization, especially actin and fibrillin family proteins that are required for plastoglobule development. This highlighted the functions of both photosynthesis and plant strengthening in the mature sheath of maize. Taken together, these data revealed metabolic changes during the maturation of the sheath tissue in the maize leaf.

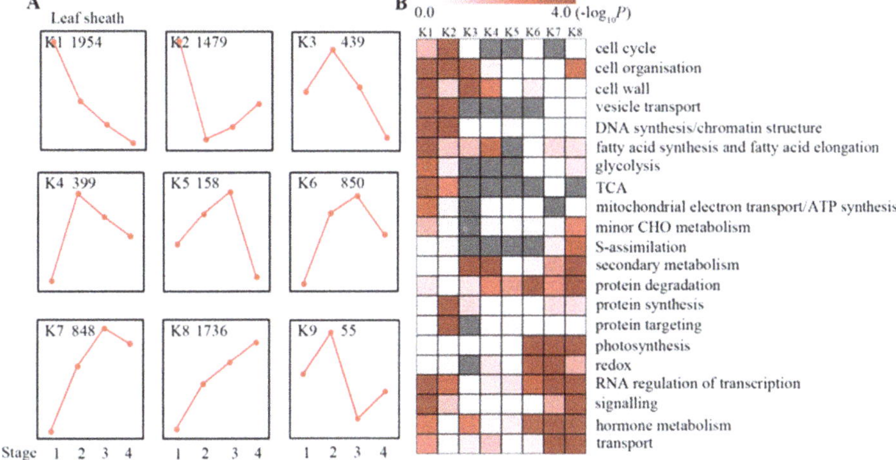

Figure 2. Transcriptome profiling during sheath maturation in maize. (**A**) K-means clustering showing the expression profile of the maize sheath transcriptome. Eight major clusters (K1–K8) were identified along the four developmental stages (stage 1 to 4) in sheath tissues from 7918 differentially expressed genes. (**B**) Functional category enrichment was calculated using the MapMan binning method among the eight major clusters in A. The shade of red represents the significance level of log 10 transformed *p*-values calculated from Fisher's exact test. Gray blocks indicate functional category enrichment was absent.

2.3. Cell Wall and Lignin Synthesis during Leaf Sheath Maturation in Maize

The cell wall plays an important role in shaping and strengthening cells. We noticed a dramatic change in the expression of cell wall-related genes during sheath maturation (Figure 3A). As shown in Figure 3A, cellulose synthases (*CESAs*)—the key genes of cellulose biosynthesis in primary and secondary cell walls—exhibited the highest expression levels in stages 1 or 2. Most of them exhibited dramatically decreased expression in stages 3 and 4, during late sheath maturation. Consistent with the expression patterns of the *CESAs*, most genes encoding CSL (cellulose synthase-like) proteins and cell wall proteins—for example, arabinogalactan proteins (AGP), leucine-rich repeat family proteins (LRR), and reversibly glycosylated polypeptide (RGP)—also showed decreased expression patterns in the late stages (S3 and S4) of sheath maturation. This indicated that the cell wall might be built in stages 1 and 2.

As a main component of the secondary cell wall, lignin plays an important role in providing strength and rigidity to support the cells and the plant body. As shown in Figure 3B (Supplemental Table S2), key enzymes related to lignin biosynthesis [18,19]—e.g., 4-coumarate-CoA ligase (4CL), cinnamoyl CoA reductase (CCR), *O*-methyltransferase (COMT), and cinnamyl alcohol dehydrogenase (CAD)—exhibited increased expression during the maturation of the sheath, which indicated the accumulation of lignin with the maturation of the leaf sheath tissues.

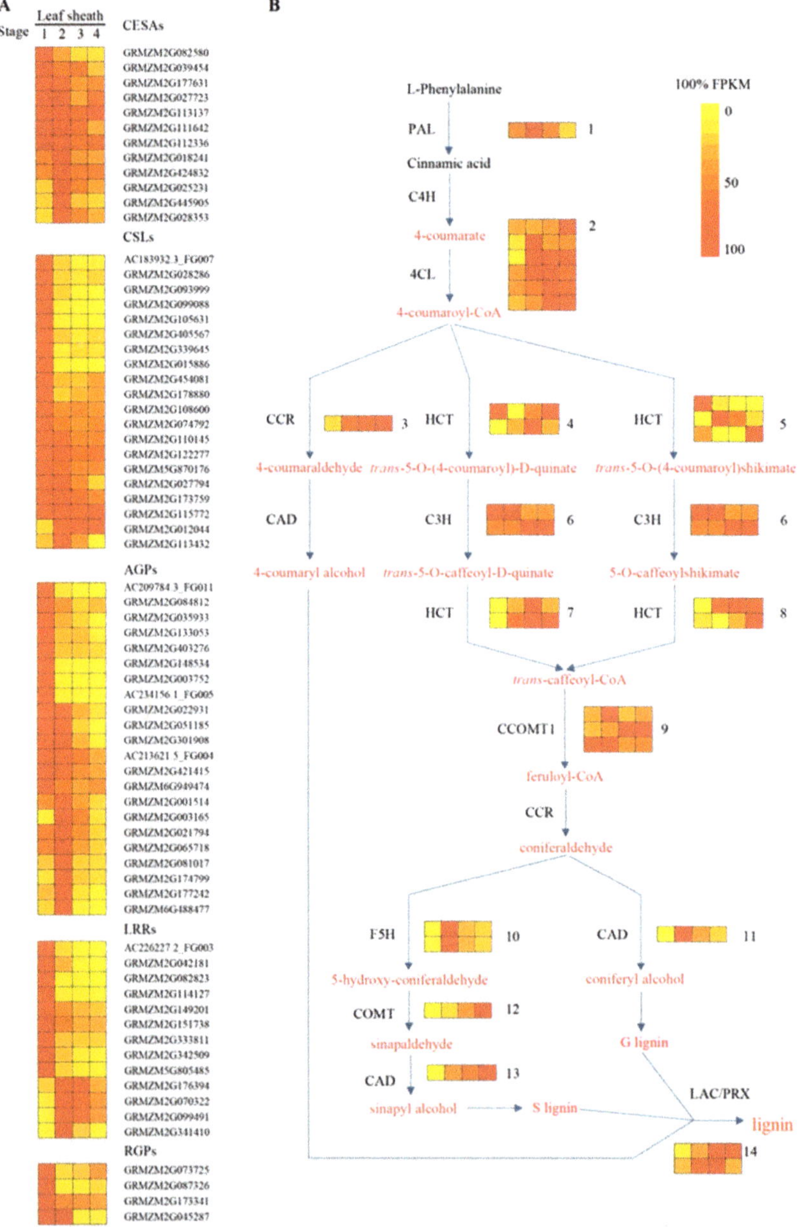

Figure 3. Heatmaps showing the expression profiles of genes related to cell wall (**A**) and lignin biosynthesis (**B**). Relative gene expression was calculated from the maximum fragments per kilobase million (FPKM) values among the maturation zones. The detailed expression patterns and identities of the genes in each of these biosynthetic pathways are shown in Supplementary Table S2.

2.4. Changes in Transcription Factors during Sheath Maturation in Maize

As key factors that regulate gene expression, transcription factors (TFs) play important roles in plant growth, development and the response to various environmental stress. We detected 456 differentially expressed TFs ($q < 0.001$) during sheath maturation and grouped them into six groups (G1 to G6, Figure 4A and Supplemental Table S3) using a hierarchical clustering (HCL) program. As shown in Figure 4A, groups G3 to G6 included most of the TFs (91%). The TFs in group G3 exhibited their highest expression levels in stage 4 during sheath maturation, and those in G4 were expressed at their peak level in stage 3. The goladen2-like (GLK) family of TFs, including *GLK1*, which regulates chloroplast development in maize [20], were exclusively enriched in groups G3 and G4 (Figure 4B). The DNA binding with one finger (DOF) family of TFs, such as *CDF3*, likely regulate photoperiod gene expression [21]. CO-like family members, such as *COL3* [22,23], regulate gene expression in photomorphogenesis and during lateral root development and function as a day length-sensitive regulator of shoot branching. The MYB family of TFs, such as *MYB4* [24,25], which respond to UV-B and *LHY* (involved in circadian rhythm [26]), were all mainly enriched in groups G3 and G4. As such, they indicated an increase in photosynthesis in stage 3 and 4 sheath tissues. Groups G5 and G6 included TFs that were highly expressed in stage 1, the earliest maturation stage that we harvested. The trihelix family members were highly enriched in both groups G5 and G6, and the basic helix-loop-helix (bHLH) family (e.g., *MUTE*) controlled meristemoid differentiation during the early stage of stomatal development [27]. The changes in the expression of TFs suggested the transcriptional regulation of gene expression during sheath maturation.

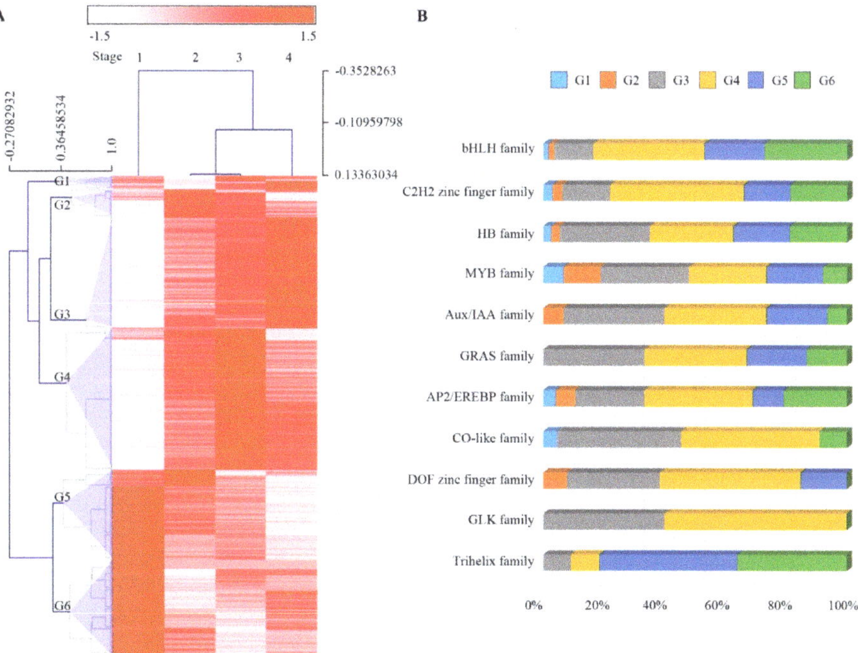

Figure 4. Dynamics of accumulated transcription factor profiles. (**A**) Hierarchical clustering (HCL) of the transcription factors related to sheath maturation in maize, six lineages (G1 to G6) were identified. The bar represents the expression of transcription factors normalized by row across four stages during sheath maturation in the HCL program. (**B**) Distribution of transcription factor families in groups G1 to G6 in the leaf sheath.

2.5. Identification of Genes Expressed at High Levels in the Leaf Sheath

To determine whether some genes were specifically expressed in the leaf sheath, we compared the sheath transcriptome with that of the blade at each stage during sheath maturation. In total, 167 genes were expressed at higher levels and 362 genes were expressed at lower levels in the sheath, than that in the blade, across the four maturation stages, with the FDR controlled at 0.001 (Supplemental Table S4).

As shown in Figure 5, of the 362 genes that were expressed at lower levels in the sheath than in the blade, 22% of them participated in the photosynthetic pathway. In contrast, of the 167 genes that were expressed at higher levels in the sheath than in the blade, none participated in photosynthesis. In addition to the photosynthetic pathway, genes involved in tetrapyrrole synthesis and redox scavenging were also specifically expressed at high levels in the leaf blade, consistent with the function of the green leaf blade to harvest light for photosynthesis, and redox reactive species generated by the light need to be controlled by enzymes such as ascorbate peroxidase. Interestingly, for genes in the functional categories of DNA synthesis/chromatin structure, cell wall development, lipid metabolism and cell organization, the enriched genes were expressed at much higher levels in the leaf sheath than in the blade. This may reflect a prolonged phase of cell growth and division in sheath tissues relative to the blade. As almost no sheath-specific genes were reported in a previous study on maize, we focused on identifying genes that were uniquely expressed at high levels in the leaf sheath.

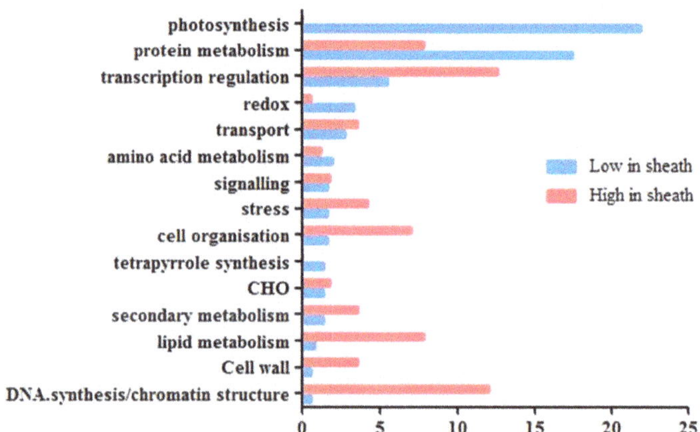

Figure 5. Functional category distribution of genes expressed at high or low levels in the sheath tissue.

As shown in Table 1, we identified 15 genes that were expressed at very low levels or not significant levels in the leaf blade, while their expression was high in the leaf sheath across the four developmental stages sampled. Among them were two genes (*GRMZM2G026556* and *GRMZM2G022606*) that are homologous to *BLADE ON PETIOLE2* (*BOP2*) in *Arabidopsis* and rice. BOP2 acts in cells adjacent to lateral organ boundaries to repress genes that confer meristem cell fate and induce genes that promote lateral organ fate and polarity in *Arabidopsis*, and plays a primary role in rice to regulate the proximal-distal axis [28,29]. *DON-glucosyltransferase 1* (*DOGT1, GRMZM2G403740*), encodes an enzyme that presumably regulates BR activity in *Arabidopsis* [30,31], and a homolog of *SMALLER WITH VARIABLE BRANCHES* (*SVB, GRMZM2G131409*), a protein with a conserved domain of unknown function (DUF538) that influences trichome development in *Arabidopsis* [32] were also expressed at higher levels in sheath relative to blade. In addition, *glycerol-3-phosphate acyltransferase* (*GPAT2, GRMZM2G033767*) and GDSL-like Lipase/Acylhydrolase superfamily protein (*GDSL-like Lipase, GRMZM5G862317*), which likely participates in lipid metabolism [33,34]; *carboxyesterase* (*CXE18, GRMZM2G104141*), which may play a role in cell elongation [35] and senescence-associated gene (*SAG12, GRMZM2G061879*), which controls nitrogen allocation during senescence in *Arabidopsis* [36], were all highly expressed in the sheath tissue. This indicated the possible role of these genes in sheath development and maturation.

Table 1. Genes expressed at high levels in the sheath tissue.

ID	Short Description	Symbol	Blade				Sheath			
			Stage				Stage			
			1	2	3	4	1	2	3	4
GRMZM2G026556	BTB/POZ domain-containing protein	BOP2	0.3	0.28	1.05	0.73	9.91	18.47	31.56	38.95
GRMZM2G022606	BTB/POZ domain-containing protein	BOP2	0.4	0.3	0.55	0.3	9.09	10.58	14.64	20.82
GRMZM2G403740	don-glucosyltransferase 1	DOGT1	0.05	0.06	0.55	0.11	0.37	2.12	3.89	13.76
GRMZM2G131409	Encodes Smaller with Variable Branches	SVB	0.58	0.52	0.33	0	0.49	15.39	16.42	8.07
GRMZM2G033767	glycerol-3-phosphate acyltransferase 2	GPAT2	0	0.55	1.25	0.6	0.83	63.1	105.91	33.77
GRMZM5G862317	GDSL-like Lipase/Acylhydrolase superfamily protein	GDSL-like Lipase	0.51	2.79	3.18	3.56	57.19	42.25	50.57	39.53
GRMZM2G104141	carboxyesterase 18	CXE18	0	0.72	2.11	0.87	0	21.06	73.57	39.17
GRMZM2G061879	senescence-associated gene 12	SAG12	1.41	6.92	7.68	1.38	1.58	51.99	136.22	196.46
GRMZM2G082227	AP2/B3-like transcriptional factor family protein NGATHA1	NGA1	1.51	2.08	2.48	2.4	7.13	6.81	14.38	16.34
GRMZM2G445684	C2H2-type zinc finger family protein No Transmitting Tract	NTT	0.22	0.17	0.13	0.16	17.68	15.79	39.79	42.81
GRMZM2G071101	C2H2-type zinc finger family protein No Transmitting Tract	NTT	0.31	1.77	0.75	0.34	10.08	19.09	45.68	42.21
GRMZM2G034113	homeobox 7	HB-7	1.67	0.91	1.97	6.79	14.24	17.7	43.08	43.7
GRMZM5G842695	MATE efflux family protein	-	0.11	0.6	0.38	0.12	0.12	3.07	5.09	6.33
GRMZM2G343291	Protein of unknown function	-	4.47	13.77	4.61	0.57	2.06	69.06	117.95	94.52
GRMZM2G136571	alpha/beta-Hydrolases superfamily protein	-	0.83	0.77	1.53	0.55	0.63	7.03	34.12	25.2
GRMZM2G412436	Protein of unknown function	-	0.98	2.39	1.31	0.17	2.25	18.56	15.16	4.33

Note: The expression of genes was represented by the FPKM value in the table.

Transcription factors may also play important roles in distinguishing sheath tissues from blade tissues. The B3 domain-containing transcription factor *NGA1*, which negatively regulates cell proliferation in *Brassica rapa* [37,38], is expressed at a high level in the sheath (*GRMZM2G082227*) tissue only, which may suggest its role in prohibiting the development of sheath tissue into blade tissue. The maize *NTT* (*NO TRANSMITTING TRACT*) homologs *GRMZM2G071101* and *GRMZM2G445684*, two members of the C2H2-type zinc finger family that determine the distal cell fate in the root and influence transmitting tract development and pollen tube growth in *Arabidopsis* [39–41], as well as the homeobox family member *GRMZM2G034113*, the homolog of which (the *AtHB7* gene) is involved in ABA signaling during water stress, were all highly expressed in the sheath tissue and exhibited quite low expression in the blade [42,43]. This suggested that these TFs may play a role in the development or tissue-specific function of the leaf sheath. To confirm the reliability of the RNA-seq data, the expression levels of these four transcription factors at the four blades and sheath maturation points discussed above were also quantified by qRT-PCR (Supplemental Figure S2, Supplemental Table S5). The consistency between the results of the two methods verified the reliability of the enriched expression of these TFs in the leaf sheath.

3. Discussion

Cereal crops, such as maize, are a primary calorie source for human and animal diets, and the most important organs for the conversion of photosynthetic energy into carbon are the leaves. Therefore, understanding leaf development in maize is important for agriculture to further increase maize yields. Several studies have uncovered photosynthetic development in the maize leaf blade [6,7]. However, very few studies have focused on leaf sheath tissues. Consequently, very few sheath-specific genes in maize have been identified and our understanding of sheath development is also limited.

Using a comprehensive transcriptome dataset, we captured the dynamic transcriptome changes that occurred during leaf sheath tissue maturation. In the earliest stage of maturation (stage 1), basic cellular structures were constructed, as the genes associated with cell organization, vesicle transport and DNA synthesis/chromatin structure were highly expressed. In stage 2, as the sheath tissue continued to expand, there was an increase in the expression of the genes required for cell wall biosynthesis, modification, and secondary metabolism. By stage 3, the sheath cells were well primed for photosynthesis, as they expressed genes required for photosynthesis and redox reactions at high levels. In addition to genes required for photosynthesis, by stage 4, there was an increased expression of genes participating in secondary metabolism and lignin biosynthesis, which suggests an increase in the strength of the sheath tissues. Therefore, these four stages of leaf sheath differentiation define distinct steps in the sheath maturation process.

We inferred from the RNA-seq data that different tissues in the maize leaf are highly specialized in their biological functions. By comparing the transcriptome of the leaf blade and sheath and identifying the genes expressed at high or very low levels in the sheath tissue across four stages of maturation, we found that increased expression of photosynthesis-related genes in the leaf blade corresponded with its exclusive function in photosynthesis. We also found increased expression of genes that participate in DNA synthesis/chromatin structure, cell wall maintenance, lipid metabolism and cell organization in the sheath tissue, which are involved in building and strengthening the tissue. Interestingly, several genes were highly expressed, specifically in the sheath tissue (e.g., the *BOP2* homologs *GRMZM2G026556* and *GRMZM2G022606*). The *BOP* gene is required for the patterning of the leaf petiole and blade in *Arabidopsis* [28], and a recent study demonstrated that it activates proximal sheath differentiation and suppresses distal blade differentiation in rice [29], indicating its important role in sheath development. More research needs to be performed to test the functions of *DOGT1* (*GRMZM2G403740*) and *SVB* (*GRMZM2G131409*) in maize, even though their *Arabidopsis* homologs affect hormone activity and influence plant development [30–32].

In our survey, we found that TFs may play important roles in both sheath development and functional specialization. During the early stage of sheath maturation, trihelix family and bHLH family members may participate in cell development, and in the late stage of maturation. When the sheath starts to carry out photosynthesis, the members of the GLK, DOF and CO-like families may play roles in light signaling and chloroplast development. Interestingly, TFs from the B3 domain (*GRMZM2G082227*), C2H2-type zinc finger (*GRMZM2G071101* and *GRMZM2G445684*) and homeobox (*GRMZM2G034113*) families were expressed at a higher level in the sheath tissue than in the blade, which may suggest a role for these transcription factors in sheath development and functional specialization. Taken together, our results provide detailed insight into dynamic changes during maize leaf sheath maturation and serve as a valuable resource for maize functional genomics study.

4. Materials and Methods

4.1. Plant Material and Sampling

Maize inbred B73 were planted in the growth chamber every day for 13 days with a light intensity of 500 μmol/m^2/s, 12:12 L/D, 31 °C L/22 °C D, and 50% relative humidity. The first planting was 13 days, and the blade and sheath of the third leaves were harvested at 2 mm sections above and below the leaf ligule from the seedlings on the 10th (stage 1), 11th (stage 2), 12th (stage 3) and 13th (stage 4) days after planting. Tissue from 10 leaves was harvested as a pool and frozen in liquid nitrogen for RNA extraction and library construction, and 3 biological replicates were harvested at each stage.

4.2. SEM Observation

Scanning electron microscopy was used to observe the cell structure in the maize leaf blade and sheath tissues. Stage 4 seedlings, including the leaf blade, ligule and sheath were fixed in 4% glutaraldehyde in 25 mM sodium phosphate buffer (pH 6.8) at 4 °C for at least 24 h. Then, the samples were rinsed by 25 mM sodium phosphate (pH 6.8) and dehydrated in a graded ethanol series at 4 °C. Thereafter, the samples were immersed in tertiary butyl alcohol and stored in a refrigerator at 4 °C. After sublimation of the frozen tertiary butyl alcohol in a vacuum, the samples were mounted on SEM stubs with double-sided tape, sputter-coated with gold, and examined under a scanning electron microscope (JSM-6610LV, JEOL, Tokyo, Japan).

4.3. RNA-Seq Library Construction and Sequencing

Total RNA was extracted from each sample using the TRIZOL reagent (Invitrogen, Carlsbad, CA, USA). The concentration of the RNA was determined using a DeNovix Spectrophotometer (DeNovix, DS-11, Wilmington, DE, USA), and the RNA quality was determined by 1% agarose gel electrophoresis. The RNA-Seq library construction method of Wang et al. was used [44]. Briefly, the total RNA was purified using the TURBO DNA-free Kit (Ambion, Austin, TX, USA) to completely remove genomic DNA contamination, after that, the poly(A) RNA was isolated from the purified total RNA using poly(T) oligonucleotide-attached magnetic beads (Invitrogen, Carlsbad, CA, USA). Following purification, the mRNA was fragmented into small pieces using divalent cations under elevated temperatures, and the cleaved RNA fragments were reverse-transcribed into first-strand cDNA using reverse transcriptase and random primers. Second-strand cDNA synthesis was performed using DNA polymerase I and RNase H, and the cDNA fragments were processed for end repair, a single adenine base was added, and sequences were ligated to the adapters. These products were then purified and enriched by PCR to create the final cDNA libraries and sequenced on the Illumina Hi-Seq 2500 for single-ended sequencing. Sequencing data have been deposited in the NCBI Sequence Read Archive under the accession number SRP133466.

4.4. Mapping of Reads

The FASTX-toolkit (http://hannonlab.cshl.edu/fastx_toolkit/) was used to remove the adapters of the raw reads with the fastx_clipper setting a parameter of "-a ADAPTER". The sequence quality was examined by FastQC (http://www.bioinformatics.babraham.ac.uk/projects/fastqc/) setting "-t 6 –noextract", and low quality reads were filtered by the FASTX-toolkit setting parameters as "q20p80" (i.e., for each read kept, 80% of bases must have sequence quality greater than 20, which indicates 1% sequencing error rate). The clean reads were mapped to the Maize B73 genome (version 3) obtained from Phytozome using Tophat v2.0.10 (http://tophat.cbcb.umd.edu/) setting -read-mismatches 2 –num-threads 6'. Count data were generated by Cuffdiff embed in Cufflinks pipeline v2.1.1 setting "–num-threads 6 -frag-bias-correct" [45]. The gene expression level was normalized as FPKM, and genes with FPKM > 1 were considered to be expressed.

4.5. Defining Differentially Expressed Genes and Cluster Analysis

Differentially expressed (DE) genes were identified by DESeq v1.34.1 [46] in Bioconductor (http://www.bioconductor.org/), based on a comparison across all the sheath samples (pairwise comparisons with stage 1 as control) or between the blade and sheath in each stage with a false discovery rate (FDR) controlled by the Benjamini and Hochberg method set at 0.001. The MapMan program (http://mapman.gabipd.org) was used to assign genes into functional categories. Clustering analysis of KMC (k-means clustering)—based on the Kendall's Tau distance metric with maximum iterations set as 50—and HCL support trees, based on Kendall's Tau distance metric with average linkage method, was performed through the MEV software (http://www.tm4.org/). Functional enrichment analysis was performed using Fisher's exact test, according to the method by Li et al. [6]. The ID of the DE genes were used to generate a text file. Then, the R software was used to complete the enrichment analysis.

4.6. Quantitative RT-PCR Analysis

cDNA was prepared using the EasyScript One-Step gDNA Removal and cDNA Synthesis SuperMix (TRAN, Beijing, China) and qRT–PCR analyses were conducted using TransStart Tip Green qPCR SuperMix (TRAN, Beijing, China) on a Step One System (Applied Biosystems, Branchburg, NJ, USA). The quantification method ($2^{-\Delta\Delta Ct}$) was used and the variation in expression was estimated using three biological replicates [47]. The maize *Ubi2* (UniProtKB/TrEMBL, Q42415, https://www.uniprot.org/statistics/TrEMBL) gene was used as an internal control to normalize the data. The PCR conditions consisted of an initial denaturation step at 94 °C for 30 s, followed by 40 cycles at 94 °C for 5 s and 60 °C for 30 s.

Supplementary Materials: The following are available online at http://www.mdpi.com/1422-0067/20/10/2472/s1. Figure S1. The original graphs showing the K-mean clusters of the differentially expressed genes (DEGs); Figure S2. qRT-PCR and RNA-seq comparison; Table S1. K-mean clusters of DEGs during sheath maturation; Table S2. Expression of genes involved in cell wall and lignin synthesis; Table S3. Hierarchical clustering (HCL) of transcription factors related to sheath maturation in maize; Table S4. Genes expressed at higher or lower levels in the sheath tissue compared with their expression in the blade tissue in maize across four stages of maturation (FDR < 0.001); Table S5. Primers used for qRT-PCR.

Author Contributions: Formal analysis, L.D., L.Q., Z.D. and R.B.; Software, X.D. and P.L.; Supervision, P.L.; Visualization, X.W.; Writing—original draft, P.L.; Writing—review & editing, Y.C., T.P.B. and P.L. All authors reviewed the manuscript.

Acknowledgments: This work is supported by National Key Research and Development Program of China 2016YFD0101003, NSFC 91435108, as well as the Taishan Pandeng program.

References

1. Duvick, D. Genetic progress in yield of United States maize (*Zea mays* L.). *Maydica* **2005**, *50*, 193.
2. Kong, F.; Zhang, T.; Liu, J.; Heng, S.; Shi, Q.; Zhang, H.; Wang, Z.; Ge, L.; Li, P.; Lu, X.; et al. Regulation of Leaf Angle by Auricle Development in Maize. *Mol. Plant* **2017**, *10*, 516–519. [CrossRef] [PubMed]
3. Lewis, M.W.; Bolduc, N.; Hake, K.; Htike, Y.; Hay, A.; Candela, H.; Hake, S. Gene regulatory interactions at lateral organ boundaries in maize. *Development* **2014**, *141*, 4590–4597. [CrossRef] [PubMed]
4. Lewis, M.W.; Hake, S. Keep on growing: Building and patterning leaves in the grasses. *Curr. Opin. Plant Biol.* **2016**, *29*, 80–86. [CrossRef] [PubMed]
5. Tian, F.; Bradbury, P.J.; Brown, P.J.; Hung, H.; Sun, Q.; Flint-Garcia, S.; Rocheford, T.R.; McMullen, M.D.; Holland, J.B.; Buckler, E.S. Genome-wide association study of leaf architecture in the maize nested association mapping population. *Nat. Genet.* **2011**, *43*, 159–162. [CrossRef] [PubMed]
6. Li, P.; Ponnala, L.; Gandotra, N.; Wang, L.; Si, Y.; Tausta, S.L.; Kebrom, T.H.; Provart, N.; Patel, R.; Myers, C.R.; et al. The developmental dynamics of the maize leaf transcriptome. *Nat. Genet.* **2010**, *42*, 1060–1067. [CrossRef] [PubMed]

7. Wang, L.; Czedik-Eysenberg, A.; Mertz, R.A.; Si, Y.; Tohge, T.; Nunes-Nesi, A.; Arrivault, S.; Dedow, L.K.; Bryant, D.W.; Zhou, W.; et al. Comparative analyses of C(4) and C(3) photosynthesis in developing leaves of maize and rice. *Nat. Biotechnol.* **2014**, *32*, 1158–1165. [CrossRef]

8. Juarez, M.T.; Kui, J.S.; Thomas, J.; Heller, B.A.; Timmermans, M.C.P. microRNA-mediated repression of rolled leaf1 specifies maize leaf polarity. *Nature* **2004**, *428*, 84–88. [CrossRef]

9. Nardmann, J. The maize duplicate genes narrow sheath1 and narrow sheath2 encode a conserved homeobox gene function in a lateral domain of shoot apical meristems. *Development* **2004**, *131*, 2827–2839. [CrossRef]

10. Moreno, M.A.; Harper, L.C.; Krueger, R.W.; Dellaporta, S.L.; Freeling, M. liguleless1 encodes a nuclear-localized protein required for induction of ligules and auricles during maize leaf organogenesis. *Genes Dev.* **1997**, *11*, 616.

11. Becraft, P.W.; Bongard-Pierce, D.K.; Sylvester, A.W.; Poethig, R.S.; Freeling, M. The liguleless-1 gene acts tissue specifically in maize leaf development. *Dev. Biol.* **1990**, *141*, 220–232. [CrossRef]

12. Sylvester, A.W.; Cande, W.Z.; Freeling, M. Division and differentiation during normal and liguleless-1 maize leaf development. *Development* **1990**, *110*, 985–1000.

13. Walsh, J.; Waters, C.A.; Freeling, M. The maize gene liguleless2 encodes a basic leucine zipper protein involved in the establishment of the leaf blade-sheath boundary. *Genes Dev.* **1998**, *12*, 208.

14. Moon, J.; Candela, H.; Hake, S. The Liguleless narrow mutation affects proximal-distal signaling and leaf growth. *Development* **2013**, *140*, 405–412. [CrossRef] [PubMed]

15. Hatfield, R.D.; Marita, J.M. Maize development: Cell wall changes in leaves and sheaths. *Am. J. Plant Sci.* **2017**, *8*, 1248. [CrossRef]

16. Johnston, R.; Wang, M.; Sun, Q.; Sylvester, A.W.; Hake, S.; Scanlon, M.J. Transcriptomic analyses indicate that maize ligule development recapitulates gene expression patterns that occur during lateral organ initiation. *Plant Cell* **2014**, *26*, 4718–4732. [CrossRef]

17. Chen, Y.; Kelly, E.E.; Masluk, R.P.; Nelson, C.L.; Cantu, D.C.; Reilly, P.J. Structural classification and properties of ketoacyl synthases. *Protein Sci.* **2011**, *20*, 1659–1667. [CrossRef]

18. Boerjan, W.; Ralph, J.; Baucher, M. Lignin biosynthesis. *Annu. Rev. Plant Biol.* **2003**, *54*, 519–546. [CrossRef]

19. Barros, J.; Serk, H.; Granlund, I.; Pesquet, E. The cell biology of lignification in higher plants. *Ann. Bot.* **2015**, *115*, 1053. [CrossRef]

20. Waters, M.T.; Moylan, E.C.; Langdale, J.A. GLK transcription factors regulate chloroplast development in a cell-autonomous manner. *Plant J. Cell Mol. Biol.* **2008**, *56*, 432–444. [CrossRef]

21. Song, Y.H.; Ito, S.; Imaizumi, T. Flowering time regulation: Photoperiod- and temperature-sensing in leaves. *Trends Plant Sci.* **2013**, *18*, 575–583. [CrossRef] [PubMed]

22. Datta, S.; Hettiarachchi, G.H.C.M.; Deng, X.-W.; Holm, M. Arabidopsis CONSTANS-LIKE3 Is a Positive Regulator of Red Light Signaling and Root Growth. *Plant Cell* **2006**, *18*, 70–84. [CrossRef] [PubMed]

23. Tripathi, P.; Carvallo, M.; Hamilton, E.E.; Preuss, S.; Kay, S.A. Arabidopsis B-BOX32 interacts with CONSTANS-LIKE3 to regulate flowering. *Proc. Natl. Acad. Sci. USA* **2017**, *114*, 172–177. [CrossRef] [PubMed]

24. Jin, H.; Cominelli, E.; Bailey, P.; Parr, A.; Mehrtens, F.; Jones, J.; Tonelli, C.; Weisshaar, B.; Martin, C. Transcriptional repression by AtMYB4 controls production of UV-protecting sunscreens in Arabidopsis. *Embo J.* **2014**, *19*, 6150–6161. [CrossRef] [PubMed]

25. Zhao, J.; Zhang, W.; Zhao, Y.; Gong, X.; Guo, L.; Zhu, G.; Wang, X.; Gong, Z.; Schumaker, K.S.; Guo, Y. SAD2, an importin -like protein, is required for UV-B response in Arabidopsis by mediating MYB4 nuclear trafficking. *Plant Cell* **2007**, *19*, 3805–3818. [CrossRef]

26. Shalit-Kaneh, A.; Kumimoto, R.W.; Filkov, V.; Harmer, S.L. Multiple feedback loops of the Arabidopsis circadian clock provide rhythmic robustness across environmental conditions. *Proc. Natl. Acad. Sci. USA* **2018**, *115*, 7147–7152. [CrossRef] [PubMed]

27. Pillitteri, L.J.; Bogenschutz, N.L.; Torii, K.U. The bHLH protein, MUTE, controls differentiation of stomata and the hydathode pore in Arabidopsis. *Plant Cell Physiol.* **2008**, *49*, 934–943.

28. Ha, C.M.; Jun, J.H.; Nam, H.G.; Fletcher, J.C. BLADE-ON-PETIOLE 1 and 2 control Arabidopsis lateral organ fate through regulation of LOB domain and adaxial-abaxial polarity genes. *Plant Cell* **2007**, *19*, 1809–1825. [CrossRef]

29. Toriba, T.; Tokunaga, H.; Shiga, T.; Nie, F.; Naramoto, S.; Honda, E.; Tanaka, K.; Taji, T.; Itoh, J.I.; Kyozuka, J. BLADE-ON-PETIOLE genes temporally and developmentally regulate the sheath to blade ratio of rice leaves. *Nat. Commun.* **2019**, *10*, 619. [CrossRef] [PubMed]

30. Poppenberger, B.; Fujioka, S.; Soeno, K.; George, G.L.; Vaistij, F.E.; Hiranuma, S.; Seto, H.; Takatsuto, S.; Adam, G.; Yoshida, S.; et al. The UGT73C5 of Arabidopsis thaliana glucosylates brassinosteroids. *Proc. Natl. Acad. Sci. USA* **2005**, *102*, 15253–15258. [CrossRef]

31. Husar, S.; Berthiller, F.; Fujioka, S.; Rozhon, W.; Khan, M.; Kalaivanan, F.; Elias, L.; Higgins, G.S.; Li, Y.; Schuhmacher, R.; et al. Overexpression of the UGT73C6 alters brassinosteroid glucoside formation in Arabidopsis thaliana. *BMC Plant Biol.* **2011**, *11*, 51. [CrossRef] [PubMed]

32. Marks, M.D.; Wenger, J.P.; Gilding, E.; Jilk, R.; Dixon, R.A. Transcriptome analysis of Arabidopsis wild-type and gl3-sst sim trichomes identifies four additional genes required for trichome development. *Mol. Plant* **2009**, *2*, 803–822. [CrossRef] [PubMed]

33. Beisson, F.; Li, Y.; Bonaventure, G.; Pollard, M.; Ohlrogge, J.B. The acyltransferase GPAT5 is required for the synthesis of suberin in seed coat and root of Arabidopsis. *Plant Cell* **2007**, *19*, 351–368. [CrossRef] [PubMed]

34. Lai, C.P.; Huang, L.M.; Chen, L.O.; Chan, M.T.; Shaw, J.F. Genome-wide analysis of GDSL-type esterases/lipases in Arabidopsis. *Plant Mol. Biol.* **2017**, *95*, 181–197. [CrossRef] [PubMed]

35. Qin, Y.; Leydon, A.R.; Manziello, A.; Pandey, R.; Mount, D.; Denic, S.; Vasic, B.; Johnson, M.A.; Palanivelu, R. Penetration of the stigma and style elicits a novel transcriptome in pollen tubes, pointing to genes critical for growth in a pistil. *PLoS Genet.* **2009**, *5*, e1000621. [CrossRef] [PubMed]

36. James, M.; Poret, M.; Masclaux-Daubresse, C.; Marmagne, A.; Coquet, L.; Jouenne, T.; Chan, P.; Trouverie, J.; Etienne, P. SAG12, a Major Cysteine Protease Involved in Nitrogen Allocation during Senescence for Seed Production in Arabidopsis thaliana. *Plant Cell Physiol.* **2018**, *59*, 2052–2063. [CrossRef] [PubMed]

37. Kwon, S.H.; Chang, S.C.; Ko, J.-H.; Song, J.T.; Kim, J.H. Overexpression of Brassica rapa NGATHA1 Gene Confers De-Etiolation Phenotype and Cytokinin Resistance on Arabidopsis thaliana. *J. Plant Biol.* **2011**, *54*, 119–125. [CrossRef]

38. Lee, B.H.; Kwon, S.H.; Lee, S.J.; Park, S.K.; Song, J.T.; Lee, S.; Lee, M.M.; Hwang, Y.S.; Kim, J.H. The Arabidopsis thaliana NGATHA transcription factors negatively regulate cell proliferation of lateral organs. *Plant Mol. Biol.* **2015**, *89*, 529–538. [CrossRef] [PubMed]

39. Crawford, B.C.; Ditta, G.; Yanofsky, M.F. The NTT gene is required for transmitting-tract development in carpels of Arabidopsis thaliana. *Curr. Biol.* **2007**, *17*, 1101–1108. [CrossRef]

40. Crawford, B.C.; Sewell, J.; Golembeski, G.; Roshan, C.; Long, J.A.; Yanofsky, M.F. Plant development. Genetic control of distal stem cell fate within root and embryonic meristems. *Science* **2015**, *347*, 655–659.

41. Marsch-Martinez, N.; Zuniga-Mayo, V.M.; Herrera-Ubaldo, H.; Ouwerkerk, P.B.; Pablo-Villa, J.; Lozano-Sotomayor, P.; Greco, R.; Ballester, P.; Balanza, V.; Kuijt, S.J.; et al. The NTT transcription factor promotes replum development in Arabidopsis fruits. *Plant J.* **2014**, *80*, 69–81. [CrossRef] [PubMed]

42. Ré, D.A.; Capella, M.; Bonaventure, G.; Chan, R.L. Arabidopsis AtHB7 and AtHB12 evolved divergently to fine tune processes associated with growth and responses to water stress. *BMC Plant Biol.* **2014**, *14*, 1–14. [CrossRef]

43. Olsson, A.; Engström, P.; Söderman, E. The homeobox genes ATHB12 and ATHB7 encode potential regulators of growth in response to water deficit in Arabidopsis. *Plant Mol. Biol.* **2004**, *55*, 663–677. [CrossRef] [PubMed]

44. Lin, W.; Si, Y.; Dedow, L.K.; Ying, S.; Peng, L.; Brutnell, T.P. A Low-Cost Library Construction Protocol and Data Analysis Pipeline for Illumina-Based Strand-Specific Multiplex RNA-Seq. *PLoS ONE* **2011**, *6*, e26426.

45. Trapnell, C.; Roberts, A.; Goff, L.; Pertea, G.; Kim, D.; Kelley, D.R.; Pimentel, H.; Salzberg, S.L.; Rinn, J.L.; Pachter, L. Differential gene and transcript expression analysis of RNA-seq experiments with TopHat and Cufflinks. *Nat. Protoc.* **2012**, *7*, 562–578. [CrossRef] [PubMed]

46. Anders, S.; Huber, W. Differential expression analysis for sequence count data. *Genome Biol.* **2010**, *11*, R106. [CrossRef] [PubMed]

47. Livak, K.J.; Schmittgen, T.D. Analysis of relative gene expression data using real-time quantitative PCR and the $2^{-\Delta\Delta CT}$ method. *Methods* **2001**, *25*, 402–408. [CrossRef] [PubMed]

International Journal of
Molecular Sciences

Article

Early Response of Radish to Heat Stress by Strand-Specific Transcriptome and miRNA Analysis

Zhuang Yang, Wen Li, Xiao Su, Pingfei Ge, Yan Zhou, Yuanyuan Hao, Huangying Shu, Chonglun Gao, Shanhan Cheng, Guopeng Zhu and Zhiwei Wang *

Hainan Key Laboratory for Sustainable Utilization of Tropical Bioresources, College of Horticulture, Hainan University, Haikou 570228, China
* Correspondence: zwwang22@163.com or wangzhiwei@hainanu.edu.cn

Received: 26 May 2019; Accepted: 4 July 2019; Published: 6 July 2019

Abstract: Radish is a crucial vegetable crop of the *Brassicaceae* family with many varieties and large cultivated area in China. Radish is a cool season crop, and there are only a few heat tolerant radish varieties in practical production with little information concerning the related genes in response to heat stress. In this work, some physiological parameter changes of young leaves under short-term heat stress were detected. Furthermore, we acquired 1802 differentially expressed mRNAs (including encoding some heat shock proteins, heat shock factor and heat shock-related transcription factors), 169 differentially expressed lncRNAs and three differentially expressed circRNAs (novel_circ_0000265, novel_circ_0000325 and novel_circ_0000315) through strand-specific RNA sequencing technology. We also found 10 differentially expressed miRNAs (ath-miR159b-3p, athmiR159c, ath-miR398a-3p, athmiR398b-3p, ath-miR165a-5p, ath-miR169g-3p, novel_86, novel_107, novel_21 and ath-miR171b-3p) by small RNA sequencing technology. Through function prediction and enrichment analysis, our results suggested that the significantly possible pathways/complexes related to heat stress in radish leaves were circadian rhythm-plant, photosynthesis—antenna proteins, photosynthesis, carbon fixation in photosynthetic organisms, arginine and proline metabolism, oxidative phosphorylation, peroxisome and plant hormone signal transduction. Besides, we identified one lncRNA–miRNA–mRNAs combination responsive to heat stress. These results will be helpful for further illustration of molecular regulation networks of how radish responds to heat stress.

Keywords: radish; heat stress; transcriptome sequencing; lncRNA; miRNA; physiological response

---

1. Introduction

Due to the significant increase in greenhouse gas emissions from human activities, especially the oxides of carbon dioxide, methane, chlorofluorocarbons and nitrogen, the global warming and the rising temperature have made plants in danger of high-temperature stress [1]. When the ambient temperature is higher than the suitable growth temperature of plants, some catalytic enzymes in the plant slowly lose their activity, resulting in structural damage of cells, which may lead to abnormal growth, flowering and seed yield reduction [2,3].

As sessile organisms, plants have evolved lots of complex regulatory mechanisms to cope with environmental stresses [4,5]. The excess production of singlet oxygen, superoxide anion radical, hydrogen peroxide and hydroxyl, also known as reactive oxygen species (ROS), are induced in plants under abiotic stress, which undermines chloroplasts and even results in cell death [6,7]. To scavenge superfluous ROS, plants have evolved antioxidative enzymes system, including superoxide dismutase (SOD) and peroxidases (POD). Overexpression of SOD in plants can protect the physiological processes of plants by clearing superoxide. The hydrogen peroxide could be eliminated by a variety of POD [8,9]. In addition, membrane lipid peroxidation occurs under heat stress (HS), and malondialdehyde (MDA) is produced. The accumulation of osmotic adjustment substances enhances tolerance to HS as one of the physiological basis of plant heat tolerance. Plant osmotic adjustment substances mainly include amino acids, soluble sugars, soluble proteins and soluble phenols [10]. These studies suggest that the changes in soluble sugar, chlorophyll, free proline and MDA contents, as well as SOD and POD activities under HS, can be used as indicators for the evaluation and screening of crop heat tolerance.

RNA sequencing (RNA-Seq) technique is an effective and vigorous method to investigate potential functional genes and regulatory networks of plant tolerance to various stresses [11,12]. For example, Li et al [5] validated that BRs played a vital role in inducing pepper tolerance to chilling stress at the transcription level. In elite rice, several heat shock factors (HSFs) were identified after heat treatment [13].

Radish (*Raphanus sativus* L.) is an important vegetable crop of the *Brassicaceae* family including many economically important species with a low temperature optimum. It is also used for medicine. Although there are many varieties of radish in China, there is a lack of good varieties suitable for summer growth, which greatly restricts the planting range of radish and the supply of radish in summer. Therefore, studying the response mechanism of radish under heat stress and breeding heat-tolerant cultivars are of great significance.

Recently, the draft genome sequences of radish have been constructed [14], and its genome characteristics and root transcriptome have been reported [15], providing an important condition for analyzing the molecular response of radish to biotic and abiotic stresses including heat shock. Wang et al. [16] identified 6600 differentially expressed genes (DEGs) and three significant pathways in relieving heat shock-induced damages and increasing thermotolerance in radish root under heat shock. Karanja et al. [17] found 172 RsNACs in the radish genome, some of which could be involved in the response to various abiotic stresses. More and more studies have shown that non-coding RNAs (ncRNAs; including lncRNAs, miRNAs and circRNAs) play important roles in organisms [18], and the expression of some genes responsive to stress is regulated by ncRNAs [19,20]. However, the expression profiles of heat-responsive mRNA and ncRNA in radish leaves remain unclear. In this work, we obtained the data of soluble sugar, chlorophyll, free proline and MDA contents as well as SOD and POD activities, and transcriptome including mRNA, non-coding RNA from radish leaves under high temperature stress. The results would provide valuable clues for deep molecular analysis of radish heat response.

2. Results

2.1. Morphological Changes Under Heat Stress

At 40 °C, all seedlings from the cultivar "Huoche" grew well at the initial stage, and then some of them shriveled, withered or died. After recovery treatment for 3 d, the damaged seedlings gradually returned to normal growth. These morphological data indicate that the cultivar "Huoche" has a certain heat tolerance (Figure 1).

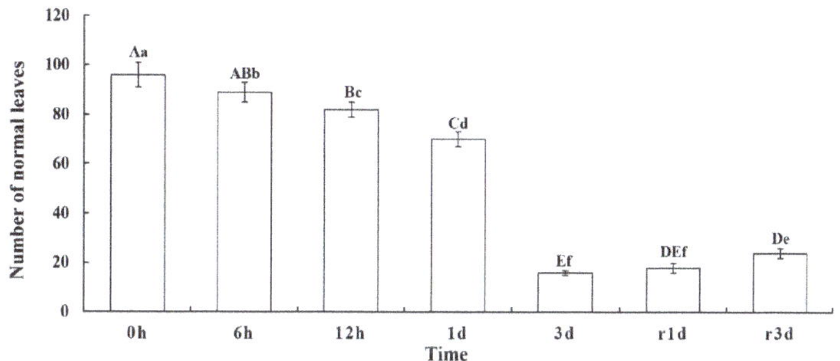

Figure 1. Morphological changes and normal leaf number of radish seedlings under heat stress and recovery treatment. (**A**) The radish seedlings under heat shock for 0 h, 6 h, 12 h, 1 d and 3 d as well as under recovery treatment for 1 d and 3 d. (**B**) The number of normal leaves in "Huoche" under heat shock for 0 h, 6 h, 12 h, 1 d and 3 d as well as under recovery treatment for 1 d and 3 d. Bars with different letters above the columns of figures indicate significant differences at p less than 0.05 (lowercase letters) or less than 0.01 (capital letters).

2.2. Physiological Response to Heat Stress in Leaves Of The Cultivar "Huoche"

POD and SOD activity increased significantly after 6 h of heat stress, and still remained high after 3 d of heat stress. The heat-induced POD and SOD activity declined to control level after recovery (Figure 2A,B). Chlorophyll, soluble sugar and MDA contents were not affected by short-term heat stress, but decreased significantly after 3 d of heat stress, even during the recovery period (Figure 2C–E). We observed that heat stress did not significantly change the content of free proline (Figure 2F). These data suggest that heat stress reduces photosynthesis, but alleviates oxidative damage by increasing antioxidant enzymes in the cultivar "Huoche".

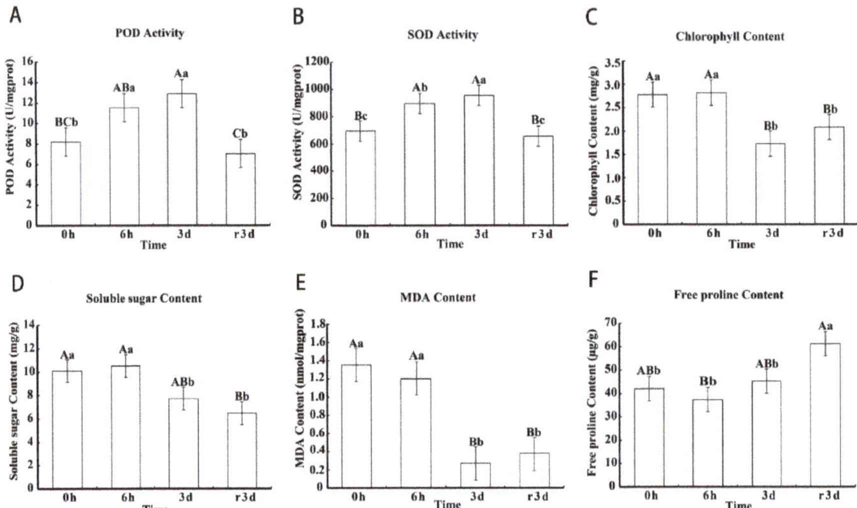

Figure 2. Physiological changes of radish seedlings under heat stress. (**A–F**) The changes in the activities of peroxidases (POD) and superoxide dismutase (SOD), the contents of chlorophyll, soluble sugar, MDA and free proline of the samples in response to heat stress (HS), respectively. Bars with different letters above the columns of figures indicate significant differences at *p* less than 0.05 (lowercase letters) or less than 0.01 (capital letters).

2.3. Mapping and Quantitative Assessment of Illumina Sequence

To comprehensively evaluate how HS influenced transcript profile, six strand-specific RNA libraries (detecting mRNA, lncRNA and circRNA) and six small RNA libraries were constructed. For strand-specific RNA sequencing, the numbers of raw reads of these six samples range from 87.47 to 103.85 million, yielding 12.62 to 15.14 G clean bases for RNA-Seq with a Q20 percentage over 98.06%, Q30 percentage over 94.83%, and a GC percentage between 42.26% and 43.25% (Table S1). For small RNA sequencing, the numbers of raw reads of these six samples range from 13.58 to 14.90 million, yielding 0.679 to 0.745G clean bases for RNA-Seq with a Q20 percentage over 97.73%, Q30 percentage over 94.87%, and a GC percentage between 49.83% and 50.13% (Table S2).

Each library was aligned with the released reference genome as mentioned above. For strand-specific RNA sequencing, over 91.82% clean reads were successfully mapped to the reference genome, in which at least 84.51% and 4.39% clean reads were uniquely mapped and multiple mapped, respectively (Table S3). For small RNA sequencing, over 84.97% reads were successfully mapped to the reference genome, in which at least 61.07% and 22.42% reads were mapped to chains with the same direction of reference sequences and chains with the opposite direction of reference sequences, respectively (Table S4).

2.4. Analysis of DE mRNAs

In QHC06vsQHC00, 1802 differentially expressed (DE) mRNAs (1040 up-regulated and 762 down-regulated) were detected (Figure 3A, Table S5). In the meanwhile, we constructed and analyzed the Venn diagram to compare overlapping relationships between two groups (Figure 3B), and the results indicated that most mRNAs were shared by QHC00 and QHC06.

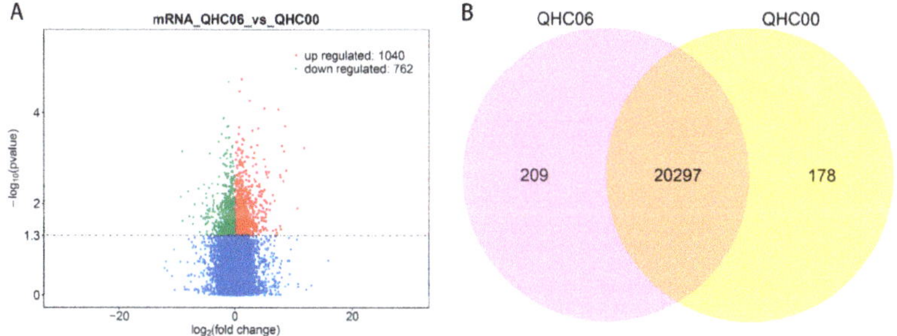

Figure 3. Analysis of mRNA between QHC06 and QHC00. (**A**) Volcano plot of mRNAs in QHC06 vs. QHC00. Significantly differentially expressed mRNAs are represented by red dots (up-regulated) and green dots (down-regulated), whereas non-differentially expressed mRNAs are represented by blue dots. (**B**) Venn diagram of the numbers of expressed mRNAs between QHC06 and QHC00.

2.5. GO and KEGG Enrichment Analysis of DE mRNA Corresponding Genes under HS

Based on gene ontology (GO) analysis of DE mRNA corresponding genes, the significantly enriched biological processes (BPs) were metabolic process (GO: 0008152), cellular process (GO: 0009987), organic substance metabolic process (GO: 0071704) and cellular metabolic process (GO: 0044237), the most significantly enriched cellular components (CCs) were cell (GO: 0005623), cell part (GO: 0044464), intracellular (GO: 0005622) and intracellular part (GO: 0044424). Structural molecule activity (GO: 0005198) and cofactor binding (GO: 0048037) were significantly enriched within the molecular function (MF) categories (Figure 4).

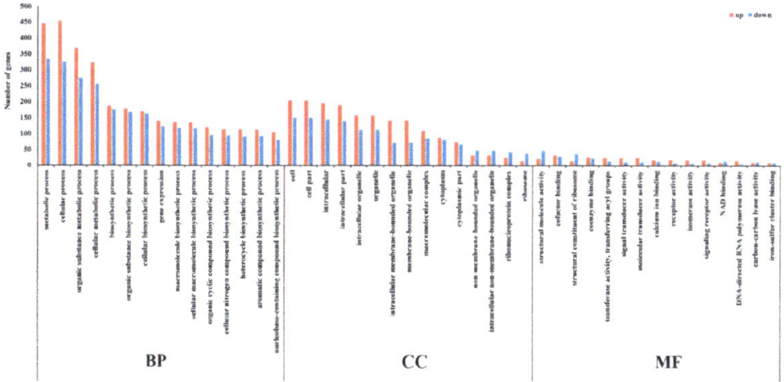

Figure 4. Gene ontology (GO) classification of differentially expressed (DE) mRNA corresponding genes in the cultivar "Huoche" under heat stress. The ordinate is the enriched GO term, and the abscissa is the number of differentially expressed genes in this term. Different colors are used to distinguish biological processes, cellular components, and molecular functions.

With the Kyoto Encyclopedia of Genes and Genome (KEGG) pathway annotation, 108 pathways/complexes were enriched (Table S6) and circadian rhythm-plant (ath04712) was notably enriched in QHC06 vs. QHC00, which might play a crucial part in the response to HS in the cultivar "Huoche" (Figure 5). In this pathway, PHYA (phytochrome A), PRR5 (pseudo-response regulator 5), PRR7 (pseudo-response regulator 7), TOC1 (pseudo-response regulator 1), CK2β (casein kinase II subunit beta) and GI (gigantea) were up-regulated, while COP1 (E3 ubiquitin-protein ligase RFWD2), CHE (transcription factor TCP21 (protein CCA1 HIKING EXPEDITION)), CCA1 (circadian clock associated 1), LHY (MYB-related transcription factor LHY), CDF1 (Dof zinc finger protein DOF5.5) and CK2α (casein kinase II subunit alpha) were down-regulated (Figure 6). Additionally, photosynthesis—antenna proteins (ath00196; Figure 7), photosynthesis (ath00195; Figure 8), carbon fixation in photosynthetic organisms (ath00710; Figure S1), arginine and proline metabolism (ath00330; Figure S2), oxidative phosphorylation (ath00190; Figure S3), peroxisome (ath04146; Figure S4) and plant hormone signal transduction (ath04075; Figure S5) could be also responsive to heat stress.

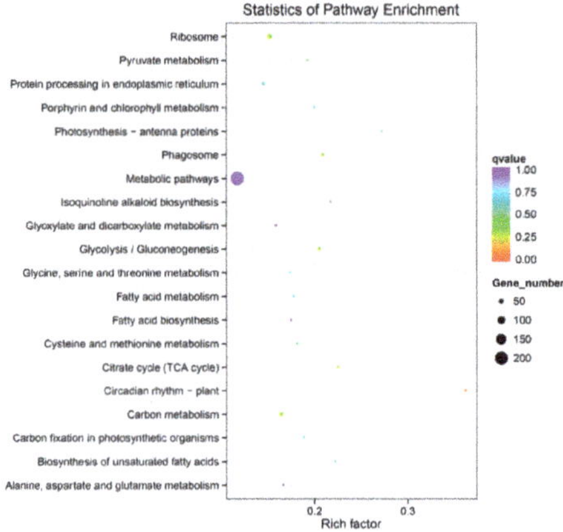

Figure 5. Kyoto Encyclopedia of Genes and Genome (KEGG) enrichment of DE mRNA corresponding genes in the cultivar "Huoche" under heat stress. The vertical axis represents the pathway name, and the horizontal axis represents the rich factor. The size of the point indicates the number of differentially expressed genes in the pathway, and the color of the point corresponds to a different q-value range.

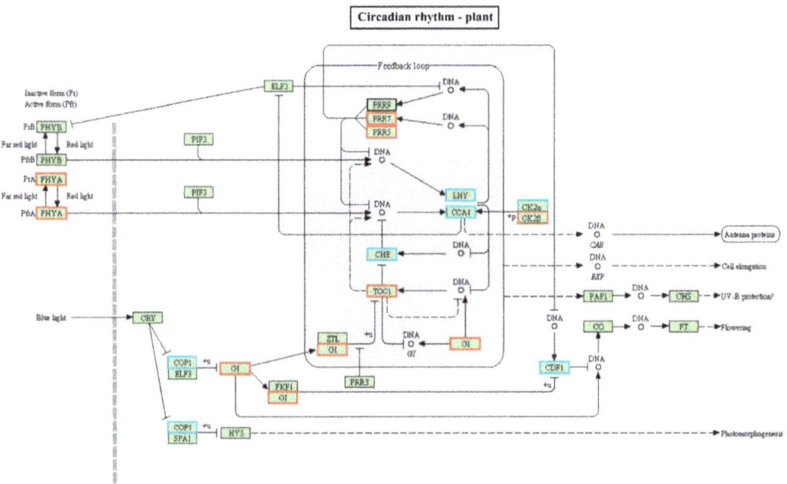

Figure 6. The "circadian rhythm-plant" pathway enriched by KEGG analysis of DE mRNA corresponding genes in the cultivar "Huoche" under heat stress. KEGG orthology (KO) nodes containing up-regulated genes are marked with red boxes, while KO nodes containing down-regulated genes are marked with blue boxes. PHYA, phytochrome A; COP1, E3 ubiquitin-protein ligase RFWD2; GI, gigantea; PRR7, pseudo-response regulator 7; PRR5, pseudo-response regulator 5; CHE, transcription factor TCP21 (protein CCA1 HIKING EXPEDITION); TOC1, pseudo-response regulator 1; LHY, MYB-related transcription factor LHY; CCA1, circadian clock associated 1; CK2α, casein kinase II subunit alpha; CK2β, casein kinase II subunit beta; CDF1, Dof zinc finger protein DOF5.5.

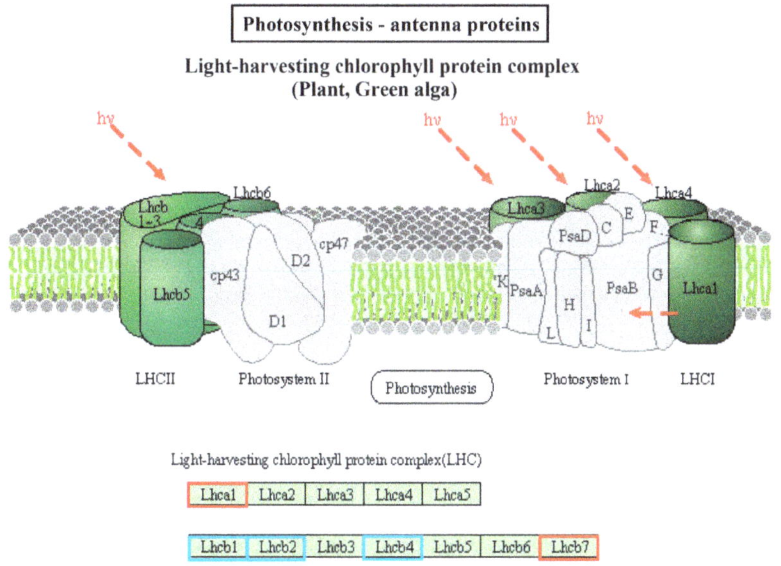

Figure 7. The "photosynthesis—antenna proteins" enriched by KEGG analysis of DE mRNA corresponding genes in the cultivar "Huoche" under heat stress. KO nodes containing up-regulated genes are marked with red boxes, while KO nodes containing down-regulated genes are marked with blue boxes. Lhca1, light-harvesting complex I chlorophyll a/b binding protein 1; Lhcb1, light-harvesting complex II chlorophyll a/b binding protein 1; Lhcb2, light-harvesting complex II chlorophyll a/b binding protein 2; Lhcb4, light-harvesting complex II chlorophyll a/b binding protein 4; Lhcb7, light-harvesting complex II chlorophyll a/b binding protein 7.

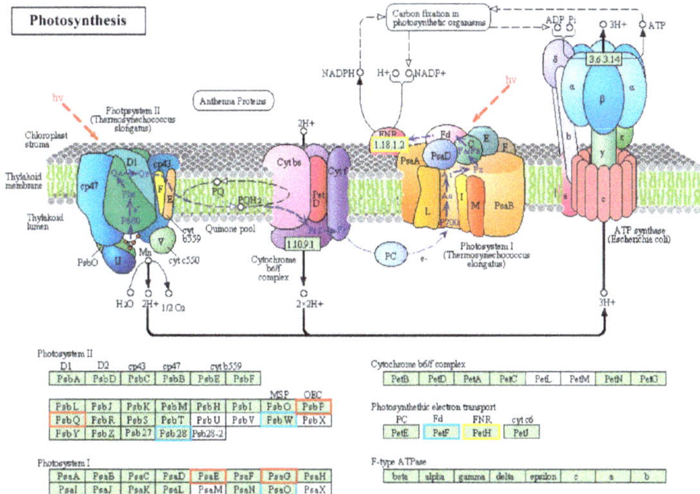

Photosystem II

D1	D2	cp43	cp47	cyt b559	
PsbA	PsbD	PsbC	PsbB	PsbE	PsbF

				MSP	OEC
PsbL	PsbJ	PsbK	PsbM	PsbH	PsbI	PsbO	PsbP
PsbQ	PsbR	PsbS	PsbT	PsbU	PsbV	PsbW	PsbX
PsbY	PsbZ	Psb27	Psb28	Psb28-2			

Photosystem I

PsaA	PsaB	PsaC	PsaD	PsaE	PsaF	PsaG	PsaH
PsaI	PsaJ	PsaK	PsaL	PsaM	PsaN	PsaO	PsaX

Cytochrome b6/f complex

PetB	PetD	PetA	PetC	PetL	PetM	PetN	PetG

Photosynthethic electron transport

PC		Fd		FNR	cyt c6	
PetE		PetF		PetH	PetJ	

F-type ATPase

beta	alpha	gamma	delta	epsilon	c	a	b

Figure 8. The "photosynthesis" pathway enriched by KEGG analysis of DE mRNA corresponding genes in the cultivar "Huoche" under heat stress. KO nodes containing up-regulated genes are marked with red boxes, while KO nodes containing down-regulated genes are marked with blue boxes. PsbP, photosystem II oxygen-evolving enhancer protein 2; PsbQ, photosystem II oxygen-evolving enhancer protein 3; PsbW, photosystem II PsbW protein; Psb28, photosystem II 13kDa protein; PsaE, photosystem I subunit IV; PsaG, photosystem I subunit V; PsaO, photosystem I subunit PsaO; PetF, ferredoxin.

2.6. DE mRNAs Encoding Transcription Factor

In this study, a total of 165 DE mRNA were predicted to associate with 54 TF families (Table S7). The top four abundant types were the Orphans family with 13 DE mRNAs, the MYB family with 11 DE mRNAs, the AP2-EREBP family with 10 DE mRNAs, and the bZIP family with 10 DE mRNAs, respectively.

2.7. Identification of DE mRNA Encoding Heat Shock Protein (HSP) and Heat Shock Factor (HSF)

In order to further excavate the DE mRNA related to HS in the cultivar "Huoche", we found that the heat stress transcription factor A-8-like (LOC108834589), heat shock factor-binding protein 1-like (LOC108833879), hsp70 nucleotide exchange factor fes1 (LOC108859078), hsp70 nucleotide exchange factor fes1-like (LOC108857317), 28 kDa heat- and acid-stable phosphoprotein (LOC108806935) and HEAT repeat-containing protein 5B%2C (LOC108817282) were differentially expressed (Table 1).

Table 1. Selected DE mRNAs related to HS in QHC06 vs. QHC00.

mRNA ID	Gene ID	Regulation	Gene Description
XM_018589933.1	108817282	up	HEAT repeat-containing protein 5B%2C
XM_018607917.1	108834589	up	heat stress transcription factor A-8-like
XM_018632926.1	108859078	up	hsp70 nucleotide exchange factor fes1
XM_018579151.1	108806935	down	28 kDa heat- and acid-stable phosphoprotein
XM_018607274.1	108833879	down	heat shock factor-binding protein 1-like
XM_018631289.1	108857317	down	hsp70 nucleotide exchange factor fes1-like

2.8. Analysis of DE lncRNAs, DE miRNAs and DE circRNAs

Compared with the control group, DE lncRNAs, miRNAs and circRNAs in the heat-treated group were displayed in the forms of a Volcano plot and Venn diagram (Figure 9A–D and Figure 10A,B). The detailed data of the up-regulated and down-regulated lncRNAs between QHC06 and QHC00 are listed in Table S8. All DE miRNAs and circRNAs are listed in Tables 2 and 3, respectively. We identified 169 DE lncRNAs (117 up-regulated and 52 down-regulated), 10 DE miRNAs (two up-regulated and eight down-regulated) and three DE circRNAs (one up-regulated and two down-regulated) between QHC06 and QHC00, respectively.

Table 2. The detailed information of DE miRNAs.

miRNA ID	QHC06 Read Count	QHC00 Read Count	log2 Fold Change	*p*-Value	*q*-Value	Regulation
ath-miR159b-3p	22,398.8097	16761.99	0.41194	0.00079	0.014687	up
ath-miR159c	6837.4573	4883.9258	0.4737	0.00233	0.030311	up
ath-miR398b-3p	284.819304	4018.3814	−2.8401	7.15E−11	9.30E−09	down
ath-miR398a-3p	16.1500394	158.0362	−2.4915	5.62E−09	3.65E−07	down
ath-miR165a-5p	18.7396158	111.39067	−2.1263	2.12E−08	9.20E−07	down
ath-miR169g-3p	2.63494616	22.542957	−2.2541	1.95E−07	6.35E−06	down
novel_86	28.5158186	74.079789	−1.2807	1.34E−06	3.49E−05	down
novel_107	6.19793394	24.899317	−1.6446	1.64E−05	0.000356	down
novel_21	778.011	2374.9094	−1.3037	0.00097	0.01577	down
ath-miR171b-3p	129.729447	229.73667	−0.76804	0.00204	0.029436	down

Table 3. The detailed information of DE circRNAs.

circRNA ID	QHC06 Read Count	QHC00 Read Count	log2 Fold Change	*p*-Value	*q*-Value	Regulation
novel_circ_0000265	202.3891602	76.29972516	1.3908	0.0015051	0.020068	up
novel_circ_0000325	7.966953075	46.82442697	−2.5937	1.09E−05	0.0004377	down
novel_circ_0000315	0	9.680244794	−6.3484	0.000834	0.016679	down

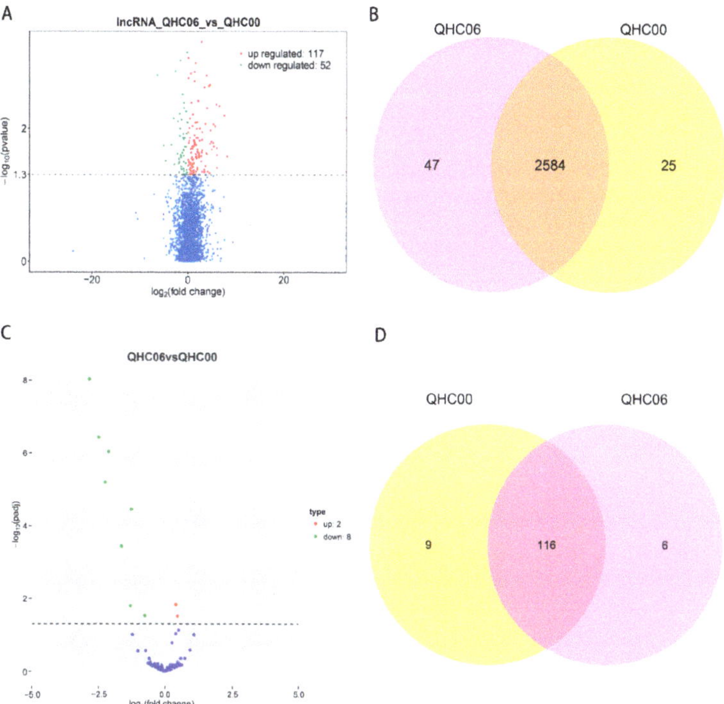

Figure 9. Analysis of lncRNA and miRNA between QHC06 and QHC00. (**A**) Volcano plot of lncRNAs in QHC06 vs. QHC00. Significantly differentially expressed lncRNAs are represented by red dots (up-regulated) and green dots (down-regulated), whereas non-differentially expressed lncRNAs are represented by blue dots. (**B**) Venn diagram of the numbers of expressed lncRNAs between QHC06 and QHC00. (**C**) Volcano plot of miRNAs in QHC06 vs. QHC00. Significantly differentially expressed miRNA are represented by red dots (up-regulated) and green dots (down-regulated), whereas non-differentially expressed genes were represented by blue dots. (**D**) Venn diagram of the numbers of expressed miRNAs between QHC06 and QHC00.

We predicted the biological function of DE lncRNA through its co-location and co-expression with protein-coding genes. Figure 10C showed the Venn diagram of the intersection analysis between DE lncRNA targeted mRNA and DE mRNA. The numbers of up-regulated targeted mRNA of up-regulated lncRNAs, down-regulated targeted mRNA of down-regulated lncRNAs, up-regulated targeted mRNA of down-regulated lncRNAs and down-regulated targeted mRNA of up-regulated lncRNAs were 172, 103, 22 and 53, respectively (Table S9). Interestingly, a heat-related gene (HEAT repeat-containing protein 5B%2C) was also found in the list of up-regulated targeted mRNAs of up-regulated lncRNAs. We took another intersection analysis between DE miRNA targeted mRNA and DE mRNA based on psRNATarget online software and identified 18 up-regulated targeted mRNAs of down-regulated miRNAs and three down-regulated targeted mRNAs of up-regulated miRNAs under HS (Figure 10D, Table 4). These results suggest that these mRNAs would be specially controlled by corresponding lncRNAs or miRNAs in the cultivar "Huoche" in response to heat stress.

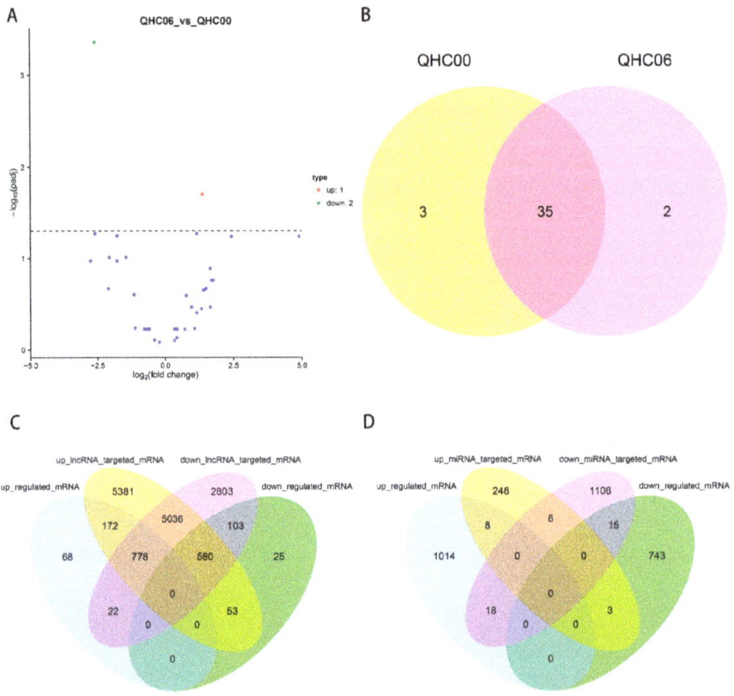

Figure 10. circRNA analysis and comprehensive analysis between QHC06 and QHC00. (**A**) Volcano plot of circRNAs in QHC06 vs. QHC00. Significantly differentially expressed circRNAs are represented by red dots (up-regulated) and green dots (down-regulated), whereas non-differentially expressed circRNAs are represented by blue dots. (**B**) Venn diagram of the numbers of expressed circRNAs between QHC06 and QHC00. (**C**)Venn diagram of the intersection analysis between DE lncRNA targeted mRNA and DE mRNA. (**D**) Venn diagram of the intersection analysis between DE miRNA targeted mRNA and DE mRNA.

Table 4. Information of DE lncRNA and its DE targeted mRNA.

DE miRNA ID	miRNA Regulation	DE Targeted mRNA ID	mRNA Regulation
ath-miR171b-3p	down	XM_018577681.1	up
ath-miR165a-5p	down	XM_018579231.1	up
novel_21	down	XM_018579451.1	up
ath-miR165a-5p	down	XM_018585227.1	up
ath-miR398b-3p, ath-miR398a-3p	down	XM_018585419.1	up
ath-miR165a-5p	down	XM_018587204.1	up
ath-miR169g-3p	down	XM_018589104.1	up
novel_107	down	XM_018590932.1	up
ath-miR169g-3p	down	XM_018596526.1	up
novel_86	down	XM_018598856.1	up
ath-miR171b-3p	down	XM_018599972.1	up
ath-miR398a-3p, ath-miR398b-3p	down	XM_018604024.1	up
novel_86	down	XM_018618012.1	up
novel_107	down	XM_018620431.1	up
ath-miR171b-3p	down	XM_018622399.1	up
ath-miR398a-3p, ath-miR398b-3p	down	XM_018622487.1	up
novel_107	down	XM_018628114.1	up
novel_86	down	XM_018635962.1	up
ath-miR159c	up	XM_018609293.1	down
ath-miR159b-3p	up	XM_018617329.1	down
ath-miR159b-3p, ath-miR159c	up	XM_018634874.1	down

2.9. Functional Prediction of DE lncRNA and DE miRNA in the Cultivar "Huoche"

On the basis of GO enrichment analysis of the targeted mRNA of DE lncRNAs, cellular process (GO: 0009987) and cellular metabolic process (GO: 0044237) were remarkably enriched among the BP categories. The most notably enriched CC was the cellular component (GO: 0005575). Among the CC categories, structural molecule activity (GO: 0005198) and structural constituent of ribosome (GO: 0003735) were significantly enriched (Figure 11). On the basis of the GO enrichment analysis of the targeted mRNA of DE miRNAs, the most remarkably enriched BPs were negative regulation of apoptotic process (GO: 0043066), negative regulation of programmed cell death (GO: 0043069) and negative regulation of cell death (GO: 0060548). Anion binding (GO: 0043168), small molecule binding (GO: 0036094), nucleotide binding (GO: 0000166) and nucleoside phosphate binding (GO: 1901265) were significantly enriched among the MF categories (Figure 12).

We utilized KEGG enrichment analysis to determine the most important biochemical metabolic pathways and signal transduction pathways involved in specific genes. The results showed that ribosome (ath03010) was the significantly enriched KEGG pathways of DE lncRNAs (Figure 13A). The notably enriched KEGG pathways of DE miRNAs were propanoate metabolism (ath00640) and phagosome (ath04145; Figure 13B).

When comprehensively analyzing KEGG pathways among DE mRNA, DE lncRNA and DE miRNA, we found that their enriched pathways had high similarity. For instance, in the "circadian rhythm-plant" pathway, the gene in CCA1 node was enriched both in KEGG pathways of DE mRNA and DE miRNAs. The corresponding genes enriched in the "carbon fixation in photosynthetic organisms", "arginine and proline metabolism", "photosynthesis", "oxidative phosphorylation", "peroxisome" and "plant hormone signal transduction" of DE mRNA were almost enriched in the KEGG analysis of DE lncRNAs. Furthermore, the genes in EC:4.1.2.13 node and in EC:3.5.3.12 node were enriched among KEGG pathways of DE mRNA, DE lncRNA and DE miRNAs in the "carbon fixation in photosynthetic organisms" and "arginine and proline metabolism" pathways, respectively. All these results showed that these pathways might have something to do with thermotolerance of radish leaves.

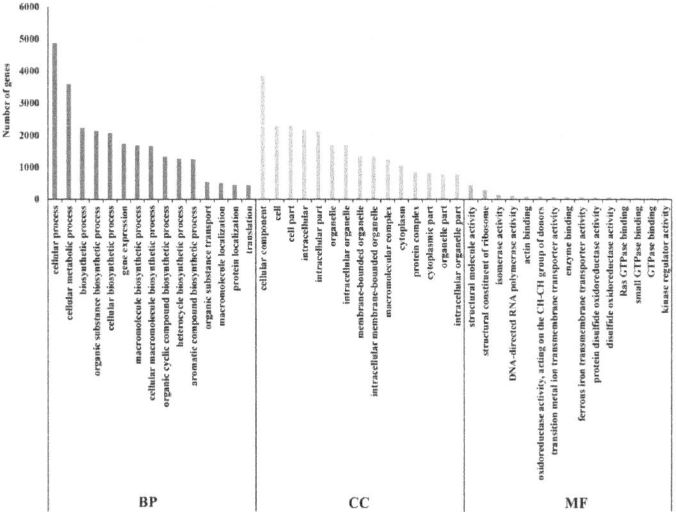

Figure 11. GO analysis of lncRNA in the cultivar "Huoche" under heat stress. The histogram of GO enrichment analysis of the targeted mRNA of DE lncRNAs. The ordinate is the enriched GO term, and the abscissa is the number of differentially expressed genes in this term. Different colors are used to distinguish biological processes, cellular components and molecular functions.

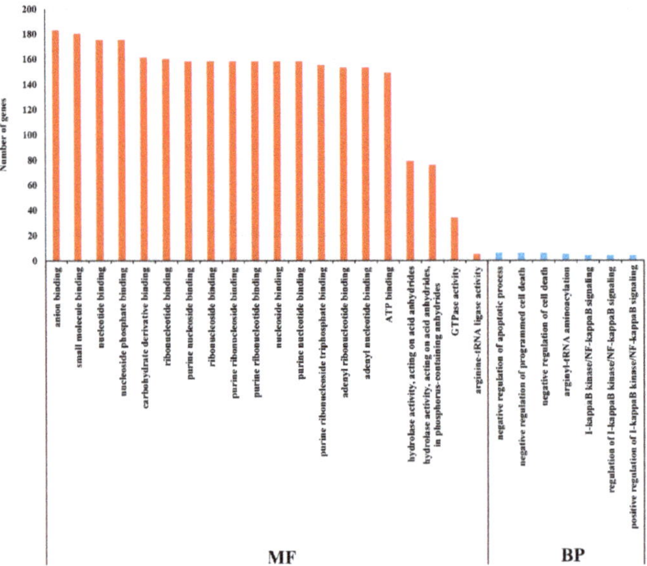

Figure 12. GO analysis of miRNAs in the cultivar "Huoche" under heat stress. The histogram of GO enrichment analysis of DE miRNA targeted mRNAs. The ordinate is the enriched GO term, and the abscissa is the number of differentially expressed genes in this term. Different colors are used to distinguish biological processes, cellular components and molecular functions.

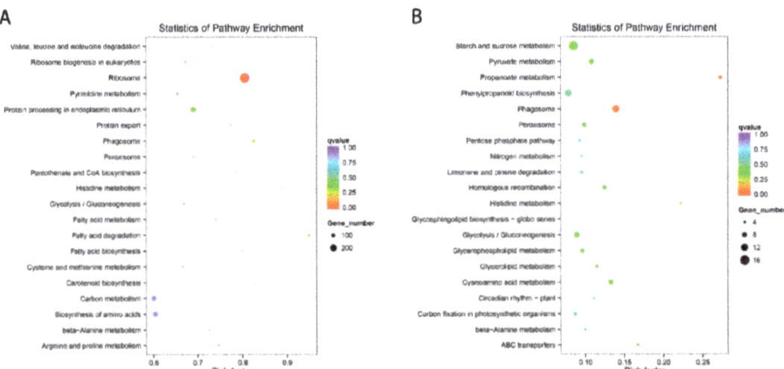

Figure 13. The enriched KEGG pathway scatterplots of non-coding RNAs (ncRNAs) in the cultivar "Huoche" under heat stress. (**A**) and (**B**) KEGG enrichment of lncRNAs and miRNAs in "Huoche" under HS. The vertical axis represents the pathway name, and the horizontal axis represents the Rich factor. The size of the point indicates the number of differentially expressed genes in the pathway, and the color of the point corresponds to a different *q*-value range.

2.10. Regulatory Network in Response to HS

After the construction of lncRNA–miRNA–mRNA combinations, we built a competing endogenous RNA (ceRNA) regulation network with XR_001945247.1 as a decoy, ath-miR165a-5p as a center, XM_018579231.1 and XM_018585227.1 as the target. Since a small number of DE circRNAs were detected, no ceRNA molecular network of circRNA–miRNA–mRNA was constructed.

3. Discussion

Heat stress, as a dominating global concern, has a profound effect on the crop's growth and developments, particularly in agronomic yield [21]. In this study, we found that most of the leaves of the cultivar the cultivar "Huoche" withered and died under the heat stress of 40 °C, and some individual plants could survive. Furthermore, we detected the changes of SOD, POD and MDA under heat stress. To identify HS-related genes, we analyzed the transcriptome of young leave of the cultivar "Huoche" under HS, 1802 DE mRNAs were gained. Furthermore, some lncRNAs and miRNAs were found to be significantly induced under HS, and a regulatory combination of lncRNA–miRNA–mRNA was suggested.

3.1. TFs Response to HS

Transcription factors (TFs) are a group of DNA-binding proteins that regulate the expression of the relative genes, which play a pivotal part in plant tolerance to abiotic stress [22,23].

Although there are little known about their specific functions, a few Orphans transcription factors may regulate the sensitivity of heat shock [24]. MYB participates in response to heat by regulating downstream relative genes in plants [25]. AP2-EREBPs not only function as key regulators in many developmental processes but also are associated with plant responses to abiotic environmental stresses [26,27]. In plants, bZIP participates in many processes, such as stress signaling and plant stress tolerance [28]. In this study, the Orphans, MYB, AP2-EREBP and bZIP families ranked in the top four among all TFs, indicating that they may play key roles in response to HS in radish (Table S6).

Phytochrome interacting factors (PIFs) with bHLH domain have been involved in the signaling mechanism of heat response [29]. Jain et al. [30] identified that HB genes played a critical role in the development and abiotic stress response in rice. Some TCP genes could be significantly up-regulated during the early duration under heat shock in soybean [31]. The expression of *SbTCP19* was restricted by heat in *Sorghum* [32]. Zhang et al. [33] concluded that heat stress could induce the expression of *OsMSR15* in rice, and two C2H2-type zinc finger motifs were the component of *OsMSR15*. Furthermore, heat stress was able to reduce the expression and enzymatic activity of cinnamate 3-hydroxylase (C3H) [34]. In our experiment, we found that bHLH, HB, TCP, C2H2 and C3H transcription factors under HS were significantly differentially regulated, suggesting their potential role of heat tolerance in the radish cultivar "Huoche" (Table S7).

NAM, ATAF and CUC (NAC) transcription factors, a large protein family, could function by helping regulate plant abiotic stress responses, and they could be applied to enhance stress tolerance in plants by genetic engineering [35]. The overexpression of NAC TF JUNGBRUNNEN1 (JUB1; ANAC042) could promote the production of heat shock proteins and improved the thermotolerance of *Arabidopsis thaliana* [36,37]. Similarly, two NAC genes were differentially expressed were detected in our study (Table S7). Chen et al. [38] found that *OsMADS87* might help to improve the thermal resilience of rice. Duan et al. [39] identified that heat stress upregulated five *BrMADS* genes in the Chinese cabbage. We also acquired one up-regulated MADS gene (agamous-like MADS-box protein AGL19%2C) in this work (Table S7).

3.2. HS-Responsive HSPs and HSFs

Heat stress induces the expression of genes encoding heat shock proteins (HSPs) activated by heat shock factors (HSFs), which interact with heat shock elements in the promoter of HSP genes [40]. HSPs are a class of molecular chaperones involved in heat stress [41], including HSP100, HSP90, HSP70, HSP60 and small HSPs (sHSPs) [42]. *HSP70* is involved in the feedback control of HS and relates to the activity of *HSFA1a* [43]. Wang et al. [16] identified four up-regulated and 13 down-regulated *HSP70* transcripts under HS in radish taproots. In this study, after 6 h under HS, one up-regulated *HSP70* component (hsp70 nucleotide exchange factor fes1) and one down-regulated *HSP70* component (hsp70 nucleotide exchange factor fes1-like) were identified in radish leaves (Table 1).

HSFs play crucial roles in heat stress response by mediating the expression of HSPs [44]. On the basis of the structural features of the oligomerization domains, plant HSFs are classified into three categories, i.e., HSFA, HSFB and HSFC [43]. In tomato, *HSFA1* serves as a master regulator of induced thermotolerance [45]. *HSFA3* is controlled by the DREB2A transcription factor and is important for the establishment of heat tolerance in *Arabidopsis* [46]. *TaHsfA6f* is a transcriptional activator regulating some heat stress protection genes in wheat [47]. In the current study, *HSFA-8-like* was up-regulated after 6 h under HS in radish (Table 1). Taken together, HSPs and HSFs play a key role in the response to high temperature, and the further research of radish *HSFA-8-like* is valuable.

3.3. HS-Induced miRNAs

In recent years, more and more ncRNAs have emerged as key regulatory molecules in response to high temperature, whose regulatory mechanisms have been revealed in some plants [48]. miRNAs that were predominantly 20–24 nucleotides function by silencing of target mRNAs [49].

Both Xin et al. [50] and Hivrale et al. [51] found that miR159 was up-regulated in response to HS. The overexpression of miR159 could inhibit MYB transcripts to reduce plant thermotolerance [48,52]. In our study, ath-miR159b-3p and ath-miR159c were also significantly induced under HS (Table 2). We also found that ath-miR169g-3p and ath-miR171b-3p were down-regulated in radish leaves under HS, similar to the results in *Populus tomentosa* [53,54], but miR169 and miR171 were up-regulated in response to HS in *Arabidopsis thaliana* [55,56], suggesting different regulation networks of these two small RNA under HS between *Arabidopsis* and radish. The miRNA miR398 is important for response to different abiotic stresses, especially heat stress. There are two essential target genes of miR398 in *Arabidopsis*, *CSDs* (closely related Cu/Zn-SODs) and *CCS1* (the copper chaperone for SOD) [57,58]. The CSDs could scavenge the superoxide radicals, and its expression is fine-tuned by the cleavage of miR398-directed mRNA. Under HS, CSD and CCS relate to the synthesis of HSF and HSP. Some studies have indicated that down-regulation of miR398 could up-regulate *CSDs* expression, which promotes the accumulation of HSF and HSP, helping plants resist heat stress [56,59]. In the present study, ath-miR398b-3p and ath-miR398a-3p were significantly down-regulated in radish leaves under HS (Table 2), similar to that in *Brassica rapa* and *Populus tomentosa* [54,60]. In addition, we identified that SOD activity in the radish was increased immediately after heat treatment, these results indicated that the *miR398-CSD/CCS* pathway could play a key role in response to HS in radish.

Based on the above data, there may be such an assumption for the response of radish leaf to heat stress: The chlorophyll content of radish leaves decreased under heat stress (Figure 2), and the expression of some genes in photosynthesis pathway and light capture protein complex was affected (Figures 7 and 8). At the same time, circadian rhythm-plant pathway was significantly affected by heat stress, and this pathway had a direct impact on the light capture protein complex (Figure 6). Antioxidant enzymes POD and SOD increased rapidly, and membrane lipid peroxidation marker MDA decreased (Figure 2), which alleviated oxidative stress. Correspondingly, some genes of peroxisome were affected (Figure S4). In addition, plant hormone signal transduction was also significantly affected by high temperature, especially the up-regulation of ABF in ABA transduction pathway (Figure S5), which can regulate stomatal closure and alleviate heat stress. These results provided basic data for further clarifying how radish seedlings respond to heat stress at the molecular level.

4. Materials and Methods

4.1. Plant Materials and HS Treatment

The seeds of the radish cultivar "Huoche" were soaked in water at 25 °C for three days. Germinated seeds were individually sown in plastic pots (35 cm × 22 cm) containing nutrient enriched peat soil in climate chamber with 16 h light at 25 °C and 8 h dark at 15 °C, 70% humidity and 4000 lx light (PQX-330B-30HM, Ningbo Life Technology Co. Ltd., Ningbo, China). Twenty-day-old seedlings were treated by heat shock at 40 °C.

After heat treatment of 0 h (Control) and 6 h, young leaves (the second and third leaf from the top) were collected for strand-specific RNA sequencing and small RNA sequencing (labeling QHC00 and QHC06). For physiological parameter measurement, radish seedlings were heat treated (16 h light at 40 °C) and 8 h dark at 30 °C) for 3 d, and then they were treated by normal temperature

(16 h light at 25 °C and 8 h dark at 15 °C) for 3 d. Three biological repeats for RNA sequencing and physiological analysis were performed, and the collected samples were frozen immediately in liquid nitrogen and stored at −80 °C for further use. QHC001, QHC002 and QHC003 represent the three biological replicates of QHC00. QHC061, QHC062 and QHC063 represent the three biological replicates of QHC06.

4.2. Morphological and Physiological Analysis

According to the manufacturer's protocols (Nanjing Jiancheng Bioengineering Institute, Nanjing, China), the activities of POD, SOD and the contents of free proline, chlorophyll, soluble sugar and MDA of the samples were detected and determined using peroxidase assay kit, copper-zinc superoxide dismutase (CuZn-SOD) assay kit, proline assay kit, chlorophyll assay kit, plant soluble sugar content test kit, malondialdehyde (MDA) assay kit (TBA method), respectively.

4.3. RNA Isolation and RNA-Seq

The frozen samples were sent to Novogene Bioinformatics Technology Co. Ltd. (Beijing, China). Total RNA of each sample was isolated with Trizol reagents under the manufacturer's instruction (Thermo Fisher Scientific, Shanghai, China). Afterward, RNA quantification and qualification were checked with the methods described by Zhang et al. [61].

mRNA, lncRNA and circRNA data were generated by strand-specific sequencing library using the rRNA-depleted RNA by NEBNext® Ultra™ Directional RNA Library Prep Kit for Illumina® with the dUTP second-strand marking (NEB, Ipswich, MA, USA). Small RNA data were generated by NEBNext® Multiplex Small RNA Library Prep Kit for Illumina® (NEB, Ipswich, MA, USA). The detailed protocols were described previously [62,63]. All data of raw reads were uploaded in NCBI Sequence Read Archive (SRA, http://www.ncbi.nlm.nih.gov/Traces/sra/; this SRA submission will be released on 26-05-2020 or upon publication) with accession numbers of SRR8980866 (sRNA-QHC001), SRR8980865 (sRNA-QHC002), SRR8980864 (sRNA-QHC003), SRR8980863 (sRNA-QHC061), SRR8980862(sRNA-QHC062), SRR8980861 (sRNA-QHC063), SRR8980860 (lncRNA-QHC001), SRR8980859 (lncRNA-QHC002), SRR8980868 (lncRNA-QHC003), SRR8980867(lncRNA-QHC061), SRR8980858 (lncRNA-QHC062) and SRR8980857 (lncRNA-QHC063).

The clean reads were aligned with the radish reference genome (Rs1.0, https://www.ncbi.nlm.nih.gov/genome/?term=Raphanus%20sativus; 29-09-2015) using Hisat2 v2. 0.4 [64].

4.4. Differential Expression Analysis

DE mRNAs and DE lncRNAs were determined with the aid of the Ballgown [65]. DESeq2 was used to identified DE miRNAs and DE circRNAs [66].

4.5. GO and KEGG Enrichment Analysis

Gene ontology (GO) annotations and functional enrichment analysis of DE mRNA corresponding genes and DE ncRNA targeted genes were performed on the GO seq R package. GO terms are distributed into biological processes (BP), cellular components (CC) and molecular functions (MF) [67].

KEGG (Kyoto Encyclopedia of Genes and Genome) is the main public database related to the pathway. It is a systematic analysis of gene function and genomic information database, which is helpful for studying genes and expressions as a whole network (https://www.genome.jp/kegg/; 20-04-2018) [68]. By means of KOBAS v2.0 software [69], we detected the statistic enrichment of DE mRNA corresponding genes and DE ncRNA targeted genes in KEGG pathways.

4.6. Analysis of Transcription Factors

Transcription factors (TFs) are a group of proteins that can specifically bind to specific sequences upstream of the 5′ end of the gene, thereby ensuring that the target gene is expressed at a specific time and space at a specific intensity. Plant transcription factor prediction was implemented by iTALK

1.2 software. The basic principle is used to identify TFs by hmmscan using the TF families and rules defined in the database [70].

4.7. Construction of the Regulation Network of Competing Endogenous RNA

RNA can be mutually regulated by competitive binding to a common microRNA response element (MRE), which constitutes competing endogenous RNA (ceRNA). The ceRNA has been found to include protein-encoding mRNAs, non-coding RNAs and pseudogene transcripts [71,72]. Based on psRobot online version software, lncRNA–miRNA–mRNA pairs and circRNA–miRNA–mRNA pairs were screened. Cytoscape 3.7.1 software was adopted to construct the regulation networks between ncRNA and mRNA in response to heat stress.

Supplementary Materials: Supplementary Materials can be found online http://www.mdpi.com/1422-0067/20/13/3321/s1. **Table S1**: Quality evaluation of sample sequencing output data for strand-specific RNA libraries. **Table S2**: Quality evaluation of sample sequencing output data for small RNA libraries. **Table S3**: Reads of strand-specific RNA libraries and reference genome comparison list. **Table S4**: Reads of small RNA libraries and reference genome comparison list. **Table S5**: The detailed information of DE mRNAs. **Table S6**: KEGG analysis results of DE mRNA corresponding genes. **Table S7**: DE mRNAs encoding transcription factor. **Table S8**: The detailed information of DE lncRNAs between QHC06 and QHC00. **Table S9**: The detailed information between DE lncRNA and its DE targeted mRNA. **Figure S1**: The "carbon fixation in photosynthetic organisms" pathway enriched by KEGG analysis of DE mRNA corresponding genes in the cultivar "Huoche" under heat stress. KO nodes containing up-regulated genes are marked with red boxes, while KO nodes containing down-regulated genes are marked with blue boxes. EC:4.1.2.13, fructose-bisphosphate aldolase, class I; EC:3.1.3.11, fructose-1,6-bisphosphatase I; EC:2.7.2.3, phosphoglycerate kinase; EC:4.1.1.39, ribulose-bisphosphate carboxylase small chain; EC:4.1.1.31, phosphoenolpyruvate carboxylase. **Figure S2**: The "arginine and proline metabolism" pathway enriched by KEGG analysis of DE mRNA corresponding genes in the cultivar "Huoche" under heat stress. KO nodes containing up-regulated genes are marked with red boxes, while KO nodes containing down-regulated genes are marked with blue boxes. EC:6.3.1.2, glutamine synthetase; EC:1.4.1.4, glutamate dehydrogenase (NADP+); EC:1.2.1.38, N-acetyl-gamma-glutamyl-phosphate reductase; EC:1.14.11.2, prolyl 4-hydroxylase; EC:4.1.1.50, S-adenosylmethionine decarboxylase; EC:3.5.3.12, agmatine deiminase; EC:2.6.1.1, aspartate aminotransferase, cytoplasmic; EC:4.1.1.19, arginine decarboxylase. **Figure S3**: The "oxidative phosphorylation" pathway enriched by KEGG analysis of DE mRNA corresponding genes in the cultivar "Huoche" under heat stress. KO nodes containing up-regulated genes are marked with red boxes, while KO nodes containing down-regulated genes are marked with blue boxes. EC:3.6.1.1, inorganic pyrophosphatase; Ndufs4, NADH dehydrogenase (ubiquinone) Fe-S protein 4; Ndufs5, NADH dehydrogenase (ubiquinone) Fe-S protein 5; Ndufs6, NADH dehydrogenase (ubiquinone) Fe-S protein 6; Ndufs7, NADH dehydrogenase (ubiquinone) Fe-S protein 7; Ndufv1, NADH dehydrogenase (ubiquinone) flavoprotein 1; Ndufa2, NADH dehydrogenase (ubiquinone) 1 alpha subcomplex subunit 2; Ndufa5, NADH dehydrogenase (ubiquinone) 1 alpha subcomplex subunit 5; Ndufa6, NADH dehydrogenase (ubiquinone) 1 alpha subcomplex subunit 6; Ndufa8, NADH dehydrogenase (ubiquinone) 1 alpha subcomplex subunit 8; COX10, heme o synthase; COX17, cytochrome c oxidase assembly protein subunit 17; D, V-type H+-transporting ATPase subunit D; F, V-type H+-transporting ATPase subunit F; d, V-type H+-transporting ATPase subunit d. Figure S4: The "peroxisome" enriched by KEGG analysis of DE mRNA corresponding genes in the cultivar "Huoche" under heat stress. KO nodes containing up-regulated genes are marked with red boxes, while KO nodes containing down-regulated genes are marked with blue boxes. PXMP2, peroxisomal membrane protein 2; MPV17, protein Mpv17; ACAA1, acetyl-CoA acyltransferase 1; FAR, alcohol-forming fatty acyl-CoA reductase; AGT, alanine-glyoxylate transaminase/serine-glyoxylate transaminase/serine-pyruvate transaminase. **Figure S5**: The "plant hormone signal transduction" pathway enriched by KEGG analysis of DE mRNA corresponding genes in the cultivar "Huoche" under heat stress. KO nodes containing up-regulated genes are marked with red boxes, while KO nodes containing down-regulated genes are marked with blue boxes. AUX1, auxin influx carrier (AUX1 LAX family); ARF, auxin response factor; GH3, auxin responsive GH3 gene family; SAUR, SAUR family protein; A-ARR, two-component response regulator ARR-A family; PYR/PYL, abscisic acid receptor PYR/PYL family; PP2C, protein phosphatase 2C; ABF, ABA responsive element binding factor; EBF1/2, EIN3-binding F-box protein; EIN3, ethylene-insensitive protein 3; BSK, BR-signaling kinase; CYCD3, cyclin D3, plant; JAZ, jasmonate ZIM domain-containing protein.

Author Contributions: Conceptualization, Z.W. and Z.Y.; methodology, Z.W. and Z.Y.; software, Z.Y.; validation, Z.Y., W.L., X.S., P.G. and Y.Z.; investigation, Z.Y., P.G., Y.Z., Y.H., H.S., C.G., S.C. and G.Z.; writing—original draft preparation, Z.Y.; writing—review and editing, Z.W., Z.Y., W.L., X.S., P.G., Y.Z., Y.H., H.S., C.G., S.C. and G.Z.; supervision, Z.W.

Funding: This research was funded by the National Natural Science Foundation of China, grant number 31660091 and 31470412.

Acknowledgments: Great thanks to anonymous reviewers for their valuable comments.

Conflicts of Interest: The authors declare no conflict of interest.

Abbreviations

ROS	Reactive oxygen species
SOD	Superoxide dismutase
POD	Peroxidases
HS	Heat stress
MDA	Malondialdehyde
RNA-Seq	RNA sequencing
DEGs	Differentially expressed genes
ncRNAs	Non-coding RNAs
r1d	Recovery treatment for 1 d
r3d	Recovery treatment for 3 d
DE	Differentially expressed
GO	Gene Ontology
BP	Biological processes
CC	Cellular components
MF	Molecular functions
KEGG	Kyoto Encyclopedia of Genes and Genome
PHYA	Phytochrome A
PRR5	Pseudo-response regulator 5
PRR7	Pseudo-response regulator 7
TOC1	Pseudo-response regulator 1
CK2β	Casein kinase II subunit beta
GI	Gigantea
COP1	E3 ubiquitin-protein ligase RFWD2
CHE	Transcription factor TCP21 (protein CCA1 HIKING EXPEDITION)
CCA1	Circadian clock associated 1
LHY	MYB-related transcription factor LHY
CDF1	Dof zinc finger protein DOF5.5
CK2α	Casein kinase II subunit alpha
KO	KEGG orthology
Lhca1	Light-harvesting complex I chlorophyll a/b binding protein 1
Lhcb1	Light-harvesting complex II chlorophyll a/b binding protein 1
Lhcb2	Light-harvesting complex II chlorophyll a/b binding protein 2
Lhcb4	Light-harvesting complex II chlorophyll a/b binding protein 4
Lhcb7	Light-harvesting complex II chlorophyll a/b binding protein 7
PsbP	Photosystem II oxygen-evolving enhancer protein 2
PsbQ	Photosystem II oxygen-evolving enhancer protein 3
PsbW	Photosystem II PsbW protein
Psb28	Photosystem II 13kDa protein
PsaE	Photosystem I subunit IV
PsaG	Photosystem I subunit V
PsaO	Photosystem I subunit PsaO
PetF	ferredoxin
TFs	Transcription factors
ceRNA	Competing endogenous RNA
HSP	Heat shock protein
HSF	Heat shock factor
PIFs	Phytochrome interacting factors
C3H	Cinnamate 3-hydroxylase
NAC	NAM, ATAF, and CUC
sHSPs	Small HSPs
CuZn-SOD	Copper-zinc superoxide dismutase
MRE	MicroRNA response element
Ndufs4	NADH dehydrogenase (ubiquinone) Fe-S protein 4

Ndufs5	NADH dehydrogenase (ubiquinone) Fe-S protein 5
Ndufs6	NADH dehydrogenase (ubiquinone) Fe-S protein 6
Ndufs7	NADH dehydrogenase (ubiquinone) Fe-S protein 7
Ndufv1	NADH dehydrogenase (ubiquinone) flavoprotein 1
Ndufa2	NADH dehydrogenase (ubiquinone) 1 alpha subcomplex subunit 2
Ndufa5	NADH dehydrogenase (ubiquinone) 1 alpha subcomplex subunit 5
Ndufa6	NADH dehydrogenase (ubiquinone) 1 alpha subcomplex subunit 6
Ndufa8	NADH dehydrogenase (ubiquinone) 1 alpha subcomplex subunit 8
COX10	Heme o synthase
COX17	Cytochrome c oxidase assembly protein subunit 17
PXMP2	Peroxisomal membrane protein 2
MPV17	Protein Mpv17
ACAA1	Acetyl-CoA acyltransferase 1
FAR	Alcohol-forming fatty acyl-CoA reductase
AGT	Alanine-glyoxylate transaminase/serine-glyoxylate transaminase/serine-pyruvate Transaminase
AUX1	Auxin influx carrier (AUX1 LAX family)
ARF	Auxin response factor
GH3	Auxin responsive GH3 gene family
SAUR	SAUR family protein
A-ARR	Two-component response regulator ARR-A family
PYR/PYL	Abscisic acid receptor PYR/PYL family
PP2C	Protein phosphatase 2C
ABF	ABA responsive element binding factor
EBF1/2	EIN3-binding F-box protein
EIN3	Ethylene-insensitive protein 3
BSK	BR-signaling kinase
CYCD3	Cyclin D3, plant
JAZ	Jasmonate ZIM domain-containing protein

References

1. Song, L.; Chow, W.S.; Sun, L.; Li, C.; Peng, C. Acclimation of photosystem II to high temperature in two *Wedelia* species from different geographical origins: Implications for biological invasions upon global warming. *J. Exp. Bot.* **2010**, *61*, 4087–4096. [CrossRef] [PubMed]

2. Miller, G.A.D.; Mittler, R.O.N. Could heat shock transcription factors function as hydrogen peroxide sensors in plants? *Ann. Bot.* **2006**, *98*, 279–288. [CrossRef] [PubMed]

3. Kotak, S.; Larkindale, J.; Lee, U.; von Koskull-Döring, P.; Vierling, E.; Scharf, K.D. Complexity of the heat stress response in plants. *Curr. Opin. Plant Biol.* **2007**, *10*, 310–316. [CrossRef] [PubMed]

4. Wang, M.; Zou, Z.; Li, Q.; Xin, H.; Zhu, X.; Chen, X.; Li, X. Heterologous expression of three *Camellia sinensis* small heat shock protein genes confers temperature stress tolerance in yeast and *Arabidopsis thaliana*. *Plant Cell Rep.* **2017**, *36*, 1125–1135. [CrossRef] [PubMed]

5. Li, J.; Yang, P.; Kang, J.; Gan, Y.; Yu, J.; Calderón-Urrea, A.; Lyu, J.; Zhang, G.; Feng, Z.; Xie, J. Transcriptome analysis of pepper (*Capsicum annuum*) revealed a role of 24-epibrassinolide in response to chilling. *Front. Plant Sci.* **2016**, *7*, 1281. [CrossRef]

6. Gill, S.S.; Tuteja, N. Reactive oxygen species and antioxidant machinery in abiotic stress tolerance in crop plants. *Plant Physiol. Biochem.* **2010**, *48*, 909–930. [CrossRef] [PubMed]

7. Anwar, A.; Yan, Y.; Liu, Y.; Li, Y.; Yu, X. 5-aminolevulinic acid improves nutrient uptake and endogenous hormone accumulation, enhancing low-temperature stress tolerance in cucumbers. *Int. J. Mol. Sci.* **2018**, *19*, 3379. [CrossRef]

8. Mittler, R. Oxidative stress, antioxidants and stress tolerance. *Trends Plant Sci.* **2002**, *7*, 405–410. [CrossRef]

9. Meloni, D.A.; Oliva, M.A.; Martinez, C.A.; Cambraia, J. Photosynthesis and activity of superoxide dismutase, peroxidase and glutathione reductase in cotton under salt stress. *Environ. Exp. Bot.* **2003**, *49*, 69–76. [CrossRef]

10. Chaitanya, K.V.; Sundar, D.; Reddy, A.R. Mulberry leaf metabolism under high temperature stress. *Biol. Plant* **2001**, *44*, 379–384. [CrossRef]

11. Marioni, J.C.; Mason, C.E.; Mane, S.M.; Stephens, M.; Gilad, Y. RNA-seq: An assessment of technical reproducibility and comparison with gene expression arrays. *Genome Res.* **2008**, *18*, 1509–1517. [CrossRef] [PubMed]

12. Wang, Z.; Gerstein, M.; Snyder, M. RNA-Seq: A revolutionary tool for transcriptomics. *Nat. Rev. Genet.* **2009**, *10*, 57. [CrossRef] [PubMed]

13. Fu, C.; Wang, F.; Liu, W.; Liu, D.; Li, J.; Zhu, M.; Liao, Y.; Liu, Z.; Huang, H.; Zeng, X.; et al. Transcriptomic analysis reveals new insights into high-temperature-dependent glume-unclosing in an elite rice male sterile line. *Front. Plant Sci.* **2017**, *8*, 112. [CrossRef] [PubMed]

14. Kitashiba, H.; Li, F.; Hirakawa, H.; Kawanabe, T.; Zou, Z.; Hasegawa, Y.; Tonosaki, K.; Shirasawa, S.; Fukushima, A.; Yokoi, S.; et al. Draft sequences of the radish (*Raphanus sativus* L.) genome. *DNA Res.* **2014**, *21*, 481–490. [CrossRef] [PubMed]

15. Mitsui, Y.; Shimomura, M.; Komatsu, K.; Namiki, N.; Shibata-Hatta, M.; Imai, M.; Katayose, Y.; Mukai, Y.; Kanamori, H.; Kurita, K.; et al. The radish genome and comprehensive gene expression profile of tuberous root formation and development. *Sci. Rep.* **2015**, *5*, 10835. [CrossRef] [PubMed]

16. Wang, R.; Mei, Y.; Xu, L.; Zhu, X.; Wang, Y.; Guo, J.; Liu, L. Genome-wide characterization of differentially expressed genes provides insights into regulatory network of heat stress response in radish (*Raphanus sativus* L.). *Funct. Integr. Genom.* **2018**, *18*, 225–239. [CrossRef] [PubMed]

17. Karanja, B.K.; Xu, L.; Wang, Y.; Muleke, E.M.m.; Jabir, B.M.; Xie, Y.; Zhu, X.; Cheng, W.; Liu, L. Genome-wide characterization and expression profiling of *NAC* transcription factor genes under abiotic stresses in radish (*Raphanus sativus* L.). *PeerJ* **2017**, *5*, 4172. [CrossRef] [PubMed]

18. Wang, Z.; Wu, Z.; Raitskin, O.; Sun, Q.; Dean, C. Antisense-mediated *FLC* transcriptional repression requires the P-TEFb transcription elongation factor. *Proc. Natl. Acad. Sci. USA* **2014**, *111*, 7468–7473. [CrossRef] [PubMed]

19. Contreras-Cubas, C.; Palomar, M.; Arteaga-Vázquez, M.; Reyes, J.L.; Covarrubias, A.A. Non-coding RNAs in the plant response to abiotic stress. *Planta* **2012**, *236*, 943–958. [CrossRef]

20. Nejat, N.; Mantri, N. Emerging roles of long non-coding RNAs in plant response to biotic and abiotic stresses. *Crit. Rev. Biotechnol.* **2018**, *38*, 93–105. [CrossRef]

21. Li, B.; Gao, K.; Ren, H.; Tang, W. Molecular mechanisms governing plant responses to high temperatures. *J. Integr. Plant Biol.* **2018**, *60*, 757–779. [CrossRef] [PubMed]

22. Nakashima, K.; Yamaguchi-Shinozaki, K.; Shinozaki, K. The transcriptional regulatory network in the drought response and its crosstalk in abiotic stress responses including drought, cold, and heat. *Front. Plant Sci.* **2014**, *5*, 170. [CrossRef] [PubMed]

23. Essebier, A.; Lamprecht, M.; Piper, M.; Bodén, M. Bioinformatics approaches to predict target genes from transcription factor binding data. *Methods* **2017**, *131*, 111–119. [CrossRef] [PubMed]

24. Jones, D.L.; Petty, J.; Hoyle, D.C.; Hayes, A.; Oliver, S.G.; Riba-Garcia, I.; Gaskell, S.J.; Stateva, L. Genome-wide analysis of the effects of heat shock on a *Saccharomyces cerevisiae* mutant with a constitutively activated cAMP-dependent pathway. *Comp. Funct. Genom.* **2004**, *5*, 419–431. [CrossRef] [PubMed]

25. Lim, C.J.; Yang, K.A.; Hong, J.K.; Choi, J.S.; Yun, D.J.; Hong, J.C.; Chung, W.S.; Lee, S.Y.; Cho, M.J.; Lim, C.O. Gene expression profiles during heat acclimation in *Arabidopsis thaliana* suspension-culture cells. *J. Plant Res.* **2006**, *119*, 373–383. [CrossRef]

26. Riechmann, J.L.; Meyerowitz, J.L. The AP2/EREBP family of plant transcription factors. *Biol. Chem.* **1998**, *379*, 633. [CrossRef] [PubMed]

27. Tang, M.; Liu, X.; Deng, H.; Shen, S. Over-expression of *JcDREB*, a putative AP2/EREBP domain-containing transcription factor gene in woody biodiesel plant *Jatropha curcas*, enhances salt and freezing tolerance in transgenic *Arabidopsis thaliana*. *Plant Sci.* **2011**, *181*, 623–631. [CrossRef]

28. Zhao, J.; Zhang, X.; Wollenweber, B.; Jiang, D.; Liu, F. Water deficits and heat shock effects on photosynthesis of a transgenic *Arabidopsis thaliana* constitutively expressing *ABP9*, a bZIP transcription factor. *J. Exp. Bot.* **2008**, *59*, 839–848. [CrossRef]

29. Bita, C.; Gerats, T. Plant tolerance to high temperature in a changing environment: Scientific fundamentals and production of heat stress-tolerant crops. *Front. Plant Sci.* **2013**, *4*, 273. [CrossRef]

30. Jain, M.; Tyagi, A.K.; Khurana, J.P. Genome-wide identification, classification, evolutionary expansion and expression analyses of homeobox genes in rice. *FEBS J.* **2008**, *275*, 2845–2861. [CrossRef]

31. Feng, Z.J.; Xu, S.C.; Liu, N.; Zhang, G.W.; Hu, Q.Z.; Gong, Y.M. Soybean TCP transcription factors: Evolution, classification, protein interaction and stress and hormone responsiveness. *Plant Physiol. Biochem.* **2018**, *127*, 129–142. [CrossRef] [PubMed]

32. Francis, A.; Dhaka, N.; Bakshi, M.; Jung, K.H.; Sharma, M.K.; Sharma, R. Comparative phylogenomic analysis provides insights into TCP gene functions in *Sorghum*. *Sci. Rep.* **2016**, *6*, 38488. [CrossRef] [PubMed]

33. Zhang, X.; Zhang, B.; Li, M.J.; Yin, X.M.; Huang, L.F.; Cui, Y.C.; Wang, M.L.; Xia, X. *OsMSR15* encoding a rice C2H2-type zinc finger protein confers enhanced drought tolerance in transgenic *Arabidopsis*. *J. Plant Biol.* **2016**, *59*, 271–281. [CrossRef]

34. Martinez, V.; Mestre, T.C.; Rubio, F.; Girones-Vilaplana, A.; Moreno, D.A.; Mittler, R.; Rivero, R.M. Accumulation of flavonols over hydroxycinnamic acids favors oxidative damage protection under abiotic stress. *Front. Plant Sci.* **2016**, *7*, 838. [CrossRef] [PubMed]

35. Shao, H.; Wang, H.; Tang, X. NAC transcription factors in plant multiple abiotic stress responses: Progress and prospects. *Front. Plant Sci.* **2015**, *6*, 902. [CrossRef]

36. Shahnejat-Bushehri, S.; Mueller-Roeber, B.; Balazadeh, S. *Arabidopsis* NAC transcription factor JUNGBRUNNEN1 affects thermomemory-associated genes and enhances heat stress tolerance in primed and unprimed conditions. *Plant Signal. Behav.* **2012**, *7*, 1518–1521. [CrossRef] [PubMed]

37. Wu, A.; Allu, A.D.; Garapati, P.; Siddiqui, H.; Dortay, H.; Zanor, M.-I.; Asensi-Fabado, M.A.; Munné-Bosch, S.; Antonio, C.; Tohge, T.; et al. *JUNGBRUNNEN1*, a reactive oxygen species–responsive NAC transcription factor, regulates longevity in *Arabidopsis*. *Plant Cell* **2012**, *24*, 482–506. [CrossRef]

38. Chen, C.; Begcy, K.; Liu, K.; Folsom, J.J.; Wang, Z.; Zhang, C.; Walia, H. Heat stress yields a unique MADS box transcription factor in determining seed size and thermal sensitivity. *Plant Physiol.* **2016**, *171*, 606–622. [CrossRef]

39. Duan, W.; Song, X.; Liu, T.; Huang, Z.; Ren, J.; Hou, X.; Li, Y. Genome-wide analysis of the MADS-box gene family in *Brassica rapa* (Chinese cabbage). *Mol. Genet. Genom.* **2015**, *290*, 239–255. [CrossRef]

40. Chauhan, H.; Khurana, N.; Agarwal, P.; Khurana, P. Heat shock factors in rice (*Oryza sativa* L.): Genome-wide expression analysis during reproductive development and abiotic stress. *Mol. Genet. Genom.* **2011**, *286*, 171. [CrossRef]

41. Wang, W.; Vinocur, B.; Shoseyov, O.; Altman, A. Role of plant heat-shock proteins and molecular chaperones in the abiotic stress response. *Trends Plant Sci.* **2004**, *9*, 244–252. [CrossRef] [PubMed]

42. Al-Whaibi, M.H. Plant heat-shock proteins: A mini review. *J. King Saud Univ. Sci.* **2011**, *23*, 139–150. [CrossRef]

43. Qu, A.L.; Ding, Y.F.; Jiang, Q.; Zhu, C. Molecular mechanisms of the plant heat stress response. *Biochem. Biophys. Res. Commun.* **2013**, *432*, 203–207. [CrossRef] [PubMed]

44. Guo, M.; Liu, J.H.; Ma, X.; Luo, D.X.; Gong, Z.H.; Lu, M.H. The plant heat stress transcription factors (hsfs): Structure, regulation, and function in response to abiotic stresses. *Front. Plant Sci.* **2016**, *7*, 114. [CrossRef] [PubMed]

45. Mishra, S.K.; Tripp, J.; Winkelhaus, S.; Tschiersch, B.; Theres, K.; Nover, L.; Scharf, K.D. In the complex family of heat stress transcription factors, HsfA1 has a unique role as master regulator of thermotolerance in tomato. *Genes Dev.* **2002**, *16*, 1555–1567. [CrossRef] [PubMed]

46. Schramm, F.; Larkindale, J.; Kiehlmann, E.; Ganguli, A.; Englich, G.; Vierling, E.; Von Koskull-Döring, P. A cascade of transcription factor DREB2A and heat stress transcription factor HsfA3 regulates the heat stress response of *Arabidopsis*. *Plant J.* **2008**, *53*, 264–274. [CrossRef]

47. Xue, G.-P.; Drenth, J.; McIntyre, C.L. TaHsfA6f is a transcriptional activator that regulates a suite of heat stress protection genes in wheat (*Triticum aestivum* L.) including previously unknown Hsf targets. *J. Exp. Bot.* **2015**, *66*, 1025–1039. [CrossRef]

48. Zhao, J.; He, Q.; Chen, G.; Wang, L.; Jin, B. Regulation of non-coding RNAs in heat stress responses of plants. *Front. Plant Sci.* **2016**, *7*, 1213. [CrossRef]

49. Rogers, K.; Chen, X. Biogenesis, turnover, and mode of action of plant microRNAs. *Plant Cell* **2013**, *25*, 2383–2399. [CrossRef]

50. Xin, M.; Wang, Y.; Yao, Y.; Xie, C.; Peng, H.; Ni, Z.; Sun, Q. Diverse set of microRNAs are responsive to powdery mildew infection and heat stress in wheat (*Triticum aestivum* L.). *BMC Plant Biol.* **2010**, *10*, 123. [CrossRef]

51. Hivrale, V.; Zheng, Y.; Puli, C.O.R.; Jagadeeswaran, G.; Gowdu, K.; Kakani, V.G.; Barakat, A.; Sunkar, R. Characterization of drought- and heat-responsive microRNAs in switchgrass. *Plant Sci.* **2016**, *242*, 214–223. [CrossRef] [PubMed]

52. Reyes, J.L.; Chua, N.H. ABA induction of miR159 controls transcript levels of two MYB factors during *Arabidopsis* seed germination. *Plant J.* **2007**, *49*, 592–606. [CrossRef] [PubMed]

53. Lu, S.; Sun, Y.H.; Chiang, V.L. Stress-responsive microRNAs in *Populus*. *Plant J.* **2008**, *55*, 131–151. [CrossRef] [PubMed]

54. Chen, L.; Ren, Y.; Zhang, Y.; Xu, J.; Sun, F.; Zhang, Z.; Wang, Y. Genome-wide identification and expression analysis of heat-responsive and novel microRNAs in *Populus tomentosa*. *Gene* **2012**, *504*, 160–165. [CrossRef] [PubMed]

55. Mahale, B.M.; Fakrudin, B.; Ghosh, S.; Krishnaraj, P.U. LNA mediated in situ hybridization of miR171 and miR397a in leaf and ambient root tissues revealed expressional homogeneity in response to shoot heat shock in *Arabidopsis thaliana*. *J. Plant Biochem. Biotechnol.* **2014**, *23*, 93–103. [CrossRef]

56. Guan, Q.; Lu, X.; Zeng, H.; Zhang, Y.; Zhu, J. Heat stress induction of *miR398* triggers a regulatory loop that is critical for thermotolerance in Arabidopsis. *Plant J.* **2013**, *74*, 840–851. [CrossRef] [PubMed]

57. Sunkar, R.; Zhu, J.K. Novel and stress-regulated microRNAs and other small RNAs from *arabidopsis*. *Plant Cell* **2004**, *16*, 2001–2019. [CrossRef]

58. Beauclair, L.; Yu, A.; Bouché, N. microRNA-directed cleavage and translational repression of the copper chaperone for superoxide dismutase mRNA in Arabidopsis. *Plant J.* **2010**, *62*, 454–462. [CrossRef]

59. Sunkar, R.; Kapoor, A.; Zhu, J.K. Posttranscriptional induction of two Cu/Zn superoxide dismutase genes in *Arabidopsis* is mediated by downregulation of miR398 and important for oxidative stress tolerance. *Plant Cell* **2006**, *18*, 2051–2065. [CrossRef]

60. Yu, X.; Wang, H.; Lu, Y.; de Ruiter, M.; Cariaso, M.; Prins, M.; van Tunen, A.; He, Y. Identification of conserved and novel microRNAs that are responsive to heat stress in *Brassica rapa*. *J. Exp. Bot.* **2012**, *63*, 1025–1038. [CrossRef]

61. Zhang, Y.; Zhao, G.; Li, Y.; Mo, N.; Zhang, J.; Liang, Y. Transcriptomic analysis implies that GA regulates sex expression via ethylene-dependent and ethylene-independent pathways in cucumber (*Cucumis sativus* L.). *Front. Plant Sci.* **2017**, *8*, 10. [CrossRef]

62. Zhou, J.; Xiong, Q.; Chen, H.; Yang, C.; Fan, Y. Identification of the spinal expression profile of non-coding RNAs involved in neuropathic pain following spared nerve injury by sequence analysis. *Front. Mol. Neurosci.* **2017**, *10*, 91. [CrossRef] [PubMed]

63. Zhang, S.; Zhu, D.; Li, H.; Li, H.; Feng, C.; Zhang, W. Characterization of circRNA-associated-ceRNA networks in a senescence-accelerated mouse prone 8 brain. *Mol. Ther.* **2017**, *25*, 2053–2061. [CrossRef] [PubMed]

64. Kim, D.; Langmead, B.; Salzberg, S.L. HISAT: A fast spliced aligner with low memory requirements. *Nat. Methods* **2015**, *12*, 357. [CrossRef] [PubMed]

65. Frazee, A.C.; Pertea, G.; Jaffe, A.E.; Langmead, B.; Salzberg, S.L.; Leek, J.T. Ballgown bridges the gap between transcriptome assembly and expression analysis. *Nat. Biotechnol.* **2015**, *33*, 243. [CrossRef]

66. Anders, S.; Huber, W. Differential expression analysis for sequence count data. *Genome Biol.* **2010**, *11*, R106. [CrossRef]

67. Young, M.D.; Wakefield, M.J.; Smyth, G.K.; Oshlack, A. Gene ontology analysis for RNA-seq: Accounting for selection bias. *Genome Biol.* **2010**, *11*, R14. [CrossRef]

68. Kanehisa, M.; Araki, M.; Goto, S.; Hattori, M.; Hirakawa, M.; Itoh, M.; Katayama, T.; Kawashima, S.; Okuda, S.; Tokimatsu, T.; et al. KEGG for linking genomes to life and the environment. *Nucleic Acids Res.* **2008**, *36*, D480–D484. [CrossRef]

69. Mao, X.; Cai, T.; Olyarchuk, J.G.; Wei, L. Automated genome annotation and pathway identification using the KEGG Orthology (KO) as a controlled vocabulary. *Bioinformatics* **2005**, *21*, 3787–3793. [CrossRef]

70. Pérez-Rodríguez, P.; Riaño-Pachón, D.M.; Corrêa, L.G.G.; Rensing, S.A.; Kersten, B.; Mueller-Roeber, B. PlnTFDB: Updated content and new features of the plant transcription factor database. *Nucleic Acids Res.* **2010**, *38*, D822–D827. [CrossRef]

71. Salmena, L.; Poliseno, L.; Tay, Y.; Kats, L.; Pandolfi, P.P. A ceRNA Hypothesis: The rosetta stone of a hidden rna language? *Cell* **2011**, *146*, 353–358. [CrossRef] [PubMed]
72. Tay, Y.; Rinn, J.; Pandolfi, P.P. The multilayered complexity of ceRNA crosstalk and competition. *Nature* **2014**, *505*, 344. [CrossRef] [PubMed]

International Journal of *Molecular Sciences*

Article

Unravelling the MicroRNA-Mediated Gene Regulation in Developing Pongamia Seeds by High-Throughput Small RNA Profiling

Ye Jin [1], Lin Liu [1], Xuehong Hao [1], David E. Harry [2], Yizhi Zheng [1], Tengbo Huang [1] and Jianzi Huang [1,*]

[1] Guangdong Key Laboratory of Plant Epigenetics, College of Life Sciences and Oceanography, Shenzhen University, Shenzhen 518060, China
[2] TerViva, Oakland, CA 94612, USA
* Correspondence: bioken2001@sina.com; Tel.: +86-755-8652-7612

Received: 29 June 2019; Accepted: 15 July 2019; Published: 17 July 2019

Abstract: Pongamia (*Millettia pinnata* syn. *Pongamia pinnata*) is a multipurpose biofuel tree which can withstand a variety of abiotic stresses. Commercial applications of Pongamia trees may substantially benefit from improvements in their oil-seed productivity, which is governed by complex regulatory mechanisms underlying seed development. MicroRNAs (miRNAs) are important molecular regulators of plant development, while relatively little is known about their roles in seed development, especially for woody plants. In this study, we identified 236 conserved miRNAs within 49 families and 143 novel miRNAs via deep sequencing of Pongamia seeds sampled at three developmental phases. For these miRNAs, 1327 target genes were computationally predicted. Furthermore, 115 differentially expressed miRNAs (DEmiRs) between successive developmental phases were sorted out. The DEmiR-targeted genes were preferentially enriched in the functional categories associated with DNA damage repair and photosynthesis. The combined analyses of expression profiles for DEmiRs and functional annotations for their target genes revealed the involvements of both conserved and novel miRNA-target modules in Pongamia seed development. Quantitative Real-Time PCR validated the expression changes of 15 DEmiRs as well as the opposite expression changes of six targets. These results provide valuable miRNA candidates for further functional characterization and breeding practice in Pongamia and other oilseed plants.

Keywords: *Millettia pinnata*; woody oilseed plants; seed development; miRNA

---

1. Introduction

Woody oilseed plants have attracted increasing attention as alternative feedstocks for biodiesel production in recent years [1]. While conventional feedstocks such as soybean (*Glycine max*), rapeseed (*Brassica napus*), and maize (*Zea mays*) are mainly herbaceous crops that need to be grown on arable lands, the woody oilseed plants are usually adapted to more diverse environments such as mountain areas or intertidal zones [2]. Pongamia is one such tree species with high yield of non-edible seed oils that are highly suitable for biodiesel preparation [3,4]. Pongamia trees can not only withstand a variety of adverse conditions like drought and high salinity [5,6], but also undergo symbiotic nitrogen fixation by their root nodules [7,8]. Hence, they can be planted on wastelands or marginal lands with limited impact on food production and reduced consumption of nitrogen fertilizers. In addition, the extracts and derivatives from various parts of the plant have shown certain pharmacological activities such as antioxidant, antimicrobial, antidiabetic, and antihyperammonemic [9–11], which may potentially contribute to increasing the added values of the biodiesel products from this species.

To establish large-scale plantations of Pongamia trees for commercial biodiesel production, one of the major requirements is the improvement in oil-seed productivity, which requires a thorough understanding of the regulatory mechanisms underlying seed development in this species. Tissue- and stage-specific gene expression patterns related to seed development have already been revealed in major oilseed crops through global transcriptome profiling [12–14]. In Pongamia, several efforts have developed a bulk of transcriptomic data as a preliminary attempt on characterizing functional genes and profiling their expression at transcriptional level [6,15–17]. These transcriptomic data, along with the reference gene sequences from soybean, have facilitated the isolation and characterization of four circadian clock genes (*ELF4*, *LCL1*, *PRR7*, and *TOC1*) and two fatty acid desaturase genes (*SAD* and *FAD2*), which have shown distinct patterns of transcriptional regulation in relation to flowering time and seed development in Pongamia [18–20].

On the other hand, microRNAs (miRNAs) are a class of non-coding small RNAs that mainly regulate gene expression at post-transcriptional level [21,22]. As the upstream regulators of protein-coding genes, miRNAs participate in a wide range of biological processes essential for plant growth and stress responses [23,24]. With regard to seed development, miRNA-mediated regulations have already been uncovered at early or late developmental stages in Arabidopsis. For instance, miR156 can target *SQUAMOSA promoter-binding protein-like* (*SPL*) *10* and *SPL11* during early embryogenesis to repress precocious accumulation of certain transcripts normally expressed in maturation phase [25], while miR160-directed repression of *Auxin Response Factor* (*ARF*) *10* plays important roles in seed germination and post-embryonic developmental programs [26]. In addition, with the rapid development of high-throughput sequencing technology, numerous conserved and novel miRNAs have also been identified in developing seeds of a number of oilseed crops, such as rapeseed [27,28], soybean [29,30], and peanut (*Arachis hypogaea*) [31,32]. These studies have implicated the regulatory roles of miRNAs in multiple steps of seed development with special interests in seed oil accumulation. Nevertheless, to the best of our knowledge, large-scale identifications of miRNAs have only been reported in a few oilseed tree species [33,34].

In our previous studies, we have characterized three major developmental phases (i.e., embryogenesis phase, seed-filling phase, and desiccation phase) of Pongamia seeds based on morphological changes and physiological events [35]. In this work, we presented the first comprehensive investigation of miRNAs in Pongamia seeds through Illumina sequencing of nine small RNA libraries from these three phases. Millions of small RNA reads were generated, leading to the discovery of hundreds of conserved and novel miRNAs, among which a subset of miRNAs were identified with differential expressions between developmental phases. Furthermore, putative target genes for these miRNAs were obtained by computational prediction. Meanwhile, expression patterns of several selected miRNAs and their predicted targets during seed development were also examined by quantitative Real-Time PCR (qRT-PCR). Lastly, the potential roles of miRNAs and their target genes in Pongamia seed development were discussed.

2. Results

2.1. Deep Sequencing of Small RNA Libraries in Developing Pongamia Seeds

To characterize miRNAs and their expression profiles in developing Pongamia seeds, we constructed nine small RNA libraries based on the seed samples collected at the three developmental phases, each with three biological replicates. A total of 283,250,572 raw reads were yielded from the nine libraries by Illumina platform (Table 1). After filtering out low-quality sequences, adapter contaminants, polyA-containing sequences, and reads smaller than 18 nt, approximately 18 to 21 million clean reads were obtained for each library. These clean reads accounted for 62.22% to 67.49% of the raw reads and represented 2.44 to 3.43 million unique sequences in each library. The length distribution of these small RNA reads showed that 24-nt RNAs were the largest population in all nine libraries from the three developmental phases, while the second most abundant group was 21-nt RNAs in the libraries from the embryogenesis phase, but was 22-nt RNAs in the libraries from both the seed-filling phase and the desiccation phase (Figure S1). Subsequently, the clean reads were searched against the Rfam and the NCBI GenBank databases to identify rRNA, tRNA, snRNA, and snoRNA sequences. According to the combined results from both databases, the percentage of these four types of non-coding small RNAs was substantially higher in libraries from the embryogenesis phase ($8.31 \pm 0.77\%$ of the clean reads) than in those from the seed-filling phase ($3.97 \pm 0.49\%$) and the desiccation phase ($5.85 \pm 0.68\%$) (Table S1). For further analysis, the clean reads matching the above four types of small RNAs were excluded. In general, both the length distribution and the relative abundance of various types of small RNAs exhibited a developmental phase-specific pattern in Pongamia seeds.

2.2. Identification of Conserved and Novel miRNAs in Developing Pongamia Seeds

To identify miRNA homologs in Pongamia seeds, we used the remaining clean reads in each library to match against the Viridiplantae mature miRNAs in the miRBase (Release 21.0). In total, 236 conserved miRNAs belonging to 49 families were identified in the nine libraries, with an average of nearly five miRNA members per family (Table S2). For 24 miRNA families, only one member was found. These conserved miRNAs were aligned to 67 plant species, among which *Glycine max*, *Medicago truncatula*, and *Populus trichocarpa* were the most frequent ones. At the embryogenesis phase, 1,723,573, 1,519,586, and 1,115,971 reads from the three biological replicates perfectly matched 174, 172, and 168 known plant mature miRNAs belonging to 38, 37, and 37 miRNA families, respectively (Table 1). At the seed-filling phase, 777,660, 823,736, and 1,207,854 reads perfectly matched 157, 165, and 183 known plant miRNAs within 38, 41, and 42 families, respectively. At the desiccation phase, 1,230,083, 1,206,643, and 1,271,965 reads perfectly matched 163, 162, and 160 known plant miRNAs within 35, 36, and 34 families, respectively. These conserved miRNAs were combined into 194, 199, and 184 nonredundant conserved miRNAs expressed at the three developmental phases, respectively (Figure 1). Taken together, the number of total reads matching the conserved miRNAs was highest at the embryogenesis phase (4,359,130), followed by those at the desiccation phase (3,708,691) and the seed-filling phase (2,809,250). MIR156 and MIR166 were the two largest miRNA families with 26 and 24 members, respectively (Table S2). MIR166 was also the most abundantly expressed family at all three phases, followed by MIR167 and MIR396 at the embryogenesis phase, while by MIR159 and MIR167 at the two later phases. There were 18, 21, and 10 conserved miRNAs exclusively expressed at each one of the three developmental phases, respectively (Figure 1). For example, mpi-miR160c-5p and mpi-miR171b-5p were only expressed at the embryogenesis phase, mpi-miR168d-3p and mpi-miR5037 only at the seed-filling phase, and mpi-miR162e-3p and mpi-miR482c-5p only at the desiccation phase. Nevertheless, most of these phase-specific conserved miRNAs were represented by less than 10 reads. On the other hand, 154 conserved miRNAs were expressed throughout all three phases (Figure 1), with their expression levels either remained stable or varied greatly between phases.

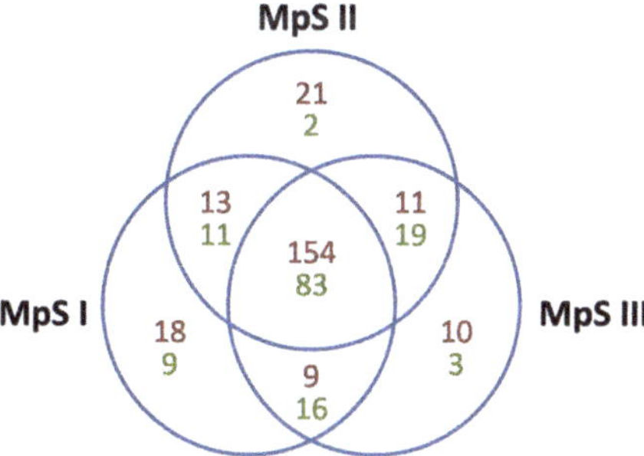

Figure 1. Venn diagram showing the distribution of miRNAs among the three developmental phases of Pongamia seeds. MpSI, the embryogenesis phase; MpSII, the seed-filling phase; MpSIII, the desiccation phase. The values in red and green indicate the numbers of conserved and novel miRNAs, respectively.

The conserved miRNAs identified by homology searching were further subjected to pre-miRNA prediction by exploring the Pongamia seed transcriptome developed by our previous study [15]. The mRNA sequences with hairpin-like structures and with more than 10 miRNA reads anchoring the 5p- and/or 3p-arm were considered as putative pre-miRNAs. As a result, 16 pre-miRNA sequences for the conserved miRNAs were identified, including eight anchored by miRNA reads in the 5p-arm, two by reads in the 3p-arm, and six by reads in both arms (Table S3). These candidate precursors had a mean length of 149 bp, a GC content of 43.78%, an MFE of −55.56 and an MFEI of −0.90, all of which were conformed to the features of miRNA biogenesis [36].

To predict novel miRNAs in Pongamia seeds, the clean reads excluding those with hits to rRNA, tRNA, snRNA, and snoRNA sequences were aligned with the mRNA sequences in the Pongamia seed transcriptome to retrieve their precursors. Only the reads matching pre-miRNA sequences with characteristic hairpin-like structures and with no homology to previously known plant miRNAs were singled out as novel miRNAs in Pongamia. Overall, there were 143 novel miRNAs identified in Pongamia seeds (Table S4). The lengths of these novel miRNAs ranged between 20 and 24 nt, with 21 nt being the most common. The majority of these novel miRNAs (135 out of 143) have a precursor anchored by sequencing reads in just one arm. Some novel miRNAs were derived from the same read tag mapping to different unigenes in the Pongamia seed transcriptome. Within the nine libraries, 87, 82, and 68 novel miRNAs were identified from the three biological replicates at the embryogenesis phase, 70, 64, and 83 novel miRNAs were identified at the seed-filling phase, and 97, 93, and 94 novel miRNAs were identified at the desiccation phase (Table 1). These novel miRNAs were combined into 119, 115, and 121 nonredundant novel miRNAs expressed at the three developmental phases, respectively (Figure 1). Compared with the conserved miRNAs, these novel miRNAs were expressed at relatively low levels, the majority of them being represented by less than 10 reads in each library. Fourteen novel miRNAs were specifically expressed at one phase, while 83 novel miRNAs were expressed throughout all three phases (Figure 1). At all three phases, mpi-nmiR0043-5p, mpi-nmiR0124-3p, mpi-nmiR0125-5p, mpi-nmiR0126-3p, and mpi-nmiR0127-5p were the five most abundant novel miRNAs (Table S4).

2.3. Differential Expression of miRNAs during Pongamia Seed Development

To investigate miRNA expression changes during Pongamia seed development, we quantified their expression levels using the transcripts per million (TPM) values, which normalized the read counts of each identified miRNA to the total read counts in each library. Based on the TPM values, we firstly performed Pearson's correlation analysis and principal component analysis to evaluate the reproducibility of expression data among the three biological replicates. Pearson's correlation analysis indicated high correlations of miRNA expression levels among replicates within the same developmental phase, with an average coefficient of 0.9991, 0.9980, and 0.9991 for the three phases, respectively (Figure S2). Principal component analysis revealed a clear assignment of three groups corresponding to the three phases (Figure S3), which also demonstrated a good reproducibility of the miRNA expression data yielded in this study.

With a minimum cutoff of 2-fold changes in average TPM values, we sorted out differentially expressed miRNAs (DEmiRs) between developmental phases of Pongamia seeds. A total of 115 DEmiRs, including 82 conserved miRNAs and 33 novel miRNAs, were identified from the two successive comparisons (Table S5). In the comparison between the embryogenesis phase and the seed-filling phase, 44 significantly up-regulated and 44 significantly down-regulated miRNAs were detected (Figure 2). In the comparison between the seed-filling phase and the desiccation phase, 19 and 39 miRNAs were observed to be significantly up-regulated and down-regulated, respectively. In other words, there were much more DEmiRs (88) between the former two phases than those (58) between the latter two phases.

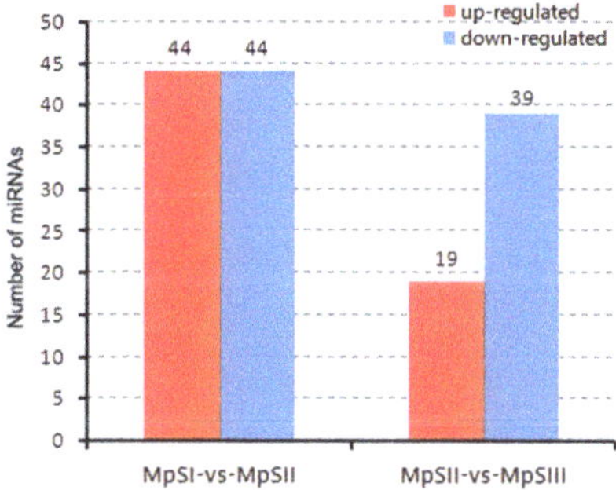

Figure 2. The numbers of DEmiRs between developmental phases of Pongamia seeds. The values in red and blue indicate the numbers of up-regulated and down-regulated miRNAs, respectively.

More specifically, five and eight DEmiRs were significantly up-regulated and down-regulated in both comparisons, respectively (Table 2). Fifteen DEmiRs were significantly up-regulated from the embryogenesis phase to the seed-filling phase, and then significantly down-regulated from the seed-filling phase to the desiccation phase, showing a bell-shape change in expression level (Table 3). On the contrary, three DEmiRs displayed a V-shape change with significant down-regulation followed by significant up-regulation in the two successive comparisons (Table 3). The remaining 84 DEmiRs showed a significant change in expression level only in one comparison. Among those conserved miRNAs, mpi-miR168d-3p and mpi-miR167f-5p showed the highest degree of up-regulation in the former and the latter comparison, respectively, whereas mpi-miR399d-3p and mpi-miR9662a-3p were the most significantly down-regulated ones in the two comparisons, respectively.

2.4. Prediction and Functional Annotation of Pongamia miRNA Targets

To gain insight into the regulatory roles of the miRNAs in Pongamia seeds, their putative targets were predicted by aligning the identified miRNA sequences with the unigenes in Pongamia seed transcriptome. As a result, 1327 targets were identified for 210 conserved miRNAs and 121 novel miRNAs through such computational screening, with an average of about four targets per miRNA molecule (Table S6). Among the conserved miRNAs, mpi-miR156u-5p (40 targets), mpi-miR156j-5p (23 targets), mpi-miR171h-3p (23 targets), mpi-miR156l-5p (22 targets), and mpi-miR396i-3p (22 targets) were the top five with the largest number of target genes in diverse functional categories, implying that they were involved in multiple processes during Pongamia seed development. On the other hand, miRNAs from different families could co-target the same gene.

For instance, Unigene22944 encoding an Auxin Signaling F-Box 2-like protein was co-targeted by four conserved miRNAs (mpi-miR1511-3p, mpi-miR393c-3p, mpi-miR5139-5p, and mpi-miR8155-3p), while Unigene17682 for a disease resistance protein was co-targeted by a conserved miRNA (mpi-miR1510-3p) and a novel miRNA (mpi-nmiR0004-3p). Only 56 targets were identified as transcription factors (TFs) belonging to 23 families, among which NF-YA, bHLH, and B3 were most abundantly represented (Figure S4). Interestingly, we also found eight target genes encoding enzymes in lipid metabolic pathways (Table 4). Among them, only one unigene encoding a mitochondrial cardiolipin synthase was targeted by a conserved miRNA, mpi-miR168f-5p, which was specifically expressed at the desiccation phase. Four unigenes encoding 3-ketoacyl-CoA synthase 4, linoleate 13S-lipoxygenase 3, and phospholipid:diacylglycerol acyltransferase 1, were targeted by the same novel miRNA, mpi-nmiR0017-3p, which expressed only at the embryogenesis and the seed-filling phases in low levels. Unigene22005 encoding a phospholipase A2 was targeted by another novel miRNA, mpi-nmiR0038-5p, with low expressions at the embryogenesis and the desiccation phases. Unigene22800 for a chloroplastic stearoyl-ACP 9-desaturase 6 and Unigene4253 for a malonyl-transferase were targeted by two DEmiRs, mpi-nmiR0028-3p and mpi-nmiR0102-5p, showing a bell-shape pattern of expression and a desiccation-phase-specific expression, respectively.

Table 1. Statistics of sequencing reads and miRNAs.

Statistical Items	MpSI-1	MpSI-2	MpSI-3	MpSII-1	MpSII-2	MpSII-3	MpSIII-1	MpSIII-2	MpSIII-3
Number of raw reads	31,711,290	31,959,711	31,763,941	27,733,326	33,988,872	31,143,787	31,167,480	32,681,945	31,100,220
Number of clean reads	21,400,622	21,401,693	21,217,533	17,612,476	21,414,373	19,377,920	20,039,686	20,872,904	19,792,612
Retention rate	67.49%	66.96%	66.80%	63.51%	63.00%	62.22%	64.30%	63.87%	63.64%
Number of unique tags	2,757,270	2,876,894	2,882,107	3,066,337	3,427,062	2,942,567	2,593,393	2,774,500	2,443,701
Number of conserved miRNAs	174	172	168	157	165	183	163	162	160
Number of novel miRNAs	87	82	68	70	64	83	97	93	94

Table 2. Pongamia miRNAs with the same tendencies of expression changes in two comparisons between developmental phases.

miRNA_ID	Reads Count									Fold Change	
	MpSI-1	MpSI-2	MpSI-3	MpSII-1	MpSII-2	MpSII-3	MpSIII-1	MpSIII-2	MpSIII-3	MpSII/MpSI	MpSIII/MpSII
mpi-miR156f-5p	0	0	0	2	2	4	12	13	12	7.65	1.74
mpi-miR482a-3p	2380	4221	2288	5869	5817	7435	39410	19325	18703	1.58	1.49
mpi-nmiR0004-3p	21	20	9	25	22	24	69	80	106	1.07	1.28
mpi-nmiR0083-3p	0	0	0	11	11	9	71	70	84	9.71	2.28
mpi-nmiR0119-5p	0	0	0	2	2	6	10	8	14	7.92	1.26
mpi-miR156a-3p	2112	3098	1195	736	356	577	316	347	424	−1.39	−1.19
mpi-miR164a-5p	141	121	130	7	20	32	11	7	7	−2.37	−1.67
mpi-miR166a-5p	3344	2597	2102	718	822	1330	437	495	803	−1.03	−1.22
mpi-miR166k-5p	408	297	290	19	21	30	7	4	9	−3.37	−2.32
mpi-miR171j-5p	66	54	35	5	5	5	0	2	0	−2.83	−3.43
mpi-miR393b-5p	384	272	219	44	84	133	35	38	42	−1.31	−1.64
mpi-miR399a-3p	5907	4743	3838	908	1026	1399	725	552	741	−1.65	−1.24
mpi-miR399b-3p	3876	4196	2814	1094	848	1048	412	207	306	−1.35	−2.25

Table 3. Pongamia miRNAs with the opposite tendencies of expression changes in two comparisons between developmental phases.

miRNA_ID	Reads Count									Fold Change	
	MpSI-1	MpSI-2	MpSI-3	MpSII-1	MpSII-2	MpSII-3	MpSIII-1	MpSIII-2	MpSIII-3	MpSII/MpSI	MpSIII/MpSII
mpi-miR156c-5p	84	111	69	327	203	197	95	68	83	2.01	−2.16
mpi-miR168d-3p	0	0	0	2	4	9	0	0	0	8.49	−8.49
mpi-miR171b-3p	96	94	52	170	226	432	219	142	198	2.22	−1.02
mpi-miR398c-3p	393	398	260	964	1092	1553	360	317	436	2.25	−2.21
mpi-miR398f-3p	10	10	6	113	142	169	9	7	13	4.52	−4.40
mpi-miR408a-3p	130	137	75	379	446	612	176	170	199	2.56	−1.91
mpi-miR408b-3p	15	15	6	68	98	147	35	45	59	3.62	−1.66
mpi-miR408d-3p	7	8	8	84	63	76	37	54	45	3.75	−1.28
mpi-miR4403-5p	3	3	3	6	5	6	2	2	2	1.39	−2.06
mpi-miR5037	0	0	0	2	2	2	0	0	0	7.32	−7.32
mpi-nmiR0028-3p	7	6	7	25	32	19	7	7	3	2.44	−2.75
mpi-nmiR0034-3p	4	5	3	6	6	9	4	2	3	1.28	−1.73
mpi-nmiR0050-5p	0	0	0	5	3	6	0	0	0	8.50	−8.50
mpi-nmiR0072-5p	14	25	12	16	28	47	2	8	5	1.25	−3.06
mpi-nmiR0103-5p	0	0	0	4	16	10	0	0	0	9.63	−9.63
mpi-miR396c-3p	1094	422	725	69	78	151	401	240	257	−2.50	1.13
mpi-miR396e-5p	9	8	4	0	0	0	5	3	5	−8.57	7.88
mpi-miR399d-3p	43	33	29	0	0	0	5	7	2	−10.94	7.99

Table 4. Pongamia miRNAs with putative targets in relation to lipid metabolism.

miRNA_ID	Target_ID	Target_Annotation	MpSI_TPM	MpSII_TPM	MpSIII_TPM
mpi-miR168f-5p	Unigene24517	cardiolipin synthase (CMP-forming), mitochondrial	0.01	0.01	0.36
mpi-nmiR0017-3p	Unigene10960	3-ketoacyl-CoA synthase 4	1.02	1.02	0.01
	Unigene15710	linoleate 13S-lipoxygenase 3-1, chloroplastic			
	Unigene26514	phospholipid:diacylglycerol acyltransferase 1-like			
	Unigene49632	3-ketoacyl-CoA synthase 4			
mpi-nmiR0028-3p	Unigene22800	stearoyl-ACP 9-desaturase 6, chloroplastic	11.54	20.81	3.09
mpi-nmiR0038-5p	Unigene22005	phospholipase A2	1.42	0.01	0.37
mpi-nmiR0102-5p	Unigene4253	malonyltransferase	0.01	0.01	2.53

Among the predicted miRNA targets, 396 genes were targeted by the DEmiRs. This subset of target genes were assigned with GO terms and KEGG pathways, and then subjected to enrichment analysis with the set of unigenes from the whole transcriptome as a background. In the category of Molecular Function, DNA polymerase activity and phosphoric diester hydrolase activity were the two most significantly enriched GO terms (Table S7). With respect to Biological Process, 48 GO terms were significantly enriched with the DEmiR-targeted genes. Among them, chromosome segregation, regulation of organelle organization, and DNA repair were the most enriched processes. As for Cellular Components, the DEmiR-targeted genes were significantly overrepresented in photosynthetic membrane, clathrin-coated vesicle membrane, and thylakoid membrane. Meanwhile, 74 DEmiR-targeted genes were assigned with 46 KEGG pathways (Table S8). Three pathways, including non-homologous end-joining, base excision repair, and photosynthesis, were significantly enriched with the DEmiR-targeted genes.

2.5. Validation of DEmiRs during Pongamia Seed Development

In order to verify the expression patterns of candidate Pongamia miRNAs obtained from the small RNA sequencing data, we conducted stem-loop qRT-PCR for 15 randomly chosen DEmiRs, including 12 conserved and three novel miRNAs (Table S9). RNAs extracted from the seeds representing the three developmental phases were used as templates. Note that the RNAs for small RNA sequencing and stem-loop qRT-PCR were separately prepared at the same time points. The stem-loop qRT-PCR results for these 15 DEmiRs were basically consistent with the sequencing results (Figure 3).

Five miRNAs (mpi-miR156c-5p, mpi-miR171b-3p, mpi-miR398c-3p, mpi-miR408a-3p, and mpi-nmiR0028-3p) showed a bell-shape pattern with their expression levels significantly up-regulated from the embryogenesis phase to the seed-filling phase, and then significantly down-regulated from the seed-filling phase to the desiccation phase. Four miRNAs (mpi-miR167c-5p, mpi-miR1507c-3p, mpi-miR2218-3p, and mpi-nmiR0004-3p) were continuously up-regulated during the two successive switches in developmental phase. Conversely, four miRNAs (mpi-miR171j-5p, mpi-miR396c-5p, mpi-miR399b-3p, and mpi-nmiR0135-5p) were down-regulated all through the three phases. The expression level of mpi-miR858a-5p did not change significantly during the first two phases, but then sharply dropped down at the desiccation phase. There was also one miRNA (mpi-miR399d-3p) displaying a near V-shape change in expression level. The linear regression analysis showed a highly significant correlation between the expression profiles revealed by small RNA sequencing and qRT-PCR results (Figure S5). In addition, we also monitored the expression profiles of target genes for these DEmiRs. The DEmiRs with more than three predicted targets were not subjected to such target expression detection. Among six tested DEmiR-targeted genes, five exhibited a V-shape or near V-shape change in expression level (Figure S6). These five target genes included the aforementioned Unigene22800 involved in lipid metabolism, Unigene22770 encoding a superoxide dismutase, Unigene25493 for a cytochrome P450 CYP72A219-like protein, and two unigenes with unknown functions (Table S10). Besides, a disease resistance protein gene (Unigene8399) was continuously down-regulated as seeds matured. Generally, the tendencies of expression changes for these target genes were opposite to those for the corresponding miRNAs, suggesting that the miRNA-mediated regulation might occur in these targets. The above results again demonstrated that our small RNA sequencing data was reliable in temporal expression analysis of Pongamia miRNAs during seed development.

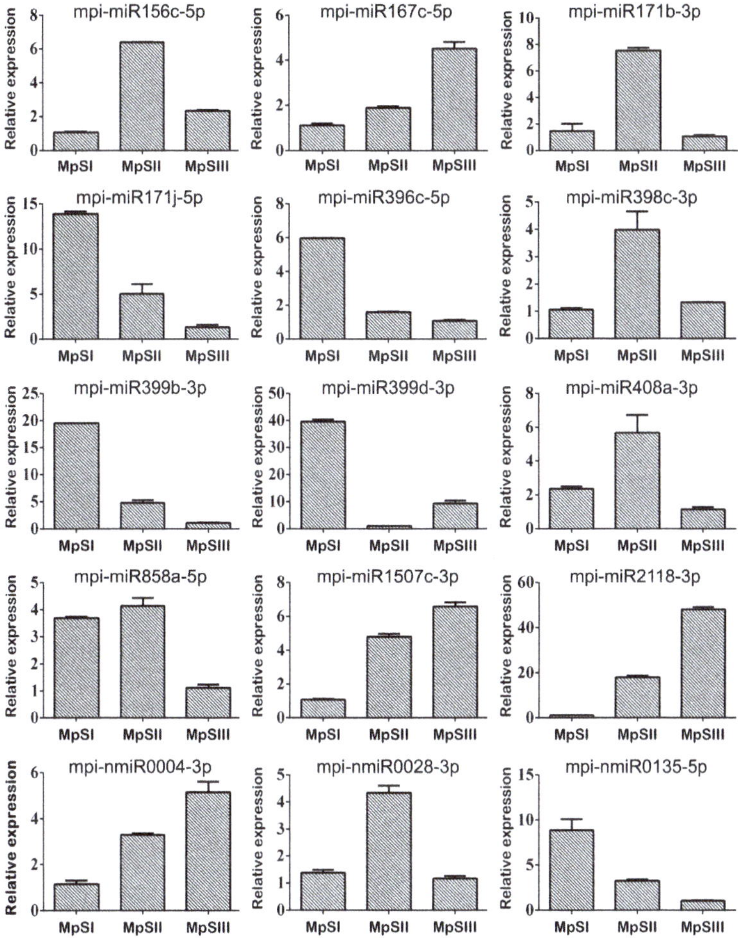

Figure 3. Relative expression levels of 15 DEmiRs evaluated by stem-loop qRT-PCR. An U6 snRNA gene of Pongamia was used as internal control. Bars represent standard deviations of three technical replicates.

3. Discussion

The seed development process of oilseed plants is instrumental in determining the oil content and quality of the end products. Despite the abundant research on molecular mechanisms regulating seed development at transcriptional level, the miRNA-mediated post-transcriptional regulation of this process remains relatively unstudied, especially in woody oilseed species. In this study, we used high throughput sequencing to characterize miRNAs and examine their expression profiles in developing seeds of Pongamia whose genome has not yet been sequenced, and obtained approximately 283 million small RNA raw reads from nine libraries representing three developmental phases. The small RNAs in Pongamia seeds displayed a wide range of variation in length with the 24-nt RNAs being the most abundant class in all nine libraries, followed by the 21-nt and 22-nt RNAs. Such length distribution pattern of small RNAs was also observed in model plants like Arabidopsis and rice (*Oryza sativa*) [37,38], as well as in some legume relatives (e.g., soybean, *Medicago truncatula*, and peanut) [30,31,39], implying that Pongamia might possess similar processing components for small RNAs biogenesis as other plant species. Moreover, the higher abundance of 24-nt small RNAs was suggested to be related to the silencing of transposons and heterochromatic repeats for ensuring normal seed formation and such abundance of 24-nt small RNAs tended to decrease as seed matured [40], which was also evidently shown in developing Pongamia seeds.

Based on the large quantity of small RNA reads, a total of 236 conserved miRNAs within 49 families and 143 novel miRNAs not found in other plants were identified in this work. The reads matching these miRNAs accounted for less than 6% of the total small RNA reads, indicating that miRNAs only contributed a small portion to small RNAs in Pongamia seeds. Generally, the conserved miRNAs were expressed at higher levels than the novel miRNAs, which was also in accordance with the former results in Arabidopsis and soybean [29,41]. Specifically, 154 conserved and 83 novel miRNAs were expressed at all three phases, implicating that they were indispensable throughout the whole developmental process of Pongamia seeds. The novel miRNAs could be Pongamia-specific miRNAs whose validity needed further confirmation. As for the conserved miRNAs, 15 out of the 49 families (MIR1507, 1510, 1511, 1514, 1515, 2089, 2118, 2119, 2218, 2643, 4403, 482, 5037, 530, and 5559) were only observed in species belonging to the order Fabales. They were classified as Fabales-specific miRNA families and suggested to be young miRNA families by a previous study [29]. Besides, there were also some families appearing in certain species within other eudicot orders (e.g., MIR5139, 8155, and 8175) or monocot orders (e.g., MIR9653, 9662, and 9778) instead of Fabales. These miRNA families probably had an ancient origin, but might have been lost or not yet been identified in other species in Fabales. In other words, both the ancient regulatory pathways and the novel and unique pathways mediated by different kinds of miRNAs might be present in Pongamia.

Identification of the DEmiRs between seed developmental phases and the corresponding target genes could provide valuable information on understanding their functions. In this work, 115 DEmiRs with differential expression in at least one comparison between two successive phases were identified. There were substantially more DEmiRs (88) between the embryogenesis phase and the seed-filling phase than those (58) between the seed-filling phase and the desiccation phase. Consistent with this tendency, more protein-coding genes were observed to be differentially expressed between the former two phases than those between the latter two phases [35]. To some extent, the general expression patterns of both DEmiRs and differentially expressed genes (DEGs) correlated with a more remarkable change in seed traits and oil content during the early developmental stages of Pongamia seeds [35,42]. Meanwhile, there were substantially more down-regulated DEmiRs (39) than up-regulated ones (19) from the seed-filling phase to the desiccation phase. Considering that miRNAs always negatively regulate protein-coding genes, the above observation also coincides with our previous findings of more up-regulated genes than down-regulated ones during the same developmental switch [35]. The DEmiRs identified in this study could potentially target 396 genes, which were preferentially enriched in certain GO terms or KEGG pathways associated with DNA repair and photosynthesis. DNA damage repair was crucial for the maintenance of genome integrity over cell division during seed development [43], while photosynthesis could provide energy and oxygen for enhancing biosynthetic fluxes to lipids in developing legume seeds [44]. Therefore, the enrichment of DEmiR-targeted genes in these two functional categories was quite reasonable.

Involvements of miRNAs in seed development have already been established in several conserved miRNA families. By binding to plant-specific transcription factor genes *SPL10* and *SPL11*, miR156 could negatively regulate phase transition and seed maturation while its expression gradually declined as seed developed [25]. Some members of the MIR156 family could also target certain genes in fatty acid biosynthetic pathway and could thus affect seed oil content of rapeseed [45]. In this study, nine MIR156 members were screened out as DEmiRs and they targeted dozens of genes with diverse functions (Table S6). Since both *SPL10* and *SPL11* transcripts were not found in the Pongamia seed transcriptome [15], none of those nine members were predicted to target any *SPL* homologs. Notably, mpi-miR156a-3p and mpi-miR156c-5p displayed significant changes in expression levels in both comparisons between seed developmental phases. The former targeted a protein S-acyltransferase (PAT) 10 for protein lipid modification and was continuously down-regulated from the embryogenesis phase to the desiccation phase, whereas the latter targeted a cytochrome P450 CYP72A219-like protein for metal binding and showed a bell-shape pattern of expression changes. The expression changes of mpi-miR156c-5p and its target gene (Unigene25493) were further validated by qRT-PCR. These two miRNAs might play critical roles at the early and mid stages of seed development in Pongamia as those reported in some other plants [25,45].

As another upstream regulator of seed development, miR160 could repress *ARF10* and *ARF17* to modulate the expression of early auxin response genes [26,46]. In this study, mpi-miR160c-5p was exclusively expressed at the embryogenesis phase and thus recognized as a DEmiR in developing Pongamia seeds. The predicted targets of this miRNA were *ARF17* and *ARF18*. This result implied that miR160 might mainly function at early developmental stages of Pongamia seeds.

The miR164-mediated regulation of the No Apical Meristem (NAC) transcription factors also played an essential role in seed development. Expression of a miR164-resistant version of Cup-Shaped Cotyledon (CUC) 1 within the NAC family could cause cotyledon orientation defects [47]. Among the three MIR164 members identified in this work, only mpi-miR164a-5p was a DEmiR with significant down-regulation from the embryogenesis phase to the desiccation phase. This miRNA was predicted to target the transcripts of a HHE cation-binding domain protein and a UPF0481 protein instead of any NAC domain-containing proteins in the Pongamia seed transcriptome, suggesting the existence of specialized regulatory components coupled with miR164 in this species, which was different from the miR164-NAC module in other plants.

The MYB transcription factors usually serve as positive regulators of ABA responses. By silencing *MYB33* and *MYB101*, miR159 could act as a negative regulator preventing the transition from seed dormancy to germination [48]. In addition, the miR159-mediated regulation of MYB33 and MYB65 expression might also be responsible for determining seed size [49]. We found four MIR159 members in the list of DEmiRs (Table S5). Except for one member showing a bell-shape expression pattern, the other three members (mpi-miR159a-5p, mpi-miR159b-3p, and mpi-miR159d-5p) all exhibited a significant down-regulated expression in favor of releasing the constraints on seed germination.

The APETALA2 (AP2) transcription factors compose a large family with a variety of regulatory functions, including the control of seed size and seed mass [50,51]. Although the *AP2* genes could be regulated by miR172 [52], there was still no evidence supporting the direct involvement of the miR172-AP2 module in seed development. Additionally, up-regulation of miR172 was shown to promote seed germination through its interaction with miR156 and the *Delay of Gemination1* (*DOG1*) gene [53]. However, mpi-miR172c-3p was significantly down-regulated as seed matured and it could target diverse transcripts including an *AP2* homolog in Pongamia (Table S6). Hence, we speculated that the miR172 might participate in the regulation of seed development via diversified mechanisms in different plant species.

In addition to the above conserved miRNAs having been reported to be relevant to seed development, there are also some other DEmiRs which may be candidate regulators for this process in Pongamia. For instance, mpi-miR482a-3p displayed a significantly increasing expression and could target several transcripts encoding a tyrosyl-DNA phosphodiesterase (TDP) 1 for DNA 3′-end processing [54]. As aforementioned, DNA damage repair was crucial for maintaining genome integrity during seed development. Another interesting candidate was mpi-nmiR0028-3p which targeted a stearoyl-ACP 9-desaturase 6 for converting stearic acid to oleic acid. This novel miRNA was also a DEmiR with a bell-shape expression pattern and could probably make an impact on fatty acid composition in developing Pongamia seeds. The contrasting expression changes of this novel miRNA and its target gene (Unigene22800) were also verified by qRT-PCR (Figure 3 and Figure S6).

4. Materials and Methods

4.1. Plant Material and Sample Collection

Three eight-year-old Pongamia trees cultivated in Shenzhen University, Shenzhen, China (22°32′ N, 113°55′ E), with the monthly average temperature ranging from about 15°C to 29°C and an average annual rainfall of approximately 2000 mm, were used as the biological replicates for seed sampling. The inflorescences on different sub-branches of each tree were labeled with tags recording their first flowering dates. Pods were harvested at 10 weeks after flowering (WAF), 20 WAF, and 30 WAF, representing the three developmental phases of Pongamia seeds as previously described [35]. At each sampling time point, the seeds were manually separated from pods, washed with distilled water, immediately frozen in liquid nitrogen, and then stored at −80°C before RNA extraction for further experiments.

4.2. RNA Isolation, Library Construction and Small RNA Sequencing

Total RNA was isolated from Pongamia seeds using a modified CTAB method [55]. At each sampling time point, the seeds from each of the three trees were separately subjected to RNA extraction. The resulting nine RNA samples were further purified with the RNeasy Plant Mini Kit (Qiagen, Hilden, Germany) following the manufacturer's instructions. RNA concentration and quality of each sample was assessed by an Agilent 2100 Bioanalyzer (Agilent Technologies, Palo Alto, CA, USA). Subsequently, the RNA molecules with different sizes were separated using polyacrylamide gel electrophoresis. The 18–30 nt small RNAs were excised and then ligated to 5′ and 3′ adapters by T4 RNA ligase in two separate steps. The ligation products were reverse transcribed and then amplified by PCR. The 140–160 bp PCR products were enriched to generate nine cDNA libraries for sequencing on the Illumina HiSeq 2500 platform (Illumina, San Diego, CA, USA) at GeneDenovo Biotechnology Co., Ltd. (GeneDenovo, Guangzhou, China). All sequence data of the nine libraries were deposited in the NCBI Sequence Read Archive (SRA) database under the accession number PRJNA550227.

4.3. Sequencing Data Processing and Identification of Conserved and Novel miRNAs

Raw reads of the nine libraries were firstly filtered by in-house Perl scripts to remove low-quality reads, reads containing 5′ adapters and polyA tails, reads without 3′ adapters, and reads shorter than 18 nt. Then, the resulting clean reads were blasted against the NCBI GenBank (http://www.ncbi.nlm.nih.gov/genbank/) database and the Rfam (http://rfam.xfam.org/) database to screen out rRNAs, tRNAs, snRNA, and snoRNAs. The remaining reads without matches to the above four types of RNAs were searched against the miRBase (21.0) (http://www.mirbase.org/) database to identify phylogenetically conserved miRNAs. Only the reads with no mismatches to those currently known plant miRNA sequences were considered as conserved miRNAs. The remaining unannotated reads were aligned with the Pongamia seed transcriptome to predict potential novel miRNAs and their hairpin precursors by the Mireap software (http://sourceforge.net/projects/mireap/) with default settings. The miRNA expression levels were calculated and normalized as transcripts per million (TPM) values based on the total number of clean reads in each library. For miRNAs that were not expressed in the samples, the expression levels were set to 0.01. To determine the statistical significance of expression difference, the fold change and the P value were calculated for each pairwise comparison between seed developmental phases. Both absolute fold change ≥ 2 and P value ≤ 0.05 were adopted as the thresholds to identify DEmiRs.

4.4. Identification and Analysis of miRNA Targets

Putative target genes of conserved and novel miRNAs were predicted by the PatMatch software (https://www.arabidopsis.org/cgi-bin/patmatch/nph-patmatch.pl), which queried the Pongamia seed transcriptome with the parameters as follows: no more than four mismatches between miRNA and target (G-U bases count as 0.5 mismatches); no more than two adjacent mismatches in the miRNA/target duplex; no adjacent mismatches in positions 2–12, no mismatches in positions 10–11, and no more than 2.5 mismatches in positions 1–12 of the miRNA/target duplex (5′ of miRNA); and the minimum free energy (MFE) of the miRNA/target duplex should be ≥ 75% compared to the MFE of the miRNA bound to its perfect complement. The predicted target genes were assigned with GO (http://www.geneontology.org/) and KEGG (http://www.genome.jp/kegg/) annotations. Specifically, for those genes targeted by DEmiRs, the hypergeometric tests were conducted to identify significantly enriched GO terms or KEGG pathways with all unigenes in the Pongamia seed transcriptome as a background. The calculated P values were gone through false discovery rate (FDR) correction, taking FDR ≤ 0.05 as a threshold.

4.5. Quantitative Real-Time PCR for miRNAs and Target Genes

To validate changes in expression levels of miRNAs during the three developmental phases of Pongamia seeds, total RNAs were isolated and enriched for small RNA fractions as above described, and then reverse transcribed by stem-loop RT-PCR [56]. The resulting cDNAs were diluted and subjected to quantitative real-time PCRs on an ABI PRISM 7300 Sequence Detection System (Applied Biosystems, Foster City, CA, USA) using the ChamQ Universal SYBR qPCR Master Mix (Vazyme Biotech, Nanjing, China). Primers for stem-loop reverse transcription and real-time PCR were separately designed for 15 randomly selected DEmiRs (Table S9). For expression profiling of target genes, total RNAs from the same samples as those for miRNA profiling were employed for first-strand cDNA synthesis using the Super Script First-Strand cDNA Synthesis Kit (Invitrogen, Carlsbad, CA, USA), followed by Real-Time PCR with specific primers for target genes (Table S10). U6 snRNA and actin were used as the internal reference for the normalization of miRNAs and target genes, respectively. All real-time PCRs were run in three technical replicates. The primer specificity was confirmed by melting curve analysis. The relative expression levels of the tested miRNAs and target genes were calculated with the $2^{-\Delta\Delta Ct}$ method.

5. Conclusions

In summary, we presented the first large-scale collection of small RNAs from Pongamia seeds by high-throughput sequencing and identified 236 conserved miRNAs belonging to 49 families as well as 143 novel miRNAs. Among them, 82 conserved miRNAs and 33 novel miRNAs with significantly differential expression between successive developmental phases were sorted out as DEmiRs. The enrichment of DEmiR-targeted genes in the functional categories related to DNA damage repair and photosynthesis suggested that the regulation of these two processes should be pivotal in controlling Pongamia seed development. Through the combined analyses of expression profiles for DEmiRs and functional annotations for their target genes, we not only found some miRNAs within the families (MIR156, MIR159, MIR160, MIR164, and MIR172) previously confirmed to be capable of regulating seed development in herbaceous model plants, but also proposed some candidates (e.g., mpi-miR482a-3p and mpi-nmiR0028-3p) with specialized regulatory mechanisms for seed development and fatty acid biosynthesis in this woody plant. These miRNA candidates deserve further functional characterization and may be potentially applied in genetic breeding of new varieties with desired seed traits for Pongamia and other oilseed plants.

Supplementary Materials: Supplementary Materials can be found at http://www.mdpi.com/1422-0067/20/14/3509/s1.

Author Contributions: Conceptualization, J.H. and Y.Z.; Methodology, Y.J. and X.H.; Validation, Y.J. and X.H.; Formal analysis, L.L., T.H. and J.H.; Resources, J.H. and Y.Z.; Writing—original draft preparation, J.H.; Writing—review and editing, D.E.H. and J.H.; Supervision, T.H. and Y.Z.; Project administration, J.H.; Funding acquisition, J.H. and Y.Z. All authors read and approved the final manuscript.

Funding: This research was funded by the National Natural Science Foundation of China (No. 31300275 to J.H. and No. 31370289 to Y.Z.), and the Guangdong Innovation Research Team Fund (No.2014ZT05S078 to L.L. and T.H.).

Acknowledgments: We are grateful to the reviewers and editors for their comments and suggestions for improving the manuscript.

Conflicts of Interest: The authors declare no conflict of interest.

Abbreviations

ARF	Auxin Response Factor
DEmiRs	Differentially Expressed miRNAs
NAC	No Apical Meristem
qRT-PCR	quantitative Real-Time PCR
SPL	SQUAMOSA promoter-binding protein-like
TPM	Transcripts Per Million
WAF	Weeks After Flowering

References

1. Wang, L.; Yu, H.; He, X.; Liu, R. Influence of fatty acid composition of woody biodiesel plants on the fuel properties. *J. Fuel Chem. Technol.* **2012**, *40*, 397–404. [CrossRef]
2. Balat, M. Potential alternatives to edible oils for biodiesel production-A review of current work. *Energy Convers. Manag.* **2011**, *52*, 1479–1492. [CrossRef]
3. Karmee, S.K.; Chadha, A. Preparation of biodiesel from crude oil of *Pongamia pinnata*. *Bioresour. Technol.* **2005**, *96*, 1425–1429. [CrossRef] [PubMed]
4. Naik, M.; Meher, L.C.; Naik, S.N.; Das, L.M. Production of biodiesel from high free fatty acid Karanja *Pongamia pinnata* oil. *Biomass Bioenerg.* **2008**, *32*, 354–357. [CrossRef]
5. Sangwan, S.; Rao, D.V.; Sharma, R.A. A review on *Pongamia pinnata* (L.) Pierre: A great versatile leguminous plant. *Nat. Sci.* **2010**, *8*, 130–139.
6. Huang, J.; Lu, X.; Yan, H.; Chen, S.; Zhang, W.; Huang, R.; Zheng, Y. Transcriptome characterization and sequencing-based identification of salt-responsive genes in *Millettia pinnata*, a semi-mangrove plant. *DNA Res.* **2012**, *19*, 195–207. [CrossRef] [PubMed]
7. Biswas, B.; Gresshoff, P.M. The role of symbiotic nitrogen fixation in sustainable production of biofuels. *Int. J. Mol. Sci.* **2014**, *15*, 7380–7397. [CrossRef]
8. Gresshoff, P.M.; Hayashi, S.; Biswas, B.; Mirzaei, S.; Indrasumunar, A.; Reid, D.; Samuel, S.; Tollenaere, A.; van Hameren, B.; Hastwell, A.; et al. The value of biodiversity in legume symbiotic nitrogen fixation and nodulation for biofuel and food production. *J. Plant Physiol.* **2015**, *172*, 128–136. [CrossRef]
9. Badole, S.L.; Bodhankar, S.L. Investigation of antihyperglycaemic activity of aqueous and petroleum ether extract of stem bark of *Pongamia pinnata* on serum glucose level in diabetic mice. *J. Ethnopharmacol.* **2009**, *123*, 115–120. [CrossRef]
10. Al Muqarrabun, L.M.R.; Ahmat, N.; Ruzaina, S.A.S.; Ismail, N.H.; Sahidin, I. Medicinal uses, phytochemistry and pharmacology of *Pongamia pinnata* (L.) Pierre: A review. *J. Ethnopharmacol.* **2013**, *150*, 395–420. [CrossRef]
11. Roy, R.; Pal, D.; Sur, S.; Mandal, S.; Saha, P.; Panda, C.K. Pongapin and Karanjin, furanoflavanoids of *Pongamia pinnata*, induce G2/M arrest and apoptosis in cervical cancer cells by differential reactive oxygen species modulation, DNA damage, and nuclear factor kappa-light-chain-enhancer of activated B cell signaling. *Phytother. Res.* **2019**, *33*, 1084–1094. [PubMed]
12. Jones, S.I.; Vodkin, L.O. Using RNA-Seq to profile soybean seed development from fertilization to maturity. *PLoS ONE* **2013**, *8*, e59270. [CrossRef] [PubMed]
13. Severin, A.J.; Woody, J.L.; Bolon, Y.; Joseph, B.; Diers, B.W.; Farme, A.D.; Muehlbauer, G.J.; Rex, T.N.; David, G.; James, E.S.; et al. RNA-Seq atlas of *Glycine max*: A guide to the soybean transcriptome. *BMC Plant Biol.* **2010**, *10*, 160. [CrossRef] [PubMed]
14. Xu, H.; Kong, X.; Chen, F.; Huang, J.; Lou, X.; Zhao, J. Transcriptome analysis of *Brassica napus* pod using RNA-Seq and identification of lipid-related candidate genes. *BMC Genom.* **2015**, *16*, 858. [CrossRef] [PubMed]
15. Huang, J.; Guo, X.; Hao, X.; Zhang, W.; Chen, S.; Huang, R.; Gresshoff, P.M.; Zheng, Y. De novo sequencing and characterization of seed transcriptome of the tree legume *Millettia pinnata* for gene discovery and SSR marker development. *Mol. Breed.* **2016**, *36*, 75. [CrossRef]

16. Wegrzyn, J.L.; Whalen, J.; Kinlaw, C.S.; Harry, D.E.; Puryear, J.; Loopstra, C.A.; Gonzalez-Ibeas, D.; Vasquez-Gross, H.A.; Famula, R.A.; Neale, D.B. Transcriptomic profile of leaf tissue from the leguminous tree, *Millettia pinnata*. *Tree Genet. Genomes* **2016**, *12*, 44. [CrossRef]

17. Sreeharsha, R.V.; Mudalkar, S.; Singha, K.T.; Reddy, A.R. Unravelling molecular mechanisms from floral initiation to lipid biosynthesis in a promising biofuel tree species, *Pongamia pinnata* using transcriptome analysis. *Sci. Rep.* **2016**, *6*, 34315. [CrossRef]

18. Moolam, A.R.; Singh, A.; Shelke, R.G.; Scott, P.T.; Gresshoff, P.M.; Rangan, L. Identification of two genes encoding microsomal oleate desaturases (FAD2) from the biodiesel plant *Pongamia pinnata* L. *Trees* **2016**, *30*, 1351–1360. [CrossRef]

19. Ramesh, A.M.; Kesari, V.; Rangan, L. Characterization of a stearoyl-acyl carrier protein desaturase gene from potential biofuel plant, *Pongamia pinnata* L. *Gene* **2014**, *542*, 113–121. [CrossRef]

20. Winarto, H.P.; Liew, L.C.; Gresshoff, P.M.; Scott, P.T.; Singh, M.B.; Bhalla, P.L. Isolation and characterization of circadian clock genes in the biofuel plant Pongamia (*Millettia pinnata*). *BioEnerg. Res.* **2015**, *8*, 760–774. [CrossRef]

21. Bartel, D.P. MicroRNAs: Genomics, biogenesis, mechanism, and function. *Cell* **2004**, *116*, 281–297. [CrossRef]

22. Filipowicz, W.; Bhattacharyya, S.N.; Sonenberg, N. Mechanisms of post-transcriptional regulation by microRNAs: Are the answers in sight? *Nat. Rev. Genet.* **2008**, *9*, 102–114. [CrossRef] [PubMed]

23. Jones-Rhoades, M.W.; Bartel, D.P.; Bartel, B. MicroRNAs and their regulatory roles in plants. *Annu. Rev. Plant Biol.* **2006**, *57*, 19–53. [CrossRef] [PubMed]

24. Sunkar, R.; Li, Y.; Jagadeeswaran, G. Functions of microRNAs in plant stress responses. *Trends Plant Sci.* **2012**, *17*, 196–203. [CrossRef] [PubMed]

25. Nodine, M.D.; Bartel, D.P. MicroRNAs prevent precocious gene expression and enable pattern formation during plant embryogenesis. *Gene Dev.* **2010**, *24*, 2678–2692. [CrossRef] [PubMed]

26. Liu, P.P.; Montgomery, T.A.; Fahlgren, N.; Kasschau, K.D.; Nonogaki, H.; Carrington, J.C. Repression of *AUXIN RESPONSE FACTOR10* by microRNA160 is critical for seed germination and post-germination stages. *Plant J.* **2007**, *52*, 133–146. [CrossRef] [PubMed]

27. Korbes, A.P.; Machado, R.; Guzman, F. Identifying conserved and novel microRNAs in developing seeds of *Brassica napus* using deep sequencing. *PLoS ONE* **2012**, *7*, e50663. [CrossRef]

28. Zhao, Y.T.; Wang, M.; Fu, S.X.; Yang, W.C.; Qi, C.K.; Wang, X.J. Small RNA profiling in two *Brassica napus* cultivars identifies microRNAs with oil production and development-correlated expression and new small RNA classes. *Plant Physiol.* **2012**, *158*, 813–823. [CrossRef]

29. Goettel, W.; Liu, Z.; Xia, J.; Zhang, W.; Zhao, P.X.; An, Y.Q. Systems and evolutionary characterization of microRNAs and their underlying regulatory networks in soybean cotyledons. *PLoS ONE* **2014**, *9*, e86153. [CrossRef]

30. Song, Q.; Liu, Y.; Hu, X.; Zhang, W.; Ma, B.; Chen, S.; Zhang, J. Identification of miRNAs and their target genes in developing soybean seeds by deep sequencing. *BMC Plant Biol.* **2011**, *11*, 5. [CrossRef]

31. Ma, X.; Zhang, X.; Zhao, K.; Li, F.; Li, K.; Ning, L.; He, J.; Xin, Z.; Yin, D. Small RNA and degradome deep sequencing reveals the roles of microRNAs in seed expansion in peanut *Arachis hypogaea* L. *Front. Plant Sci.* **2018**, *9*, 349. [CrossRef] [PubMed]

32. Zhao, C.; Xia, H.; Frazier, T.P.; Yao, Y.; Bi, Y.; Li, A.; Li, M.; Li, C.; Zhang, B.; Wang, X. Deep sequencing identifies novel and conserved microRNAs in peanuts *Arachis hypogaea* L. *BMC Plant Biol.* **2010**, *10*, 3. [CrossRef] [PubMed]

33. Galli, V.; Guzman, F.; de Oliveira, L.F.V.; Loss-Morais, G.; Körbes, A.P.; Silva, S.D.A.; Margis-Pinheiro, M.M.A.N.; Margis, R. Identifying microRNAs and transcript targets in *Jatropha* seeds. *PLoS ONE* **2014**, *9*, e83727. [CrossRef] [PubMed]

34. Yin, D.; Li, S.; Shu, Q.; Gu, Z.; Wu, Q.; Feng, C.; Xu, W.; Wang, L. Identification of microRNAs and long non-coding RNAs involved in fatty acid biosynthesis in tree peony seeds. *Gene* **2018**, *666*, 72–82. [CrossRef] [PubMed]

35. Huang, J.; Hao, X.; Jin, Y.; Guo, X.; Shao, Q.; Kumar, K.S.; Ahlawat, Y.K.; Harry, D.E.; Joshi, C.P.; Zheng, Y. Temporal transcriptome profiling of developing seeds reveals a concerted gene regulation in relation to oil accumulation in Pongamia (*Millettia pinnata*). *BMC Plant Biol.* **2018**, *18*, 140. [CrossRef] [PubMed]

36. Meyers, B.C.; Axtell, M.J.; Bartel, B.; Bartel, D.P.; Baulcombe, D.; Bowman, J.L.; Cao, X.; Carrington, J.C.; Chen, X.; Green, P.J.; et al. Criteria for annotation of plant microRNAs. *Plant Cell* **2008**, *20*, 3186–3190. [CrossRef] [PubMed]

37. Rajagopalan, R.; Vaucheret, H.; Trejo, J.; Bartel, D.P. A diverse and evolutionarily fluid set of microRNAs in *Arabidopsis thaliana*. *Gene. Dev.* **2006**, *20*, 3407–3425. [CrossRef]

38. Wu, Y.; Yang, L.; Yu, M.; Wang, J. Identification and expression analysis of microRNAs during ovule development in rice *Oryza sativa* by deep sequencing. *Plant Cell Rep.* **2017**, *36*, 1815–1827. [CrossRef]

39. Szittya, G.; Moxon, S.; Santos, D.M.; Jing, R.; Fevereiro, M.P.S.; Moulton, V.; Dalmay, T. High-throughput sequencing of *Medicago truncatula* short RNAs identifies eight new miRNA families. *BMC Genom.* **2008**, *9*, 593. [CrossRef]

40. Zabala, G.; Campos, E.; Varala, K.K.; Bloomfield, S.; Jones, S.I.; Win, H.; Tuteja, J.H.; Calla, B.; Clough, S.J.; Hudson, M.; et al. Divergent patterns of endogenous small RNA populations from seed and vegetative tissues of *Glycine max*. *BMC Plant Biol.* **2012**, *12*, 177. [CrossRef]

41. Fahlgren, N.; Jogdeo, S.; Kasschau, K.D.; Sullivan, C.M.; Chapman, E.J.; Laubinger, S.; Smith, L.M.; Dasenko, M.; Givan, S.A.; Weigel, D.; et al. MicroRNA gene evolution in *Arabidopsis lyrata* and *Arabidopsis thaliana*. *Plant Cell* **2010**, *22*, 1074–1089. [CrossRef] [PubMed]

42. Sharma, S.S.; Islam, M.A.; Negi, M.S.; Tripathi, S.B. Changes in oil content and fatty acid profiles during seed development in *Pongamia pinnata*(L.) Pierre. *Indian J. Plant Physiol.* **2015**, *20*, 281–284. [CrossRef]

43. Diaz, M.; Pecinkova, P.; Nowicka, A.; Baroux, C.; Sakamoto, T.; Gandha, P.Y.; Jeřábková, H.; Matsunaga, S.; Grossniklaus, U.; Pecinka, A. The SMC5/6 complex subunit NSE4A is involved in DNA damage repair and seed development. *Plant Cell* **2019**, *31*, 1579–1597. [CrossRef] [PubMed]

44. Yang, S.; Miao, L.; He, J.; Zhang, K.; Li, Y.; Gai, J. Dynamic transcriptome changes related to oil accumulation in developing soybean seeds. *Int. J. Mol. Sci.* **2019**, *20*, 2202. [CrossRef] [PubMed]

45. Wang, J.; Jian, H.; Wang, T.; Wei, L.; Li, J.; Li, C.; Li, L. Identification of microRNAs actively involved in fatty acid biosynthesis in developing *Brassica napus* seeds using high-throughput sequencing. *Front. Plant Sci.* **2016**, *7*, 1570. [CrossRef] [PubMed]

46. Mallory, A.C.; Bartel, D.P.; Bartel, B. MicroRNA-directed regulation of Arabidopsis *AUXIN RESPONSE FACTOR17* is essential for proper development and modulates expression of early auxin response genes. *Plant Cell* **2005**, *17*, 1360–1375. [CrossRef] [PubMed]

47. Mallory, A.C.; Dugas, D.V.; Bartel, D.P.; Bartel, B. MicroRNA regulation of NAC-domain targets is required for proper formation and separation of adjacent embryonic, vegetative, and floral organs. *Curr. Biol.* **2004**, *14*, 1035–1046. [CrossRef]

48. Reyes, J.L.; Chua, N. ABA induction of miR159 controls transcript levels of two MYB factors during Arabidopsis seed germination. *Plant J.* **2007**, *49*, 592–606. [CrossRef] [PubMed]

49. Allen, R.S.; Li, J.; Stahle, M.I.; Dubroué, A.; Gubler, F.; Millar, A.A. Genetic analysis reveals functional redundancy and the major target genes of the Arabidopsis miR159 family. *Proc. Natl. Acad. Sci. USA* **2007**, *104*, 16371–16376. [CrossRef] [PubMed]

50. Ohto, M.; Fischer, R.L.; Goldberg, R.B.; Nakamura, K.; Harada, J.J. Control of seed mass by APETALA2. *Proc. Natl. Acad. Sci. USA* **2005**, *102*, 3123–3128. [CrossRef]

51. Ohto, M.; Floyd, S.K.; Fischer, R.L.; Goldberg, R.B.; Harada, J.J. Effects of APETALA2 on embryo, endosperm, and seed coat development determine seed size in Arabidopsis. *Sex. Plant Reprod.* **2009**, *22*, 277–289. [CrossRef] [PubMed]

52. Aukerman, M.J.; Sakai, H. Regulation of flowering time and floral organ identity by a microRNA and its APETALA2-Like target genes. *Plant Cell* **2003**, *15*, 2730–2741. [CrossRef] [PubMed]

53. Huo, H.; Wei, S.; Bradford, K.J. *DELAY OF GERMINATION1* (*DOG1*) Regulates both seed dormancy and flowering time through microRNA pathways. *Proc. Natl. Acad. Sci. USA* **2016**, *113*, E2199–E2206. [CrossRef] [PubMed]

54. Pommier, Y.; Huang, S.N.; Gao, R.; Das, B.B.; Murai, J.; Marchand, C. Tyrosyl-DNA-phosphodiesterases (TDP1 and TDP2). *DNA Repair* **2014**, *19*, 114–129. [CrossRef] [PubMed]

55. Fu, X.; Deng, S.; Su, G.; Zeng, Q.; Shi, S. Isolating high-quality RNA from mangroves without liquid nitrogen. *Plant Mol. Biol. Rep.* **2004**, *22*, 197. [CrossRef]
56. Varkonyi-Gasic, E.; Wu, R.; Marion, W.; Walton, E.F.; Hellens, R.P. Protocol: A highly sensitive RT-PCR method for detection and quantification of microRNAs. *Plant Methods* **2007**, *3*, 12. [CrossRef]

International Journal of
Molecular Sciences

Article

Nitrogen Fertilizer Induced Alterations in The Root Proteome of Two Rice Cultivars

Jichao Tang [1], Zhigui Sun [1], Qinghua Chen [1], Rebecca Njeri Damaris [2], Bilin Lu [1,*] and Zhengrong Hu [2,*]

[1]   Hubei Collaborative Innovation Center for Grain Industry, Agricultural College, Yangtze University, Jingzhou 434025, China
[2]   State Key Laboratory of Biocatalysis and Enzyme Engineering, School of Life Sciences, Hubei University, Wuhan 430062, China
*   Correspondence: blin9921@sina.com (B.L.); huzhengrong1001@hubu.edu.cn (Z.H.)

Received: 4 June 2019; Accepted: 24 July 2019; Published: 26 July 2019

Abstract: Nitrogen (N) is an essential nutrient for plants and a key limiting factor of crop production. However, excessive application of N fertilizers and the low nitrogen use efficiency (NUE) have brought in severe damage to the environment. Therefore, improving NUE is urgent and critical for the reductions of N fertilizer pollution and production cost. In the present study, we investigated the effects of N nutrition on the growth and yield of the two rice (*Oryza sativa* L.) cultivars, conventional rice Huanghuazhan and indica hybrid rice Quanliangyou 681, which were grown at three levels of N fertilizer (including 135, 180 and 225 kg/hm^2, labeled as N9, N12, N15, respectively). Then, a proteomic approach was employed in the roots of the two rice cultivars treated with N fertilizer at the level of N15. A total of 6728 proteins were identified, among which 6093 proteins were quantified, and 511 differentially expressed proteins were found in the two rice cultivars after N fertilizer treatment. These differentially expressed proteins were mainly involved in ammonium assimilation, amino acid metabolism, carbohydrate metabolism, lipid metabolism, signal transduction, energy production/regulation, material transport, and stress/defense response. Together, this study provides new insights into the regulatory mechanism of nitrogen fertilization in cereal crops.

Keywords: nitrogen fertilizer; rice; proteome; cultivars; nitrogen use efficiency (NUE)

---

1. Introduction

Nitrogen (N), an essential plant nutrient, is a key factor limiting the crop growth and productivity [1,2]. The amount of applied fertilizer in the world has increased 10-fold in the last half century, to meet the food demand of the growing world population. This increasing trend is predicted to be even greater in this century [3]. However, less than half of the field applied nitrogen fertilizer is absorbed and utilized by plants, while most of it is dissipated in the atmosphere, and leached into groundwater, lakes and rivers, which induce increasingly severe pollutions to the environment [4]. Thus, the improvement of N use efficiency (NUE) is an urgent task for agricultural sustainability and environmental protection. Rice (*Oryza sativa* L.), cultivated in more than 100 countries, is the major food source of at least half of the world's population [5]. Therefore, proper management of N fertilizer is pivotal for improvement of rice yield.

Plants have developed an intricate N detection system, in which N status is constantly sensed and the plants continuously adapt to the changes by regulating gene expression, enzyme activity, and substance metabolism [6,7]. In plants, inorganic nitrogen is firstly reduced to ammonia (NH$_3$) and then incorporated into organic compounds [8,9]. Afterwards, ammonia is assimilated into these amino acids that serve as vital nitrogen carriers in plants, including glutamine, glutamate, asparagine, and aspartate [10]. The major enzymes, responsible for the biosynthesis of these nitrogen-carrying amino acids, are glutamine synthetase (GS), glutamate dehydrogenase (GDH), glutamate synthase (GOGAT), aspartate aminotransferase (AspAT), as well as asparagine synthetase (AS) [10]. GS is a key enzyme for ammonia assimilation, which has high affinity for ammonia and ability to incorporate ammonia efficiently into organic combination (compounds). The glutamate synthase cycle has been well established in that NH_3 is converted into organic compounds via its assimilation by GS, and this is the major route of nitrogen assimilation in plants [11]. In addition, N can be transferred into and out of proteins in different organs, and be moved between different organs in plants, which are achieved via the activity of transaminases, glutamine-amide transferases and GS [10]. However, excessive ammonium ion (NH4$^+$) is one of the inorganic pollutants in aqueous solution, which could induce eutrophication and impair self-purification of water [12].

Numerous researches have been conducted to illuminate the regulatory mechanisms involved in NUE of crop plants [13]. Genetic engineering methods have become the research focus for improvement of biological N fixation [1,14–16]. However, these techniques are not very effective in improving NUE of crop plants. Furthermore, manipulating genes involved in the nitrogen uptake and assimilation have not yielded any obvious improvement of nitrogen utilization capacity of plant [1,14]. For example, overexpression of *nitrite reductase* (*NiR*) genes, which encodes the enzyme catalyzing the first and the rate-limiting step of N assimilation, did not cause any obvious change of NUE in plants [14].

Recently, the proteomic approach has become an effective tool in the study of nitrogen regulation. Ding et al. (2011) identified twelve protein spots by two-dimensional gel electrophoresis (2-DE), providing a better understanding of the mechanism underlying rice response to low nitrogen stress [17]. Besides, Liao et al. (2012) demonstrated that 47 differentially expressed proteins were identified in young ears of maize treated with different levels of nitrogen fertilizer, and the regulatory process was associated with hormonal metabolism [2]. Chandna and Ahmad (2015) investigated the effects of the nitrogen nutrition on protein expression pattern of leaves in low-N sensitive and low-N tolerant wheat (*Triticum aestivum* L.) varieties, and found proteins related to photosynthesis, glycolysis, nitrogenmetabolism, sulphur metabolism and defense [18]. Hakeem et al. (2012) analyzed the leaf proteins expression patterns of high and low N-responsive contrasting rice genotypes grown at three levels of N fertilizer, and determined that the differentially expressed proteins were involved in the energy production/regulation and metabolism [1].

Root system is essential and critical for plant growth, due to its involvement in nutrients and water absorption, as well as plant hormones synthesis [19,20]. Development of root system is remarkably affected by nitrogen supply and its distribution in the soil [21]. Liu et al. (2018) reported that five elite super hybrid rice cultivars displayed greater root traits and higher grain yield under N treatment at level of 300 kg/hm^2 [22]. Total root number and total root length of these cultivars reached maximum at 55 days after transplanting. However, the molecular mechanism underlying N fertilizer response of root system is still unclear. In the present study, the agronomic traits of two rice cultivars were determined, conventional rice Huanghuazhan and indica hybrid rice Quanliangyou 681, grown at three levels of nitrogen fertilizer (including 135, 180 and 225 kg/hm^2, labeled as N9, N12, N15, respectively). Then, the root proteins of the two rice cultivars under N nutrition (at the level of N15) were identified by using tandem mass tags (TMT)-labelling proteomic platform. This work will provide new insights into the N fertilizer response of rice root system, and the identified key proteins can be used for developing new strategies to improve NUE of rice.

2. Results

2.1. Determination of Agronomic Traits

Accumulation of dry matter is essential for crop yield formation, and dry weight is an extensively used parameter for assessing the growth conditions of plants. As shown in Table 1, the shoot dry weight increased with the growth of plants, which was also significantly affected by nitrogen fertilizer. For Huanghuazhan cultivar, three levels of N fertilizer treatments increased the values of dry weight at the four stages, and the positive effects of N12 and N15 were relatively outstanding. Similar results were observed in Quanliangyou 681, and N15 displayed the most remarkable impact. When compared, the difference in dry weight of the two rice cultivars showed that the values for Quanliangyou 681 were significantly higher than that of Huanghuazhan at a given level of N fertilizer at any stage. Cereal yield, a key agronomic trait for agricultural production, was also measured in this study. As shown in Figure 1, three levels of nitrogen fertilizer treatments increased the yields of Huanghuazhan cultivar in undifferentiated ways. Differentially, the production of Quanliangyou 681 improved with the increase of nitrogen fertilizer concentration, and the effect of N15 was the most notable. Furthermore, the yield of Quanliangyou 681 was significantly higher than that of Huanghuazhan with or without the treatment of nitrogen fertilizer (at the same level). These results suggested that the three levels of N fertilizer treatments could improve the growth and yields of the two rice cultivars, especially for Quanliangyou 681; and N12 and N15 could be used as the optimal concentration of nitrogen fertilizer treatment for Huanghuazhan and Quanliangyou 681, respectively.

Table 1. Dry weight (t/hm^2) of rice cultivars under Nitrogen (N) treatment at different levels in different stages.

Cultivar	Level	Tillering Stage	Jointing Stage	Full Heading Stage	Mature Stage
Huanghuazhan	N0	0.57 ± 0.06c	2.34 ± 0.17c	8.70 ± 0.37c	11.86 ± 0.19b
	N9	1.21 ± 0.12b	3.63 ± 0.04b	10.29 ± 0.29c	16.68 ± 0.26a
	N12	1.57 ± 0.15a	4.38 ± 0.17a	12.58 ± 0.45b	17.06 ± 0.35a
	N15	1.78 ± 0.15a	4.61 ± 0.20a	14.94 ± 0.54a	16.86 ± 0.27a
Quanliangyou 681	N0	0.72 ± 0.07b	3.07 ± 0.14d *	9.94 ± 0.46c *	14.77 ± 0.16b *
	N9	2.15 ± 0.02a *	4.25 ± 0.26c *	14.76 ± 0.16b *	18.21 ± 0.29a *
	N12	2.39 ± 0.19a *	4.91 ± 0.07b *	14.59 ± 0.062b *	18.10 ± 0.23a *
	N15	2.48 ± 0.25a *	5.76 ± 0.17a *	16.26 ± 0.13a *	19.23 ± 1.15a *

Note: Huanghuazhan and Quanliangyou 681 rice seedlings were treated with three levels of nitrogen fertilizer (including 135, 180 and 225 kg/hm^2, labeled as N9, N12, N15, respectively), and the untreated group was considered as the control (labeled as N0). Different letters (a, b, c) indicate significant differences among different nitrogen fertilizer levels for a given cultivar in a given stage based on Duncan's multiple range tests ($p < 0.05$). Star (*) represents statistical significance between the two cultivars at a given nitrogen fertilizer level in a given stage based on Independent-samples *t* test ($p < 0.05$).

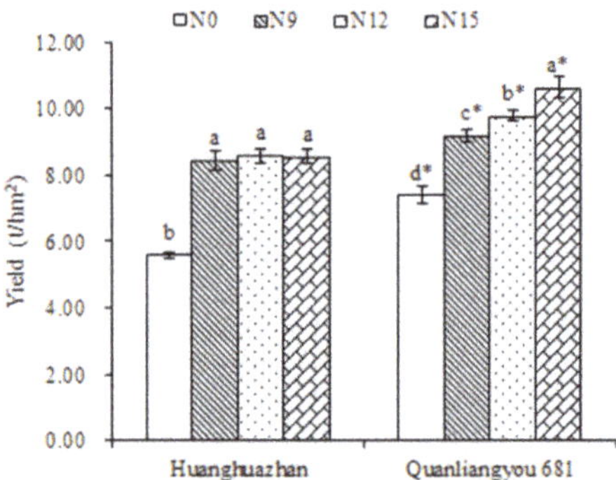

Figure 1. Grain yields of the two rice cultivars treated with N fertilizer at three levels. Seeds of two rice varieties, conventional rice Huanghuazhan and indica hybrid rice Quanliangyou 681, were soaked, pre-germinated for 48 h, and sown in soil. Seedlings of the two rice varieties were treated with nitrogen fertilizer at three levels (including 135, 180 and 225 kg/hm^2, labeled as N9, N12, N15, respectively), and the untreated group was considered as the control (labeled as N0). Different letters (a, b, c) indicate significant differences among different nitrogen fertilizer levels for a given cultivar based on Duncan's multiple range tests ($p < 0.05$). Star (*) represents statistical significance between the two cultivars at a given nitrogen fertilizer level based on Independent-samples *t* test ($p < 0.05$).

2.2. Proteome-Wide Analysis of Differentially Expressed Protein of The Two Rice Cultivars Treated With Nitrogen Fertilizer

In total, 6728 proteins were identified by searching the rice database, with 6093 protein being quantified. All the annotation and quantification information of these proteins are presented in Table S1. The quantitative repeatability of proteins was evaluated by principal component analysis (PCA) and relative standard deviation (RSD) statistical analysis. Good quantitative repeatability of proteins was determined by the small RSD values and high aggregation degree between repeated samples (Figure S1).

In this present study, proteins with the threshold change fold >2 or <0.5, and p value < 0.05 were considered as up-regulated and down-regulated proteins, respectively. As shown in Figure 2, the number of differentially expressed proteins and an overlap of these proteins were summarized. In total, 511 differentially expressed proteins in the two rice cultivars after application of N fertilizer were identified, and all the protein information is provided in Table S2. N15 treatment induced 109 differentially expressed proteins (93 up-regulated and 16 down-regulated) in Quanliangyou 681, while 283 differentially expressed proteins (160 up-regulated and 123 down-regulated) in Huanghuazhan (Figure 2A). Under control condition, 68 up-regulated and 108 down-regulated proteins were identified in Huanghuazhan, when compared with Quanliangyou 681. After application of nitrogen, the expression levels of 63 proteins were increased and 115 proteins decreased in Huanghuazhan when compared with Quanliangyou 681 (Figure 2A). Figure 2B showed that there were 40 differentially expressed proteins shared by the comparisons of 681_N15 vs. 681_cK, and H_N15 vs. H_CK. Interestingly, 43 proteins revealed differentially expression abundance in two sample comparisons (H_CK vs. 681_CK, H_N15 vs. 681_N15) were regulated by rice cultivar difference, regardless of N fertilizer treatment.

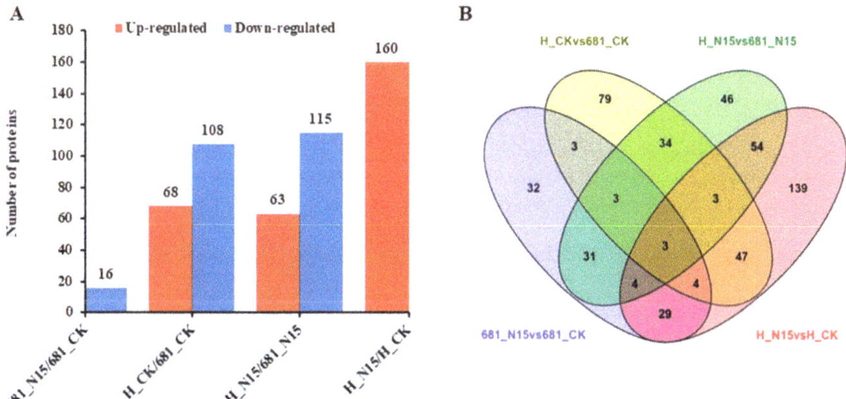

Figure 2. Histogram (**A**) and Venn diagram (**B**) of the number distribution of differentially expressed proteins in different comparison groups. Proteomics analysis was performed on the roots of the two rice cultivars at booting stage with/without treatment of N15 (225 kg/hm^2 nitrogen fertilizer). N15 represented rice under nitrogen fertilizer treatment with the concentration of 225 kg/hm^2, while control check (CK) represented the control without nitrogen fertilizer treatment; 681 and H represented Quanliangyou 681 and Huanghuazhan cultivars, respectively. Proteins with threshold change fold > 2 and p value < 0.05 were regarded as up-regulated, while quantitative ratio < 0.5 and p value < 0.05 were considered as down-regulated.

2.3. Validation of Differentially Expressed Proteins by Parallel Reaction Monitoring

To validate the reliability of TMT-labeling proteomic results, parallel reaction monitoring (PRM) analysis was carried out. In detail, we selected 16 differentially expressed proteins due to their functional significance concluded from proteome analysis, and the ratio of protein abundance varied in a wide range. Then, we analyzed the expression levels of these 16 proteins among the four regimes (H_CK; 681_CK; H_N15; 681_N15). Consequently, a total of 64 quantitative values produced by PRM analysis were obtained. Correlation analysis was carried out of the 64 PRM values with their corresponding protein expression level generated by TMT labeling proteomic platform. As shown in Figure 3, the results generated by PRM analysis showed a strong correlation with the data produced by TMT labeling detection system (Pearson correlation coefficients r^2 = 0.93). In addition, the distributions of peptide fragment ion peak area of all the 16 proteins is provided in the Figure S3.

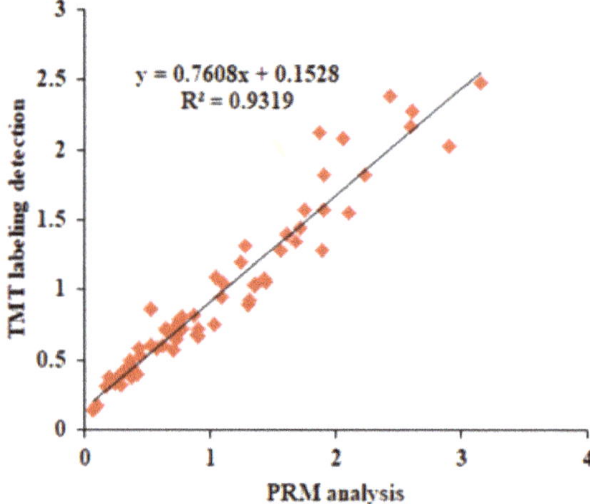

Figure 3. Validation of differentially expressed proteins by parallel reaction monitoring (PRM). Correlation of fold change analyzed by tandem mass tags (TMT) labelling detection platform (*y* axis) with the data obtained by PRM reaction (*x* axis).

2.4. Gene Ontology Classification Analysis of Differentially Expressed Proteins

To further explore the cellular functions of these differentially expressed proteins regulated by N fertilizer in rice, Gene Ontology (GO) classification analysis was performed. The top three cellular components were cell, membrane and macromolecular complex related proteins in Huanghuazhan, while extracellular region, cell, membrane related proteins in Quanliangyou 681 (Figure 4A). In the ontology of molecular function, binding and catalytic activity related proteins were the overwhelming preponderant in the two cultivars (Figure 4B). As for biological processes classification, metabolic process, single-organism process and cellular process related proteins were the dominant components in the two rice cultivars after N fertilizer treatment (Figure 4C).

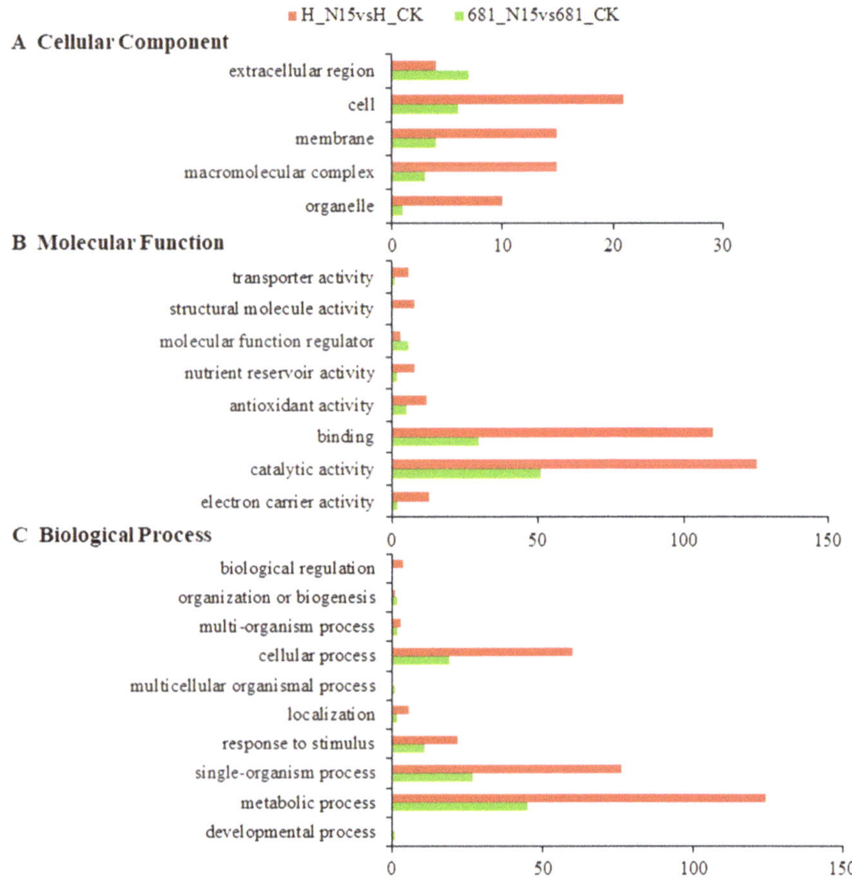

Figure 4. Statistical distribution chart of differentially expressed proteins in the two rice cultivars roots under each Gene Ontology (GO) category (2nd Level). **A.** Cellular Component; **B.** Biological Process; **C.** Molecular Function. N15 represented rice under nitrogen fertilizer treatment with the concentration of while 225 kg/hm^2, while control check (CK) represented the control without nitrogen fertilizer treatment; 681 and H represented Quanliangyou 681 and Huanghuazhan cultivars, respectively.

2.5. Clustering Analysis of The Differentially Expressed Proteins

To better understand the involved pathways of these differentially expressed proteins, clustering analyses based on GO enrichments and Encyclopedia of Genes and Genomes (KEGG) pathway enrichments were conducted. Several biological processes altered in the two rice cultivars after application of nitrogen fertilizer, such as defense response to bacterium, fungus, and other organism; photosynthetic electron transport chain, as well as hexose metabolic process (Figure S2). As shown in Figure 5, the differentially expressed proteins in the comparison of 681_N15 vs. 681_CK were mainly enriched in valine, leucine and isoleucine degradation, amino sugar and nucleotide sugar metabolism, alpha-linolenic acid metabolism, galactose metabolism, cysteine and methionine metabolism. As for Huanghuazhan, the proteins regulated by nitrogen fertilizer were enriched in diverse pathways, such as photosynthesis, diterpenoid biosynthesis, alanine, aspartate and glutamate metabolism, linoleic acid metabolism, carbon fixation in photosynthetic organisms, nitrogen metabolism, arginine biosynthesis, and tryptophan metabolism. Under control condition, the differentially expressed proteins between the two rice cultivars were largely enriched in several pathways, including ribosome, photosynthesis-antenna proteins, plant-pathogen interaction, MAPK signaling pathway-plant. However, after treatment of the N fertilizer, the pathways enriched by differentially expressed proteins were changed, which mainly involved in biosynthesis of secondary metabolites, glycine, serine and threonine metabolism, phenylpropanoid biosynthesis, terpenoid backbone biosynthesis.

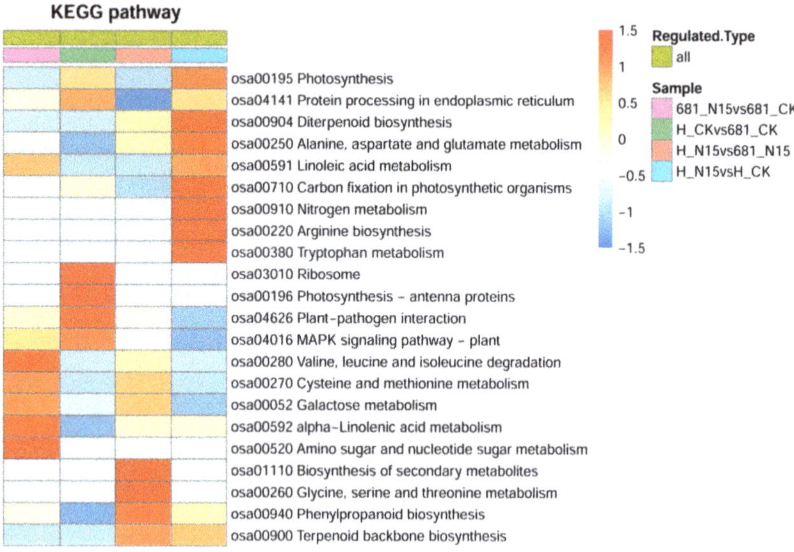

Figure 5. Heatmap for cluster analysis of the enrichment patterns of Encyclopedia of Genes and Genomes (KEGG) pathways. N15 represented rice under nitrogen fertilizer treatment with the concentration of while 225 kg/hm^2, while control check (CK) represented the control without nitrogen fertilizer treatment; 681 and H represented Quanliangyou 681 and Huanghuazhan cultivars, respectively.

To obtain a better understanding of the possible regulatory mechanism of nitrogen fertilizer in different rice cultivars, several typical groups of differentially expressed proteins were summarized (Tables 2–4, and Tables S3–S5). These proteins are involved in ammonium assimilation, signal transduction, substance metabolism, energy metabolism, material transport, and stress/defense response.

As shown in Table 2, several differentially expressed proteins involved in ammonium assimilation were identified in the comparisons of H_N15 vs. H_CK, and 681_N15 vs. 681_CK. As for Huanghuazhan cultivar, two glutamine synthetases (B8AVN6; B8AJN3) and one glutamate synthase (B8AWG6) were down-regulated, but one glutathione synthetase (B8AZE9) and one glutamate carboxypeptidase (B8A9D2) were found as up-regulated. Differentially, in Quanliangyou 681 cultivar, there was only one up-regulated glutamine synthetase (B8AQE3) identified.

Table 2. The differentially expressed proteins involved in ammonium assimilation in the comparisons of H_N15 vs. H_CK, and 681_N15 vs. 681_CK.

Protein Accession	Protein Description	Gene Name	H_N15/H_CK	681_N15/681_CK
B8AQE3	Glutamine synthetase	*OsI_13264*	1.359	3.216
B8AVN6	Glutamine synthetase	*OsI_17756*	0.486	0.782
B8AJN3	Glutamine synthetase	*OsI_10575*	0.259	0.605
B8AZE9	Glutathione synthetase	*OsI_18978*	2.226	1.734
B8AWG6	glutamate synthase 2 [NADH], chloroplastic	*OsI_20922*	0.308	0.605
B8A9D2	probable glutamate carboxypeptidase 2 isoform X1	*OsI_03680*	2.028	1.108
B8AGA6	glutathionyl-hydroquinone reductase YqjG isoform X2	*OsI_06963*	2.115	1.618

Note: In this present study, proteins with the threshold change fold >2 or <0.5, and *p* value < 0.05 were considered as up-regulated and down-regulated proteins, respectively. Black color represented up-regulated protein, and gray color indicated down-regulated protein. N15 represented rice under nitrogen fertilizer treatment with the concentration of 225 kg/hm^2, while control check (CK) represented the control without N fertilizer treatment; 681 and H represented Quanliangyou 681 and Huanghuazhan cultivar, respectively.

In the present study, members of proteins associated with signal transduction were identified in in the comparisons of H_N15 vs. H_CK, and 681_N15 vs. 681_CK (Table 3). Two up-regulated receptor-like kinases (A2Z8U1; A2Y361) and two 1-aminocyclopropane-1-carboxylate oxidases (A2ZC94; B8AY41) were up-regulated, while one serine/threonine-protein kinase STY46 (B8B2K2) and one aspartokinase 1 (B8ANP3) were down-regulated by nitrogen fertilizer in Huanghuazhan cultivar. For Quanliangyou 681, all the proteins involved in signal transduction were up-regulated by N nutrition, including one cysteine-rich receptor-like protein kinase (B8B6Z1), two 1-aminocyclopropane-1-carboxylates (A2ZC94; B8AY41), one serine/threonine-protein kinase STY46 (B8AGJ3), and one phosphoenolpyruvate carboxykinase (A2XEP1); except aspartokinase 1 (B8ANP3).

Table 3. The differentially expressed proteins involved in signal transduction in the comparisons of H_N15 vs. H_CK, and 681_N15 vs. 681_CK.

Protein Accession	Protein Description	Gene Name	H_N15/H_CK	681_N15/681_CK
A2Z8U1	putative receptor-like protein kinase	*OsI_34134*	3.803	1.294
A2Y361	receptor-like protein kinase FERONIA	*OsI_19449*	2.090	1.301
B8B6Z1	cysteine-rich receptor-like protein kinase 25	*OsI_26338*	1.814	2.017
A2ZC94	1-aminocyclopropane-1-carboxylate oxidase	*OsI_35404*	2.959	3.840
B8AY41	1-aminocyclopropane-1-carboxylate oxidase	*OsI_18467*	2.102	2.676
B8AGJ3	serine/threonine-protein kinase STY46	*OsI_05618*	1.331	2.621
A2XEP1	phosphoenolpyruvate carboxykinase [ATP]	*OsI_10803*	1.220	2.310
B8B2K2	serine/threonine-protein kinase STY46	*OsI_24506*	0.447	0.745
B8ANP3	aspartokinase 1, chloroplastic isoform X1	*OsI_14341*	0.402	0.359

Note: In this present study, proteins with the threshold change fold >2 or <0.5, and *p* value < 0.05 were considered as up-regulated and down-regulated proteins, respectively. Black color represented up-regulated protein, and gray color indicated down-regulated protein. N15 represented rice under nitrogen fertilizer treatment with the concentration of 225 kg/hm^2, while control check (CK) represented the control without N fertilizer treatment; 681 and H represented Quanliangyou 681 and Huanghuazhan cultivar, respectively.

Moreover, many differentially expressed proteins involved in energy metabolism and material transport were also found in the comparisons of H_N15 vs. H_CK, and 681_N15 vs. 681_CK. As shown in Table 4, one NAD(P)H-dependent oxidoreductase 2 (B8BFL3), and one glutamate dehydrogenase (A2XW22) were up-regulated in Quanliangyou 681 after treatment of N15, and similar results were observed in the category of material transport (one sugar carrier protein C, A2YML7; and one calmodulin-like protein, B8ACJ8). Differently, more proteins were found in Huanghuazhan cultivar with involvement in energy metabolism and material transport, such as NAD(P)H-dependent oxidoreductase 1/2, ferredoxin, ferredoxin–NADP reductase, some dehydrogenases, ABC transporter G family members, pyrophosphate-energized vacuolar membranes, potassium channel KAT2 isoform X1, transmembrane protein, etc.

Table 4. The differentially expressed proteins involved in energy metabolism and material transport in the comparisons of H_N15 vs. H_CK, and 681_N15 vs. 681_CK.

Protein Accession	Protein Description	Gene Name	H_N15/H_CK	681_N15/681_CK
	Energy Metabolism			
A2Z4G9	probable NAD(P)H-dependent oxidoreductase 1	*OsI_32550*	2.723	1.387
B8BFL3	probable NAD(P)H-dependent oxidoreductase 2	*OsI_32552*	2.242	2.204
A2XNS0	Ferredoxin	*OsI_15810*	2.574	0.622
A2YI62	Ferredoxin–NADP reductase	*OsI_24896*	2.283	1.372
A2ZCK1	Alcohol dehydrogenase family-3	*OSI9Ba083C*	4.092	1.227
B8ADR5	cytokinin dehydrogenase 2	*OsI_00771*	3.210	1.205
A2YBK1	aldehyde dehydrogenase family 2 member B7, mitochondrial	*OsI_22484*	2.081	1.288
A2XW22	Glutamate dehydrogenase 2, mitochondrial	*GDH2*	1.416	2.212
B8AF09	Glyceraldehyde-3-phosphate dehydrogenase	*OsI_07948*	0.402	0.643
A2Y8B2	L-lactate dehydrogenase	*OsI_21291*	0.371	0.569
A2XU83	Glyceraldehyde-3-phosphate dehydrogenase	*OsI_16160*	0.361	1.098
	Material Transport			
B8BLE1	ABC transporter G family member 48	OsI_36727	2.274	1.518
B8B6Q3	ABC transporter G family member 43	OsI_26239	2.152	1.900
A2WPG7	pyrophosphate-energized vacuolar membrane proton pump	OsI_01741	2.025	1.262
A2XAP0	pyrophosphate-energized vacuolar membrane proton pump	OsI_09320	0.390	0.685
B8ACQ3	potassium channel KAT2 isoform X1	OsI_00861	0.486	0.783
B8BBQ9	dipeptide transport ATP-binding protein	OsI_28206	0.445	0.747
A2YP92	"transmembrane protein, putative	OsI_27085	0.432	1.146
A2YVV8	putative copper transporter 5.2	OsI_29464	0.474	0.892
A2YML7	sugar carrier protein C	OsI_26468	1.572	2.925
B8ACJ8	calmodulin-like protein	OsI_01318	1.446	5.881

Note: In this present study, proteins with the threshold change fold >2 or <0.5, and *p* value < 0.05 were considered as up-regulated and down-regulated proteins, respectively. Black color represented up-regulated protein, and gray color indicated down-regulated protein. N15 represented rice under nitrogen fertilizer treatment with the concentration of 225 kg/hm^2, while control check (CK) represented the control without N fertilizer treatment; 681 and H represented Quanliangyou 681 and Huanghuazhan cultivar, respectively.

Many proteins associated with substance metabolism were identified in the comparisons of H_N15 vs. H_CK, and 681_N15 vs. 681_CK (Table S3), which included aminotransferase, methyltransferase, aspartic proteinase nepenthesin, glycosyltransferase, glucosidase, pyrophosphate–fructose 6-phosphate, lipoxygenase, etc. In the category of substance metabolism, the N fertilizer regulated proteins were more in the comparison of H_N15 vs. H_CK than that of 681_N15 vs. 681_CK. Although the number of differentially expressed proteins was smaller in Quanliangyou 681, all the proteins were up-regulated by N fertilizer in this cultivar, expect one O-methyltransferase 2 (Table S3).

Proteins related to stress/defense response were also found to be regulated by N fertilizer, such as glutathione S-transferase, peroxidase, dehydrin, late embryogenesis abundant protein (LEA), momilactone A synthase, and chitinase, etc. Similarly, the differentially expressed proteins in the comparison of 681_N15 vs. 681_CK were less than that of H_N15 vs. H_CK, but most of them were up-regulated by N nutrition (Table S4). Furthermore, the differentially expressed proteins in the comparisons of H_CK vs. 681_CK, and H_N15 vs. 681_N15 were also summarized in the above classifications (Table S5). And the results suggested that the most proteins were down-regulated in Huanghuazhan compared with Quanliangyou 681, regardless of nitrogen fertilizer or not. Based on these results, we speculated that application of nitrogen fertilizer could influence a wide variety of biological processes in plants, and the expression levels of these proteins are relatively higher for Quanliangyou 681.

3. Discussion

To improve cereal production, application of inorganic nitrogenous fertilizers is one of necessary input. The amount of N fertilizer applied in agriculture has increased at an amazing speed in the last 50 years [12]. Consequently, the use of nitrogen fertilizers has already shown severe damage to the environment [1]. Therefore, it is urgent and crucial for the agricultural industry to analyze the molecular mechanism underlying NUE of crop plants. In the present study, we designed three levels for nitrogen fertilizer treatment of the two rice cultivars, including 135, 180 and 225 kg/hm^2, labeled as N9, N12, N15, respectively. The results showed that three levels of N treatments increased the shoot dry weight and yields of the two rice cultivars, especially for Quanliangyou 681; N12 and N15 could be used as optimal concentration of N fertilizer application for Huanghuazhan and Quanliangyou 681, respectively (Table 1 and Figure 1).

Although previous studies have involved proteomic analysis of N fertilizer response in rice, most of them are concentrated on leaf proteins and are conducted by two-dimensional gel electrophoresis (2-DE) [1,3,23,24]. Studies focused on the changes of root proteome in response to N nutrition are limited. Ding et al. (2011) performed a proteomic study to investigate the response of rice root to low N stress, and they identified twelve proteins that were involved in tricarboxylic acid cycle, adenylate metabolism, phenylpropanoid metabolism, and protein degradation [17]. Besides, to analyze proteins regulated by nitrogen and cytokinin in rice roots, Ding et al. (2012) carried out a comparative proteomic analysis using two-dimensional polyacrylamide gel electrophoresis. Twenty-eight proteins were successfully identified, which were categorized into classes related to energy, metabolism, disease/defense, protein degradation, signal transduction, transposons, and unclear classification [25]. However, the number of identified differentially expressed proteins is relatively limited when using two-dimensional gel electrophoresis approach.

In the present study, a proteomic analysis was performed using TMT labeling detection platform, to explore the possible mechanism underlying N fertilizer response of rice root system. A total of 6093 proteins were quantified, and 511 differentially expressed proteins were identified (Figure 2 and Tables S1 and S2). Go analysis showed that metabolic process related proteins were the dominant group in the ontology of biological process (Figure S2). Consistently, the KEGG pathway analysis indicated that the proteins involved in amino acid metabolism, nitrogen metabolism and glycometabolism were significantly enriched in the two rice cultivars after treatment of N15 (Figure 5). These results determined that diverse biological processes may be involved in N fertilizer response of rice root system, which are summarized as follows.

3.1. Ammonium Assimilation

Plants take up inorganic nitrogen in the form of nitrate and ammonium from the soil and dinitrogen in the atmosphere. Ammonium is the final form of inorganic nitrogen, which is derived from primary nitrate reduction catalyzed by nitrate reductase and ferredoxin (Fd)-nitrite reductase, as well as absorption from soil and symbiotic dinitrogen fixation in root nodules of leguminous plants. Ammonium is then assimilated into glutamine and glutamate via glutamine synthetase (GS) and glutamate synthase (GOGAT) through GS/GOGAT cycle [26,27]. The glutamate (Glu) with ATP-dependent NH4$^+$ are converted into glutamine (Gln) by GS, then the amide group of Gln is transferred to α-ketoglutarate (2-OG) to yield Glu [28]. It has been reported that the entire N in plant is channeled through the GS catalyzed reactions [28]. N atom can pass through the reaction catalyzed by GS many times, from uptake from the soil, assimilation and remobilization to final deposition in a storage protein of seed [29]. During this process, Gln serves as one important nitrogen donors for the biosynthesis of organic nitrogenous compounds (eg., nucleotides, amino acids and chlorophyll). Therefore, GS is a pivotal factor that controls nitrogen assimilation in plant. In this study, three differentially expressed GSs were identified, including two down-regulated (B8AJN3 and B8AVN6) in Huanghuazhan and one up-regulated (B8AQE3) in Quanliangyou 681 after treatment with N fertilizer (Table 2). Besides, one glutamate synthase 2 (B8AWG6) was decreased in Huanghuazhan, while one glutamate dehydrogenase (A2XW22) was increased in Quanliangyou 681. These results indicated that the roles of nitrogen fertilizer and the mode of nitrogen assimilation might be different between the two cultivars. Accordantly, Hakeem et al. (2012) revealed that the expression intensity of GS enhanced in rice genotype 'Munga Phool' (low nitrogen efficiency) after treatments with KNO_3 at levels of 10 mM and 25 mM, but the intensity in 'Rai Sudha' (high nitrogen efficiency) decreased [1]. The vital role of GS has been highlighted in productivity of maize kernel in the study of Hirel and Lea (2001) by using a quantitative genetic approach [28]. One QTL for thousand kernels weight was identified coincident with a GS (Gln1–4) locus, while two QTLs for thousand kernel weight and yield were coincident with another GS (Gln1–3) locus. Furthermore, the positive correlation between kernel yield and GS activity has been revealed in previous study [30]. Herein, we deduced that the higher level of GS expression in Quanliangyou 681 could enhance the incorporation of $NH4^+$ into organic compounds, which is beneficial to increase grain production.

3.2. Signal Transduction

It is very common for living organisms to perceive signal through cell-surface receptors [28]. In plants, several different types of cell-surface receptors are determined to perceive diverse signals and stimuli from the environment [31–33]. Plant receptor-like kinases (RLKs) are transmembrane proteins, defined by the presence of a signal sequence: an amino-terminal extracellular domain with a transmembrane region, and a carboxyl-terminal intracellular kinase domain [34,35]. RLKs control an extensive range of biological processes, including development, hormone perception, disease resistance and self-incompatibility. In our study, there were two receptor-like protein kinases up-regulated by N fertilizer application in Huanghuazhan (A2Z8U1; A2Y361), and one up-regulated in Quanliangyou 681 (B8B6Z1), indicating the enhancement of signal transduction by N fertilizer (Table 3). Besides, the expression levels of one serine/threonine-protein kinases STY46 (B8AGJ3) was increased in Quanliangyou 681, while that of (B8B2K2) was decreased in Huanghuazhan cultivar, after N fertilizer treatment, suggesting the difference in N-regulated signal transduction between the two cultivars. The synthesis of ethylene, one of classical plant hormones, is from methionine through S-adenosyl-L-methionine and 1-aminocyclopropane-1-carboxylic acid (ACC), which is catalyzed by ACC synthase and ACC oxidase, respectively [36]. Interestingly, two 1-aminocyclopropane-1-carboxylate oxidases (A2ZC94; B8AY41) were up-regulated in the two rice cultivars, which implied that ethylene signaling may be affected by N fertilizer. Based on these results, we can speculate that signal transduction may be involved in the nitrogen nutrition response of plants.

The 14-3-3 proteins play crucial roles in a variety of important physiological pathways regulated by phosphorylation. These proteins complete signal transduction by binding to the phosphorylated target, which accomplish an alteration of structure that regulates activity [33]. There are a wide range of processes controlled by 14-3-3s, including the fundamental nitrogen and carbon assimilation, starch synthase, Glu synthase, ATP synthase, ascorbate peroxidase, and methyl transferase [37,38]. In the present study, application of N fertilizer had no significant effect on the expression of 14-3-3 protein in the two rice cultivars; whereas the amount of five 14-3-3 proteins (A2XL95; A2XUA6; A2X6G8; A2YWB4; A2YVG3) were down-regulated in the comparison of H_CK vs. 681_CK (Table S5). This result showed that the expression of 14-3-3 proteins may be of constitutive nature in the two rice cultivars, and the higher levels of these proteins in Quanliangyou 681 may be contributed to enhance signal transduction.

3.3. Energy Metabolism and Material Transport

Respiration is the process of organic matter oxygenolysis and energy production. Most oxidative degradation of respiratory substrates occurs through dehydrogenation reactions. Hydrogen acceptors, such as NAD+, FAD or NADP+, will be transferred into NADH, FADH2 or NADPH after accepting hydrogen ion and electron, which promotes synthesis of ATP to provide energy for organism [39]. As shown in Table 4, most of the proteins related to respiration were up-regulated by nitrogen fertilizer in

Huanghuazhan cultivar, including NAD(P)H-dependent oxidoreductase (A2Z4G9; B8BFL3), ferredoxin (A2XNS0), dehydrogenase (A2ZCK1; B8ADR5; A2YBK1; A2XW22). As for Quanliangyou 681, one NAD(P)H-dependent oxidoreductase 2 (B8BFL3) and one glutamate dehydrogenase 2 (A2XW22) were also up-regulated after treatment of N fertilizer. These results indicated that application of nitrogen fertilizer could enhance energy metabolism in the two cultivars.

Membrane bound H^+ pumps are proposed to constitute the primary transducers on the basis of the chemiosmotic hypothesis, which make living cells to interconvert light, chemical, and electrical energy [40]. Through the transmembrane electrochemical gradients, H^+ pumps provide energy for the transport of other solutes, or transduce the H^+ electrochemical gradient generated by membrane-linked anisotropic redox reactions to the synthesis of ATP [40]. The energy-dependent transport of solutes through vacuolar membrane of plants is driven by two kinds of H^+ pumps: vacuolar (V type) H^+-ATPase and H^+-translocating (pyrophosphate-energized) inorganic pyrophosphatase (H^+-PPase) [41]. Both H^+-ATPase and H^+-PPase are abundant and ubiquitous in the vacuolar membranes of plant cells, and they make an essential contribution to the H^+-electrochemical potential difference. The ATP-binding cassette (ABC) transporter superfamily is the largest transporter gene family. The proteins bind ATP and utilize the energy to drive the transport of diverse molecules, including sugars, amino acids, peptides, proteins, metal ions, and plenty of hydrophobic compounds and metabolites across the plasma membrane and intracellular membranes [42–44]. In our study, N nutrition increased the expression levels of two ABC transporter G family members (B8BLE1; B8B6Q3) and one pyrophosphate-energized vacuolar membrane proton pump (A2WPG7) in Huanghuazhan, but no obvious change was observed in Quanliangyou 681. However, one sugar carrier protein C (A2YML7) and one calmodulin-like protein (B8ACJ8) were identified as up-regulated by N fertilizer in Quanliangyou 681(Table 4). These above results illustrated that energy metabolism and matter transport in plant cells might be associated with nitrogen nutrition response of plant root system.

3.4. Substance Metabolism

Aminotransferases have been extensively studied since their discovery over 80 years ago, owing to their vital functions in the synthesis of amino acids by catalyzing the process of amino transfer [45,46]. In the present study, three down-regulated (B8BGM4; B8APP5; B8AZ97) aminotransferases were identified in Huanghuazhan, and one up-regulated (A2XE78) was found in Quanliangyou 681, after application of nitrogen fertilizer (Table S3). As a form of inorganic nitrogen, ammonium is derived from several metabolic pathways and assimilated into glutamine, glutamate, asparagine and carbamoylphosphate [27,47–49]. Specifically, the asparagine synthesis is catalyzed by asparagine synthetase through amidation of aspartate by using either glutamine or ammonium as an amino donor, which is catalyzed by ammonia-dependent or glutamine-dependent asparagine synthetase [50,51]. For most non-leguminous plants, glutamine and asparagine function as the major nitrogen transport and storage compounds from source to sink organs [52]. Due to the high nitrogen/carbon ratio and stability, asparagine is considered as an optimal nitrogen transport and reserve compound. Interestingly, the expression level of one glutamine-dependent asparagine synthetase (B8ALL8) was significantly increased by N fertilizer in Quanliangyou 681, while no obvious change was observed in Huanghuazhan cultivar. This result revealed the difference in nitrogen metabolism between the two rice cultivars, and the probable superiority of nitrogen transport and reserve for Quanliangyou 681.

Glycosidase, also known as glycosidic hydrolase, is a major category of hydrolases belonging to the glycosylase family [53]. They are the enzymes hydrolyzing glycosidic bonds, and playing a critical role in carbohydrate metabolism. Glycosyltransferases are divided into 47 different families based on both PSI-BLAST sequence analysis and substrate/product stereochemistry [54]. The glycosyltransferase reaction is the process involving the transfer of a monosaccharide from an activated sugar donor to a saccharide, protein, DNA, lipid or small molecule acceptor [55,56]. Glycosyltransferase-mediated reaction is proposed to proceed through an oxocarbenium-ion-like transition state, which is similar to glycosidase reactions [57,58]. In our study, two glycosyltransferase (A2Z7U4; A2WUT6) and 6 glycosidases (B8B1F4; B8B1F5; B8B6Z6; A2WYZ7; A2WYX6; A2WYX5) were significantly up-regulated in Huanghuazhan, while only two glycosidases (A2WYZ7; A2WYX6) were increased in Quanliangyou 681 cultivar, after treatment of N fertilizer (Table S3). These results indicated that the enhancements of substance metabolism induced by N fertilizer are different between the two rice cultivars, and Huanghuazhan seems to be more sensitive to nitrogen nutrition.

3.5. Stress and Defense Response

Peroxidase (POD) is one key enzyme for the oxidative detoxification, due to its high affinity for H_2O_2 [59]. It has been determined that de-novo synthesis of the enzymatic protein or up-regulation of related gene, are contributed to the increase of antioxidant enzyme activity [60,61]. In our study, many PODs were changed after N fertilizer treatment in the two rice cultivars, with 5 up-regulated in Quanliangyou 681; while 4 up-regulated and 6 down-regulated in Haunghuazhan (Table S4). Late embryogenesis abundant protein (LEA) protein has been proposed to function as an antioxidant and a membrane stabilizer during water stress, which plays an important role in plant abiotic stress tolerance [62]. LEA proteins are divided into several different groups, among which one group (LEA II) constituting of dehydrins (DHNs). DHNs have been reported to function as membrane stabilizers during freezing induced dehydration [63], functioning as possible osmoregulatory substance [64] or as radical scavengers [65]. Overexpression of *LEA/DHN* enhanced the stress tolerance in transgenic plants [66,67]. In the present study, three LEA proteins (A2YTZ6; A2Y720; A2WU85) and one dehydrin (A2ZDX4) were significantly increased in Quanliangyou 681 after N fertilizer treatment, but no obvious difference was observed in Huanghuazhan cultivar (Table S4). Glutathione S-transferase acts as a detoxification enzyme that catalyzes the reduction of tripeptide glutathione (GSH; g-Glu- Cys-Gly) into multitudinous hydrophobic and electrophilic substrates [68]. It has been determined that GSTs are associated with high tolerance to abiotic stress in plant [68,69]. In this present study, 5 up-regulated (A2Z9M2; A2WZ34; A2Z9L3; A2Z9J9; A2Z6F5) and 2 down-regulated (A2XK19; A2WQ51) GSTs were found in Huanghuazhan after nitrogen fertilizer application (Table S4). However, only one GST (A2WQ51) changed (down-regulated) in Quanliangyou 681 treated with N nutrition. These results indicated that the effects of nitrogen nutrition on antioxidant system are different between the two cultivars. Although the antioxidant enzymes in Hunaghuazhan seemed to be more sensitive to N fertilizer, their expression levels were relatively higher for Quanliangyou 681 (Table S5, see the comparison of H_15 vs. 681_N15). Above these results, we speculated that the higher stress tolerance in Quanliangyou 681 may be beneficial for root absorbing the water and nutrition, consequently improving the growth and grain production.

Momilactone A and momilactone B are important phytoalexins in rice, which play vital roles in the rice defense system against pathogens and insects [70–72]. Chitinase also plays an important role in cell defense, by attacking on chitin molecules that are the main structural component in fungal cell wall and insects' skeleton [18,73]. Besides this, dramatic increase in chitinase level by numerous abiotic agents (ethylene, salicylic acid, salt solutions, ozone, UV light) and by biotic factors (fungi, bacteria, viruses, viroids, fungal cell wall components and oligosaccharides) also proved their role in plant defense response [74,75]. In our present study, the expression levels of several momilactone A synthases (B8B5L2; B8B5L4; A2YPN5; A2YPP1; A2Z1W4; B8B5L1) and chitinases (A2Y4F6; A2Z7A3; A2Y4F5) were significantly increased by N fertilizer application in Huanghuazhan cultivar, whereas no obvious change was observed in Quanliangyou 681 (Table S4). This result indicated that the defense system could be improved by N nutrition in Huanghuazhan cultivar.

4. Conclusions

This study highlighted the differentially expressed proteins in the two rice cultivars' response to N nutrition. These proteins were mainly involved in ammonium assimilation, substance metabolism, signal transduction, energy metabolism, material transport, and stress/defense response. Although the number of N fertilizer-regulated proteins was more for Huanghuazhan cultivar, the expression levels of most key proteins were relatively higher in Quanliangyou 681. Therefore, we speculated that Huanghuazhan is more sensitive to N nutrition, while the higher yield improved by N fertilizer of Quanliangyou 681 may be owing to the relatively higher background levels of these responsive proteins. These proteins provide a better understanding of nitrogen regulatory functions in cereal crops, which is necessary for precise identification of potential molecular protein markers to assist the breeding for high NUE cultivars.

5. Materials and Methods

5.1. Plant Materials and Growth Conditions

Seeds of two rice cultivars, conventional rice Huanghuazhan and indica hybrid rice Quanliangyou 681, were used in this study, which were provided from Hubei Seed Industry Group and Hubei Winall Hi-tech Seed Company, respectively. This present experiment was performed at Lake Tai farm in Jinzhou city during the rice-growing season (May–October, 2017). The soil fertility was medium, and the previous crop was wheat. Seedlings of the two rice cultivars were treated with nitrogen fertilizer, at three levels of 135, 180 and 225 kg/hm^2 (labeled as N9, N12, N15, respectively), and the untreated group was considered as the control (labeled as N0). For N fertilizer treatment, basal fertilizer was compound fertilizer (15:15:15); tillering fertilizer and panicle fertilizer were urea, and the proportion of the three N fertilizers was 6:2:2. Phosphate fertilizer was one-off basal application, and potash fertilizer was used as basal and panicle fertilizer with the ratio of 7:3. Totally, the ratio of fertilizing amounts of N, P, K was 2:1:2. The rice seeds were soaked, pre-germinated for 48h, and sown in soil. Every group had three biological replicates, and each of them was grown within an independent zone (30 m^2), and different zones were isolated by ridges, with independent managements of water and fertilizer.

5.2. Measurement of Rice Dry Weight and Grain Yield

The plant samples for dry weight measurement were collected at four key stages, including tillering, jointing, full heading and mature period. For one zone, five holes of plants were collected without roots. The samples were dried at 105 °C for 30 mins followed by 80 °C for 48 h, the dried samples were used for measurement of shoot dry weight after being allowed to cool to room temperature. After harvesting, the grain yields of the two rice cultivars were determined, respectively.

5.3. Protein Extraction

Four regimes were designed in the proteomic study, and each regime had three biological replicates. The groups of H_CK, and 681_CK represented as Huanghuazhan or Quanliangyou 681 under control condition; while H_N15 and 681_N15 indicated as Huanghuazhan or Quanlaingyou 681 under N treatment, respectively. The root samples in rice booting stage were collected respectively, and frozen in liquid nitrogen, then kept at −80 °C. Protein extraction was determined by the method described by Hu et al. (2017) and Zhou et al. (2018) with some modifications [76,77]. The sample was ground into powder, then transferred into extraction buffer (containing 10 mM dithiothretiol, 1% protease inhibitor and 2 mM EDTA), and sonicated three times by using a high intensity ultrasonic processor (Scientz). Isometric Tris-saturated phenol (pH 8.0) was added, and the mixture was centrifuged at 5000× *g* at 4 °C for 10 min, the supernatant was transferred into a new centrifuge tube. Proteins were added five volumes of ammonium acetate/ methanol and precipitated for 12 h. The remaining precipitate was washed with ice-cold methanol and acetone, successively. Finally, the protein was redissolved in 8 M urea. The protein concentration was measured by the method described by Walker (2009) [78], by using BCA kit according to the manufacturer's instructions.

5.4. Trypsin Digestion

Trypsin digestion was determined according to the method described in our previous study with slight modifications [76]. In detail, the protein solution was added with 5 mM dithiothreitol (DDT) and reduced at 56 °C for 30 min. Then the solution was alkylated with 11 mM iodoacetamide for 15 min at room temperature in darkness. The protein sample was diluted by adding 100 mM triethylammonium bicarbonate (TEAB) to insure the urea concentration less than 2M. Finally, trypsin was added at the ratio of 1:50 (trypsin-to-protein mass ratio) for the first digestion at 37 °C overnight, and 1:100 for a second 4 h-digestion.

5.5. TMT Labeling

The tryptic peptide was desalted by Strata X C18 column (Phenomenex, Torrance, CA, USA) and vacuum freeze-dried. Peptide was reconstituted in 0.5 M TEAB and labeled by using TMT kit according to the manufacturer's protocol. In brief, one-unit TMT reagent was dissolved in acetonitrile, and mixed with peptides, then incubated for 2 h at room temperature. The labeled peptide was pooled, desalted and vacuum freeze-dried.

5.6. HPLC Fractionation

The labeled peptides were fractionated by high pH reverse-phase HPLC with Agilent 300Extend C18 column (5 μm particles, 4.6 mm ID, 250 mm length). Briefly, peptides were firstly separated with a gradient of 8–32% acetonitrile (pH 9.0) over 60 min into 60 fractions. The peptides were combined into 18 fractions (final fraction 1 = 1, 19, 37, 55; final fraction 2 = 2, 20, 38, 56; final fraction 3 = 3, 21, 39, 57, … … … ; final fraction 18 = 18, 36, 54); afterwards, they were dried by vacuum centrifuging.

5.7. LC-MS/MS Analysis

The peptides were dissolved in solvent A, separated through the EASY-nLC 1000 UPLC system. The solvent A was composed of 0.1% formic acid and 2% acetonitrile, while solvent B contained 0.1% formic acid and 90% acetonitrile. The gradient was comprised of an increase from 9% to 25% solvent B over 26 min, 25% to 36% in 26–34 min, and climbing to 80% in 3 min and then holding at 80% for the last 3 min, all at a constant flow rate of 700 nL/min. The peptides, separated by UPLC system, were subjected into NSI ion source and analyzed with application of tandem mass spectrometry (MS/MS) in Q ExactiveTM Plus (Thermo, Bremen, Germany). The electrospray voltage applied was 2.0 kV. The m/z scan range was 350 to 1550 for full scan, and intact peptides were detected in the Orbitrap at a resolution of 60,000. Peptides were then selected for MS/MS using NCE setting as 100 and the fragments were detected in the Orbitrap at a resolution of 15,000. Data was collected by using a data-dependent procedure (DDA) that alternated between one MS scan followed by 20 MS/MS scans with 30s dynamic exclusion. The parameters of automatic gain control (AGC), signal threshold and maximum time of injection were set as 5E4, 5000 ions/s and 200 ms, respectively.

5.8. Database Search

The MS/MS data were further processed by using Maxquant search engine (v.1.5.2.8). Tandem mass spectra were searched against UniProt Oryza sativa_india database (37,383 sequences) coupled with reverse decoy database. Trypsin/P was designated as the specific cleavage enzyme allowing up to 2 missing cleavages. In addition, the mass tolerance for precursor ions was set as 20 ppm in First search and 5 ppm in Main search, and the mass error was set as 0.02 Da for fragment ions. Carbamidomethyl on cysteine was designated as fixed modification, while oxidation on methionine was specified as variable modifications. The quantitative method was set as TMT-6plex, and the false discovery rate (FDR) for the identification of protein and propensity score matching (PSM) were both adjusted to <1%.

5.9. Bioinformatics Methods

The bioinformatics analysis methods were according to the detailed description in the previous studies [76,77]. Totally, in this present study we performed the protein annotation, including Gene Ontology (GO) annotation, domain annotation, Encyclopedia of Genes and Genomes (KEGG) pathway annotation; functional enrichment analysis, including enrichment of GO analysis, pathway analysis and protein domain analysis; as well as enrichment-based clustering analysis.

5.10. Parallel Reaction Monitoring (PRM) Validation of Differentially Expressed Proteins

To validate the reliability of the results produced by TMT labeling proteomic platform, parallel reaction monitoring (PRM) analysis was carried out, which has emerged as an alternative method of targeted proteins/peptides quantification [79,80]. A total of 16 differentially expressed proteins were selected for the project of PRM, due to their functional significance concluded from proteome analysis, and the ratio of protein abundance varied in a wide range. The experimental operation was as the following: protein extraction and trypsin digestion were consistent with the above TMT-labelling detection proteomics analysis. As for separation of peptide fragment by an EASY-nLC 1000 UPLC system, the tryptic peptides were dissolved in 0.1% formic acid (solvent A). The gradient was comprised of an increase from 4% to 16% solvent B (0.1% formic acid and 90% acetonitrile) over 38 min, 16% to 30% in 14 min and climbing to 80% in 4 min then holding at 80% for the last 4 min, with a constant flow rate of 500 nL/min.

The peptides were subjected to NSI source followed by MS/MS in Q ExactiveTM Plus (Thermo, Bremen, Germany). The electrospray voltage was set as 2.0 kV. The m/z scan range applied was 400 to 1100 for full scan, and intact peptides were detected in the Orbitrap at a resolution of 70,000. Peptides were then selected for MS/MS with NCE setting as 27, and the fragments were detected in the Orbitrap with resolution of 17,500. A data-independent procedure was applied that alternated between one MS scan followed by 20 MS/MS scans. Automatic gain control (AGC) was set at 3E6 and 1E5 for full MS and MS/MS, respectively, while the maxumum IT was set at 50 ms for full MS and 300 ms for MS/MS. The isolation window for MS/MS was set at 1.6 *m/z*.

The resulting MS data was processed by using Skyline (v.3.6, Seattle, WA, USA). Parameters for peptide: enzyme was set as Trypsin [KR/P], and max missed cleavage set as 0. The peptide length was set as 7–25, and alkylation on cysteine was specified as fixed modification. Transition settings: precursor charges were set as 2 and 3; ion charges were set as 1; ion types were set as b, y. The fragment ions were selected from ion 3 to the last one, and the ion match tolerance was set as 0.02 Da.

5.11. Statistical Analysis

All experiments performed in this study were repeated three times. For analysis of agronomic characters, the values were shown as mean ± SD of three replicates, and statistical analyses were conducted by one-way analysis of variance (ANOVA) concatenated with independent-samples t test. For enrichments of GO analysis and KEGG pathway analysis, a two-tailed Fisher's exact test was employed to test the enrichment of the differentially expressed protein against all identified proteins.

Supplementary Materials: Supplementary materials can be found at http://www.mdpi.com/1422-0067/20/15/3674/s1.

Author Contributions: B.L. conceived and designed the research framework; J.T., Z.S. and Q.C. performed the experiments; Z.H. analyzed the data and wrote the manuscript; R.N.D. revised the manuscript; Z.H. and B.L. supervised the work and finalized this manuscript. All authors read and approved the manuscript.

Funding: This work was supported the National Key Research and Development Program of China (No. 2017YFD030140404).

Acknowledgments: We thank Jingjie PTM BioLab Co. Ltd. (Hangzhou, China) for the mass spectrometry analysis.

References

1. Hakeem, K.R.; Chandna, R.; Ahmad, A.; Qureshi, M.I.; Iqbal, M. Proteomic analysis for low and high nitrogen-responsive proteins in the leaves of rice genotypes grown at three nitrogen levels. *Appl. Biochem. Biotechnol.* **2012**, *168*, 834–850. [CrossRef] [PubMed]

2. Liao, C.; Peng, Y.; Ma, W.; Liu, R.; Li, C.; Li, X. Proteomic analysis revealed nitrogen-mediated metabolic, developmental, and hormonal regulation of maize (*Zea mays* L.) ear growth. *J. Exp. Bot.* **2012**, *63*, 5275–5288. [CrossRef] [PubMed]

3. Song, C.; Zeng, F.; Feibo, W.; Ma, W.; Zhang, G. Proteomic analysis of nitrogen stress-responsive proteins in two rice cultivars differing in N utilization efficiency. *J. Integr. Omics* **2010**, *1*, 78–87.

4. Frink, C.R.; Waggoner, P.E.; Ausubel, J.H. Nitrogen fertilizer: Retrospect and prospect. *Proc. Natl. Acad. Sci. USA* **1999**, *96*, 1175–1180. [CrossRef] [PubMed]

5. Muthayya, S.; Sugimoto, J.D.; Montgomery, S.; Maberly, G.F. An overview of global rice production, supply, trade, and consumption. *Ann. N Y Acad. Sci.* **2014**, *1324*, 7–14. [CrossRef] [PubMed]

6. Sakakibara, H. Nitrate-specific and cytokinin-mediated nitrogen signaling pathways in plants. *J. Plant Res.* **2003**, *116*, 253–257. [CrossRef] [PubMed]

7. Sakakibara, H.; Takei, K.; Hirose, N. Interactions between nitrogen and cytokinin in the regulation of metabolism and development. *Trends Plant Sci.* **2006**, *11*, 440–448. [CrossRef] [PubMed]

8. Crawford, N.M.; Arst, H.N., Jr. The molecular genetics of nitrate assimilation in fungi and plants. *Annu. Rev. Genet.* **1993**, *27*, 115–146. [CrossRef]

9. Hoff, T.; Truong, H.N.; Caboche, M. The use of mutants and transgenic plants to study nitrate assimilation. *Plant Cell Environ.* **1994**, *17*, 489–506. [CrossRef]

10. Lam, H.M.; Coschigano, K.T.; Oliveira, I.C.; Melo-Oliveira, R.; Coruzzi, G.M. The molecular-genetics of nitrogen assimilation into amino acids in higher plants. *Annu. Rev. Plant Biol.* **1996**, *47*, 569–593. [CrossRef]

11. Miflin, B.J.; Lea, P.J. Ammonia assimilation. In *Amino Acids and Derivatives*; Academic Press: Cambridge, MA, USA, 1980; pp. 169–202.

12. Pham, T.D.; Do, T.T.; Doan, T.H.Y.; Nguyen, T.A.H.; Mai, T.D.; Kobayashi, M.; Adachi, Y. Adsorptive removal of ammonium ion from aqueous solution using surfactant-modified alumina. *Environ. Chem.* **2017**, *14*, 327–337. [CrossRef]

13. Shrawat, A.K.; Good, A.G. Genetic Engineering Approaches to Improving Nitrogen use Efficiency. ISB News Report. May 2008. Available online: http://www.isb.vt.edu/news/2008/artspdf/may0801.pdf (accessed on 28 November 2011).

14. Djennane, S.; Chauvin, J.E.; Meyer, C. Glasshouse behaviour of eight transgenic potato clones with a modified nitrate reductase expression under two fertilization regimes. *J. Exp. Bot.* **2002**, *53*, 1037–1045. [CrossRef] [PubMed]

15. Lian, X.; Wang, S.; Zhang, J.; Feng, Q.; Zhang, L.; Fan, D.; Li, X.; Yuan, D.; Han, B.; Zhang, Q. Expression profiles of 10,422 genes at early stage of low nitrogen stress in rice assayed using a cDNA microarray. *Plant Mol. Biol.* **2006**, *60*, 617–631. [CrossRef] [PubMed]

16. Wang, X.; Li, Y.; Fang, G.; Zhao, Q.; Zeng, Q.; Li, X.; Gong, H.; Li, Y. Nitrite promotes the growth and decreases the lignin content of indica rice calli: A comprehensive transcriptome analysis of nitrite-responsive genes during in vitro culture of rice. *PLoS ONE* **2014**, *9*, e95105. [CrossRef] [PubMed]

17. Ding, C.; You, J.; Liu, Z.; Rehmani, M.I.; Wang, S.; Li, G.; Wang, Q.; Ding, Y. Proteomic analysis of low nitrogen stress-responsive proteins in roots of rice. *Plant Mol. Biol. Rep.* **2011**, *29*, 618–625. [CrossRef]

18. Chandna, R.; Ahmad, A. Nitrogen stress-induced alterations in the leaf proteome of two wheat varieties grown at different nitrogen levels. *Physiol. Mol. Biol. Plants* **2015**, *21*, 19–33. [CrossRef] [PubMed]

19. Yang, C.; Yang, L.; Yang, Y.; Ouyang, Z. Rice root growth and nutrient uptake as influenced by organic manure in continuously and alternately flooded paddy soils. *Agric. Water Manag.* **2004**, *70*, 67–81. [CrossRef]

20. Yoshida, T.; Ancajas, R.R. Nitrogen Fixation by Bacteria in the Root Zone of Rice. *Soil Sci. Soc. Am. J.* **1971**, *35*, 156–158. [CrossRef]

21. Forde, B.; Lorenzo, H. The nutritional control of root development. *Plant Soil* **2001**, *232*, 51–68. [CrossRef]

22. Liu, K.; He, A.; Ye, C.; Liu, S.; Lu, J.; Gao, M.; Fan, Y.; Lu, B.; Tian, X.; Zhang, Y. Root morphological traits and spatial distribution under different nitrogen treatments and their relationship with grain yield in super hybrid rice. *Sci. Rep. (UK)* **2018**, *8*, 131. [CrossRef]

23. Kim, D.Y.; Shibato, J.; Kim, D.W.; Oh, M.K.; Kim, M.K.; Shim, I.S.; Iwahashi, H.; Masuo, Y.; Rakwal, R. Gel-based proteomics approach for detecting low nitrogen-responsive proteins in cultivated rice species. *Physiol. Mol. Biol. Plants* **2009**, *15*, 31–41. [CrossRef] [PubMed]

24. Kim, S.G.; Wang, Y.; Wu, J.; Kang, K.Y.; Kim, S.T. Physiological and proteomic analysis of young rice leaves grown under nitrogen-starvation conditions. *Plant Biotechnol. Rep.* **2011**, *5*, 309. [CrossRef]

25. Ding, C.; You, J.; Wang, S.; Liu, Z.; Li, G.; Wang, Q.; Ding, Y. A proteomic approach to analyze nitrogen-and cytokinin-responsive proteins in rice roots. *Mol. Biol. Rep.* **2012**, *39*, 1617–1626. [CrossRef] [PubMed]

26. Cai, H.; Zhou, Y.; Xiao, J.; Li, X.; Zhang, Q.; Lian, X. Overexpressed glutamine synthetase gene modifies nitrogen metabolism and abiotic stress responses in rice. *Plant Cell Rep.* **2009**, *28*, 527–537. [CrossRef] [PubMed]

27. Gaufichon, L.; Reisdorf-Cren, M.; Rothstein, S.J.; Chardon, F.; Suzuki, A. Biological functions of asparagine synthetase in plants. *Plant Sci.* **2010**, *179*, 141–153. [CrossRef]

28. Hirel, B.; Lea, P.J. Ammonia assimilation. In *Plant Nitrogen*; Springer: Berlin/Heidelberg, Germany, 2001; pp. 79–99.

29. Coque, M.; Gallais, A. Genomic regions involved in response to grain yield selection at high and low nitrogen fertilization in maize. *Theor. Appl. Genet.* **2006**, *112*, 1205–1220. [CrossRef] [PubMed]

30. Gallais, A.; Hirel, B. An approach to the genetics of nitrogen use efficiency in maize. *J. Exp. Bot.* **2004**, *55*, 295–306. [CrossRef] [PubMed]

31. Shiu, S.H.; Bleecker, A.B. Receptor-like kinases from Arabidopsis form a monophyletic gene family related to animal receptor kinases. *Proc. Natl. Acad. Sci. USA* **2001**, *98*, 10763–10768. [CrossRef] [PubMed]

32. Bleecker, A.B.; Kende, H. Ethylene: A gaseous signal molecule in plants. *Annu. Rev. Cell Dev. Biol.* **2000**, *16*, 1–18. [CrossRef]

33. Inoue, T.; Higuchi, M.; Hashimoto, Y.; Seki, M.; Kobayashi, M.; Kato, T.; Tabata, S.; Shinozaki, K.; Kakimoto, T. Identification of CRE1 as a cytokinin receptor from Arabidopsis. *Nature* **2001**, *409*, 1060. [CrossRef]

34. Walker, J.C. Structure and function of the receptor-like protein kinases of higher plants. *Plant Mol. Biol.* **1994**, *26*, 1599–1609. [CrossRef] [PubMed]

35. Torii, K.U. Receptor kinase activation and signal transduction in plants: An emerging picture. *Curr. Opin. Plant Biol.* **2000**, *3*, 361–367. [CrossRef]

36. Yang, S.F.; Hoffman, N.E. Ethylene biosynthesis and its regulation in higher plants. *Annu. Rev. Plant Physiol.* **1984**, *35*, 155–189. [CrossRef]

37. Sehnke, P.C.; Chung, H.J.; Wu, K.; Ferl, R.J. Regulation of starch accumulation by granule-associated plant 14-3-3 proteins. *Proc. Natl. Acad. Sci. USA* **2001**, *98*, 765–770. [CrossRef] [PubMed]

38. Finnie, C.; Borch, J.; Collinge, D.B. 14-3-3 proteins: Eukaryotic regulatory proteins with many functions. *Plant Mol. Biol.* **1999**, *40*, 545–554. [CrossRef]

39. Buchanan, B.B.; Gruissem, W.; Jones, R.L. (Eds.) *Biochemistry and Molecular Biology of Plants*, 2nd ed.; John Wiley & Sons: Hoboken, NJ, USA, 2015.

40. Mitchell, P. Chemiosmotic coupling in oxidative and photosynthetic phosphorylation. *Biol. Rev.* **1996**, *41*, 445–501. [CrossRef]

41. Sarafian, V.; Kim, Y.; Poole, R.J.; Rea, P.A. Molecular cloning and sequence of cDNA encoding the pyrophosphate-energized vacuolar membrane proton pump of Arabidopsis thaliana. *Proc. Natl. Acad. Sci. USA* **1992**, *89*, 1775–1779. [CrossRef]

42. Higgins, C.F. ABC transporters: From microorganisms to man. *Annu. Rev. Cell Biol.* **1992**, *8*, 67–113. [CrossRef]

43. Dean, M.; Allikmets, R. Evolution of ATP-binding cassette transporter genes. *Curr. Opin. Genet. Dev.* **1995**, *5*, 779–785. [CrossRef]

44. Dean, M.; Hamon, Y.; Chimini, G. The human ATP-binding cassette (ABC) transporter superfamily. *J. Lipid Res.* **2001**, *42*, 1007–1017. [CrossRef]

45. Christen, P.; Metzler, D.E. *Transaminases*; John Wiley & Sons: Hoboken, NJ, USA, 1985.

46. Calton, G.J.; Wood, L.L.; Updike, M.H.; Lantz, L.; Hamman, J.P. The Production of L–Phenylalanine by Polyazetidine Immobilized Microbes. *Bio/technology* **1986**, *4*, 317. [CrossRef]

47. Masclaux-Daubresse, C.; Reisdorf-Cren, M.; Pageau, K.; Lelandais, M.; Grandjean, O.; Kronenberger, J.; Valadier, M.H.; Feraud, M.; Jouglet, T.; Suzuki, A. Glutamine synthetase-glutamate synthase pathway and glutamate dehydrogenase play distinct roles in the sink-source nitrogen cycle in tobacco. *Plant Physiol.* **2006**, *140*, 444–456. [CrossRef] [PubMed]

48. Purnell, M.P.; Botella, J.R. Tobacco isoenzyme 1 of NAD (H)-dependent glutamate dehydrogenase catabolizes glutamate in vivo. *Plant Physiol.* **2007**, *143*, 530–539. [CrossRef] [PubMed]

49. Miyashita, Y.; Good, A.G. NAD (H)-dependent glutamate dehydrogenase is essential for the survival of Arabidopsis thaliana during dark-induced carbon starvation. *J. Exp. Bot.* **2008**, *59*, 667–680. [CrossRef] [PubMed]

50. Larsen, T.M.; Boehlein, S.K.; Schuster, S.M.; Richards, N.G.; Thoden, J.B.; Holden, H.M.; Rayment, I. Three-dimensional structure of Escherichia coli asparagine synthetase B: A short journey from substrate to product. *Biochemistry* **1999**, *38*, 16146–16157. [CrossRef] [PubMed]

51. Scofield, M.A.; Lewis, W.S.; Schuster, S.M. Nucleotide sequence of Escherichia coli asnB and deduced amino acid sequence of asparagine synthetase B. *J. Biol. Chem.* **1990**, *265*, 12895–12902. [PubMed]

52. Lea, P.J.; Ireland, R.J. *The Enzymes of Glutamine. Glutamate, Asparagine, and Aspartate Metabolism. Plant Amino Acids*; CRC Press: Boca Raton, FL, USA, 1998; pp. 63–124.

53. Babcock, G.D.; Esen, A. Substrate specificity of maize β-glucosidase. *Plant Sci.* **1994**, *101*, 31–39. [CrossRef]

54. Campbell, J.A.; Davies, G.J.; Bulone, V.; Henrissat, B. A classification of nucleotide-diphospho-sugar glycosyltransferases based on amino acid sequence similarities. *Biochem. J.* **1997**, *326*, 929. [CrossRef] [PubMed]

55. Radominsk-Pandya, A.; Czernik, P.J.; Little, J.M.; Battaglia, E.; Mackenzie, P.I. Structural and functional studies of UDP-glucuronosyltransferases. *Drug Metab. Rev.* **1999**, *31*, 817–899. [CrossRef]

56. Ünligil, U.M.; Rini, J.M. Glycosyltransferase structure and mechanism. *Curr. Opin. Struct. Biol.* **2000**, *10*, 510–517.

57. Murray, B.W.; Wittmann, V.; Burkart, M.D.; Hung, S.C.; Wong, C.H. Mechanism of human α-1, 3-fucosyltransferase V: Glycosidic cleavage occurs prior to nucleophilic attack. *Biochemistry* **1997**, *36*, 823–831. [CrossRef] [PubMed]

58. Kim, S.C.; Singh, A.N.; Raushel, F.M. Analysis of the galactosyltransferase reaction by positional isotope exchange and secondary deuterium isotope effects. *Arch. Biochem. Biophys.* **1998**, *267*, 54–59. [CrossRef]

59. Zhang, S.; Zhang, H.; Qin, R.; Jiang, W.; Liu, D. Cadmium induction of lipid peroxidation and effects on root tip cells and antioxidant enzyme activities in Vicia faba L. *Ecotoxicology* **2009**, *18*, 814–823. [CrossRef] [PubMed]

60. Verma, S.; Dubey, R.S. Lead toxicity induces lipid peroxidation and alters the activities of antioxidant enzymes in growing rice plants. *Plant Sci.* **2003**, *164*, 645–655. [CrossRef]

61. Alvarez, M.E.; Lamb, C. Oxidative burst-mediated defense responses in plant disease resistance. *Cold Spring Harbor Monogr. Arch.* **1997**, *34*, 815–839.

62. Tunnacliffe, A.; Wise, M.J. The continuing conundrum of the LEA proteins. *Naturwissenschaften* **2007**, *94*, 791–812. [CrossRef]

63. Close, T.J. Dehydrins: Emergence of a biochemical role of a family of plant dehydration proteins. *Physiol. Plantarum* **1996**, *97*, 795–803. [CrossRef]

64. Nylander, M.; Svensson, J.; Palva, E.T.; Welin, B.V. Stress-induced accumulation and tissue-specific localization of dehydrins in Arabidopsis thaliana. *Plant Mol. Biol.* **2001**, *45*, 263–279. [CrossRef]

65. Hara, M.; Terashima, S.; Fukaya, T.; Kuboi, T. Enhancement of cold tolerance and inhibition of lipid peroxidation by citrus dehydrin in transgenic tobacco. *Planta* **2003**, *217*, 290–298.

66. Sivamani, E.; Bahieldin, A.; Wraith, J.M.; Al-Niemi, T.; Dyer, W.E.; Ho, T.H.D.; Qu, R. Improved biomass productivity and water use efficiency under water deficit conditions in transgenic wheat constitutively expressing the barley HVA1 gene. *Plant Sci.* **2000**, *155*, 1–9. [CrossRef]

67. Xu, D.; Duan, X.; Wang, B.; Hong, B.; Ho, T.H.D.; Wu, R. Expression of a late embryogenesis abundant protein gene, HVA1, from barley confers tolerance to water deficit and salt stress in transgenic rice. *Plant Physiol.* **1996**, *110*, 249–257. [CrossRef] [PubMed]

68. Dixon, D.P.; Hawkins, T.; Hussey, P.J.; Edwards, R. Enzyme activities and subcellular localization of members of the Arabidopsis glutathione transferase superfamily. *J. Exp. Bot.* **2009**, *60*, 1207–1218. [CrossRef] [PubMed]

69. Yu, T.A.O.; Li, Y.S.; Chen, X.F.; Hu, J.; Chang, X.U.N.; Zhu, Y.G. Transgenic tobacco plants overexpressing cotton glutathione S-transferase (GST) show enhanced resistance to methyl viologen. *J. Plant Physiol.* **2003**, *160*, 1305–1311. [CrossRef] [PubMed]

70. Agrawal, G.K.; Rakwal, R.; Tamogami, S.; Yonekura, M.; Kubo, A.; Saji, H. Chitosan activates defense/stress response (s) in the leaves of Oryza sativa seedlings. *Plant Physiol. Biochem.* **2002**, *40*, 1061–1069. [CrossRef]

71. Jung, Y.H.; Lee, J.H.; Agrawal, G.K.; Rakwal, R.; Kim, J.A.; Shim, J.K.; Lee, S.K.; Jeon, J.S.; Koh, H.J.; Lee, Y.H.; et al. The rice (Oryza sativa) blast lesion mimic mutant, blm, may confer resistance to blast pathogens by triggering multiple defense-associated signaling pathways. *Plant Physiol. Biochem.* **2005**, *43*, 397–496. [CrossRef] [PubMed]

72. Nojiri, H.; Sugimori, M.; Yamane, H.; Nishimura, Y.; Yamada, A.; Shibuya, N.; Kodama, O.; Murofushi, N.; Omori, T. Involvement of jasmonic acid in elicitor-induced phytoalexin production in suspension-cultured rice cells. *Plant Physiol.* **1996**, *110*, 387–392. [CrossRef]

73. Sharma, N.; Sharma, K.P.; Gaur, R.K.; Gupta, V.K. Role of chitinase in plant defense. *Asian J. Biochem.* **2011**, *6*, 29–37. [CrossRef]

74. Gupta, V.; Misra, A.; Gupta, A.; Pandey, B.; Gaur, R. Rapd-Pcr of Trichoderma Isolates and In Vitro Antagonism Against Fusarium Wilt Pathogens of Psidium Guajaval. *J. Plant Prot. Res.* **2010**, *50*, 256–262. [CrossRef]

75. Punja, Z.K.; Zhang, Y.Y. Plant chitinases and their roles in resistance to fungal diseases. *J. Nematol.* **1993**, *25*, 526.

76. Hu, Z.; Liu, A.; Bi, A.; Amombo, E.; Gitau, M.M.; Huang, X.; Liang, C.; Fu, J. Identification of differentially expressed proteins in bermudagrass response to cold stress in the presence of ethylene. *Environ. Exp. Bot.* **2017**, *139*, 67–78. [CrossRef]

77. Zhou, H.; Finkemeier, I.; Guan, W.; Tossounian, M.A.; Wei, B.; Young, D.; Huang, J.; Messens, J.; Yang, X.; Zhu, J.; et al. Oxidative stress-triggered interactions between the succinyl-and acetyl-proteomes of rice leaves. *Plant Cell Environ.* **2018**, *41*, 1139–1153. [CrossRef] [PubMed]

78. Walker, J.M. The bicinchoninic acid (BCA) assay for protein quantitation. In *The Protein Protocols Handbook*; Humana Press: Totowa, NJ, USA, 2009; pp. 11–15.

79. Peterson, A.C.; Russell, J.D.; Bailey, D.J.; Westphall, M.S.; Coon, J.J. Parallel reaction monitoring for high resolution and high mass accuracy quantitative, targeted proteomics. *Mol. Cell Proteomics* **2012**, *11*, 1475–1488. [CrossRef] [PubMed]

80. Rauniyar, N. Parallel reaction monitoring: a targeted experiment performed using high resolution and high mass accuracy mass spectrometry. *Int. J. Mol. Sci.* **2015**, *16*, 28566–28581. [CrossRef] [PubMed]

International Journal of
Molecular Sciences

Article

Identification and Analysis of Micro-Exon Genes in the Rice Genome

Qi Song [1,2], Fang Lv [2], Muhammad Tahir ul Qamar [1,2], Feng Xing [1,2], Run Zhou [1,2], Huan Li [1,2] and Ling-Ling Chen [1,2,*]

[1] National Key Laboratory of Crop Genetic Improvement, Huazhong Agricultural University, Wuhan 430070, China; aries981@webmail.hzau.edu.cn (Q.S.); m.tahirulqamar@webmail.hzau.edu.cn (M.T.u.Q.); xfengr@mail.hzau.edu.cn (F.X.); 18162856086@163.com (R.Z.); lihuan2729@163.com (H.L.)
[2] Hubei Key Laboratory of Agricultural Bioinformatics, College of Informatics, Huazhong Agricultural University, Wuhan 430070, China; lvfang2009@163.com
* Correspondence: llchen@mail.hzau.edu.cn; Tel.: +86-27-8728-0877

Received: 29 April 2019; Accepted: 29 May 2019; Published: 31 May 2019

Abstract: Micro-exons are a kind of exons with lengths no more than 51 nucleotides. They are generally ignored in genome annotation due to the short length, whereas recent studies indicate that they have special splicing properties and important functions. Considering that there has been no genome-wide study of micro-exons in plants up to now, we screened and analyzed genes containing micro-exons in two *indica* rice varieties in this study. According to the annotation of Zhenshan 97 (ZS97) and Minghui 63 (MH63), ~23% of genes possess micro-exons. We then identified micro-exons from RNA-seq data and found that >65% micro-exons had been annotated and most of novel micro-exons were located in gene regions. About 60% micro-exons were constitutively spliced, and the others were alternatively spliced in different tissues. Besides, we observed that approximately 54% of genes harboring micro-exons tended to be ancient genes, and 13% were *Oryza* genus-specific. Micro-exon genes were highly conserved in *Oryza* genus with consistent domains. In particular, the predicted protein structures showed that alternative splicing of in-frame micro-exons led to a local structural recombination, which might affect some core structure of domains, and alternative splicing of frame-shifting micro-exons usually resulted in premature termination of translation by introducing a stop codon or missing functional domains. Overall, our study provided the genome-wide distribution, evolutionary conservation, and potential functions of micro-exons in rice.

Keywords: micro-exons; constitutive splicing; alternative splicing; ancient genes; domain

1. Introduction

Micro-exons are a set of small exons with lengths no more than 51 nucleotides. Previous studies identified some micro-exon genes from mammals, insects, and plants [1–3]. Although micro-exon genes distribute widely in various species, systematic recognition of them has not been performed in a long period. In initial studies, the identification of micro-exons was not simple due to its mismatch until canonical and noncanonical splice sites were considered in alignment corrections [4]. After that, many alignment tools were improved to be more sensitive to micro-exons, and some pipelines were developed specially for micro-exons identification [1,5].

Alternative splicing is a process in which precursor mRNA adopts different splice sites and contributes to the expansion of sequence diversity and function [6,7]. In general, micro-exons whose lengths are not multiples of three nucleotides could cause frame-shifting in splicing, which may trigger nonsense-mediated decay [8]. Therefore, some alternative isoforms appear to be unstable or have no function. In addition, several studies have revealed that the level of micro-exon splicing is related to physiological disease in human; for example, mis-regulation of alternative splicing for neural micro-exons mediated by nSR100 is associated with autism [1,9]. Micro-exons are also good objects to investigate mechanism of splicing. Some studies have revealed there are sets of regulatory sequences in micro-exons and flanking introns, which are called exonic and intronic splicing enhancers (ESE and ISE). These motifs are recognized and bound by RNA-binding proteins (RBP) and contribute to splicing efficiency, consistent with the fact that most micro-exons are included in transcripts [5,10,11].

Micro-exons could function in various ways, including alternative splicing, such as degradation of the transcripts via nonsense-mediated decay (NMD), altering protein domain architecture, and introducing novel post-translational modification sites [12]. There have been several studies revealing that mis-regulation of splicing level could relate to several diseases in humans, such as autism spectrum disorders and schizophrenia [1,13], while few comprehensive studies have been performed in plants. However, we have observed that quite a few micro-exons were located in protein domains of some rice genes affecting foundational functions, such as *OsHAP3* and *RSR1*, which can regulate chloroplast biogenesis and starch biosynthesis, respectively [14,15]. Thus, it is crucial to determine detailed properties and roles of micro-exons in rice.

Nowadays, two important *indica* rice lines, ZS97 and MH63, have been assembled and annotated, providing a good basis for genome-wide micro-exon detection [16]. We extracted micro-exons from the genome annotations and compared micro-exons identified from alignments between RNA-seq data and genomes. Based on transcriptome data in different tissues and conditions, we aimed to explore different splicing types of micro-exons and the conservation of micro-exon genes in related species, and revealed the changes of protein structures caused by micro-exon splicing. We found that micro-exons accounted for 6.3% of all exons annotated by genomic information, and the lengths of 40.7% micro-exons were multiples of three nucleotides. In addition, more than half of micro-exon genes were evolutionarily ancient genes, and less than one-sixth were evolutionarily young genes, which only occurred in *Oryza* species. Gene ontology (GO) analysis showed that micro-exon genes varied in multiple metabolic and biosynthetic processes and enriched ribonucleotide binding. Further studies illustrated that micro-exons tended to be located in protein domains, such as protein kinase domain, RNA binding domains, etc. Alternative splicing of micro-exons could result in local structural recombination, while frame-shifting micro-exons might lead to premature termination of translation. Our study systematically analyzed the distribution, evolution, and functional information of micro-exons in rice genome.

2. Results

2.1. Identification and Inclusion Ratio of Micro-Exons in ZS97 and MH63

There were 16,508 and 17,517 micro-exons in the genome annotation of ZS97 and MH63, respectively, which accounted for ~6.3% of all the annotated exons. About 23% of the annotated genes contained micro-exons, implying that micro-exons were widely distributed in rice genomes. Among these annotated micro-exons, more than half (8816 in ZS97 and 9399 in MH63) were internal micro-exons, and the first or last micro-exons tended to be shorter than internal micro-exons (Figure 1A). Next, we employed RNA-seq data from four tissues (seedling shoot, root, flag leaf, and young panicle) of ZS97 and MH63 in four grown conditions (high/low temperature and long/short daytime) to identify micro-exons. In total, 7645 and 8137 internal micro-exons were identified in ZS97 and MH63, respectively. Compared with the above internal micro-exons identified in genome annotation, 2517 and 2824 were newly identified micro-exons in ZS97 and MH63, respectively (Figure 1B). Among these newly identified micro-exons, 1861 and 2079 were located in annotated gene regions, indicating that they might be un-annotated alternative splicing micro-exons. In total, the lengths of 40.7% micro-exons were multiples of three nucleotides, which were beneficial for the stability of open reading frames (ORFs) [1,5], whereas the others, which were not multiples of three nucleotides, could result in frame-shifting and encode different amino acid sequences.

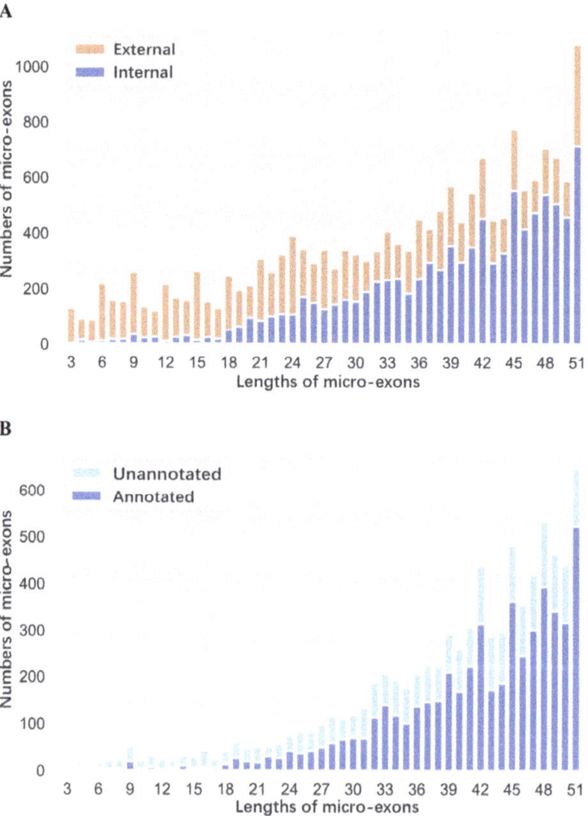

Figure 1. The length distribution of micro-exons in MH63. (**A**) The length distribution of external and internal micro-exons based on MH63 genome annotation. (**B**) The distribution of lengths for the annotated and unannotated micro-exons, which were identified from RNA-seq data in four tissues.

2.2. Evolutionary Age of Genes Containing Micro-Exons

We investigated the evolutionary age of genes containing micro-exons by a previously reported approach [17,18]. The protein sequences of micro-exon genes were aligned to the non-redundant (NR) protein database in 13 taxonomic levels (details in the methods section). Then 5123 micro-exons of 3565 genes in MH63 were employed to construct their phylostratigraphic profiles (Figure 2A). We assigned a phylostratum (PS) value for each gene and defined the genes from PS1 to PS3 as "old genes", while genes from PS11 to PS13 were defined as "young genes". In total, 54.2% and 13.2% micro-exon genes were divided into old and young genes, respectively. After that, we compared the coding sequence lengths between young and old genes and observed that the average coding length of young genes was longer than that of old genes (Figure 2B, Wilcoxon rank-sum test, *p*-value < 0.001), but the lengths of their micro-exons were similar (Figure 2C). The average gene expression level of old micro-exon genes was much higher than that of young micro-exon genes in all the flag leaf, panicle, and shoot and root tissues (Figure 2D, Wilcoxon rank-sum test, *p*-value < 0.001), indicating that old micro-exon genes might perform more fundamental or essential functions than young micro-exon genes.

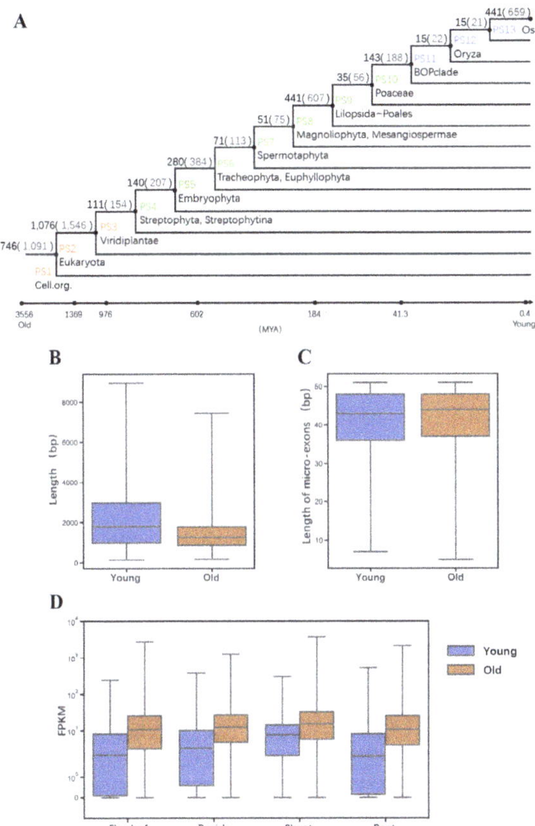

Figure 2. The evolutionary ages of micro-exons genes. (**A**) The phylogenetic tree shows the genes containing micro-exons in different evolutionary times, from phylostratum (PS)1 (single-cell organisms) to PS13 (*O. sativa*). There are two numbers on each branch: the numbers of genes corresponding to each level are outside the brackets, while the numbers of identified micro-exons are in brackets. The genes of PS1–3 are assigned as "old genes" while the genes of PS11–13 are "young genes". (**B**) The comparison of coding sequence (CDS) lengths between the young genes and old genes. (**C**) The comparison of micro-exon lengths between the young genes and old genes. (**D**) The comparison of gene expression in four tissues between the young genes and old genes. FPKM represents pair-end fragments per kilobase of exon model per million mapped fragments. The blue color represents young genes and the orange color represents old genes.

2.3. Conservation of Domains and Micro-Exons

In order to study the characteristics of micro-exons, we explored the conservation of micro-exons in angiosperm. McScanX [19] was employed to gain the collinear gene pairs among MH63 and ZS97, seven other representative *Oryza* species, *Zea mays*, and *Arabidopsis thaliana*. It was revealed that most micro-exons were highly conserved in *Oryza* species, owning the same open read frames in coding sequence and encoding the same amino acids. When we compared micro-exons in MH63 with those in *Zea mays*, there were 2509 micro-exons conserved in the open read frame and length, and about half of them had nucleotide substitutions, implying that micro-exons were relatively conserved in monocotyledons. However, when comparing MH63 with dicotyledon (*Arabidopsis thaliana*), only 197 micro-exons were conserved (Table 1), indicating that the micro-exon genes were highly divergent in monocotyledon and dicotyledon. Based on the above evolutionary age analyses, we speculated that micro-exon genes might originate from evolutionary old genes, and had independent differentiation and function in monocotyledon and dicotyledon.

Table 1. Conservation of genes containing micro-exons.

Species (Paired with MH63)	Total Gene Pairs	Gene Pairs with Micro-Exons	Gene Pairs with Conserved Micro-Exons
Arabidopsis thaliana	1828	253	214 (197)
Oryza barthii	27,306	3821	3183 (3470)
Oryza glaberrima	25,755	3559	2949 (3571)
Oryza glumipatula	27,314	3731	3111 (3434)
Oryza meridionalis	21,571	3156	2691 (2975)
Oryza nivara	27,192	3638	3089 (3422)
Oryza rufipogon	29,323	3864	3208 (3523)
Oryza sativa	26,237	3875	3217 (3561)
Zea mays	21,953	3524	2386 (2509)
ZS97	38,649	3747	3161 (4251)

To further investigate the protein functions related to micro-exons, we detected phosphorylation sites and domains in the micro-exon genes. Only 44 phosphorylation sites were identified in 30 micro-exons, indicating that only a few micro-exons may be involved in the function of cell signal transduction. On the other hand, about 58% micro-exons were located in domains. We then compared the enriched domains containing whole/part of micro-exons and their upstream/downstream domains, which did not include micro-exons. Table 2 showed the top ten domains containing micro-exons and the neighboring domains not containing micro-exons. Protein kinase domain (PF00069), RNA recognition motif (PF00076) and WD domain, and G-beta repeat (PF00400) were commonly enriched in micro-exon containing domains and their upstream/downstream domains. AP2 domain (PF00847), K-box region (PF01486), glycosyl hydrolase family 1 (PF00232), and protein tyrosine kinase (PF07714) were only enriched in domains containing micro-exons. The above results indicated that micro-exons might be related with functions of protein kinase and RNA recognition. Additionally, we found that the proportions of consistent domains (>95%) in gene pairs with conserved micro-exons in different species were higher than the ratios of non-conserved micro-exons (Figure 3A). Figure 3B–E showed four types of micro-exon related domains in collinear gene pairs between ZS97 and MH63. The first type was that micro-exons in collinear gene pairs had the same gene structure and were located in the same protein domain (Figure 3B); the second type was that the gene structures of collinear gene pairs were different, and only the micro-exon in one genome was located in the protein domain and the other genome was not (Figure 3C); and the third and fourth types were that the homolog sequences of micro-exons in one genome were contained in a longer exon in the other genomes, which formed similar or different protein domains (Figure 3D–E).

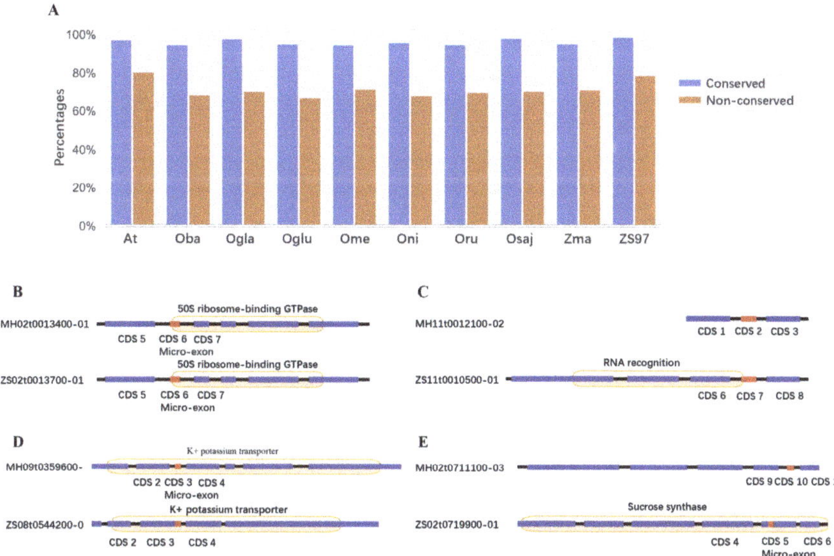

Figure 3. The proportion of conservation for micro-exons and domains in gene pairs. (**A**) The percentage of consistent domains between the conserved and non-conserved micro-exons. (**B–E**) Several examples of consistent and inconsistent domains with conserved and non-conserved micro-exons in ZS97 and MH63.

Table 2. The top10 domains including and excluding micro-exons.

Rank	Domains Including Micro-Exons	Domains Excluding Micro-Exons
1	Protein kinase domain (PF00069, 66)	WD domain, G-beta repeat (PF00400, 163)
2	RNA recognition motif (PF00076, 42)	RNA recognition motif (PF00076, 78)
3	AP2 domain (PF00847, 36)	Protein kinase domain (PF00069, 60)
4	K-box region (PF01486, 28)	IQ calmodulin-binding motif (PF00612, 46)
5	Glycosyl hydrolase family 1 (PF00232, 24)	PPR repeat family (PF13041, 44)
6	Protein tyrosine kinase (PF07714, 23)	PPR repeat (PF01535, 39)
7	Myb-like DNA-binding domain (PF00249, 20)	SRF-type transcription factor (PF00319, 32)
8	Serine carboxypeptidase (PF00450, 20)	C2 domain (PF00168, 28)
9	WD domain, G-beta repeat (PF00400, 17)	Helicase conserved C-terminal domain (PF00271, 28)
10	Major facilitator superfamily (PF07690, 16)	Gelsolin repeat (PF00626, 25)

2.4. Quantification of Micro-Exon Usage and Gene Ontology Enrichment Analysis

To measure the constitutive or alternative splicing ratio of micro-exons, percent spliced-in (PSI) values in all samples were calculated by using the reads mapped to splicing junctions. In total, 4921 and 5162 micro-exons in ZS97 and MH63, respectively, had PSI values > 0.1 in at least one RNA-seq dataset. More than 70% of micro-exons tended to be constitutively spliced (CS; PSI ≥ 0.9) in all samples, which was consistent with the results in humans [5], and the other micro-exons were alternatively spliced (AS) in different tissues (Figure 4A). The PSI values in the same tissue (flag leaf/root/panicle/shoot) under different conditions were similar, suggesting that temperature and light had little effect on gene splicing.

Hundreds of micro-exons had various splicing types in different tissues, but only one micro-exon was AS across all samples in MH63. As seen in Figure 4B, 3679 micro-exons were expressed in the four tissues, and 838 micro-exons were expressed in two or three tissues. In addition, 41, 110, 47, and 296 micro-exons were specifically expressed in flag leaf, root, shoot, and panicle, respectively. Then we performed GO analysis of these tissue-specific micro-exon genes and observed that the enriched functions were highly divergent in different tissues (Figure 4C). The common functions in the four tissues mainly enriched in the 'cellular aromatic compound metabolic process' and 'heterocycle metabolic process'. Furthermore, the 'organic substance metabolic process' was enriched in flag leaf and shoot, while the 'cellular nitrogen compound metabolic/biosynthetic process' and 'aromatics compound/heterocycle/nucleobase-containing compound biosynthetic process' were enriched in the panicle, root, and shoot. However, the 'macromolecule metabolic process', 'protein metabolic process', and 'phosphorus metabolic process' were only enriched in the flag leaf; the 'reproductive process', 'RNA biosynthetic process', and 'regulation of RNA metabolic process' were merely enriched in the panicle; the 'regulation of (cellular) biosynthetic process', 'multicellular organismal process', and 'developmental process' were only enriched in the root, indicating that these specifically expressed micro-exon genes played different roles in the four tissues.

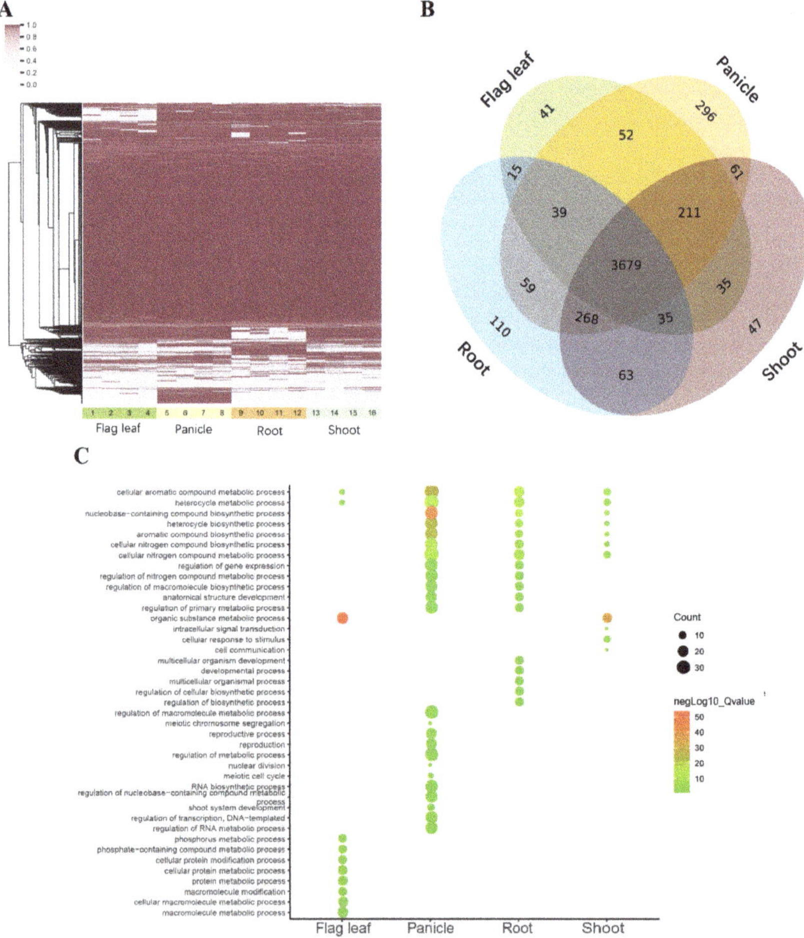

Figure 4. The percent spliced-in (PSI) values and gene ontology (GO) enrichment analyses in four MH63 tissues. (**A**) PSI of 5162 micro-exons in 16 samples. 1–4: flag leaf, 5–8: panicle, 9–12: root, 13–16: shoot in order of high temperature long daytime, high temperature short daytime, low temperature long daytime, and low temperature short daytime. (**B**) Micro-exons with PSI ≥10% in the four tissues. (**C**) GO analysis of tissue-specific micro-exon genes.

2.5. Structure and Functional Analyses of Alternative Spliced Micro-Exons

To explore the influence of AS micro-exons in protein structure and function, we predicted and analyzed the three-dimensional (3D) structures of some representative micro-exon genes. Generally, AS micro-exons with a multiple of three nucleotides only changed the local sequence and structure, and the global protein structure changed little, whereas AS micro-exons with lengths that were not multiple of three could cause frame-shifting and great changes in amino acid sequences and 3D structures [20]. Figure 5A–D showed four examples of 3D structures containing AS micro-exons. As seen in Figure 5A, the MH05t0027800-01 protein (rice starch regulator 1, *RSR1*) contained two Apetala 2 (AP2) domains; one AP2 domain included a 45-nt micro-exon with three DNA binding sites [15,21], and another downstream AP2 domain had a similar 3D structure. Kim et al. reported that the AP2 domain had an 18 amino acid core region that formed an amphipathic α-helix, which was completely overlapped with the micro-exon [21]. It was interesting that the 45-nt micro-exon was CS in flag leave and shoot but AS in panicle and root. Considering that *RSR1* acts as a transcription factor in starch synthesis [21], this variation might have different effects on starch synthesis in different tissues. Figure 5B showed the structure of the MH01t0092800-26 protein (*OsGI*) with or without a micro-exon. When the micro-exon was not included, it only affected the local structure but did not change the overall structure of the protein. A previous study revealed that *OsGI* was expressed in the biological clock and had different effects on flowering time in long days or short days [22]. The micro-exon was AS in flag leaf and CS in root and shoot, indicating its dynamic effect in different tissues and conditions. Besides, the *OsGI* gene had another micro-exon and various isoforms. Figure 5C showed the structures of MH03t0653800-01 protein (*OsSUV3*) containing or not containing a 46-nt AS micro-exon, whose function is related to high salt resistance [23]. The exclusion of the micro-exon led to the conformation change of the *OsSUV3* protein C-terminal due to the 46-nt micro-exon located at the back end of the gene. Figure 5D illustrated the structure of the rice heterochronic gene MH07t0148200-01 (SUPERNUMERARY BRACT, *SNB*), whose conformation is much different from the structure not containing a 31-nt AS micro-exon. It was interesting that *SNB* genes also had two AP2 containing micro-exons. Previous studies reported that mutants of the rice *SNB* gene with an insertion in the terminal region could cause significant sterility [24]. Taken together, most micro-exons with lengths that were not multiples of three nucleotides could change the open reading frame and cause overall structure changes after missing micro-exons.

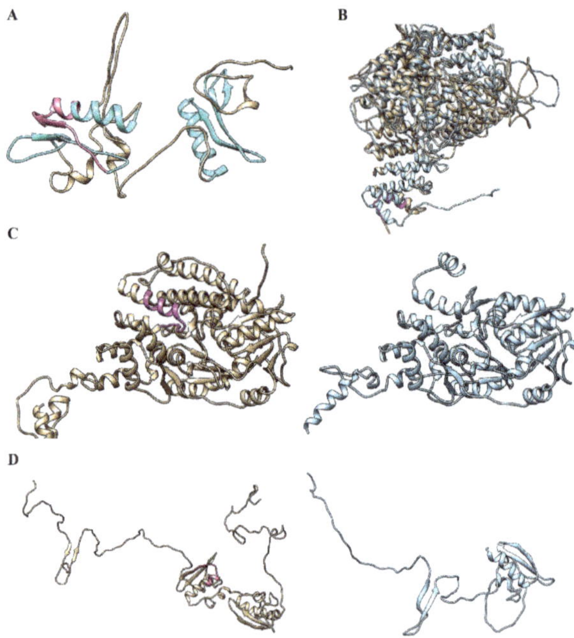

Figure 5. Structure comparison of proteins harboring micro-exons. (**A**) Structure of MH05t0027800-01 (*RSR1*). Two AP2 domains are presented in cyan and micro-exon is in pink. (**B–D**) Structural comparison of the proteins containing micro-exon (tan) or excluding micro-exon (cyan). (**B**) is MH01t0092800-26 (*OsGI*) with a 42-nt micro-exon, (**C**) is MH03t0653800-01 (*OsSUV3*) with a 46-nt micro-exon, and (**D**) is MH07t0148200-01 (*SNB*) with a 31-nt micro-exon. Micro-exons are highlighted in magenta.

3. Discussion

Previous studies have reported some micro-exons and their roles in various species, but there is no systematic study of micro-exons in plants. In this study, we used RNA-seq data from various rice tissues to identify thousands of micro-exons. About one third of them were not annotated but located in the region of the annotated genes. While more than half of the micro-exon genes were ancient genes, the other micro-exons were conserved in *Oryza* species and other monocotyledons while not conserved in dicotyledons, implying that these genes might have an ancient origin but differ in later evolutionary processes. Moreover, most micro-exons were located in protein domains and generally tended to be CS. All these results suggested that the micro-exon genes may play crucial roles in primary functions of the organism, and stable splicing of micro-exons was essential for maintaining their functions.

The results of GO analysis indicated that the micro-exon genes were enriched in many biological processes, and they were enriched in ribonucleotide binding in molecular function. We found that a large number of domains containing micro-exons were related to DNA or RNA binding, such as AP2 and RNA recognition motif, which may be a key role of micro-exons in molecular functions. It has been reported that splicing mis-regulation of brain-expressed micro-exons led to brain-related diseases [1]; therefore, many micro-exon genes have important functions. Furthermore, the tissue-specific protein contained disordered regions and conserved binding motifs [25]. Although most micro-exons were widely expressed in the four tissues, genes containing tissue-specific micro-exons also had distinctive functions in ZS97 and MH63.

Previous studies reported three effects of micro-exons: leading to premature stop codons, changing the protein structure, and creating sites for post-translational modification [12]. Only a few phosphorylation sites were found in the micro-exons of ZS97 and MH63, implying that phosphorylation only represented a few cases of post-translational modification. On the other hand, most micro-exons were contained in domains, and some of them harbored functional sites. Besides, the secondary structure of micro-exons usually consisted of several types, including α-helix, β-strand, and coil. There were two AP2 domains in *RSR1* [21], and a 45-nt AS micro-exon lied in one of the domains, harboring three DNA binding sites. There is no doubt that micro-exon exclusion will affect gene function. As *RSR1* is a transcription factor, the splicing level of this micro-exon may affect the expression of downstream genes. Additionally, inclusion or exclusion of the micro-exon could easily cause premature terminal of translation, especially for the lengths that were not multiples of three nucleotides. In some cases, we observed many shortened sequences by alternative splicing, and the encoded protein functions can be influenced distinctly.

This study provided an overview on micro-exons in ZS97 and MH63, and demonstrated the relationship of protein structure, function, and micro-exon splicing. The splicing mis-regulation of micro-exons leads to disease in human [1,5]. In this case, a high proportion of micro-exons were CS, and they had precise and stable splicing process and conserved domains, providing a guarantee of regular metabolic process. In addition, the predicted domains and structures in this study illustrated that the changes among proteins were caused by AS micro-exon inclusion and exclusion to some extent. However, the predicted structures cannot completely explain the variation, and for many genes, their 3D structures were hard to predict. Despite the lack of software and structure data, the exact splicing of micro-exons in transcripts and proteins cannot be obtained only by RNA-seq data. Generally, the genes with micro-exons have various isoforms in the expression process. For instance, a 34-nt micro-exon located in *OsTrx1* is a histone H3K9 methyltransferase gene, but the micro-exon is not annotated in the annotation information. With advanced techniques, complete sequences and structures of the isoforms can be obtained through full-length transcript data and protein data, and then the precise roles of micro-exons in proteins revealed.

4. Materials and Methods

4.1. Materials and Data Preparation

To identify micro-exons in ZS97 and MH63, RNA-sequencing was performed from four tissues (flag leaf, panicles, seedling shoot, and roots) in four different conditions (high temperature and long day, high temperature and short day, low temperature and long day, and low temperature and short day. High temperature: 28–32 °C; low temperature: 22–25 °C; long day: 14 h light and 10 h dark; short day: 10 h light and 14 h dark). Besides, each sample had two biological replicates (Illumina HiSeq2000 platform). Finally, a total of 948 and 988 million strand-specific paired-end 101-nt reads were obtained from ZS97 and MH63 (an average of 29.6 and 30.9 million per sample), respectively. Genome data and annotation information of ZS97 and MH63 were downloaded from the website (http://rice.hzau.edu.cn;versionRS1; accessed on 4 July 2016). Genome and annotation information of other plant species were retrieved from Ensembl Plants (http://plants.ensembl.org/; accessed on 12 April 2018).

4.2. Identification of Micro-Exons

Firstly, we extracted all micro-exons whose lengths were in the range of 3 to 51 nt from genome annotations. The micro-exon, as the first or last exon in genes, was examined as to whether it was coding sequences or untranslated regions. Then to identify the micro-exons in ZS97 and MH63 based on RNA-seq data, all the reads were mapped to their reference genomes using HISAT2 [26]. A strand-specific parameter was set in the alignments (–rna-strandness RF) and unique mapping reads with less than 2 mismatches were retained. For each mapping, we reserved at least 6-nt on both sides to completely align the micro-exons. In order to detect as many micro-exons as possible, the insertions supported by more than 10 reads in all samples were identified as candidate internal micro-exons. The candidate micro-exons were divided into annotated micro-exons and novel micro-exons according to the annotation. Compared with annotated genes, we detected the locations of the novel micro-exons. The distribution of length of novel and annotated micro-exons were measured and displayed, and the whole identification process is showed in Figure S1A.

4.3. Evolutionary Age of Genes Containing Micro-Exons

Previous methods described that the protein sequences from the NR protein database were attributed to 13 taxonomic levels (PS1: Cellular organisms; PS2: Eukaryota; PS3: Viridiplantae; PS4: Streptophyta, Streptophytina; PS5: Embryophyta; PS6: Tracheophyta, Euphyllophyta; PS7: Spermatophyta; PS8: Magnoliophyta, Mesangiospermae; PS9: Liliopsida, Petrosaviidae, Commelinids, Poales; PS10: Poaceae; PS11: BOP clade; PS12: Oryzoideae, Oryzeae, Oryza; and PS13: *O. sativa*) [17]. The protein sequences containing micro-exons from MH63 were aligned to the 13 levels of non-redundant databases using BLASTP with *e*-values $\leq 1 \times 10^{-5}$, identity ≥ 0.3, and coverage ≥ 0.8. The age of a gene was assigned the taxonomic level of its alignment, and the gene that failed to be aligned with all databases was categorized to PS13 (*O. sativa*). Then the micro-exons were divided as their corresponding types of genes. The genes of PS1–3 were assigned as "old genes", while the genes of PS11–13 were "young genes". Next, the length of CDS and micro-exons of the old and young genes were calculated and compared. Furthermore, the gene expression levels of the old and young genes in four tissues were displayed.

4.4. Conservation of Micro-Exons and Domains

The protein sequences of MH63 were aligned to the other species by BLASTP [27] with default parameters, and the collinear gene pairs were obtained using MCScanX [19]. Then the alignments of micro-exon regions were applied to measure the conservation of micro-exons. The micro-exons in the alignments, which had the same phase and length of CDS in both MH63 and another species, were considered as conserved ones. In addition, Interproscan [28] was used to predict domains in all the sequences. For each micro-exon, we compared the domains of micro-exons in MH63 with the ones in other species. At last, we also examined phosphorylation sites in the micro-exons via the Plant Protein Phosphorylation DataBase (P3DB, http://www.p3db.org/).

4.5. Percent Splice-In (PSI) Index of Micro-Exons

To measure the usages of micro-exons, the results of HISAT2 were used to establish a non-redundant transcript annotation with StringTie [29]. For each micro-exon gene, the longest transcript containing micro-exons was searched by the annotation information. Then we constructed transcripts that contained micro-exons and flanked 100-nt to measure PSI for each micro-exon. We also constructed a dataset without micro-exons but flanked 100-nt. The transcripts would be combined into one in the cases where the distance among multiple micro-exons were within 100-nt. After that, the RNA-seq data was mapped onto the constructed transcripts using BWA [30] (single-end; options mem) with no more than 2 mismatches. As shown in Figure S1B, R_L and R_R represented the number of reads supporting the left and right junction for each micro-exon, respectively, while $R_{skipped}$ was the number of reads spanning the junction where micro-exons were skipped. The counted reads should span the junction at least 3-nt. With these quantities, the number of reads supporting micro-exon inclusion, R_{tot}, was computed as follows:

$$R_{tot} = 2 \min\{R_L, R_R\} \tag{1}$$

Previous studies demonstrated that Equation (1) can avoid cases in which alternative 5' or 3' splice site biases the estimated micro-exon usage [5]. In addition, the sum of R_L and R_R (or $R_{skipped}$) should be no less than 10, otherwise the PSI value should be considered missing. The PSI value was computed as follows:

$$PSI = R_{tot} / (R_{tot} + R_{skipped}) \tag{2}$$

We combined the PSI values of two biological replicates by the following rules: if a micro-exon had PSI value only in one biological replicate, the value was used; otherwise, if both replicates had PSI values and their difference was less than 10%, we used the mean; other cases were assigned to missing.

Generally, micro-exons with PSI values in the range of 10–90% were categorized as AS micro-exons, and >90% of PSI were assigned as CS. For each tissue, micro-exons with PSI $\geq$ 10% in at least two samples were collected, and then the micro-exons and relevant genes were compared and displayed in four tissues.

4.6. Functional Annotation of Micro-Exon Genes

To detect the function of micro-exon genes, micro-exons with PSI $\geq$ 10% in at least two samples were collected and the genes containing tissue-specific micro-exons were retained. Then AgriGO [31] was performed to do GO enrichment analysis with the reference *Oryza sativa* subsp. *indica*. Furthermore, the genes containing AS micro-exons in each tissue were also analyzed.

4.7. Structures and Domains of Micro-Exons

To investigate the structure of micro-exons, the protein sequences of micro-exon genes were extracted from annotation information. The longest sequences were selected if micro-exon genes had several transcripts. Then the sequences with inclusion and exclusion of micro-exons were constructed. In addition, the unannotated sequences from transcript assembly and prediction were used as supplementary data. After that, the protein structure of some genes were predicted using RaptorX [32], and the domains were predicted with Interproscan [28]. Chimera [33] was employed to display and compare the protein structures including and excluding micro-exons.

Supplementary Materials: Supplementary materials can be found at http://www.mdpi.com/1422-0067/20/11/2685/s1.

Author Contributions: Q.S., F.L., F.X., H.L. and R.Z. contributed to the acquisition and analysis of the data, and wrote the draft of the manuscript. M.T.u.Q. predicted the 3D structures, L.-L.C. contributed to the conception and design of the analysis, supervision and analysis of the data, and improving the manuscript. All authors read and approved the final manuscript.

Funding: This work was supported by the National Natural Science Foundation of China (31571351, 31871269) and the Fundamental Research Funds for the Central Universities (2662017PY043).

Acknowledgments: Authors would like to thank the editors as well as the reviewers who have given us useful suggestions for improvements.

Conflicts of Interest: The authors declare no conflict of interest.

Abbreviations

ESE	Exonic splicing enhancers
ISE	Intronic splicing enhancers
RBP	RNA-binding proteins
GO	Gene ontology
ORF	Open reading frames
NR	Non-redundant
PS	Phylostratum
PSI	Percent spliced-in
AP2	Apetala 2
SNB	Supernumerary bract
P3DB	Plant Protein Phosphorylation DataBase
AS	Alternatively spliced
CS	Constitutively spliced

References

1. Irimia, M.; Weatheritt, R.J.; Ellis, J.D.; Parikshak, N.N.; Gonatopoulos-Pournatzis, T.; Babor, M.; Quesnel-Vallières, M.; Tapial, J.; Raj, B.; O'Hanlon, D.; et al. A Highly Conserved Program of Neuronal Microexons Is Misregulated in Autistic Brains. *Cell* **2014**, *159*, 1511–1523. [CrossRef] [PubMed]
2. McAllister, L. Alternative Splicing of Micro-Exons Creates Multiple Forms of the Insect Cell Adhesion Molecule Fasciclin I. *J. Neurosci.* **1992**, *12*, 895–905. [CrossRef]
3. Bournay, A.-S.; Hedley, P.E.; Maddison, A.; Waugh, R.; Machray, G.C. Exon Skipping Induced by Cold Stress in a Potato Invertase Gene Transcript. *Nucleic Acids Res.* **1996**, *24*, 2347–2351. [CrossRef]
4. Volfovsky, N. Computational Discovery of Internal Micro-Exons. *Genome Res.* **2003**, *13*, 1216–1221. [CrossRef]
5. Li, Y.I.; Sanchez-Pulido, L.; Haerty, W.; Ponting, C.P. RBFOX and PTBP1 proteins regulate the alternative splicing of micro-exons in human brain transcripts. *Genome Res.* **2015**, *25*, 1–13. [CrossRef] [PubMed]
6. Romero, P.R.; Zaidi, S.; Fang, Y.Y.; Uversky, V.N.; Radivojac, P.; Oldfield, C.J.; Cortese, M.S.; Sickmeier, M.; Legall, T.; Obradovic, Z. Alternative splicing in concert with protein intrinsic disorder enables increased functional diversity in multicellular organisms. *Proc. Natl. Acad. Sci. USA* **2006**, *103*, 8390–8395. [CrossRef]
7. Black, D.L. Mechanisms of Alternative Pre-Messenger RNA Splicing. *Annu. Rev. Biochem.* **2003**, *72*, 291–336. [CrossRef]
8. Lykke-Andersen, S.; Jensen, T.H. Nonsense-mediated mRNA decay: An intricate machinery that shapes transcriptomes. *Nat. Rev. Mol. Cell Biol.* **2015**, *16*, 665–677. [CrossRef]
9. Quesnel-Vallières, M.; Irimia, M.; Cordes, S.P.; Blencowe, B.J. Essential roles for the splicing regulator nSR100/SRRM4 during nervous system development. *Genes Dev.* **2015**, *29*, 746–759. [CrossRef]
10. Simpson, C.G.; Hedley, P.E.; Watters, J.A.; Clark, G.P.; Mcquade, C.; Machray, G.C.; Brown, J.W. Requirements for mini-exon inclusion in potato invertase mRNAs provides evidence for exon-scanning interactions in plants. *RNA* **2000**, *6*, 422. [CrossRef] [PubMed]
11. Simpson, C.G.; Thow, G.; Clark, G.P.; Jennings, S.N.; Watters, J.A.; Brown, J.W. Mutational analysis of a plant branchpoint and polypyrimidine tract required for constitutive splicing of a mini-exon. *RNA* **2002**, *8*, 47. [CrossRef] [PubMed]
12. Ustianenko, D.; Weyn-Vanhentenryck, S.M.; Zhang, C. Microexons: Discovery, regulation, and function: Microexons: Discovery, regulation, and function. *Wiley Interdiscip. Rev. RNA* **2017**, *8*, e1418. [CrossRef] [PubMed]
13. Huntsman, M.M.; Tran, B.-V.; Potkin, S.G.; Bunney, W.E.; Jones, E.G. Altered ratios of alternatively spliced long and short 2 subunit mRNAs of the -amino butyrate type A receptor in prefrontal cortex of schizophrenics. *Proc. Natl. Acad. Sci. USA* **1998**, *95*, 15066–15071. [CrossRef]
14. Miyoshi, K.; Ito, Y.; Serizawa, A.; Kurata, N. OsHAP3 genes regulate chloroplast biogenesis in rice. *Plant J.* **2010**, *36*, 532–540. [CrossRef]
15. Fu, F.-F.; Xue, H.-W. Coexpression Analysis Identifies Rice Starch Regulator1, a Rice AP2/EREBP Family Transcription Factor, as a Novel Rice Starch Biosynthesis Regulator. *PLANT Physiol.* **2010**, *154*, 927–938. [CrossRef]
16. Zhang, J.; Chen, L.-L.; Xing, F.; Kudrna, D.A.; Yao, W.; Copetti, D.; Mu, T.; Li, W.; Song, J.-M.; Xie, W.; et al. Extensive sequence divergence between the reference genomes of two elite *indica* rice varieties Zhenshan 97 and Minghui 63. *Proc. Natl. Acad. Sci. USA* **2016**, *113*, E5163–E5171. [CrossRef] [PubMed]
17. Wang, W.; Mauleon, R.; Hu, Z.; Chebotarov, D.; Tai, S.; Wu, Z.; Li, M.; Zheng, T.; Fuentes, R.R.; Zhang, F.; et al. Genomic variation in 3,010 diverse accessions of Asian cultivated rice. *Nature* **2018**, *557*, 43–49. [CrossRef]
18. Domazet-Lošo, T.; Brajković, J.; Tautz, D. A phylostratigraphy approach to uncover the genomic history of major adaptations in metazoan lineages. *Trends Genet.* **2007**, *23*, 533–539. [CrossRef]
19. Wang, Y.; Tang, H.; DeBarry, J.D.; Tan, X.; Li, J.; Wang, X.; Lee, T.-H.; Jin, H.; Marler, B.; Guo, H.; et al. MCScanX: A toolkit for detection and evolutionary analysis of gene synteny and collinearity. *Nucleic Acids Res.* **2012**, *40*, e49. [CrossRef]
20. Wang, P.; Yan, B.; Guo, J.; Hicks, C.; Xu, Y. Structural genomics analysis of alternative splicing and application to isoform structure modeling. *Proc. Natl. Acad. Sci. USA* **2005**, *102*, 18920–18925. [CrossRef]
21. Kim, S.; Soltis, P.S.; Wall, K.; Soltis, D.E. Phylogeny and Domain Evolution in the APETALA2-like Gene Family. *Mol. Biol. Evol.* **2006**, *23*, 107–120. [CrossRef]

22. Fowler, S. GIGANTEA: A circadian clock-controlled gene that regulates photoperiodic flowering in Arabidopsis and encodes a protein with several possible membrane-spanning domains. *EMBO J.* **1999**, *18*, 4679–4688. [CrossRef]

23. Tuteja, N.; Tarique, M.; Tuteja, R. Rice SUV3 is a bidirectional helicase that binds both DNA and RNA. *BMC Plant Biol.* **2014**, *14*. [CrossRef]

24. Lee, D.-Y.; An, G. Two AP2 family genes, SUPERNUMERARY BRACT (SNB) and OsINDETERMINATE SPIKELET 1 (OsIDS1), synergistically control inflorescence architecture and floral meristem establishment in rice: SNB and OsIDS1 control rice inflorescence architecture and floral meristem. *Plant J.* **2012**, *69*, 445–461.

25. Buljan, M.; Chalancon, G.; Eustermann, S.; Wagner, G.P.; Fuxreiter, M.; Bateman, A.; Babu, M.M. Tissue-specific splicing of disordered segments that embed binding motifs rewires protein interaction networks. *Mol. Cell* **2012**, *46*, 871–883. [CrossRef]

26. Pertea, M.; Kim, D.; Pertea, G.M.; Leek, J.T.; Salzberg, S.L. Transcript-level expression analysis of RNA-seq experiments with HISAT, StringTie and Ballgown. *Nat. Protoc.* **2016**, *11*, 1650. [CrossRef]

27. AltschuP, S.F.; Gish, W.; Miller, W.; Myers, E.W.; Lipman, D.J. Basic Local Alignment Search Tool. *J. Mol. Biol.* **1990**, *215*, 403–410. [CrossRef]

28. Jones, P.; Binns, D.; Chang, H.-Y.; Fraser, M.; Li, W.; McAnulla, C.; McWilliam, H.; Maslen, J.; Mitchell, A.; Nuka, G.; et al. InterProScan 5: Genome-scale protein function classification. *Bioinformatics* **2014**, *30*, 1236–1240. [CrossRef]

29. Pertea, M.; Pertea, G.M.; Antonescu, C.M.; Chang, T.-C.; Mendell, J.T.; Salzberg, S.L. StringTie enables improved reconstruction of a transcriptome from RNA-seq reads. *Nat. Biotechnol.* **2015**, *33*, 290–295. [CrossRef]

30. Li, H.; Durbin, R.; Li, H.; Durbin, R. Fast and accurate short read alignment with Burrows-Wheeler transform. Bioinformatics 25, 1754-1760. *Bioinformatics* **2009**, *25*, 1754–1760. [CrossRef]

31. Du, Z.; Zhou, X.; Ling, Y.; Zhang, Z.; Su, Z. agriGO: A GO analysis toolkit for the agricultural community. *Nucleic Acids Res.* **2010**, *38*, W64–W70. [CrossRef] [PubMed]

32. Källberg, M.; Wang, H.; Wang, S.; Peng, J.; Wang, Z. Template-based protein structure modeling using the RaptorX web server. *Nat. Protoc.* **2012**, *7*, 1511–1522. [CrossRef] [PubMed]

33. Pettersen, E.F.; Goddard, T.D.; Huang, C.C.; Couch, G.S.; Greenblatt, D.M.; Meng, E.C.; Ferrin, T.E. UCSF Chimera–a visualization system for exploratory research and analysis. *J. Comput. Chem.* **2004**, *25*, 1605–1612. [CrossRef] [PubMed]

International Journal of
Molecular Sciences

Article

Overexpression of a *S*-Adenosylmethionine Decarboxylase from Sugar Beet M14 Increased *Araidopsis* Salt Tolerance

Meichao Ji [1,2,†], Kun Wang [1,†], Lin Wang [1], Sixue Chen [1,3], Haiying Li [1,2], Chunquan Ma [1,2,*] and Yuguang Wang [1,4,*]

[1] Engineering Research Center of Agricultural Microbiology Technology, Ministry of Education, Heilongjiang University, Harbin 150080, China; jimeichao25@gmail.com (M.J.); laissezfairewk@gmail.com (K.W.); wanglin1995@gmail.com (L.W.); schen@ufl.edu (S.C.); 1999020@hlju.edu.cn (H.L.)

[2] Key Laboratory of Molecular Biology, College of Heilongjiang Province, College of Life Sciences, Heilongjiang University, Harbin 150080, China

[3] Department of Biology, Genetics Institute, Plant Molecular and Cellular Biology Program, University of Florida, Gainesville, FL 32610, USA

[4] Key Laboratory of Sugar Beet Genetic Breeding of Heilongjiang Province, Heilongjiang University, Harbin 150080, China

* Correspondence: chqm@hlju.edu.cn (C.M.); wangyuguang@hlju.edu.cn (Y.W.); Tel.: +86-045186604372 (C.M.); +86-045186609753 (Y.W.)

† These authors contributed equally to this work.

Received: 10 February 2019; Accepted: 17 April 2019; Published: 23 April 2019

Abstract: Polyamines play an important role in plant growth and development, and response to abiotic stresses. Previously, differentially expressed proteins in sugar beet M14 (*Bv*M14) under salt stress were identified by iTRAQ-based quantitative proteomics. One of the proteins was an S-adenosylmethionine decarboxylase (SAMDC), a key rate-limiting enzyme involved in the biosynthesis of polyamines. In this study, the *BvM14-SAMDC* gene was cloned from the sugar beet M14. The full-length *BvM14-SAMDC* was 1960 bp, and its ORF contained 1119 bp encoding the SAMDC of 372 amino acids. In addition, we expressed the coding sequence of *BvM14-SAMDC* in *Escherichia coli* and purified the ~40 kD BvM14-SAMDC with high enzymatic activity. Quantitative real-time PCR analysis revealed that the *BvM14-SAMDC* was up-regulated in the *Bv*M14 roots and leaves under salt stress. To investigate the functions of the *BvM14-SAMDC*, it was constitutively expressed in *Arabidopsis thaliana*. The transgenic plants exhibited greater salt stress tolerance, as evidenced by longer root length and higher fresh weight and chlorophyll content than wild type (WT) under salt treatment. The levels of spermidine (Spd) and spermin (Spm) concentrations were increased in the transgenic plants as compared with the WT. Furthermore, the overexpression plants showed higher activities of antioxidant enzymes and decreased cell membrane damage. Compared with WT, they also had low expression levels of *RbohD* and *RbohF*, which are involved in reactive oxygen species (ROS) production. Together, these results suggest that the BvM14-SAMDC mediated biosynthesis of Spm and Spd contributes to plant salt stress tolerance through enhancing antioxidant enzymes and decreasing ROS generation.

Keywords: sugar beet; salt stress; *S*-adenosylmethionine decarboxylase; ROS; antioxidant enzyme

---

1. Introduction

Plant growth and development was frequently affected by adverse environment factors (e.g., cold, salt, alkali and drought), and salt stress is a major factor limiting crop production [1]. When plants are exposed to salinity environment, the processes of protein synthesis and photosynthesis can be inhibited by excessive accumulation of salt in plants [2]. Furthermore, under salt stress conditions, excessive reactive oxygen species (ROS) were accumulated in the plant, and high levels of ROS can cause membrane lipid peroxidation and metabolic perturbation [3]. In order to survive the salinity, plants have evolved a series of complex tolerance mechanisms to prevent damage caused by salt stress from morphological, physiological and biochemical levels [4–6].

Polyamines (PAs) play an important role in regulating plant morphology, growth, embryonic development and response to stress conditions [7–9]. In higher plants, putrescine (Put), spermine (Spm) and spermidine (Spd) are the main types of polyamines. In the process of PA synthesis, Put is firstly synthesized from arginine catalyzed by arginine decarboxylase, agmatine iminohydrolase and N-carbamoylputrescine amidohydrolase. Then through spermidine synthase and spermine synthase, Put is converted to Spm and Spd with decarboxylated S-adenosylmethionine (dcSAM) as a donor [10,11]. dcSAM is synthesized from the decarboxylation of S-adenosylmethionine (SAM) in a reaction catalyzed by SAM decarboxylase (SAMDC) [12].

Previous studies reported that polyamines were involved in plant responses to abiotic stresses such as cold, drought and salinity [13,14]. Several enzymes involved in polyamine synthesis were shown to be important in plant stress tolerance [15,16]. For example, transgenic plants overexpression of *SAM* synthase *(SAMS)* from tomato showed enhanced alkali stress tolerance and maintained nutrient acquisition under the alkali stress [17]. Similarly, overexpressing a sugar beet *BvM14-SAMS2* in *Arabidopsis* increased salt and H_2O_2 tolerance of the plants [18]. SAMDC is a key enzyme in the synthesis of Spm and Spd. Down-regulation of *SAMDC* expression could decrease the Spm and Spd contents and lead to sensitivity to biotic and abiotic stresses [19,20]. In addition, overexpression of *SAMDC* was shown to play an important role in improving plant stress tolerance [21]. However, the mechanisms underlying the SAMDC regulation of plant tolerance to biotic and abiotic stresses are still to be investigated.

Sugar beet monosomic addition line M14 was obtained through hybridization between sugar beet (*Beta vulgaris* L.) and wild sugar beet (*Beta corolliflora* Zoss.) [22,23]. The M14 line exhibits several interesting characteristics, such as salt and cold tolerance, as well as apomixes. These traits may be attributed to the retention of the 9th wild sugar beet chromosome in the M14 [24]. Previously, we used iTRAQ-based proteomics to profile protein changes in the M14 line under salt stress [25] and found a *BvM14-SAMDC* was strongly induced by salt stress. In the present study, we aim to gain insight into the mechanisms of BvM14-SAMDC mediated salt stress tolerance by overexpressing the *BvM14-SAMDC* in *Arabidopsis*. To understand the *BvM14-SAMDC* functions, polyamine contents, antioxidant enzyme activities and expression levels of antioxidant and ROS related genes were analyzed in the transgenic plants. Our results demonstrated that BvM14-SAMDC enhances salt stress tolerance through elevating the antioxidant system and suppressing ROS generation.

2. Results

2.1. Isolation of BvM14- SAMDC and Sequence Analysis

In order to acquire the sequence information of *BvM14-SAMDC*, a unigene with significant homology to plant *SAMDC* was found in our sugar beet transcriptome data set. Based on the unigene sequence and a RACE method, a full-length cDNA of *BvM14-SAMDC* were amplified by RT-PCR. The full-length *BvM14-SAMDC* is comprised of 1,960 bp with an open reading frame of 1119 bp nucleotides (Figure S1). The putative BvM14-SAMDC protein contains 372 amino acids with a predicted molecular mass of 40.7 kDa and a pI of 4.65. A large number of plant SAMDC homologs are present in the NCBI non-redundant database by BLASTP search using the BvM14-SAMDC sequence as a query. The BvM14-SAMDC protein shared 69–85% amino acid sequence identity with other plant SAMDCs.

2.2. Prokaryotic Expression and Enzymatic Activity Assay of Recombinant BvM14-SAMDC

To test whether the isolated *BvM14-SAMDC* codes for a functional protein, the open reading frame was cloned under the control of the T7 promoter in a pET28a vector, and transformed into *E.coli* cells. Following induction with 0.1 mM IPTG, a ~40 kDa protein band could be observed on an SDS gel of protein extracts from the *E.coli* cells transformed with *BvM14-SAMDC*, but not with the empty vector (Figure 1a). After purification of the protein extract through a nickel column, the ~40 kDa BvM14-SAMDC could be identified on Western with an anti-His-tag antibody (Figure 1b). The purified protein is active with a specific activity of 2.11 units/mg. This result demonstrates that the *BvM14-SAMDC* gene encodes the right protein with SAMDC enzymatic activity.

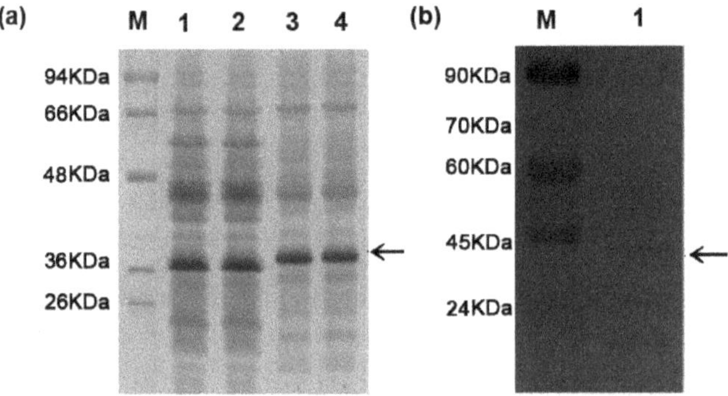

Figure 1. *Ecoli* expression of the recombinant sugar beet M14 S-adenosylmethionine decarboxylase (BvM14-SAMDC) protein. (**a**) Sodium dodecyl sulfate polyacrylamide gel electrophoresis (SDS-PAGE) analysis of bacterial expression of BvM14-SAMDC. Total soluble protein fractions with the empty vector after 4 h isopropyl β-ᴅ-thiogalactoside (IPTG) induction (lanes 1 and 2); and total soluble protein fractions with pET28a-*BvM14-SAMDC* after 4 h IPTG induction (lanes 3 and 4). (**b**) Immunoblot identification of the purified recombinant BvM14-SAMDC using anti-His antibody. Arrows indicate the BvM14-SAMDC position.

2.3. Expression Patterns of BvM14-SAMDC under Salt Stress

To analyze tissue expression patterns of *BvM14-SAMDC*, real-time quantitative PCR was conducted using RNA extracted from in roots, stems, flowers and leaves of the M14 line. As shown in Figure 2a, *BvM14-SAMDC* was ubiquitously expressed in different organs, with roots showing higher levels than other organs (Figure 2a). Furthermore, to determine the response of *BvM14-SAMDC* to salt stress, the expression patterns of *BvM14-SAMDC* were analyzed in roots and leaves under 400 mM NaCl treatment and sampled at different time points. After salt stress, the increasing *BvM14-SAMDC* transcript level appeared much earlier in the M14 roots than in leaves (Figure 2b,c). The maximum levels of *BvM14-SAMDC* expression were observed at 6 and 12 h in roots and leaves, respectively, under salt stress. The results showed that *BvM14-SAMDC* transcription was significantly induced in roots and leaves by salt stress. Therefore, it is reasonable to speculate that *BvM14-SAMDC* participates in regulating plant salt stress tolerance.

2.4. Overexpression of BvM14-SAMDC Increased Salt Stress Tolerance in Arabidopsis

To study the functions of *BvM14-SAMDC* under salinity stress, we used *Agrobacterium tumefaciens* mediated transformation to produce transgenic Arabidopsis plants over-expressing the *BvM14-SAMDC* gene (Figure S2a). BLASTp analysis of BvM14-SAMDC protein sequences in the *Arabidopsis thaliana* database showed that the BvM14-SAMDC had the highest sequence similarity (77%) to an Arabidopsis SAMDC1 protein (AtSAMDC1). Therefore, a T-DNA insertion mutant of *AtSAMDC1* was selected and verified (Figure S2b,c). The expression level of *AtSAMDC1* was decreased by 72.5% in the *atsamdc1* mutant (KD) compared with WT, which was caused by the T-DNA insertion in the promoter region (Figure S2d,e). Furthermore, we transformed the *atsamdc1* mutant line with the *BvM14-SAMDC* overexpression construct to generate homozygous T3 complementation transgenic lines (CO1 and CO2) (Figure S2f).

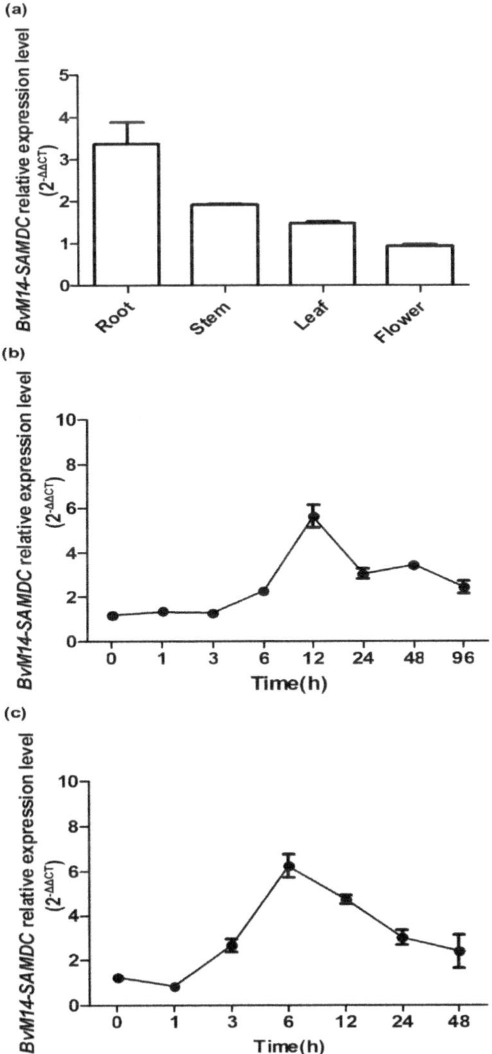

Figure 2. *BvM14-SAMDC* expression patterns in different organs and in response to salt stress treatments. (a) BvM14-SAMDC expression profiles in different organs. (b) leaves and (c) roots of the M14 plants treated with 400 mM NaCl for different time periods. Data are the means of three biological replicates with standard deviation (SD) bars.

Following salt stress, plant growth was significantly affected in WT and all transgenic lines (Figure 3a). However, the root length, fresh weight and chlorophyll content were higher in transgenic plants overexpressing *BvM14-SAMDC* under salt stress compared with WT or *atsamdc1* (Figure 3b–e). Furthermore, we found *atsamdc1* were more sensitive to salt stress than the complementation lines (CO1 and CO2) or WT. Moreover, relative electrical conductivity was found to be much higher in the KD lines than the WT, complementation lines (CO1 and CO2) under salt stress conditions (Figure 3e). The relative electrical conductivity of KD was 1.44-fold higher than that of the WT under salt stress. Usually, the relative conductivity is an important index to evaluate cell membrane permeability. The lower value in the OX lines under salinity stress indicated that overexpressing *BvM14-SAMDC* could alleviate cell membrane damage in the plants. Collectively, these results demonstrated that *BvM14-SAMDC* is involved in regulating plants salt stress response, and plants overexpressing *BvM14-SAMDC* are more tolerant to salt stress.

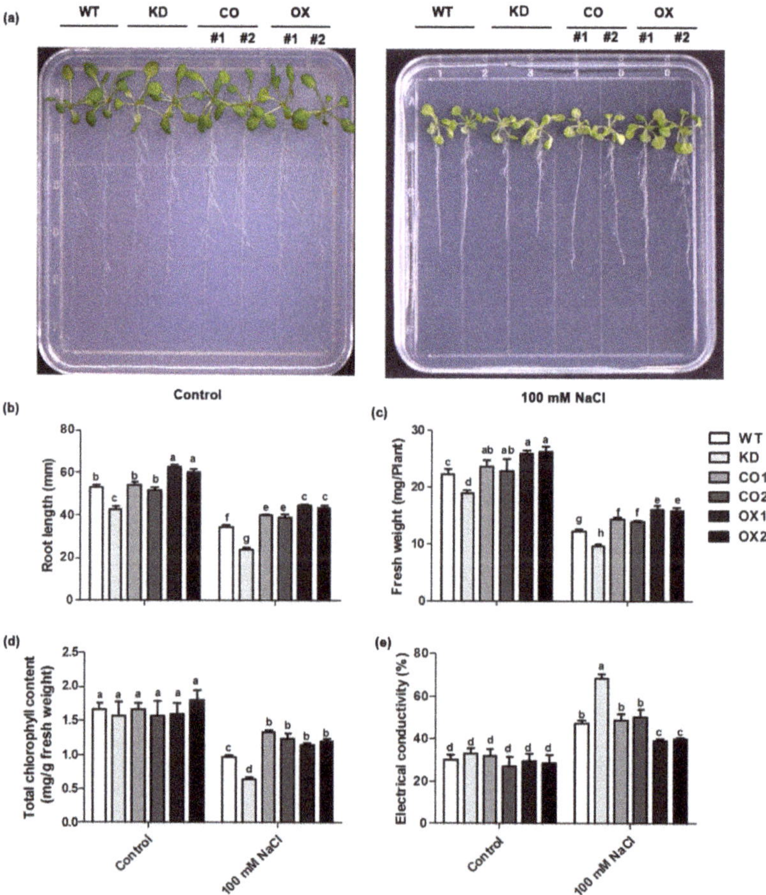

Figure 3. Effect of salt stress on seedling growth phenotype, root length, fresh weight, chlorophyll and electrical conductivity in wild type (WT), *BvM14-SAMDC*-overexpressed in wild type *Arabidopsis* (OX), *atsamdc1* mutant (KD) and transgenic *BvM14-SAMDC* in the mutant seedlings (CO) leaves. (**a**) 8-day-old WT and transgenic seedlings, grown on MS medium, were transferred to new MS solid agar plates supplemented with 0 or 100 mM NaCl, followed by growth for 10 days; (**b**) root length; (**c**) fresh weight; (**d**) chlorophyll level; and (**e**) electrical conductivity in control and 100 mM NaCl treated seedlings. Different letters indicate significantly different at *P* < 0.05. Three biological replicates were performed.

2.5. Overexpression of BvM14-SAMDC Enhanced Antioxidative Activities in Arabidopsis

To determine how overexpression of *BvM14-SAMDC* affects the plant salt stress tolerance, malondialdehyde (MDA) content and antioxidant enzyme activities were determined in WT and the transgenic lines (Figure 4). Under stress conditions, MDA content was increased in all the genotypes (Figure 4a). However, the MDA level was dramatically higher in the KD mutant plants than transgenic or WT plants. In addition, *BvM14-SAMDC* overexpression plants (OX) showed lower MDA content than the WT and KD line under salt stress. Our results indicate that overexpressing *BvM14-SAMDC* could reduce the level of lipid peroxidation, which is tightly regulated by enzymes involved in ROS-detoxifying pathways. Therefore, we examined the activities of superoxide dismutase (SOD), catalase (CAT) and peroxidase (POD) (Figure 4b–d). After salt stress, the activities of SOD increased by 23.9%, 5.2% and 30.9% in WT, KD and OX1 lines, respectively. Moreover, the activities of SOD, CAT and POD in the leaves or roots of OX lines were significantly higher than WT or the mutant under control and salt stress conditions (Figure 4b–d and Figure S3). Overall, these results suggest that reduction of MDA can be attributed to the increased activities of ROS-detoxifying enzymes in the transgenic plants.

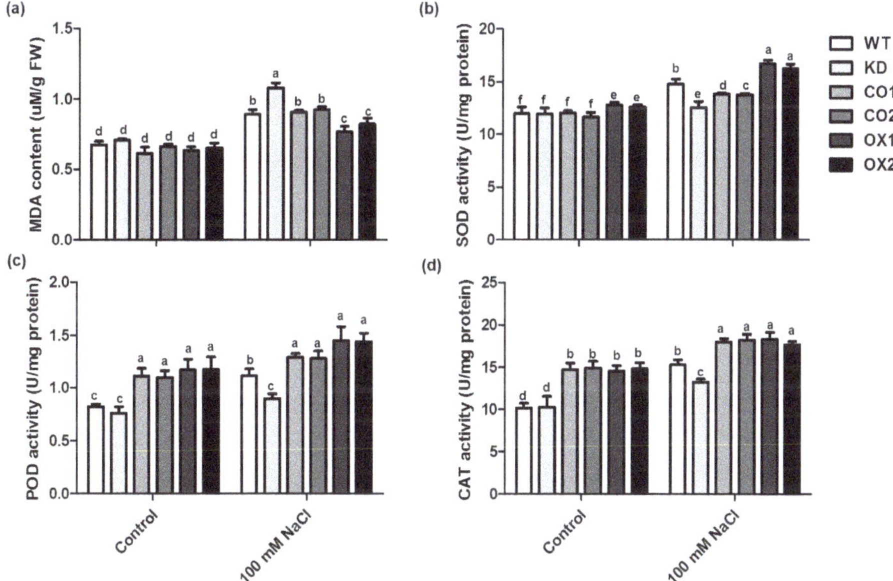

Figure 4. Effects of salt stress on antioxidant enzyme system in the leaves of wild type (WT), *BvM14-SAMDC*-overexpression in WT *Arabidopsis* (OX), *atsamdc1* mutant (KD) and transgenic *BvM14-SAMDC* in the mutant seedlings (CO). (**a**) leaf malondialdehyde (MDA) content; (**b–d**) antioxidant enzyme activities under control and salt stress (100 mM NaCl) conditions. One unit of CAT activity was defined as the amount of enzyme required for 1 μmol of H_2O_2 decomposed within 1 min. One unit of SOD activity was defined as the amount of enzyme required for inhibition of photochemical reduction of r-nitro blue tetrazolium chloride (NBT) by 50%. One unit of POD activity was defined as the amount of enzyme required for oxidation of 5 μmol of guaiacol within 1 min. Different letters indicate significantly different at $P < 0.05$. Three biological replicates were performed.

2.6. Overexpression of BvM14-SAMDC Enhanced PA Levels and PAO Activity in Arabidopsis

As to the overall SAMDC activities in *Arabidopsis*, the SAMDC activity of mutant line (KD) was 21.2% and 18.1% lower than that of WT under control and salt stress conditions, respectively. In contrast, the complementation line (CO) exhibited higher SAMDC activity than the wild type

(Figure 5a). As expected, the SAMDC activity in OX1 line was 1.46 times higher than that of WT under control conditions. Since SAMDC plays an important role in PA synthesis, polyamines were measured in the transgenic plants in comparison with WT and the mutant lines (Figure 5b–d). Our result indicates that the Put level showed no difference between the WT and transgenic plants, but the Spd and Spm levels were significantly elevated in the transgenic plants. Furthermore, Spd and Spm contents in the mutant line were dramatically lower than the transgenic or WT plants. Clearly, the synthesis of Spd and Spm was improved by overexpressing the *BvM14-SAMDC*.

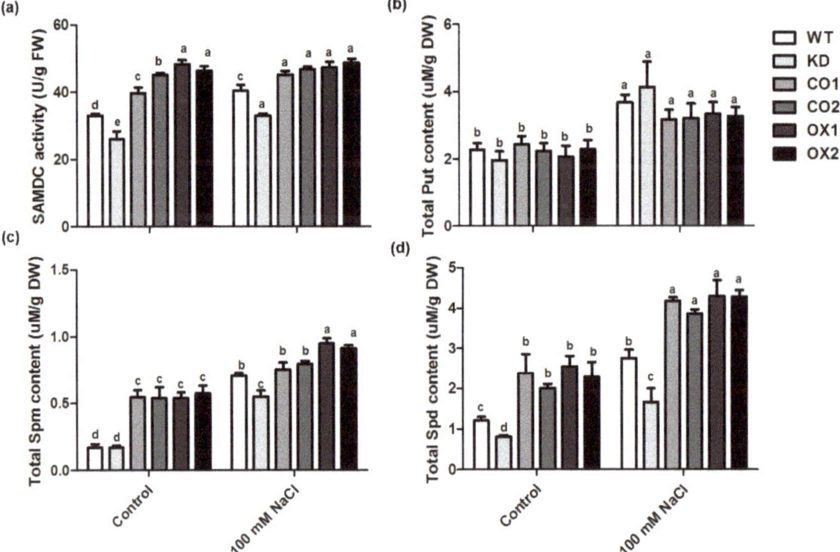

Figure 5. Effects of salt stress on polyamine metabolism in the leaves of wild type (WT), *BvM14-SAMDC*-overexpression in WT *Arabidopsis* (OX), *atsamdc1* mutant (KD) and transgenic *BvM14-SAMDC* in the mutant seedlings (CO). (**a**) SAMDC activity; (**b**) levels of putrescine (Put); (**c**) levels of spermine (Spm); and (**d**) levels of spermidine (Spd) under control and salt stress (100 mM NaCl) conditions. One unit of SAMDC activity was defined as the amount of enzyme required for catalyzing 2 μmol SAM within 1 min. Different letters indicate significantly different at $P < 0.05$. Three biological replicates were performed.

Previously, it is reported that PA played a key role in regulating H_2O_2 homeostasis through Spd and Spm catabolized by polyamine oxidase (PAO) to generate H_2O_2 in plants [11]. H_2O_2 is also an important signaling molecule to induce stress/defense responses. In the present study, higher activity of PAO was found in the transgenic plants than WT (Figure 6a and Figure S4a), and the H_2O_2 content was lower in the overexpression transgenic seedlings than in WT (Figure 6b). The generation of ROS may be attributed to the expressions of respiratory burst oxidase homolog genes (*RbohD* and *RbohF*), which were significantly enhanced in the leaves or roots of WT under salt stress (Figure 6c,d and Figure S4b,c). Salt stress-enhanced the expression of *RbohD* and *RbohF* transcription was notably inhibited in the leaves and roots of two transgenic lines compared to WT plant (Figure 6c,d and Figure S4b,c). Therefore, these results indicate that PA could alleviate ROS damage in the transgenic plants by decreasing the expression of *RbohD* and *RbohF*.

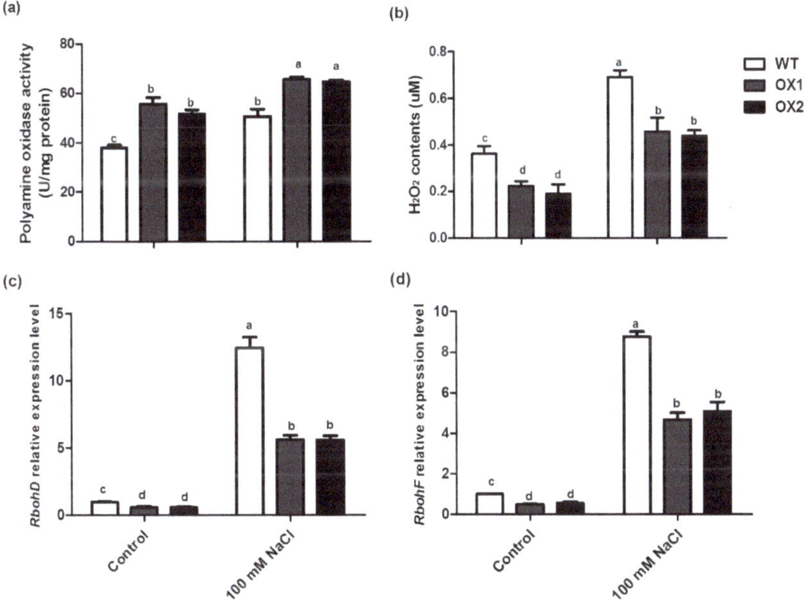

Figure 6. Effects of salt stress on polyamine oxidase (PAO) activity, H_2O_2 content and mRNA levels of *RbohD* and *RbohF* in the leaves of wild type (WT) and *BvM14-SAMDC*-overexpression in WT *Arabidopsis* (OX). (**a**) PAO activity; (**b**) H_2O_2 content; (**c**) mRNA levels of *RbohD*; and (**d**) mRNA levels of *RbohF* under control and salt stress (100 mM NaCl) conditions. One unit of PAO activity was defined as the amount of enzyme required for catalyzing 1 mmol of Spd or Spm oxidation within 1 min. Different letters indicate significantly different at $P < 0.05$. Three biological replicates were performed.

3. Discussion

Plant PAs, namely spm, spd and put, play essential roles in many different developmental and physiological processes. In particular, plant PAs are involved in regulating plant resistance to various stresses [26,27]. As the key enzyme in the synthesis of Spm and Spd biosynthesis, SAMDCs play an important role in plant stress tolerance [28,29]. In our study, *BvM14-SAMDC* expression varied in different organs of sugar beet, and it might be associated with growth and developmental processes in the sugar beet M14. In addition, *BvM14-SAMDC* expression was significantly induced by salt stresses in roots and leaves. The increasing expression of *BvM14-SAMDC* appeared much earlier in the M14 roots than in leaves, suggesting that the response of PA metabolism to salt stress is rapid in roots compared with leaves. Usually, the photosynthetic pigment is closely associated with plant salinity tolerance [30]. The OX lines showed increases in the total chlorophyll content compared to WT and mutant plants under salt stress (Figure 3d). These results predict salt stress tolerance of the *BvM14-SAMDC* overexpression plant and the sensitivity of the mutant line to salt stress.

Other studies have demonstrated the functions of PAs as free radical scavengers under various stress conditions, and they inhibit lipid peroxidation [31,32]. Here the levels of diamine (Put) and triamine (Spd and Spm) were increased under salt stress conditions in wild type and the transgenic plants (Figure 5). This result is consistent with a previous report [33]. Treatment of *Solanum chilense* plants with 125 mM NaCl increased the production of Spm, involving in salt resistance in *S. chilense* [33]. Therefore, PAs play an important role in regulating plant salt stress tolerance. Moreover, the levels of Spd and Spm were much higher in the *BvM14-SAMDC* overexpression lines under both control and salt stress conditions, and there was no significant difference in put content between WT and the transgenic plants (Figure 5b). This situation will lead to a high ratio of (Spd + Spm)/Put in OX transgenic plans. It is reported that Spd and Spm play a key role in regulating the thylakoid membrane integrity, and a major function of put is to regulate cell membrane depolarization [34]. In addition, another study also showed that overexpression of a *SAMS* in tomato (another key enzyme in PA biosynthesis) caused a high ratio of (Spd + Spm)/Put and alkali stress tolerance through up-regulating *SlSAMDC* and *SlSPDS* [17]. They speculated that the high levels of *SlSAMDC* and *SlSPDS* could increase the conversion of Put to Spm and Spd. Therefore, the increased *BvM14-SAMDC* level in the transgenic plants may be involved in increasing the Spd and Spm levels for enhanced salt stress tolerance.

Previous studies proposed that SAMDC regulates the plant response to multiple abiotic stresses such as salinity, drought, cold and high temperature. These studies suggest that overexpression of SAMDC lead to polyamine accumulation, which can confer tolerance to both biotic and abiotic stresses in plants [35–38]. However, the mechanism of SAMDC gene improving plant stress resistance is not very clear. In the present study, compared to WT, higher activities of antioxidant enzymes were detected in the *BvM14-SAMDC* overexpression lines under both salt stress and control conditions. Furthermore, the MDA and H_2O_2 levels in the transgenic plants were significantly lower than in WT (Figures 4a and 6b). It is reported that tobacco plants with decreased *SAMDC* expression showed low PA biosynthesis, and accumulated significantly more H_2O_2 [19]. In our study, although the elevated PAO activity was observed in the transgenic plants, H_2O_2 content was reduced (Figure 6b). We speculated that ROS production induced by PAO activity was effectively cleared by the ROS-detoxifying enzymes in the *BvM14-SAMDC* OX lines. In addition, the expression of the two plasma membrane localized NADPH-oxidases, *RbohD* and *RbohF* is often associated with ROS generation [39,40]. Therefore, our study showed that expression of *RbohD* and *RbohF* was dramatically enhanced in the WT under salt stress, and the increased expression was blocked by overexpressing the *BvM14-SAMDC* in the transgenic plants (Figure 6c,d and Figure S4b,c). These results show that *BvM14-SAMDC* regulates salt stress tolerance by improving antioxidant activity and reducing ROS production (Figure 7). Similarly, our previous study also found the activities of CAT and POD were higher in the sugar beet *SAMS2* overexpression plants than in WT, and the *SAMS2* plants had a low accumulation of H_2O_2 [18]. Whether SAMS2, like SAMDC, and improve plant stress resistance through reducing ROS production remains to be further studied.

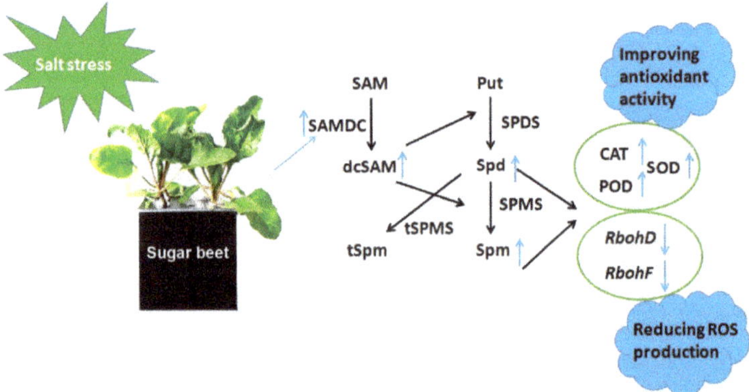

Figure 7. Overview diagram showing how the *Bv*M14-SAMDC functions in mediating plant salt stress tolerance. Salt stress can turn on the expression of BvM14-SAMDC, which plays an important role in the biosynthesis of PAs. PAs may inhibit the expression of ROS-biosynthetic enzymes and activate ROS-detoxifying enzymes, leading to reduced ROS levels and enhanced salt stress tolerance phenotype. SAMDC, S-adenosylmethionine decarboxylase; SPDS, spermidine synthase; SPMS, spermine synthase; tSPMS, thermospermine synthase; dcSAM, decarboxylated S-adenosylmethionine; Put, putrescine; SAM, S-adenosylmethionine; Spd, spermidine; Spm, spermine.

4. Materials and Methods

4.1. Plant Materials and Treatments

Sugar beet M14 was grown in a greenhouse of Heilongjiang University. After the M14 seeds were sown in vermiculite for 7 day, the seedlings were transferred to Hoagland nutrient solution with a 14 h light/10 h dark cycle, a 400 μmol m^{-2} s^{-1} light intensity, and 25 °C/21 °C day/night temperature. The 14 day old seedlings were subjected to 400 mM NaCl treatment. Sugar beet seedlings were also grown in pots with black soil from Heilongjiang University, and the organs of root, stem, leaf and flower were harvested after 3 months [41]. The *Arabidopsis* seeds were surface sterilized with the mixed solutions of NaClO (0.5%) and Triton X-100 (0.01%) for 8 min followed by washing with sterilized distilled water three times. The sterilized seeds were first incubated on Petri dishes containing Murashige and Skoog (MS) agar (0.8%) medium (containing 3% sucrose and 0.25% phyta-gel, pH 5.8) at 4 °C for 3 day before germination, and Petri dishes were sealed using parafilm. Then the seeds were germinated at 22 °C under 14 h at 300 μmol m^{-2} s^{-1} light intensity light and 10 h darkness, and parafilm was removed from the dishes. The 8-day-old seedlings were transferred to MS medium containing 100mM NaCl for salt treatment 10 day, and each Petri dish contained eight seedlings.

4.2. Cloning the Ful- Length cDNA of BvM14-SAMDC Gene

A unigene of *BvM14-SAMDC* was used for designing primers (sense primer: 5′-TCTGCTGCT TACTCAAACTGCG- 3′ and antisense primer: 5′-TATTCCAACACGGGACACTGA- 3′) to amplify the full-length cDNA sequence of the *BvM14-SAMDC* gene using reverse transcription (RT)-PCR through a RACE method [41].

4.3. Real-time Quantitative PCR

Real-time quantitative PCR (qRT-PCR) was carried out to determine the gene expression in the M14 line and Arabidopsis. Total RNA was extracted using a Trizol reagent (Invitrogen, Carlsbad, CA, USA), and cDNAs were synthesized from 0.5 μg total RNA using SuperScript-RNase H-reverse transcriptase (Clontech, Foster, CA, USA). With 18S rRNA as the internal control, samples from different organs were amplified using a Bio-Rad Thermocycler system combined with SYBR-Green fluorescent dye (TaKaRa, Dalian, China) according to the manufacturer's instructions. PCR reaction was carried out in 10 μL volumes using the following amplification protocol: 94 °C for 5 min; 94 °C for 35 s, 55 °C for 30 s, and 72 °C for 90 s; and 72 °C for 5 min, 45 cycles. The primers used for qRT-PCR analysis are listed in Table S1.

4.4. Heterologous Expression of BvM14-SAMDC in E.coli

The open reading frame (ORF) sequence of the *BvM14-SAMDC* gene was amplified with primers (5′-ATGACGGTTCCCATGGTTGG-3′ and 5′-ATTCATTTCTTCTTCTTTTTT-3′) and ligated into a pMD18-T vector. Then, the construction vector was cut with *EcoR* I and *Sal* I, the cut fragment was ligated with a pET28a carrier cut with *EcoR* I and *Sal* I. The construct was introduced into *E. coli* BL21 (DE3) for protein expression after induction with 0.5 mM IPTG. Cells were harvested after 4 h for total protein extracts from the induced *E. coli* transformed with pET28a-*BvM14-SAMDC*. The total protein extracts from the induced *E. coli* transformed with pET28a-*BvM14-SAMDC* were analyzed using 12 % SDS-PAGE and detected with anti-histidine monoclonal antibody linked to horseradish peroxidase (1:10,000) according to the manufacturer's instructions (Pierce, Carlsbad, CA, USA).

4.5. Arabidopsis Thaliana Mutant Screening

Arabidopsis T-DNA insertion lines were identified for *AtSAMDC1* mutation by genotyping PCR. Specific primers for the left and right borders of the T-DNA and for *AtSAMDC1* (F1: 5-ATT GAAGACGTTCGTCCAAAC-3, R1: 5-CTCGCCTTGTTGTGTGAGCGACAG-3, T1: 5-TAGCATCTGA

ATTTCATAACCAATCTCGATACAC-3) were used to identify the mutant lines. PCR reaction was carried out in 30 μL volumes using the following amplification protocol: 94 °C for 5 min; 94 °C for 30 s, 60 °C for 30 s, and 72 °C for 90 s; and 72 °C for 5min, 30 cycles.

4.6. Generation of 35S:BvM14-SAMDC Construct and Transformation of Arabidopsis

The ORF sequence of the *BvM14-SAMDC* was amplified using primers (5′-GTT CTTTTAA GCAATCTAG-3′ and 5′-ATTCATTTCTTCTTCTTTTTT-3′) and inserted into the pUCm-T vector. Then the construct was digested with using *EcoR* I and *BamH* I and inserted into a pCAMBIA1305 vector. *Agrobacterium tumefaciens* strain EHA105 containing pCAMBIA1305-*BvM14-SAMDC* was infiltrated into *Arabidopsis* using a floral dip method as previously described [42]. The presence of the transgene was confirmed by PCR and quantitative PCR.

4.7. Determination of Physiological Indicators

Chlorophyll (chl) content was measured according to our previous reported by homogenizing leaf samples (0.5 g) with 10 mL acetone (80% *v/v*) followed by centrifuging at 9000 g for 8 min. Absorbance was measured with a UV-visspectro photometer at 663 and 645 nm for chl a and chl b content, respectively. The activities of SOD and CAT were measured according to previously reported methods [17,43,44]. For the extraction of antioxidant enzyme, 1 g of sugar beet tissues were homogenised in 3 mL of chilled buffer containing 50 mM phosphate buffer (pH = 7.8), 2 mM EDTA, 1 mM DTT, 1 mM PMSF, 0.5% (*v/v*) Triton X-100 and 10% (*w/v*) PVP-40, and the homogenate was centrifuged at 20,000 g for 30 min. For measurement of SOD, 1mL of reaction mixture was prepared in 50 mM K-P buffer (pH 7.8) containing 2 μM riboflavin, 75 μM nitrotetrazolium blue (NBT), 100 μM EDTA, 13 mM DL-methionine and 60 μL of enzyme extract and the absorbance was taken at 560 nm. CAT enzymatic activity was calculated using the system reported by Aebi. The decline of H_2O_2 absorbance was determined in 2 mL of reaction buffer containing 100 mM K-P buffer (pH 7.0), 20 mM H_2O_2 and 20 μL enzyme extract. POD activity was measured by the increase in absorbance at 470 nm due to guaiacol oxidation

For MDA extraction, fresh leaf samples (0.5 g) were homogenized with 0.1% trichloroacetic acid (TCA). The homogenate was then centrifuged at 15,000 g for 10 min. An aliquot (1 mL) of the supernatant was mixed to 4 mL of 20% TCA prepared in 0.5% thiobarbituric acid (TBA) and incubated at 90 °C for 30 min in a shaking water bath. The reaction was stopped in an ice bath. The samples were then centrifuged at 10,000 g for 5 min, and the absorbance of the supernatant was measured at 532 nm.

For measurement of H_2O_2, fresh leaves (0.3 g) were frozen in liquid nitrogen and ground to powder in a mortar with a pestle, together with 5 mL of 5% TCA and 0.15 g activated charcoal. The mixture was centrifuged at 10,000 g for 20 min at 4 °C. The supernatant was adjusted to pH 8.4 with 17 M ammonia solution and then filtered. The filtrate was divided into aliquots of 1 mL. To one of these, the blank was added 8 μg of catalase and then kept at room temperatures for 10 min. To both aliquots with and without catalase, 1 mL of colorimetric reagent was added. The reaction solution was incubated for 10 min at 30 °C. Absorbance at 505 nm was determined spectrophotometrically.

Polyamine levels were measured according to the procedure of Guo et al [9]. Leaves (0.5 g) were extracted in 5 mL of 5% (*v/v*) cold perchloric acid (PCA) and incubated on ice for 1 h. The homogenate was centrifuged at 20,000 g for 30 min. Aliquots (0.5 mL) of supernatant were mixed with 1 mL of 2 M NaOH and 7 mL of benzoyl chloride and incubated at 37 °C for 20 min in dark for benzoylation. The benzylated polyamines were extracted to diethyl ether and resuspended in 1 mL of mobile phase solution before HPLC analysis. Polyamine levels were calculated based on standard curves in combination with a recovery of the extraction procedure. The activity of SAMDC was assayed by HPLC using an enzyme activity standard curve [9].

Polyamine oxidase was extracted in 0.1 M phosphate buffer (PH 7.0), and the reaction was initiated and incubated at 30 °C for 30 min after addition of 20 mL of 20 mM Spd or Spd into the reaction mixture (3 mL) that consisted of 2.5 mL of 0.1 M phosphate buffer, 0.1 mL of horseradish peroxidase,

0.2 mL of coloring solution (25 mL *N*, *N*-dimethylaniline and 10 mg 4-aminoantipyrine were dissolved in 100 mL of 0.1 M phosphate buffer, pH 7.0) and 0.2 mL enzyme extract or inactivated enzyme (by heating the enzyme for 20 min in a boiling water bath). Absorbance at 550 nm was recorded.

5. Conclusions

In the present study, we discovered that overexpression of *BvM14-SAMDC* in *Arabidopsis* seedlings conferred salt stress tolerance. Furthermore, the overexpression lines (OX) reduced ROS accumulation, had decreased expression of *RbohD* and *RbohF*, and increased antioxidant enzyme activities. This work clearly showed that a key enzyme SAMDC in the biosynthesis of PAs can enhance salt stress tolerance through reducing ROS levels caused by inhibiting the expression of ROS-biosynthetic enzymes and activation of ROS-detoxifying enzymes. How does salt stress turn on the expression of *BvM14-SAMDC* is a very interesting question to be addressed in the near future.

Supplementary Materials: Supplementary Materials can be found at http://www.mdpi.com/1422-0067/20/8/1990/s1.

Author Contributions: Y.W., H.L., S.C. and C.M. compiled and edited the manuscript. Y.W. and C.M. are the principal investigator of the project and conceived the overall concept of the study. Y.W., C.M., M.J., L.W. and K.W. participated in the sampling, carried out the experiments and statistical analysis. All authors read and approved the final manuscript.

Funding: Research was supported by the National Natural Science Foundation of China (Project 31701487: Analysis of transcription factor BvBHLH93 function and salt tolerance regulation mechanism in sugar beet T510), the Common College Science and Technology Innovation Team of Heilongjiang Province (2014TD004), Basic Research Work Program of Heilongjiang Provincial Higher Education Institutions (KJCXYB201706) and Youth Innovative Talents Training Program of Heilongjiang Regular Universities. We also appreciated Maoqian Wang for her project help (the provincial universities basal research project at Heilongjiang University, KJCXYB201714).

Conflicts of Interest: The authors declare no conflict of interest.

References

1. Munns, R.; Tester, M. Mechanisms of salinity tolerance. *Annu. Rev. Plant Biol.* **2008**, *59*, 651–681. [CrossRef] [PubMed]
2. Kumar, M.; Yusuf, M.A.; Nigam, M.; Kumar, M. An update on genetic modification of chickpea for increased yield and stress tolerance. *Mol. Biotechnol.* **2018**, *60*, 651–663. [CrossRef]
3. Liang, W.; Ma, X.; Wan, P.; Liu, L. Plant salt-tolerance mechanism: A review. *Biochem. Biophys. Res. Commun.* **2018**, *495*, 286–291. [CrossRef]
4. Yang, Y.; Guo, Y. Elucidating the molecular mechanisms mediating plant salt-stress responses. *New Phytol.* **2018**, *217*, 523–539. [CrossRef]
5. Shah, N.; Anwar, S.; Xu, J.; Hou, Z.; Salah, A.; Khan, S.; Gong, J.; Shang, Z.; Qian, L.; Zhang, C. The response of transgenic *Brassica* species to salt stress: A review. *Biotechnol. Lett.* **2018**, *40*, 1159–1165. [CrossRef]
6. Wu, B.; Munkhtuya, Y.; Li, J.; Hu, Y.; Zhang, Q.; Zhang, Z. Comparative transcriptional profiling and physiological responses of two contrasting oat genotypes under salt stress. *Sci. Rep.* **2018**, *8*, 16248. [CrossRef]
7. Kusano, T.; Berberich, T.; Tateda, C.; Takahashi, Y. Polyamines: Essential factors for growth and survival. *Planta* **2008**, *228*, 367–381. [CrossRef]
8. Liu, J.H.; Wang, W.; Wu, H.; Gong, X.; Moriguchi, T. Polyamines function in stress tolerance: From synthesis to regulation. *Front. Plant Sci.* **2015**, *6*, 827. [CrossRef]
9. Guo, J.; Wang, S.; Yu, X.; Dong, R.; Li, Y.; Mei, X.; Shen, Y. Polyamines regulate strawberry fruit ripening by abscisic acid, auxin, and ethylene. *Plant Physiol.* **2018**, *177*, 339–351. [CrossRef]
10. Mehta, R.A.; Cassol, T.; Li, N.; Ali, N.; Handa, A.K.; Mattoo, A.K. Engineered polyamine accumulation in tomato enhances phytonutrient content, juice quality, and vine life. *Nat. Biotechnol.* **2002**, *20*, 613–618. [CrossRef] [PubMed]
11. Angelini, R.; Cona, A.; Federico, R.; Fincato, P.; Tavladoraki, P.; Tisi, A. Plant amine oxidases "on the move": An update. *Plant Physiol. Biochem.* **2010**, *48*, 560–564. [CrossRef] [PubMed]
12. Roje, S. S-Adenosyl-L-methionine: Beyond the universal methyl group donor. *Phytochemistry* **2006**, *67*, 1686–1698. [CrossRef]

13. Romero, F.M.; Maiale, S.J.; Rossi, F.R.; Marina, M.; Ruíz, O.A.; Gárriz, A. Polyamine metabolism responses to biotic and abiotic stress. *Methods Mol. Biol.* **2018**, *1694*, 37–49. [CrossRef]

14. Shu, S.; Yuan, Y.; Chen, J.; Sun, J.; Zhang, W.; Tang, Y.; Zhong, M.; Guo, S. The role of putrescine in the regulation of proteins and fatty acids of thylakoid membranes under salt stress. *Sci. Rep.* **2015**, *5*, 14390. [CrossRef] [PubMed]

15. Kim, S.H.; Kim, S.H.; Palaniyandi, S.A.; Yang, S.H.; Suh, J.W. Expression of potato S-adenosyl-L-methionine synthase (*SbSAMS*) gene altered developmental characteristics and stress responses in transgenic Arabidopsis plants. *Plant Physiol. Biochem.* **2015**, *87*, 84–91. [CrossRef]

16. Tanou, G.; Ziogas, V.; Belghazi, M.; Christou, A.; Filippou, P.; Job, D.; Fotopoulos, V.; Molassiotis, A. Polyamines reprogram oxidative and nitrosative status and the proteome of citrus plants exposed to salinity stress. *Plant Cell Environ.* **2014**, *37*, 864–885. [CrossRef] [PubMed]

17. Gong, B.; Li, X.; VandenLangenberg, K.M.; Wen, D.; Sun, S.; Wei, M.; Li, Y.; Yang, F.; Shi, Q.; Wang, X. Overexpression of S-adenosyl-L-methionine synthetase increased tomato tolerance to alkalistress through polyamine metabolism. *Plant Biotechnol. J.* **2014**, *12*, 694–708. [CrossRef] [PubMed]

18. Ma, C.; Wang, Y.; Gu, D.; Nan, J.; Chen, S.; Li, H. Overexpression of *S-Adenosyl-l-Methionine Synthetase 2* from sugar beet M14 increased *Arabidopsis* tolerance to salt and oxidative stress. *Int. J. Mol. Sci.* **2017**, *18*, 847. [CrossRef] [PubMed]

19. Mellidou, I.; Moschou, P.N.; Ioannidis, N.E.; Pankou, C.; Gėmes, K.; Valassakis, C.; Andronis, E.A.; Beris, D.; Haralampidis, K.; Roussis, A.; et al. Silencing S-Adenosyl-L-Methionine Decarboxylase (SAMDC) in *Nicotiana tabacum* points at a polyamine-dependent trade-off between growth and tolerance responses. *Front. Plant Sci.* **2016**, *7*, 379. [CrossRef]

20. Majumdar, R.; Lebar, M.; Mack, B.; Minocha, R.; Minocha, S.; Carter-Wientjes, C.; Sickler, C.; Rajasekaran, K.; Cary, J.W. The *Aspergillus flavus* spermidine synthase (*spds*) gene, is required for normal development, aflatoxin productionand pathogenesis during infection of maize kernels. *Front. Plant Sci.* **2018**, *9*, 317. [CrossRef]

21. Mo, H.J.; Sun, Y.X.; Zhu, X.L.; Wang, X.F.; Zhang, Y.; Yang, J.; Yan, G.J.; Ma, Z.Y. Cotton S-adenosylmethionine decarboxylase-mediated spermine biosynthesis is required for salicylic acid-and leucine-correlated signaling in the defense response to *Verticillium dahliae*. *Planta* **2016**, *243*, 1023–1039. [CrossRef]

22. Ge, Y.; He, G.; Wang, Z.; Guo, D.; Qin, R.; Li, R. GISH and BAC-FISH study of apomicitic Beta M14. *Sci. China Ser. C Life Sci.* **2007**, *37*, 209–216.

23. Guo, D.; Kang, C.; Liu, L.; Li, Y. Study of apomixis in theallotriploid beet (VVC). *Agric. Sci. China* **1999**, *32*, 1–5.

24. Yang, L.; Ma, C.; Wang, L.; Chen, S.; Li, H. Salt stress induced proteome and transcriptome changes in sugar beet monosomic addition line M14. *J. Plant Physiol.* **2012**, *169*, 839–850. [CrossRef] [PubMed]

25. Yang, L.; Zhang, Y.; Zhu, N.; Koh, J.; Ma, C.; Pan, Y.; Yu, B.; Chen, S.; Li, H. Proteomic analysis of salt tolerance in sugar beet monosomic addition line M14. *J. Proteome Res.* **2013**, *12*, 4931–4950. [CrossRef]

26. Yin, J.; Jia, J.; Lian, Z.; Hu, Y.; Guo, J.; Huo, H.; Zhu, Y.; Gong, H. Silicon enhances the salt tolerance of cucumber through increasing polyamine accumulation and decreasing oxidative damage. *Ecotoxicol. Environ. Saf.* **2019**, *169*, 8–17. [CrossRef]

27. Murray Stewart, T.; Dunston, T.T.; Woster, P.M.; Casero, R.A., Jr. Polyamine catabolism and oxidative damage. *J. Biol. Chem.* **2018**, *293*, 18736–18745. [CrossRef] [PubMed]

28. Wi, S.J.; Kim, S.J.; Kim, W.T.; Park, K.Y. Constitutive S-adenosylmethionine decarboxylase gene expression increases drought tolerance through inhibition of reactive oxygen species accumulation in Arabidopsis. *Planta* **2014**, *239*, 979–988. [CrossRef] [PubMed]

29. Marco, F.; Busó, E.; Carrasco, P. Overexpression of *SAMDC1* gene in *Arabidopsis* increases expression of defense related genes as well as resistance to *Pseudomonas syringae* and *Hyaloperonospora arabidopsidis*. *Front. Plant Sci.* **2014**, *5*, 115. [CrossRef]

30. Ibrahim, W.; Qiu, C.W.; Zhang, C.; Cao, F.; Shuijin, Z.; Wu, F. Comparative physiological analysis in the tolerance to salinity and drought individual and combination in two cotton genotypes with contrasting salt tolerance. *Physiol. Plant.* **2019**, *165*, 155–168. [CrossRef]

31. Groppa, M.D.; Benavides, M.P. Polyamines and abiotic stress: Recent advances. *Amino Acids* **2008**, *34*, 35–45. [CrossRef]

32. Cona, A.; Rea, G.; Botta, M.; Corelli, F.; Federico, R.; Angelini, R. Flavin-containing polyamine oxidase is a hydrogen peroxide source in the oxidative response to the protein phosphatase inhibitor cantharidin in *Zea mays* L. *J. Exp. Bot.* **2006**, *57*, 2277–2289. [CrossRef] [PubMed]
33. Gharbi, E.; Martínez, J.P.; Benahmed, H.; Fauconnier, M.L.; Lutts, S.; Quinet, M. Salicylic acid differently impacts ethylene and polyamine synthesis in the glycophyte *Solanum lycopersicum* and the wild-related halophyte *Solanum chilense* exposed to mild salt stress. *Physiol. Plant.* **2016**, *158*, 152–167. [CrossRef]
34. Botella, M.A.; Xu, Y.; Prabha, T.N.; Zhao, Y.; Narasimhan, M.L.; Wilson, K.A.; Nielsen, S.S.; Bressan, R.A.; Hasegawa, P.M. Differential expression of soybean cysteine proteinase inhibitor genes during development and in response to wounding and methyl jasmonate. *Plant Physiol.* **1996**, *112*, 1201–1210. [CrossRef]
35. Hazarika, P.; Rajam, M.V. Biotic and abiotic stress tolerance in transgenic tomatoes by constitutive expression of S-adenosylmethionine decarboxylase gene. *Physiol. Mol. Biol. Plants* **2011**, *17*, 115–128. [CrossRef] [PubMed]
36. Wi, S.J.; Kim, W.T.; Park, K.Y. Overexpression of carnation S-adenosylmethionine decarboxylase gene generates a broad-spectrum tolerance to abiotic stresses in transgenic tobacco plants. *Plant Cell Rep.* **2006**, *25*, 1111–1121. [CrossRef]
37. Waie, B.; Rajam, M.V. Effect of increased polyamine biosynthesis on stress responses in transgenic tobacco by introduction of human S-adenosylmethionine gene. *Plant Sci.* **2003**, *164*, 727–734. [CrossRef]
38. Roy, M.; Wu, R. Overexpression of S-adenosylmethionine decarboxylase gene in rice increases polyamine level and enhances sodium chloride-stress tolerance. *Plant Sci.* **2002**, *163*, 987–992. [CrossRef]
39. Yang, L.; Ye, C.; Zhao, Y.; Cheng, X.; Wang, Y.; Jiang, Y.Q.; Yang, B. An oilseed rape WRKY-type transcription factor regulates ROS accumulation and leaf senescence in *Nicotiana benthamiana* and Arabidopsis through modulating transcription of *RbohD* and *RbohF*. *Planta* **2018**, *247*, 1323–1338. [CrossRef]
40. Morales, J.; Kadota, Y.; Zipfel, C.; Molina, A.; Torres, M.A. The Arabidopsis NADPH oxidases *RbohD* and *RbohF* display differential expression patterns and contributions during plant immunity. *J. Exp. Bot.* **2016**, *67*, 1663–1676. [CrossRef] [PubMed]
41. Wang, Y.; Zhan, Y.; Wu, C.; Gong, S.; Zhu, N.; Chen, S.; Li, H. Cloning of a cystatin gene from sugar beet M14 that can enhance plant salt tolerance. *Plant Sci.* **2012**, *191–192*, 93–99. [CrossRef]
42. Clough, S.J.; Bent, A.F. Floral dip: A simplified method for Agrobacterium mediated transformation of *Arabidopsis thaliana*. *Plant J.* **1998**, *16*, 735–743. [CrossRef]
43. Aebi, H. CAT in vitro. *Meth. Enzymol.* **1984**, *105*, 121–126. [CrossRef]
44. Wang, Y.; Stevanato, P.; Yu, L.; Zhao, H.; Sun, X.; Sun, F.; Li, J.; Geng, G. The physiological and metabolic changes in sugar beet seedlings under different levels of salt stress. *J. Plant Res.* **2017**, *130*, 1079–1093. [CrossRef] [PubMed]

International Journal of
Molecular Sciences

Article

A Cotton (*Gossypium hirsutum*) *Myo*-Inositol-1-Phosphate Synthase (*GhMIPS1D*) Gene Promotes Root Cell Elongation in *Arabidopsis*

Rendi Ma [1,†], Wangyang Song [1,†], Fei Wang [1,†], Aiping Cao [1], Shuangquan Xie [1], Xifeng Chen [1], Xiang Jin [1,2,*] and Hongbin Li [1,*]

1   College of Life Sciences, Key Laboratory of Xinjiang Phytomedicine Resource and Utilization of Ministry of Education, Shihezi University, Shihezi 832003, China; mrendi118@163.com (R.M.); swywinner@163.com (W.S.); wangfshzu@163.com (F.W.); aiping9smile@sina.com (A.C.); xiesq@shzu.edu.cn (S.X.); cxf_cc@shzu.edu.cn (X.C.)
2   Ministry of Education Key Laboratory for Ecology of Tropical Islands, College of Life Sciences, Hainan Normal University, Haikou 571158, China
*   Correspondence: jinx@hainnu.edu.cn (X.J.); lihb@shzu.edu.cn (H.L.); Tel.: +86-898-66961060 (X.J.); +86-993-2057912 (H.L.)
†   These authors contributed equally to this work.

Received: 20 January 2019; Accepted: 26 February 2019; Published: 11 March 2019

Abstract: *Myo*-inositol-1-phosphate synthase (MIPS, EC 5.5.1.4) plays important roles in plant growth and development, stress responses, and cellular signal transduction. *MIPS* genes were found preferably expressed during fiber cell initiation and early fast elongation in upland cotton (*Gossypium hirsutum*), however, current understanding of the function and regulatory mechanism of *MIPS* genes to involve in cotton fiber cell growth is limited. Here, by genome-wide analysis, we identified four *GhMIPS* genes anchoring onto four chromosomes in *G. hirsutum* and analyzed their phylogenetic relationship, evolutionary dynamics, gene structure and motif distribution, which indicates that *MIPS* genes are highly conserved from prokaryotes to green plants, with further exon-intron structure analysis showing more diverse in *Brassicales* plants. Of the four *GhMIPS* members, based on the significant accumulated expression of *GhMIPS1D* at the early stage of fiber fast elongating development, thereby, the *GhMIPS1D* was selected to investigate the function of participating in plant development and cell growth, with ectopic expression in the loss-of-function *Arabidopsis mips1* mutants. The results showed that *GhMIPS1D* is a functional gene to fully compensate the abnormal phenotypes of the deformed cotyledon, dwarfed plants, increased inflorescence branches, and reduced primary root lengths in *Arabidopsis mips1* mutants. Furthermore, shortened root cells were recovered and normal root cells were significantly promoted by ectopic expression of *GhMIPS1D* in *Arabidopsis mips1* mutant and wild-type plants respectively. These results serve as a foundation for understanding the *MIPS* family genes in cotton, and suggest that *GhMIPS1D* may function as a positive regulator for plant cell elongation.

Keywords: MIPS; exon-intron structure diversity; *Gossypium hirsutum*; loss-of-function mutant; root cell elongation

---

1. Introduction

Myo-inositol-1-phosphate synthase (MIPS; EC 5.5.1.4) is the rate-limiting enzyme to control *myo*-inositol (Ins) biosynthesis by converting D-glucose 6-phosphate (G6P) to inositol phosphate, followed by dephosphorylation reaction that is catalyzed by *myo*-inositol monophosphatase (IMP) [1]. The *MIPS* genes are widely spread in numerous organisms, from cyanobacteria to eubacteria and archaea, and ultimately to higher eukaryotes, such as higher plants and humans [2]. Numbers of *MIPS* have been identified in higher plants, including *Phaseolus vulgaris* [3], *Oryza sativa* [4], *Citrus paradise* [5], *Arabidopsis thaliana* [6], *Zea mays* [7], *Brassica napus* [8], *Glycine max* [9], and *Sesamum indicum* [10]. Protein sequence analysis of MIPS indicated that the eukaryotic MIPS family is homogenous, but different from the prokaryotic MIPS proteins [11]. This may be explained by the monophyletic origin of the eukaryotic *MIPS* genes. All eukaryotic MIPS sequences (around 510 amino acids) have shown regions of high conservation even at the nucleotide level [12], indicating that MIPS is very important for biological processes, such as embryogenesis and seed formation. There are four highly conserved amino acid sequences in MIPS proteins: GWGGNNG (domain 1), LWTANTERY (domain 2), NGSPQNTFVPGL (domain 3) and SYNHLGNNDG (domain 4) [1]. It is reported that these domains are involved in MIPS protein binding for catalyzing enzyme reaction and are essential for MIPS functions [2].

By catalyzing the biosynthesis of Ins and its derivates, including Ins polyphosphates (IPs), phospholipid phosphatidylinositol (PtdIns) and phosphoinositide phosphates (PtdInsPs), MIPS performs crucial diverse roles in biotic- and abiotic-stress responses, plant growth and organ development, and cell division [13]. The first MIPS was reported from *Archaeoglobus fulgidus* and might function under high temperatures [14]. In *Arabidopsis*, three genes encoding MIPS proteins (MIPS1–3) have been identified and appeared to have undergone functional divergence to some extent. Murphy et al. [15] found that loss-of-function *Arabidopsis mips2* mutants showed enhanced susceptibility to diverse viral, bacterial, and fungal pathogens, whereas *mips1* mutants showed no such effect. Furthermore, *mips1* mutants exhibit spontaneous cell death and enhanced resistance to the oomycete pathogen [16]. The *misp1* mutants were also reported to be sensitive to strong light stress [17]. The *mips1 mips2* double mutant and *mips1 mips2 mips3* triple mutant were embryo lethal, whereas the single *mips* mutants showed no obvious phenotype [18]. Severe reduction of MIPS activity inhibited plant growth significantly [19]. Lower IP_6 generated by suppressing *MIPS* using RNA interference (RNAi)-mediated approach in potato led to the alteration of leaf morphology, induction of leaf senescence, and reduction of tuber yield [7].

Allotetraploid cotton *Gossypium hirsutum* is the most widely planted cotton species, which provide more than 90% of natural fiber for the textile industry. The yield of cotton fiber is greatly influenced by the fiber initiation and early stage of fiber elongation. Ins and its derivates also play important role in cell growth by involving in cell wall synthesis, membrane trafficking, and signal transduction. The oxidation product of Ins, D-glucuronic acid (GlcA), is used for cell wall pectic noncellulosic compounds and, in some organisms, for the synthesis of ascorbic acid (AsA) [20–24]. Pectin precursors were reported to be essential for cotton fiber elongation in tetraploid cotton *G. hirsutum* [25]. Our previous works presented that AsA and its oxidation metabolism catalyzed by ascorbate peroxidase (GhAPX) were important for cotton fiber development in *G. hirsutum* [26,27]. PtdIns and PtdInsP were significantly accumulated in fast elongating fibers and promoted fiber cell elongation in vitro [28]. *GhMIPS* genes were preferably expressed during fiber initiation and the early stage of fiber elongation development. Nonetheless, current understanding of cotton *MIPS* family genes and their functions involving in plant growth and development and fiber cell growth is limited. In this work, we identified four *MIPS* gene family members in *G. hirsutum* through a genome-wide investigation and characterized the detailed information of phylogenetic relationship, evolutionary dynamics, exon-intron structure, and conserved motif distribution. Furthermore, after consideration of the expression levels of the four *GhMIPS* genes during fiber development, we selected the *GhMIPS1D* to further investigate its functional roles in participating in plant development and cell growth by ectopic expression of *GhMIPS1D* in *Arabidopsis mips1* mutants. The results indicated that *GhMIPS1D* could rescue the mutant abnormal phenotypes and ectopic expression of *GhMIPS1D* in *mips1* mutant and wild-type (WT) *Arabidopsis* plants promoted the root cell elongation significantly.

2. Results

2.1. Characterization, Phylogenetic Relationship and Motif Distribution Analyses of GhMIPS Gene Family Members in G. hirsutum

Cotton *MIPS* family genes were characterized by submitting the *Arabidopsis* MIPS protein sequences against *G. hirsutum* genome (allotetroploid, AADD) database derived from Phytozome (v12.1.6), followed by conserved domain recognition using InterProScan. Four *GhMIPS* members were determined: *GhMIPS1A*, *GhMIPS2A*, *GhMIPS1D* and *GhMIPS2D* (the sub-genomes were indicated by A and D). The *GhMIPS1A* located in sub-genome A chromosome 2 while *GhMIPS1D* located in sub-genome D chromosome 3, indicating homologous recombination events might occur during the individual evolution of A and D sub-genomes (Supplementary Figure S1). Due to the highly conserved gene sequences and structures, all *GhMIPS* homologues were considered as duplicates and the duplication relationships were indicated by red lines in Supplementary Figure S1. The *Ka/Ks* analysis showed that all *GhMIPS* members are under purifying selection (Supplementary Table S1). Expression patterns of *GhMIPS* family genes showed that *GhMIPS1A* and *GhMIPS1D* were the most highly expressed members in cotton fibers, with the highest expression level at 0 day post anthesis (DPA) (no fiber growth) and 3 DPA (fiber initiation growth) (Supplementary Figure S2), implying that *GhMIPS1D* may perform potential important role in fiber growth and development.

Phylogenetic tree of 70 MIPS homologues of 36 different species from prokaryotes to higher plants was constructed to investigate the evolutionary characters of the *MIPS* gene family. In general, seven prokaryotes, seven animal, six fungus and fifty plant MIPS proteins were included for the phylogenic analysis. Detailed information of MIPS used in this study is provided (Supplementary Table S2), containing organism names, accession numbers and sequences. The *MIPS* gene family is a very small one, usually including one or two members in most plant species, except for *Z. mays* (diploid, monocotyledons); *Brassica rapa*, *Brassica oleracea* and *Glysine max* (diploid, dicotyledons); *G. hirsutum* and *Gossypium barbadense* (tetraploid, dicotyledons); which have four *MIPS* members each. In addition to *B. rapa* and *B. oleracea*, the model plant *A. thaliana* has three *MIPS* members, indicating that this protein family has expanded in most *Brassicales* plants (Figure 1).

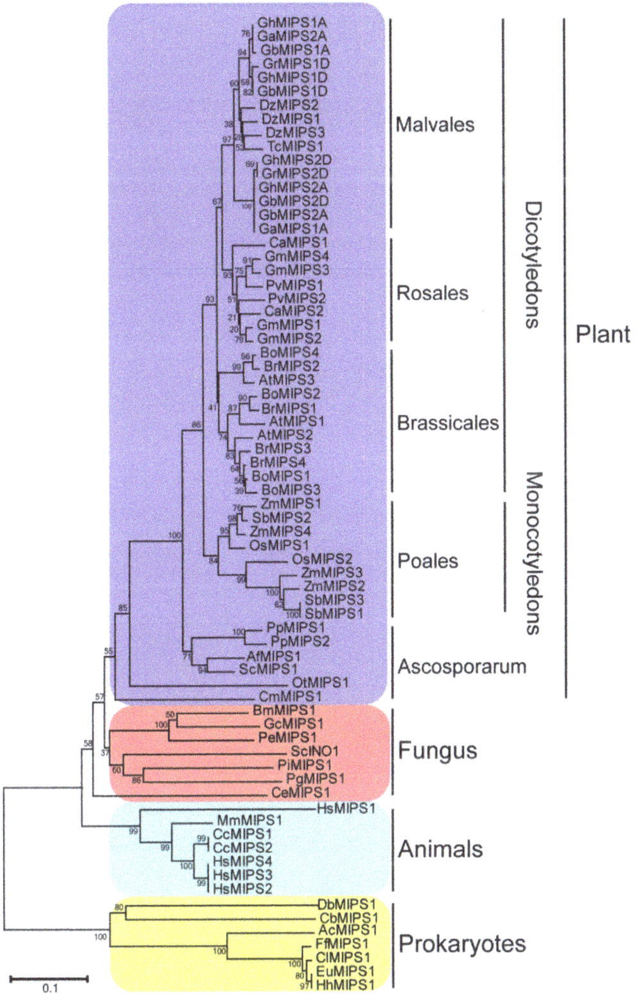

Figure 1. Phylogenetic tree of 70 MIPS homologues of 36 different species from prokaryotes to higher plants. See Supplementary Table S2 for detailed information of organism and protein sequences. The phylogenetic tree was constructed by MEGA5.0 using neighbor-joining method with bootstrap tests 1000.

GhMIPS1D and other nine typical plant MIPS proteins were used to perform multiple sequence alignment and conserved domain analysis, showing that MIPS proteins are quite conserved in plants with amino acid identity over 80%. Similar with the other nine plant MIPS proteins, cotton GhMIPS1D possessed two NAD(P)-binding domains distributing at the N and C terminus respectively, and one *myo*-inositol-1-phosphate synthase domain locating at the middle region, as well as four conserved domains that are responsible for MIPS protein binding and are essential for MIPS function exertion (Figure 2). Note that in higher plants, all MIPS proteins have around 510 amino acids, while CmMIPS1 of simple single cell algae (*Cyanidioschyzon merolae*) has 530 amino acids (Figure 2). Motif distribution analysis recognized ten conserved motifs that are organized similarly in all plant MIPS proteins (Figure 3A and see Supplementary Figure S3 for detailed sequences for motifs). These data suggest that MIPS is a highly conserved protein family in different plants, from ascosporarum to higher plants.

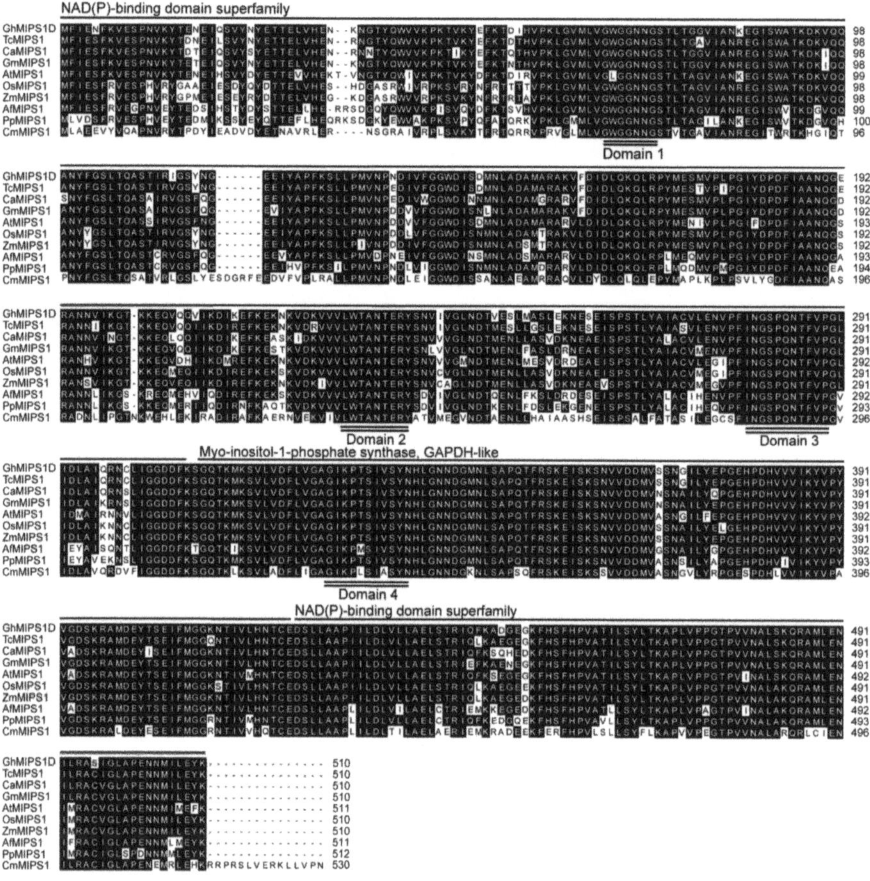

Figure 2. Multiple sequence alignment of ten representative MIPS proteins. See Supplementary Table S2 for organism and sequence information. Conserved domains are indicated with double underline (Domain 1–4).

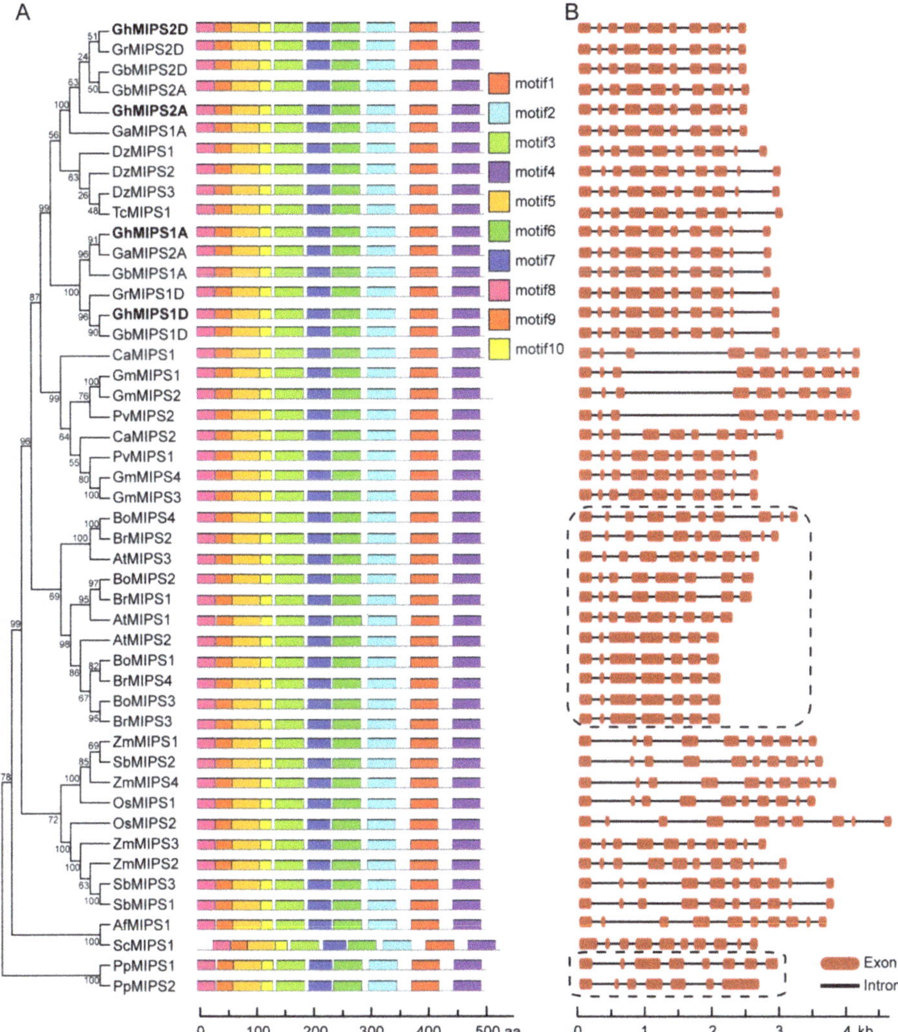

Figure 3. Conserved motif and exon-intron dynamic analyses of 48 plant *MIPS* genes. (**A**) Conserved motifs of protein sequences of 48 plant MIPS proteins. Ten conserved motifs were recognized and represented in different colors. All MIPS proteins are arranged according to their phylogenetic relationships. (**B**) The exon-intron structures of 48 plant *MIPS* genes. Two frames represent *Bryophytes* plants (*Physcomitrella patens*, seven and eight exons) and *Brassicales* plants (from seven to ten exons), which possess usual exon-intron structures. The detailed information for organism names, protein sequences and accession numbers are available in Supplementary Table S2.

2.2. Exon-Intron Structure Evolution Analysis of MIPS Gene Family

To further study the evolutionary relationship of *MIPS* family members, exon-intron structure dynamic analysis of four cotton *GhMIPS* members and other 44 plant *MIPS* genes was performed. Only one exon was found in single-cell algae *C. merolae* and *Ostreococcus tauri*. However, *Physcomitrella patens* *PpMIPS1* and *PpMIPS2* possess eight and seven exons respectively, indicating that gene family expansion and gene structure variation occurred after the terrestrial plant has been evolved (Figure 3B).

For higher plants, all MIPS proteins have ten exons except *B. rapa*, *B. oleracea* and *A. thaliana*, suggesting that *Brassicales* plants have undergone some special evolutionary events (Figure 3B, framed).

According to the phylogenetic tree and the exon-intron structures, an evolution model of the exon-intron structures for plant *MIPS* gene family was illustrated. The single-exon structure had been split into seven and eight exons shortly after the appearance of land plants. Then the last exon of the seven-exon structure homologues split into four new exons (event I, Figure 4A), while the 3rd and the last exons of the eight-exon structure members split into two new exons each (event II and event III, Figure 4A). Red arrows in Figure 4A indicate the exon split sites and black arrows indicate the amino acid sequences corresponding to the final exon-intron structures. Our data show that the formation of the ten-exon structure of *MIPS* genes is no later than arise of ferns (*Azolla filiculoides* and *Salvinia cucullata*).

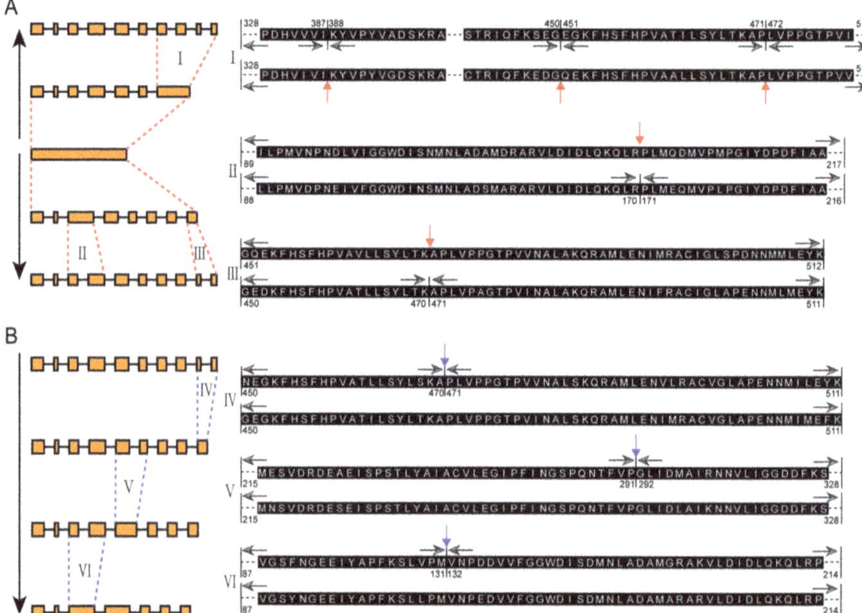

Figure 4. Illustration of the evolution model of *MIPS* exon-intron structures. (**A**) The evolution model of exon split events. *MIPS* possesses one entire exon in single cell algae. One evolution event has been recognized in *Bryophytes* plants, leading to a seven-exon *PpMIPS2* (up-arrow) and an eight-exon *PpMIPS1* (down-arrow). Then some of the exons have been split into more exons, forming the predominant exon-intron structure in higher plants (events I, II and III). The exon split sites are indicated by red arrows and amino acid numbers. (**B**) The evolution model of exon merge events in *Brassicales* plants. Three exon merge events have been identified, indicating there were three independent evolutionary events during the evolution of *Brassicales* plants. The exon-merge sites are indicated by blue arrows.

Notably, three exons merge events were recognized in *Brassicales* plants, leading to the formation of nine-exon (*AtMIPS1*), eight-exon (*BoMIPS2* and *BrMIPS1*) and seven-exon (*AtMIPS2*, *BoMIPS1*, *BoMIPS3*, *BrMIPS3* and *BrMIPS4*) *MIPS* family members, respectively (Figure 4B, and Figure 3B). Interestingly, the three exons merge events seem to occur in order (from event IV to event VI), indicating the *Brassicales* plants *MIPS* might have undergone three special evolutionary events.

2.3. Functional Complementary Analysis of GhMIPS1D in the Loss-of-Function Arabidopsis mips1 Mutant

In the light of Ins and its derivates important functions in plant development and cell growth [13], as well as the accumulated expression of *GhMIPS1D* during cotton fiber development (Supplementary Figure S2), thus, the overexpression vector *35S::GhMIPS1D-GFP* generated by cloning the *GhMIPS1D* into the modified *pCAMBIA2300-GFP* vector, was transformed into the loss-of-function *mips1* mutant and WT *Arabidopsis* plants, to investigate the *GhMIPS1D* in vivo functions. The *35S::GhMIPS1D-GFP* vector was introduced into onion epidermal cells to determine the subcellular localization of GhMIPS1D. The results showed that GhMIPS1D located in the nucleus, plasma membrane and endomembranes (Supplementary Figure S4), suggesting its likely functions in membrane trafficking and signal transduction, and regulation of gene expression.

The loss-of-function *Arabidopsis mips1* mutants were selected to perform genetic functional analysis of *GhMIPS1D*. The expression levels of *GhMIPS1D* in the transgenic *Arabidopsis* lines were examined (Supplementary Figure S5). As is reported, the T-DNA insertion *Arabidopsis* mutant lines exhibit abnormal cotyledon development (Figure 5B), shorter plants (Figure 5F) and increased inflorescence branch numbers (Figure 5G), and sensitivity to strong light stress (Supplementary Figure S6). The transgenic *mips1/GhMIPS1D* lines, obtained by transforming *35S::GhMIPS1D-GFP* vector into *Arabidopsis mips1* mutants, indicated stable expression levels of *GhMIPS1D* and rescued the aberrant phenotypes of *Arabidopsis mips1* mutants (Figure 5C–G). In addition, the *Arabidopsis* mutant lines expressing *GhMIPS1D* also showed increased tolerance to light stress under 220 μmol·m^{-2}s^{-1} (Supplementary Figure S6). These data suggest that *GhMIPS1D* is the predominantly expressed *MIPS* gene in G. *hirsutum* and is an active gene effectively to rescue the abnormal defects of growth and development in *Arabidopsis mips1* mutants.

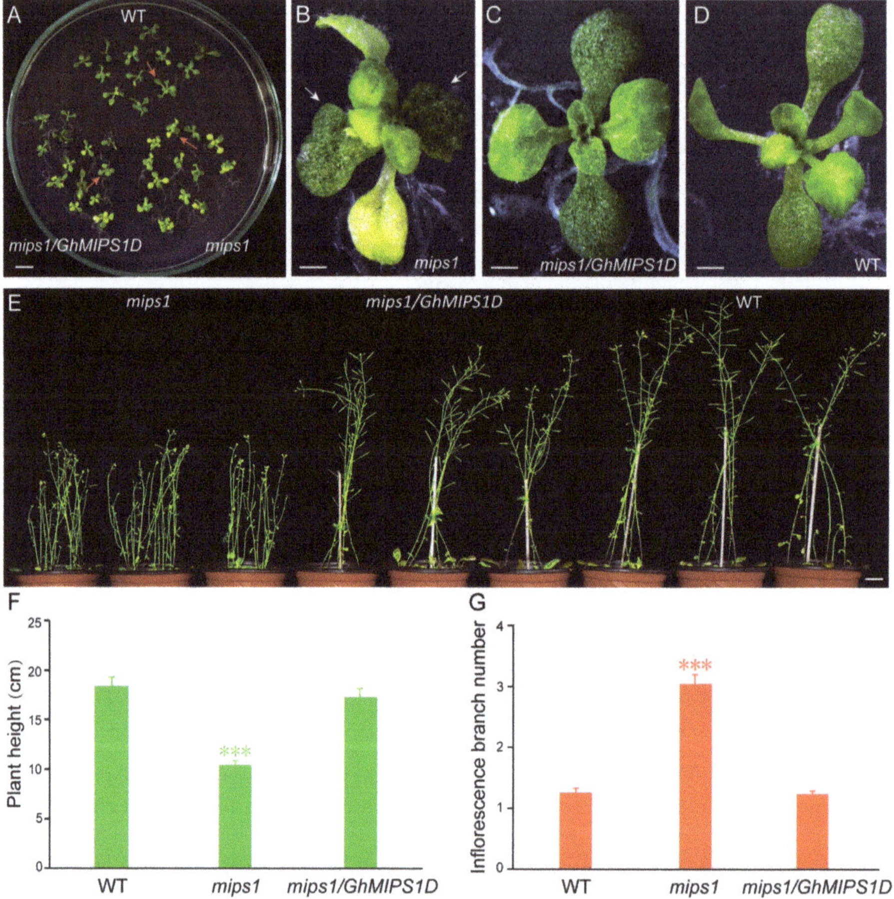

Figure 5. Ectopic expression of *GhMIPS1D* in *Arabidopsis* rescues the abnormal defects of *mips1* mutants. (**A**) The *mips1* mutant (*mips1*), transgenic (*mips1/GhMIPS1D*) and wild-type (WT) *Arabidopsis* plants grew on 0.5 × Murashige and Skoog (MS) medium for two weeks. Bar = 1 cm. (**B–D**), the enlarged image of indicated plants (red arrows in (**A**)). White arrows indicate the abnormal cotyledon in *mips1* mutants. Bars = 1 mm. (**E**) Phenotypes of *Arabidopsis* plants grew for ten weeks. Significant shorter plants and more inflorescence branches are observed in *mips1* mutants, but not in *mips1/GhMIPS1D* transgenic lines and WT plants. Bar = 1 cm. The plant height (**F**) and the number of inflorescence branches (**G**) are measured ($n = 30$). ***, $p < 0.001$ compared to WT.

2.4. Ectopic Expression of GhMIPS1D Promotes Root Cell Elongation in Arabidopsis

Considering that the Ins derivates play essential roles in root development and root hair formation [29,30], and *GhMIPS1D* gene highly expressed during the stages of fiber initiation and early elongation (Supplementary Figure S2), which indicates its potential important function for plant cell elongation. Thus, we further measured the length of primary roots and root cells in *Arabidopsis* lines ectopically expressing *GhMIPS1D*. The transgenic *mips1/GhMIPS1D* plant lines exhibited normal root length similar to that in WT, suggesting that *GhMIPS1D* is a functional gene to compensate the root elongation development defect of *Arabidopsis mips1* mutants (Figure 6A,B). In addition, the transgenic plants ectopically expressing *GhMIPS1D* in WT *Arabidopsis* demonstrated significantly increased primary root length (Figure 6A,B). Further confocal microscopy detection displayed that the primary root cell lengths are increased from 124.02 ± 10.01 μm (WT) to 152.14 ± 16.58 μm (*35S::GhMIPS1D*), indicating that *GhMIPS1D* is essential for plant cell elongation (Figure 6C,D). Totally, the genetically functional complementary analyses suggest that *GhMIPS1D* is a fully functional *MIPS* and plays as a key positive regulator to involve in cell elongation.

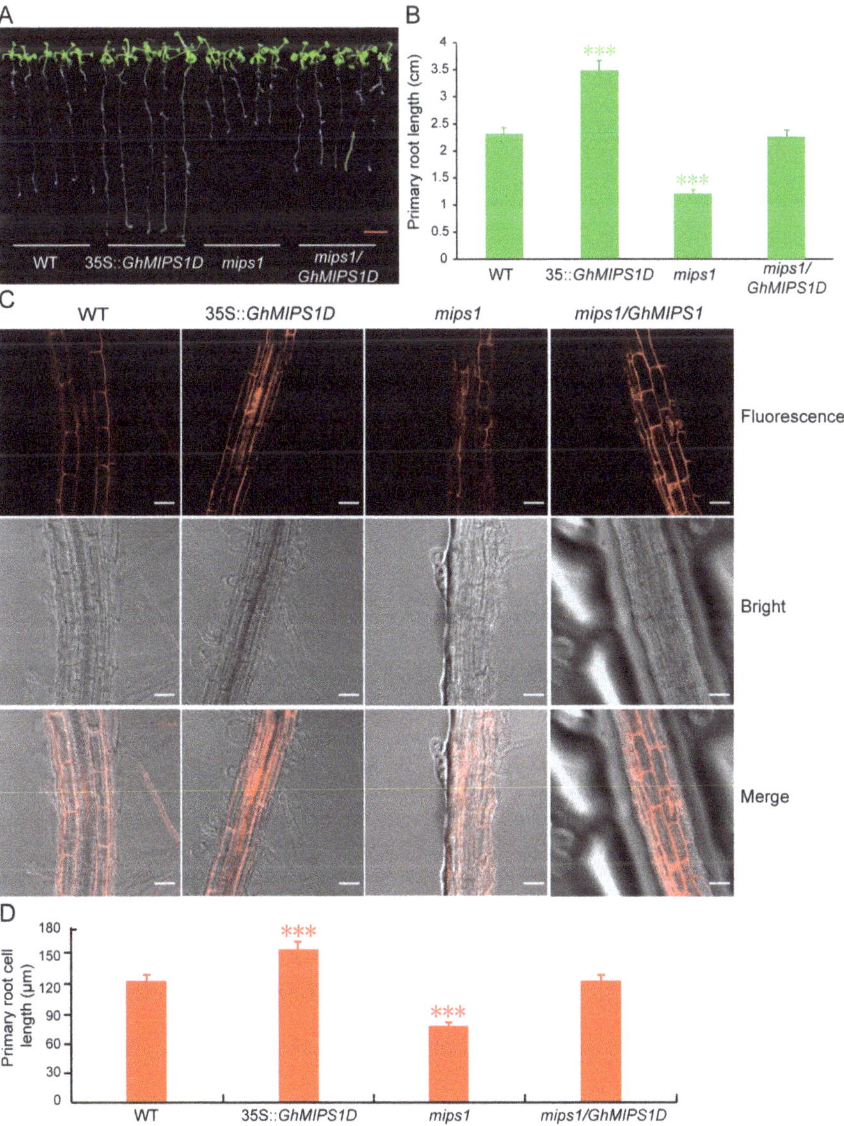

Figure 6. Ectopic expression of *GhMIPS1D* promotes *Arabidopsis* root cell elongation. (**A**) The primary root phenotypes of wild-type (WT), transgenic lines ectopically expressing *GhMIPS1D* (*35S::GhMIPS1D*), *mips1* mutants (*mips1*) and transgenic lines expressing *GhMIPS1D* in *mips1* mutants (*mips1/35S::GhMIPS1D*). Red bar = 0.5 cm. (**B**) The primary root lengths are measured (*n* = 10). (**C**) Fluorescence images of *Arabidopsis* root cells under confocal microscopy. Bars = 50 μm. (**D**) The root cell lengths are measured (*n* = 50). The results show that ectopic expression of *GhMIPS1D* significantly induces the cell length in *Arabidopsis* roots. ***, *p* < 0.001 compared to WT.

3. Discussion

Lots of *MIPS* genes have been identified in numerous living organisms, such as bacteria, fungi, animals, and higher plants, etc. [31]. Generally, there are one or two *MIPS* members in most plant species, with the exception of three members in *A. thaliana*, and of four members in *Z. mays*, *B. rapa*, *B. oleracea* and *G. max*, respectively. In this study, by genome-wide analysis, we identified four *G. hirsutum MIPS* family genes that are distributed across different chromosomes (Supplementary Figure S1). Phylogenetic tree of 70 MIPS from 36 species was constructed and used for evolutionary analysis, benefiting from the dramatically increased genome database. The protein sequences of eukaryotic MIPS exhibit high conservation, supporting the explanation that eukaryotic *MIPS* genes have a monophyletic origin (Figure 2). However, the single-cell algae *C. merolae* and *Ostreococcus tauri* possess MIPS proteins of 533 and 535 amino acids, while all higher plants have MIPS proteins of around 510 amino acids (from 509 to 511), displaying that sequence lost events had occurred after plant landing. Only two higher plant MIPS proteins possess amino acids much longer than 510 amino acids: *S. cucullate* MIPS1 (537 amino acids) and *G. max* MIPS2 (526 amino acids) (Supplementary Table S2), indicating special evolutionary events in these two species. Many studies reported the conservation of MIPS protein sequences, but few noticed the exon-intron structure evolution patterns of *MIPS* genes [32,33]. The exon-intron structure analysis presented here suggested that the *MIPS* gene structures were highly conserved in higher plants except for *Brassicales*, suggesting that *MIPS* family genes underwent more complicated evolutionary events during the *Brassicales* plants evolution (Figures 2 and 3). This may explain the fact that *Arabidopsis MIPS* family members show function diversity [17,18,31,34].

MIPS is the key biosynthetic enzyme for the formation of Ins and its derivates, including PtdIns and PtdInsP, which have been proven as a crucial regulator to control plant growth and organ development. Of the four *GhMIPS* members, *GhMIPS1D*, showing the predominantly expression during the early stages of fiber development (Supplementary Figure S2), was selected to further investigate cotton *MIPS* functions of participating in plant growth and development. After ectopic expression of *GhMIPS1D* in *Arabidopsis* loss-of-function *mips1* mutants, the aberrant phenotypes of cotyledon abnormality, shorter plants, increased inflorescence branches and sensitivity to light in the *mips1* mutants were rescued significantly (Figure 5 and Supplementary Figure S6), implying the cotton *MIPS* important roles in the involvement of plant growth and development as the pivotal enzyme to produce Ins and its derivates that are important signal molecules of the cell. Knockout of the soybean *GmMIPS1* affected the early development of the embryo and resulted in the termination of seed mature that might be caused by the reduction of IP_6 [35]. *Arabidopsis MIPS1* is essential for seed development, and the loss of *MIPS1* leads to the reduction of AsA and PtdIns and thus produces the irregular phenotypes of smaller plants, curly leaves and generation of lesions [18]. Ins produced by MIPS is crucial for *Arabidopsis* embryogenesis through regulating the synthesis of PtdIns and phosphatidylinositides and thereby involving in endomembrane structure trafficking and PIN1-mediated auxin singaling pathway [17]. The functions of the other *GhMIPS* genes should also be verified in the future for a better understanding of the diverse roles of Ins pathway in cotton. Besides, *MIPS* has also been reported to function in salt and drought tolerance, indicating that the *MIPS* gene family play various roles in plant development [36,37]. However, there still are unsolved problems should be addressed in the future, such as the mechanism leading to increased inflorescence branch numbers in *Arabidopsis misp1* mutants, as well as the possible phenotypes of loss-of-function cotton *mips1* and/or *mips2* mutants.

Due to the highly conserved protein sequences of GhMIPS1D and AtMIPS1, it is not surprising that *GhMIPS1D* could rescue the *Arabidopsis mips1* mutant phenotypes. As is well known, MIPS proteins from diverse species share highly conserved core catalytic domains, which endow the activity to catalyze rate-limiting redox reaction to generate the Ins formation from G6P [31]. The *G. hirsutum* MIPS possessed several typical domains that serve as binding and catalysis of MIPS proteins (Figure 2, Supplementary Figure S3). Since the first MIPS was reported from *Archaeoglobus fulgidus*, many MIPS proteins have been investigated from different organisms. The MIPS in higher organisms usually showed cytosolic or organellar location. In this study, subcellular distribution analysis indicated the location of GhMIPS1D in the nucleus, plasma membrane and endomembranes (Supplementary Figure S4), suggesting that cotton MIPS may perform a diverse role in different compartments to participate in plant growth and development. By suppressing the spreading of heterochromatin, MIPS can bind to its promoter to induce its own expression, providing the illustration of the regulatory mechanism at the transcriptional level [38].

MIPS catalyzed Ins biosynthesis provides the important supply for synthetizing its derivates containing PtdIns, PtdInsPs and InsPs that are vital signal molecules to affect cell growth by participating in cellular signal transduction [39]. In this study, *GhMIPS1D* indicated preferential expression during fiber initial growth stage (Supplementary Figure S2), and ectopic expression of *GhMIPS1D* in *Arabidopsis* rescued the shorter primary root lengths in *mips1* mutants and promoted the root cell elongation in WT, respectively (Figure 6). Severe reduction of IP$_6$ content in *Arabidopsis* *ipk1-1* mutants significantly inhibits the root hair elongation [40]. The decrease of PtdInsP$_2$ content in *Arabidopsis* results in suppression of pollen tube elongation by affecting apical pectin deposition and membrane trafficking [41]. Although many Ins derivates have been reported to be involved in cell growth, whereas, this is the first time to prove that *MIPS* family members could promote the root cell elongation in *Arabidopsis* (Figure 6). This phenomenon is reasonable for that previous work showed that pectin and AsA play important roles in cotton fiber elongation [25,26]. *MIPS* has been reported to increase the synthesis of pectin precursor of UDP-D-glucuronic acid (UDP-GlcA) and AsA [42]. UDP-GlcA, PtdIns and PtdInsP that are important Ins derivates have been studied to involve in fiber elongation growth as important components of the cell wall and cell membranes [25,28].

De novo synthesis of Ins is required for the correct transport and localization of auxin during embryo pattern formation in *Arabidopsis* [17,18]. The functions of *MIPS* during cotton fiber development has not been well studied before. Many studies have reported that auxin signaling is essential for the initiation and early elongation of cotton fiber [43,44]. In the present study, *GhMIPS1D* was highly expressed at the early stage of cotton fiber development (Supplementary Figure S2), providing the potential possible direct or indirect link between MIPS catalyzed synthesis of Ins and its derivates and auxin signal transduction. *Arabidopsis mips1* mutants exhibit a significant reduction of PtdIns content and decrease of auxin polar transport rate that regulates the cellular auxin distribution [45]. Further investigation on *MIPS* and phytohormone crosstalk will provide a better understanding of functions of *MIPS* during cotton fiber development in the future.

In conclusion, we identified four *MIPS* family gene members in *G. hirsutum*, the phylogenic relationship, multiple sequence alignment, and motif distribution analyses showed that the *MIPS* gene family is highly conserved. Further genetically functional complementary analysis showed that *GhMIPS1D* is the predominantly expressed *MIPS* gene in cotton fibers, and ectopic expression of *GhMIPS1D* in *Arabidopsis mips1* mutant rescues the phenotypes of abnormal cotyledon development, increased inflorescence branch numbers and light stress sensitivity. Moreover, ectopic expression of *GhMIPS1D* in WT *Arabidopsis* promotes root cell elongation significantly. Our results provide a genome-wide analysis of the *MIPS* gene family in *G. hirsutum* and suggest that the *GhMIPS1D* is a positive regulator involving in plant cell elongation.

4. Materials and Methods

4.1. Sequence Acquirement and Chromosomal Distribution of GhMIPS Genes

The prokaryotic MIPS protein sequences were retrieved from NCBI (database downloaded on 31 December 2018) by online BLASTP. The eukaryotic MIPS protein sequences were determined using local BLASTP program by submitting *Arabidopsis* MIPS to the genome database for each organism. The genome data were downloaded from the online database Phytozome (v12.1.6). The candidate MIPS proteins were confirmed by screening the conserved domains using InterProScan. The genomic location of *GhMIPS* genes was performed using local BLASTN against the *G. hirsutum* genome database. Mapinspect was used to draw the location of *MIPS* genes in different chromosomes.

4.2. Phylogenic and Evolutionary Analyses

Multiple sequence alignment was performed using ClustalW and the phylogenetic tree was constructed by MEGA5.0 [46] using the neighbor-joining method with bootstrap tests 1000. The evolutionary analyses were performed as previously described [47,48]. Briefly, the exon-intron distribution analysis was performed using the gene structure display server (GSDS, http://gsds. cbi.pku.edu.cn/index.php). Conserved motifs were recognized by MEME online software (http: //meme-suite.org/tools/meme). The *Ka/Ks* was calculated by Dnasp (v6) software [49].

4.3. Plant Materials

The *Arabidopsis mips1* mutant (SALK_023626) was obtained from Lijia Qu, State Key Laboratory of Protein and Plant Gene Research, College of Life Sciences, Peking University. All *Arabidopsis* plants were grown on 0.5 × MS medium as previously described [17].

4.4. Construction of Vectors, Subcellular Localization Analysis and Ectopic Expression of GhMIPS1D in Arabidopsis

The full-length of the coding sequences of *GhMIPS1D* gene was amplified using primers in Supplementary Table S3, and then cloned into the modified pCAMBIA2300-GFP vector using *KpnI* and *XbaI* to generate *35S::GhMIPS1D-GFP*, which was used for further ectopic expression and subcellular localization analyses. The successfully constructed vectors were then transformed into *Agrobacterium tumefaciens* stain GV3101.

The subcellular localization analysis was performed using onion epidermal cells and *A. tumefaciens* containing *35S::GhMIPS1D-GFP* vector. Briefly, the inside epidermal layer of the onion was separated. After soaking in 75% ethanol for 10 min and washing 3–4 times with sterile water, the inside epidermis was cut and co-cultivated with *A. tumefaciens* containing *35S::GhMIPS1D-GFP* vector on 1/2 MS at 28 °C under darkness. After a sub-culturing for 24 h, a confocal laser-scanning microscope (Zeiss LSM510, Oberkochen, Germany) was used to detect the GFP signals with an activation wavelength of 488 nm.

Wild-type (WT) and *mips1* mutant *Arabidopsis* plants were transformed using *A. tumefaciens* stain GV3101 containing *35S::GhMIPS1D-GFP* vector. The functional analysis of transgenic *Arabidopsis* was performed as reported before [50].

4.5. Statistical Analysis

All statistics were performed by One-way ANOVA followed by Bonferroni test using SigmaStat software (Version 4.0) (Starcom Information Technology Ltd, Bangalore, India). *** represents significant difference at $p < 0.001$ level.

Supplementary Materials: Supplementary materials can be found at http://www.mdpi.com/1422-0067/20/5/ 1224/s1.

Author Contributions: Conceptualization, X.J. and H.L.; Methodology, X.J.; Software, R.M., W.S. and F.W.; Validation, R.M., W.S. and F.W.; Investigation, A.C., S.X. and X.C.; Resources, H.L.; Data Curation, X.J.; Writing—Original Draft Preparation, W.S. and X.J.; Writing—Review and Editing, F.W. and H.L.; Visualization, W.S., F.W., R.M. and A.C.; Supervision, X.J. and H.L.; Project Administration, R.M., W.S. and X.J.; Funding Acquisition, H.L.

Funding: This work was supported by grants of the National Natural Science Foundation of China (Grant number 31660408, 31260339) and Scientific and Technological Achievement Transformation Project of Bingtuan (Grant number 2016AC017).

Acknowledgments: The authors thank Lijia Qu from State Key Laboratory of Protein and Plant Gene Research, College of Life Sciences, Peking University for kindly providing the Arabidopsis mips1 mutant. We also thank Xianzhong Huang from the College of Life Sciences, Shihezi University for providing the modified vector pCAMBIA2300-GFP.

Conflicts of Interest: The authors declare no conflicts of interest.

Abbreviations

APX	Ascorbate peroxidase
AsA	Ascorbic acid
BLAST	Basic local alignment search tool
DPA	Day post anthesis
G6P	D-glucose 6-phosphate
GlcA	D-glucuronic acid
IMP	*Myo*-inositol monophosphatase
Ins	*Myo*-inositol
IP	*Myo*-inositol polyphosphate
MEGA	Molecular evolutionary genetics analysis
MIPS	*Myo*-inositol-1-phosphate synthase
MS	Murashige and Skoog
PtdIns	phospholipid phosphatidylinositol
PtdInsP	phosphoinositide phosphate
RNAi	RNA interference
UDP-GlcA	UDP-D-glucuronic acid
WT	Wild-type

References

1. Majumder, A.L.; Chatterjee, A.; Dastidar, K.G.; Majee, M. Diversification and evolution of L-*myo*-inositol 1-phosphate synthase. *FEBS Lett.* **2003**, *553*, 3–10. [CrossRef]
2. Majumder, A.L.; Johnson, M.D.; Henry, S.A. 1-L-*myo*-inositol-1-phosphate synthase. *Biochim. Biophys. Acta* **1997**, *1348*, 245–256. [CrossRef]
3. Johnson, M.D.; Wang, X. Differentially expressed forms of 1-L-*myo*-inositolphosphate synthase (EC5.5.1.4) in *Phaseolus vulgaris. J. Biol. Chem.* **1996**, *271*, 17215–17218. [CrossRef] [PubMed]
4. Hait, N.C.; Chaudhuri, A.R.; Das, A.; Bhattacharyya, S.; Majumder, A.L. Processing and activation of chloroplast L-*myo*-inositol 1-phosphate synthase from *Oryza sativa* requires signals from both light and salt. *Plant Sci.* **2002**, *162*, 559–568. [CrossRef]
5. Abu-abied, M.; Holland, D. The gene c-ino1 from *Citrus paradise* is highly homologous to tur1 and ino1 from the yeast and *Spirodela* encoding for *myo*-inositol phosphate synthase. *Plant Physiol.* **1994**, *106*, 1689. [CrossRef] [PubMed]
6. Johnson, M.D.; Sussex, I.M. 1-L-*myo*-inositol 1-phosphate synthase from *Arabidopsis thaliana. Plant Physiol.* **1995**, *107*, 613–619. [CrossRef] [PubMed]
7. Keller, R.; Brearley, C.A.; Trethewey, R.N.; Muller-Rober, B. Reduced inositol content and altered morphology in transgenic potato plants inhibited for 1-D-*myo*-inositol 3-phosphate synthase. *Plant J.* **1998**, *16*, 403–410. [CrossRef]
8. Larson, S.R.; Raboy, V. Linkage mapping of maize and barley *myo*-inositol 1-phosphate synthase DNA sequences: Correspondence with low phytic acid mutation. *Theor. Appl. Genet.* **1999**, *99*, 27–36. [CrossRef]
9. Iqbal, M.; Afzal, A.; Yaegashi, S.; Ruben, E.; Triwitayakorn, K.; Njiti, V.; Ahsan, R.; Wood, A.; Lightfoot, D. A pyramid of loci for partial resistance to *Fusarium solani f.* sp. glycines maintains *myo*-inositol-1-phosphate synthase expression in soybean roots. *Theor. Appl. Genet.* **2002**, *105*, 1115–1123. [CrossRef] [PubMed]
10. Chun, J.A.; Jin, U.H.; Lee, J.W.; Yi, Y.B.; Hyung, N.I.; Kang, M.H.; Pyee, J.H.; Suh, M.C.; Kang, C.W.; Seo, H.Y.; et al. Isolation and characterization of a *myo*-inositol 1-phosphate synthase cDNA from developing sesame (*Sesamum indicum* L.) seeds: Functional and differential expression, and salt-induced transcription during germination. *Planta* **2003**, *216*, 874–880. [PubMed]
11. Bachhawat, N.; Mande, S.C. Complex evolution of the inositol-1-phosphate synthase gene among archaea and eubacteria. *Trends Genet.* **2000**, *16*, 111–113. [CrossRef]

12. Hegeman, C.E.; Good, L.L.; Grabau, E.A. Expression of D-*myo*-inositol-3-phosphate synthase in soybean implication for phytic acid biosynthesis. *Plant Physiol.* **2001**, *125*, 1941–1948. [CrossRef] [PubMed]

13. Gillaspy, G.E. The cellular language of *myo*-inositol signaling. *New Phytol.* **2011**, *192*, 823–839. [CrossRef] [PubMed]

14. Chen, L.J.; Zhou, C.; Yang, H.Y.; Roberts, M.F. Inositol 1-phosphate synthase from *Archaeoglobus fulgidus* is a class II aldolase. *Biochemistry* **2000**, *39*, 12415–12423. [CrossRef] [PubMed]

15. Murphy, A.M.; Otto, B.; Brearley, C.A.; Carr, J.P.; Hanke, D.E. A role for inositol hexakisphosphate in the maintenance of basal resistance to plant pathogens. *Plant J.* **2008**, *56*, 638–652. [CrossRef] [PubMed]

16. Meng, P.H.; Raynaud, C.; Tcherkez, G.; Blanchet, S.; Massoud, K.; Domenichini, S.; Herry, Y.; Soubigou-Taconnat, L.; Lelarge-Trouverie, C.; Saindrenan, P.; et al. Crosstalks between *myo*-inositol metabolism, programmed cell death and basal immunity in *Arabidopsis*. *PLoS ONE* **2009**, *4*, e7364. [CrossRef] [PubMed]

17. Luo, Y.; Qin, G.J.; Zhang, J.; Liang, Y.; Song, Y.Q.; Zhao, M.P.; Tsuge, T.; Aoyama, T.; Liu, J.J.; Gu, H.Y.; et al. D-*myo*-inositol-3-phosphate affects phosphatidylinositol-mediated endomembrane function in *Arabidopsis* and is essential for auxin-regulated embryogenesis. *Plant Cell* **2011**, *23*, 1352–1372. [CrossRef] [PubMed]

18. Donahue, J.L.; Alford, S.R.; Torabinejad, J.; Kerwin, R.E.; Nourbakhsh, A.; Ray, W.K.; Hernick, M.; Huang, X.; Lyons, B.M.; Hein, P.P.; et al. The *Arabidopsis myo*-inositol 1-phosphate synthase1 gene is required for *myo*-inositol synthesis and suppression of cell death. *Plant Cell* **2010**, *22*, 888–903. [CrossRef] [PubMed]

19. Raboy, V. Seeds for a better future: 'Low phytate' grains help to overcome malnutrition and reduce pollution. *Trends Plant Sci.* **2001**, *6*, 458–462. [CrossRef]

20. Baig, M.M.; Kelly, S.; Loewus, F. L-ascorbic acid biosynthesis in higher plants from L-gulono-1, 4-lactone and L-galactono-1, 4-lactone. *Plant Physiol.* **1970**, *46*, 277–280. [CrossRef] [PubMed]

21. Allison, J.H.; Stewart, M.A. *Myo*-inositol and ascorbic acid in developing rat brain. *J. Neurochem.* **1973**, *20*, 1785–1788. [CrossRef] [PubMed]

22. Banhegyi, G.; Braun, L.; Csala, M.; Puskás, F.; Mandl, J. Ascorbate metabolism and its regulation in animals. *Free Radic. Biol. Med.* **1997**, *23*, 793–803. [CrossRef]

23. Loewus, F. Inositol metabolism and cell wall formation in plants. *Fed. Proc.* **1965**, *24*, 855–862. [PubMed]

24. Loewus, F.A. Inositol and plant cell wall polysaccharide biogenesis. *Subcell. Biochem.* **2006**, *39*, 21–45. [PubMed]

25. Pang, C.Y.; Wang, H.; Pang, Y.; Xu, C.; Jiao, Y.; Qin, Y.M.; Western, T.L.; Yu, S.X.; Zhu, Y.X. Comparative proteomics indicates that biosynthesis of pectic precursors is important for cotton fiber and *Arabidopsis* root hair elongation. *Mol. Cell Proteom.* **2010**, *9*, 2019–2033. [CrossRef] [PubMed]

26. Li, H.B.; Qin, Y.M.; Pang, Y.; Song, W.Q.; Mei, W.Q.; Zhu, Y.X. A cotton ascorbate peroxidase is involved in hydrogen peroxide homeostasis during fibre cell development. *New Phytol.* **2007**, *175*, 462–471. [CrossRef] [PubMed]

27. Tao, C.C.; Jin, X.; Zhu, L.P.; Xie, Q.L.; Wang, X.C.; Li, H.B. Genome-wide investigation and expression profiling of APX gene family in *Gossypium hirsutum* provide new insights in redox homeostasis maintenance during different fiber development stages. *Mol. Genet. Genom.* **2018**, *293*, 685–697. [CrossRef] [PubMed]

28. Liu, G.J.; Xiao, G.H.; Liu, N.J.; Liu, D.; Chen, P.S.; Qin, Y.M.; Zhu, Y.X. Targeted Lipidomics Studies Reveal that Linolenic Acid Promotes Cotton Fiber Elongation by Activating Phosphatidylinositol and Phosphatidylinositol Monophosphate Biosynthesis. *Mol. Plant* **2015**, *8*, 911–921. [CrossRef] [PubMed]

29. Kusano, H.; Testerink, C.; Vermeer, J.E.; Tsuge, T.; Shimada, H.; Oka, A.; Munnik, T.; Aoyama, T. The *Arabidopsis* Phosphatidylinositol Phosphate 5-Kinase PIP5K3 is a key regulator of root hair tip growth. *Plant Cell* **2008**, *20*, 367–380. [CrossRef] [PubMed]

30. Stenzel, I.; Ischebeck, T.; König, S.; Hołubowska, A.; Sporysz, M.; Hause, B.; Heilmann, I. The type B phosphatidylinositol-4-phosphate 5-kinase 3 is essential for root hair formation in *Arabidopsis thaliana*. *Plant Cell* **2008**, *20*, 124–141. [CrossRef] [PubMed]

31. Abid, G.; Silue, S.; Muhovski, Y.; Jacquemin, J.M.; Toussaint, A.; Baudoin, J.P. Role of *myo*-inositol phosphate synthase and sucrose synthase genes in plant seed development. *Gene* **2009**, *439*, 1–10. [CrossRef] [PubMed]

32. Cui, M.; Liang, D.; Wu, S.; Ma, F.W.; Lei, Y.S. Isolation and developmental expression analysis of L-*myo*-inositol-1-1phosphate synthase in four *Actinidia* species. *Plant Physiol. Biochem.* **2013**, *73*, 351–358. [CrossRef] [PubMed]

33. Basak, P.; Maitra-Majee, S.; Das, J.K.; Mukherjee, A.; Ghosh Dastidar, S.; Pal Choudhury, P.; Lahiri Majumder, A. An evolutionary analysis identifies a conserved pentapeptide stretch containing

the two essential lysine residues for rice L-*myo*-inositol 1-phosphate synthase catalytic activity. *PLoS ONE* **2017**, *12*, e0185351. [CrossRef] [PubMed]

34. Fleet, C.M.; Yen, J.Y.; Hill, E.A.; Gillaspy, G.E. Co-suppression of *AtMIPS* demonstrates cooperation of *MIPS1*, *MIPS2* and *MIPS3* in maintaining *myo*-inositol synthesis. *Plant Mol. Biol.* **2018**, *97*, 253–263. [CrossRef] [PubMed]

35. Nunes, A.C.; Vianna, G.R.; Cuneo, F.; Amaya-Farfán, J.; de Capdeville, G.; Rech, E.L.; Aragão, F.J. RNAi-mediated silencing of the *myo*-inositol-1-phosphate synthase gene (*GmMIPS1*) in transgenic soybean inhibited seed development and reduced phytate content. *Planta* **2006**, *224*, 125–132. [CrossRef] [PubMed]

36. Zhai, H.; Wang, F.B.; Si, Z.Z.; Huo, J.X.; Xing, L.; An, Y.Y.; He, S.Z.; Liu, Q.C. A *myo*-inositol-1-phosphate synthase gene, *IbMIPS1*, enhances salt and drought tolerance and stem nematode resistance in transgenic sweet potato. *Plant Biotechnol. J.* **2016**, *14*, 592–602. [CrossRef] [PubMed]

37. Tan, J.L.; Wang, C.Y.; Xiang, B.; Han, R.H.; Guo, Z.F. Hydrogen peroxide and nitric oxide mediated cold- and dehydration-induced *myo*-inositol phosphate synthase that confers multiple resistances to abiotic stresses. *Plant Cell Environ.* **2013**, *36*, 288–299. [CrossRef] [PubMed]

38. Latrasse, D.; Jégu, T.; Meng, P.H.; Mazubert, C.; Hudik, E.; Delarue, M.; Charon, C.; Crespi, M.; Hirt, H.; Raynaud, C.; et al. Dual function of MIPS1 as a metabolic enzyme and transcriptional regulator. *Nucleic Acids Res.* **2013**, *41*, 2907–2917. [CrossRef] [PubMed]

39. Valluru, R.; Van den Ende, W. *Myo*-inositol and beyond-emerging networks under stress. *Plant Sci.* **2011**, *181*, 387–400. [CrossRef] [PubMed]

40. Stevenson-Paulik, J.; Bastidas, R.J.; Chiou, S.T.; Frye, R.A.; York, J.D. Generation of phytate-free seeds in *Arabidopsis* through disruption of inositol polyphosphate kinases. *Proc. Natl. Acad. Sci. USA* **2005**, *102*, 12612–12617. [CrossRef] [PubMed]

41. Ischebeck, T.; Stenzel, I.; Heilmann, I. Type B phosphatidylinositol-4-phosphate 5-kinases mediate *Arabidopsis* and *Nicotiana tabacum* pollen tube growth by regulating apical pectin secretion. *Plant Cell* **2008**, *20*, 3312–3330. [CrossRef] [PubMed]

42. Kusuda, H.; Koga, W.; Kusano, M.; Oikawa, A.; Saito, K.; Hirai, M.Y.; Yoshida, K.T. Ectopic expression of *myo*-inositol 3-phosphate synthase induces a wide range of metabolic changes and confers salt tolerance in rice. *Plant Sci.* **2015**, *232*, 49–56. [CrossRef] [PubMed]

43. Zhang, M.; Zheng, J.Y.; Long, H.; Xiao, Y.H.; Yan, X.Y.; Pei, Y. Auxin regulates cotton fiber initiation via GhPIN-mediated auxin transport. *Plant Cell Physiol.* **2017**, *58*, 385–397. [CrossRef] [PubMed]

44. Samuel, Y.S.; Cheung, F.; Lee, J.J.; Ha, M.; Wei, N.E.; Sze, S.H.; Stelly, D.M.; Thaxton, P.; Triplett, B.; Town, C.D.; et al. Accumulation of genome-specific transcripts, transcription factors and phytohormonal regulators during early stages of fiber cell development in allotetraploid cotton. *Plant J.* **2006**, *47*, 761–765. [CrossRef] [PubMed]

45. Chen, H.; Xiong, L.M. *myo*-Inositol-1-phosphate synthase is required for polar auxin transport and organ development. *J. Biol. Chem.* **2010**, *285*, 24238–24247. [CrossRef] [PubMed]

46. Tamura, K.; Peterson, D.; Peterson, N.; Stecher, G.; Nei, M.; Kumar, S. MEGA5: Molecular evolutionary genetics analysis using maximum likelihood, evolutionary distance, and maximum parsimony methods. *Mol. Biol. Evol.* **2011**, *28*, 2731–2739. [CrossRef] [PubMed]

47. Jin, X.; Zhu, L.P.; Yao, Q.; Meng, X.R.; Ding, G.H.; Wang, D.; Xie, Q.L.; Tong, Z.; Tao, C.C.; Yu, L.; et al. Expression profiling of mitogen-activated protein kinase genes reveals their evolutionary and functional diversity in different rubber tree (*Hevea brasiliensis*) cultivars. *Genes* **2017**, *8*, 261. [CrossRef] [PubMed]

48. Zhu, L.P.; Jin, X.; Xie, Q.L.; Yao, Q.; Wang, X.C.; Li, H.B. Calcium-dependant protein kinase family genes involved in ethylene-induced natural rubber production in different *Hevea brasiliensis* cultivars. *Int. J. Mol. Sci.* **2018**, *19*, 947. [CrossRef] [PubMed]

49. Rozas, J.; Ferrer-Mata, A.; Sánchez-DelBarrio, J.C.; Guirao-Rico, S.; Librado, P.; Ramos-Onsins, S.E.; Sánchez-Gracia, A. DnaSP 6: DNA sequence polymorphism analysis of large datasets. *Mol. Biol. Evol.* **2017**, *34*, 3299–3302. [CrossRef] [PubMed]
50. Li, R.; Xin, S.; Tao, C.C.; Jin, X.; Li, H.B. Cotton ascorbate oxidase promotes cell growth in cultured tobacco bright yellow-2 cells through generation of apoplast oxidation. *Int. J. Mol. Sci.* **2017**, *18*, 1346. [CrossRef] [PubMed]

International Journal of
Molecular Sciences

MDPI

Review

The Function of Inositol Phosphatases in Plant Tolerance to Abiotic Stress

Qi Jia [1,2,*], Defeng Kong [1], Qinghua Li [3], Song Sun [1], Junliang Song [1], Yebao Zhu [4], Kangjing Liang [1], Qingming Ke [3], Wenxiong Lin [1,2] and Jinwen Huang [1,2,*]

[1]   Key Laboratory for Genetics Breeding and Multiple Utilization of Crops, Ministry of Education/College of Crop Science, Fujian Agriculture and Forestry University, Fuzhou 350002, China
[2]   Key Laboratory of Crop Ecology and Molecular Physiology (Fujian Agriculture and Forestry University), Fujian Province University, Fuzhou 350002, China
[3]   Putian Institute of Agricultural Sciences, Putian 351144, China
[4]   Rice Research Institute, Fujian Academy of Agricultural Sciences, Fuzhou 350018, China
*   Correspondence: jiaqi@fafu.edu.cn (Q.J.); hqliuyh@163.com (J.H.)

Received: 30 June 2019; Accepted: 13 August 2019; Published: 16 August 2019

Abstract: Inositol signaling is believed to play a crucial role in various aspects of plant growth and adaptation. As an important component in biosynthesis and degradation of *myo*-inositol and its derivatives, inositol phosphatases could hydrolyze the phosphate of the inositol ring, thus affecting inositol signaling. Until now, more than 30 members of inositol phosphatases have been identified in plants, which are classified intofive families, including inositol polyphosphate 5-phosphatases (5PTases), suppressor of actin (SAC) phosphatases, SAL1 phosphatases, inositol monophosphatase (IMP), and phosphatase and tensin homologue deleted on chromosome 10 (PTEN)-related phosphatases. The current knowledge was revised here in relation to their substrates and function in response to abiotic stress. The potential mechanisms were also concluded with the focus on their activities of inositol phosphatases. The general working model might be that inositol phosphatases would degrade the $Ins(1,4,5)P_3$ or phosphoinositides, subsequently resulting in altering Ca^{2+} release, abscisic acid (ABA) signaling, vesicle trafficking or other cellular processes.

Keywords: inositol; phosphatidylinositol; phosphatase; stress; signaling pathway

---

1. Introduction

Myo-inositol (Inositol, Ins) and its derivative metabolites are ubiquitous in all eukaryotes as both lipids and soluble compounds playing important roles in stress responses, development, and many other processes [1,2]. Upon environmental stresses, some of them are vital in various signal transduction in plants, especially inositol(1,4,5)triphosphate $(Ins(1,4,5)P_3)$ and phosphatidylinositol(4,5)bisphosphate $(PtdIns(4,5)P_2)$ [3–7]. They pass the cellular messages via addition or removal of lipids or phosphates to Ins and its derivatives, which could be mediated by synthases, kinases, phospholipases, and phosphatases [8,9]. Thus, those related enzymes are crucial in the regulation of these signaling pathways. In comparison to the other well-studied enzymes, limited information has been reviewed for the phosphatases in the Ins and phosphatidylinositol (PtdIns) signaling in plants. Here, we focus on these phosphatases and their function in abiotic tolerance.

2. The Biosynthesis and Degradation of Inositol and Its Derivatives

Inositol could be synthesized from glycolytic glucose-6-phosphate (Glc6P) or be regenerated from various phosphate forms of inositol, which is produced during the metabolism of phosphoinositides. As shown in Figure 1, Glc6P is catalyzed to *myo*-inositol-3-phosphate (Ins3P) by *myo*-inositol-3-phosphate synthase (MIPS). Subsequently, Ins3P is dephosphorylated by inositol monophosphatase (IMP) to form inositol. IMP is also responsible for the dephosphorylation of *myo*-inositol-4-phosphate (Ins4P) [10,11]. Free inositol could be linked to glycerophospholipid to generate the basic inositol containing phospholipid, phosphatidylinositol (PtdIns), by phosphatidylinositol synthase (PIS) [5]. The hydroxyl groups of PtdIns could be phosphorylated at the positions 3, 4, and 5 of the lipid head group sequentially by a series of PtdIns kinases. Unlike the animals, plants have evolved only five phosphorylated isomers, including three PtdIns monophosphates (PtdIns3P, PtdIns4P, PtdIns5P) and two PtdIns bisphosphates (PtdIns(3,5)P_2, PtdIns(4,5)P_2). The other two, PtdIns(3,4)P_2 and PtdIns(3,4,5)P_3, identified in animals, have not been found in plants [4,12].

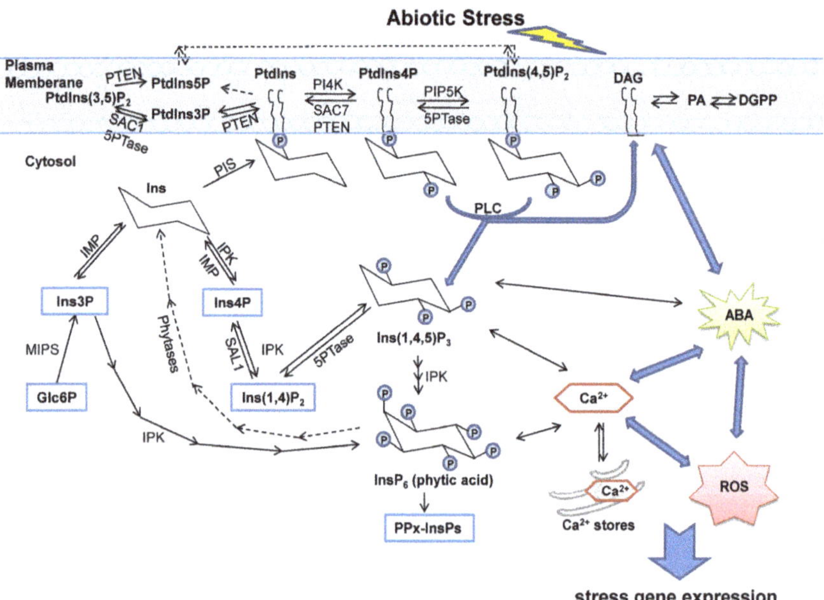

Figure 1. Schematic representation of inositol phosphatases in the plant inositol (Ins) signaling pathways under stress. It illustrated the network of the inositol phosphate (IP) and phosphoinositide (PI) signaling pathway, together with the stress responding processes, such as the ABA pathway, Ca^{2+} release, and ROS generation. The dashed arrows indicated the putative pathways. Ins is soluble, whereas phosphatidylinositol (PtdIns) is bound to the membrane. In the Ins signaling pathways, inositol(1,4,5)trisphosphate (Ins(1,4,5)P_3, IP_3), phytic acid (InsP_6), diacylglycerol (DAG), and phosphatidic acid (PA) are all signaling molecules. ABA—abscisic acid, DGPP—diacylglycerolpyrophosphate, Glc6P—glucose-6-phosphate, IMP—inositol monophosphatase, IPK—inositol polyphosphate multi kinase, MIPS—*myo*-inositol-3-phosphate synthase, P—phosphate, PIP5K—PtIns4P 5-kinase, PI4K—phosphatidylinositol 4-kinase, PIS—phosphatidylinositol synthase, PKC—protein kinase C, PLC—phospholipase C, PPx-InsPs—pyrophosphates, PTEN—phosphatase and tensin homologue deleted on chromosome 10, PtdIns—phosphatidylinositol, ROS—reactive oxygen species, SAC—suppressor of actin, 5PTases—inositol polyphosphate 5-phosphatases.

On the other hand, PtdIns4P and PtdIns(4,5)P_2 can be hydrolyzed into diacylglycerol (DAG) and the corresponding phosphoinositide phosphates (PtdInsPs) by phospholipase C (PLC) (Figure 1) [13]. DAG and inositol-1,4,5-trisphosphate (Ins(1,4,5)P_3, also abbreviated as IP_3 in this article) are believed as second messages for various signal transduction. In brief, the membrane-localized DAG activates the protein kinase C (PKC) and the soluble InsP_3 diffuses in cytosol to release Ca^{2+} from intracellular stores via a ligand-gated Ca^{2+} channel [5,14]. DAG can also be used to generate phosphatidic acid (PA), which is also an important signaling molecule [6]. Those inositol polyphosphates can be further phosphorylated by inositol polyphosphate multi kinases (IPKs) and stored as phytic acid (inositol-1,2,3,4,5,6-hexakisphosphate, $InsP_6$) in seeds and other storage tissues [4,15]. $InsP_6$ has been identified as a signaling molecular to regulate Ca^{2+} release as well [6]. Moreover, $InsP_6$ could be converted to pyrophosphates, denoted as PPx-InsPs [16]. Notably under abiotic stress, there are crosstalks between the Ins signaling pathway and phytohormones, especially abscisic acid (ABA) [4,6].

3. Phosphatases in Inositol Signaling Pathways

Among the processes of inositol phosphate (IP) and the phosphoinositide (PI) signaling pathway, dephosphorylation is catalyzed by specific inositol phosphatases on various substrates (Figure 1). Until now, dozens of enzymes have been identified, including inositol polyphosphate 5-phosphatases (5PTases), suppressor of actin (SAC) phosphatases, SAL1 phosphatase/FIERY1 (FRY1) and its homologs, inositol monophosphatase (IMP), and phosphatase and tensin homologue deleted on chromosome 10 (PTEN)-related phosphatases (Figure 2). Most knowledge of them was obtained from the studies in the model plant *Arabidopsis thaliana*. These plant inositol phosphatases have a broad function in development and adaptation by altering the IP or PI signaling pathways. The general information of those *Arabidopsis* proteins was listed in Table 1. Interestingly, one certain inositol phosphatase could hydrolyze several substrates, even both inositol phosphate and phosphoinositide. One substrate could be degraded by more than one enzyme as well, suggesting their redundant roles in multiple aspects of life processes.

The 5PTases family is the biggest family of the mentioned inositol phosphatases, containing 15 members in *Arabidopsis*, 21 in rice, and 39 in soybean [17]. 5PTases hydrolyze the phosphate bond on the 5-position of the inositol ring from both inositol phosphate and phosphoinositide with the conserved inositol polyphosphate phosphatase catalytic (IPPc) domain. Due to the substrate specificity, mammalian 5PTases have been classified into four groups [18]. Group I, 5PTases hydrolyze only the water-soluble inositol polyphosphates ($Ins(1,4,5)P_3$ and $Ins(1,3,4,5)P_4$); group II the water-soluble inositol polyphosphates and the membrane-bound phosphoinositide; group III, $Ins(1,3,4,5)P_4$ and $PtdIns(3,4,5)P_3$ with a 3-position phosphate group; and group IV only phosphoinositide. Similar as the mammalian counterparts, plant 5PTases also have various substrate specificities. The substrates have been identified by biochemical evidences for twelve of the fifteen *Arabidopsis* 5PTases, including Group I, Group II, and Group IV 5PTases (Table 1). Since several 5PTases could hydrolyze $Ins(1,4,5)P_3$ to prevent its accumulation, it is believed to terminate the corresponding $Ins(1,4,5)P_3$ pathway and alter abscisic acid (ABA) signaling, Ca^{2+} release, and reactive oxygen species (ROS) production [19–21].

The SAC phosphatases are polyphosphoinositide phosphatases, containing the enzymatic SAC domain [22]. There are nine members in *Arabidopsis* [23]. Most *Arabidopsis* SAC phosphatases have a ubiquitous expression pattern, except for AtSAC6 which is only expressed in flowers under normal growth condition. Their expression was not altered by treatment with phytohormones (auxin, cytokinin, GA, and ABA) [23]. When two-week-old seedlings were treated with various stresses (dark, cold, salt, and wounding), only *AtSAC6* has been identified to be induced by salt stress, indicating it would be involved in salt response [23]. Besides, the *sac9* mutants exhibit a constitutive stress response with highly up-regulated stress-induced genes and over-accumulation of ROS [24]. Though there is limited knowledge on their substrate specificity, SAC phosphatases have been found to affect the accumulation of some certain phosphatidylinositol phosphates, such as $PtdIns(4,5)P_2$, $PtdIns(3,5)P_2$, and PtdIns4P, in addition to having a possible role in vesicle trafficking [24–26].

Table 1. Phosphatases of the inositol signaling pathway in *Arabidopsis thaliana*.

Name	Gene ID	Substrates	Cellular Localization	Expression Patterns	Function	References
		5PTase—hydrolyze inositol-5-phosphate				
At5TPase1	At1G34120	Group II Ins(1,4,5)P$_3$, Ins(1,3,4,5)P$_4$, PtdIns(4,5)P$_2$	-	leaf, flower, bolt, seedling	alter ABA and light signaling, stomatal opening, seedling development	[19,21,27,28]
At5TPase2	At4G18010	Group II Ins(1,4,5)P$_3$, Ins(1,3,4,5)P$_4$, PtdIns(4,5)P$_2$	-	leaf, flower, bolt, seedling	alter ABA signaling, seedling development	[20,27,28]
At5TPase3	At1G71710	Group II PtdIns(4,5)P$_2$, PtdIns(3,4,5)P$_3$, Ins(1,4,5)P$_3$, Ins(1,3,4,5)P$_4$,	-	-	-	[29]
At5TPase4	At3G63240	Group IV PtdIns(4,5)P$_2$	-	-	-	[29]
At5TPase5/ MRH3/BST1	At5G65090	-	-	-	root hair development	[30,31]
At5TPase6/CVP2	At1G05470	Group IV PtdIns(4,5)P$_2$, PtdIns(3,4,5)P$_3$	-	vascular system	foliar vein patterning, root branching	[32–34]
At5TPase7/ CVL1	At2G32010	Group IV PtdIns(4,5)P$_2$, PtdIns(3,4,5)P$_3$	plasma membrane, nuclear speckles	vascular system	foliar vein patterning, root branching, salt tolerance, and ROS production	[33–35]
At5TPase8	At2G37440	-	-	-	-	-
At5TPase9	At2G01900	Group IV PtdIns(4,5)P$_2$, PtdIns(3,4,5)P$_3$	-	root	salt tolerance and ROS production endocytosis	[36]
At5TPase10	At5G04980	-	-	-	-	-
At5TPase11	At1G47510	Group IV PtdIns(4,5)P$_2$, PtdIns(3,5)P$_2$, PtdIns(3,4,5)P$_3$	cell surface or plasma membrane	flower, leaf, root, silique, bolt, seedling	seedling development	[27,37]
At5TPase12	At2G43900	Group I Ins(1,4,5)P$_3$	-	pollen grain, leaf and flower (mostly); root, stem and young seedling (weakly)	pollen dormancy/germination	[38,39]

Table 1. *Cont.*

Name	Gene ID	Substrates	Cellular Localization	Expression Patterns	Function	References
At5TPase13	At1G05630	Group I Ins(1,4,5)P$_3$,	nucleus	young seedlings, flowers	cotyledon vein development, alter auxin, ABA, sugar and PHOTOTROPIN1 signaling, root gravitropism, vesicle trafficking	[38–43]
At5TPase14	At2G31830	Group II PtdIns(4,5)P$_2$, PtdIns(3,4,5)P$_3$, Ins(1,4,5)P$_3$	-	pollen grain	-	[38,39]
At5TPase15/ FRA3	At1G65580	Group II PtdIns(4,5)P$_2$, PtdIns(3,4,5)P$_3$, Ins(1,4,5)P$_3$	-	seedling, stem, root, flower, mature leaf (weak)	secondary wall synthesis and actin organization	[44]
SAC—hydrolyze phosphatidylinositol phosphates						
SAC1/FRA7	At1G22620	PtdIns(3,5)P$_2$	Golgi	ubiquitous, predominant in vascular tissues and fibers of stems	cell morphogenesis, cell wall synthesis, actin organization	[23,45]
SAC2	At3G14205	-	tonoplast	ubiquitous	vacuolar function	[23,26]
SAC3	At3G43220	-	tonoplast	ubiquitous	vacuolar function	[23,26]
SAC4	At5G20840	-	tonoplast	ubiquitous	vacuolar function	[23,26]
SAC5	At1G17340	-	tonoplast	ubiquitous	vacuolar function	[23,26]
SAC6/SAC1b	At5G66020	-	endoplasmic reticulum	pollen grain	embryo development	[22,23]
SAC7/SAC1c/ RHD4	At3G51460	PtdIns4P	endoplasmic reticulum	most tissues (strong)	embryo development, root hair development	[22,23,25]
SAC8/AtSAC1a	At3G51830	-	endoplasmic reticulum	hypocotyls of seedlings, pollen grain, most tissues (week),	embryo development	[22,23]

Table 1. *Cont.*

Name	Gene ID	Substrates	Cellular Localization	Expression Patterns	Function	References
SAC9	At3G59770	-	-	root (strong), leaf and shoot (weak)	cell wall formation, stress response	[23,24,46]
SAL—hydrolyze inositol-1-phosphate						
AtSAL1/ AtFIERY1 (AtFRY1)/ HOS2/RON1	At5G63980	Ins(1,4)P2, Ins(1,3,4)P3, PAP, PAPS	chloroplast, mitochondria	vascular tissue	alter ABA, auxin and stress signaling (cold, drought, salt, lithium, high light, cadmium), venation patterning	[47–55]
AtSAL2	At5G64000	Ins(1,4)P$_2$, PAP,	-	-	-	[56]
IMP—hydrolyze inositol-3-phosphate, inositol-4-phosphate						
IMP/VTC4	At3G02870	Ins3P, Ins1P, L-Galactose-1-P	cytosol	photosynthetictissuesbiosynthesis, alter cold, salt and ABA responses	seed development, ascorbate responses	[11,57–59]
IMPL1	At1G31190	Ins3P, Ins1P, Ins2P, L-Galactose-1-P	chloroplast	ubiquitous	seed development	[11,57,59]
IMPL2	At4G39120	Histidinol 1-P	chloroplast	root (strong), hypocotyl (weak)	seed development, histidinebiosynthesis	[11,57,59]
PTEN—hydrolyze inositol-3-phosphate						
PTEN1	At5G39400	PtdIns(3,4,5)P$_3$, phosphotyrosin	vesicles, autophagic body	pollen grain	pollen development	[60,61]
PTEN2a	At3G19420	PtdIns3P, PtdIns(3,4)P$_2$, PtdIns(3,5)P$_2$, PtdIns4P, PtdIns(3,4,5)P$_3$, phosphotyrosin	-	seedling, leaf, flower, silique	-	[62]
PTEN2b	At3G50110	PtdIns3P, phosphotyrosin	-	seedling, leaf, flower, silique	-	[62]

BST1, BRISTLED1; CVL1, CVP2-like1; CVP2, cotyledon vascular pattern2; FRA3, fragile fiber 3; FRA7, fragile fiber 7; HOS2, high expression of osmotic stress-regulated gene expression 2; IMP, *myo*-inositol monophosphatase; Ins, inositol, MRH3, root hair morphogenesis 3; P, phosphate; PAP, 3′-phosphoadenosine 5′-phosphate; PAPS, 2′-PAP and 3′-phosphoadenosine 5′-phosphosulfate; PTEN, phosphatase and tensin homologue deleted on chromosome 10, PtdIns, phosphatidylinositol, RHD4, root hair defective 4; RON1, rotunda 1; SAC, suppressor of actin, VTC4, vitamin C 4, 5PTases, inositol polyphosphate 5-phosphatases.

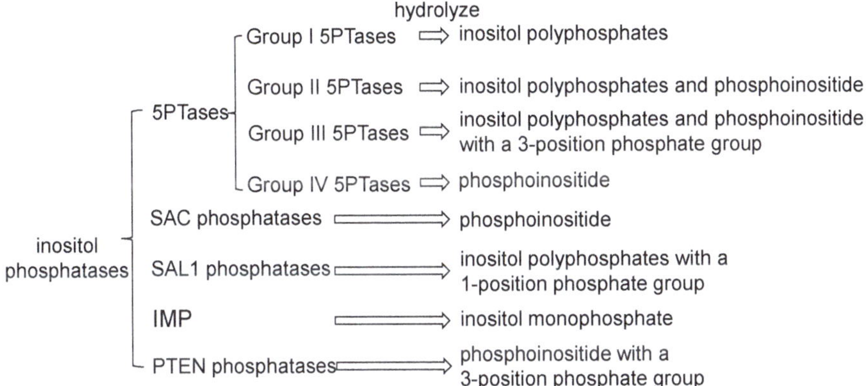

Figure 2. Inositol phosphatases and their inositol-related substrates overviewed in this study. IMP—inositol monophosphatase, PTEN—phosphatase and tensin homologue deleted on chromosome 10, SAC—suppressor of actin, 5PTases—inositol polyphosphate 5-phosphatases.

Comparing to 5PTases and SAC phosphatases, there are fewer members in the SAL, IMP, and PTEN families and most of them behave as bifunctional enzymes (Table 1). AtSAL1 and AtSAL2 exhibit the activities of not only inositol polyphosphate 1-phosphatase but also 3′(2′),5′-bisphosphate nucleotidase [47,56]. The other SAL1 homologues without inositol phosphatases are not listed here. AtSAL1 has been identified as an important player in response to various stresses, probably through both enzyme activities [48,49,51,53,63–65]. Three IMP members have been identified in *Arabidopsis* [11]. IMP and inositol monophosphatase-like 1 (IMPL1) exhibit bifunctional activities affecting both inositol and ascorbate synthesis pathways, whereas IMPL2 is a histidinol-phosphate phosphatase affecting histone biosynthesis pathways [57,66]. The IMPs from other plants have been shown to play a role in stress tolerance [67–69], which we will discuss later. PTEN members are also dual phosphatases for protein and phosphoinositide phosphates [62]. The transcript and protein analyses showed that *AtPTEN2a* and *AtPTEN2b* were up-regulated at transcriptional level, but not at protein level under salt and osmotic stress [62], suggesting their potential roles in plant adaptation to stress. But no further evidence has been reported yet.

4. Function of Inositol Phosphatases under Abiotic Stress

4.1. 5PTases and Plant Responses to Abiotic Stress

The capacity of 5PTases hydrolyzing IP_3 is believed to be vital in the termination of IP_3, consequently altering Ca^{2+} oscillations, ABA signaling, and other stress-related pathways. The transgenic *Arabidopsis* plants overexpressing mammalian type I (group I) inositol polyphosphate 5-phosphatase (InsP 5-ptase) exhibited increased drought tolerance with less water loss [70]. The contents of IP_3 and IP_6 were decreased in the transgenic lines as expected, thus attenuating ABA induction and Ca^{2+} signal transduction. The stomata were less responsive to the inhibition of opening by ABA and more sensitive to ABA-induced closure. Furthermore, the microarray data showed that *dehydration-responsive element-binding protein 2A* (DREB2A), encoding a drought-inducible ABA-independent transcription factor, and the DREB2A-regulated genes were induced in the InsP 5-ptase transgenic plants, suggesting the drought tolerance is mediated via the DREB2A-dependent pathway [70].

For plant 5PTases, it is common to take a role in the degradation process of inositol phosphate or phosphoinositide, terminating the IP_3 signaling, thus altering of ABA pathway and Ca^{2+} release, which is believed to be vital in stress tolerance [19,21,34,39]. However, only three of the 15 At5PTases have been identified to play important roles in abiotic stress with genetic and biochemical evidences until now. At5PTase7 and At5PTase9 function in salt tolerance, and At5PTase13 in low nutrient and sugar stress [35,36,41].

The T-DNA insertion mutants of *At5PTase7* or *At5PTase9* increased salt sensitivity and the overexpression plants increased salt tolerance [35,36]. Mutation in either *At5PTase7* or *At5PTase9* reduced ROS production in the *Arabidopsis* roots after 10 to 15 min after salt treatment. Additionally, the expression of salt-responsive genes, such as *RD29A* and *RD22*, was not induced highly in both mutants as in the wild-type under salt stress [35,36]. It suggested that the defect in *At5PTase7* or *At5PTase9* would disturb ROS production and salt-responsive gene expression, probably hampering the subsequent rescue signal transduction. Interestingly, the *At5PTase9* mutants appeared to have a better ability to resistant osmotic stress. Meanwhile, the *At5PTase9* mutants decreased Ca^{2+} influx and fluid-phase endocytosis [36]. Though the At5PTase7 and At5PTase9 isomers take non-redundant roles in regulating plant responses to salt stress, they share the same substrates, membrane-bound phosphoinositide, indicating that phosphoinositide would be important in salt tolerance [36].

At5TPase13 is one of the four At5TPases (At5TPase12-15), which contain the plant specific WD40 repeats [38,44]. The T-DNA insertion mutants of *At5TPase13* showed a reduction of root growth under limited nutrient conditions and germination rates in response to sugar stress, along with ABA insensitivity [41]. The yeast two-hybrid analyses suggested that its WD40 repeat domain interacts with the sucrose nonfermenting-1-related kinase (SnRK1.1), which is an energy/stress sensor [41]. The genetic and biochemical evidences indicated that At5TPase13 acts as a positive regulator of SnRK1.1 under low-nutrient or low-sugar conditions, as a negative regulator under severe starvation conditions through affecting the proteasomal degradation of SnRK1.1. Strangely, the *At5ptase13* mutants accumulate less IP_3 in response to sugar stress [41]. Again, At5PTase13 could alter cytosolic Ca^{2+} to regulate PHOYOTROPIN1 signaling under blue light [40].

Besides, several transcriptional analyses showed that the expression of multiple At5PTases is greatly up- or down-regulated in response to a series of abiotic stresses, such as cold, osmotic, salt, drought, oxidative, and heat [35,36,71]. Considering the general function of the known 5PTases in the inositol pathway, Ca^{2+} signaling, ABA responses, ROS generation, vesicle trafficking, and possible connection with other phytohormones [43,71], it could imply their potential roles in plant responses to abiotic stress.

4.2. SAL1 and Plant Responses to Abiotic Stress

AtSAL1, identified as a homologue of the yeast HAL2 in *Arabidopsis* and also well-known as FIERY1 (FRY1), has dual enzymatic activity of inositol phosphatase and nucleotidase, which play a role in both inositol signaling and nucleotide metabolite [47,48]. AtSAL1 functions broadly in responses to abiotic stresses, including salt, cold, lithium, drought, cadmium, high light, and oxidative, probably with the contributions of both enzymatic activity [48,49,51–53,63,72,73]. Here we will focus on its activity of inositol polyphosphate 1-phosphatase. Remarkably, it can hydrolyze the signaling molecular IP_3, thus affecting the subsequent steps in a similar pattern of 5PTases, which we have discussed above.

It seems the effects of AtSAL1 on stress responses are controversial. Ectopic expression of *AtSAL1* could increase lithium tolerance in yeast by modifying Na^+ and Li^+ effluxes [47]. Ectopic expression of its homologue in soybean, *GmSAL1*, could alleviate salinity stress in tobacco BY-2 cells [74]. Mutation in *AtSAL1* would cause more sensitivity to salt, osmotic, and cold stress in *Arabidopsis* [48,72]. However, another *Atsal1* mutant, *hos2* with a single amino acid substitution exhibited as more resistant to lithium and salt stress [72]. Moreover, overexpression of *AtSAL1* or ectopic expression of *GmSAL1* could not enhance salt tolerance in *Arabidopsis* [49,74]. Loss function in *AtSAL1* would enhance drought and cadmium resistance in *Arabidopsis*, suggesting it would be a negative regulator of stress tolerance [51,63]. Expressing the modified *SAL1*, by inserting the META motif from black yeast *Aureobasidium pullulans*, *ApHal2*, improved salt and drought tolerance in *Arabidopsis* [73]. It seems the presence of the META motif should be responsible for its ability on the stress tolerance, but the mechanism is still obscure.

The molecular mechanism of AtSAL1 in stress responses seems to be complicated for its multiple effects in various cellular processes. First, AtSAL1 would regulate stress tolerance and ABA responses via IP$_3$ signaling. The *Atsal1* mutants increase IP$_3$ accumulation and the expression levels of ABA and stress genes, including *RD29A*, cold-specific *CRT-binding factor 2* (*CBF2*), and *CBF3* [48]. On the contrary, ectopic expression of *GmSAL1* leads to a reduction of IP$_3$ accumulation and suppression of the ABA-induced stomatal closure [74]. Furthermore, it also showed AtSAL1 could regulate Ca^{2+} release and modulate the auxin pathway by IP$_3$ signaling in plant development [54,55]. It seems a similar consequence of AtSAL1 in the IP$_3$ signaling as for 5PTases. Maybe further investigation will supply evidence that AtSAL1 takes a role in Ca^{2+} release and its downstream signaling in response to abiotic stress as well. Secondly, AtSAL1 also regulates the ion homeostasis via the IP$_3$ pathway. Ectopic expression of *AtSAL1* could modify Na$^+$ and Li$^+$ effluxes in yeast for lithium and salt tolerance [47]. *GmSAL1*-transgenic BY-2 cells could compartmentalize more Na$^+$ in vacuolar for protection from salt stress [74]. Additionally, AtSAL1 takes a role more likely as a phosphoadenosine phosphatase under drought, high light, and oxidative stress, for only 3′-phosphoadenosine 5′-phosphate (PAP), not IP$_3$, accumulated in the *Atsal1* mutants, when suffering stresses [49,52,53,65]. The genetic evidences indicated that PAP accumulation could also affect the ABA pathway, relying on, rather, the negative regulator ABH1 in the branched ABA pathway, than ABI1 in the core ABA pathway [49]. AtSAL1 could protect 5′ to 3′ exoribonucleases (XRNs) by degrading PAP and subsequently modulate the expression of the corresponding nuclear genes, supposed as the chloroplast retrograde pathway [52,53,65]. Besides, the *AtSAL1*-deficient mutants have been found to attenuate endoplasmic reticulum (ER) stress under cadmium stress [63]. But no exploration has been made to determine its connection with the IP3 signaling or SAL1-PAP pathway. This would provide a new insight on the mechanism of AtSAL1 in various stress tolerance [63].

4.3. IMPs and Plant Responses to Abiotic Stress

IMPs were first identified in tomato to play a role in inositol synthesis with high sensitivity to lithium [10]. Their homologues in *Arabidopsis* have also been characterized as multi-functional enzymes involved in inositol, ascorbate, and histone biosynthesis [57,59,66], so do their homologues in other plants, such as rice (*Oryza sativa* L.), chickpea (*Cicer arietinum* L.), soybean (*Glycine max*), barley (*Hordeum vulgare*), and *Medicago truncatula* [68,69,75]. The genetic studies showed that IMPs play a role in seed development in *Arabidopsis* [11]. Chickpea IMP could also influence seed size/weight [76]. But few explorations have been made with *Arabidopsis* IMP on stress tolerance yet. Only some authors have tried assays in chickpea and rice suggesting that IMPs also function in response to abiotic stress [67–69]. But it is still unclear how IMPs influence the inositol pathway to confer stress.

Biochemical evidence demonstrated that CaIMP contains the same enzyme activity as *Arabidopsis* IMP and IMP activity is increased in chickpea seedlings under abiotic stresses, including salt, cold, heat, dehydration, and paraquat. It is consistent with the results of the transcript analyses by qRT-PCR, which showed that *CaIMP* is induced under abiotic stress and ABA treatment [69]. The *CaIMP*-transgenic *Arabidopsis* plants exhibited enhanced tolerance to abiotic stress, whereas the *IMP*-deficient *Arabidopsis* mutants increased the sensitivity to stress during seed germination and seedling growth. The inositol content and ascorbate content of the *CaIMP*-overexpressing lines are higher than the wild-type and the vector control, suggesting CaIMP would improve the plant tolerance to stress through both metabolic pathways [69]. Association analyses performed with 60 chickpea germplasm accessions showed that NCPGR90, a simple sequence repeat marker for phytic acid content and drought tolerance, is located to the 5′UTR of *CaIMP* [68]. The transcript lengths of *CaIMP* are different between the drought-tolerant and drought-susceptible accessions, suggesting this variation might regulate phytic acid contents in plants, thus conferring drought tolerance in chickpea [68]. In another study, this variation also causes the differential protein level and enzymatic activity of CaIMP [76].

Rice *OsIMP* is significantly upregulated by cold and ABA treatment by transcript analyses [67]. The promoter analyses on sequence also identified several important stress-responding cis-acting elements, including ABRE-element (abscisic acid responsiveness), LTR (low-temperature responsiveness), TCA-element (salicylic acid responsiveness), GARE-motif (gibberellins responsive), and MBS (MYB binding site). Ectopic expression of *OsIMP* in tobacco improved cold tolerance. The transgenic plants contained more inositol content, less hydrogen peroxide (H_2O_2), and less malondialdehyde (MDA), with increased antioxidant enzyme activities under normal and cold stress conditions [67]. It suggested that the accumulation of inositol by expressing *OsIMP* would modulate the antioxidant enzymes' activities to conquer cold stress.

5. Conclusions

Substantial evidences demonstrate inositol phosphates, phosphoinositides, and the related inositol signaling play a crucial role in various life processes of development and environmental adaptation in plants [1,4,6,7,12]. When plants suffer abiotic stress from the environment, a membrane receptor would accept the stimulus and the membrane-associated phosphoinositides would pass the cellular message by producing second messages, lipid-bound DAG, and soluble IP_3. Components involved in the inositol pathways have been noted for their general roles in stress tolerance. This article focused on the knowledge about inositol phosphatases, which are considered to be more important in the degradation pathway of IP_3 signaling, and their function in plant responses to abiotic stress.

Around 30 members of inositol phosphatases from five families have been identified. Their functions and mechanisms are still largely unknown. Biochemical and physiological data, especially those from analytical techniques, have delineated their substrates and the affecting signals. Moreover, the genetic evidences elucidate the genes' function and how to pass the signals. In general, loss-in-function of inositol phosphatases usually cause the accumulation of IP_3 or phosphoinositides, thus facilitating Ca^{2+} release from cellular stores and affecting ABA or other phytohormones' pathways. For their effects on lipid-bound phosphoinositides, several enzymes have been proved to be involved in vesicle trafficking. For most of the inositol phosphatases, the existed evidences could only support part of the model. There are also some other puzzles. Since phytic acid ($InsP_6$) could also serve as a signaling molecule to regulate Ca^{2+} release [6], what is the role of inositol phosphatases in this process? There are multiple genes in the same family, especially 5PTases and SAC phosphatases. How do plants coordinate their function? Most of the knowledge about these enzymes is obtained from the mode plant *Arabidopsis*. Study from other plants is relatively rare. Do these inositol phosphatases take a universal role in all plants under abiotic stress? Hopefully, more exploration will expand our understanding about inositol phosphatases.

Author Contributions: Conceptualization, Q.J., J.H. and W.L.; invesitigation, Q.J., Q.L., Y.Z. and Q.K.; resources, Q.J., D.K., S.S., J.S.; writing—original draft preparation, Q.J.; writing—review and editing, Q.J., D.K. and Q.L.; visualization, Q.J.; supervison, Q.J. and J.H.; project administration, K.L., Q.K. and W.L.; funding acquisition, Q.J.

Funding: This work is funded by the National Natural Science Foundation of China (31501232), the China Postdoctoral Science Foundation (2014T70603, 2013M540527), the Fujian Agriculture and Forestry University (FAFU) Science Fund for Distinguished Young Scholars (xjq201629), and the FAFU Science Grant for Innovation (CXZX2018119).

Conflicts of Interest: The authors declare no conflicts of interest.

Abbreviations

ABA	abscisic acid
ABRE	abscisic acid responsiveness
BST1	BRISTLED1
CBF	CRT-binding factor
CVL1	CVP2-like1
CVP2	cotyledon vascular pattern2
DAG	diacylglycerol
DGPP	diacylglycerol pyrophosphate
DREB2A	dehydration-responsive element-binding protein 2A

ER	endoplasmic reticulum
FRA3	fragile fiber 3
FRA7	fragile fiber 7
FRY1	FIERY1
Glc6P	glucose-6-phosphate
GARE	gibberellins responsive
HOS2	high expression of osmotic stress-regulated gene expression 2
IMP	*myo*-inositol monophosphatase
IMPL	inositol monophosphatase-like
Ins	inositol
InsP 5-ptase	inositol polyphosphate 5-phosphatase
IP	inositol phosphate
IP$_3$	Inositol(1,4,5)trisphosphate
IPK	inositol polyphosphate multi kinase
IPPc	inositol polyphosphate phosphatase catalytic
LTR	low-temperature responsiveness
MBS	MYB binding site
MDA	malondialdehyde
MRH3	root hair morphogenesis 3
MIPS	*myo*-inositol-3-phosphate synthase
P	phosphate
PA	phosphatidic acid
PAP	3′-phosphoadenosine 5′-phosphate
PAPS	2′-PAP and 3′-phosphoadenosine 5′-phosphosulfate
PI	phosphoinositide
PIP5K	PtIns4P 5-kinase
PIP4K	PtIns4P 4-kinase
PIS	phosphatidylinositol synthase
PKC	protein kinase C
PLC	phospholipase C
PPx-InsPs	pyrophosphates
PTEN	phosphatase and tensin homologue deleted on chromosome 10
PtdIns	phosphatidylinositol
RHD4	root hair defective 4
RON1	rotunda 1
ROS	reactive oxygen species
SAC	suppressor of actin
SnRK1.1	sucrose nonfermenting-1-related kinase
TCA	salicylic acid responsiveness
VTC4	vitamin C 4
XRN	5′ to 3′ exoribonuclease
5PTases	inositol polyphosphate 5-phosphatases

References

1. Valluru, R.; Van den Ende, W. *myo*-inositol and beyond—Emerging networks under stress. *Plant Sci.* **2011**, *181*, 387–400. [CrossRef] [PubMed]

2. Michell, R.H. Inositol derivatives: Evolution and functions. *Nat. Rev. Mol. Cell Biol.* **2008**, *9*, 151–161. [CrossRef] [PubMed]

3. Heilmann, M.; Heilmann, I. Plant phosphoinositides—Complex networks controlling growth and adaptation. *Biochim. Biophys. Acta Mol. Cell Biol. Lipids* **2015**, *1851*, 759–769. [CrossRef] [PubMed]

4. Gillaspy, G.E. The cellular language of *myo*-inositol signaling. *New Phytol.* **2011**, *192*, 823–839. [CrossRef] [PubMed]

5. Munnik, T.; Vermeer, J.E.M. Osmotic stress-induced phosphoinositide and inositol phosphate signalling in plants. *Plant Cell Environ.* **2010**, *33*, 655–669. [CrossRef] [PubMed]

6. Hou, Q.; Ufer, G.; Bartels, D. Lipid signalling in plant responses to abiotic stress. *Plant Cell Environ.* **2016**, *39*, 1029–1048. [CrossRef] [PubMed]

7. Xue, H.-W.; Chen, X.; Mei, Y. Function and regulation of phospholipid signalling in plants. *Biochem. J.* **2009**, *421*, 145–156. [CrossRef]

8. Michell, R.H. Inositol and its derivatives: Their evolution and functions. *Adv. Enzyme Regul.* **2011**, *51*, 84–90. [CrossRef]

9. Xiong, L.; Schumaker, K.S.; Zhu, J.-K. Cell signaling during cold, drought, and salt stress. *Plant Cell* **2002**, *14*, S165–S183. [CrossRef]

10. Gillaspy, G.E.; Keddie, J.S.; Oda, K.; Gruissem, W. Plant inositol monophosphatase is a lithium-sensitive enzyme encoded by a multigene family. *Plant Cell* **1995**, *7*, 2175–2185.

11. Sato, Y.; Yazawa, K.; Yoshida, S.; Tamaoki, M.; Nakajima, N.; Iwai, H.; Ishii, T.; Satoh, S. Expression and functions of *myo*-inositol monophosphatase family genes in seed development of *Arabidopsis*. *J. Plant Res.* **2011**, *124*, 385–394. [CrossRef] [PubMed]

12. Heilmann, I. Phosphoinositide signaling in plant development. *Development* **2016**, *143*, 2044–2055. [CrossRef] [PubMed]

13. Hong, Y.; Zhao, J.; Guo, L.; Kim, S.-C.; Deng, X.; Wang, G.; Zhang, G.; Li, M.; Wang, X. Plant phospholipases D and C and their diverse functions in stress responses. *Prog. Lipid Res.* **2016**, *62*, 55–74. [CrossRef] [PubMed]

14. Van Leeuwen, W.; Ökrész, L.; Bögre, L.; Munnik, T. Learning the lipid language of plant signalling. *Trends Plant Sci.* **2004**, *9*, 378–384. [CrossRef] [PubMed]

15. Xia, H.J.; Yang, G. Inositol 1,4,5-trisphosphate 3-kinases: Functions and regulations. *Cell Res.* **2005**, *15*, 83–91. [CrossRef] [PubMed]

16. Williams, S.P.; Gillaspy, G.E.; Perera, I.Y. Biosynthesis and possible functions of inositol pyrophosphates in plants. *Front. Plant Sci.* **2015**, *6*, 67. [CrossRef] [PubMed]

17. Zhang, Z.; Li, Y.; Luo, Z.; Kong, S.; Zhao, Y.; Zhang, C.; Zhang, W.; Yuan, H.; Cheng, L. Expansion and Functional Divergence of Inositol Polyphosphate 5-Phosphatases in Angiosperms. *Genes* **2019**, *10*, 393. [CrossRef] [PubMed]

18. Majerus, P.W.; Kisseleva, M.V.; Norris, F.A. The Role of Phosphatases in Inositol Signaling Reactions. *J. Biol. Chem.* **1999**, *274*, 10669–10672. [CrossRef] [PubMed]

19. Berdy, S.E.; Kudla, J.; Gruissem, W.; Gillaspy, G.E. Molecular characterization of At5PTase1, an inositol phosphatase capable of terminating inositol trisphosphate signaling. *Plant Physiol.* **2001**, *126*, 801–810. [CrossRef]

20. Sanchez, J.-P.; Chua, N.-H. *Arabidopsis* PLC1 Is Required for Secondary Responses to Abscisic Acid Signals. *Plant Cell* **2001**, *13*, 1143–1154. [CrossRef]

21. Burnette, R.N. An *Arabidopsis* Inositol 5-Phosphatase Gain-of-Function Alters Abscisic Acid Signaling. *Plant Physiol.* **2003**, *132*, 1011–1019. [CrossRef] [PubMed]

22. Despres, B.; Bouissonnié, F.; Wu, H.-J.; Gomord, V.; Guilleminot, J.; Grellet, F.; Berger, F.; Delseny, M.; Devic, M. Three SAC1-like genes show overlapping patterns of expression in *Arabidopsis* but are remarkably silent during embryo development. *Plant J.* **2003**, *34*, 293–306. [CrossRef] [PubMed]

23. Zhong, R.; Ye, Z.-H. The SAC Domain-Containing Protein Gene Family in *Arabidopsis*. *Plant Physiol.* **2003**, *132*, 544–555. [CrossRef] [PubMed]

24. Williams, M.E.; Torabinejad, J.; Cohick, E.; Parker, K.; Drake, E.J.; Thompson, J.E.; Hortter, M.; Dewald, D.B. Mutations in the *Arabidopsis* phosphoinositide phosphatase gene SAC9 lead to overaccumulation of PtdIns(4,5)P_2 and constitutive expression of the stress-response pathway. *Plant Physiol.* **2005**, *138*, 686–700. [CrossRef] [PubMed]

25. Thole, J.M.; Vermeer, J.E.M.; Zhang, Y.; Gadella, T.W.J.; Nielsen, E. Root hair defective4 encodes a phosphatidylinositol-4-phosphate phosphatase required for proper root hair development in *Arabidopsis thaliana*. *Plant Cell* **2008**, *20*, 381–395. [CrossRef] [PubMed]

26. Nováková, P.; Hirsch, S.; Feraru, E.; Tejos, R.; van Wijk, R.; Viaene, T.; Heilmann, M.; Lerche, J.; De Rycke, R.; Feraru, M.I.; et al. SAC phosphoinositide phosphatases at the tonoplast mediate vacuolar function in *Arabidopsis*. *Proc. Natl. Acad. Sci. USA* **2014**, *111*, 2818–2823. [CrossRef] [PubMed]

27. Ercetin, M.E.; Gillaspy, G.E. Molecular characterization of an *Arabidopsis* gene encoding a phospholipid-specific inositol polyphosphate 5-phosphatase. *Plant Physiol.* **2004**, *135*, 938–946. [CrossRef] [PubMed]

28. Gunesekera, B.; Torabinejad, J.; Robinson, J.; Gillaspy, G.E. Inositol Polyphosphate 5-Phosphatases 1 and 2 Are Required for Regulating Seedling Growth. *Plant Physiol.* **2007**, *143*, 1408–1417. [CrossRef]

29. Ercetin, M.E. Molecular Characterization and Loss-of-Function Analysis of an *Arabidopsis thaliana* Gene Encoding a Phospholipid-Specific Inositol Polyphosphate 5-Phosphatase. Ph.D. Thesis, Virginia Polytechnic Institute and State University, Blacksburg, VA, USA, 19 May 2005.

30. Jones, M.A.; Raymond, M.J.; Smirnoff, N. Analysis of the root-hair morphogenesis transcriptome reveals the molecular identity of six genes with roles in root-hair development in *Arabidopsis*. *Plant J.* **2006**, *45*, 83–100. [CrossRef]

31. Parker, J.S.; Cavell, A.C.; Dolan, L.; Roberts, K.; Grierson, C.S. Genetic Interactions during Root Hair Morphogenesis in *Arabidopsis*. *Plant Cell* **2000**, *12*, 1961–1974. [CrossRef]

32. Carland, F.M.; Nelson, T. COTYLEDON VASCULAR PATTERN2–Mediated Inositol (1,4,5) Triphosphate Signal Transduction Is Essential for Closed Venation Patterns of *Arabidopsis* Foliar Organs. *Plant Cell* **2004**, *16*, 1263–1275. [CrossRef] [PubMed]

33. Carland, F.; Nelson, T. CVP2- and CVL1-mediated phosphoinositide signaling as a regulator of the ARF GAP SFC/VAN3 in establishment of foliar vein patterns. *Plant J.* **2009**, *59*, 895–907. [CrossRef] [PubMed]

34. Rodriguez-Villalon, A.; Gujas, B.; van Wijk, R.; Munnik, T.; Hardtke, C.S. Primary root protophloem differentiation requires balanced phosphatidylinositol-4,5-biphosphate levels and systemically affects root branching. *Development* **2015**, *142*, 1437–1446. [CrossRef] [PubMed]

35. Kaye, Y.; Golani, Y.; Singer, Y.; Leshem, Y.; Cohen, G.; Ercetin, M.; Gillaspy, G.; Levine, A. Inositol Polyphosphate 5-Phosphatase7 Regulates the Production of Reactive Oxygen Species and Salt Tolerance in *Arabidopsis*. *Plant Physiol.* **2011**, *157*, 229–241. [CrossRef] [PubMed]

36. Golani, Y.; Kaye, Y.; Gilhar, O.; Ercetin, M.; Gillaspy, G.; Levine, A. Inositol Polyphosphate Phosphatidylinositol 5-Phosphatase9 (At5PTase9) Controls Plant Salt Tolerance by Regulating Endocytosis. *Mol. Plant* **2013**, *6*, 1781–1794. [CrossRef] [PubMed]

37. Ercetin, M.E.; Ananieva, E.A.; Safaee, N.M.; Torabinejad, J.; Robinson, J.Y.; Gillaspy, G.E. A phosphatidylinositol phosphate-specific myo-inositol polyphosphate 5-phosphatase required for seedling growth. *Plant Mol. Biol.* **2008**, *67*, 375–388. [CrossRef]

38. Zhong, R.; Ye, Z.-H. Molecular and biochemical characterization of three WD-repeat-domain-containing inositol polyphosphate 5-phosphatases in *Arabidopsis* thaliana. *Plant Cell Physiol.* **2004**, *45*, 1720–1728. [CrossRef]

39. Wang, Y.; Chu, Y.-J.; Xue, H.-W. Inositol polyphosphate 5-phosphatase-controlled Ins(1,4,5)P3/Ca^{2+} is crucial for maintaining pollen dormancy and regulating early germination of pollen. *Development* **2012**, *139*, 2221–2233. [CrossRef]

40. Chen, X.; Lin, W.-H.; Wang, Y.; Luan, S.; Xue, H.-W. An Inositol Polyphosphate 5-Phosphatase Functions in PHOTOTROPIN1 Signaling in Arabidopis by Altering Cytosolic Ca^{2+}. *Plant Cell* **2008**, *20*, 353–366. [CrossRef]

41. Ananieva, E.A.; Gillaspy, G.E.; Ely, A.; Burnette, R.N.; Erickson, F.L. Interaction of the WD40 Domain of a *myo*inositol Polyphosphate 5-Phosphatase with SnRK1 Links Inositol, Sugar, and Stress Signaling. *Plant Physiol.* **2008**, *148*, 1868–1882. [CrossRef]

42. Wang, Y.; Lin, W.-H.; Chen, X.; Xue, H.-W. The role of *Arabidopsis* 5PTase13 in root gravitropism through modulation of vesicle trafficking. *Cell Res.* **2009**, *19*, 1191–1204. [CrossRef] [PubMed]

43. Lin, W.-H.; Wang, Y.; Mueller-Roeber, B.; Brearley, C.A.; Xu, Z.-H.; Xue, H.-W. At5PTase13 Modulates Cotyledon Vein Development through Regulating Auxin Homeostasis. *Plant Physiol.* **2005**, *139*, 1677–1691. [CrossRef] [PubMed]

44. Zhong, R.; Burk, D.H.; Morrison, W.H.; Ye, Z.-H. *FRAGILE FIBER3*, an *Arabidopsis* gene encoding a type II inositol polyphosphate 5-phosphatase, is required for secondary wall synthesis and actin organization in fiber cells. *Plant Cell* **2004**, *16*, 3242–3259. [CrossRef] [PubMed]

45. Zhong, R.; Burk, D.H.; Nairn, C.J.; Wood-Jones, A.; Morrison, W.H.; Ye, Z.-H. Mutation of SAC1, an *Arabidopsis* SAC domain phosphoinositide phosphatase, causes alterations in cell morphogenesis, cell wall synthesis, and actin organization. *Plant Cell* **2005**, *17*, 1449–1466. [CrossRef] [PubMed]

46. Vollmer, A.H.; Youssef, N.N.; DeWald, D.B. Unique cell wall abnormalities in the putative phosphoinositide phosphatase mutant AtSAC9. *Planta* **2011**, *234*, 993–1005. [CrossRef] [PubMed]

47. Quintero, F.J.; Garciadeblás, B.; Rodríguez-Navarro, A. The *SAL1* gene of *Arabidopsis*, encoding an enzyme with 3′(2′),5′-bisphosphate nucleotidase and inositol polyphosphate 1-phosphatase activities, increases salt tolerance in yeast. *Plant Cell* **1996**, *8*, 529–537. [PubMed]

48. Xiong, L.; Lee, B.; Ishitani, M.; Lee, H.; Zhang, C.; Zhu, J.K. FIERY1 encoding an inositol polyphosphate 1-phosphatase is a negative regulator of abscisic acid and stress signaling in Arabidopsis. *Genes Dev.* **2001**, *15*, 1971–1984. [CrossRef]

49. Chen, H.; Zhang, B.; Hicks, L.M.; Xiong, L. A Nucleotide Metabolite Controls Stress-Responsive Gene Expression and Plant Development. *PLoS ONE* **2011**, *6*, e26661. [CrossRef]

50. Kim, B.-H.; von Arnim, A.G. *FIERY1* regulates light-mediated repression of cell elongation and flowering time via its 3′(2′),5′-bisphosphate nucleotidase activity. *Plant J.* **2009**, *58*, 208–219. [CrossRef]

51. Wilson, P.B.; Estavillo, G.M.; Field, K.J.; Pornsiriwong, W.; Carroll, A.J.; Howell, K.A.; Woo, N.S.; Lake, J.A.; Smith, S.M.; Harvey Millar, A.; et al. The nucleotidase/phosphatase SAL1 is a negative regulator of drought tolerance in *Arabidopsis*. *Plant J.* **2009**, *58*, 299–317. [CrossRef]

52. Estavillo, G.M.; Crisp, P.A.; Pornsiriwong, W.; Wirtz, M.; Collinge, D.; Carrie, C.; Giraud, E.; Whelan, J.; David, P.; Javot, H.; et al. Evidence for a SAL1-PAP Chloroplast Retrograde Pathway That Functions in Drought and High Light Signaling in *Arabidopsis*. *Plant Cell* **2011**, *23*, 3992–4012. [CrossRef] [PubMed]

53. Chan, K.X.; Mabbitt, P.D.; Phua, S.Y.; Mueller, J.W.; Nisar, N.; Gigolashvili, T.; Stroeher, E.; Grassl, J.; Arlt, W.; Estavillo, G.M.; et al. Sensing and signaling of oxidative stress in chloroplasts by inactivation of the SAL1 phosphoadenosine phosphatase. *Proc. Natl. Acad. Sci. USA* **2016**, *113*, E4567–E4576. [CrossRef] [PubMed]

54. Robles, P.; Fleury, D.; Candela, H.; Cnops, G.; Alonso-Peral, M.M.; Anami, S.; Falcone, A.; Caldana, C.; Willmitzer, L.; Ponce, M.R.; et al. The *RON1/FRY1/SAL1* gene is required for leaf morphogenesis and venation patterning in *Arabidopsis*. *Plant Physiol.* **2010**, *152*, 1357–1372. [CrossRef] [PubMed]

55. Zhang, J.; Vanneste, S.; Brewer, P.B.; Michniewicz, M.; Grones, P.; Kleine-Vehn, J.; Löfke, C.; Teichmann, T.; Bielach, A.; Cannoot, B.; et al. Inositol trisphosphate-induced Ca^{2+} signaling modulates auxin transport and PIN polarity. *Dev. Cell* **2011**, *20*, 855–866. [CrossRef] [PubMed]

56. Gil-Mascarell, R.; López-Coronado, J.M.; Bellés, J.M.; Serrano, R.; Rodríguez, P.L. The *Arabidopsis* HAL2-like gene family includes a novel sodium-sensitive phosphatase. *Plant J.* **1999**, *17*, 373–383. [CrossRef] [PubMed]

57. Torabinejad, J.; Donahue, J.L.; Gunesekera, B.N.; Allen-Daniels, M.J.; Gillaspy, G.E. VTC4 is a bifunctional enzyme that affects myoinositol and ascorbate biosynthesis in plants. *Plant Physiol.* **2009**, *150*, 951–961. [CrossRef] [PubMed]

58. Conklin, P.L.; Gatzek, S.; Wheeler, G.L.; Dowdle, J.; Raymond, M.J.; Rolinski, S.; Isupov, M.; Littlechild, J.A.; Smirnoff, N. *Arabidopsis thaliana* VTC4 encodes L-galactose-1-P phosphatase, a plant ascorbic acid biosynthetic enzyme. *J. Biol. Chem.* **2006**, *281*, 15662–15670. [CrossRef] [PubMed]

59. Nourbakhsh, A.; Collakova, E.; Gillaspy, G.E. Characterization of the inositol monophosphatase gene family in *Arabidopsis*. *Front. Plant Sci.* **2015**, *5*, 725. [CrossRef]

60. Zhang, Y.; Li, S.; Zhou, L.-Z.; Fox, E.; Pao, J.; Sun, W.; Zhou, C.; McCormick, S. Overexpression of *Arabidopsis thaliana PTEN* caused accumulation of autophagic bodies in pollen tubes by disrupting phosphatidylinositol 3-phosphate dynamics. *Plant J.* **2011**, *68*, 1081–1092. [CrossRef]

61. Gupta, R.; Ting, J.T.L.; Sokolov, L.N.; Johnson, S.A.; Luan, S. A tumor suppressor homolog, *AtPTEN1*, is essential for pollen development in *Arabidopsis*. *Plant Cell* **2002**, *14*, 2495–2507. [CrossRef]

62. Pribat, A.; Sormani, R.; Rousseau-Gueutin, M.; Julkowska, M.M.; Testerink, C.; Joubès, J.; Castroviejo, M.; Laguerre, M.; Meyer, C.; Germain, V.; et al. A novel class of PTEN protein in *Arabidopsis* displays unusual phosphoinositide phosphatase activity and efficiently binds phosphatidic acid. *Biochem. J.* **2012**, *441*, 161–171. [CrossRef] [PubMed]

63. Xi, H.; Xu, H.; Xu, W.; He, Z.; Xu, W.; Ma, M. A SAL1 Loss-of-Function *Arabidopsis* Mutant Exhibits Enhanced Cadmium Tolerance in Association with Alleviation of Endoplasmic Reticulum Stress. *Plant Cell Physiol.* **2016**, *57*, 1210–1219. [CrossRef] [PubMed]

64. Litthauer, S.; Chan, K.X.; Jones, M.A. 3′-Phosphoadenosine 5′-Phosphate Accumulation Delays the Circadian System. *Plant Physiol.* **2018**, *176*, 3120–3135. [CrossRef] [PubMed]

65. Phua, S.Y.; Yan, D.; Chan, K.X.; Estavillo, G.M.; Nambara, E.; Pogson, B.J. The *Arabidopsis* SAL1-PAP Pathway: A Case Study for Integrating Chloroplast Retrograde, Light and Hormonal Signaling in Modulating Plant Growth and Development? *Front. Plant Sci.* **2018**, *9*, 1171. [CrossRef] [PubMed]

66. Petersen, L.N.; Marineo, S.; Mandala, S.; Davids, F.; Sewell, B.T.; Ingle, R.A. The Missing Link in Plant Histidine Biosynthesis: *Arabidopsis myo*inositol monophosphatase-like2 Encodes a Functional Histidinol-Phosphate Phosphatase. *Plant Physiol.* **2010**, *152*, 1186–1196. [CrossRef] [PubMed]

67. Zhang, R.-X.; Qin, L.-J.; Zhao, D.-G. Overexpression of the *OsIMP* Gene Increases the Accumulation of Inositol and Confers Enhanced Cold Tolerance in Tobacco through Modulation of the Antioxidant Enzymes' Activities. *Genes* **2017**, *8*, 179. [CrossRef] [PubMed]

68. Joshi-Saha, A.; Reddy, K.S. Repeat length variation in the 5'UTR of *myo*-inositol monophosphatase gene is related to phytic acid content and contributes to drought tolerance in chickpea (*Cicer arietinum* L.). *J. Exp. Bot.* **2015**, *66*, 5683–5690. [CrossRef]

69. Saxena, S.C.; Salvi, P.; Kaur, H.; Verma, P.; Petla, B.P.; Rao, V.; Kamble, N.; Majee, M. Differentially expressed *myo*-inositol monophosphatase gene (*CaIMP*) in chickpea (*Cicer arietinum* L.) encodes a lithium-sensitive phosphatase enzyme with broad substrate specificity and improves seed germination and seedling growth under abiotic stresses. *J. Exp. Bot.* **2013**, *64*, 5623–5639. [CrossRef]

70. Perera, I.Y.; Hung, C.-Y.; Moore, C.D.; Stevenson-Paulik, J.; Boss, W.F. Transgenic *Arabidopsis* Plants Expressing the Type 1 Inositol 5-Phosphatase Exhibit Increased Drought Tolerance and Altered Abscisic Acid Signaling. *Plant Cell* **2008**, *20*, 2876–2893. [CrossRef]

71. Lin, W.H.; Ye, R.; Ma, H.; Xu, Z.H.; Xue, H.W. DNA chip-based expression profile analysis indicates involvement of the phosphatidylinositol signaling pathway in multiple plant responses to hormone and abiotic treatments. *Cell Res.* **2004**, *14*, 34–45. [CrossRef]

72. Xiong, L.; Lee, H.; Huang, R.; Zhu, J.-K. A single amino acid substitution in the *Arabidopsis* FIERY1/HOS2 protein confers cold signaling specificity and lithium tolerance. *Plant J.* **2004**, *40*, 536–545. [CrossRef] [PubMed]

73. Gašparič, M.B.; Lenassi, M.; Gostinčar, C.; Rotter, A.; Plemenitaš, A.; Gunde-Cimerman, N.; Gruden, K.; Žel, J. Insertion of a Specific Fungal 3′-phosphoadenosine-5′-phosphatase Motif into a Plant Homologue Improves Halotolerance and Drought Tolerance of Plants. *PLoS ONE* **2013**, *8*, e81872. [CrossRef] [PubMed]

74. Ku, Y.-S.; Koo, N.S.-C.; Li, F.W.-Y.; Li, M.-W.; Wang, H.; Tsai, S.-N.; Sun, F.; Lim, B.L.; Ko, W.-H.; Lam, H.-M. GmSAL1 hydrolyzes inositol-1,4,5-trisphosphate and regulates stomatal closure in detached leaves and ion compartmentalization in plant cells. *PLoS ONE* **2013**, *8*, e78181. [CrossRef] [PubMed]

75. Ruszkowski, M.; Dauter, Z. Structural Studies of *Medicago truncatula* Histidinol Phosphate Phosphatase from Inositol Monophosphatase Superfamily Reveal Details of Penultimate Step of Histidine Biosynthesis in Plants. *J. Biol. Chem.* **2016**, *291*, 9960–9973. [CrossRef] [PubMed]

76. Dwivedi, V.; Parida, S.K.; Chattopadhyay, D. A repeat length variation in *myo*-inositol monophosphatase gene contributes to seed size trait in chickpea. *Sci. Rep.* **2017**, *7*, 4764. [CrossRef] [PubMed]

MDPI
St. Alban-Anlage 66
4052 Basel
Switzerland
Tel. +41 61 683 77 34
Fax +41 61 302 89 18
www.mdpi.com

International Journal of Molecular Sciences Editorial Office
E-mail: ijms@mdpi.com
www.mdpi.com/journal/ijms

9 783039 219605